*Jürgen Gmehling, Bärbel Kolbe,
Michael Kleiber, and Jürgen Rarey*

Chemical Thermodynamics

Related Titles

Jess, Andreas/Wasserscheid, Peter/Kragl, Udo

Chemical Technology -
An Integral Textbook

2012
ISBN: 978-3-527-30446-2

Anwar, Sara/Carroll, John J.

Carbon Dioxide Thermodynamic Properties Handbook

2011
ISBN-13: 978-1-118-01298-7

Sandler, Stanley I.

An Introduction to Applied Statistical Thermodynamics

2010
ISBN-13: 978-0-470-91347-5

Engell, Sebastian (ed.)

Logistic Optimization of Chemical Production Processes

2008
ISBN-13: 978-3-527-30830-9

Sandler, Stanley I.

Chemical and Engineering Thermodynamics

2008
ISBN-13: 978-0-471-66181-8

Bertau, Martin/Mosekilde, Erik/Westerhoff, Hans V. (eds.)

Biosimulation in Drug Development

2007
ISBN-13: 978-3-527-31699-1

*Jürgen Gmehling, Bärbel Kolbe, Michael Kleiber,
and Jürgen Rarey*

Chemical Thermodynamics

for Process Simulation

WILEY-VCH

WILEY-VCH Verlag GmbH & Co. KGaA

The Authors

Prof. Dr. Jürgen Gmehling
Carl von Ossietzky Univ. Oldenburg
Inst. für Reine & Angew. Chem.
Technische Chemie
26111 Oldenburg
Germany

Dr.-Ing. Bärbel Kolbe
ThyssenKrupp Uhde GmbH
Friedrich-Uhde-Str. 15
44141 Dortmund
Germany

Dr.-Ing. Michael Kleiber
ThyssenKrupp Uhde GmbH
Friedrich-Uhde-Str. 2
65812 Bad Soden
Germany

Dr. Jürgen Rarey
Carl von Ossietzky Univ. Oldenburg
Inst. für Reine & Angew. Chem.
Technische Chemie
26111 Oldenburg
Germany

1st Reprint 2012
2nd Reprint 2013
3rd Reprint 2013
4th Reprint 2015

All books published by **Wiley-VCH** are carefully produced. Nevertheless, authors, editors, and publisher do not warrant the information contained in these books, including this book, to be free of errors. Readers are advised to keep in mind that statements, data, illustrations, procedural details or other items may inadvertently be inaccurate.

Library of Congress Card No.: applied for

British Library Cataloguing-in-Publication Data
A catalogue record for this book is available from the British Library.

Bibliographic information published by the Deutsche Nationalbibliothek
The Deutsche Nationalbibliothek lists this publication in the Deutsche Nationalbibliografie; detailed bibliographic data are available on the Internet at <http://dnb.d-nb.de>.

© 2012 Wiley-VCH Verlag & Co. KGaA, Boschstr. 12, 69469 Weinheim, Germany

All rights reserved (including those of translation into other languages). No part of this book may be reproduced in any form – by photoprinting, microfilm, or any other means – nor transmitted or translated into a machine language without written permission from the publishers. Registered names, trademarks, etc. used in this book, even when not specifically marked as such, are not to be considered unprotected by law.

Cover Design WMX-Design, Bruno Winkler, Heidelberg
Typesetting Laserwords Private Limited, Chennai, India
Printing and Binding betz-druck GmbH, Darmstadt

Printed in the Federal Republic of Germany
Printed on acid-free paper

Print ISBN: 978-3-527-31277-1

Contents

Authors *XIII*
Preface *XV*
List of Symbols *XIX*

1	**Introduction** *1*	
2	***PvT* Behavior of Pure Components** *5*	
2.1	General Description *5*	
2.2	Caloric Properties *10*	
2.3	Ideal Gases *14*	
2.4	Real Fluids *16*	
2.4.1	Auxiliary Functions *16*	
2.4.2	Residual Functions *17*	
2.4.3	Fugacity and Fugacity Coefficient *19*	
2.4.4	Phase Equilibria *23*	
2.5	Equations of State *27*	
2.5.1	Virial Equation *27*	
2.5.2	High Precision Equations of State *32*	
2.5.3	Cubic Equations of State *40*	
2.5.4	Generalized Equations of State and Corresponding States Principle *45*	
2.5.5	Advanced Cubic Equations of State *52*	
	Additional Problems *58*	
	References *61*	
3	**Correlation and Estimation of Pure Component Properties** *65*	
3.1	Characteristic Physical Property Constants *65*	
3.1.1	Critical Data *66*	
3.1.2	Acentric Factor *71*	
3.1.3	Normal Boiling Point *72*	
3.1.4	Melting Point and Enthalpy of Fusion *74*	
3.1.5	Standard Enthalpy and Standard Gibbs Energy of Formation *77*	
3.2	Temperature-Dependent Properties *80*	

3.2.1	Vapor Pressure	82
3.2.2	Liquid Density	94
3.2.3	Enthalpy of Vaporization	97
3.2.4	Ideal Gas Heat Capacity	102
3.2.5	Liquid Heat Capacity	109
3.2.6	Speed of Sound	113
3.3	Correlation and Estimation of Transport Properties	114
3.3.1	Liquid Viscosity	114
3.3.2	Vapor Viscosity	120
3.3.3	Liquid Thermal Conductivity	125
3.3.4	Vapor Thermal Conductivity	130
3.3.5	Surface Tension	133
3.3.6	Diffusion Coefficients	136
	Additional Problems	141
	References	143
4	**Properties of Mixtures**	**147**
4.1	Property Changes of Mixing	148
4.2	Partial Molar Properties	149
4.3	Gibbs–Duhem Equation	153
4.4	Ideal Mixture of Ideal Gases	154
4.5	Ideal Mixture of Real Fluids	156
4.6	Excess Properties	157
4.7	Fugacity in Mixtures	159
4.7.1	Fugacity of an Ideal Mixture	159
4.7.2	Phase Equilibrium	160
4.8	Activity and Activity Coefficient	161
4.9	Application of Equations of State to Mixtures	162
4.9.1	Virial Equation	163
4.9.2	Cubic Equations of State	164
	Additional Problems	174
	References	175
5	**Phase Equilibria in Fluid Systems**	**177**
5.1	Thermodynamic Fundamentals	186
5.2	Application of Activity Coefficient Models	193
5.3	Calculation of Vapor–Liquid Equilibria Using g^E-Models	197
5.4	Fitting of g^E-Model Parameters	216
5.4.1	Check of VLE Data for Thermodynamic Consistency	221
5.4.2	Recommended g^E-Model Parameters	231
5.5	Calculation of Vapor–Liquid Equilibria Using Equations of State	235
5.5.1	Fitting of Binary Parameters of Cubic Equations of State	240
5.6	Conditions for the Occurrence of Azeotropic Behavior	248
5.7	Solubility of Gases in Liquids	259
5.7.1	Calculation of Gas Solubilities Using Henry Constants	261

5.7.2	Calculation of Gas Solubilities Using Equations of State	*270*
5.7.3	Prediction of Gas Solubilities	*271*
5.8	Liquid–Liquid Equilibria	*273*
5.8.1	Temperature Dependence of Ternary LLE	*286*
5.8.2	Pressure Dependence of LLE	*288*
5.9	Predictive Models	*289*
5.9.1	Regular Solution Theory	*290*
5.9.2	Group Contribution Methods	*292*
5.9.3	UNIFAC Method	*293*
5.9.3.1	Modified UNIFAC (Dortmund)	*300*
5.9.3.2	Weaknesses of the Group Contribution Methods UNIFAC and Modified UNIFAC	*309*
5.9.4	Predictive Soave–Redlich–Kwong (PSRK) Equation of State	*312*
5.9.5	VTPR Group Contribution Equation of State	*317*
	Additional Problems	*326*
	References	*330*

6 Caloric Properties *333*

6.1	Caloric Equations of State	*333*
6.1.1	Internal Energy and Enthalpy	*333*
6.1.2	Entropy	*336*
6.1.3	Helmholtz Energy and Gibbs Energy	*337*
6.2	Enthalpy Description in Process Simulation Programs	*339*
6.2.1	Route A: Vapor as Starting Phase	*340*
6.2.2	Route B: Liquid as Starting Phase	*344*
6.2.3	Route C: Equation of State	*346*
6.3	Caloric Properties in Chemical Reactions	*354*
6.4	The *G*-Minimization Technique	*361*
	Additional Problems	*364*
	References	*364*

7 Electrolyte Solutions *365*

7.1	Introduction	*365*
7.2	Thermodynamics of Electrolyte Solutions	*369*
7.3	Activity Coefficient Models for Electrolyte Solutions	*374*
7.3.1	Debye–Hückel Limiting Law	*374*
7.3.2	Bromley Extension	*376*
7.3.3	Pitzer Model	*377*
7.3.4	Electrolyte-NRTL Model by Chen	*378*
7.3.5	LIQUAC Model	*387*
7.3.6	MSA Model	*396*
7.4	Dissociation Equilibria	*396*
7.5	Influence of Salts on the Vapor-Liquid Equilibrium Behavior	*398*
7.6	Complex Electrolyte Systems	*400*

Additional Problems *401*
References *402*

8   Solid–Liquid Equilibria *405*
8.1   Thermodynamic Relations for the Calculation of Solid–Liquid Equilibria *408*
8.1.1   Solid–Liquid Equilibria of Simple Eutectic Systems *410*
8.1.1.1 Freezing Point Depression *417*
8.1.2   Solid–Liquid Equilibria of Systems with Solid Solutions *419*
8.1.2.1 Ideal Systems *419*
8.1.2.2 Solid–Liquid Equilibria for Nonideal Systems *420*
8.1.3   Solid–Liquid Equilibria with Intermolecular Compound Formation in the Solid State *424*
8.1.4   Pressure Dependence of Solid–Liquid Equilibria *427*
8.2   Salt Solubility *427*
8.3   Solubility of Solids in Supercritical Fluids *432*
        Additional Problems *434*
        References *437*

9   Membrane Processes *439*
9.1   Osmosis *439*
9.2   Pervaporation *443*
        Additional Problems *444*
        References *444*

10   Polymer Thermodynamics *445*
10.1   Introduction *445*
10.2 g^E-models *451*
10.3   Equations of State *462*
10.4   Influence of Polydispersity *479*
        Additional Problems *482*
        References *484*

11   Applications of Thermodynamics in Separation Technology *487*
11.1   Verification of Model Parameters Prior to Process Simulation *492*
11.1.1 Verification of Pure Component Parameters *492*
11.1.2 Verification of g^E-Model Parameters *493*
11.2   Investigation of Azeotropic Points in Multicomponent Systems *501*
11.3   Residue Curves, Distillation Boundaries, and Distillation Regions *503*
11.4   Selection of Entrainers for Azeotropic and Extractive Distillation *511*
11.5   Selection of Solvents for Other Separation Processes *518*
11.6   Examination of the Applicability of Extractive Distillation for the Separation of Aliphatics from Aromatics *519*

　　　　　　Additional Problems　522
　　　　　　References　523

12　　　　**Enthalpy of Reaction and Chemical Equilibria**　525
12.1　　　　Enthalpy of Reaction　526
12.1.1　　　Temperature Dependence　527
12.1.2　　　Consideration of the Real Gas Behavior on the Enthalpy of Reaction　529
12.2　　　　Chemical Equilibrium　531
12.3　　　　Multiple Chemical Reaction Equilibria　551
12.3.1　　　Relaxation Method　552
12.3.2　　　Gibbs Energy Minimization　556
　　　　　　Additional Problems　563
　　　　　　References　565

13　　　　**Special Applications**　567
13.1　　　　Formaldehyde Solutions　567
13.2　　　　Vapor Phase Association　573
　　　　　　Additional Problems　587
　　　　　　References　589

14　　　　**Practical Applications**　591
14.1　　　　Flash　591
14.2　　　　Joule–Thomson Effect　593
14.3　　　　Adiabatic Compression and Expansion　595
14.4　　　　Pressure Relief　600
14.5　　　　Limitations of Equilibrium Thermodynamics　606
　　　　　　Additional Problems　608
　　　　　　References　610

15　　　　**Introduction to the Collection of Example Problems**　613
15.1　　　　Mathcad Examples　613
15.2　　　　Examples Using the Dortmund Data Bank (DDB) and the Integrated Software Package DDBSP　615
15.3　　　　Examples Using Microsoft Excel and Microsoft Office VBA　616

Appendix A Pure Component Parameters　619

Appendix B Coefficients for High Precision Equations of State　641

Appendix C Useful Derivations　645
A1.　　　　Relationship between $(\partial s/\partial T)_P$ and $(\partial s/\partial T)_v$　646
A2.　　　　Expressions for $(\partial u/\partial v)_T$ and $(\partial s/\partial v)_T$　646
A3.　　　　c_P and c_v as Derivatives of the Specific Entropy　647
A4.　　　　Relationship between c_P and c_v　648

A5.	Expression for $(\partial h/\partial P)_T$	649
A6.	Expression for $(\partial s/\partial P)_T$	650
A7.	Expression for $[\partial(g/RT)/\partial T]_P$ and van't Hoff Equation	651
A8.	General Expression for c_v	651
A9.	Expression for $(\partial P/\partial v)_T$	652
A10.	Cardano's Formula	652
B1.	Derivation of the Kelvin Equation	653
B2.	Equivalence of Chemical Potential μ and Gibbs Energy g for a Pure Substance	654
B3.	Phase Equilibrium Condition for a Pure Substance	655
B4.	Relationship between Partial Molar Property and State Variable (Euler Theorem)	657
B5.	Chemical Potential in Mixtures	658
B6.	Relationship between Second Virial Coefficients of Leiden and Berlin Form	659
B7.	Derivation of Expressions for the Speed of Sound for Ideal and Real Gases	659
B8.	Activity of the Solvent in an Electrolyte Solution	661
B9.	Temperature Dependence of the Azeotropic Composition	662
C1.	$(s-s^{id})_{T,P}$	664
C2.	$(h-h^{id})_{T,P}$	665
C3.	$(g-g^{id})_{T,P}$	665
D1.	Fugacity Coefficient for a Pressure-Explicit Equation of State	665
D2.	Fugacity Coefficient of the Virial Equation (Leiden Form)	666
D3.	Fugacity Coefficient of the Virial Equation (Berlin Form)	668
D4.	Fugacity Coefficient of the Soave–Redlich–Kwong Equation of State	669
D5.	Fugacity Coefficient of the PSRK Equation of State	671
E1.	Derivation of the Wilson Equation	675
E2.	Notation of the Wilson, NRTL, and UNIQUAC Equations in Process Simulation Programs	678
E3.	Inability of the Wilson Equation to Describe a Miscibility Gap	679
F1.	$(h-h^{id})$ for Soave–Redlich–Kwong Equation of State	681
F2.	$(s-s^{id})$ for Soave–Redlich–Kwong Equation of State	683
F3.	$(g-g^{id})$ for Soave–Redlich–Kwong Equation of State	683
F4.	Antiderivatives of c_p^{id} Correlations	683
G1.	Speed of Sound as Maximum Velocity in an Adiabatic Pipe with Constant Cross-Flow Area	685
G2.	Maximum Mass Flux of an Ideal Gas	685
	References	687

Appendix D Standard Thermodynamic Properties for Selected Electrolyte Compounds 689

Appendix E Regression Technique for Pure Component Data 691

Appendix F Regression Techniques for Binary Parameters *695*
        References *709*

Appendix G Ideal Gas Heat Capacity Polynomial Coefficients for Selected Compounds *711*

Appendix H UNIFAC Parameters *713*

Appendix I Modified UNIFAC Parameters *715*

Appendix J PSRK Parameters *721*

Appendix K VTPR Parameters *725*
        References *727*

    Index *729*

Authors

Jürgen Gmehling I finished an apprenticeship as a laboratory technician and studied Chemical Engineering at the Technical College in Essen before studying Chemistry leading to the degree of Diplom-Chemiker in 1970 and my doctoral degree from the University of Dortmund in Inorganic Chemistry in 1973. After graduation I worked in the research group of Prof. Ulfert Onken (Chair of Chemical Engineering) at the University of Dortmund. My research activities were directed to applied thermodynamics, in particular the development of group contribution methods and the synthesis of separation processes. From 1977–1978 I spent 15 months with Prof. J.M. Prausnitz at the Department of Chemical Engineering in Berkeley, California.

In 1989 I joined the faculty of the University of Oldenburg as Professor of Chemical Engineering. The research activities of my group are mainly directed to the computer-aided synthesis, design and optimization of chemical processes. Our research results, in particular the software products, such as the Dortmund Data Bank, the group contribution methods for nonelectrolyte and electrolyte systems (UNIFAC, modified UNIFAC, PSRK, VTPR, LIQUAC, LIFAC, COSMO-RS(Ol)) and the sophisticated software packages for process developments are used worldwide by a large number of chemical engineers in industry during their daily work. The importance of the group contribution methods developed by us is demonstrated by the fact that the systematic further development of these methods is supported by a consortium of more than 50 companies for more than 15 years.

I am also president and CEO of the company "DDB Software and Separation Technology (DDBST), Oldenburg, Germany (www.ddbst.com)" founded by myself and two coworkers in 1989. This company is responsible for the Dortmund Data Bank (DDB), the largest worldwide factual data bank for thermophysical properties, and at the same time engaged in consulting and developing software for the synthesis and design of separation processes. In collaboration with Dr. K. Fischer I founded the "Laboratory for Thermophysical Properties (LTP GmbH (www.ltp-oldenburg.de))" as an "An-Institute" at the Carl von Ossietzky University of Oldenburg in 1999, a company that is engaged in the measurement of thermophysical data (pure component properties, phase equilibria, excess properties, transport properties, reaction rates, etc.) over wide temperature and pressure ranges.

For my research activities I have received various awards, e.g. the "Arnold-Eucken Prize" in 1982 from GVC (Gesellschaft für Verfahrenstechnik und Chemieingenieurwesen), the "Rossini Lecture Award 2008" from the International Association of Chemical Thermodynamics and 2010 the "Gmelin-Beilstein Denkmünze" from GDCh (Gesellschaft Deutscher Chemiker).

Bärbel Kolbe After graduating in Chemical Engineering I finished my thesis on the description and measurement of the properties of liquid mixtures at the University of Dortmund in the research group of Jürgen Gmehling in 1983. I continued to work with Jürgen Gmehling for another 3 years. During this time I participated in the publication of the Dechema Chemistry Data Series on VLE and the first edition of this book in the German language was being written.

I have been working for more than twenty years as a senior process engineer first for the Krupp Koppers GmbH and, since 1997, for ThyssenKrupp Uhde. Within the research and development department of ThyssenKrupp Uhde my main focuses are thermophysical properties, thermal separation technology and new processes.

Michael Kleiber After having graduated in mechanical engineering, I worked as a scientific assistant and finished my thesis at the Technical University of Brunswick in 1994 with a supplement to the UNIFAC method for halogenated hydrocarbons. In the following years, I worked for the former Hoechst AG and their legal successors in a wide variety of tasks in the fields of process development, process simulation and engineering calculations. Meanwhile, I moved to ThyssenKrupp Uhde as a colleague of Bärbel Kolbe as Chief Development Engineer. I am a member of the German Board of Thermodynamics and have made contributions to several process engineering standard books like VDI Heat Atlas, Winnacker-Küchler and Ullmann's Encyclopedia of Industrial Chemistry.

Jürgen Rarey I studied and graduated in Chemistry and did my PhD in Chemical Engineering. I built my first computer in the 3rd year of study and developed some interfacing to lab equipment and control software. I became fascinated by the possibilities of simulation and learned the importance of the correct description of basic phenomena to the outcome of simulations (garbage in – garbage out).

In 1989, near the end of my PhD study, I moved to Oldenburg in northern Germany. Since April 1989 I have held a permanent position at the University of Oldenburg in the group of Prof. Gmehling.

In the same year I cofounded DDBST GmbH (www.ddbst.com). This company today is the most well-known provider of thermophysical property data and estimation methods for process simulation and further applications in safety, environmental protection, etc.

Since the 1980s I have taught many courses on applied thermodynamics for chemical process simulation for external participants from industry both in Oldenburg and in-house for companies in Europe, the US, Middle East, South Africa, Japan, etc. (alone or together with Prof. Gmehling).

An additional role in my professional life is a Honorary Professor position in Durban, South Africa. Currently, my group there consists of 4 MSc students, all highly motivated. We developed and published estimation methods for a number of important properties like vapor pressure, liquid viscosity, water and alkane solubility, etc. The first of our methods on normal boiling point estimation is already generally regarded as the primary and best method available.

Preface

More than 20 years ago, the first edition of the textbook *Thermodynamics* was published in the German language.

Its target was to demonstrate the basic principles, how thermodynamics can contribute to solve manifold kinds of problems in gas, oil and chemical processing, pharmaceutical and food production, in environmental industry, in plant design by engineering companies, and also for institutions dealing with hazardous materials like the fire brigade, transport companies or the Technical Supervisory Associations. For all these purposes, it is often decisive to have a profound knowledge of the thermophysical properties, transport properties, phase equilibria, and chemical equilibria. Therefore, a large part of the first edition and also of this completely new edition is dedicated to the evaluation of these quantities. The mentioned properties are also helpful in the evaluation of nonequilibrium properties such as kinetic data and reaction rates, which are not subject of this book.

Databases filing published experimental physical properties and phase equilibrium data are a prerequisite for developing thermodynamic models and for determining reliable model parameters, which describe the problem to be solved with adequate accuracy. A long way has been covered since the beginning of the professional filing of phase equilibrium data. Starting with a few hundred compounds in 1973, pure component and mixture properties for more than 33 000 components can now be found in the Dortmund Data Bank (DDB). A great step forward in modeling was the further development of the solution of group concept, which makes the prediction of, for example, phase equilibria possible. A lot of experimental work was performed to systematically fill the gaps where no data for the determination of group interaction parameters were available. Together with the fast developing computer technology and on the basis of professional databases like the DDB, process simulators nowadays allow rapid calculation of phase equilibria, transport properties, caloric data, the various thermophysical properties, and chemical equilibria. Even the thermodynamics of large industrial processes is routinely modeled using commercial process simulators. While a large variety of models and model options can be selected by a simple mouse-click, the task of the engineer or chemist remains to choose the most appropriate model, and one should be aware of its accuracy, its possible limitations, and the quality of the model parameters for the system of interest. A thorough understanding of thermodynamics is still obligatory; otherwise misconceptions of processes or design errors are the consequences.

The new edition of the textbook, now written in the English language, is called *Chemical Thermodynamics for Process Simulation*. It specifically targets readers working in the fields of process development, process synthesis, or process optimization and therefore presents the fundamentals of thermodynamics not only for students but also on the level required for experienced process engineers. The most important models that are applied in process industry are thoroughly explained, as well as their adjustment with the help of factual databases (data regression). Cubic equations of state with g^E mixing rules present a great step forward toward a universal model for both subcritical and supercritical systems and are therefore emphasized.

In addition, models for special substances like carboxylic acids, hydrogen fluoride, formaldehyde, electrolytes, and polymers are introduced and the capabilities of high-precision equations of state and various predictive methods are explained. Recommendations for the parameter fitting procedure and numerous hints to avoid pitfalls during process simulation are given. Because of the space limitation in the book we were not able to cover the whole range of thermodynamics, for example, adsorption has been left out completely as it cannot be presented within a short chapter.

The English language was chosen to extend the readership to students and engineers from all over the world. Although none of the authors is a native speaker, we found it even more convenient to describe the particular issues in the language generally used in scientific publications.

The team of four authors with considerably different backgrounds reflects the importance of thermodynamics in both academia and industrial application. The authors present their biography and special research interests on separate pages following this preface.

In contrast to other textbooks on thermodynamics, we assume that the readers are familiar with the fundamentals of classical thermodynamics, that means the definitions of quantities like pressure, temperature, internal energy, enthalpy, entropy, and the three laws of thermodynamics, which are very well explained in other textbooks. We therefore restricted ourselves to only a brief introduction and devoted more space to the description of the real behavior of the pure compounds and their mixtures. The ideal gas law is mainly used as a reference state; for application examples, the real behavior of gases and liquids is calculated with modern g^E models, equations of state, and group contribution methods.

Of course, by taking into account the real behavior the solution of the examples becomes much more complex, but at the same time they are closer to industrial practice. For a textbook, there is a difficulty to describe the typical iterative procedures in phase equilibrium and process calculations. In order to achieve a better understanding, we decided to provide MathCAD-sheets and DDBST programs so that the reader has the chance to reproduce the examples on his own. MathCAD was chosen because of its convenient way to write equations in close-to-textbook form and without cryptic variable names. We prefer SI units but do not stick to them obsessively. In the examples and diagrams we used the most convenient units. We think that the parallel use of various units will remain the status quo for the time being, and engineers and chemists should be able to cope with this situation. We are aware that the current value of the gas constant is $R = 8.31447$ J/(mol K). However, still many applications are based on the old value $R = 8.31433$ J/(mol K). Luckily, except for the high-precision equations of state this distinction is by far beyond the accuracy scope of our calculations.

For a complete understanding, mathematical derivations can often not be avoided or are even necessary for the understanding. If they interrupt the flow of the presentation, we have moved them to a special chapter in the Appendix, so that the reader can follow the main ideas more easily. Of course, no textbook can cover all possible and interesting derivations, but we hope that the reader will gain a feeling for the methodology in thermodynamics and is able to carry out similar derivations on his own.

We hope that this book closes a gap between scientific development and its application in industry. We are grateful to all the people who gave us valuable support and advice during the compilation of the manuscript. None of the authors was capable to write an adequate chapter on polymer thermodynamics. Therefore we are especially obliged to Prof. Dr. Sabine Enders. She wrote an excellent chapter fully in line with the targets and structure of this book. Many other people gave valuable advice. We are thankful to Prof. Dr. Wolfgang Wagner, Prof. Dr. Hans Hasse, Prof. Dr. Josef Novak, Prof. Dr. Roland Span, Todd Willman, Dr. Michael Sakuth, Ingo Schillgalies, Jens Otten, Dr. André Mohs, Dr. Bastian Schmid, Dr. Jens Ahlers, Dr. Silke Nebig, Dr. Torben Laursen, Dr. Heiner Landeck, Prof. Dr. Ravi Prasad Andra, Dr. Michael Benje, the colleagues and coworkers from DDBST GmbH and the research group at the Carl-von-Ossietzky-University of Oldenburg, who provided many impressive figures of the book. Furthermore, we are deeply thankful to our families for supporting us during all the time.

Jürgen Gmehling
Bärbel Kolbe
Michael Kleiber
Jürgen Rarey

List of Symbols

a	attractive parameter in cubic equations of state	J m^3 mol^{-2}
a	specific Helmholtz energy	J mol^{-1}, J kg^{-1}
a_i	activity of component i	
$a_{ij}, b_{ij}, c_{ij}, d_{ij}$	binary or group interaction parameter in local composition models (Wilson, NRTL, UNIQUAC, UNIFAC ...)	
A	Helmholtz energy	J
A	area	m^2
A_m	parameter in Debye–Hückel equation	kg$^{0.5}$ mol$^{0.5}$
$a, b, c, \ldots$	constants in pure component property correlations	
$A, B, C \ldots$	constants in pure component property correlations	
A_n, B_n	parameters for the description of association reactions of degree n	
A_ϕ	parameter in Pitzer–Debye–Hückel term	
b	repulsive parameter in equations of state	m^3 mol^{-1}
B	second virial coefficient	m^3 mol^{-1}
B_{ij}	parameter in Pitzer equation	
B_{ij}	cross second virial coefficient	m^3 mol^{-1}
c	volume concentration	mol m^{-3}
c_P	specific isobaric heat capacity	J mol^{-1} K^{-1}, J kg^{-1} K^{-1}
c_V	specific isochoric heat capacity	J mol^{-1} K^{-1}, J kg^{-1} K^{-1}
c_σ	specific liquid heat of vaporization along the saturation line	J mol^{-1} K^{-1}, J kg^{-1} K^{-1}
C	third virial coefficient	(m^3)^2 mol^{-2}
d	droplet diameter	m
d_i	segment diameter (PC-SAFT)	m
D_{ij}	diffusion coefficient of component i in component j	m^2 s^{-1}
e	elementary charge; $e = 1.602189 \cdot 10^{-19}$ C	C
E	Ackermann correction factor	
f_i	fugacity of component i	Pa
F	objective function	
F	Faraday's constant; $F = 96484.56$ C/mol	C mol^{-1}
F_i	surface area/mole fraction of component i (UNIQUAC, UNIFAC)	–
$F(r)$	integral distribution function	

List of Symbols

F_{ij}	force between two ions i and j	N
g	specific Gibbs energy	J mol^{-1}
Δg	Gibbs energy of mixing	J mol^{-1}
Δg_{ij}	interaction parameter of the NRTL equation	K
Δg_R^0	standard Gibbs energy of reaction	J mol^{-1}
G	Gibbs energy	J
h	Planck's constant; $h = 6.6242 \cdot 10^{-34}$ Js	Js
h	specific enthalpy	J mol^{-1}, J kg^{-1}
Δh_f^0	standard enthalpy of formation	J mol^{-1}
Δg_f^0	standard Gibbs energy of formation	J mol^{-1}
Δh_R^0	standard enthalpy of reaction	J mol^{-1}
Δh_m	specific enthalpy of fusion	J mol^{-1}, J kg^{-1}
$h - h^{id}$	specific isothermal enthalpy difference between actual and ideal gas state, calculated with an EOS	J mol^{-1}
$\Delta h_i^{\infty L}$	specific enthalpy of solution of Henry component i	J mol^{-1}
Δh_{sol}	enthalpy of solution	J mol^{-1}, J kg^{-1}
Δh_v	specific enthalpy of vaporization	J mol^{-1}, J kg^{-1}
H	enthalpy	J
H_{ij}	Henry constant of component i in solvent j	Pa
I	ionic strength	mol kg^{-1}
I_x	molar ionic strength	mol mol^{-1}
J_i	flux through membrane of component i	kg s^{-1} m^{-2}
k	Boltzmann's constant; k = $1.38048 \cdot 10^{-23}$ J/K	J/K
k_{ij}	binary interaction parameter in cubic equations of state	
K	chemical equilibrium constant	
K	liquid-liquid distribution coefficient	
K_{cry}	cryoscopic constant	K kg mol^{-1}
K_{sp}	solubility product	
K_n	chemical equilibrium constant for association of degree n	
K_{in}	chemical equilibrium constant for association of degree n, component i	
K_{Mij}	chemical equilibrium constant for mixed association, components i and j	
K_i	K-factor for component i ($K_i = y_i/x_i$)	
l	membrane thickness	m
L	amount of liquid	mol, kg
m	arbitrary specific thermodynamic function	
m	mass	kg
m_i	molality of component i	mol kg^{-1}
$\overline{m}_i$	partial molar property	
$\dot{m}$	mass flow	kg s^{-1}
M	molar mass	g mol^{-1}
$\overline{M}$	moment of distribution function	
$\Delta m, \Delta M$	property change of mixing	
n	number of components	
n	number of data points	
$\dot{n}$	mole flow	mol s^{-1}
n_A	number of atoms in a molecule	
N_A	Avogadro's number; $N_A = 6.023 \cdot 10^{23}$	
N_{th}	number of theoretical stages	

Symbol	Description	Units
n_f	number of degrees of freedom	
n_i	number of moles of component i	mol
n_T	total number of moles	mol
N	total number of species	mol
Nu	Nußelt number	
p_i	partial pressure of component i	Pa
P	Parachor	
P	total pressure	Pa
P_i	permeability	kg s^{-1} m^{-1} Pa^{-1}
p^G	vapor pressure around a droplet	Pa
P^*	apparent permeability	kg s^{-1} m^{-2} Pa^{-1}
P_i^s	vapor pressure of component i	Pa
Poy_i	Poynting factor of component i	
Pr	Prandtl number	
q_i	relative van der Waals surface area of component i	
q	charge	C
q	vapor fraction	
q	specific heat	J mol^{-1}, J kg^{-1}
$\dot{q}$	specific heat flux	W mol^{-1}, W kg^{-1}
Q	heat	J
$\dot{Q}$	heat flow	W
Q_k	relative van der Waals surface area of group k	
r_i	relative van der Waals volume of component i	
r_i	ionic radius	
r_i	segment number	
r_{ij}	distance between two ions i and j	m
R	universal gas constant; $R = 8.314471$ J/mol K $= 1.98721$ cal/mol K	J mol^{-1} K^{-1}
Re	Reynolds number	
R_k	relative van der Waals volume of group k	
s	specific entropy	J mol^{-1} K^{-1}, J kg^{-1} K^{-1}
s_{abs}	absolute specific entropy	J mol^{-1} K^{-1}, J kg^{-1} K^{-1}
Δs^{id}	specific entropy of mixing	J mol^{-1} K^{-1}, J kg^{-1} K^{-1}
Δs_R^0	standard entropy of reaction	J mol^{-1} K^{-1}, J kg^{-1} K^{-1}
S	entropy	J K^{-1}
S_{12}	selectivity	
T	absolute temperature	K
u	specific internal energy	J mol^{-1}, J kg^{-1}
Δu_{ij}	interaction parameter of the NRTL equation	K
u	internal energy	J
V	amount of vapor	mol, kg
v	specific volume	m^3 mol^{-1}, m^3 kg^{-1}
v^*	characteristic volume	m^3 mol^{-1}
V	volume	m^3
V_i	volume fraction/mole fraction of component i (UNIQUAC, UNIFAC)	

w	specific work	J mol^{-1}, J kg^{-1}
w	velocity	m s^{-1}
w^*	speed of sound	m s^{-1}
W_t	technical work	J
w_t	specific technical work	J mol^{-1}, J kg^{-1}
w	weighting factor in objective functions	
$W(r)$	distribution function	
w_i	weighting factor of data point i	
w_i	weight fraction of component i	
x_i	mole fraction of component i in the liquid phase	
x'_i	mole fraction of component i on salt-free basis	
X	group mole fraction	
X	chemical conversion	
y	mole fraction in the vapor phase	
z	compressibility factor; $z = Pv/RT$	
z	length	m
z_i	charge of ion i	C
z_i	mole fraction	
z_n	true mole fraction of associate of degree n	
z_{in}	true mole fraction of associate of degree n, component i	
z_{Mij}	true mole fraction of mixed associate, components i and j	
$\bar{\bar{z}}$	segment molar quantity	

Greek Symbols

α_{ij}	separation factor	
α_{ij}	nonrandomness parameter in the NRTL equation	
α	thermal expansion coefficient	K^{-1}
α	degree of dissociation	
α	heat transfer coefficient	W m^2 K^{-1}
α	function in cubic EOS	
α	reduced Helmholtz energy in high precision EOS	
α, β, γ	constants in pure component property correlations	
γ_i	activity coefficient of component i	
β	parameter in Bromley equation	kg mol^{-1}
β	mass transfer coefficient	m s^{-1}
Γ_k	group activity coefficient	
δ_i	solubility parameter of component i	(J m^{-3})$^{0.5}$
δ_{ij}	excess virial coefficient	m^3 mol^{-1}
Δ	difference value of a thermodynamic property	
Δ_b	group increment for normal boiling point	
$\Delta_A, \Delta_B, \Delta_C, \Delta_D$	group constants for c_p^{id}	
$\Delta_{Born} g^E$	Born term for regarding the dielectricity constant of the solvent	J mol^{-1}
Δ_G	group increment for standard Gibbs energy of formation	
Δ_H	group increment for standard enthalpy of formation	
Δ_T	group increment for critical temperature	
Δ_P	group increment for critical pressure	
Δ_V	group increment for critical volume	
ε	depth of potential well (PC-SAFT)	

Symbol	Description	Units
ε	stop criterion	
ε	performance number	
ε	relative dielectricity constant	
ε_0	vacuum dielectricity constant; $\varepsilon_0 = 8.854188 \cdot 10^{-12}$ C^2 N^{-1} m^{-2}	C^2 N^{-1} m^{-2}
ζ	local volume fraction	
ζ	local volume fraction	
η	dynamic viscosity	Pa s
η	efficiency	
ϑ	Celsius temperature	°C
Θ_k	surface area fraction of group k (UNIFAC)	
Θ_{ij}	local concentration of species i around species j	
χ	isothermal compressibility factor	Pa^{-1}
κ	isentropic exponent; $\kappa = c_p^{id}/c_v^{id}$	
λ	reduced well width (PC-SAFT)	
λ	thermal conductivity	W K^{-1} m^{-1}
λ_{ij}	parameter in Pitzer model	
$\Delta\lambda_{ij}$	interaction parameter of the Wilson equation	K
Λ_{ij}	Wilson parameter	
μ	dipole moment	Debye
μ_i	chemical potential of component i	J mol^{-1}
μ_{ijk}	parameter in Pitzer equation	
ν	kinematic viscosity	m^2 s^{-1}
ν_0	frequency	s^{-1}
ν_k	number of structural groups of type k	
ν_i	stoichiometric coefficient of component i	
ξ	local volume fraction	
Π	osmotic pressure	Pa
π	number of phases	
ρ	density	mol m^{-3}, kg m^{-3}
σ	hard sphere segment diameter (PC-SAFT)	
σ	standard deviation	
σ	surface tension	N m^{-1}
σ	symmetry number	
τ	shear stress	N m^{-2}
τ_{ij}	binary interaction parameter in the UNIQUAC model	
Φ	osmotic coefficient	
Φ_i	volume fraction of component i	
ϕ_i	factor defined in Eq. (5.15)	
φ_{el}	electric potential	V
φ_i	fugacity coefficient of component i	
χ	Flory-Huggins parameter	
Ψ	UNIFAC parameter	
ω	acentric factor	

Special symbols

$-$	partial property	
∞	value at infinite dilution	

Subscripts

$\pm$	entire electrolyte
0	reference state
1,2,3,4 ...	process steps
a	anion
amb	ambient
az	value at the azeotropic point
b	value at the normal boiling point
c	critical property
c	cation
calc	calculated value
Comp	compressor
CWR	cooling water return
CWS	cooling water supply
D	value for the dimer
DH	Debye–Hückel
dil	dilution
elec	electrolyte
exp	experimental value
eut	value at the eutectic point
f	value for the formation reaction
G	gas phase
HF	hemiformal
$i, j, \ldots$	component
inv	inversion temperature
LC	local composition model
LR	long range
m	value at the melting point
max	maximum occuring value
mix	mixture
M	value for the monomer
MR	middle range
MSA	Mean Spherical Approximation method
P	at constant pressure
R	value for the chemical reaction
r	reduced property
ref	reference state
rev	for reversible processes
solv	solvent
SR	short range
tot	total
tr	value at the triple point
trans	transition point
trs	transition
true	true value in maximum likelihood method
v	vaporization
V	at constant volume

Superscripts

*	electrolyte reference state
assoc	association term

rep	caused by repulsive forces
att	caused by attractive forces
$\alpha, \beta, \varphi, \pi, ', ''$	phases
C	combinatorial part
disp	dispersion
E	excess property
hs	hard sphere
hc	hard chain
id	in the ideal gas state
L	liquid phase
m	molality scale
0	standard state
R	residual part
(r)	reference fluid
o	at zero pressure
pure	pure component
S	salt free
s	saturation state
S	solid phase
S	solvent free basis
subl	sublimation
tot	totally dissociated
V	vapor phase
(0)	simple fluid
(1)	deviation from the behavior of simple fluids (corresponding state principle)
∞	at infinite dilution
I, II, III ...	phases

Mathematics

ln logarithm basis e
log logarithm basis 10

Conversion factors

$1\,\text{kPa} = 0.009869\,\text{atm} = 0.01\,\text{bar} = 7.50062\,\text{Torr} = 10^3\,\text{N/m}^2 = 1000\,\text{Pa}$
$1\,\text{J} = 1\,\text{kg m}^2/\text{s}^2 = 1\,\text{Nm} = 0.238846\,\text{cal}$
$1\,\text{Debye} = 3.336 \cdot 10^{-30}\,\text{C m}$

1
Introduction

Process simulation is a tool for the development, design and optimization of processes in the chemical, petrochemical, pharmaceutical, energy producing, gas processing, environmental, and food industry. It provides a representation of the particular basic operations of the process using mathematical models for the different unit operations, ensuring that the mass and energy balances are maintained. Nowadays, simulation models are of extraordinary importance for scientific and technical developments and even for economic and political decisions, as it is, for example, the case in the climate modeling. Unfortunately, the wide and productive use of process simulation in industry is currently being limited by a lack of understanding of thermodynamics and its application in process simulation. This book is dedicated to provide this knowledge in a convenient way.

The development of process simulation started in the 1960s, when appropriate hardware and software became available and could connect the remarkable knowledge about thermophysical properties, phase equilibria, reaction equilibria, reaction kinetics, and the particular unit operations. A number of comprehensive simulation programs have been developed, commercial ones (ASPEN Plus, ChemCAD, HySys, Pro/II, ProSim, and SYSTEM 7) as well as in-house simulators in large companies, for example, VTPlan (Bayer AG) or ChemaSim (BASF), not to mention the large number of in-house tools that cover the particular calculation tasks of small companies working in process engineering. Nevertheless, all the simulators have in common that they are only as good as the models and the corresponding model parameters available. Most of the process simulators provide a default databank for the user, but finally the user himself is responsible for the parameters used. In this context, it is very useful that since the 1970s databanks for pure component and mixture data (e.g., DIPPR, DDB) have been compiled. Today even the data formats are well established so that they can easily be interpreted by computer programs. These activities belong to the main fundamentals of modern process simulation. Furthermore, these systematic data collections made it possible to develop prediction methods like ASOG, UNIFAC, or PSRK, which enabled the process engineers to adjust reasonable parameters even for systems where no data are available. One of the most important purposes of this book is to enable the user to choose the model and the model parameters in an appropriate way.

Chemical Thermodynamics: for Process Simulation, First Edition.
Jürgen Gmehling, Bärbel Kolbe, Michael Kleiber, and Jürgen Rarey.
© 2012 Wiley-VCH Verlag GmbH & Co. KGaA. Published 2012 by Wiley-VCH Verlag GmbH & Co. KGaA.

Various degrees of effort can be applied in process simulation. A simple split balance can give a first overview of the process without introducing any physical relationships into the calculation. The user just defines split factors to decide which way the particular components take. In a medium level of complexity, shortcut methods are used to characterize the various process operations. The rigorous simulation with its full complexity can be considered as the most common case. The particular unit operations (reactors, columns, heat exchangers, flash vessels, compressors, valves, pumps, etc.) are represented with their correct physical background and a model for the thermophysical properties.

Different physical modes are sometimes available for the same unit operation. A distillation column can, for example, be modeled on the basis of theoretical stages or using a rate-based model, taking into account the mass transfer on the column internals. A simulation of this kind can be used to extract the data for the design of the process equipment or to optimize the process itself. During recent years, dynamic simulation has become more and more important. In this context, "dynamic" means that the particular input data can be varied with time so that the time-dependent behavior of the plant can be modeled and the efficiency of the process control can be evaluated.

For both steady-state and dynamic simulation, the correct representation of thermophysical properties, phase equilibria, mass transfer, and chemical reactions mainly determines the quality of the simulation. As those are strongly influenced by thermodynamic relationships, a correct application of chemical thermodynamics is an indispensable requirement for a successful process simulation. This includes the handling of raw data to obtain pure component and mixture parameters with a sufficient quality as well as the choice of an appropriate model and the ability to assess the capabilities of the particular models. Poor thermophysical property data or data of insufficient availability, inappropriate models and model parameters and a missing perception about the sensitivities are among the most common mistakes in process simulation and can lead to a wrong equipment design. Creative application of thermodynamic knowledge can result in the development of feasible and elegant process alternatives, which makes thermodynamics an essential tool for business success in the industries mentioned above.

However, one must be aware that there are a lot of pitfalls beyond thermodynamics. Unknown components, foam formation, slow mass transfer, fouling layers, decomposition, or side reactions might lead to unrealistic results. The occurrence of solids in general is always a challenge, where only small scale-ups are possible ($\sim 1:10$) in contrast to fluid processes, where a scale-up of $1:1000$ is nothing unusual. For crystallization, the kinetics of crystal growth is often more important than the phase equilibrium itself. Nevertheless, even under these conditions simulation can yield a valuable contribution for understanding the principles of a process.

Nowadays, process simulations are the basis for the design of plants and the evaluation of investment and operation costs, as well as for follow-up tasks like process safety analysis, emission lists or performance evaluation. For process development and optimization purposes, they can effectively be used to compare various options and select the most promising one, which, however, should in general be verified experimentally. Therefore, a state-of-the-art process simulation can make a considerable contribution for both plant contractors and operating companies in reducing costs.

The structure of this book is as follows:

In Chapter 2, the particular phenomena of the behavior of fluids are introduced and explained for pure fluids. The main types of equations of state and their abilities are described. In Chapter 3, the various quantities of interest in process simulation, their correlation and prediction are discussed in detail.

In Chapter 4, the important terms for the description of the thermodynamics of mixtures are explained, including the g^E mixing rules, which have become the main progress of modern equations of state for mixtures.

Chapter 5 gives a comprehensive overview on the most important models and routes for phase equilibrium calculation, including sophisticated phenomena like the pressure dependence of liquid–liquid equilibria. The abilities and weaknesses of both g^E models and equations of state are thoroughly discussed. A special focus is dedicated to the predictive methods for the calculation of phase equilibria, applying the UNIFAC group contribution method and its derivatives, that is, the Mod. UNIFAC method and the PSRK and VTPR group contribution equations of state. Furthermore, in Chapter 6 the calculation of caloric properties and the way they are treated in process simulation programs are explained.

While Chapter 5 deals with models which are applicable to a wide variety of non-electrolyte systems, separate chapters have been composed where systems are described which require specialized models. These are electrolytes (Chapter 7), polymers (Chapter 10) and systems where chemical reactions and phase equilibrium calculations are closely linked, for example, aqueous formaldehyde solutions and substances showing vapor phase association (Chapter 13). Special phase equilibria like solid–liquid equilibria and osmosis are discussed in Chapters 8 and 9, respectively.

Certain techniques for the application of thermodynamics in separation technology are introduced in Chapter 11, for example, the concept of residue curve maps, a general procedure for the choice of suitable solvents for the separation of azeotropic systems, the verification of model parameters prior to process simulation and the identification of separation problems.

Chapter 12 gives an extensive coverage on the thermodynamics of chemical reactions, which emphasizes the importance of the real mixture behavior on the description of reaction equilibria and the enthalpies of reaction as well as solvent effects on chemical equilibrium conversion.

Chapter 14 concludes the book with the explanation of some standard operations frequently used in process simulation and design.

The book is supplemented by a large Appendix, listing parameters for the calculation of thermophysical properties and increasing the understanding by the accurate derivation of some relationships. To enable the reader to set up his own models, special care was taken to explain the regression procedures and philosophies in the Appendix, both for pure component properties and for the adjustment of binary parameters.

All the topics are illustrated with examples that are closely related to practical process simulation problems. At the end of each chapter, additional calculation examples are given to enable the reader to extend his comprehension. An introduction to a larger number of problems can be found in Chapter 15. These problems and their solutions can be downloaded from the site *www.ddbst.com*. The problems partially require the use of the software Mathcad® and the Dortmund Data Bank Software Package – Explorer Version. Both packages can be downloaded from the Internet. The DDBSP Explorer Version is free to use, whereas Mathcad® is only available for free during a tryout period of 30 days. Mathcad® files enable the users to perform the iterative calculations themselves and get a feeling for their complexity. Often, typical pitfalls in process simulation are covered in the examples, which should be helpful for the reader to avoid them in advance.

Special care was taken to deliver a complete logical representation of chemical thermodynamics. As far as possible, derivations for the particular relationships have been provided to ensure a proper understanding and to avoid mistakes in their application. Many of these derivations have been put into the Appendix in order not to interrupt the flow of the text.

2
PvT Behavior of Pure Components

2.1
General Description

If the relationship between the pressure P, the molar volume v, the absolute temperature T and, additionally, the ideal gas specific heat capacity c_P^{id} of a pure substance are known, all thermodynamic properties of this substance can be calculated. The typical *PvT* behavior is shown in Figure 2.1 in a three-dimensional diagram. All thermodynamically stable states are represented by the surface. Depending on the values of the state variables P, v, T the substance exists as a solid (S), liquid (L), or a vapor phase[1] (V) or as a combination of two or three phases. They can be characterized as follows.

A solid has a clearly defined shape. It is composed of molecules (or, respectively, ions in the case of electrolytes and metals), which stay at one place in a crystal lattice. These molecules can vibrate.

A liquid has a definite volume but an undefined shape. It will adopt the shape of the container it is filled in, but, depending on its amount, it will fill only a part of the container. The molecules in liquids can vibrate and change their place. There is more distance between them than in solids.[2]

A vapor has neither a definite volume nor a shape. It will fill the container in which it is located in completely. The distance between the molecules is by far larger than in the liquid. Therefore, its density is much lower.[3]

Liquid and vapor phases have in common that they deform and move when a shear stress is applied; these states are often referenced together as fluid phases. The different regions are separated by lines which mark the phase transitions. Homogeneous (one phase) and heterogeneous regions, where two or even three phases coexist (e.g., at the triple point), can be distinguished.

---

1) There is a slight difference between a vapor and gas. A vapor can condense when it is cooled down or compressed. A gas far away from its condensation conditions remains gaseous during the process regarded, for example, air at environmental conditions.

2) Water and ice are a well-known exception.
3) In fact, exceptions exist. At high pressures, for instance in the system CO_2–water, the vapor phase has a higher density than the liquid so that the liquid is actually the upper phase.

Chemical Thermodynamics: for Process Simulation, First Edition.
Jürgen Gmehling, Bärbel Kolbe, Michael Kleiber, and Jürgen Rarey.
© 2012 Wiley-VCH Verlag GmbH & Co. KGaA. Published 2012 by Wiley-VCH Verlag GmbH & Co. KGaA.

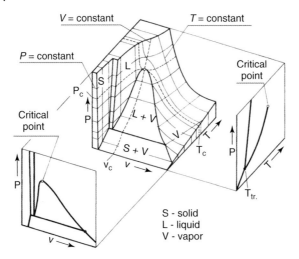

Figure 2.1 PvT-diagram.

The states and state transitions can be shown more clearly by looking at the projection onto one plane, for example, a Pv- or PT-diagram. At point A in the Pv-diagram (Figure 2.2), the regarded substance exists solely as vapor. When the vapor is compressed at constant temperature, the first liquid drop appears at point B on the dew-point curve.[4] The liquid fraction increases at constant pressure, when the total volume is reduced until the boiling point curve is reached, where the substance is completely liquefied. Between boiling point curve and dew-point curve the vapor phase and the liquid phase coexist in equilibrium. The amount of liquid and vapor, respectively, can be calculated by a mass balance or can be determined graphically by the so-called law of opposite lever arms. Further reduction of the volume can only be achieved by a strong pressure increase because of the very low compressibility of liquids. With ongoing compression, the solid line is reached, where the substance begins to crystallize. The pressure remains constant again, until the melting line is reached and the substance is transformed completely into the solid state. Boiling point curve and dew-point curve meet at the critical point. Above the critical point no coexistence of liquid and vapor is possible. The critical isotherm ($T = T_c$) shows a saddle point at the critical point, which means mathematically

$$\left(\frac{\partial P}{\partial v}\right)_{T_c} = 0 \quad \text{and} \quad \left(\frac{\partial^2 P}{\partial v^2}\right)_{T_c} = 0 \tag{2.1}$$

In order to understand the critical behavior and the role of the critical volume imagine three drums with the same volume at the same temperature and pressure.

---

4) Due to the elevated vapor pressure of small droplets (see Section 3.2.1 and Appendix C, B1), supersaturation and usually active surfaces are required to facilitate condensation.

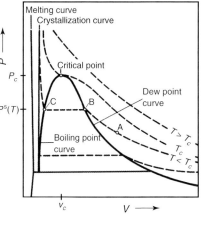

Figure 2.2 Pv-diagram.

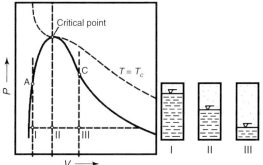

Figure 2.3 Isochoric changes of state in the Pv-diagram.

Each drum is filled with a pure substance, but the liquid levels are different (see Figure 2.3).

The first drum contains a large quantity of liquid, that is, the average volume of the two phases is smaller than the critical volume. The liquid level of the second drum is adjusted in a way that the average molar volume is identical with the critical molar volume. Consequently, the third drum contains only a small amount of liquid ($v > v_c$).

Imagine now that the first drum is heated: the pressure rises along a vertical line in the Pv-diagram, because the total volume remains constant. At the same time the liquid fraction increases because of the increasing molar volume of the liquid (lower density at higher temperatures) until point A on the boiling point curve is reached. At this point, the drum is filled completely with liquid. Further heating leads to a strong pressure increase, no phase transition takes place. In process simulators, the state of fluids at temperatures above T_c is defined as "vapor," without assigning a physical meaning to it as mentioned above. In physics, this state is often called "*fluid*" [1], as it makes no sense any more to distinguish between vapor and liquid.

When the second drum is heated, the liquid molar volume increases again, and at the same time the vapor molar volume decreases together with the pressure increase. Closely below the critical point, the phenomenon of the critical opalescence

$T = T_c - 0.5$ K $\qquad T = T_c = 305.41$ K $\qquad T = T_c + 0.5$ K

Figure 2.4 Critical opalescence of ethane.

as shown in Figure 2.4 occurs; As the enthalpy of vaporization is extremely low, parts of the substance permanently change their state, and cords are formed. When reaching the critical point, the phase boundary disappears, and vapor and liquid become identical. A thick nontransparent fog is formed, looking white or, from the top view, purple or even black. At this point, no distinction can be made whether the content of the drum is a liquid or a vapor. At the critical point and at temperatures above, the state is again "fluid," whereas a process simulator defines it as "vapor" or, respectively, "gaseous."

When the third drum is heated, all the liquid is vaporized completely because of the low liquid level when reaching the dew-point curve (point C). The fluid remains vaporous or – by definition above the critical isotherm – gaseous at further heating.

The critical point plays a major role in the history of gas liquefaction. Since the beginning of the nineteenth century, many scientists tried to liquefy gases by means of high pressures and low temperatures. Carbon dioxide and ammonia could be liquefied at ambient temperature just by pressure increase, but other gases like nitrogen, oxygen, carbon monoxide, methane, or hydrogen remained gaseous even at pressures up to 1200 bar and temperatures down to $-110\,°$C. Thus, it was believed that it is impossible to condense these gases, and they were called *"permanent gases."* In 1863, the Irish physicist Thomas Andrews (1813–1885) examined carbon dioxide, and he found out that there is a point where the difference between vapor and liquid vanishes, the so-called critical point. Above the temperature of its critical point, a gas cannot be liquefied at any pressure, whereas relatively low pressures are sufficient for temperatures below the critical temperature. This was the basis for the famous thesis [2] of Johannes Diderik van der Waals (1837–1923) in 1873, where he set up his equation of state. For the first time, a plausible explanation for the various phenomena like condensation, evaporation, phase equilibrium, and the behavior of gases in the supercritical state

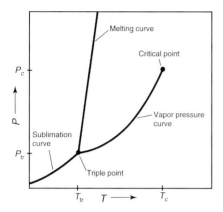

Figure 2.5 PT-diagram.

was given. Still, almost 15 years passed by until Louis Paul Cailletet (1832–1913) managed to liquefy oxygen and other permanent gases in 1877. Using the Linde process [3], it became possible to reach temperatures below 80 K. In 1898 James Dewar (1842–1923) liquefied hydrogen at 20.4 K, and Heike Kamerlingh Onnes (1853–1926) could liquefy helium at 4.2 K in 1908. Finally, all the "permanent gases" fitted into the pattern of substances with a critical point.

Another important projection of the PvT-diagram is the PT-graph (see Figure 2.5). In this projection, the dew-point line coincides with the boiling point in the vapor pressure curve. Similarly, solidus and liquidus curve coincide in the melting curve. The phase transition between the solid state and the gaseous state is described by the sublimation curve. Vapor pressure curve, melting curve, and sublimation curve meet at the triple point, where the three phases vapor, liquid, and solid coexist in equilibrium. The triple point of water is very well known and can be reproduced in a so-called triple point cell. It is used as a fix point[5] of the International Temperature Scale ITS-90 [4] ($T_{tr} = 273.16$ K or $\vartheta_{tr} = 0.01\,°C$, $P_{tr} = 611.657 \pm 0.01$ Pa). The vapor pressure curve ends at the critical point; no liquid exists above the critical temperature T_c.

It should be mentioned at this point that water has an unusual solid–liquid transfer, as the solid (ice) has a larger specific volume than the corresponding liquid. Thus, the melting curve has a negative slope; pressure increase favors the phase with the lower volume. This has numerous practical consequences. Vessels completely filled with water can burst when the temperature is lowered below 0 °C. Icebergs swim on the ocean. Ice skaters slide on a film of liquid water as the ice is melted by the pressure applied by the skates; however, this popular explanation is currently under discussion, as this effect alone is not sufficient [5]. The frictional heat and some special properties of ice surfaces play a major role as well [6].

---

5) In the ITS-90, the normal boiling point of water is no more a fix point. In fact, the normal boiling point of water is now determined to be 99.975 °C instead of 100 °C.

2.2
Caloric Properties

From mechanics, the principle of energy conservation is well known. It can be extended to the first law of thermodynamics, where the term "energy" is generalized and applied to phenomena which are related to "heat." Both "heat" and "energy" are difficult to define in a physically exact way.

In thermodynamics, one can distinguish between intensive and extensive properties. "Extensive" means that a quantity is proportional to the size of the system, which can be characterized by its mass (m) or its number of moles (n_T) in the system. An example is the volume of a system, whereas its temperature and pressure are intensive variables, as they do not depend on the size of the system.

As well, energy is also an extensive state variable of a system. The first law of thermodynamics states that the energy content of a system can only be changed by transport of energy across the system boundaries; therefore, energy cannot be generated or destroyed.

Kinetic energy or potential energy, which are known from mechanics, are contributions to the total energy of the system, the rest is called the *internal energy* U *of the system*. As for any extensive property, a related specific property

$$u = \frac{U}{m} \quad \text{or, respectively,} \quad u = \frac{U}{n_T}$$

can be defined.[6]

At this stage, there is no need to further define the character of the internal energy. In fact, it is essentially determined by different kinds of movement of the molecules (translation, rotation, vibration) and the forces between them. More details are given in Sections 3.2.4 and 3.2.5.

If a system is in equilibrium with a pressure field at its boundaries, it is useful to introduce the enthalpy as another extensive state variable. The easiest case for its application is the expansion of a gas against a piston when the system is heated up ($Q_{12} > 0$). Figure 2.6 shows two arrangements; one with a fixed (a) and one with a flexible piston (b) as system boundary.

Let 1 be the start and 2 the end of the procedure, the energy balance of case (a) is

$$Q_{12} = U_2 - U_1 \tag{2.2}$$

as the kinetic and the potential energy are not affected. For case (b), it is necessary to consider the change in the potential energy of the constant pressure field outside, when the piston is moved upward due to the expansion of the gas:

$$\begin{aligned} Q_{12} &= U_2 - U_1 + P(V_2 - V_1) \\ &= (U_2 + P_2 V_2) - (U_1 + P_1 V_1) \\ &= H_2 - H_1 \end{aligned} \tag{2.3}$$

---

[6] Usually, the symbols for extensive properties are capital letters (e.g., U, V). The corresponding specific or molar quantities are written with small letters.

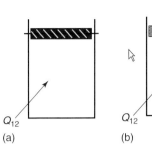
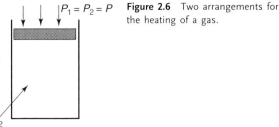

Figure 2.6 Two arrangements for the heating of a gas.

The combination $(U + PV)$ frequently occurs in process calculations and can be considered as the most widely used quantity in energy balances, especially for the energy balance of mass flows, which is the standard case in process calculations and where, different to this example, $P_1 \neq P_2$. It is called *enthalpy (H)*. The enthalpy and the related molar or specific enthalpy are defined by

$$H = U + PV \tag{2.4}$$

and

$$h = u + Pv \tag{2.5}$$

respectively. The introduction of the enthalpy is described in more detail in [3].

The entropy S can be defined by its differential in closed systems (i.e., without mass transfer):

$$dS = \frac{dU + PdV}{T} \tag{2.6}$$

The entropy is probably the quantity in thermodynamics which is most difficult to understand. It is strongly related to the second law of thermodynamics, which states that in a closed adiabatic system, where neither heat nor mass can pass the system boundaries, the entropy does not decrease. Detailed explanations can be found in [7] and [3]. An interesting new explanation has been elaborated by Thess [5], based on the work of Lieb and Yngvason [8].

Furthermore, two other combined caloric quantities can be defined, that is, the Helmholtz energy

$$A \equiv U - TS \tag{2.7}$$

and the Gibbs energy

$$G \equiv H - TS \tag{2.8}$$

Their extraordinary importance will become obvious in the following chapters.

It should be clearly pointed out that there is no absolute value for the specific caloric quantities u, h, s, a, and g. Therefore, single values for caloric properties without the definition of a reference point are meaningless; only differences between caloric properties can be interpreted. Any table for caloric properties should have defined a reference point where the particular caloric property is set to zero. As long as only pure components are involved, it is sufficient if the reference point is chosen in an arbitrary way, for example, $h = 0$ for the boiling liquid at $\vartheta = 0\,°C$.

For process simulators, a more sophisticated way is chosen which makes sure that the caloric properties are consistent even if chemical reactions occur. This is the case if the standard enthalpy of formation Δh_f^0 is taken as the reference point. This is explained in detail in the Section 3.1.5 and Chapters 6 and 12. The standard enthalpy of formation refers to the standard conditions $T_0 = 298.15\,K$ and $P_0 = 101325\,Pa$ in the state of ideal gases (see Section 2.3).[7] Therefore, the reference point for the specific enthalpy for a pure component is

$$h_{\text{ref}}(T_0 = 298.15\,K, P_0 = 101\,325\,Pa, \text{ideal gas}) = \Delta h_f^0 \tag{2.9}$$

The standard enthalpy of formation is listed in many standard reference books, for example, in [9, 10], or [11]. For a number of substances, it is also given in Appendix A.

There is no need to set up a separate convention for the internal energy u via

$$u = h - Pv \tag{2.10}$$

The third law of thermodynamics states that the entropy of a pure crystalline substance is zero at $T = 0\,K$. This defines the reference point for the specific entropy in process simulators and makes it consistent with respect to chemical reactions. Entropies using this reference point are called *absolute entropies*, which are indicated by the subscript "abs." For convenience, in the reference tables the standard entropies, that is, the absolute entropies at standard conditions, are listed:

$$s_{\text{standard}} = s_{\text{abs}}(T_0, P_0, \text{ideal gas}) \tag{2.11}$$

The combination of Δh_f^0 and s_{abs} makes it possible to set up a consistent reference point for the Gibbs energy,[8] the Gibbs energy of formation:

$$g(T_0, P_0, \text{ideal gas}) = \Delta g_f^0 \tag{2.12}$$

The Gibbs energies of formation of substances are usually listed together with the standard enthalpies of formation. As

$$a = g - Pv \tag{2.13}$$

no special convention for the reference point of the Helmholtz energy a is necessary.

From the definitions (Eq. (2.4), (2.7), and (2.8)), the total differentials can be derived. Together with Eq. (2.6), the so-called fundamental equations are obtained:

$$dU = TdS - PdV \tag{2.14}$$

7) The standard enthalpy of formation is the enthalpy of reaction, when the particular compound is formed from the chemical elements in the stable form at T_0, P_0.

8) The relationship between Δg_f^0 and s_{abs} is illustrated in Section 6.3.

2.2 Caloric Properties

$$dH = TdS + VdP \tag{2.15}$$

$$dA = -PdV - SdT \tag{2.16}$$

$$dG = VdP - SdT \tag{2.17}$$

From these equations, one can see that the following relations are valid:

$$T = \left(\frac{\partial U}{\partial S}\right)_V = \left(\frac{\partial H}{\partial S}\right)_P \tag{2.18}$$

$$P = -\left(\frac{\partial U}{\partial V}\right)_S = -\left(\frac{\partial A}{\partial V}\right)_T \tag{2.19}$$

$$V = \left(\frac{\partial H}{\partial P}\right)_S = \left(\frac{\partial G}{\partial P}\right)_T \tag{2.20}$$

$$S = -\left(\frac{\partial G}{\partial T}\right)_P = -\left(\frac{\partial A}{\partial T}\right)_V \tag{2.21}$$

Also, the second derivatives of the fundamental equations give useful relationships. According to the theorem of Schwarz, the mixed partial derivatives of continuous functions are independent of the order of differentiation. For example, Eq. (2.14) yields

$$\left(\frac{\partial U}{\partial S}\right)_V = T \qquad \left(\frac{\partial U}{\partial V}\right)_S = -P$$

$$\left(\frac{\partial^2 U}{\partial S \partial V}\right) = \left(\frac{\partial T}{\partial V}\right)_S \qquad \left(\frac{\partial^2 U}{\partial V \partial S}\right) = -\left(\frac{\partial P}{\partial S}\right)_V$$

and therefore

$$\left(\frac{\partial T}{\partial V}\right)_S = -\left(\frac{\partial P}{\partial S}\right)_V \tag{2.22}$$

Analogously, one can obtain from Eqs. (2.15) to (2.17)

$$\left(\frac{\partial T}{\partial P}\right)_S = \left(\frac{\partial V}{\partial S}\right)_P \tag{2.23}$$

$$\left(\frac{\partial S}{\partial V}\right)_T = \left(\frac{\partial P}{\partial T}\right)_V \tag{2.24}$$

$$\left(\frac{\partial V}{\partial T}\right)_P = -\left(\frac{\partial S}{\partial P}\right)_T \tag{2.25}$$

Equations (2.22–2.25) are the so-called Maxwell relations.

The total differential of the enthalpy of a pure substance as a function of temperature and pressure is

$$dh = \left(\frac{\partial h}{\partial T}\right)_P dT + \left(\frac{\partial h}{\partial P}\right)_T dP \tag{2.26}$$

The temperature dependence of the enthalpy is expressed by the specific heat capacity at constant pressure c_P. The specific heat capacity at constant pressure is defined as

$$c_P \equiv \left(\frac{\partial h}{\partial T}\right)_P \tag{2.27}$$

In a similar way, the temperature dependence of the internal energy is expressed by the specific heat capacity at constant volume c_v

$$c_v \equiv \left(\frac{\partial u}{\partial T}\right)_v \tag{2.28}$$

The expression $(\partial h/\partial P)_T$ can be written as

$$\left(\frac{\partial h}{\partial P}\right)_T = -T\left(\frac{\partial v}{\partial T}\right)_P + v \tag{2.29}$$

using the Gibbs fundamental equation and one of the Maxwell relations.[9]

2.3
Ideal Gases

The PvT behavior of a pure substance can be described by so-called equations of state (EOS). In general, an equation of state is a relationship between P, v, and T. They can be formulated in different ways, for example, volume-explicit ($v = f(T, P)$) or, most commonly, pressure-explicit ($P = f(T, v)$). An equation of state combined with the general thermodynamic correlations offers the possibility to calculate all thermodynamic properties of the substance. The simplest equation of state for describing the PvT behavior of gases is the ideal gas law:

$$Pv = RT \quad \text{or} \quad PV = n_T RT \tag{2.30}$$

with R being the universal gas constant ($R = 8.31433$ J/(mol K))[10] and n_T being the total number of moles.

The ideal gas is a fictitious model substance. The molecules are regarded as having no proprietary volume and exerting no intermolecular forces. In reality, there is no substance which fulfills these conditions, but the model of the ideal gas plays an important role as a starting point for the description of the PvT behavior of gases. Real substances behave very similar to ideal gases when the pressure approaches values of zero ($v \to \infty$), because the molecular volume and the molecular interactions can be neglected at this state.

---

9) A derivation of Eq. (2.29) is given in Appendix C, A5.
10) The currently acknowledged value for the gas constant is $R = 8.31447$ J/(mol K); however, most of the calculations are based on the old value, and the difference is usually negligible.

The thermodynamic correlations as a function of temperature, pressure, and volume are very simple for an ideal gas, as shown below.

The total differential of the enthalpy of a pure substance as a function of temperature and pressure is

$$dh = \left(\frac{\partial h}{\partial T}\right)_P dT + \left(\frac{\partial h}{\partial P}\right)_T dP \qquad (2.31)$$

The temperature dependence of the enthalpy is expressed by the specific heat capacity at constant pressure c_P. The specific heat capacity at constant pressure is defined as

$$c_P \equiv \left(\frac{\partial h}{\partial T}\right)_P \qquad (2.32)$$

In a similar way, the temperature dependence of the internal energy is expressed by the specific heat capacity at constant volume c_v

$$c_v \equiv \left(\frac{\partial u}{\partial T}\right)_v \qquad (2.33)$$

The expression $(\partial h/\partial P)_T$ can be written as

$$\left(\frac{\partial h}{\partial P}\right)_T = -T\left(\frac{\partial v}{\partial T}\right)_P + v \qquad (2.34)$$

using the Gibbs fundamental equation and one of the Maxwell relations.

This expression can be calculated easily using a volume-explicit equation of state. The ideal gas equation can be written in a volume-explicit way, that is,

$$v = \frac{RT}{P} \qquad (2.35)$$

Applying the equation of state for the ideal gas results in

$$T\left(\frac{\partial v}{\partial T}\right)_P = \frac{RT}{P} = v \qquad (2.36)$$

Therefore, the pressure dependence of the enthalpy of an ideal gas according to Eq. (2.29) results in

$$\left(\frac{\partial h^{id}}{\partial P}\right)_T = -v + v = 0 \qquad (2.37)$$

This relation shows clearly that the enthalpy of an ideal gas is not a function of the pressure. Generally, the total differential of the enthalpy of a gas as a function of temperature and pressure can be written as (see Eq. (2.29))

$$dh = c_P dT + \left[v - T\left(\frac{\partial v}{\partial T}\right)_P\right] dP \qquad (2.38)$$

Considering Eq. (2.37), this relation simplifies for the ideal gas to

$$dh^{id} = c_P^{id} dT \qquad (2.39)$$

2 PvT Behavior of Pure Components

Table 2.1 The specific state functions u, h, s, g, and a as a function of T, P, and v.

General	Ideal gas
$du = c_v dT + \left[T\left(\frac{\partial P}{\partial T}\right)_v - P\right]dv$	$du^{id} = c_v^{id} dT$
$dh = c_p dT + \left[v - T\left(\frac{\partial v}{\partial T}\right)_P\right]dP$	$dh^{id} = c_p^{id} dT$
$ds = \frac{c_p}{T} dT - \left(\frac{\partial v}{\partial T}\right)_P dP$	$ds^{id} = \frac{c_p^{id}}{T} dT - \frac{R}{P} dP$
$dg = -sdT + vdP$	$dg^{id} = -sdT + \frac{RT}{P} dP$
$da = -sdT - Pdv$	$da^{id} = -sdT - RT\frac{dv}{v}$

Analogously, the equations for the specific internal energy (u), the entropy (s), the Gibbs energy (g), and the Helmholtz energy (a) can be derived. All these equations are summarized in Table 2.1. Furthermore, for the specific heat capacities of an ideal gas it can be shown that

$$\left(\frac{\partial c_p^{id}}{\partial P}\right)_T = 0 \quad \text{and} \quad \left(\frac{\partial c_v^{id}}{\partial v}\right)_T = 0 \qquad (2.40)$$

$$c_p^{id} - c_v^{id} = R \qquad (2.41)$$

These equations show that for an ideal gas the internal energy u, the enthalpy h, and the specific heat capacities at constant pressure c_p^{id} and at constant volume c_v^{id} are only functions of the temperature and independent of pressure or volume. On the other hand, the specific entropy (s), the specific Gibbs energy (g), and the specific Helmholtz energy (a) are functions of temperature and pressure or volume even for an ideal gas.

2.4
Real Fluids

2.4.1
Auxiliary Functions

In order to describe the properties of real fluids, it is common to define certain auxiliary functions for the representation of the real behavior. One important auxiliary function is the compressibility factor z:

$$z \equiv \frac{Pv}{RT} \qquad (2.42)$$

The compressibility factor z can be applied for both gases and liquids. From the definition above, its value for the ideal gas is equal to

$$z^{id} = 1 \qquad (2.43)$$

The compressibility factor can also be interpreted as the ratio of the molar volume of a real fluid to the molar volume of the corresponding ideal gas at the same temperature and pressure:

$$z = \left(\frac{v}{v^{id}}\right)_{T,P} \tag{2.44}$$

When the pressure approaches zero, the compressibility factor of all substances approaches unity:

$$\lim_{P \to 0} z = 1 \tag{2.45}$$

Other common auxiliary functions are

- the thermal expansion coefficient α:

$$\alpha \equiv \frac{1}{v}\left(\frac{\partial v}{\partial T}\right)_P \tag{2.46}$$

- the isothermal compressibility factor κ:

$$\kappa \equiv -\frac{1}{v}\left(\frac{\partial v}{\partial P}\right)_T \tag{2.47}$$

which should not be mistaken with the compressibility factor defined in Eq. (2.42).

The auxiliary functions α and κ can also be employed to express the differentials in Table 2.1.

$$dh = c_P dT + v(1 - \alpha T)dP \tag{2.48}$$

With the help of the Maxwell relations, an equation similar to Eq. (2.41) can be derived:

$$c_P - c_V = T\frac{v\alpha^2}{\kappa} \tag{2.49}$$

The thermal expansion coefficient α and the compressibility factor κ are well suited for the description of liquids, as the volume changes of liquids with changes of the temperature or the pressure are relatively small. As a consequence, α and κ can be assumed to be constant over a limited temperature and pressure range.

2.4.2
Residual Functions

The so-called residual functions are a further option to describe the properties of real fluids. Residual functions represent the difference between the thermodynamic property of a real fluid and the corresponding ideal gas at the same temperature and pressure:

$$(m - m^{id})_{T,P}$$

The residual functions are a measure for the intermolecular forces. In the following section, the residual functions of volume, entropy, enthalpy, and Gibbs energy are examined in detail.

Combining Eqs. (2.42) and (2.43), the residual function of the volume can easily be expressed by the compressibility factor:

$$(v - v^{id}) = \frac{RT}{P}(z - 1) \qquad (2.50)$$

An expression for the entropy residual function can be derived as follows, regarding that all gases behave as ideal gases for $P \to 0$:

$$(s - s^{id})_{T,P} = (s - s^{id})_{T,P \to 0} + \int_0^P \left[\left(\frac{\partial s}{\partial P}\right)_T - \left(\frac{\partial s^{id}}{\partial P}\right)_T \right] dP \qquad (2.51)$$

The residual function at zero pressure is zero according to the ideal gas law. With the Maxwell relation[11]

$$\left(\frac{\partial s}{\partial P}\right)_T = -\left(\frac{\partial v}{\partial T}\right)_P \qquad (2.52)$$

and the ideal gas law (see Table 2.1)

$$\left(\frac{\partial s^{id}}{\partial P}\right)_T = -\frac{R}{P} \qquad (2.53)$$

we get

$$(s - s^{id})_{T,P} = \int_0^P \left[-\left(\frac{\partial v}{\partial T}\right)_P + \frac{R}{P} \right] dP \qquad (2.54)$$

This integral can be calculated directly by a volume-explicit equation of state, which is suitable to describe the PvT behavior of a pure substance. In a very similar way, the relationship for $(h - h^{id})$ can be derived:

$$(h - h^{id})_{T,P} = \int_0^P \left[v - T\left(\frac{\partial v}{\partial T}\right)_P \right] dP \qquad (2.55)$$

The residual function for the Gibbs energy is calculated from the above functions for the enthalpy and entropy

$$(g - g^{id})_{T,P} = (h - h^{id})_{T,P} - T(s - s^{id})_{T,P} \qquad (2.56)$$

From this equation, it follows that

$$(g - g^{id})_{T,P} = \int_0^P \left[v - T\left(\frac{\partial v}{\partial T}\right)_P + T\left(\frac{\partial v}{\partial T}\right)_P - \frac{RT}{P} \right] dP \qquad (2.57)$$

$$(g - g^{id})_{T,P} = \int_0^P \left[v - \frac{RT}{P} \right] dP \qquad (2.58)$$

Equation (2.58) is illustrated with a Pv diagram in Figure 2.7.

11) Derivation in Appendix C, A6.

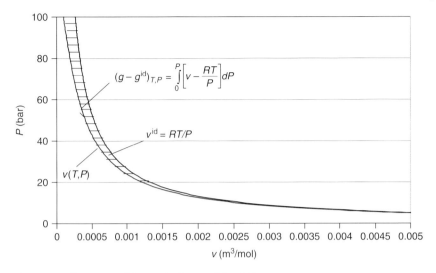

Figure 2.7 Illustration of the residual part of the Gibbs energy using the isotherm of CO_2 at $T = 320\,K$ in the Pv diagram.

The residual functions as a function of volume can be derived in a similar way as needed for a pressure-explicit equation of state. Table 2.2 summarizes the different residual functions for pressure- and volume-explicit EOS.

Besides the expression "residual function," the term "*departure function*" is frequently used. "*Departure functions*" are defined as the difference between the property of a real fluid at the temperature T and the pressure P and the property of the corresponding ideal gas at the temperature T and a reference pressure P^0:

$$[m(T, P) - m^{id}(T, P^0)].$$

In most cases, the reference pressure is set to $P^0 = 1$ atm.

2.4.3
Fugacity and Fugacity Coefficient

A special form for the description of the Gibbs energy residual function proved to be convenient. Starting point is again the equation for the ideal gas (Table 2.1):

$$dg^{id} = -s dT + \frac{RT}{P} dP \qquad (2.59)$$

Integration at constant temperature between P^0 and P gives

$$g^{id}(T, P) - g^{id}(T, P^0) = RT \ln \frac{P}{P^0} \quad (T = \text{const.}) \qquad (2.60)$$

Table 2.2 Residual properties for pressure- and volume-explicit equations of state (derivations for the integrals can be found in Appendix C).

	Volume-explicit EOS	Pressure-explicit EOS
$(u - u^{id})$	$\int_0^P \left[v - T\left(\frac{\partial v}{\partial T}\right)_P \right] dP - Pv + RT$	$\int_\infty^v \left[T\left(\frac{\partial P}{\partial T}\right)_v - P \right] dv$
$(h - h^{id})$	$\int_0^P \left[v - T\left(\frac{\partial v}{\partial T}\right)_P \right] dP$	$\int_\infty^v \left[T\left(\frac{\partial P}{\partial T}\right)_v - P \right] dv + Pv - RT$
$(s - s^{id})$	$\int_0^P \left[-\left(\frac{\partial v}{\partial T}\right)_P + \frac{R}{P} \right] dP$	$\int_\infty^v \left[\left(\frac{\partial P}{\partial T}\right)_v - \frac{R}{v} \right] dv + R \ln z$
$(a - a^{id})$	$\int_0^P \left(v - \frac{RT}{P} \right) dP - Pv + RT$	$-RT \ln z + \int_\infty^v \left(\frac{RT}{v} - P \right) dv$
$(g - g^{id})$ or $RT \ln \varphi$	$\int_0^P \left(v - \frac{RT}{P} \right) dP$	$RT(z - 1 - \ln z) + \int_\infty^v \left(\frac{RT}{v} - P \right) dv$
$(c_v - c_v^{id})$	$T \frac{\left(\frac{\partial v}{\partial T}\right)_P^2}{\left(\frac{\partial v}{\partial P}\right)_T} - T \int_0^P \left(\frac{\partial^2 v}{\partial T^2}\right)_P dP + R$	$T \int_\infty^v \left(\frac{\partial^2 P}{\partial T^2}\right)_v dv$
$(c_P - c_P^{id})$	$-T \int_0^P \left(\frac{\partial^2 v}{\partial T^2}\right)_P dP$	$T \int_\infty^v \left(\frac{\partial^2 P}{\partial T^2}\right)_v dv - T \frac{\left(\frac{\partial P}{\partial T}\right)_v^2}{\left(\frac{\partial P}{\partial v}\right)_T} - R$

Lewis introduced the auxiliary property[12] "fugacity" (lat: fugare = to flee) in order to be able to apply the above-mentioned simple equation in a similar form to real fluids:

$$g(T, P) - g(T, P^0) = RT \ln \frac{f}{f^0} \quad (T = \text{const.}) \quad (2.61)$$

The index ^{0} refers to the ideal gas in an arbitrarily chosen reference state; different reference states in the solid or liquid phase are possible.

Equation (2.61) is only the first part of the definition for the fugacity. Additionally, the two expressions should become identical when the pressure approaches zero, as in this case real fluids behave like ideal gases:

$$\lim_{P \to 0} \frac{f}{P} = 1 \quad (2.62)$$

From this point of view, the fugacity can be regarded as a "corrected pressure." In the limiting case of ideal gas behavior, pressure and fugacity are identical. In the

12) Auxiliary properties are calculated properties in contrast to physical properties, which are directly accessible via measurements.

case of a real gas the fugacity equals the pressure as a first approximation; for a more accurate description the fugacity has to be corrected by taking into account the intermolecular forces. The relationship between fugacity and pressure becomes even more evident by the introduction of the fugacity coefficient:

$$\varphi \equiv \frac{f}{P} \tag{2.63}$$

This definition implies that the fugacity coefficient for an ideal gas[13] is always 1.

Combining the above equations leads to the residual function of the Gibbs energy.

$$\left(g - g^{id}\right)_{T,P} = \left(g - g^{id}\right)_{T,P^0} + RT \ln \frac{f}{f^0} \frac{P^0}{P} \tag{2.64}$$

In accordance with the definitions above, the limiting case $[P^0 \to 0]$ leads to $[f^0/P^0 \to 1]$ and $[(g - g^{id})^0_{T,P} \to 0]$. Taking these relations into account, the residual Gibbs energy results in

$$(g - g^{id})_{T,P} = RT \ln \frac{f}{P} = RT \ln \varphi \tag{2.65}$$

The Gibbs energy, the fugacity and the fugacity coefficient, respectively, are very important functions for the calculation of phase equilibria as will be shown later. The temperature dependence of the fugacity and fugacity coefficient can be derived from the van't Hoff equation[14]

$$\left(\frac{\partial g/RT}{\partial T}\right)_P = -\frac{h}{RT^2} \tag{2.66}$$

Combining Eqs. (2.65) and (2.66) leads to a relationship for the temperature dependence of the fugacity coefficient:

$$\left(\frac{\partial \ln \varphi}{\partial T}\right)_P = -\frac{(h - h^{id})}{RT^2} \tag{2.67}$$

Similarly, equations for the pressure dependence of the fugacity coefficient can be derived from the pressure dependence of the Gibbs energy

$$\left(\frac{\partial \ln \varphi}{\partial P}\right)_T = \frac{1}{RT}\left(v - \frac{RT}{P}\right) \tag{2.68}$$

As

$$d \ln \varphi = d \ln \frac{f}{P} = \frac{P}{f} \frac{Pdf - fdP}{P^2} = \frac{df}{f} - \frac{dP}{P} \tag{2.69}$$

Eqs. (2.67) and (2.68) can also be written as

$$\left(\frac{\partial f/f}{\partial T}\right)_P = -\frac{(h - h^{id})}{RT^2} \tag{2.70}$$

---

13) For mixtures, the pressure has to be replaced by the partial pressure of the regarded components (see Section 4.7).

14) Derivation in Appendix C, A7.

and

$$\left(\frac{\partial f/f}{\partial P}\right)_T = \frac{v}{RT} \tag{2.71}$$

Example 2.1

Most of the EOS are available in the pressure-explicit form, which means

$$P = P(T, v) \quad \text{or} \quad z = \frac{Pv}{RT} = z(T, v)$$

Derive an expression for the fugacity coefficient from a pressure-explicit equation of state.

Solution

The derivation is possible by substituting the pressure P by means of the compressibility factor z:

$$P = \frac{zRT}{v}$$

$$\left(\frac{\partial P}{\partial v}\right)_T = \frac{RT}{v^2}\left[v\left(\frac{\partial z}{\partial v}\right)_T - z\right]$$

The above equation together with Eq. (2.68) results in

$$\left(\frac{\partial \ln \varphi}{\partial v}\right)_T = \left(\frac{\partial \ln \varphi}{\partial P}\right)_T \cdot \left(\frac{\partial P}{\partial v}\right)_T$$

$$= \frac{1}{RT}\left(v - \frac{RT}{P}\right)\frac{RT}{v^2}\left[v\left(\frac{\partial z}{\partial v}\right)_T - z\right]$$

Rearrangement gives

$$\left(\frac{\partial \ln \varphi}{\partial v}\right)_T = \left(1 - \frac{1}{z}\right)\left(\frac{\partial z}{\partial v}\right)_T + \frac{1}{v}(1 - z).$$

With

$$\frac{1}{z}\left(\frac{\partial z}{\partial v}\right)_T = \left(\frac{\partial \ln z}{\partial v}\right)_T \quad \text{and} \quad \frac{z}{v} = \frac{P}{RT}$$

the equation gets the form

$$\left(\frac{\partial \ln \varphi}{\partial v}\right)_T = \left(\frac{\partial z}{\partial v}\right)_T - \left(\frac{\partial \ln z}{\partial v}\right)_T + \left(\frac{1}{v} - \frac{P}{RT}\right)$$

Integration gives

$$\ln \varphi = \int_{\infty}^{v} \left[\left(\frac{\partial z}{\partial v}\right)_T - \left(\frac{\partial \ln z}{\partial v}\right)_T \right] dv + \int_{\infty}^{v} \left(\frac{1}{v} - \frac{P}{RT}\right) dv$$

$$= (z - \ln z) \Big|_{\infty}^{v} + \frac{1}{RT} \int_{\infty}^{v} \left(\frac{RT}{v} - P\right) dv$$

$$= z - 1 - \ln z + \frac{1}{RT} \int_{\infty}^{v} \left(\frac{RT}{v} - P\right) dv \qquad (2.72)$$

This equation can be used to calculate the fugacity coefficient with the help of a pressure-explicit equation of state. As $RT\ln \varphi$ is by definition equivalent to the residual Gibbs energy $(g - g^{id})_{T,P}$, the above equation is equivalent to the corresponding equation in Table 2.2.

2.4.4
Phase Equilibria

On the border lines in Figure 2.5, two phases can coexist. These two-phase regions have an outstanding importance in technical applications, as they are the prerequisite of most of the thermal separation processes. The following considerations refer to the vapor–liquid equilibrium but also apply to the other phase equilibrium lines of pure substances.

In the two-phase region, vapor and liquid coexist, and vapor and liquid have the same temperature (thermal equilibrium) and pressure (mechanical equilibrium). When they are in equilibrium with each other, vapor and liquid are called *saturated vapor* and *saturated liquid*, respectively. If a saturated liquid is further heated at constant pressure, the temperature does not rise any more but stays constant. Instead, vapor is generated until all liquid is vaporized. Similarly, if saturated vapor is cooled down, the temperature stays constant, and the vapor condenses and forms a saturated liquid. Figure 2.8 illustrates the well-known behavior of water at $P = 1.013$ bar, when it is heated up from $\vartheta_1 = 50\,°C$ to $\vartheta_2 = 150\,°C$.

For the two-phase region, where vapor and liquid coexist, it can be shown that the phase equilibrium condition is given by

$$g^L = g^V \qquad (2.73)$$

where the superscripts L and V denote boiling liquid and saturated vapor, respectively.

In case of a pure substance, the chemical potential μ is equivalent to the Gibbs energy g. The derivations of both the phase equilibrium condition (Eq. (2.73)) and the equivalence of μ and g for a pure substance are given in Appendix C, B2 and B3. According to Eqs. (2.65) and (2.73) also yields

$$f^L = f^V \qquad (2.74)$$

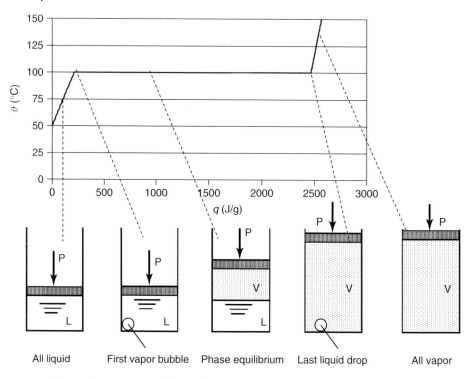

Figure 2.8 Temperature change of water between $\vartheta_1 = 50\,°C$ and $\vartheta_2 = 150\,°C$ at $P = 1.013$ bar with respect to the heat added.

and, respectively, for a pure substance

$$\varphi^L = \varphi^V \tag{2.75}$$

The state of a pure substance in the two-phase region is not sufficiently characterized by temperature and pressure, as both do not change during evaporation. The system can consist of arbitrary amounts of liquid and vapor phase. Therefore, for a complete determination of the state the vapor fraction[15] has to be given additionally. It is defined by

$$q = \frac{n^V}{n^L + n^V} \tag{2.76}$$

For a boiling liquid is $q = 0$, as no vapor exists, for a saturated vapor is $q = 1$, since the substance is completely vaporized.

---

15) In process simulators, the molar vapor fraction is often abbreviated with VFRAC.

It is easy to show [3] that the quantities v, h, and s can be calculated as the sum of saturated liquid and saturated vapor:

$$v = q\,v^V + (1-q)v^L$$
$$h = q\,h^V + (1-q)h^L$$
$$s = q\,s^V + (1-q)s^L \tag{2.77}$$

Example 2.2

To improve the heat transfer conditions and to minimize the demands on the construction materials in a steam header, boiler feed water with $P_{BFW} = 30$ bar, $\vartheta_{BFW} = 150\,°C$ has to be injected into superheated steam ($\dot{m}_{SS} = 20\,t/h$) with $\vartheta_{SS} = 400\,°C$, $P = 20$ bar to reduce the superheating to 5 K. Determine the appropriate amount of boiler feed water.[16] Use a high-precision equation of state (Appendix B) for the determination of the enthalpies, which should match the values obtained with the steam table [12] sufficiently well.

Solution

The saturation temperature of water at $P = 20$ bar is $\vartheta_S = 212.38\,°C$. The final temperature of the superheated steam should therefore be $\vartheta_{5K} = 217.38\,°C$, which corresponds to a specific enthalpy of $h_{5K} = 2813.8\,J/g$.

The energy balance is given by

$$\dot{m}_{SS}\,h_{SS} + \dot{m}_{BFW}\,h_{BFW} = \dot{m}_{5K}\,h_{5K} = (\dot{m}_{SS} + \dot{m}_{BFW})h_{5K}$$

With

$$h_{SS}(400\,°C, 20\,\text{bar}) = 3248.34\,J/g$$

$$h_{BFW}(150\,°C, 30\,\text{bar}) = 633.74\,J/g$$

we get

$$\dot{m}_{BFW} = \dot{m}_{SS}\,\frac{h_{SS} - h_{5K}}{h_{5K} - h_{BFW}} = 20\,t/h\,\frac{3248.34 - 2813.8}{2813.8 - 633.74} = 3.99\,t/h$$

Thus, approx. 20% of the superheated steam can be additionally generated by the desuperheating with boiler feed water.

Example 2.3

The condensate of the high-pressure steam of Example 2.2 shall be used to generate low-pressure steam at $P = 3$ bar by flashing it into a vessel. Calculate the amount

---

16) A small superheating is necessary to avoid condensation already in the line.

of steam generated, assuming that the condensate is saturated and no pressure drop occurs.

Solution

Using the steam table [12], the enthalpy of the condensate at $P = 20$ bar can be evaluated to be h^L (20 bar, sat.) $= 908.5$ J/g.

At $P = 3$ bar, the enthalpies of saturated vapor and saturated liquid can be determined to be

$$h^L \text{ (3 bar, sat.)} = 561.43 \text{ J/g}$$

$$h^v \text{ (3 bar, sat.)} = 2724.88 \text{ J/g}$$

According to Eq. (2.77), the vapor fraction can be calculated to be

$$q = \frac{h^L \text{ (20 bar,sat.)} - h^L \text{ (3 bar,sat.)}}{h^v \text{ (3 bar,sat.)} - h^L \text{ (3 bar,sat.)}} = \frac{908.5 - 561.43}{2724.88 - 561.43} = 0.1604$$

corresponding to a low-pressure steam generation of

$$\dot{m}_{LPS} = 20 \, t/h \cdot 0.16 = 3.21 \, t/h$$

According to the vapor pressure curve $P^s = P^s(T)$, there is a unique pressure for each temperature where a vapor–liquid equilibrium for a pure substance is possible, as long as both temperature and pressure are below their critical values. Similarly, there is a functional relationship between the enthalpy of vaporization and the boiling temperature

$$\Delta h_v = \Delta h_v(T) \tag{2.78}$$

The Clausius–Clapeyron equation gives a well-founded relationship between the enthalpy of vaporization and the vapor pressure curve.

From the phase equilibrium condition

$$g^L = g^v \tag{2.79}$$

and the definition of the Gibbs energy g

$$g = h - Ts \tag{2.80}$$

we get

$$dg^L = dg^v \tag{2.81}$$

and

$$h^v - h^L = T(s^v - s^L) \tag{2.82}$$

As

$$dg = dh - Tds - sdT = vdP - sdT \tag{2.83}$$

we obtain for the case of phase equilibrium between vapor and liquid

$$v^L dP^s - s^L dT = v^V dP^s - s^V dT \tag{2.84}$$

or

$$\frac{dP^s}{dT} = \frac{s^V - s^L}{v^V - v^L} = \frac{1}{T}\frac{h^V - h^L}{v^V - v^L} \tag{2.85}$$

Solving for Δh_v, one obtains the Clausius–Clapeyron equation

$$\Delta h_v = T\left(v^V - v^L\right)\frac{dP^s}{dT} \tag{2.86}$$

The Clausius–Clapeyron equation is an accurate relationship between vapor pressure and enthalpy of vaporization and widely used for the estimation of enthalpies of vaporization from vapor pressure data[17] (see Section 3.2.3).

2.5
Equations of State

As already mentioned, the *PvT* behavior of a substance can be described mathematically by equations of state (EOS). The most simple equation of state is the ideal gas law. The ideal gas law is a simplified model without taking into account the individual properties of a substance such as the intrinsic volume of the molecules and the attractive and repulsive forces between the molecules.

The equation of state for the ideal gas is

$$Pv = RT \quad \text{or} \quad PV = n_T RT \tag{2.30}$$

Several extensions of the ideal gas equation, which is always reproduced at $P \to 0$, are possible. They are explained in the following chapters.

2.5.1
Virial Equation

The virial equation can be derived with the help of statistical mechanics or regarded as simple expansions of the compressibility factor z in terms of pressure P (Berlin form):

$$z = 1 + B'(T)P + C'(T)P^2 + \cdots \tag{2.87}$$

or density $\rho = 1/v$ (Leiden form)

$$z = 1 + B(T)\rho + C(T)\rho^2 + \cdots \tag{2.88}$$

where the ideal gas state at $P \to 0$ or, respectively, $\rho \to 0$ is used as a reference point for the series.

---

17) The Clausius–Clapeyron equation can also be applied to melting or sublimation processes.

By definition, the virial coefficients $B, C, \ldots$ or $B', C', \ldots$ of the pure substances depend solely on temperature. The values of the coefficients are very different for each of the forms, but they are convertible into each other, for example, the second virial coefficients may be converted as

$$B' = \frac{B}{RT} \tag{2.89}$$

This is exactly true only for a polynomial with an unlimited number of terms (unlimited series). Equation (2.89) is derived in Appendix C, B6.

The main advantage of the virial equation is its theoretical background, as the virial coefficients can be connected to the potential functions of the intermolecular forces. The main disadvantage is that the virial equation is only valid for gases with low or moderate densities. If the equation is truncated after the second term, a rule of thumb says that the density should be $\rho < 1/2\rho_c$. If it is truncated after the third term, $\rho < 3/4\rho_c$ should be maintained. This limitation is necessary, because the fourth and all higher virial coefficients are unknown in nearly all practical cases. Even values for the third virial coefficient can hardly be found in the literature. The virial equation can be understood as a Taylor series of the compressibility factor z as a function of the density ρ at the point ρ_0. The reference point ρ_0 is chosen as the density at zero pressure, where the ideal gas law is valid:

$$z = z^0 + \left(\frac{\partial z}{\partial \rho}\right)_{T,\rho^0} (\rho - \rho^0) + \frac{1}{2}\left(\frac{\partial^2 z}{\partial \rho^2}\right)_{T,\rho^0} (\rho - \rho^0)^2 + \cdots \tag{2.90}$$

With

$$z^0 = \frac{P^0}{\rho^0 RT} = 1 \quad \text{at} \quad P^0 \to 0 \quad \text{and} \quad \rho^0 \to 0$$

the definition of the virial coefficients B and C from Eq. (2.90) is obtained as

$$B = \left(\frac{\partial z}{\partial \rho}\right)_{T,\rho=0} \tag{2.91}$$

$$C = \frac{1}{2}\left(\frac{\partial^2 z}{\partial \rho^2}\right)_{T,\rho=0} \tag{2.92}$$

The virial coefficients can be determined from experimental PvT data. When doing so, the theoretical background of the virial coefficients B and C as the first terms of an infinite series should be taken into account. A simple least square fit is not advisable. The correct interpretation and determination of the virial coefficients can be derived from Eq. (2.90) as

$$B = \lim_{\rho \to 0} \left(\frac{z-1}{\rho}\right) \tag{2.93}$$

and

$$C = \lim_{\rho \to 0} \left(\frac{\partial (z-1)/\rho}{\partial \rho}\right) \quad \text{or} \quad C = \lim_{\rho \to 0} \left(\frac{z-1-B\rho}{\rho^2}\right) \tag{2.94}$$

A plot of the expression

$$\frac{z-1}{\rho} = \frac{P/\rho RT - 1}{\rho} = \left(\frac{Pv}{RT} - 1\right) v$$

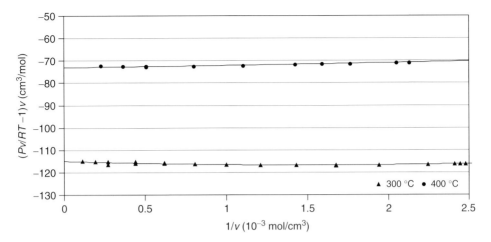

Figure 2.9 PvT data of water for the determination of the second virial coefficient data from [9].

which can be calculated from experimental PvT data, versus the density ρ yields the second virial coefficient B as the ordinate value at the density $\rho = 0$.

This procedure is shown for water in Figure 2.9 for two different temperatures. The experimental PvT data for water are taken from the literature.

The third virial coefficient can be determined in two different ways (see Eq. (2.94)): as one possibility the tangent at $\rho = 0$ of the graph in Figure 2.9 is determined. As the more exact and robust method the term

$$\left(\frac{z-1-B\rho}{\rho^2}\right) = \frac{P/\rho RT - 1 - B\rho}{\rho^2} = \left(\frac{Pv}{RT} - 1 - \frac{B}{v}\right) v^2$$

is plotted as a function of ρ and the second virial coefficient C is determined as the ordinate value at $\rho = 0$. In principle these methods may also be used to determine the virial coefficients of higher order, but in most cases the accuracy of the experimental data is not sufficient even for the determination of the third virial coefficient.

As already mentioned, the virial coefficients are temperature dependent. The temperature dependence of the second virial coefficient B is shown in Figure 2.10 for nitrogen as an example. At low temperatures, the second virial coefficient B is usually negative, which means that the molar volume of the real gas is smaller than the corresponding molar volume of the ideal gas due to attractive interactions. With rising temperatures the second virial coefficient becomes less negative until reaching a value of zero at the so-called Boyle temperature. The Boyle temperatures of nitrogen and CO_2 are approx. 330 K and 770 K, respectively. As a rule of thumb, the value of the Boyle temperature is about three times the value of the critical temperature. At temperatures above the Boyle temperature B becomes positive, but the slope of B with respect to temperature is flat. Many second virial coefficients which have been determined from experimental data are collected in [13] and stored in data banks like DDB [9].

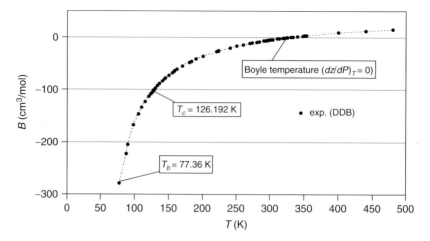

Figure 2.10 Temperature dependence of the second virial coefficient of nitrogen.

Theoretically, the second and the third virial coefficient can be calculated from models for the potential functions and with the help of the statistical mechanics. However, this procedure is rarely used in practical applications and tends to be extremely complex and even approximate for nonspherical molecules.

For nonassociating molecules, the methods of Tsonopoulos [14] and Hayden and O'Connell [15] yield pretty good results for the estimation of the second virial coefficient. The first method works with the three-parameter corresponding-states principle (Section 2.4.4); the second method takes into account several molecular effects. In case of vapor phase association (Section 13.2), the model of Hayden–O'Connell is extended by a chemical equilibrium term, whose parameters have to be fitted to experimental data.

The method of Tsonopoulos is based on an equation of the form

$$z = 1 + \left(\frac{BP_c}{RT_c}\right)\frac{P_r}{T_r} = z^{(0)} + \omega z^{(1)} \qquad (2.95)$$

In this relation, $z^{(0)}$ is a term describing the compressibility factor of a simple fluid and $z^{(1)}$ is a term taking into account the deviation of a real fluid from the behavior of the simple fluid. Additionally, Eq. (2.95) contains the so-called acentric factor ω, which will be explained in detail later (Section 2.4.4). As B is only a function of temperature, Eq. (2.95) can also be written as

$$\frac{BP_c}{RT_c} = b^{(0)}(T_r) + \omega b^{(1)}(T_r) \qquad (2.96)$$

Tsonopoulos [14] gives the following empirical correlations for $b^{(0)}(T_r)$ and $b^{(1)}(T_r)$:

$$b^{(0)}(T_r) = 0.1445 - \frac{0.330}{T_r} - \frac{0.1385}{T_r^2} - \frac{0.0121}{T_r^3} - \frac{0.000607}{T_r^8} \qquad (2.97)$$

$$b^{(1)}(T_r) = 0.0637 + \frac{0.331}{T_r^2} - \frac{0.423}{T_r^3} - \frac{0.008}{T_r^8} \qquad (2.98)$$

Example 2.4

Calculate the second virial coefficient of N_2 at 150 K using the Tsonopoulos method
Physical properties of N_2:

$$T_c = 126.15 \text{ K}, \; P_c = 33.94 \text{ bar}, \; \omega = 0.045$$

Solution

Inserting the reduced temperature

$$T_r = 150/126.15 = 1.18906$$

into Eqs. (2.97) and (2.98) gives

$$b^{(0)} = 0.1445 - \frac{0.330}{1.18906} - \frac{0.1385}{1.18906^2} - \frac{0.0121}{1.18906^3} - \frac{0.000607}{1.18906^8} = -0.23834$$

$$b^{(1)} = 0.0637 + \frac{0.331}{1.18906^2} - \frac{0.423}{1.18906^3} - \frac{0.008}{1.18906^8} = 0.0442$$

The second virial coefficient results in

$$B = \frac{8.3143 \cdot 126.15 \cdot 10}{33.94}[-0.23834 + 0.045 \cdot 0.0442] \frac{\text{cm}^3}{\text{mol}} = -73.04 \frac{\text{cm}^3}{\text{mol}}$$

This value is in good agreement with the experimental results, as can be seen in Figure 2.10.

The virial equation can not only describe the PvT behavior of gases, but also the different residual functions. Both forms (Leiden form and Berlin form) are well-suited for this purpose, but most of the thermodynamic functions can be derived easier using the Berlin form, because this form is volume explicit. Volume explicit means that the equation can be solved explicitly for v as a function of T and P. As an example, the calculations of the fugacity coefficient and of the residual function of the enthalpy are shown, using the virial equation in the Berlin form truncated after the second virial coefficient B.

With

$$\left(h - h^{\text{id}}\right)_{T,P} = \int_0^P \left[v - T\left(\frac{\partial v}{\partial T}\right)_P\right] dP \qquad (2.99)$$

and

$$z = \frac{Pv}{RT} = 1 + \frac{B(T)P}{RT} \qquad (2.100)$$

we get

$$v = \frac{RT}{P} + B(T) \qquad (2.101)$$

and

$$\left(\frac{\partial v}{\partial T}\right)_P = \frac{R}{P} + \frac{dB(T)}{dT} \qquad (2.102)$$

Inserting Eqs. (2.101) and (2.102) into Eq. (2.99) gives

$$\left(h - h^{id}\right)_{T,P} = \int_0^P \left(B - T\frac{dB}{dT}\right) dP \qquad (2.103)$$

As the expression in parentheses does not depend on pressure, the equation can be integrated easily:

$$\left(h - h^{id}\right)_{T,P} = \left(B - T\frac{dB}{dT}\right) P \qquad (2.104)$$

The expression for the fugacity coefficient is derived as follows:

$$\ln \varphi = \frac{\left(g - g^{id}\right)_{T,P}}{RT} \qquad (2.105)$$

$$\ln \varphi = \int_0^P \left(\frac{v}{RT} - \frac{1}{P}\right) dP \qquad (2.106)$$

$$\ln \varphi = \int_0^P \left(\frac{z}{P} - \frac{1}{P}\right) dP = \int_0^P \left(\frac{1}{P} + \frac{B}{RT} - \frac{1}{P}\right) dP \qquad (2.107)$$

which leads to the very simple equation

$$\ln \varphi = \frac{BP}{RT} \qquad (2.108)$$

2.5.2
High Precision Equations of State

Sometimes, cases occur where a much higher accuracy of the physical properties is decisive for the simulation, that is, the power plant process, the heat pump process, or the pressure drop calculation in a large pipeline. As already mentioned, the virial equation truncated after the second or third term is only able to describe the gas phase with a limited accuracy. For the description of the complete PvT behavior including the two-phase region several extensions of the virial equation were suggested. All these extensions have been derived empirically and contain a large number of parameters, which have to be fitted to experimental data. A large data base is necessary to obtain reliable parameters. One of the first approaches to EOS with higher accuracies has been made by Benedict, Webb, and Rubin (BWR) in 1940 [16], who used eight parameters. Their equation allows reliable calculations of PvT-data for nonpolar gases and liquids up to densities of about 1.8 ρ_c. Bender [17] has extended the BWR equation to 20 parameters. With this large number of parameters it became possible to describe the experimental data for certain substances over a large density range in an excellent way.

In the 1990s, the so-called high-precision reference equations have been developed [18]. Their significant improvement was possible due to progress in

Table 2.3 Typical accuracy demands for technical equations of state [29].

	$\rho(P,T)$ (%)	$w^*(P,T)$ (%)	$c_P(P,T)$ (%)	$P^s(T)$ (%)	$\rho'(T)$ (%)	$\rho''(T)$ (%)
$P < 30$ MPa	0.2	1–2	1–2	0.2	0.2	0.2
$P > 30$ MPa	0.5	2	2	–	–	–

measurement techniques and the development of mathematical algorithms for optimizing the structure of EOS. With respect to the accuracy of the calculated properties, their extrapolation behavior, and their reliability in regions where data are scarce, these equations define the state-of-the-art representation of thermal and caloric properties and their particular derivatives in the whole fluid range [19]. There is also an important demand to get reliable results for derived caloric properties (c_P, c_v) even if the equation can only be fitted to data for thermal properties. The main disadvantages of the reference equations are their extraordinary complexity and limited availability. There are only few substances for which the data situation justifies the development of a reference equation of state, for example, water [20], methane [21], argon [22], carbon dioxide [23], nitrogen [24, 25], ethane [26], n-butane [27], isobutane [27], and ethylene [28]. A further serious drawback is that the concept can hardly be applied to mixtures.

Technical high-precision EOS are a remarkable compromise between keeping the accuracy and gain in simplicity. They can also be applied to substances where less extensive and less accurate experimental data are available. Furthermore, these equations should enable the user to extrapolate safely to the extreme conditions often encountered in industrial processes. For example, in the LDPE[18] process ethylene is compressed to approx. 3000 bar, and it is necessary for the simulation of the process and the design of the equipment to have a reliable tool for the determination of the thermal and caloric properties.

Table 2.3 illustrates the accuracy demand for technical EOS; especially, the derived properties like the specific heat capacity c_P and the speed of sound w^* (Section 3.2.6) have to be reproduced very well.

Like reference equations, state-of-the-art technical EOS are formulated in terms of the Helmholtz energy, which is split into an ideal gas part and a residual part. Instead of the specific volume, the density is used as a variable:

$$\frac{a(T,\rho)}{RT} = \frac{a^{id}(T,\rho) + a^R(T,\rho)}{RT} = \alpha^{id}(\tau,\delta) + \alpha^R(\tau,\delta) \tag{2.109}$$

with $\tau = T_c/T$ and $\delta = \rho/\rho_c$. The ideal gas part can easily be evaluated if a sufficiently accurate correlation for c_P^{id} is available. With $c_v^{id} = c_P^{id} - R$, we obtain

---

18) LDPE: Low-density polyethylene.

$$a^{\text{id}}(T, v) = u^{\text{id}}(T) - Ts^{\text{id}}(T, v)$$

$$= \int_{T_0}^{T} c_v^{\text{id}} dT - T \int_{T_0}^{T} c_v^{\text{id}} \frac{dT}{T} - RT \ln \frac{v}{v_0} + a(T_0, v_0) \quad (2.110)$$

or

$$a^{\text{id}}(T, \rho) = \int_{T_0}^{T} c_v^{\text{id}} dT - T \int_{T_0}^{T} c_v^{\text{id}} \frac{dT}{T} + RT \ln \frac{\rho}{\rho_0} + a(T_0, \rho_0) \quad (2.111)$$

For the description of c_p^{id} or, respectively, c_v^{id}, highly flexible and accurate equations are necessary, as discussed in Section 3.2.4. From c_p^{id} or c_v^{id}, the various caloric properties can be obtained by integration (see Section 6.1). As reference points, the high-precision EOS use $T_{\text{ref}} = 298.15$ K and $P_{\text{ref}} = 101\,325$ Pa in the ideal gas state, where h_{ref} and s_{ref} are set to 0, even if this reference point is fictitious and the fluid regarded is in the liquid state.[19]

The more difficult part is to set up a correlation for the residual part of the Helmholtz energy $\alpha^R(\tau, \delta)$. For this purpose, a bank of terms that could be useful for a high-precision equation of state has been established [29]. The general setup for the fitting procedure contains 583 terms. It is expressed as

$$\alpha^R(\tau, \delta) = \sum_{i=1}^{8} \sum_{j=-8}^{12} n_{i,j} \delta^i \tau^{j/8} + \sum_{i=1}^{5} \sum_{j=-8}^{24} n_{i,j} \delta^i \tau^{j/8} \exp(-\delta)$$

$$+ \sum_{i=1}^{5} \sum_{j=16}^{56} n_{i,j} \delta^i \tau^{j/8} \exp(-\delta^2) + \sum_{i=2}^{4} \sum_{j=24}^{38} n_{i,j} \delta^i \tau^{j/2} \exp(-\delta^3)$$

(2.112)

With an evolutionary algorithm [30, 31] it is possible to select the most significant terms, where the coefficients are fitted to the available experimental data, which are weighted according to their demanded uncertainties. Details are given in [24, 28].

For technical EOS, the number of terms used from Eq. (2.112) has been restricted to 12 to keep the complexity low. Two different runs have been made for nonpolar and for polar fluids. As a result, the following functional forms have been found.

For nonpolar fluids [32, 33] (methane, ethane, propane, n-butane, n-pentane, n-hexane, n-heptane, n-octane, argon, oxygen, nitrogen, ethylene, isobutane, cyclohexane, sulfur hexafluoride, carbon monoxide, carbonyl sulfide, n-decane, hydrogen sulfide, isopentane, neopentane, isohexane, krypton, n-nonane, toluene,

---

[19] As explained in Section 2.2, for applications where chemical reactions are involved, it is advantageous that the reference points are set to the standard enthalpy of formation and to the absolute entropy, which is done in process simulators (Section 6.2). The high-precision equations of state refer only to pure components where chemical reactions are not involved; therefore, the arbitrary choice to set them to 0 at standard conditions can be accepted.

2.5 Equations of State

xenon, and R116):

$$\alpha^R(\tau,\delta) = n_1\delta\tau^{0.25} + n_2\delta\tau^{1.125} + n_3\delta\tau^{1.5} + n_4\delta^2\tau^{1.375} + n_5\delta^3\tau^{0.25} + n_6\delta^7\tau^{0.875}$$
$$+ n_7\delta^2\tau^{0.625}e^{-\delta} + n_8\delta^5\tau^{1.75}e^{-\delta} + n_9\delta\tau^{3.625}e^{-\delta^2} + n_{10}\delta^4\tau^{3.625}e^{-\delta^2}$$
$$+ n_{11}\delta^3\tau^{14.5}e^{-\delta^3} + n_{12}\delta^4\tau^{12}e^{-\delta^3} \quad (2.113)$$

For polar fluids [33, 34] (R11, R12, R22, R32, R113, R123, R125, R134a, R143a, R152a, carbon dioxide, ammonia, acetone, nitrous oxide, sulfur dioxide [35], R141b, R142b, R218, and R245fa):

$$\alpha^R(\tau,\delta) = n_1\delta\tau^{0.25} + n_2\delta\tau^{1.25} + n_3\delta\tau^{1.5} + n_4\delta^3\tau^{0.25} + n_5\delta^7\tau^{0.875} + n_6\delta\tau^{2.375}e^{-\delta}$$
$$+ n_7\delta^2\tau^2 e^{-\delta} + n_8\delta^5\tau^{2.125}e^{-\delta} + n_9\delta\tau^{3.5}e^{-\delta^2} + n_{10}\delta\tau^{6.5}e^{-\delta^2}$$
$$+ n_{11}\delta^4\tau^{4.75}e^{-\delta^2} + n_{12}\delta^2\tau^{12.5}e^{-\delta^3} \quad (2.114)$$

The coefficients for the particular substances are given in Appendix B. For R41 [33], dimethyl ether [36], and the four butenes [37] (1-butene, isobutene, *cis*-2-butene, and *trans*-2-butene) analogous equations with different exponents are available. Similar equations for strongly polar and associating fluids are currently under development [38].

The advantage of the Helmholtz formulation is that all thermal and caloric properties can be calculated by means of simple derivatives of the Helmholtz energy a with respect to τ and δ. For instance, the pressure can be obtained via

$$P(T,\rho) = -(\partial a/\partial v)_T = \rho RT\left[1 + \delta\left(\frac{\partial \alpha^R}{\partial \delta}\right)_\tau\right] \quad (2.115)$$

For pure components, the fugacity coefficient can be derived from the Gibbs energy:

$$\ln\varphi = \frac{g(T,P) - g^{id}(T,P)}{RT} \quad (2.116)$$

From the equation of state, there is an easy access to the difference $a(T,v) - a^{id}(T,v)$. With $g = a + Pv$, one can write

$$g(T,v) - g^{id}(T,v) = a(T,v) - a^{id}(T,v) + Pv - RT \quad (2.117)$$

or [33, 37]

$$g(T,\rho) = RT + RT\alpha^{id} + RT\alpha^R + RT\delta\left(\frac{\partial\alpha^R}{\partial\delta}\right)_\tau \quad (2.118)$$

To have a better accessibility to the function $g(T,v)$, it can be related to $g(T,P)$ via

$$\left[g(T,P) - g^{id}(T,P)\right] - \left[g(T,v(P)) - g^{id}(T,v(P))\right]$$
$$= g^{id}(T,v(P)) - g^{id}(T,P) \quad (2.119)$$

and

$$g^{id}(T,v(P)) - g^{id}(T,P) = \left[h^{id}(T) - Ts^{id}(T,v(P))\right] - \left[h^{id}(T) - Ts^{id}(T,P)\right]$$
$$= T\left[s^{id}(T,P) - s^{id}(T,v(P))\right] \quad (2.120)$$

as the enthalpy of an ideal gas does not depend on the pressure. Setting up the entropies in the ideal gas state, one obtains

$$s^{id}(T,P) - s^{id}(T,v(P)) = \left[s_0 + \int_{T_0}^{T} c_p^{id} \frac{dT}{T} - R\ln\frac{P}{P_0} \right]$$

$$- \left[s_0 + \int_{T_0}^{T} c_v^{id} \frac{dT}{T} + R\ln\frac{v(P)}{v_0} \right]$$

$$= R\ln\frac{T}{T_0} - R\ln\frac{P}{P_0} - R\ln\frac{v(P)}{v_0}$$

$$= R\ln\frac{RT}{Pv(P)} \frac{P_0 v_0}{RT_0}$$

$$= -R\ln z \qquad (2.121)$$

as P_0, T_0, and v_0 refer to ideal gas conditions.

Summarizing Eqs. (2.116–2.120), the fugacity coefficient can be expressed as

$$\ln\varphi = \frac{a(T,v) - a^{id}(T,v) + Pv - RT - RT\ln z}{RT} = \alpha^R(\tau,\delta) + z - 1 - \ln z \qquad (2.122)$$

Other relevant thermodynamic quantities can be derived from an equation of state in the Helmholtz energy formulation very easily. The resulting formulas are [33, 37]

$$s(T,\rho) = -\left(\frac{\partial a}{\partial T}\right)_v = R\tau\left[\left(\frac{\partial \alpha^{id}}{\partial \tau}\right)_\delta + \left(\frac{\partial \alpha^R}{\partial \tau}\right)_\delta\right] - R\alpha^{id} - R\alpha^R \qquad (2.123)$$

$$u(T,\rho) = a + Ts = RT\tau\left[\left(\frac{\partial \alpha^{id}}{\partial \tau}\right)_\delta + \left(\frac{\partial \alpha^R}{\partial \tau}\right)_\delta\right] \qquad (2.124)$$

$$h(T,\rho) = u + Pv = RT\tau\left[\left(\frac{\partial \alpha^{id}}{\partial \tau}\right)_\delta + \left(\frac{\partial \alpha^R}{\partial \tau}\right)_\delta\right] + RT\delta\left(\frac{\partial \alpha^R}{\partial \delta}\right)_\tau + RT \qquad (2.125)$$

$$c_v(T,\rho) = \left(\frac{\partial u}{\partial T}\right)_v = -R\tau^2\left[\left(\frac{\partial^2 \alpha^{id}}{\partial \tau^2}\right)_\delta + \left(\frac{\partial^2 \alpha^R}{\partial \tau^2}\right)_\delta\right] \qquad (2.126)$$

$$c_P(T,\rho) = c_v(T,\rho) + \frac{R\left[1 + \delta\left(\frac{\partial \alpha^R}{\partial \delta}\right)_\tau - \delta\tau\left(\frac{\partial^2 \alpha^R}{\partial \delta \partial \tau}\right)\right]^2}{\left[1 + 2\delta\left(\frac{\partial \alpha^R}{\partial \delta}\right)_\tau + \delta^2\left(\frac{\partial^2 \alpha^R}{\partial \delta^2}\right)_\tau\right]} \qquad (2.127)$$

A lot of discussions took place about the significance of the obtained equations. Definitely, the parameters have no physical meaning. Moreover, if terms with a theoretical background are added to the bank of terms, they are simply sorted out by the algorithm. Nevertheless, the technical equations show a remarkably accurate extrapolation behavior, and quantities such as caloric properties, speed of sound, or isochoric heat capacity in the critical region, which are generally difficult to describe, are well reproduced.

High-precision EOS provide a valuable tool, as they are currently the only way to carry out reliable calculations at extreme conditions. Moreover, these equations can serve as a reference to judge the errors of less accurate equations, especially in the case of caloric properties. Of course, their restriction to pure components means that they cannot be used in processes where reactions and separations take place. On the other hand, most of the processes have large utility units (steam, cooling water, nitrogen for inertization, ammonia or other refrigerants for cooling) and units for the conditioning of the pure component raw materials, which operate actually with pure components. For this purpose, high-precision EOS are an effective tool and can often cover large parts of processes.

Attempts have been made to apply the optimization algorithm to mixtures. However, there are few mixtures where the enormous effort can be justified, and a final structure has not been found yet. A multicomponent mixture where an accurate description is urgently required is natural gas. In this case, an equation is needed for the description of pipeline transport, storage, processing, and usage as fuel. In the GERG-equation, more than 100 000 data points have been considered to describe the properties and phase equilibria of any possible subsystem of altogether 21 natural gas components in the temperature range 90–450 K at pressures up to 35 MPa [39]. Approximately 800 parameters have been fitted. The equation comprises as well the high-precision equations of all the 21 pure components. Another system where a high-precision equation for mixtures would make sense is air for air fractionation.

Example 2.5

At 293.15 K, ethylene is already in a supercritical state ($T_c = 282.35$ K). How much ethylene is in a pressurized bottle (50 dm³), if the pressure on the manometer is $P = 20$ MPa?

Solution

Using the high-precision equation of state, the density is calculated to be $\rho = 410.5$ kg/m³. An experimental value of 410.839 kg/m³ has been reported [40], indicating that the relative error is only 0.08%. Thus, the content is calculated to be 20.525 kg.

Although this case is very difficult for a cubic equation of state, as it is a supercritical state in the vicinity of the critical temperature, the Peng–Robinson equation (Section 2.5.3) behaves quite well and yields $\rho = 430.3$ kg/m³. The error is only 4.6%. Of course, an approximation with the ideal gas law gives a terrible result in this case ($\rho = 230.2$ kg/m³, 44% error).

Example 2.6

Ethylene from a pipeline (50 t/h, 70 bar, 5 °C) passes a heat exchanger and is heated up to 45 °C. The pressure drop is negligible. As usual, in the data sheet of the heat exchanger the c_P values at inlet and outlet are given:

$$c_P(5\,°C, 70\,\text{bar}) = 3.984\,\text{J/g K}$$
$$c_P(45\,°C, 70\,\text{bar}) = 3.783\,\text{J/g K}$$

Calculate the heat to be exchanged

1) by estimation with an average c_P and
2) using a high-precision equation of state [41].

Which solution shall be preferred?

Solution

1) Ethylene is supercritical as $P > P_c$ ($P_c = 50.4\,\text{bar}$). With c_P values at 5 °C and 45 °C at 70 bar $c_{P,\text{av}}$ can be determined to be

$$c_{P,\text{av}} = (3.984 + 3.783)\,\text{J/(g K)} \cdot 0.5 = 3.8835\,\text{J/(g K)}$$
$$\Rightarrow \dot{Q} = \dot{m}\,c_{P,\text{av}}\,\Delta T = 50\,t/h \cdot 3.8835\,\text{J/(g K)} \cdot 40\,\text{K} = 2.157\,\text{MW}$$

2) The enthalpies at the particular temperatures are
$$h(5\,°C, 70\,\text{bar}) = -356.825\,\text{J/g}$$
$$h(45\,°C, 70\,\text{bar}) = -108.907\,\text{J/g}$$
$$\Rightarrow \dot{Q} = \dot{m}\Delta h = 50t/h \cdot (-108.907 + 356.825)\,\text{J/g} = 3.443\,\text{MW}.$$

There is a large difference between the two results. Although the c_P values are correct, approach (1) does not yield a reasonable result. The problem is that c_P has a sharp maximum in the vicinity of the critical point, as Figure 2.11 illustrates. This

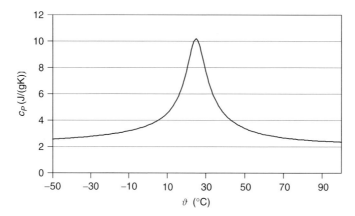

Figure 2.11 Specific heat capacity c_P of ethylene at $P = 70$ bar as a function of temperature.

example refers to an unfortunate case; however, it actually occurred in a practical application and caused a lot of confusion, as it is a common procedure to take the c_P values at inlet and outlet for the calculation of the heat duty and for the design of the heat exchanger.

Example 2.7

Ideal gases are characterized by the following conditions:

$$z = \frac{Pv}{RT} = 1 \quad (\partial T/\partial P)_h = 0 \quad \left(\frac{\partial (Pv)}{\partial P}\right)_T = 0$$

Nevertheless, these conditions can also be fulfilled at high pressures, giving the so-called ideal curves Zeno line ($z = 1$), Joule–Thomson inversion curve (($\partial T/\partial P)_h = 0$) and Boyle curve (($\partial (Pv)/\partial P)_T = 0$) (Figure 2.12). They are important tools for the assessment of the applicability of EOS; moreover, the Joule–Thomson inversion curve has a technical significance, as it is decisive for the liquefaction of gases like oxygen, nitrogen, hydrogen, or helium [3]. Calculate these lines for nitrogen using a high-precision EOS.

Solution

Using Eq. (2.113) and the coefficients from Appendix B, we get the following results:

It can be seen that the three lines differ significantly. If one of these conditions is fulfilled in the high-pressure region, it cannot be concluded at all that the ideal gas equation of state is applicable.

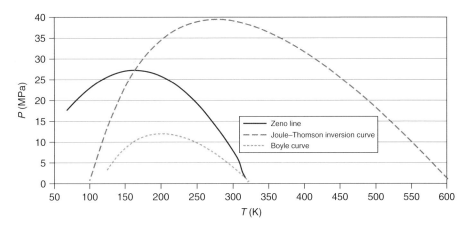

Figure 2.12 Zeno line, Joule–Thomson inversion curve and Boyle curve for nitrogen.

2.5.3
Cubic Equations of State

The so-called cubic EOS are very popular in process simulation due to their robustness and their simple applicability to mixtures. They are based on the more than 100 years old van der Waals equation of state [2]. Historically, the van der Waals equation of state played an important role, as this equation made it possible to explain condensation, vaporization, the two-phase region, and the critical phenomena by molecular concepts for the first time. All modifications of the van der Waals equation of state are based on the assumption of additivity of separate contributions to the resulting compressibility factor:

$$z = z^{rep} + z^{att} \tag{2.128}$$

The first term z^{rep} considers the intrinsic volume of the molecule to account for the volume of the closest packing. The second term z^{att} represents the attractive intermolecular forces. The van der Waals equation of state has the form

$$z = \frac{v}{v-b} - \frac{a}{RTv} \tag{2.129}$$

or

$$P = \frac{RT}{v-b} - \frac{a}{v^2} \tag{2.130}$$

Among the numerous published forms the Redlich–Kwong equation [42], the Soave–Redlich–Kwong equation [43], and the Peng–Robinson equation [44] are the best known and most successful modifications of the van der Waals equation of state. All the different modifications and extensions of the van der Waals equation of state can be written in a general form of a cubic equation of state. The individual equations are then regarded as special cases of this general formulation. This general form can be written as

$$z = \frac{v}{v-b} - \frac{\Theta}{(v^2 + \delta v + \varepsilon)} \frac{v}{RT} \tag{2.131}$$

or as a pressure-explicit equation

$$P = \frac{RT}{v-b} - \frac{\Theta}{v^2 + \delta v + \varepsilon} \tag{2.132}$$

Table 2.4 summarizes the values for Θ, δ, and ε for the most common versions of cubic EOS.

A typical example for a plot of the pressure P versus the molar volume v at constant temperature ($T < T_c$) as derived from a cubic equation of state is shown in Figure 2.13.

The different branches of the graph result from the different zero roots in the denominator. Only the very right branch has a physical meaning and is used for the description of the thermodynamic behavior of fluids. As the constant b is a measure for the intrinsic volume of the molecules, the molar volume must always be larger than the value of the constant b.

2.5 Equations of State

Table 2.4 Forms of the cubic equations of state.

Equation	Θ	δ	ε
van der Waals	a	0	0
Redlich–Kwong	$\dfrac{a}{T^{1/2}}$	b	0
Soave–Redlich–Kwong	$a(T)$	b	0
Peng–Robinson	$a(T)$	$2b$	$-b^2$

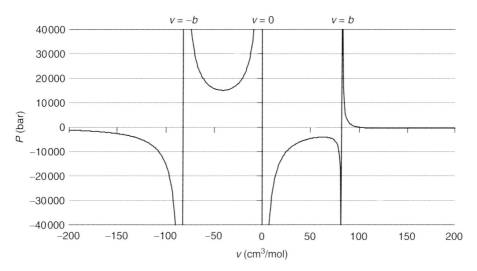

Figure 2.13 Mathematical representation of the calculated PvT behavior with the Redlich–Kwong equation of state, exemplified by the 300 K isotherm of benzene.

From the mathematical point of view, the van der Waals equation of state is a cubic polynomial of the molar volume, given by[20]

$$v^3 - \left(b + \frac{RT}{P}\right)v^2 + \frac{a}{P}v - \frac{ab}{P} = 0 \qquad (2.133)$$

The cubic form of an equation of state is the simplest form which enables the description of the PvT behavior of gases and liquids and thus the representation of the vapor–liquid equilibrium with only one model. At constant temperature and at a given pressure this equation has three solutions. These solutions may be – depending on the values of temperature and pressure – all of real type or of mixed real and complex type. Figure 2.14 shows an isotherm in the Pv-diagram, calculated with the Soave–Redlich–Kwong equation for ethanol at 473.15 K. The chosen temperature is lower than the critical temperature of ethanol ($T_c = 516.2$ K),

---

20) It is thoroughly described below how a and b can be calculated.

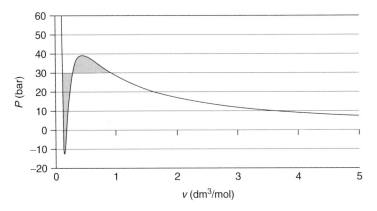

Figure 2.14 Isotherm for ethanol at $\vartheta = 200\,°C$ in the Pv-diagram, calculated with the Soave–Redlich–Kwong equation of state.

consequently three real solutions exist for a given temperature within a certain pressure range. The left and the right solutions represent the molar volumes of the liquid and the vapor at the saturation vapor pressure P^s; the intermediate solution has no physical meaning. If the solutions are of mixed real and complex type, only the real one corresponds to a solution of the equation of state with physical meaning.

The solutions of cubic EOS can be obtained iteratively or analytically using Cardano's formula (see Appendix C, A10 and Example 2.9). As it does not contain any iteration, Cardano's formula is very convenient to use. However, it has been shown that it is by far not the fastest option [45].

The Gibbs phase rule [46] states that at a certain temperature the phase equilibrium between the liquid phase and the vapor phase for a pure substance is reached at one corresponding pressure – the saturated vapor pressure. Using a cubic equation of state, the saturated vapor pressure is calculated as follows.

When the phase equilibrium is reached, the fugacities of the vapor and the liquid are identical:

$$f^L(T, P^s) = f^v(T, P^s) \tag{2.134}$$

The fugacities of the two phases can be calculated with any equation of state which is valid for both the vapor and liquid phase. The relationship between pressure and fugacity has already been derived in Section 2.4.3:

$$\left(\frac{\partial f/f}{\partial P}\right)_T = \frac{v}{RT} \tag{2.135}$$

which is equivalent to

$$RT\left(\frac{df}{f}\right)_T = v\,dP = d(Pv) - P\,dv \tag{2.136}$$

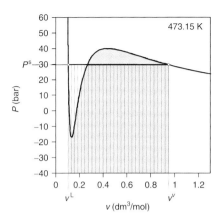

Figure 2.15 Maxwell's area construction to determine the vapor pressure, exemplified by the isotherm at $T = 473.15$ K for ethanol (Soave–Redlich–Kwong equation).

Integration from $v = v^L$ to $v = v^V$ at $P = P^s(T)$ yields

$$RT \ln \frac{f^V(T, P^s)}{f^L(T, P^s)} = P^s(v^V - v^L) - \int_{v_L}^{v^V} P dv \qquad (2.137)$$

Considering Eq. (2.134), one obtains

$$P^s(v^V - v^L) = \int_{v^L}^{v^V} P dv \qquad (2.138)$$

Figure 2.15 makes clear that the two terms

$$P^s(v^V - v^L) \quad \text{and} \quad \int_{v^L}^{v^V} P dv$$

are equal when the two areas (the gray area and the hatched area) have the same size. Thus, it is possible to find the saturation pressure by varying the pressure until the two marked areas are equal. The described graphic solution for finding the saturation pressure is called *Maxwell's equal area construction*. The saturation pressure $P^s(T)$ can be calculated for any isotherm $T < T_c$. As can be seen in Figure 2.14, the isotherms calculated from an equation of state may show negative values for the pressure. This is a merely mathematical phenomenon which does not have an influence on the calculation; in equilibrium, only the points L and V linked by the horizontal line between these points exist. In this example (ethanol at $T = 473.15$ K), the saturation pressure is calculated to be $P^s = 29.7$ bar.

In principle, the parameters a and b are specific for each substance. A fit to experimental data (e.g., vapor pressure data, PvT data, enthalpies of vaporization) is one option to determine the numerical values of the parameters. The prerequisite for a good fit is an extensive base of experimental data over a wide temperature and pressure range. The more popular approach for the determination of the parameters a and b is a generalization of the equation of state based on the critical data. This will be shown here for the Redlich–Kwong equation as an example. The Redlich–Kwong equation has the form

$$P = \frac{RT}{v-b} - \frac{a}{T^{1/2} v(v+b)} \tag{2.139}$$

or

$$z = \frac{v}{v-b} - \frac{a}{RT^{3/2}(v+b)} \tag{2.140}$$

According to Eq. (2.1) the first and the second derivative of the pressure with respect to the molar volume must be zero at the critical point.

Application to the Redlich–Kwong equation results in

$$\left(\frac{\partial P}{\partial v}\right)_T = -\frac{RT}{(v-b)^2} + \frac{a}{T^{1/2}} \frac{2v+b}{v^2(v+b)^2} = 0 \tag{2.141}$$

$$\left(\frac{\partial^2 P}{\partial v^2}\right)_T = \frac{2RT}{(v-b)^3} + \frac{2a}{T^{1/2}} \frac{v(v+b) - (2v+b)^2}{v^3(v+b)^3} = 0 \tag{2.142}$$

and thus for the critical point

$$\frac{RT_c}{(v_c - b)^2} = \frac{a}{T_c^{1/2}} \frac{2v_c + b}{v_c^2 (v_c + b)^2} \tag{2.143}$$

$$\frac{2RT_c}{(v_c - b)^3} = \frac{2a}{T_c^{1/2}} \cdot \frac{(2v_c + b)^2 - v_c(v_c + b)}{v_c^3 (v_c + b)^3} \tag{2.144}$$

The two unknowns a and b are determined by solving these two equations. Combination and transformation gives a relation for b as a function of the critical volume

$$b^3 + 3v_c b^2 + 3v_c^2 b - v_c^3 = 0 \tag{2.145}$$

This cubic equation has one real solution

$$b = (2^{1/3} - 1) v_c = 0.25992 v_c \tag{2.146}$$

The parameter a is now obtained by inserting this result into Eq. (2.143)

$$a = \frac{1}{3(2^{1/3} - 1)} RT_c^{3/2} v_c = 1.28244 RT_c^{3/2} v_c \tag{2.147}$$

The compressibility factor at the critical point z_c can be obtained by inserting these constants a and b into the original form of the Redlich–Kwong equation:

$$z_c = \frac{v_c}{v_c - b} - \frac{a}{RT_c^{3/2}(v_c + b)} \tag{2.148}$$

$$z_c = \frac{1}{(2 - 2^{1/3})} - \frac{1}{3 \cdot 2^{1/3}(2^{1/3} - 1)} = \frac{1}{3} \tag{2.149}$$

Simultaneously, the above equation represents a relationship between the different critical properties:

$$v_c = \frac{RT_c}{3P_c} \tag{2.150}$$

Using Eq. (2.150) together with Eqs. (2.146) and (2.147), the parameters a and b can be expressed as a function of the critical pressure and the critical temperature:

$$a = \frac{R^2 T_c^{5/2}}{9\left(2^{2/3} - 1\right) P_c} = 0.42748 \frac{R^2 T_c^{5/2}}{P_c} \tag{2.151}$$

$$b = \frac{1}{3}\left(2^{1/3} - 1\right) \frac{RT_c}{P_c} = 0.08664 \frac{RT_c}{P_c} \tag{2.152}$$

The values of the parameters a and b can be very different depending on how they have been determined, either from T_c and v_c (Eqs. (2.146) and (2.147)) or from T_c and P_c (Eqs. (2.151) and (2.152)). In reality, the experimental critical data do not fulfill either Eqs. (2.149) or (2.150), which is the reason for this discrepancy. In fact, the experimental compressibility factor at the critical point z_c never matches the value 1/3 as the Redlich–Kwong equation of state demands. In most practical cases, Eqs. (2.151) and (2.152) are referred to for determining the parameters a and b.

2.5.4
Generalized Equations of State and Corresponding States Principle

The state variables P, v, and T may also be written in a dimensionless or so-called reduced form, which means that they are divided by their critical values:

$$P_r \equiv \frac{P}{P_c}, \quad v_r \equiv \frac{v}{v_c} \quad \text{and} \quad T_r \equiv \frac{T}{T_c} \tag{2.153}$$

Combining the expressions for the parameters a and b of the Redlich–Kwong equation in Eqs. (2.146) and (2.147) with the definition of the reduced properties leads to a dimensionless form of the Redlich–Kwong equation with the compressibility factor z as a function of the reduced temperature T_r and reduced molar volume v_r:

$$z = \frac{v_r}{v_r - b/v_c} - \frac{a/\left(T_c^{3/2} v_c R\right)}{T_r^{3/2}\left(v_r + b/v_c\right)} \tag{2.154}$$

$$z = \frac{v_r}{v_r - 0.25992} - \frac{1.28244}{T_r^{3/2}\left(v_r + 0.25992\right)} \tag{2.155}$$

In a similar way, the van der Waals equation of state Eq. (2.130) can be written just by means of the reduced quantities:

$$P_r = \frac{8T_r}{3v_r - 1} - \frac{3}{v_r^2} \tag{2.156}$$

These forms of a generalized equation of state only require the critical temperature and the critical pressure as substance-specific parameters. Therefore, these correlations are an example for the so-called two-parameter corresponding-states principle, which means that the compressibility factor and thus the related thermodynamic properties for all substances should be equal at the same values of their reduced properties. As an example, the reduced vapor pressure as a function of the reduced temperature should have the same value for all substances, provided that the regarded equation of state can reproduce the PvT behavior of the substance on the basis of the critical data. In reality, the two-parameter corresponding-states principle is only well-suited to reflect the properties of simple, almost spherical, nonpolar molecules (noble gases as Ar, Kr, Xe). For all other molecules, the correlations based on the two-parameter corresponding-states principle reveal considerable deviations. To overcome these limitations, a third parameter was introduced, which is characteristic for a particular substance. The most popular third parameter is the so-called acentric factor, which was introduced by Pitzer:

$$z = f(T_r, P_r, \omega)$$

with

$$\omega = \left[\log P_r^s(\text{Ar, Kr, Xe}) - \log P_r^s\right]_{T_r=0.7}$$

$$\omega = -1 - \log\left(P_r^s\right)_{T_r=0.7} \tag{2.157}$$

The introduction of the acentric factor ω as defined above adds the information on the reduced vapor pressure at the reduced temperature $T_r = 0.7$ to the equation of state. The acentric factor can be illustrated in a $\log P_r^s - 1/T_r$ diagram as shown in Figure 2.16. The diagram shows the vapor pressure curve of the simple fluids (Ar, Kr, Xe) and the vapor pressure curve of ethanol, both in their reduced form. The

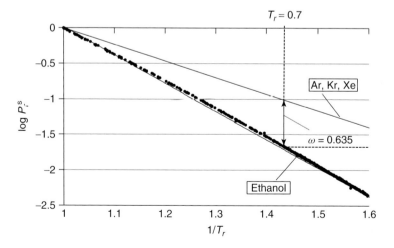

Figure 2.16 Definition of the acentric factor, exemplified for ethanol. Experimental vapor pressure data from [9].

distance between both curves at $T_r = 0.7$ corresponds to the value of the acentric factor as defined by Eq. (2.157).

A modification of the corresponding-state principle by introducing a parameter related to the vapor pressure curve is reasonable, because experimental vapor pressure data as a function of temperature are easy to retrieve. Furthermore, the vapor–liquid equilibrium is a very sensitive indicator for deviations from the simple corresponding-state principle. The value $T_r = 0.7$ was chosen because this temperature is not far away from the normal boiling point for most substances. Additionally, the reduced vapor pressure at $T_r = 0.7$ of the simple fluids has the value $P_r^s = 0.1$ (log $P_r^s = -1$). As a consequence, the acentric factor of simple fluids is 0 and the three-parameter correlation simplifies to the two-parameter correlation.

Pitzer suggested the following approach for the representation of the compressibility factor as a function of T_r, P_r, and ω:

$$z = z^{(0)}(T_r, P_r) + \omega z^{(1)}(T_r, P_r) \tag{2.158}$$

The term $z^{(0)}$ is the contribution of a simple fluid to the compressibility factor. This term can be derived from PvT-data of simple fluids. The term $z^{(1)}$ is the generalized term for molecules with behavior different from simple fluids. Values for $z^{(0)}$ and for $z^{(1)}$ as a function of T_r and P_r are tabulated in [10], for example. The values were, for example, determined by the method of Lee and Kesler [47], who calculated the compressibility factor of a certain substance from the compressibility factor of a simple fluid (argon, krypton, xenon) $z^{(0)}$ as defined above, and the compressibility factor of a reference fluid (n-octane) $z^{(r)}$:

$$z = z^{(0)} + \frac{\omega}{\omega^{(r)}} \left(z^{(r)} - z^{(0)} \right) \tag{2.159}$$

with $z^{(r)}$ and $z^{(0)}$ as functions of P_r and T_r.

The functions $z^{(r)}$ and $z^{(0)}$ are expressed by a modified BWR-equation (see Section 2.4.2).

The three-parameter corresponding-state principle is applicable to many substances; only for strongly polar or associating substances large deviations between the theoretical and experimental values can occur.

In chemical industry, the cubic equations of Soave–Redlich–Kwong [43] and Peng and Robinson [44], have become the most popular EOS. In both equations, the acentric factor is introduced into the equation in a similar way, that is, by modifying the parameter a, with a so-called α-function depending on the temperature and the acentric factor. The quality of the reproduction of the PvT behavior and the vapor pressure were significantly improved, even for polar substances or larger molecules. The α-function is generalized, meaning that besides the critical data the acentric factor is the only information referring to the substance regarded.

The formulations for the compressibility factor and for the pressure are

Soave–Redlich–Kwong

$$z = \frac{v}{v - b} - \frac{a(T)}{RT(v + b)} \tag{2.160}$$

$$P = \frac{RT}{v-b} - \frac{a(T)}{v(v+b)} \tag{2.161}$$

$$a(T) = 0.42748 \frac{R^2 T_c^2}{P_c} \alpha(T) \tag{2.162}$$

with

$$\alpha(T) = \left[1 + \left(0.48 + 1.574\omega - 0.176\omega^2\right)\left(1 - T_r^{0.5}\right)\right]^2 \tag{2.163}$$

$$b = 0.08664 \frac{RT_c}{P_c} \tag{2.164}$$

Peng–Robinson

$$z = \frac{v}{v-b} - \frac{a(T)v}{RT\left[v(v+b) + b(v-b)\right]} \tag{2.165}$$

$$P = \frac{RT}{v-b} - \frac{a(T)}{v(v+b) + b(v-b)} \tag{2.166}$$

$$a(T) = 0.45724 \frac{R^2 T_c^2}{P_c} \alpha(T) \tag{2.167}$$

with

$$\alpha(T) = \left[1 + \left(0.37464 + 1.54226\omega - 0.26992\omega^2\right)\left(1 - T_r^{0.5}\right)\right]^2 \tag{2.168}$$

$$b = 0.0778 \frac{RT_c}{P_c} \tag{2.169}$$

For $T = T_c (T_r = 1)$, $\alpha(T)$ becomes unity, that is, the constants a and b can again be directly determined from the critical data, as it is the case for the original Redlich–Kwong equation.

Example 2.8

A pressurized vessel ($V = 0.5 \, \text{m}^3$) contains 5 kg steam (H_2O) at $\vartheta = 220\,°C$. Calculate the pressure in the vessel

1) according to the ideal gas law,
2) using the virial equation,
3) using the Soave–Redlich–Kwong equation, and
4) using a high-precision equation of state.

Molar mass:	$M = 18.015 \, \text{g/mol}$
Second virial coefficient:	$B = -293 \, \text{cm}^3/\text{mol}$
Critical temperature:	$T_c = 647.096 \, \text{K}$
Critical pressure:	$P_c = 220.64 \cdot 10^5 \, \text{Pa}$ (220.64 bar)
Acentric factor:	$\omega = 0.3443$

2.5 Equations of State

Solution

First, the molar volume of the steam is calculated:

$$v = \frac{V}{n_T} \quad n_T = \frac{m}{M}$$

$$v = \frac{0.5 \text{ m}^3 \cdot 18.015 \cdot 10^{-3} \text{ kg/mol}}{5 \text{ kg}} = 1.8015 \text{ dm}^3/\text{mol}$$

Knowing the molar volume and the temperature, the pressure in the vessel can be calculated with the particular EOS.

1) ideal gas law:

$$P = \frac{RT}{v} = \frac{8.3143 \text{ J} \cdot 493.15 \text{ K mol}}{\text{mol K} \cdot 1.8015 \cdot 10^{-3} \text{ m}^3} = 22.76 \text{ bar}$$

2) Virial equation (Leiden form):

$$\frac{Pv}{RT} = z = 1 + \frac{B}{v}$$

$$z = 1 + \frac{(-293 \cdot 10^{-6})}{1.8015 \cdot 10^{-3}} = 0.8374$$

$$P = z\frac{RT}{v} = 0.8374 \cdot 22.76 \text{ bar} = 19.06 \text{ bar}$$

3) Soave–Redlich–Kwong equation:

$$T_r = \frac{493.15}{647.096} = 0.7621$$

$$P = \frac{RT}{v-b} - \frac{a(T)}{v(v+b)}$$

$$a(T) = a\alpha(T)$$

$$a = 0.42748 \cdot \frac{R^2 T_c^2}{P_c} = 0.42748 \cdot \frac{8.3143^2 \cdot 647.096^2}{220.64 \cdot 10^5} \frac{\text{J}^2\text{m}^2}{\text{mol}^2\text{N}}$$

$$= 0.5608 \text{ Nm}^4/\text{mol}^2$$

$$b = 0.8664 \cdot \frac{RT_c}{P_c} = 2.1126 \cdot 10^{-5} \text{ m}^3/\text{mol}$$

$$\alpha(T) = \left[1 + (0.48 + 1.574 \cdot 0.3443 - 0.176 \cdot 0.3443^2)\right.$$
$$\left.(1 - 0.7621^{0.5})\right]^2 = 1.2705$$

$$a(T) = a\alpha(T) = 0.5608 \cdot 1.2705 \text{ Nm}^4/\text{mol}^2 = 0.7125 \text{ Nm}^4/\text{mol}^2$$

$$P = \frac{8.3143 \cdot 493.15}{1.8015 \cdot 10^{-3} - 2.1126 \cdot 10^{-5}} \text{ Pa}$$
$$- \frac{0.7125}{1.8015 \cdot 10^{-3}(1.8015 \cdot 10^{-3} + 2.1126 \cdot 10^{-5})} \text{ Pa}$$
$$= 23.03 \text{ bar} - 2.1699 \text{ bar} = 20.86 \text{ bar}$$

4) High-precision equation of state.

Using a high-precision equation of state for water [20] in the FLUIDCAL system [41], the result is

$$P = P(493.15 \text{ K}; 10 \text{ kg/m}^3) = 20.38 \text{ bar}$$

This result can be taken as the reference. It becomes obvious that the ideal gas law and the virial equation are not appropriate to be used at ~20 bar. The error of the Soave–Redlich–Kwong equation (2.4%) might be acceptable; however, the equation has some difficulties with water as a strongly polar substance. Usually, even better results can be expected. The Soave–Redlich–Kwong equation is generalized, that is, T_c, P_c, and ω are the only input that is used, whereas the second virial coefficient had been adjusted to vapor densities.

Example 2.9

Calculate the density of propylene at $T = 300$ K and $P = 10$ bar using the Peng–Robinson equation of state. The vapor pressure of propylene is $P^s(300 \text{ K}) = 12.1$ bar.

Solution

The Peng–Robinson equation of state is given by Eq. (2.165)

$$z = \frac{v}{v-b} - \frac{a(T)v}{RT[v(v+b)+b(v-b)]} \quad (2.165)$$

It can be rewritten as [44]

$$z^3 + \left(\frac{bP}{RT}-1\right)z^2 + \left(\frac{aP}{R^2T^2} - 3\left(\frac{bP}{RT}\right)^2 - 2\frac{bP}{RT}\right)z$$

$$+ \left[\left(\frac{bP}{RT}\right)^3 + \left(\frac{bP}{RT}\right)^2 - \frac{aP}{R^2T^2}\frac{bP}{RT}\right] = 0 \quad (2.170)$$

The advantage of this formulation is that all coefficients and z itself remain dimensionless. The particular terms occurring in this equation are

$$b = 0.0778 \frac{RT_c}{P_c} = 0.0778 \frac{8.3143 \cdot 365.57}{46.646 \cdot 10^5} \frac{\text{J K}}{\text{mol K Pa}} \frac{\text{Pa}}{\text{J/m}^3}$$

$$= 5.0695 \cdot 10^{-5} \frac{\text{m}^3}{\text{mol}}$$

$$\alpha(300 \text{ K}) = \left[1 + (0.37464 + 1.54226\omega - 0.26992\omega^2)(1 - T_r^{0.5})\right]^2$$

$$= \left[1 + (0.37464 + 1.54226 \cdot 0.1408 - 0.26992 \cdot 0.1408^2)\right.$$

$$\left.(1 - (300/365.57)^{0.5})\right]^2$$

$$= 1.113426$$

2.5 Equations of State

$$a(300\text{ K}) = 0.45724 \frac{R^2 T_c^2}{P_c} \alpha(300\text{K})$$

$$= 0.45724 \frac{8.3143^2 \cdot 365.57^2}{46.646 \cdot 10^5} \cdot \frac{J^2 K^2}{mol^2 K^2} \cdot \frac{m^3}{J} \cdot 1.113426$$

$$= 1.008286 \frac{J\, m^3}{mol^2}$$

$$\frac{bP}{RT} = \frac{5.0695 \cdot 10^{-5} \cdot 10 \cdot 10^5}{8.3143 \cdot 300} \frac{m^3}{mol\, m^3} \frac{J}{J\, K} \frac{mol\, K}{} = 0.020324$$

$$\frac{aP}{R^2 T^2} = \frac{1.008286 \cdot 10 \cdot 10^5}{8.3143^2 \cdot 300^2} \frac{J\, m^3}{mol^2\, m^3} \frac{J}{J^2 K^2} \frac{mol^2 K^2}{} = 0.162065$$

For the application of Cardano's formula (see Appendix C, A10), Eq. (2.170) is written as

$$z^3 + Uz^2 + Sz + T = 0$$

with

$$U = \frac{bP}{RT} - 1 = 0.020324 - 1 = -0.979676$$

$$S = \frac{aP}{R^2 T^2} - 3\left(\frac{bP}{RT}\right)^2 - 2\frac{bP}{RT} = 0.162065 - 3 \cdot 0.020324^2 - 2 \cdot 0.020324$$
$$= 0.120178$$

$$T = \left(\frac{bP}{RT}\right)^3 + \left(\frac{bP}{RT}\right)^2 - \frac{aP}{R^2 T^2}\frac{bP}{RT} = 0.020324^3$$
$$+ 0.020324^2 - 0.162065 \cdot 0.020324 = -0.00287235$$

The terms in Cardano's formula are

$$P = \frac{3S - U^2}{3} = \frac{3 \cdot 0.120178 - 0.979676^2}{3} = -0.199744$$

$$Q = \frac{2U^3}{27} - \frac{US}{3} + T = -0.033276$$

giving the discriminant

$$D = \left(\frac{P}{3}\right)^3 + \left(\frac{Q}{2}\right)^2 = -1.83358 \cdot 10^{-5} < 0$$

This means that three real solutions are obtained. With the abbreviations

$$\Theta = \sqrt{-\frac{P^3}{27}} = 0.01718 \quad \text{and} \quad \Phi = \arccos\left(\frac{-Q}{2\Theta}\right) = 0.2519$$

where the angle is in the circular measure, the solutions are

$$z_1 = 2\Theta^{1/3} \cos\left(\frac{\Phi}{3}\right) - \frac{U}{3} = 0.84081$$

$$z_2 = 2\Theta^{1/3} \cos\left(\frac{\Phi}{3} + \frac{2\pi}{3}\right) - \frac{U}{3} = 0.03195$$

$$z_3 = 2\Theta^{1/3} \cos\left(\frac{\Phi}{3} + \frac{4\pi}{3}\right) - \frac{U}{3} = 0.10692$$

The largest value of the three solutions (1) corresponds to a vapor compressibility factor, and the lowest value (2) corresponds to a liquid one, whereas the middle value (3) has no physical meaning. Taking into account that $P = 10$ bar is lower than $P^s(300\,K) = 12.1$ bar, the vapor solution is relevant. The solution is

$$v(300\,K, 10\,bar) = \frac{z_1 RT}{P} = \frac{0.84081 \cdot 8.3143 \cdot 300}{10 \cdot 10^5} \frac{J\,K}{mol\,K} \frac{m^3}{J}$$

$$= 2.0972 \cdot 10^{-3} \frac{m^3}{mol}$$

and

$$\rho(300\,K, 10\,bar) = \frac{1}{v} = \frac{1}{2.0972 \cdot 10^{-3}} \frac{mol}{m^3} \cdot 42.081 \frac{g}{mol} = 20.065 \frac{kg}{m^3}$$

The value obtained with a high-precision equation of state is $\rho(300\,K, 10\,bar) = 20.06\,kg/m^3$. This illustrates the extraordinary quality of the Peng–Robinson equation of state for nonpolar substances.

2.5.5
Advanced Cubic Equations of State

The most widely used generalized cubic EOS, Soave–Redlich–Kwong [43] and Peng and Robinson [44], yield remarkably good results for nonpolar substances at elevated pressures. Nevertheless, they still show weaknesses that restrict their applicability to certain areas:

- the vapor pressures are often well reproduced but cannot be adjusted according to the accuracy demand of the calculation;
- the calculated liquid densities are often drastically wrong so that they are replaced by the results of density correlations (see Chapter 3) in the calculation routes of process simulators;
- the accuracy of the liquid heat capacities is usually not sufficient for process simulation.

Especially for polar substances these disadvantages become more and more obvious, not to mention the weaknesses in the description of mixtures (see Chapter 4).

On the other hand, the ability to handle supercritical components makes it worth to seek for more capable cubic EOS. A detailed analysis of the existing options has been performed by Ahlers et al. [48].

An approach to overcome the above mentioned difficulties was the PSRK (Predictive Soave–Redlich–Kwong) equation developed by Holderbaum and Gmehling in 1991 [49, 50], based on the Soave–Redlich–Kwong equation [43]. Its main progress is the use of g^E mixing rules (see Chapter 4) for a significant improvement of the description of mixtures with polar compounds. For the improvement of the pure component vapor pressures, an α-function, individual for every substance, was introduced. In the PSRK equation, the Mathias–Copeman approach [51] is used, which is a polynomial extension of the Soave α-function (Eq. (2.163)):

$$\alpha(T_r) = \left[1 + c_1\left(1 - \sqrt{T_r}\right) + c_2\left(1 - \sqrt{T_r}\right)^2 + c_3\left(1 - \sqrt{T_r}\right)^3\right]^2 \quad T_r \leq 1 \tag{2.171}$$

For the supercritical state, the parameters c_2 and c_3 are set to 0 to obtain a more reliable extrapolation, yielding again the original Soave α-function:

$$\alpha(T_r) = \left[1 + c_1\left(1 - \sqrt{T_r}\right)\right]^2 \quad T_r \geq 1 \tag{2.172}$$

The Mathias–Copeman parameters c_1, c_2, and c_3 are obtained by adjusting them to experimental vapor pressure data. If they are not available, the original Soave α-function (Eq. (2.163)) can further be used.[21] However, the original Soave α-function shows a qualitatively wrong extrapolation to high reduced temperatures that cannot be corrected by the Mathias–Copeman term. An appropriate α-function should fulfill the following criteria:

- At the critical point, the constraint is $\alpha = 1$.
- As the α-function represents the attractive forces between the molecules, its value should be positive.
- α should decrease with increasing temperature. For very high temperatures, it should approach 0 asymptotically.
- The α-function should be continuous. The parameters should not vanish if it is differentiated twice, as the correct representation of liquid heat capacities becomes more and more important.

According to an analysis of Noll [52], the Twu-α-function [53]

$$\alpha(T_r) = T_r^{N(M-1)} \exp\left[L\left(1 - T_r^{MN}\right)\right] \quad T_r \leq 1 \tag{2.173}$$

with the adjustable parameters L, M, and N can be regarded as appropriate for the subcritical region. For the supercritical state, an auxiliary function containing the acentric factor

$$\alpha(T_r) = \alpha^{(0)} + \omega(\alpha^{(1)} - \alpha^{(0)}) \quad T_r \geq 1 \tag{2.174}$$

---

21) The acentric factor ω can be estimated as described in Section 3.1.2, depending on the particular data situation.

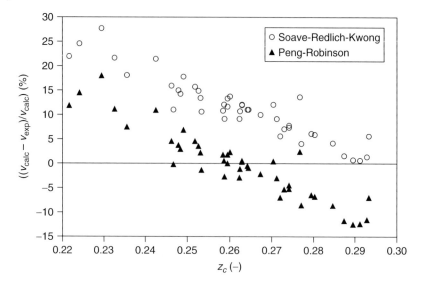

Figure 2.17 Relative deviations between experimental and calculated liquid molar volumes at $T_r = 0.7$ as a function of the critical compressibility factor z_c. Each triangle or circle represents a substance.

was first introduced. $\alpha^{(1)}$ and $\alpha^{(0)}$ are based on parameters that have been obtained by the adjustment of gas solubility data of nitrogen and hydrogen in long-chain alkanes:

$$\alpha^{(0)} = T_r^{-0.792614} \exp\left[0.401219 \cdot \left(1 - T_r^{-0.992614}\right)\right] \qquad (2.175)$$

$$\alpha^{(1)} = T_r^{-1.984712} \exp\left[0.024955 \cdot \left(1 - T_r^{-9.984712}\right)\right] \qquad (2.176)$$

Meanwhile, the latest results report that it is justified to use Eq. (2.173) in the supercritical region as well, using the parameters adjusted in the subcritical region [54].

All cubic EOS discussed perform poorly concerning the liquid density, where the parameter a or, respectively, the α-function have a negligible influence. Figure 2.17 illustrates the deviations of the liquid molar volume at $T_r = 0.7$ for the Soave–Redlich–Kwong and the Peng–Robinson equation of state. The 40 substances regarded are characterized by their critical compressibility z_c.

Only for substances with a high critical compressibility ($z_c = 0.275$–0.295, for example, hydrogen, nitrogen, and noble gases), the SRK equation of state has advantages in comparison to the Peng–Robinson equation of state. Even for simple nonpolar fluids such as alkanes and aromatic hydrocarbons there are more than 10% deviation from the correct value. The error becomes even larger for polar compounds such as ketones, alcohols, and halogenated hydrocarbons. Although the deviations for the Peng–Robinson equation are significantly lower, the results are still not acceptable for process simulation purposes.

An improvement can be achieved with the volume translation concept introduced by Peneloux et al. [55]. The idea is that the specific volume calculated by the equation of state is corrected by addition of a constant parameter c. The volume translation has no effect on the vapor–liquid equilibrium calculation, as both the liquid and the vapor volume are simultaneously translated by a constant value. The procedure has also little effect on the calculated vapor volumes, as c is in the order of magnitude of a liquid volume far away from the critical temperature.

The successor of the PSRK group contribution equation of state is the volume-translated Peng–Robinson equation (VTPR), based on the Peng–Robinson equation, which usually performs slightly better than the Soave–Redlich–Kwong equation of state. It makes use of the Twu-α-function, the volume translation and the g^E mixing rule (see Chapter 4) for the description of mixture properties [48]. By introducing the volume translation parameter c, the Peng–Robinson equation of state Eq. (2.166) is transformed into

$$P = \frac{RT}{(v+c-b)} - \frac{a(T)}{(v+c)(v+c+b)+b(v+c-b)} \qquad (2.177)$$

c can be obtained by adjustment to experimental liquid densities at $T_r = 0.7$:

$$c = v_{\text{calc}} - v_{\text{exp}} \qquad (2.178)$$

Alternatively, the generalized function

$$c = -0.252 \frac{R \cdot T_c}{P_c} (1.5448 z_c - 0.4024) \qquad (2.179)$$

can be used. In case they are not available, the required critical data can be estimated (see Section 3.1.1). For polar components and long chain alkanes, Eq. (2.179) does not perform very well.

With an adjusted c parameter, the deviations in the liquid density for $T_r < 0.8$ are in the range 1–2%. In the vicinity of the critical point ($T_r = 0.8\ldots 1.0$), larger errors are obtained, revealing a basic weakness in principle of cubic EOS. In process simulation, it should still be considered to override the results for the liquid density in the summary table with an appropriate correlation according to Section 3.2.2. For extreme pressures (e.g. >1000 bar), the application of a high-precision equation of state has to be preferred.

It is interesting to note that the introduction of a temperature-dependent volume translation has failed [56]. Although the agreement between experimental and calculated values improved, intersections of isotherms for $T_r > 1$ in the Pv-diagram have been detected. A temperature-dependent volume translation can only be used for special applications where the occurrence of extrapolation failures like that is not of importance.

For the application to mixtures, the volume translation parameter c can be calculated by a simple linear mixing rule:

$$c = \sum_i x_i \cdot c_i \qquad (2.180)$$

whereas for the parameters a and b the mixing rules

$$b = \sum_i \sum_j x_i x_j \left(\frac{b_{ii}^{3/4} + b_{jj}^{3/4}}{2} \right)^{4/3} \qquad (2.181)$$

and

$$\frac{a}{b} = \sum_i x_i \frac{a_{ii}}{b_{ii}} \cdot \frac{g_{res}^E}{0.53087} \qquad (2.182)$$

are used. g_{res}^E is explained in Chapter 4.

Using Eq. (2.180), the absolute values of mixture densities can be reproduced sufficiently well. The excess volume v^E of liquids (see Chapter 4) is usually not met; however, this is acceptable for process simulation applications.

Concerning the caloric properties, the enthalpy of vaporization is linked to the vapor pressure curve via the Clausius–Clapeyron equation (Eq. (2.86)). Therefore, the quality is mainly determined by the ability of the α-function in representing the vapor pressure data, which is usually sufficient (see above). The enthalpy of vaporization can easily be calculated with a cubic equation of state by subtracting the residual parts of the enthalpy of vapor and liquid in the saturation state

$$\Delta h_v(T) = \left[h^v(T, P^s) - h^{id}(T, P=0) \right] - \left[h^L(T, P^s) - h^{id}(T, P=0) \right]$$
$$= h^v(T, P^s) - h^L(T, P^s) \qquad (2.183)$$

Figure 2.18 is a typical diagram used for the illustration of the enthalpy of a pure substance in process simulation. Essentially, there are three lines.

- The enthalpy of the ideal gas, which appears to be a straight line as for most substances the specific heat capacity c_P^{id} is usually not a strong function of temperature in the region regarded here.
- The enthalpy on the dew-point line, which is more or less identical to the ideal gas enthalpy at low pressures but differs significantly at higher saturation pressures.
- The enthalpy on the saturated liquid line, which differs from the enthalpy on the dew-point line by the enthalpy of vaporization. It meets the enthalpy on the dew-point line at the critical point with a vertical tangent. The change of the liquid enthalpy with pressure is usually negligible and plays a minor role in process simulation. It is not regarded in this diagram.

As a reference point, the standard enthalpy of formation at $T = 298.15$ K and $P = 101\,325$ Pa, ideal gas state is used. It is equal to the enthalpy at $T = 298.15$ K and $P = 0$, as the enthalpy of ideal gases does not depend on the pressure.

Using an equation of state, the enthalpy for both vapor and liquid is calculated by subtracting the residual part determined with the equation of state from the ideal gas enthalpy. Further explanations concerning the enthalpy diagram are given in Section 6.2.

An analytical expression for c_P^L using the VTPR equation of state has been derived by Coniglio et al. [57], where the particular derivatives can be found in

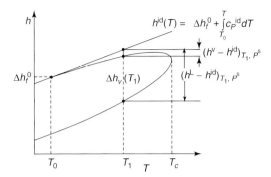

Figure 2.18 Calculation route for the enthalpy of vaporization.

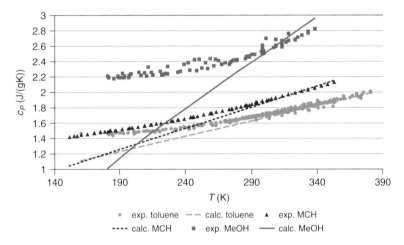

Figure 2.19 Representation of the liquid heat capacities of toluene, methanol, and methyl cyclohexane with the VTPR equation without simultaneous fitting to P^s and c_P^L.

the paper of Gasem *et al.* [58]. As $\partial^2 c/\partial T^2$ vanishes, the volume translation has no influence on c_P.

Like many models used in process simulation (see Section 6.2), the cubic equations in general have weaknesses in the correct representation of c_P^L. Some examples for the VTPR equation are given in Figure 2.19.

The reason for this behavior is explained in Section 6.2 and in [59]; the reproduction of the slope of the enthalpy of vaporization, which is directly related to the difference $(c_P^L - c_P^{id})$, is not accurate enough. The problem has been solved by adjusting the parameters of the Twu-α-function simultaneously to both vapor pressure and liquid heat capacity data [60]. By varying the weighting factors of both data types, it is usually possible to find a set of Twu-α-parameters that represents both vapor pressures and liquid heat capacities [61] with the required accuracy.

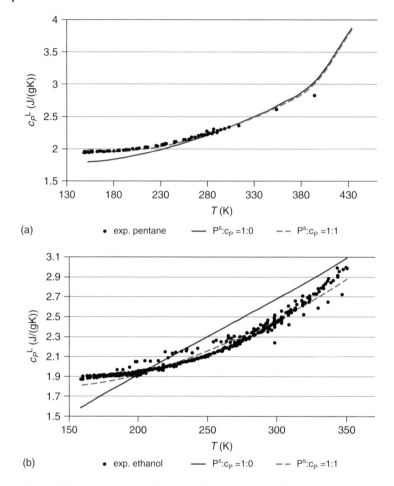

Figure 2.20 Representation of the liquid heat capacities of n-pentane (a) and ethanol (b) with the VTPR equation after simultaneous fitting to vapor pressure and heat capacity data. Experimental data from [9].

Examples are given in Figure 2.20. It should be mentioned that of course a reliable correlation for c_P^{id} is necessary.

Additional Problems

P2.1 Calculate the compressibility factor and the molar volume of methanol vapor at 200 °C and 10 bar
   a. using the ideal gas law
   b. with the virial equation truncated after the third virial coefficient.
   (virial coefficients: $B = -219$ cm^3/mol; $C = -17300$ cm^6/mol^2.)

Additional Problems

P2.2 A container with a volume of $V = 0.1$ m^3 is filled with $m = 10$ kg ethylene at a temperature of $T = 300$ K. What will be the pressure and compressibility factor of the gas? Use the virial equation truncated after the second virial coefficient and the Peng-Robinson equation of state to describe the PVT-behavior. (virial coefficient: $B = -138$ cm^3/mol, all other properties are given in Appendix A).

P2.3 Derive an expression for the second virial coefficient based on the van der Waals equation of state.

P2.4 Derive an expression for the volume dependence of the internal energy U at constant temperature based on the van der Waals equation of state.

P2.5 Calculate the pressure of 1 mol CO$_2$ in a container of 2.5 dm^3 at 40 °C via the
   a. ideal gas law
   b. virial equation
   c. Redlich-Kwong-equation.
   (virial coefficient: $B = -110$ cm^3/mol; for all other required properties see Appendix A).

P2.6 Derive the expressions for the residual functions $(h - h^{id})$, $(s - s^{id})$, $(g - g^{id})$ using the Redlich-Kwong equation of state.

P2.7 Calculate the change in enthalpy of 1 mol of liquid ethanol during isothermal compression from 1 to 100 bar at a temperature of 25 °C. The compressibility coefficient ($1.14 \cdot 10^{-4}$ bar^{-1}), the thermal expansion coefficient ($1.1 \cdot 10^{-3}$ K^{-1}) and the molar volume (58.04 cm^3/mol) are regarded as constant.

P2.8 Search for experimental second virial coefficients for nitrogen in the DDB using the DDBSP Explorer Version. Compare the values to the estimation results from the Tsonopoulos method via DDBSP-Predict. Estimate the Boyle temperature from the experimental findings.

P2.9 Retrieve the vapor pressure and the liquid density data for methane from the DDBSP Explorer Version and export the values to Excel. Implement a liquid vapor pressure curve calculation for the van der Waals equation of state in Excel-VBA and compare the results along the vapor-liquid coexistence curve to the experimental data.

P2.10 Estimate the acentric factor of methane, propane, pentane, and heptane using the critical data and normal boiling temperatures given in Appendix A and discuss the results.

P2.11 Calculate the vapor pressures of benzene between 280 K and 540 K using the Soave-Redlich-Kwong equation of state with critical data and acentric factor given in Appendix A. Compare the slope of the vapor pressure curve in the log(P^s) vs. 1/T diagram with the slope calculated via the vapor pressure correlation also given in Appendix A.

P2.12 In a heat exchanger, gaseous propylene is cooled down from $\vartheta_1 = 90$ °C, $P_1 = 20$ bar to $\vartheta_2 = 60$ °C. The pressure drop across the heat exchanger is $\Delta P = 2$ bar. How much cooling water is necessary? The supply and return temperatures of the cooling water are $\vartheta_{CWS} = 30$ °C and $\vartheta_{CWR} = 40$ °C, respectively. Use the Peng-Robinson equation of state for propylene and the function given for c_P^L in Appendix A for water.

P2.13 A closed vessel is filled completely with liquid water at $\vartheta_1 = 25\,°C$, $P_1 = 1$ bar. Due to solar radiation, it is heated up to $\vartheta_2 = 60\,°C$. Which pressure P_2 is built up? Use a high-precision equation of state. Calculate the transferred heat.

P2.14 An ideal gas is heated up from T_1 to T_2 in a heat exchanger. The pressure drop is $\Delta P = P_2 - P_1 > 0$. Why can the duty be calculated with the isobaric heat capacity by $q_{12} = c_P^{id} \cdot (T_2 - T_1)$, although a pressure drop occurs?

P2.15 Show that the relationship $c_P = -T \left(\frac{\partial^2 g}{\partial T^2} \right)_P$ is correct.

P2.16 A vessel is filled with nitrogen at $\vartheta_1 = 20\,°C$ and $P_1 = 1$ bar. With the help of a high precision equation of state,
   a. calculate the duty if the drum is heated up isobarically to $\vartheta_2 = 100\,°C$.
   b. calculate the duty if the drum is heated up isochorically to $\vartheta_2 = 100\,°C$.
   Interpret the results.

P2.17 Show that the relationship $\left(\frac{\partial g}{\partial T} \right)_P = -s$ is correct.

P2.18 Calculate the vapor pressure of acetone at $T_1 = 260$ K, $T_2 = 360$ K, and $T_3 = 450$ K using
   a. the vapor pressure equation listed in Appendix A
   b. the high-precision equation of state
   c. the Peng-Robinson equation of state
   d. the Redlich-Kwong-Soave equation of state
   e. the Redlich-Kwong equation of state
   f. the PSRK equation of state using the generalized SRK α-function
   g. the VTPR equation of state.
   What are the conclusions of the results?
   (All required parameters can be found in the Appendices A, B, J, K and the textbook page on www.ddbst.com.)

P2.19 Calculate the saturated density of liquid methanol at $T_1 = 300$ K and $T_2 = 430$ K using
   a. the density equation listed in Appendix A
   b. the high-precision equation of state
   c. the Peng-Robinson equation of state
   d. the Redlich-Kwong-Soave equation of state
   e. the Redlich-Kwong equation of state
   f. the PSRK equation of state using the generalized SRK α-function
   g. the VTPR equation of state.
   What are the conclusions of the results?
   (All required parameters can be found in the Appendices A, B, J, K and the textbook page on www.ddbst.com.)

P2.20 Calculate the compressibility factor z of gaseous propylene at $P_1 = 2$ bar and $P_2 = 10$ bar at $T = 293.15$ K, using the Tsonopoulos and the Peng-Robinson equations of state. Check the results with a high-precision equation of state. All required parameters can be found in the Appendices A and B.

P2.21 Make a list of all the input parameters necessary for the calculation of the enthalpy of the saturated vapor of a pure substance if
a. the Peng-Robinson equation of state
b. the vapor pressure equation in combination with the Peng-Robinson equation of state
c. the VTPR equation of state.
is used. The vapor pressure itself shall not be an input parameter.

P2.22 Calculate
a. the fugacity f^s in the saturation state
b. the molar volumes v^L and v^V in the saturation state
c. the enthalpy of vaporization.
for n-butane, benzene, and water at $\vartheta_1 = 30\,°C$, $\vartheta_2 = 80\,°C$, and $\vartheta_3 = 130\,°C$ using the Soave-Redlich-Kwong equation of state. All required parameters can be found in Appendix A.

P2.23 In Appendix A, the standard Gibbs energy of formation at $\vartheta = 25\,°C$ and $P = 1$ atm of water in the ideal gas state is reported to be $\Delta g_f^0 = -228590\,J/mol$. Estimate the Gibbs energy of formation for the liquid phase at $\vartheta = 25\,°C$ using the vapor pressure equation given in Appendix A.

References

1. Löffler, H.J. (1969) *Thermodynamik, Grundlagen und Anwendungen auf reine Stoffe*, Band **1**, Springer, Berlin.
2. van der Waals, J.D. (1873) Over de Continuiteit van den Gas- en Vloeistoftoestand. Thesis. Leiden.
3. Baehr, H.D. and Kabelac, S. (2006) *Thermodynamik*, Springer, Berlin.
4. Blanke, W. (1989) *PTB Mitt.*, **99**, 411–418.
5. Thess, A. (2010) *The Entropy Principle*, Springer, Berlin.
6. Rosenberg, R. (2005) *Phys. Today*, **12**, 50–55.
7. Falk, G. and Ruppel, W. (1976) *Energie und Entropie*, Springer, Berlin.
8. Lieb, E.H. and Yngvason, J. (1999) *Phys. Rep.*, **310**, 1–96.
9. Dortmund Data Bank www.ddbst.com.
10. Reid, R.C., Prausnitz, J.M., and Poling, B.E. (1987) *The Properties of Gases and Liquids*, McGraw-Hill, New York.
11. VDI-Gesellschaft Verfahrenstechnik und Chemieingenieurwesen (ed.) (2010) *VDI Heat Atlas*, 2nd edn, Springer, Berlin.
12. Wagner, W. and Overhoff, U. (2006) *Extended IAPWS-IF97 Steam Tables*, Springer, Berlin.
13. Dymond, J.H. and Smith, E.B. (1969) *The Virial Coefficients of Gases*, Clarendon Press, Oxford.
14. Tsonopoulos, C. (1974) *AIChE J.*, **20**, 263–272.
15. Hayden, J.G. and O'Connell, J.P. (1975) *Ind. Eng. Chem. Proc. Des. Dev.*, **14**, 209–216.
16. Benedict, M., Webb, G.B., and Rubin, L.C. (1940) *J. Chem. Phys.*, **8**, 334–345.
17. Bender, E. (1973) *The Calculation of Phase Equilibria from a Thermal Equation of State Applied to the Pure Fluids Argon, Nitrogen, Oxygen, their Mixtures*, C.F. Müller, Karlsruhe.
18. Span, R. (2000) *Multiparameter Equations of State – an Accurate Source of Thermodynamic Property Data*, Springer, Berlin.
19. Span, R., Wagner, W., Lemmon, E.W., and Jacobsen, R.T. (2001) *Fluid Phase Equilib.*, **183–184**, 1–20.

20. Wagner, W. and Pruß, A. (2002) *J. Phys. Chem. Ref. Data*, **31**, 387–535.
21. Setzmann, U. and Wagner, W. (1991) *J. Phys. Chem. Ref. Data*, **20**, 1061–1155.
22. Tegeler, C., Span, R., and Wagner, W. (1999) *J. Phys. Chem. Ref. Data*, **28**, 779–850.
23. Span, R. and Wagner, W. (1996) *J. Phys. Chem. Ref. Data*, **25**, 1509–1596.
24. Span, R., Lemmon, E.W., Jacobsen, R.T., and Wagner, W. (1998) *Int. J. Thermophys.*, **19**, 1121–1132.
25. Span, R., Lemmon, E.W., Jacobsen, R.T., Wagner, W., and Yokozeki, A. (2000) *J. Phys. Chem. Ref. Data*, **29**, 1361–1433.
26. Bücker, D. and Wagner, W. (2006) *J. Phys. Chem. Ref. Data*, **35**, 205–266.
27. Bücker, D. and Wagner, W. (2006) *J. Phys. Chem. Ref. Data*, **35**, 929–1019.
28. Smukala, J., Span, R., and Wagner, W. (2000) *J. Phys. Chem. Ref. Data*, **29**, 1053–1121.
29. Span, R. and Wagner, W. (2003) *Int. J. Thermophys.*, **24**, 1–39.
30. Setzmann, U. and Wagner, W. (1989) *Int. J. Themophys.*, **10**, 1103–1126.
31. Span, R., Collmann, H.-J., and Wagner, W. (1998) *Int. J. Thermophys.*, **19**, 491–500.
32. Span, R. and Wagner, W. (2003) *Int. J. Thermophys.*, **24**, 41–109.
33. Lemmon, E.W. and Span, R. (2006) *J. Chem. Eng. Data*, **24**, 41–109.
34. Span, R. and Wagner, W. (2003) *Int. J. Thermophys.*, **24**, 111–162.
35. Ihmels, E.C. and Lemmon, E.W. (2003) *Fluid Phase Equilib.*, **207**, 111–130.
36. Ihmels, E.C. and Lemmon, E.W. (2007) *Fluid Phase Equilib.*, **260**, 36–48.
37. Lemmon, E.W. and Ihmels, E.C. (2005) *Fluid Phase Equilib.*, **228–229**, 173–187.
38. Piazza, L., Herres, G., and Span, R. (2006) 16th Symposium of Thermophysical Properties, Boulder, July 30–August 4, 2006.
39. Kunz, O., Kilmeck, R., Wagner, W., and Jaeschke, M. (2007) *The GERG-2004 Wide-Range Reference Equation of State for Natural Gases*, GERG Technical Monograph TM15, Fortschr.-Ber. VDI, Reihe 6, Nr. **557**, VDI-Verlag, Düsseldorf.
40. Achtermann, H.J., Baehr, H.D., and Bose, T.K. (1989) *J. Chem. Thermodyn.*, **21**, 1023–1043.
41. Wagner, W. (2005) *FLUIDCAL, Software for the Calculation of Thermodynamic and Transport Properties of Several Fluids*, Bochum, Ruhr-Universität Bochum.
42. Redlich, O. and Kwong, J.N.S. (1949) *Chem. Rev.*, **44**, 233–244.
43. Soave, G. (1972) *Chem. Eng. Sci.*, **27**, 1197–1203.
44. Peng, D.Y. and Robinson, D.B. (1976) *Ind. Eng. Chem. Fundam.*, **15**, 59–64.
45. Deiters, U.K. (2002) *AIChE J.*, **48**, 882–886.
46. Moore, W.J. and Hummel, D.O. (1986) *Physikalische Chemie*, 4th edn, Walter de Gruyter, Berlin.
47. Lee, B.I. and Kesler, M.G. (1975) *AIChE J.*, **21**, 510–527.
48. Ahlers, J., Yamaguchi, T., and Gmehling, J. (2004) *Ind. Eng. Chem. Res.*, **43**, 6569–6576.
49. Holderbaum, T. and Gmehling, J. (1991) *Fluid Phase Equilib.*, **70**, 251–265.
50. Horstmann, S., Fischer, K., and Gmehling, J. (2000) *Fluid Phase Equilib.*, **167**, 173–186.
51. Mathias, P.M. and Copeman, T.W. (1983) *Fluid Phase Equilib.*, **13**, 91–108.
52. Noll, O. (1998) Analyse der Beschreibung thermodynamischer Eigenschaften von Reinstoffen und Mischungen unter Verwendung kubischer Zustandsgleichungen. Thesis. Oldenburg.
53. Twu, C.H., Bluck, D., Cunningham, J.R., and Coon, J.E. (1991) *Fluid Phase Equilib.*, **69**, 33–50.
54. AIF Projekt (IGF-Project No 15345N) : Weiterentwicklung einer universellen Gruppenbeitragszustandsgleichung "VTPR", Oldenburg 2003-2010.
55. Peneloux, A., Rauzy, E., and Freze, R. (1982) *Fluid Phase Equilib.*, **8**, 7–23.
56. Pfohl, O. (1999) *Fluid Phase Equilib.*, **163**, 157–159.
57. Coniglio, L., Rauzy, E., Peneloux, A., and Neau, E. (2002) *Fluid Phase Equilib.*, **200**, 375–398.
58. Gasem, K.A.M., Gao, W., Pan, Z., and Robinson, R.L. Jr. (2001) *Fluid Phase Equilib.*, **181**, 113–125.

59. Kleiber, M. (2003) *Ind. Eng. Chem. Res.*, **42**, 2007–2014.
60. Diedrichs, A. (2005) Optimierung eines dynamischen Differenzkalorimeters zur experimentellen Bestimmung von Wärmekapazitäten, Thesis. Oldenburg.
61. Diedrichs, A., Rarey, J., and Gmehling, J. (2006) *Fluid Phase Equilib.*, **248**, 56–69.

3
Correlation and Estimation of Pure Component Properties

Equations of state valid both for vapor and liquid phase as well as for pure components and mixtures provide all the necessary volumetric and caloric properties which are needed for process simulation except c_p^{id}, where a correlation has to be provided. In case of the real liquid mixture behavior, the activity coefficient approach provides an alternative option which is still applied in most cases in chemical industry. This method is explained in detail in Chapter 5. It is more appropriate for low-pressure applications. Concerning the physical property calculation, it is more accurate, as all the properties are represented by separate correlations. The following section explains the correlation and estimation methods for the particular properties.

Commercially available compilations of pure component characteristic physical property constants and temperature-dependent properties[1] for convenient use in process simulators have been developed by AIChE (DIPPR), DDBST, DECHEMA, NIST (TRC), and PPDS. They should be checked first before estimating thermophysical properties. DDBST has developed a comprehensive pure component thermophysical property estimation software (Artist), where many of the methods discussed in this chapter are implemented.

In this chapter, the empirical mixing rules for the particular pure component properties are also mentioned, as well as the diffusion coefficients.

3.1
Characteristic Physical Property Constants

For the description of thermophysical properties, estimation methods based on the three parameter corresponding states principle mostly employ the critical temperature (T_c), the critical pressure (P_c), and the acentric factor (ω) as characteristic

---

[1] In fact, the particular quantities are pressure-dependent as well. As the vapor density is usually calculated with an equation of state, it can be omitted in this chapter. For the other quantities, the pressure dependence is less distinctive than the temperature dependence and accounted for with correction terms. Properties which are linked to the saturation line (e.g., vapor pressure, enthalpy of vaporization) are only temperature-dependent.

Chemical Thermodynamics: for Process Simulation, First Edition.
Jürgen Gmehling, Bärbel Kolbe, Michael Kleiber, and Jürgen Rarey.
© 2012 Wiley-VCH Verlag GmbH & Co. KGaA. Published 2012 by Wiley-VCH Verlag GmbH & Co. KGaA.

constants of a substance. Especially, these quantities are the basis for cubic equations of state. As the results of calculations significantly depend on the quality of the input data, it is worth to determine them carefully.

However, especially the experimental determination of critical constants usually requires a high experimental effort. Moreover, a lot of substances used in chemical industry decompose at high temperatures before the critical point is reached. Thus, these quantities are often not available from experiment. Therefore, reliable estimation methods are necessary, as the estimation of many other properties (density, thermal conductivity, surface tension, heat capacity, vapor pressure, and enthalpy of vaporization) is based on the critical data and needs reliable input data. In the following section, the currently most frequently used estimation methods for T_c, P_c, and ω are introduced, as well as the ones for the normal boiling point, critical volume, and melting point.

In combination with correlation and estimation methods for the temperature-dependent thermophysical properties, a system of methods can be established that finds for any combination of given data the optimum method to estimate unknown data. As an extreme case, one can generate all the data just from the structural formula as the only available information. This might make sense in case of a not important side component; however, it should be avoided if possible as the errors propagate more and more. The strategy to obtain reliable data should be to use as much information as possible from data banks or experiments.

3.1.1
Critical Data

Critical temperatures, critical pressures, and critical volumes can be estimated with group contribution methods. The most common ones are from Lydersen [1], Ambrose [2, 3], Joback and Reid [4], and Constantinou and Gani [5]. They are all explained in [6]. A recent development by Rarey and Nannoolal [7, 8] seems to be the most accurate one, but it is more difficult to apply.

The earliest estimation method for the critical temperature is the Guldberg rule, which was established already in 1887:

$$\frac{T_b}{T_c} = \frac{2}{3} \qquad (3.1)$$

Although it is very simple, it can be regarded as the basis of other methods like the Joback method, where the ratio between the normal boiling point and critical point in the vicinity of 2/3 can be found as well.

The Joback method needs only the structural formula as input information except for the estimation of the critical temperature, where additionally the normal boiling point is required. In case it is unknown, it can be estimated with the Joback method as well. The Joback relationships are

$$\frac{T_c}{K} = \frac{T_b}{K}\left[0.584 + 0.965\sum\Delta_T - \left(\sum\Delta_T\right)^2\right]^{-1} \qquad (3.2)$$

$$\frac{P_c}{\text{bar}} = \left(0.113 + 0.0032 n_A - \sum \Delta_P\right)^{-2} \qquad (3.3)$$

$$\frac{v_c}{\text{cm}^3/\text{mol}} = 17.5 + \sum \Delta_v \qquad (3.4)$$

The group contributions for Δ_T, Δ_P, and Δ_v can be taken from Table 3.1, as well as Δ_b, Δ_H, and Δ_G, which are discussed later. n_A is the number of atoms in a molecule including H-atoms. The group incrementation of a molecule is simple, as the increments are directly listed.

For the critical temperature, the quality of the estimation is usually quite good; usually, the correct value is met except for a few K. Poling et al. [6] report an average error for the Joback method of 1.1% of the absolute critical temperature, if the normal boiling point is known. Only 1% of the tested substances had an error larger than 5%. If the normal boiling point is not known, the Constantinou and Gani method [5] can be regarded as an alternative. The average error of the Joback method for the critical pressure is 4.6% [6]. As the case may be, these errors can be transferred to the estimation of vapor pressures (Section 3.2.1) or to the accuracy of cubic equations of state. The critical volume is less important. It is used for the estimation of the liquid density with the COSTALD method (Section 3.2.2) or for the VTPR equation (Section 2.5.5), if no reference point is available. The average error for the estimation of critical volumes with the Joback method is 3.1%.

The average errors for the Joback method reported by Nannoolal [8] are slightly larger: 1.41% for the critical temperature, 7.11% for the critical pressure, and 3.73% for the critical volume. A detailed review can be found in [7, 8].

In all cases, the statements for the average errors refer to molecules with more than three carbon atoms, as for simpler substances experimental data are usually available and estimations do not make sense. As well, the Joback method gives unreasonable results for large molecules with more than 20 C-atoms [7, 8].

If T_c, P_c, and v_c are available, it is strongly recommended to check their consistency with the critical compressibility factor

$$z_c = \frac{P_c v_c}{R T_c} \qquad (3.5)$$

z_c is usually in the range $0.21 < z_c < 0.29$.

Example 3.1

Estimate the critical data of m-xylene using the Joback method (Figure 3.1). The normal boiling point is given to be $T_b = 412.25$ K.

Solution

The group incrementation of m-xylene is as follows:

    4 = CH– (ring) groups
    2 = C< (ring) groups
    2 –CH$_3$ groups

Table 3.1 Group contributions for the Joback method.

Structural group	Δ_T	Δ_P	Δ_V	Δ_b	Δ_H	Δ_G
$-CH_3$	0.0141	−0.0012	65	23.58	−76.45	−43.96
$>CH_2$	0.0189	0.0000	56	22.88	−20.64	8.42
$>CH-$	0.0164	0.0020	41	21.74	29.89	58.36
$>C<$	0.0067	0.0043	27	18.25	82.23	116.02
$=CH_2$	0.0113	−0.0028	56	18.18	−9.63	3.77
$=CH-$	0.0129	−0.0006	46	24.96	37.97	48.53
$=C<$	0.0117	0.0011	38	24.14	83.99	92.36
$=C=$	0.0026	0.0028	36	26.15	142.14	136.70
$\equiv CH$	0.0027	−0.0008	46	9.20	79.30	77.71
$\equiv C-$	0.0020	0.0016	37	27.38	115.51	109.82
$-CH_2-$ (ring)	0.0100	0.0025	48	27.15	−26.80	−3.68
$>CH-$ (ring)	0.0122	0.0004	38	21.78	8.67	40.99
$>C<$ (ring)	0.0042	0.0061	27	21.32	79.72	87.88
$=CH-$ (ring)	0.0082	0.0011	41	26.73	2.09	11.30
$=C<$ (ring)	0.0143	0.0008	32	31.01	46.43	54.05
$-F$	0.0111	−0.0057	27	−0.03	−251.92	−247.19
$-Cl$	0.0105	−0.0049	58	38.13	−71.55	−64.31
$-Br$	0.0133	0.0057	71	66.86	−29.48	−38.06
$-I$	0.0068	−0.0034	97	93.84	21.06	5.74
$-OH$ (alcohols)	0.0741	0.0112	28	92.88	−208.04	−189.20
$-OH$ (phenols)	0.0240	0.0184	−25	76.34	−221.65	−197.37
$-O-$	0.0168	0.0015	18	22.42	−132.22	−105.00
$-O-$ (ring)	0.0098	0.0048	13	31.22	−138.16	−98.22
$>C=O$	0.0380	0.0031	62	76.75	−133.22	−120.50
$>C=O$ (ring)	0.0284	0.0028	55	94.97	−164.50	−126.27
$-CH=O$	0.0379	0.0030	82	72.24	−162.03	−143.48
$-COOH$	0.0791	0.0077	89	169.09	−426.72	−387.87
$-COO-$	0.0481	0.0005	82	81.10	−337.92	−301.95
$=O$	0.0143	0.0101	36	−10.50	−247.61	−250.83
$-NH_2$	0.0243	0.0109	38	73.23	−22.02	14.07
$>NH$	0.0295	0.0077	35	50.17	53.47	89.39
$>NH$ (ring)	0.0130	0.0114	29	52.82	31.65	75.61
$>N-$	0.0169	0.0074	9	11.74	123.34	163.16
$-N=$	0.0255	−0.0099	–	74.60	23.61	–
$-N=$ (ring)	0.0085	0.0076	34	57.55	55.52	79.93
$=NH$	–	–	–	–	93.70	119.66
$-CN$	0.0496	−0.0101	91	125.66	88.43	89.22
$-NO_2$	0.0437	0.0064	91	152.54	−66.57	−16.83
$-SH$	0.0031	0.0084	63	63.56	−17.33	−22.99
$-S-$	0.0119	0.0049	54	68.78	41.87	33.12
$-S-$ (ring)	0.0019	0.0051	38	52.10	39.10	27.76

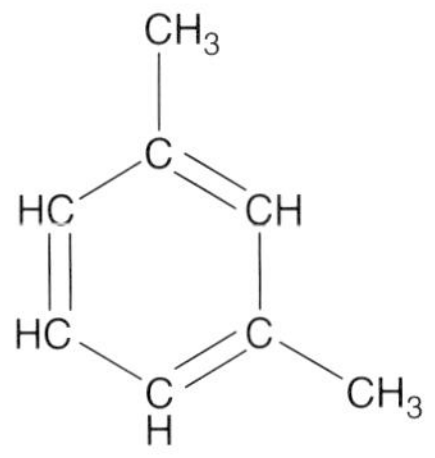

Figure 3.1 Structural formula of m-xylene.

The particular contributions add up to

$$\sum \Delta_T = 4 \cdot (0.0082) + 2 \cdot (0.0143) + 2 \cdot (0.0141) = 0.0896$$
$$\sum \Delta_P = 4 \cdot (0.0011) + 2 \cdot (0.0008) + 2 \cdot (-0.0012) = 0.0036$$
$$\sum \Delta_v = 4 \cdot (41) + 2 \cdot (32) + 2 \cdot (65) = 358$$

The results for the critical data are

$$T_c = 412.25 \text{ K}/(0.584 + 0.965 \cdot 0.0896 - 0.0896^2) = 622.32 \text{ K}$$
$$P_c = (0.113 + 0.0032 \cdot 18 - 0.0036)^{-2} \text{ bar} = 35.86 \text{ bar}$$
$$v_c = (17.5 + 358) \text{ cm}^3/\text{mol} = 0.3755 \text{ m}^3/\text{kmol}$$

The experimental values [9] for m-xylene are

$T_c = 617.05$ K
$P_c = 35.41$ bar
$v_c = 0.3750 \text{ m}^3/\text{kmol}$

The agreement between estimated and experimental values is remarkably satisfactory. The check of the critical compressibility factor using the estimated values yields

$$z_c = \frac{35.86 \cdot 10^5 \cdot 0.3755 \cdot 10^{-3}}{8.3143 \cdot 622.32} = 0.260$$

which is well within the usual range between 0.21 and 0.29.

It should be noted that the properties of m-xylene were most likely used to regress the group parameters of the various methods.

Recently, a new method has been developed by Rarey and Nannoolal [7, 8]. Based on the Dortmund Data Bank, the method has made use of all the data currently available. It has overcome the difficulties of other methods for large molecules with more than 20 carbon atoms. The calculation equations are

$$T_c = T_b \left(0.699 + \frac{1}{0.9889 + \left(\sum_i v_i \Delta_{T,i} + GI \right)^{0.8607}} \right) \tag{3.6}$$

$$\frac{P_c}{\text{kPa}} = \frac{\left(\dfrac{M}{\text{g/mol}}\right)^{-0.14041}}{\left(0.00939 + \sum_i v_i \Delta_{P,i} + \text{GI}\right)^2} \tag{3.7}$$

$$\frac{V_c}{\text{cm}^3/\text{mol}} = \frac{\sum_i v_i \Delta_{v,i} + \text{GI}}{n_A^{-0.2266}} + 86.1539 \tag{3.8}$$

v_i is the number of occurrences of group i in the molecule. n_A is the number of atoms without hydrogen. The particular group contributions are given on www.ddbst.com. The group assignment philosophy is a little bit different from other methods.

The number of atoms is evaluated without counting the hydrogen atoms. The groups are sometimes ambiguous; therefore, a priority number is assigned to the groups. The lower the priority number, the higher is its priority.

GI refers to group interaction parameters which have to be taken into account when more than one hydrogen-bonding group is present in the molecule. In these cases, other models often give large deviations. The use of the group interactions is obligatory. The group interaction contribution has to be divided by the number of atoms:

$$\text{GI} = \frac{1}{n_A} \sum_{i=1}^{m} \sum_{j=1}^{m} \frac{C_{ij}}{m-1} \tag{3.9}$$

with

C_{ij} group interaction contribution between group i and group j ($C_{ii} = 0$),
n_A number of atoms except hydrogen,
m total number of interaction groups in the molecule.

Different from other group contribution methods, the combination of carbon and hydrogen atoms is not decisive for the group assignment. In contrast, the number of other strongly electronegative atoms already bound to a carbon is taken into account as this more closely correlates with the estimated properties. For example, in the Joback method a –CHF– segment would be incremented as >CH– and F, whereas the Rarey/Nannoolal method can also treat it as >C< and F. The priority decides which group is finally used.

Example 3.2

Estimate the critical data of R227ea (CF$_3$–CHF–CF$_3$) with the Rarey/Nannoolal method. The normal boiling point is given by $T_b = 256.73$ K [10]. The molar mass is $M = 170.027$ g/mol. The number of atoms in the molecules is 10, where hydrogen is not counted.

Solution

The group assignment can be performed as follows:

Group ID	Group name	Frequency	Contribution T_c	Contribution P_c	Contribution V_c
21	F connected to C substituted with one F	6	0.0181302	0.00034933	3.3646
19	F connected to nonaromatic C	1	0.0156068	0.00007328	−5.0331
7	C in a chain connected to at least one F	3	0.0528003	0.00034310	33.7577
124	Component has one H	1	−0.0027180	0.00028103	−6.1909
121	C with three halogens	2	−0.0013023	0.00004387	1.5807
			0.2774663	0.00356733	113.3981

giving the following results for the critical values:

$$T_c = 256.73 \text{ K} \left(0.699 + \frac{1}{0.9889 + 0.2774663^{0.8607}} \right) = 373.855 \text{ K}$$

$$P_c = \frac{170.027^{-0.14041}}{(0.00939 + 0.00356733)^2} \text{ kPa} = 2.896 \text{ MPa}$$

$$v_c = \left(\frac{113.3981}{10^{-0.2266}} + 86.1539 \right) \text{ cm}^3/\text{mol} = 277.23 \text{ cm}^3/\text{mol}$$

The reported values [10] are $T_c = 375.95$ K, $P_c = 2.999$ MPa, and $v_c = 293.035$ cm^3/mol. With the Joback method, one would have obtained $T_c = 379.7$ K, $P_c = 3.17$ MPa, and $v_c = 301.5$ cm^3/mol.

3.1.2
Acentric Factor

As discussed in Section 2.5.4, the simple two-parameter corresponding states principle indicates that a generalized equation of state for all substances can be created using only two specific parameters, for example, T_c and P_c. The success of this approach is restricted to simple, spherical molecules like Ar, Kr, Xe, or CH$_4$, where vapor pressure and compressibility factor can be reasonably described. For other molecules, the simple two-parameter corresponding states principle leads to significant errors. A large improvement has been achieved with the introduction of a third parameter which describes the vapor pressure curve (extended three-parameter principle of corresponding states). The most common parameter of this kind is the so-called acentric factor, which is defined as

$$\omega = -1 - \log \left(\frac{P^s}{P_c} \right)_{T/T_c=0.7} \tag{2.157}$$

The physical meaning of this parameter is the description of orientation-dependent intermolecular interactions. The parameter is used in many correlations for the estimation of thermophysical properties, especially in cubic equations of state. The definition makes sense, as vapor pressures as a function of temperature are easily accessible and characteristic for a substance. The reference temperature $T = 0.7 \cdot T_c$ has been chosen because it is often close to the normal boiling point. Therefore, an estimation of ω would essentially be similar to an estimation of the vapor pressure curve or, respectively, the normal boiling point.

Although estimation methods for the acentric factor are available (e.g., group contribution method of Constantinou and Gani [11]), this concept is not recommended, as an estimation of ω would only be a redundancy of the normal boiling point estimation. To avoid inconsistencies, it is recommended that ω is always calculated by its defining equation 2.157. If the vapor pressure curve is unknown, the normal boiling point and the critical point can be estimated, and the Hoffmann–Florin equation or the Rarey/Moller method (Section 3.2.1) can be used to calculate the vapor pressure curve.

The acentric factor increases with the size of the molecule, but only in extreme cases values >1 can be obtained, for example, for hydrocarbons with a molar mass >300. Helium ($\omega = -0.39$) and hydrogen ($\omega = -0.216$) have negative acentric factors as so-called quantum gases. Methane and the noble gases argon, krypton, xenon, and neon have acentric factors close to 0. Otherwise, $\omega < 0$ can be ruled out. If such a value is evaluated, something is wrong with the vapor pressure curve or the critical point.

3.1.3
Normal Boiling Point

The normal boiling point is an easily accessible physical property and has been measured for a large number of substances. In case that it is not available, the normal boiling point can be estimated with group contribution methods, for example, Joback and Reid [4] and Constantinou and Gani [5], analogously to the estimation of the critical point. The estimation formula for the Joback method is

$$T_b/K = 198 + \sum \Delta_b \tag{3.10}$$

The particular group contributions can be taken from Table 3.1. The accuracy is much lower than for the estimation of the critical data.

There is an opportunity of taking a known substance with a similar structure as reference. Because of the high uncertainty, for the normal boiling point estimation this procedure is recommended to avoid possible large errors.

Example 3.3

Estimate the normal boiling point of pentylcyclohexane (Figure 3.2).

1) using Joback's method,
2) using Joback's method with methylcyclohexane as reference substance.

Solution

1) Joback's method
   The group incrementation for pentylcyclohexane is

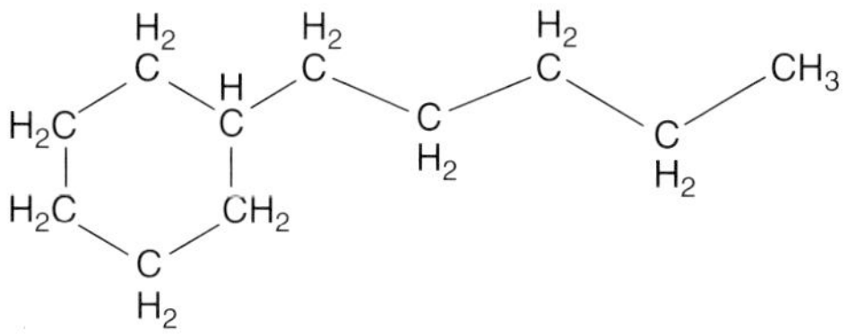

Figure 3.2 Structural formula of pentylcyclohexane.

    5 · CH_2 (ring)
    1 · CH (ring)
    1 · CH_3
    4 · CH_2

The group contributions add up to

$$\sum \Delta_b = 5 \cdot (27.15) + 1 \cdot (21.78) + 1 \cdot (23.58) + 4 \cdot (22.88) = 272.63$$

The result for the normal boiling point estimation is

$$T_b = (198 + 272.63) \text{ K} = 470.63 \text{ K}$$

2) Joback's method with methylcyclohexane as reference substance (Figure 3.3)
The normal boiling point for methylcyclohexane is $T_b = 373.95$ K.
The sum of group contributions for methylcyclohexane can be calculated backward:

$$\sum \Delta_{b,\text{methylcyclohexane}} = 373.95 - 198 = 175.95$$

The difference of the structural formulas of pentylcyclohexane and methylcyclohexane are the four aliphatic CH_2-groups. Therefore, the sum of group contributions for pentylcyclohexane can be determined to be

$$\sum \Delta_b = 175.95 + 4 \cdot (22.88) = 267.47$$

The corresponding estimation for the normal boiling point is

$$T_b = (198 + 267.47) \text{ K} = 465.47 \text{ K}$$

The experimental value for pentylcyclohexane is $T_b = 476.75$ K. The Joback method provides an excellent result in this case. The use of a reference substance is not a guarantee to obtain a more accurate result but helps to avoid large errors.

A new equation for estimating the normal boiling point has been developed by Rarey and Nannoolal [12]:

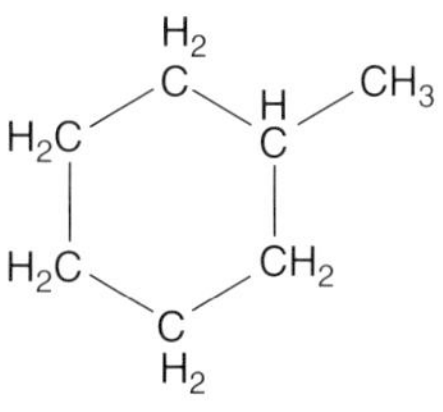

Figure 3.3 Structural formula of methylcyclohexane.

$$\frac{T_b}{K} = \frac{\sum_i v_i \Delta_{bi} + GI}{n_A^{0.6587} + 1.6902} + 84.3359 \qquad (3.11)$$

with

$$GI = \frac{1}{n_A} \sum_{i=1}^{m} \sum_{j=1}^{m} C_{ij} \qquad (3.12)$$

and

C_{ij} group interaction contribution between group i and group j ($C_{ii} = 0$),
n_A number of atoms except hydrogen.

v_i is the number of occurrences of group i in the molecule. The values for the particular group contributions are listed on www.ddbst.com. The results for the representation of the data sets are reported to be superior to those obtained with other methods [12]. An evaluation with more than 2000 substances yielded average deviations of 16.5 K for the Joback method and only 1.5 K for the Rarey/Nannoolal method, which demonstrates the superiority of the latter one.

Example 3.4

Estimate the normal boiling point of pentylcyclohexane (Figure 3.2) with the Rarey/Nannoolal method.

Solution

The group contributions are

Group ID	Group name	Frequency	Contribution T_b	Sum
9	CH$_2$ in a ring	5	239.4957	1197.4785
10	CH in a ring	1	222.1163	222.1163
4	CH$_2$ in a chain	4	239.4511	957.8044
1	CH$_3$ not connected to N,O,F,Cl	1	177.3066	177.3066
				2554.7058

giving

$$T_b = \frac{2554.7058 \text{ K}}{11^{0.6587} + 1.6902} + 84.3359 \text{ K} = 474.8 \text{ K}$$

which is an almost perfect result.

3.1.4
Melting Point and Enthalpy of Fusion

Melting points are listed for a large number of substances in data banks and handbooks. Their estimation is hardly possible, as they are influenced by the crystal structure, enthalpy of fusion, which represents the molecular interactions, and by the entropy of fusion, which depends on the molecular symmetry. Therefore, the application of simple group contribution methods [4, 5] is not very successful, as the information about the symmetry of a molecule is lost after the incrementation.

The case is similar for the enthalpy of fusion. To a large extent, it depends on the structure of the crystal that is formed. In principle, the Clausius–Clapeyron equation can be applied; however, the information about the pressure dependence of the melting point is usually not available. The so-called Walden rule [13] gives at least an estimation for aromatic compounds like benzene or naphthalene:

$$\frac{\Delta h_m(T_m)}{J/mol} \frac{1}{T_m/K} \approx 54.4 \tag{3.13}$$

As the fusion process is accompanied by with a change in volume, the melting point is a function of pressure. Very high pressures can change the melting point significantly. Analogously to Eq. (2.86), the Clausius–Clapeyron equation can be applied:

$$\Delta h_m = T_m \frac{dP_m}{dT_m} (v^L - v^S) \tag{3.14}$$

Setting a reference melting point $T_{m,ref}$ at $P_{m,ref}$ and substituting

$$\Delta v_m = v^L - v^S \tag{3.15}$$

and

$$\frac{dP_m}{dT_m} \approx \frac{P_m - P_{m,ref}}{T_m - T_{m,ref}} \tag{3.16}$$

one gets after rearrangement

$$T_m = \frac{T_{m,ref}}{1 - (P_m - P_{m,ref}) \frac{\Delta v_m}{\Delta h_m}} \tag{3.17}$$

The approximation $\ln(1+x) \approx x$ for small values of x leads to

$$T_m = T_{m,ref} \exp\left[(P_m - P_{m,ref}) \frac{\Delta v_m}{\Delta h_m}\right] \tag{3.18}$$

Using Eq. (3.18), only trends can be predicted. The necessary input information about Δh_m and Δv_m with respect to temperature is usually missing. If data are available, the Simon–Glatzel equation [14]

$$T_m = T_{m,ref} \left(\frac{P_m/MPa}{a} + 1\right)^{1/c} \tag{3.19}$$

can successfully be applied for correlation, where a temperature close to the triple point is used for $T_{m,ref}$ and the corresponding pressure has been neglected. For substances like water, where the melting point decreases with increasing pressure, the Simon–Glatzel equation is not appropriate. With an extension made by Kechin [15], this difficulty can be overcome:

$$T_m = T_{m,ref} \left(\frac{P_m/MPa}{a_1} + 1\right)^{a_2} \exp\left(-a_3 \frac{P_m}{MPa}\right) \tag{3.20}$$

Table 3.2 Parameters for the Kechin equation.

Substance	$T_{m,ref}$ (K)	a_1	a_2	a_3
Water	273.25	550.1	0.4593	0.00111
Methanol	175.17	358.6	0.3082	0
Ethanol	158.37	549.4	0.4667	0
Cyclohexane	279.55	280.0	0.5476	0
Benzene	278.24	347.9	0.3689	0
Cyclohexanone	242.4	228.9	0.3257	0
Cyclopentanol	256.0	380.5	0.5208	0

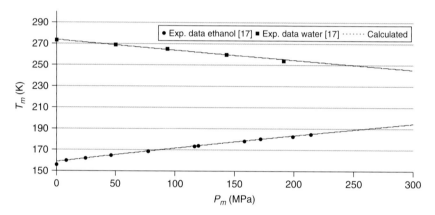

Figure 3.4 Description of the pressure dependence of the melting point using the Simon–Glatzel equation.

For $a_3 = 0$, Eq. (3.20) reduces to the Simon–Glatzel equation. For a few substances, the parameters a_1, a_2, and a_3 are listed in Table 3.2.

In Figure 3.4, the pressure dependence of the experimental and correlated melting points for water and ethanol is plotted.

Example 3.5

In a television broadcast [18] it was reported that at the bottom of the Mariana Trench, the deepest part of the world's oceans with a depth of 11 000 m, the temperature is $\vartheta = -2\,°C$, which is below the well-known melting point of $0\,°C$. Check whether this is plausible or not. The differences in the properties between pure water and sea water shall be neglected. For the liquid density of water, $\rho^L = 1000\,kg/m^3$ shall be used. The pressure is $P_{sea\,level} = 1$ bar.

Solution

Due to hydrostatics, the pressure at the bottom of the Mariana Trench is

$$P = P_{\text{sea level}} + \rho^L g h = 1 \text{ bar} + 1000 \, \frac{\text{kg}}{\text{m}^3} \cdot 9.81 \, \frac{\text{m}}{\text{s}^2} \cdot 11000 \text{ m} = 1 \text{ bar}$$
$$+ 107910000 \text{ Pa} = 1080.1 \text{ bar}$$

According to the Simon–Glatzel equation, the melting point of pure water at this pressure is

$$T_m = T_{m,\text{ref}} \left(\frac{P_m/\text{MPa}}{a_1} + 1 \right)^{a_2} \exp\left(-a_3 \frac{P_m}{\text{MPa}} \right)$$

$$= 273.25 \text{ K} \left(\frac{108.01}{550.1} + 1 \right)^{0.4593} \exp(-0.00111 \cdot 108.01) = 263.18 \text{ K}$$

This means that the melting point decreases by almost 10 K due to the elevated pressure; therefore, the temperature of $\vartheta = -2\,°C$ at the bottom of the Mariana Trench is possible. Due to the salts dissolved in the sea water, the melting point is further lowered (cryoscopic constant, Section 8.1.1.1), but the main effect is caused by the pressure elevation in this case.

At very high pressures, the pressure dependence of the melting point of water is extremely complicated, as a lot of different crystal structures occur [19]. The lowest melting point observed is $\vartheta_m = -23\,°C$ at 2100 bar, whereas very high melting points like $\vartheta_m = 175\,°C$ at 36500 bar are also possible. Correlations for the melting pressure curves of the different modifications of ice are given in [20].

3.1.5
Standard Enthalpy and Standard Gibbs Energy of Formation

Standard enthalpies of formation Δh_f^0 and standard Gibbs energies of formation Δg_f^0 are important for the calculation of enthalpies of reaction and chemical equilibria. For their estimation, the standard state at $T_0 = 298.15$ K and $P_0 = 101325$ Pa in the ideal gas state is used.[2] In process simulation programs, standard enthalpies and standard Gibbs energies of formation in the ideal gas state are usually taken as reference points for enthalpy calculation so that enthalpy and Gibbs energy differences are consistent with respect to chemical reactions.

For performing estimations, the group contribution method of Joback and Reid [4] can again be used. The corresponding formulas are

$$\frac{\Delta h_f^0}{\text{kJ/mol}} = 68.29 + \sum \Delta_H \tag{3.21}$$

---

2) In thermodynamics, also other standard states for the liquid and the solid phase are possible. In process simulators, usually the ideal gas state is used.

$$\frac{\Delta g_f^0}{\text{kJ/mol}} = 53.88 + \sum \Delta_G \qquad (3.22)$$

The group contribution parameters can be taken from Table 3.1. A more sophisticated method has been proposed by Domalski and Hearing [21, 22]. As well, the Benson method is widely used and further developed with industrial support [23, 24].

Example 3.6

Estimate the standard enthalpy of formation and the standard Gibbs energy of formation of ethyl acetate (CH_3–COO–C_2H_5) with the Joback method.

Solution

The group incrementation of ethyl acetate is given by

2 · CH_3
1 · CH_2
1 · COO

$$\sum \Delta_H = 2 \cdot (-76.45) + 1 \cdot (-20.64) + 1 \cdot (-337.92) = -511.46$$

$$\sum \Delta_G = 2 \cdot (-43.96) + 1 \cdot (8.42) + 1 \cdot (-307.95) = -381.45$$

One obtains

$$\frac{\Delta h_f^0}{\text{kJ/mol}} = 68.29 - 511.46 = -443.17$$

$$\frac{\Delta g_f^0}{\text{kJ/mol}} = 53.88 - 381.45 = -327.57$$

From data tables, we get $\Delta h_f^0 = -442.91$ kJ/mol and $\Delta g_f^0 = -327.39$ kJ/mol. The excellent agreement can be attributed to the reaction equilibria of esterification reactions of ethyl acetate, which are very well known and which might have given a large contribution to the parameter database.

The problem of differences between large numbers has always to be taken into account. Relatively small errors in determining Δh_f^0 and especially Δg_f^0 can lead to significant errors in estimating the enthalpy of reaction or the equilibrium constant, respectively. Therefore, the results of estimation methods should be handled with care. The following example gives an impression about the sensitivity of Δh_f^0 and Δg_f^0.

Example 3.7

Estimate the enthalpy of reaction and the equilibrium concentration y_1 of the theoretically possible isomerization reaction in the gas phase

$$n\text{-butane} \rightleftharpoons \text{isobutane}$$

at $\vartheta = 25\,°C$ and $P = 1.01325$ bar. The standard enthalpies of formation and the standard Gibbs energies of formation are

	Δh_f^0 (J/mol)	Δg_f^0 (J/mol)
Isobutane (1)	−134510	−20878
n-Butane (2)	−126150	−17154

As well, calculate the range of possible results if each of these given quantities has an error of ±1%.

Solution

As the reaction conditions are equal to the standard conditions, the given values can directly be adopted (see Chapter 12):

$$\Delta h_R^0 = (-134\,510 + 126\,150)\ \text{J/mol} = -8360\ \text{J/mol}$$

$$\Delta g_R^0 = (-20\,878 + 17\,154)\ \text{J/mol} = -3724\ \text{J/mol}$$

$$\Rightarrow K = \exp\left(-\frac{\Delta g_R^0}{RT}\right) = \exp\left(\frac{3724\ \text{J/mol}}{8.3143\ \frac{\text{J}}{\text{mol K}} \cdot 298.15\ \text{K}}\right) = 4.4919$$

$$K = \frac{y_1}{1 - y_1} \Rightarrow y_1 = \frac{K}{1 + K} = 0.8179$$

The influence of the errors is highest if both errors go into different directions:

$$\Delta h_{R,\max}^0 = (-135\,855 + 124\,888)\ \text{J/mol} = -10\,967\ \text{J/mol}$$

$$\Delta h_{R,\min}^0 = (-133\,165 + 127\,411)\ \text{J/mol} = -5754\ \text{J/mol}$$

$$\Delta g_{R,\max}^0 = (-21\,087 + 16\,982)\ \text{J/mol} = -4105\ \text{J/mol}$$

$$\Rightarrow K_{\max} = \exp\left(-\frac{\Delta g_{R,\max}^0}{RT}\right) = 5.2367$$

$$K_{\max} = \frac{y_{1,\max}}{1 - y_{1,\max}} \Rightarrow y_{1,\max} = \frac{K_{\max}}{1 + K_{\max}} = 0.8397$$

$$\Delta g_{R,\min}^0 = (-20669 + 17326)\ \text{J/mol} = -3343\ \text{J/mol}$$

$$\Rightarrow K_{\min} = \exp\left(-\frac{\Delta g_{R,\min}^0}{RT}\right) = 3.853$$

$$K_{\min} = \frac{y_{1,\min}}{1 - y_{1,\min}} \Rightarrow y_{1,\min} = \frac{K_{\min}}{1 + K_{\min}} = 0.7939$$

Thus, in this case 1% error in Δh_f^0 can lead to more than 30% error in the enthalpy of reaction. The error for Δg_f^0 looks less dramatic; however, the equilibrium concentration of n-butane can vary between approx. 16–21%, which is quite significant. Fortunately, due to the RT term in the denominator the influence of the error decreases with increasing temperature, and chemical reactions usually take place at higher temperatures.

3.2
Temperature-Dependent Properties

For the particular temperature-dependent thermodynamic properties, a number of special equations has been developed which are able to describe the various properties in the interesting temperature range, that is, from the triple point to the critical point for a liquid property and in an arbitrarily chosen temperature range (e.g., -200 to $1500\,°C$) for a vapor property. All of them have a number of adjustable parameters that can be fitted to experimental data, usually in a way that the objective function F, the sum of squares of the relative deviations between calculated and experimental data, becomes a minimum:

$$F = \sum_i \left(\frac{m_{i,\text{calc}} - m_{i,\text{exp}}}{m_{i,\text{exp}}} \right)^2 \overset{!}{\rightarrow} \min \qquad (3.23)$$

If a very large temperature range is covered and errors introduced by the uncertainty of the temperature as the independent property become significant, the maximum-likelihood function [25] can be applied:

$$F = \sum_i \left[\left(\frac{T_{\text{true},i} - T_{\text{exp},i}}{\sigma_T} \right)^2 + \left(\frac{m_{\text{true},i} - m_{\text{exp},i}}{\sigma_m} \right)^2 \right] \qquad (3.24)$$

where σ_m and σ_T are the standard deviations of the property m and the temperature T. Usually, the experimental uncertainty is inserted. For each data point, a "true" value of the temperature T_{true} is assumed. These true values are regarded as adjustable parameters. The "true" values m_{true} of the property m are calculated with the true values of the temperature as argument. This fitting procedure is much more complicated, as regarding the true values of the temperature as adjustable increases the number of values to be optimized by the fitting routine drastically. Nevertheless, the maximum-likelihood procedure can be regarded as a method without arbitrary mixing of data with different quality or different order of magnitude.

An equation fitted like that is called a *correlation*. Its quality is determined by the value of the obtained minimum, which is usually converted into an average deviation. The lesser the value of the objective function is, the higher the quality of the correlation can be regarded. On the other hand, the number of adjusted parameters must be taken into account. The user has to take care that all the fitted coefficients are significant. This means that if a coefficient is set to a standard value (usually 0), the fitting procedure gives a significantly worse value for the objective function.

Furthermore, the extrapolation behavior of a correlation should always be checked. It should not show unreasonable values when it is applied outside

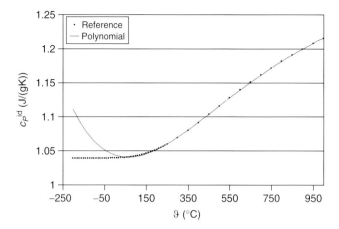

Figure 3.5 Qualitatively wrong representation of c_p^{id} of nitrogen at low temperatures with a polynomial. (Reference data from [10].)

the temperature range where it is fitted. In a parameter data bank, the valid temperature range should be stored as well.

Inaccurate extrapolations must be accepted as long as no better information is available. Things become worse if a qualitatively wrong extrapolation behavior is calculated. Generally, it can be distinguished between two types of extrapolation errors. First, an extrapolation can fail because the property itself does not exist outside the given temperature range. For instance, the vapor pressure curve ends at the critical point. Values beyond this point do not exist. Equations like the Wagner equation (see Eq. (3.37)) contain terms like $(1 - T_r)^{1.5}$, which are not defined at temperatures $T > T_c$. Therefore, the equation causes a runtime error, which is physically correct but not convenient for process simulation, where evaluations like that might take place during multiply nested iterations.

The second cause for an extrapolation to fail is that due to the structure of the equation a qualitatively wrong behavior is obtained. An example is the correlation of c_p^{id} with a polynomial (Figure 3.5).

The solution is that special extrapolation functions are defined which take over at the limiting temperature with the same value and the same slope. These extrapolation functions ensure that the behavior beyond the valid temperature range is at least qualitatively correct. It must be emphasized that this does not mean that the accuracy is good. A collection of well-proven extrapolation functions is given in [26].

The various pure component property correlations can be applied to mixtures with simple mixing rules, although sometimes with limited success. Therefore, the corresponding mixing rules are given at the end of the particular chapters. In case of vapor pressure and enthalpy of vaporization, the appropriate mixing rule is provided by the g^E-model used.

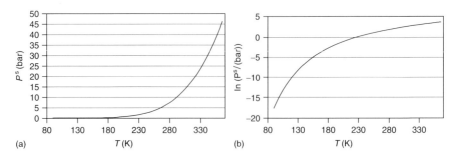

Figure 3.6 Vapor pressure of propylene as a function of temperature.

3.2.1
Vapor Pressure

For many tasks in process simulation, the vapor pressure is the most important quantity. It is often decisive for the determination of the number of theoretical stages of distillation columns and the calculation of temperature profiles. In environmental protection, the vapor pressure is important to determine the load of a component in exhaust air and to evaluate the options for an exhaust air treatment. Furthermore, a vapor pressure curve that has been carefully evaluated is also useful for the estimation of other thermophysical properties, especially the enthalpy of vaporization.

The vapor pressure is an exponential function of temperature, starting at the triple point and ending at the critical point. It comprises several orders of magnitude; therefore, a graphical representation usually represents only part of its characteristics. Figure 3.6 shows two diagrams with linear and with logarithmic axes for the vapor pressure of propylene. The linear diagram makes it impossible to identify even the qualitative behavior at low temperatures, whereas in the logarithmic diagram only the order of magnitude can be identified on the axis. At least, logarithmic diagrams allow for the comparison between vapor pressures of different substances. On the other hand, many process calculations require extremely accurate vapor pressure curves, for example, the separation of isomers by distillation. For the visualization of a fit of a vapor pressure curve, special techniques must be applied (see Appendix E), the simple graphical comparison between experimental and calculated data in a diagram is not possible.

Fortunately, nowadays a large amount of experimental data for common industrial components is available together with tabulated correlation parameters from different sources, for example, DDB [17], DIPPR [9], PPDS [27], or VDI Heat Atlas [28], and there is usually no need to estimate the vapor pressure curve of key components in process simulation. Due to the predominant influence of the vapor pressure on the vapor–liquid separation factor, one would also not rely on estimated data for this purpose.

One way to develop a correlation for the vapor pressure is to apply the integration of the Clausius–Clapeyron equation using simplifying assumptions:

$$\frac{dP^s}{dT} = \frac{\Delta h_v}{T(v^v - v^L)} \tag{3.25}$$

Neglecting v^L in comparison with v^v ($v^v \gg v^L$), the ideal gas approximation

$$v^v - v^L \approx \frac{RT}{P^s} \tag{3.26}$$

can be used, which is valid for low pressures and nonassociating substances. One can obtain the differential equation

$$\frac{dP^s}{P^s} = \frac{\Delta h_v}{R} \frac{dT}{T^2} \tag{3.27}$$

Assuming that the enthalpy of vaporization is independent of temperature, the differential equation can easily be integrated to

$$\ln \frac{P^s}{P_0} = A + \frac{B}{T} \tag{3.28}$$

with A and B as component specific parameters. An arbitrary pressure unit P_0 is introduced into the equation to ensure that the argument of the logarithm is dimensionless. A change of the pressure unit would only affect the parameter A.

However, this simple equation can only reproduce the vapor pressure curve with sufficient accuracy in a very small temperature range. There are a lot of approaches that introduce additional adjustable parameters. For example, assuming that the enthalpy of vaporization is a linear function of the temperature, the Kirchhoff equation

$$\ln \frac{P^s}{P_0} = A + \frac{B}{T} + C \ln \frac{T}{K} \tag{3.29}$$

is obtained. A very popular modification of Eq. (3.28) is the Antoine equation

$$\ln \frac{P^s}{P_0} = A + \frac{B}{T + C} \tag{3.30}$$

which, however, has been found empirically and cannot be derived directly from the Clausius–Clapeyron equation. The Antoine equation has proved to be successful as long as it is used for the temperature range where it was validated. It should be avoided to fit the Antoine parameters over a wide temperature range. The common practice to fit two Antoine equations for a substance, that is, one for temperature below and one for temperatures above the normal boiling point, is not acceptable for process simulation, as these equations do usually not have the same value and the same slope at the point where the change-over between the two equations takes place. The equation is more successful if the application range is restricted, for example, to the low pressure range. If parameters for Eq. (3.30) are given, one has to take care which units are used for temperature and pressure.[3]

[3] A lot of notations for the Antoine equation are used, making its application often a guessing game if the form used is not outlined clearly. Temperature and pressure units, the signs of the parameters and the basis of the logarithm can be varied.

There are certain restrictions for the parameters of the Antoine equation. Its temperature derivative for the vapor pressure is given by

$$\frac{dP^s}{dT} = P^s \frac{-B}{(T+C)^2} \tag{3.31}$$

As $dP^s/dT > 0$, this leads to the constraint

$$B < 0 \tag{3.32}$$

Applying the Antoine equation (3.30) to the Clausius–Clapeyron equation, the allowable range for the parameter C can be derived. Using Eq. (3.31) and the approximation (3.26), the relationship

$$\Delta h_v = \frac{-RT^2 B}{(T+C)^2} \tag{3.33}$$

is obtained. The temperature derivative of the enthalpy of vaporization is always negative for nonassociating substances. From Eq. (3.33), we obtain

$$\frac{d\Delta h_v}{dT} = \frac{-2BCRT}{(T+C)^3} < 0 \tag{3.34}$$

As $B < 0$ and $T > 0$, the only way to fulfill this constraint is

$$-T < C < 0 \tag{3.35}$$

At $T = -C$, the Antoine equation has a pole; therefore, the application limit of the Antoine equation should be well above this temperature.

For process simulation applications, a high accuracy is often required, especially when components with similar vapor pressures have to be separated in distillation columns. When a parameter database is set up, for example, the one in a commercial process simulation program, it cannot be known in advance at which conditions exact vapor pressures will be required. The advantage of the Antoine equation is its simplicity; in contrast to many other vapor pressure equations it can easily be converted into a temperature-explicit equation. However, it cannot be applied for the whole temperature range from the triple point to the critical point. In these cases, more capable vapor pressure correlations are recommended, that is, the extended Antoine equation and the Wagner equation.

In principle, the extended Antoine equation is a collection of various useful terms. It comprises the Antoine equation (3.30) itself as well as the Kirchhoff equation (3.29):

$$\ln \frac{P^s}{P_0} = A + \frac{B}{T+C} + DT + E \ln T + F T^G \tag{3.36}$$

It is rarely applied with all the parameters being active. Its advantage is that the user of a simulation program can handle many different equations with one single formula code.

A very useful correlation is the Wagner equation

$$\ln \frac{P^s}{P_c} = \frac{1}{T_r} \left[A(1-T_r) + B(1-T_r)^{1.5} + C(1-T_r)^3 + D(1-T_r)^6 \right] \tag{3.37}$$

where $T_r = T/T_c$. It can correlate the whole vapor pressure curve from the triple point to the critical point with an excellent accuracy. The Wagner equation has been developed by identifying the most important terms in Eq. (3.37) with a structural optimization method [29]. The extraordinary capability of the equation has also been demonstrated by Moller [30], who showed that the Wagner equation can reproduce the difficult term $\Delta h_V/(v^V - v^L)$ reasonably well. Equation (3.37) is called the *3–6-form*, where the numbers refer to the exponents of the last two terms. Some authors [6] prefer the 2.5–5-form, which is reported to be slightly more accurate:

$$\ln \frac{P^s}{P_c} = \frac{1}{T_r} \left[A(1-T_r) + B(1-T_r)^{1.5} + C(1-T_r)^{2.5} + D(1-T_r)^5 \right] \quad (3.38)$$

For the application of the Wagner equation, accurate critical data are required. As long as the experimental data points involved are far away from the critical point (e.g., only points below atmospheric pressure), estimated critical data are usually sufficient. As the critical point is automatically met due to the structure of the equation, the Wagner equation extrapolates reasonably to higher temperatures, even if the critical point is only estimated. However, like all vapor pressure equations it does not extrapolate reliably to lower temperatures.

Sometimes, users of process simulation programs calculate vapor pressures beyond the critical point, although it is physically meaningless. If the Wagner equation is applied above the critical temperature, it will yield a mathematical error. Therefore, the simulation program must provide an extrapolation function that continues the vapor pressure line with the same slope [26].

For the particular parameters of the Wagner equation, the following ranges of values are reasonable for both forms:

$$A = -9 \ldots -5$$
$$B = -10 \ldots 10$$
$$C = -10 \ldots 10$$
$$D = -20 \ldots 20 \quad (3.39)$$

If these ranges are exceeded, one should carefully check the critical data used and the experimental data points for possible outliers. Coefficients for the Wagner equations (3.37) and (3.38) can be found in [6, 31, 32], or [28].

For a vapor pressure correlation, average deviations should be well below 0.5%. Data points with correlation deviations larger than 1% should be rejected, as long as there are enough other values available. Exceptions can be made for vapor pressures below 1 mbar, as the accuracy of the measurements is lower in that range. The structure of the deviations should always be carefully interpreted (see Appendix E). For the parameters of the Wagner and the Antoine equation, some more restrictions are given by McGarry [31] and Chase [33]. Especially the paper of McGarry [31] has proved to be a useful source of reliable vapor pressure curves over decades. He has described and applied several empirical constraints in the fitting procedure of the Wagner equation and gives the parameters for more than 250 substances, most of them having proved to be very accurate. Some exceptions can be detected using the criterion given in Eq. (3.39).

Despite this high accuracy demand for vapor pressures, there is also a need for good estimation methods. These requirements refer mostly to medium and low pressures for molecules with a limited complexity, as small molecules usually have well-established vapor pressure equations, whereas large molecules usually have a volatility which is so low that a purification by distillation is not possible.

The estimation of vapor pressures is one of most difficult problems in thermodynamics. Due to the exponential relationship between vapor pressure and temperature, a high accuracy must not be expected. Deviations in the range of 5–10% have to be tolerated. Thus, estimated vapor pressure correlations should not be used for a main substance in a distillation column to evaluate the final design; however, they can be very useful to decide about the behavior of side components without additional measurements.

Therefore, the average deviations are less important for a vapor pressure estimation method than the number of substances where the method fails completely. In this area, a simple estimation procedure based on the vapor pressure equation of Hoffmann and Florin [34] has proved to be more appropriate than various methods from the literature [35–37] that are based on corresponding states or group contributions. The equation of Hoffmann–Florin is based on the charts developed by Cox [38]. Cox charts are constructed in a way that for some reference fluids the scale of the abscissa is adjusted to make the vapor pressure curve in a log P^s–T-diagram a straight line. The Hoffmann–Florin equation has two adjustable parameters α and β:

$$\ln \frac{P^s}{P_0} = \alpha + \beta f(T) \tag{3.40}$$

with

$$f(T) = \frac{1}{T/K} - 7.9151 \cdot 10^{-3} + 2.6726 \cdot 10^{-3} \log \frac{T}{K} - 0.8625 \cdot 10^{-6} \frac{T}{K} \tag{3.41}$$

The equation can be fitted to two or more known data points for the vapor pressure. It is not a very exact correlation but it shows a good extrapolation behavior, especially superior to Eq. (3.28). It is important that the distance between the two data points (T_1, P_1^s) and (T_2, P_2^s) is large enough so that significant parameters are obtained. α and β can then be calculated via

$$\alpha = \ln \frac{P_1^s}{P_0} - \ln \frac{P_1^s}{P_2^s} \cdot \frac{f(T_1)}{f(T_1) - f(T_2)} \tag{3.42}$$

and

$$\beta = \frac{\ln \frac{P_1^s}{P_2^s}}{f(T_1) - f(T_2)} \tag{3.43}$$

The Hoffmann–Florin equation can be transformed to the extended Antoine equation (3.36) by setting

$$A = \alpha - 7.9151 \cdot 10^{-3}\beta$$
$$B = \beta$$
$$D = -0.8625 \cdot 10^{-6}\beta$$
$$E = \frac{2.6726 \cdot 10^{-3}}{\ln 10}\beta$$
$$C = F = G = 0$$

If only one or no vapor pressure point is available, the normal boiling point, the critical point or both can be estimated, and the adjustable parameters α and β can be fitted to these artificial data points, giving a procedure for the estimation of vapor pressures.

Example 3.8

Estimate the vapor pressure of chloroform at $\vartheta_1 = -41.7\,°C$, $\vartheta_2 = 4.5\,°C$, and $\vartheta_3 = 120.1\,°C$ without using any experimental data point.

Solution

Using Joback's method, we obtain

$T_b = 334.13$ K (true value: 334.25 K)
$T_c = 532.11$ K (true value: 536.45 K)
$P_c = 4.98$ MPa (true value: 5.55 MPa).

With

$T_1 = 334.13$ K
$P_1^s = 0.101325$ MPa
$T_2 = 532.11$ K
$P_2^s = 4.98$ MPa.

one obtains by setting $P_0 = 1$ Pa

$\alpha = 19.5596$
$\beta = -5233.61$.

The results are

$P^s(-41.7\,°C) = 5.69$ hPa (true value: 5 hPa)
$P^s(4.5\,°C) = 99.94$ hPa (true value: 100 hPa)
$P^s(120.1\,°C) = 0.518$ MPa (true value: 0.5 MPa).

In most cases, the quality of the normal boiling point or, respectively, another given data point is decisive for the quality of the estimation. The comparably large error in the critical pressure estimation is often acceptable.

Example 3.9

Estimate the vapor pressure of ethylene glycol monoacetate ($CH_3-COO-CH_2-CH_2-OH$) at $\vartheta_1 = 101.27\,°C$. At $\vartheta_2 = 175.63\,°C$, the vapor pressure is reported to be $P^s(\vartheta_2) = 727.9\,hPa$ [39].

Solution

Using Joback's method, we obtain

$$T_c = T_b \cdot 1.3856$$
$$P_c = 4.415\,MPa$$

In an iterative procedure, T_b is estimated to calculate the critical temperature. Using T_b and T_c as the given reference points, α and β are determined with Eqs. (3.42) and (3.43). With Eqs. (3.40) and (3.41) it is checked whether $P^s(\vartheta_2)$ can be reproduced. As long as it is not met, the procedure is repeated with another T_b as starting value. Using $P_0 = 1\,Pa$, the final result is

$T_b = 459.33\,K$
$T_c = 636.45\,K$
$\alpha = 21.264$
$\beta = -9930.45$.

Using Eq. (3.40), one obtains $P^s(\vartheta_1) = 38.35\,hPa$. The measured value is 37.9 hPa [39].

Example 3.10

Estimate the vapor pressure of RE-218 ($CF_3-O-CF_2-CF_3$) at $\vartheta_1 = 50.3\,°C$ and at $\vartheta_2 = -45\,°C$. The normal boiling point is $\vartheta_b = -23.7\,°C$ [40].

Solution

Using Joback's method, one obtains

$\sum \Delta_T = 0.1257 \rightarrow T_c = 361.78\,K$ (experimental value: 356.85 K [40])
$\sum \Delta_P = -0.0312 \rightarrow P_c = 2.999\,MPa$ (experimental value: 2.315 MPa [41])
$\alpha = 20.0309$
$\beta = -3722.423$

The result is

$P^s\,(50.3\,°C) = 12.71\,bar$ (experimental value: 11.319 bar [41])
$P^s\,(-45\,°C) = 0.3455\,bar$ (experimental value: 0.372 bar [40])

Although the critical pressure is again estimated very poorly, the result remains acceptable. Using the experimental values for the critical point, the results would be

$$P^s(50.3\,°C) = 11.24\,bar$$
$$P^s(-45\,°C) = 0.3641\,bar$$

From Moller et al. [42], a new group contribution estimation method with the normal boiling point as reference has been developed. Their vapor pressure equation refers to the Thomson rule [43], meaning that the C parameter of the Antoine equation (3.30) is correlated with the normal boiling point. Furthermore, a logarithmic term is supplemented to account for alcohols and carboxylic acids:

$$\ln \frac{P^s}{101.325\,\text{kPa}} = B' \frac{T/K - T_b/K}{T/K + 2.65 - \frac{(T_b/K)^{1.485}}{135}} + D' \ln \frac{T}{T_b} \qquad (3.44)$$

where B' and D' are determined by

$$B' = 9.42208 + \sum_i v_i \Delta B_i + n_A \sum_j v_j \Delta B_j + \sum_k \Delta B_k + \frac{1}{2} \sum_i \sum_j GI_{ij} \qquad (3.45)$$

$$D' = D + \frac{1}{n_A} \sum_i v_i \Delta E_i \qquad (3.46)$$

The particular contributions can be explained as follows:

ΔB_i are the group contributions of the structural groups which are size-independent. v_i is the frequency of group i in the molecule. If the group is assigned with a size dependence (index j), additional group constant contributions ΔB_k have to be added, and the group contributions ΔB_j are multiplied with the number of atoms n_A in the molecule except hydrogen. The GI_{ij} are the group interaction contributions, where $GI_{ij} = GI_{ji}$ and the interaction of a group with itself is zero.

If the group assignment is ambiguous, the priority number decides which group has to be applied; the group with the lower priority number has to be preferred. D' denotes a correction term for aliphatic alcohols and carboxylic acids, consisting of group contributions ΔE_i and a constant value D. n_A is the number of atoms except hydrogen. The groups are listed on www.ddbst.com, where also further information about the groups, that is, the size dependence and the priority, is given.

If, instead of the normal boiling point, another reference vapor pressure is available, T_b can be adjusted in a way that the reference point is met. If no reference point is available, T_b can be estimated. In this case, the accuracy of Eq. (3.44) is, of course, significantly lowered as mentioned above.

For this method, the authors report an average error of 5% for the more than 2300 substances used in the database [42].

However, it must be emphasized that the estimation of vapor pressures without at least two reliable data points can only be regarded as an estimation of the order of magnitude of the vapor pressure to get an orientation about the volatility of the compound.

Example 3.11

Estimate the vapor pressure of n-propyl benzene at $\vartheta_1 = 100\,°C$ and $\vartheta_2 = 200\,°C$ using the Rarey/Moller method.

The structural formula of n-propyl benzene is given in Figure 3.7.

The normal boiling point is $\vartheta_b = 159.22\,°C$ [10].

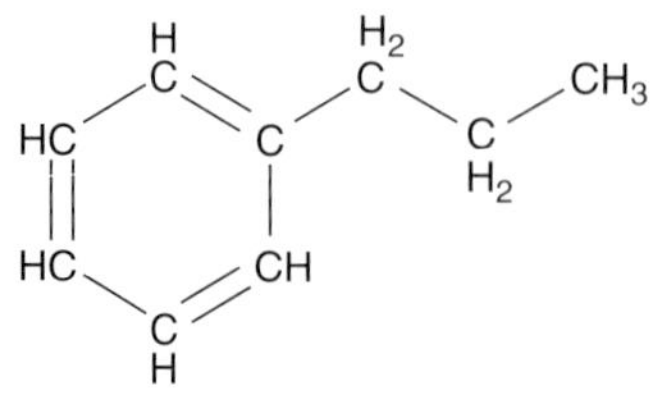

Figure 3.7 Structural formula of n-propyl benzene.

Solution

The structure increments can be set up as follows:

Increment number	Group name	Frequency	ΔB_i	$\sum \Delta B_i$
4	CH_2 in a chain	2	0.07545	0.1509
1	CH_3	1	−0.00227	−0.00227
17	C in an aromatic ring	1	0.11192	0.11192
16	CH in an aromatic ring	5	0.01653	0.08265

				0.3432

$B' = 9.42208 + 0.3432 = 9.76528$

$D' = 0$

With $T_b = 432.37$ K, Eq. (3.44) yields P^s (373.15 K) = 0.1616 bar and P^s (473.15 K) = 2.6452 bar. The reference values [10] are P^s_{ref} (373.15 K) = 0.1665 bar and P^s_{ref} (473.15 K) = 2.5921 bar, respectively.

Example 3.12

Repeat Example 3.9 using the Rarey/Moller method.

Solution

The following group contributions can be identified:

Increment number	Description	Frequency	n_A	ΔB_i	$\sum \Delta B_i$
1	methyl group	1	−	−0.00227	−0.00227
54	ester in a chain	1	−	0.55698	0.55698
7	CH_2 attached to electronegative atom	2	−	0.11758	0.23516
153	OH group, large molecule, size dependent	1	7	−0.04696	−0.32872
176	long OH group constant	1	−	5.21138	5.21138
207	alcohol–ester interaction	2 · 0.5	−	−1.38632	−1.38632

					4.28621

Therefore, we get
$$B' = 9.42208 + 4.28621 = 13.70829$$

For the logarithmic correction term, one gets
$$D' = -4.798 + \frac{1}{7} \cdot 6.578 = -3.8583$$

giving the vapor pressure curve
$$\ln \frac{P^s}{101.325 \text{ kPa}} = 13.70829 \frac{T/K - T_b/K}{T/K + 2.65 - \frac{(T_b/K)^{1.485}}{135}} - 3.8583 \ln \frac{T}{T_b}$$

Using the reference point P^s (175.63 °C) = 727.9 hPa, we can adjust T_b in an iterative procedure to be $T_b = 460.955$ K. Then the vapor pressure at $\vartheta_1 = 101.27$ °C can be determined to be $P^s(101.27\,°C) = 49.36$ hPa, whereas the correct value is 37.9 hPa [39].

Example 3.13

Repeat Example 3.10 using the Rarey/Moller method.

Solution

The following group contributions can be identified:

Increment number	Description	Frequency	ΔB_i	$\sum \Delta B_i$
9	C attached to electronegative atom	3	−0.0896	−0.2688
39	F attached to C with two other halogen atoms	2	0.09402	0.18804
38	F attached to C with one other halogen atom	1	0.1054	0.1054
51	ether oxygen	1	0.15049	0.15049
150	no hydrogen	1	−0.19373	−0.19373
			$\sum$	−0.0186

Therefore, we get
$$B' = 9.42208 - 0.0186 = 9.40348$$

and
$$\ln \frac{P^s}{101.325 \text{ kPa}} = 9.40348 \frac{T/K - 249.45}{T/K + 2.65 - \frac{249.45^{1.485}}{135}} = 9.40348 \frac{T/K - 249.45}{T/K - 24.215}$$

The results are

P^s (50.3 °C) = 10.367 bar (experimental value: 11.319 bar [41])
P^s (−45 °C) = 0.3795 bar (experimental value: 0.372 bar [40]).

Estimation methods for the vapor pressure are still not reliable enough to apply them in process simulations for components whose behavior is decisive. Vapor pressure estimations can be used for components where the order of magnitude of the vapor pressure is sufficient (e.g., to decide whether the component occurs at the top or at the bottom of a column). Even in this case, it depends on the frequency it is used. For the simulation and optimization of a process for basic chemicals it is strongly recommended to have vapor pressure curves based on data points for all the components involved, whereas for the batch production of fine chemicals with a tight timeframe or during process evaluation an estimation might be acceptable.

Especially in the case of the vapor pressure the average deviation for the reproduction of the database is only a rough indicator for the quality of an estimation method. Managable deviations in a prediction can easily be handled. Much more interesting is the number of substances where the prediction completely fails. It makes hardly any difference whether the deviation is 2 or 3% in case of a good estimation, or whether it is 60 or 150% in case of a bad estimation; instead, it is decisive not to use a bad estimation, as it can lead to a serious failure of a process simulation even if it is applied to a trace component.

It is worth mentioning that for small droplets in the nanometer region the vapor pressure is a strong function of the droplet size. It is described by the Kelvin equation [44]

$$\ln \frac{P^G}{P^s} = \frac{4\sigma}{\rho^L RTd} \qquad (3.47)$$

where P^G is the vapor pressure of a droplet with the diameter d and P^s is the normal vapor pressure.[4] Table 3.3 gives an impression of the vapor pressure elevation of water droplets at $\vartheta = 20\,°C$, where the effect is relatively large due to the high surface tension σ of water.

Table 3.3 Change in vapor pressure of water droplets as a function of the droplet size.

d (nm)	P^G/P^s
1000	1.002
100	1.022
10	1.244
1	8.904
0.5	79.29
0.4	236.59
0.3	1463.3
0.2	55,976
0.1	$3.13 \cdot 10^9$

4) A derivation of Eq. (3.47) is given in Appendix C.

There is a remarkable consequence of Eq. (3.47). If there are no droplets in the vapor phase, condensation would not be possible, as a droplet with $d = 0$ ends up with an infinite saturation pressure, resulting in immediate evaporation. Only due to statistical fluctuations or seed particles (dust, microorganisms) [44, 45] droplet formation can happen at all. Once a droplet has been formed, a small growth in size makes the saturation pressure fall rapidly, leading to an extremely fast growth of the droplet. For evaporation, the situation is vice versa, so that a small decrease of the droplet size leads to rapid evaporation. Thus, the Kelvin equation describes an unstable equilibrium, where small disturbances lead to disappearance or growth of the droplets.

Example 3.14

Estimate the vapor pressure elevation of a 1 nm droplet of n-hexane at $\vartheta = 50\,°C$. The thermophysical properties can be taken from Appendix A:

$\sigma = 15.24$ mN/m
$\rho^L = 631.1$ kg/m^3
$M = 86.178$ g/mol
$P^s = 0.5409$ bar.

Solution

$$\ln \frac{P^G}{P^s} = \frac{4\sigma}{\rho^L RTd} = \frac{4 \cdot 15.24 \cdot 10^{-3}}{631.1 \cdot 8.3143 \cdot 323.15 \cdot 10^{-9}} \frac{\text{N}}{\text{m}} \frac{\text{m}^3}{\text{kg}} \frac{\text{mol K}}{\text{J}}$$

$$\cdot \frac{1}{K} \frac{1}{m} \frac{86.178 \cdot 10^{-3} \text{ kg}}{\text{mol}} = 3.098$$

$$P^G = P^s \cdot e^{3.098} = 11.98 \text{ bar}$$

Two substances frequently occur in process calculations as solids: water and carbon dioxide. It is useful to know the relationships of their sublimation pressures. They are

$$\ln \frac{P^{\text{subl}}_{H_2O}}{611.657 \text{ Pa}} = -13.928169 \cdot \left(1 - \left(\frac{T}{273.16 K}\right)^{-1.5}\right)$$

$$+ 34.7078238 \cdot \left(1 - \left(\frac{T}{273.16 K}\right)^{-1.25}\right) \tag{3.48}$$

valid between 190 K < T < 273.16 K [20], and

$$\ln \frac{P^{\text{subl}}_{CO_2}}{\text{Pa}} = 22.9862 - \frac{2979.9116}{T/K} + 0.73009 \ln \frac{T}{K} \tag{3.49}$$

valid between 154.15 K < T < 216.15 K [17].

3.2.2
Liquid Density

The density of a pure substance or a mixture is obviously a fundamental quantity in any process calculation. In addition to the simple determination of the volume demand or, respectively, the weight of an amount of substance or a stream, it is for instance required for pressure drop calculations, column hydraulics, calculation of reaction forces of streams in pipes with bends and many more. The density of vapor or gaseous substances is strongly pressure-dependent and always determined with equations of state (see Chapter 2), whereas it is a good approximation to treat the density of liquids at moderate pressures only as a function of the temperature. The typical shape of a liquid density plot is shown in Figure 3.8.

Liquid densities ρ_L are mostly correlated by the Rackett equation, which can generally reproduce the density on the whole saturation line within 0.2–0.3% except the area close to the critical point:

$$\rho^L = \frac{A}{B^{1+(1-T/C)^D}} \quad (3.50)$$

An even more precise equation is the one suggested by PPDS.[5]

$$\frac{\rho^L}{\text{kg/m}^3} = \frac{\rho_c}{\text{kg/m}^3} + A\left(1 - \frac{T}{T_c}\right)^{0.35} + B\left(1 - \frac{T}{T_c}\right)^{2/3} + C\left(1 - \frac{T}{T_c}\right)$$
$$+ D\left(1 - \frac{T}{T_c}\right)^{4/3} \quad (3.51)$$

The accuracy difference between Eqs. (3.50) and (3.51) is illustrated with the deviation plot shown in Figure 3.9. Coefficients for Eq. (3.51) for a large number of substances are given in [28].

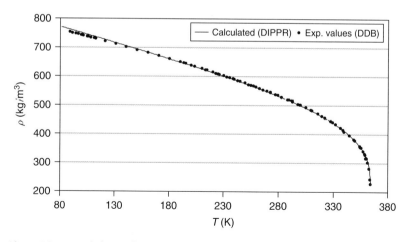

Figure 3.8 Typical shape of a saturated liquid density plot (substance: propylene).

5) Physical Property Data Service.

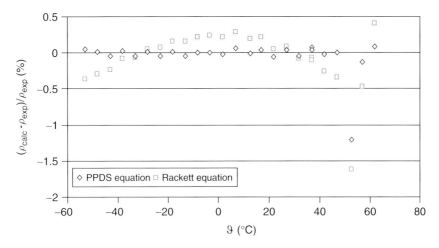

Figure 3.9 Density deviation plot for R125 (pentafluoroethane) for the PPDS and Rackett equation. (Data from Widiatmo et al. [46].)

For both correlations (Eqs. (3.50) and (3.51)) there are some weaknesses when the density of water is described. The well-known maximum of the liquid density at 4 °C, which is not very important from the technical point of view, and the fact that the decrease of the liquid density with increase of temperature is extremely flat (only 1% between 10 and 50 °C, whereas the density of benzene with a comparable melting point decreases by 5%) makes correlation difficult. It is recommended to use a simple polynomial

$$\frac{\rho^L_{\text{water}}}{\text{kg/m}^3} = 18.015 \cdot \sum_{i=0}^{5} A_i \left(\frac{T}{K}\right)^i \tag{3.52}$$

with

$A_0 = -13.418392 \qquad A_1 = 0.6884103$
$A_2 = -2.44970115 \cdot 10^{-3} \qquad A_3 = 3.7060667 \cdot 10^{-6}$
$A_4 = -2.11062995 \cdot 10^{-9} \qquad A_5 = -1.12273895 \cdot 10^{-13}$

For the estimation of liquid densities or specific volumes, respectively, the COSTALD method [16] has proved to be reliable. It can be applied in the range $0.25 < T_r < 1$. The calculation equations are

$$v_s = \left(\rho^L\right)^{-1} = v \cdot V_R^{(0)} \left(1 - \omega V_R^{(\delta)}\right) \tag{3.53}$$

$$V_R^{(0)} = 1 + a(1 - T_r)^{1/3} + b(1 - T_r)^{2/3} + c(1 - T_r) + d(1 - T_r)^{4/3} \tag{3.54}$$

$$V_R^{(\delta)} = \left(e + fT_r + gT_r^2 + hT_r^3\right) / (T_r - 1.00001) \tag{3.55}$$

The coefficients in Eqs. (3.53–3.55) are

$a = -1.52816$	$b = 1.43907$	$c = -0.81446$	$d = 0.190454$
$e = -0.296123$	$f = 0.386914$	$g = -0.0427458$	$h = -0.0480645$

The characteristic volume v^* is an adjustable parameter that can be fitted to one or more experimental data points. If no data are available, v^* can be replaced by v_c, which results in a loss of accuracy but is often acceptable.

Example 3.15

Estimate the liquid density of *n*-hexane at $T = 293.15$ K using the COSTALD method.

The following values are given:

$T_c = 507.8$ K; $v^* = v_c = 386.8$ cm^3/mol; $\omega = 0.3002$; $M = 86.178$ g/mol

Solution

Applying Eqs. (3.53–3.55), we obtain

$V_R^{(0)} = 0.3798$

$V_R^{(\delta)} = 0.2277$

$v_s = 136.85$ cm^3/mol $\Rightarrow \rho^L = 629.7$ kg/m^3

The experimental value is $\rho^L = 659.4$ kg/m^3.

In process simulation, it is fully sufficient that the liquid density always refers to the saturation line, the pressure effect on the density is neglected. The pressure dependence of the density is necessary e.g. for safety considerations like the calculation of the pressure build-up in a closed volume or for the determination of the speed of sound. For the determination of the pressure effect, there are a number of correlations listed in [6]. The Tait equation

$$v(T, P) = v(T, P_0) \left[1 - C \ln \frac{B + P}{B + P_0} \right] \tag{3.56}$$

can be applied at pressures up to 150 MPa and higher. P_0 is a reference pressure where the density is known, usually the saturation pressure $P^s(T)$. B and C are substance-specific parameters which should be fitted to experimental data. If it is justified by the data available, they should be made temperature-dependent [47]. Several estimation methods have been proposed, for example, by Thomson et al. [48]:

$$C = 0.0861488 + 0.0344483\,\omega \tag{3.57}$$

$$\frac{B}{P_c} = -1 - 9.070217\,(1 - T_r)^{1/3} + 62.45326\,(1 - T_r)^{2/3} - 135.1102\,(1 - T_r)$$
$$+ (1 - T_r)^{4/3} \exp\left(4.79594 + 0.250047\,\omega + 1.14188\,\omega^2\right) \tag{3.58}$$

However, it is recommended that predictions in this area should be taken more or less as a guess. For higher accuracy, high-precision equations of state (see Section 2.5.2) are recommended if available.

For mixtures, the liquid density can be derived by a linear mixing rule of the specific volume, where the excess volume (see Chapter 4) is neglected:

$$v_{mix}^L = \sum_i x_i v_i^L \tag{3.59}$$

which is equivalent to

$$\rho_{mix}^L = \left(\sum_i \frac{x_i}{\rho_i^L} \right)^{-1} \tag{3.60}$$

The unit of the density must refer to the kind of concentration used, that is, either the mole fraction (mol/m³) or the mass fraction (kg/m³).

The errors caused by this simplification are usually small. The largest error has been reported for the ethanol–water mixture, where the contribution of the excess volume can arise up to 3.5% [17]. It is worth mentioning that Eq. (3.60) yields better results than a linear mixing rule for the density, which is often applied due to its simplicity, especially when the pure component densities are significantly different.

Example 3.16

Estimate the density of a binary mixture consisting of 40 wt% n-hexane (1) and 60 wt% chloroform (2) at $\vartheta = 50\,°C$. The pure component densities are referred to be [49]:

$$\rho_1 = 631.1 \text{ kg/m}^3, \quad \rho_2 = 1437.5 \text{ kg/m}^3.$$

Solution

Applying Eq. (3.60) leads to

$$\rho_{mix} = \left(\frac{0.4}{631.1} + \frac{0.6}{1437.5} \right)^{-1} \frac{\text{kg}}{\text{m}^3} = 951.3 \text{ kg/m}^3$$

A linear mixing rule for the density would yield

$$\rho_{mix} = (0.4 \cdot 631.1 + 0.6 \cdot 1437.5) \frac{\text{kg}}{\text{m}^3} = 1114.9 \text{ kg/m}^3$$

Figure 3.10 illustrates the mixing rule for the density and shows the large errors for a system where the pure component densities differ significantly.

3.2.3
Enthalpy of Vaporization

The operation costs of distillation processes are mainly influenced by the enthalpy of vaporization and, to a smaller extent, by the specific isobaric heat capacity of liquids. Therefore, its importance can easily be understood.

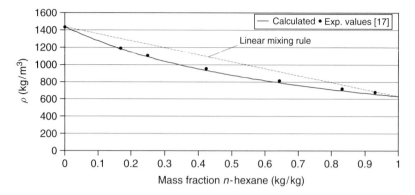

Figure 3.10 Mixture density for the system n-hexane-chloroform at $\vartheta = 50°C$.

Figure 3.12 shows some typical enthalpies of vaporization as a function of temperature. For substances which are not strongly associating in the vapor phase, the data follow a monotonically declining function with a relatively flat slope. With increasing temperature, the slope becomes larger. At the critical point, where vapor and liquid become identical, the enthalpy of vaporization becomes 0. Enthalpies of vaporization can differ by almost one order of magnitude. Water has the largest one (~2200 J/g at 100 °C), whereas halogenated hydrocarbons often show enthalpies of vaporization below 200 J/g even in the flat region. For substances which form associates in the vapor phase (e.g., formic acid in Figure 3.12), the course of the enthalpy of vaporization with respect to temperature is different and shows a more or less clearly defined maximum (see Section 13.2).

For non-associating substances, enthalpies of vaporization can be very well correlated with the extended Watson equation

$$\Delta h_v = A \left(1 - T_r\right)^{B+CT_r+DT_r^2+ET_r^3} \tag{3.61}$$

The order of magnitude of the average deviation should be ~0.5%. For temperatures near the critical point, the values for the enthalpy of vaporization are small and the slopes are large; therefore, higher deviations can be accepted, and it is recommended to leave this region out of the database for regression. The parameters C, D, and E should only be fitted if many data with a high quality are available, otherwise, they should be set to 0. In that case, B is often close to $B = 0.38$.

In many cases, the PPDS equation

$$\Delta h_v = RT_c \left(A\tau^{1/3} + B\tau^{2/3} + C\tau + D\tau^2 + E\tau^6\right) \tag{3.62}$$

with

$$\tau = 1 - T/T_c \tag{3.63}$$

can yield impressive results, especially when high-precision data have to be reproduced or in the case of associating substances (see Section 13.2).

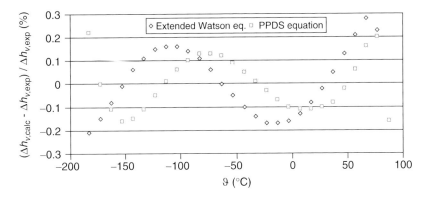

Figure 3.11 Deviation plots for the enthalpy of vaporization of propane using the extended Watson equation and the PPDS equation. Reference data from high-precision equation of state [10].

Figure 3.11 shows an example for a comparison between Eqs. (3.61) and (3.63). Usually, as in this case, Eq. (3.63) performs slightly better. Coefficients for Eq. (3.63) for a large number of substances are given in [28].

Again, users of a simulation program could use enthalpies of vaporization for supercritical components, which is physically unrealistic but unavoidable in practical applications. For that case, it is useful to define an extrapolation function to avoid a mathematical error.

There is a peculiarity for the estimation of enthalpies of vaporization: the Clausius–Clapeyron equation

$$\Delta h_v = T\left(v^v - v^l\right)\frac{dP^s}{dT} \tag{3.64}$$

is an exact thermodynamic relationship of the enthalpy of vaporization with the easier accessible quantities vapor pressure and specific liquid and vapor volume at the saturation state.

Equation (3.62) is often simplified by using the ideal gas law for v^v and neglecting the liquid volume v^l ($v^v \gg v^l$), giving

$$\Delta h_v = \frac{RT^2}{P^s}\frac{dP^s}{dT} \tag{3.65}$$

This approximation is often useful for estimation enthalpies of vaporization in the vacuum region. For pressures greater than approx. 1 bar, a qualitatively wrong curvature is obtained; thus, Eq. (3.65) cannot be recommended for general use. For v^v, a cubic equation of state is more appropriate, for example, the Peng–Robinson equation of state, which is appropriate for the calculation of saturated vapor densities from the triple point to temperatures close to the critical point.

In Figure 3.12, the curves for the enthalpies of vaporization for water, methanol, diethyl ether, and sulfur dioxide are in fact based on calculated values. For sulfur

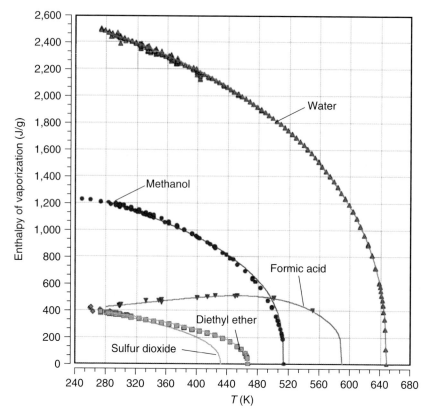

Figure 3.12 Enthalpies of vaporization as a function of temperature for some selected compounds. (Data from [17].)

dioxide, only a few data points in the vicinity of the normal boiling point are available. Nevertheless, with the help of good PvT and vapor pressure data it is possible to set up a function for the enthalpy of vaporization with good accuracy. For formic acid, the application of the Clausius–Clapeyron equation (3.62) using the association model for v^v (see Section 13.2) only yields acceptable results in the low-pressure region, where the limiting assumption of an ideal gas mixture of the associates is valid. In this case, accurate experimental data for the enthalpy of vaporization are essential.

The application of the Clausius–Clapeyron equation should be restricted to a certain temperature range. The first term dP^s/dT can be considered as quite exact, as a reliable vapor pressure equation is a necessary requirement for process simulation. The limitation is that the common vapor pressure equations like Wagner and Antoine badly extrapolate to low temperatures. Furthermore, even if measured data at low temperatures are available, the relative errors are quite high. As a rule of thumb, it is recommended not to use the Clausius–Clapeyron equation

Table 3.4 Impression on the quality of cubic equations of state for the prediction of saturated vapor volumes (data taken from Smith and Srivastava [50, 51]).

Substance	T (K)	v^v exp. (cm³/mol)	v^v SRK (cm³/mol)	Dev. SRK (%)	v^v PR (cm³/mol)	Dev. PR (%)
Dichloromethane	384	3729	3773	1.2	3748	0.5
R22	286	2771	2809	1.4	2787	0.6
R152a	307	2817	2876	2.1	2850	1.2
Methanol	399	3967	4043	1.9	4025	1.5
Propanol-1	441	3687	3773	2.3	3703	0.4
Methane	145	1239	1249	0.8	1237	0.2
n-Hexane	432	3157	3251	3.0	3200	1.4
Water	443.15	4375	4460	1.9	4452	1.8
o-Xylene	515	4570	4682	2.5	4630	1.3

for vapor pressures $P^s < 1$ mbar. Enthalpies of vaporization at low temperatures, where the assumption of an ideal gas is justified, should rather be derived from reliable values at higher temperature and the specific heat capacities for liquid and ideal gas:

$$\Delta h_v(T) = \Delta h_v(T_0) - \left(c_P^L - c_P^{id}\right)(T - T_0) \tag{3.66}$$

For the accuracy of the Clausius–Clapeyron equation at higher temperatures, the term $(v^v - v^L)$ is decisive. v^L can easily be calculated from a density correlation that is usually available. It is negligible in comparison to v^v except in the region near the critical point, where, unfortunately, the quality of the correlations is lower. In most cases, v^v is again determined by a generalized cubic equation of state, usually the Peng–Robinson or Soave–Redlich–Kwong (SRK) equation of state. Table 3.4 gives an impression about the accuracy of saturated vapor volumes at vapor pressures $P^s \approx 8$ bar. It should not be taken as a comprehensive statistical analysis, but it demonstrates that an error of approx. 1–2% for v^v is a reasonable assumption. In the critical region, the calculation of v^v with a cubic equation of state gives poor results. Considering that both v^v and v^L are much less accurate and that the difference between v^v and v^L rapidly decreases, it is not recommended to apply the Clausius–Clapeyron equation in the critical region. As a rule of thumb, it should be avoided to use it at temperatures above $T > T_c - 30$ K.

With these two restrictions, it can be assumed that an error of 1–2% coming from v^v is usually obtained for the calculated enthalpy of vaporization. The values outside the application range of the Clausius–Clapeyron equation can be estimated by fitting Eqs. (3.61) or (3.63) to data generated with the Clausius–Clapeyron equation and extrapolating it.

Example 3.17

Estimate the enthalpy of vaporization of acetone at $T = 273.15$ K. The given data are

$$\rho^L = 812.89 \text{kg/m}^3, \ T_c = 508.1 \text{ K}, \ P_c = 4.6924 \text{ MPa}, \ \omega = 0.3064,$$
$$M = 58.08 \text{ g/mol};$$

Wagner coefficients (Eq. (3.38)):

$$A = -7.670734, \ B = 1.965917, \ C = -2.445437, \ D = -2.899873$$

Solution

Using the Wagner coefficients, at $T = 273.15$ K the vapor pressure can be determined to be $P^s = 93.0$ hPa. The specific volumes can be calculated to be
$v^L = 7.145 \cdot 10^{-5}$ m^3/mol (see Appendix A)
$v^V = 0.2453$ m^3/mol (from Peng–Robinson equation of state)
The derivative of the Wagner equation with respect to temperature is

$$\frac{dP^s}{dT} = -\frac{P^s}{T}\left[\ln\frac{P^s}{P_c} + A + 1.5B(1-T_r)^{0.5} + 2.5C(1-T_r)^{1.5} + 5D(1-T_r)^4\right]$$
(3.67)

The result is $dP^s/dT = 492.84$ Pa/K, giving

$$\Delta h_v = 273.15 \text{K} \cdot (0.2453 - 7.145 \cdot 10^{-5})\frac{\text{m}^3}{\text{mol}} \cdot 492.84 \frac{\text{Pa}}{\text{K}} = 33012 \text{ J/mol}$$
$$= 568.4 \text{ J/g}$$

The experimental value is reported to be 558.9 J/g. The deviation is 1.7%.

3.2.4
Ideal Gas Heat Capacity

The specific heat capacity of an ideal gas is the basic quantity for the enthalpy calculation, as it is independent from molecular interactions. It is also possible to define a real gas heat capacity, but for process calculations it is more convenient to account for the real gas effects with the enthalpy description of the equation of state used (see Section 6.2). In process calculations, the specific heat capacity of ideal gases mainly determines the duty of gas heat exchangers, and it has an influence on the heat transfer coefficient as well.

The specific heat capacity of an ideal gas is defined as the heat per amount of substance in the ideal gas state necessary to obtain a certain temperature change. It must be distinguished between the specific isobaric heat capacity c_p^{id} (at constant pressure) and the specific isochoric heat capacity c_v^{id} (at constant volume). For ideal gases, both quantities are related [52] via

$$c_p^{id} = c_v^{id} + R \tag{3.68}$$

The specific isochoric heat capacity c_v^{id} consists of four parts, which refer to the kinetic energy of the molecules, the rotation energy of the molecules, the energy of the valence and deformation vibrations in the molecules, and the anharmonicity correction:

$$c_v^{id} = c_v^{id,kin} + c_v^{id,rot} + c_v^{id,vibr} + c_v^{id,anharm} \tag{3.69}$$

The term *"degree of freedom"* is important to understand how c_v^{id} depends on temperature. Each fully activated degree of freedom contributes $1/2 R$ to c_v^{id}. The kinetic energy part is pretty simple. The three translation coordinates can be regarded as the degrees of freedom. They are fully activated just above 0 K, therefore, the contribution to c_v^{id} is

$$c_v^{id,kin} = \frac{3}{2} R \tag{3.70}$$

over the whole temperature range. As long as one-atomic molecules like He, Ne, or Ar are regarded, no rotation degrees of freedom can be activated, as, according to quantum mechanics, the moment of inertia of a single atom is too small. Furthermore, a single atom cannot vibrate against itself; therefore, in this case the kinetic part is the only contribution to c_v^{id}:

$$c_v^{id} = c_v^{id,kin} = \frac{3}{2} R \tag{3.71}$$

Considering two-atomic molecules, the rotation parts can be activated. There are three rotation degrees of freedom according to the three coordinates for the orientation of the rotation axis. They are fully activated when the absolute temperature exceeds a few K. However, the third coordinate, that is, the connection line between the two atoms, can again not be activated due to quantum mechanics because of the small moment of inertia. Thus, the rotation contribution of the rotation energy to c_v^{id} is

$$c_v^{id,rot,2\text{-atomic}} = \frac{2}{2} R \tag{3.72}$$

Equation (3.72) is also valid for linear molecules like CO_2, where the rotation around the common axis of the atoms cannot be activated. For polyatomic nonlinear molecules, the third rotation coordinate can be activated. In this case, the rotation part of c_v^{id} is

$$c_v^{id,rot,polyatomic} = \frac{3}{2} R \tag{3.73}$$

The vibration contribution to c_v^{id} is the most complicated one. The number of basic frequencies in a molecule with n_A atoms is

$$N_{vibr} = 3n_A - 3 - N_{rot} \tag{3.74}$$

where N_{rot} is the number of rotation degrees of freedom. Each basic frequency has two degrees of freedom (potential and kinetic energy). The vibration degrees of freedom are gradually activated with rising temperature. Using statistical thermodynamics [44, 52], it can be shown that the contribution of the basic frequency ν_0 is

$$c_v^{\text{id,vibr},v_0} = R \frac{\left(\frac{h\nu_0}{2kT}\right)^2}{\sinh^2\left(\frac{h\nu_0}{2kT}\right)} \tag{3.75}$$

which is the so-called Planck–Einstein function. For many molecules, the basic frequencies can be obtained by spectroscopic data or quantum chemical calculations.

For very high temperatures, it must be taken into account that rotations and vibrations are not independent from each other. The effect results in an additional positive contribution to c_v^{id}, the so-called correction for anharmonic vibration.

The typical curvature of c_p^{id} as a function of temperature is shown in Figure 3.13. On the left-hand side at low temperatures, only the constant part for the translation and rotation energy is active. With increasing temperature, the function increases monotonically, until all the possible vibrations in the molecule are completely activated. For one-atomic gases, c_p^{id} is a constant, as only the translation degrees of freedom can be activated. Then, c_p^{id} becomes almost constant again. At high temperatures, dissociation of the molecule might occur, resulting again in a rise of c_p^{id}. In this context, this effect should be assigned to a chemical reaction instead of a pure component property.

The ideal gas heat capacity can be quite well derived by the interpretation of spectroscopic data. But in the recent years, the measurement of the speed of sound has more and more become the favorite method [53–55]. The relationship between c_p^{id} and the speed of sound w^* of an ideal gas is (see Section 3.2.6, Eq. 3.99 and

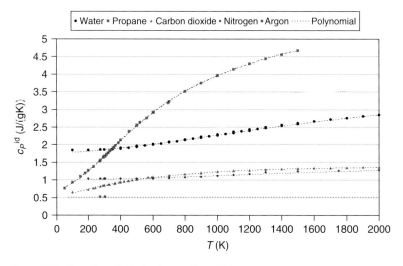

Figure 3.13 Specific isobaric heat capacities of ideal gases as functions of temperature. (Data from [17].)

Appendix C, B7):

$$c_p^{id} = \frac{R(w^*)^2}{(w^*)^2 - RT} \qquad (3.76)$$

The following example explains the strong link between the speed of sound and the ideal gas heat capacity.

Example 3.18

Determine the specific isobaric ideal gas heat capacity of methane ($M = 16.043$ g/mol) at $\vartheta = 20\,°C$, $P = 0.1$ bar. The speed of sound at these conditions is $w^* = 445.33$ m/s. Determine the sensitivity of the specific heat capacity to the speed of sound.

Solution

First, the gas constant has to be turned into mass units:

$$R = 8.3143 \frac{J}{\text{mol K}} \frac{\text{mol}}{16.043\,\text{g}} \frac{1000\,\text{g}}{\text{kg}} = 518.251 \frac{J}{\text{kgK}}$$

$$c_p^{id} = \frac{R(w^*)^2}{(w^*)^2 - RT} = \frac{518.251\,\text{J/(kgK)} \cdot 445.33^2\,\text{m}^2/\text{s}^2}{445.33^2\,\text{m}^2/\text{s}^2 - 518.251\,\text{J/(kgK)} \cdot 293.15\,\text{K}} = 2215.4 \frac{J}{\text{kgK}}$$

The value obtained from the equation in Appendix A is $c_p^{id} = 2207.3$ J/kgK.

For the determination of the sensitivity, the derivation of Eq. (3.76) is set up:

$$\frac{dc_p^{id}}{dw^*} = \frac{2Rw^*\left((w^*)^2 - RT\right) - R(w^*)^2 \cdot 2w^*}{((w^*)^2 - RT)^2} = -\frac{2R^2 w^* T}{(w^{*2} - RT)^2}$$

At this point, one obtains

$$\frac{dc_p^{id}}{dw^*} = -\frac{2R^2 w^* T}{((w^*)^2 - RT)^2} = -\frac{2 \cdot 518.251^2 (\text{J/(kgK)})^2 \cdot 445.33\,\text{m/s} \cdot 293.15\,\text{K}}{(445.33^2\,\text{m}^2/\text{s}^2 - 518.251\,\text{J/(kgK)} \cdot 293.15\text{K})^2}$$

$$= -0.0326 \frac{\text{J/(gK)}}{\text{m/s}}$$

This is a relatively high sensitivity, as an error of 1% in the speed of sound (4.45 m/s) would correspond to an error of 0.145 J/(gK), that is,

$$\frac{\Delta c_p^{id}}{c_p^{id}} = \frac{0.145\,\text{J/(gK)}}{2215.4\,\text{J/(kgK)}} = 6.5\%$$

However, the accuracy in the measurement of the speed of sound is better than 0.1% [56], and therefore c_p^{id} can be determined very reliably.

The measurement of the speed of sound is also used to evaluate the exact value of the universal gas constant R. As $c_p^{id} = 2.5R$ for one-atomic gases according to Eqs. (3.68) and (3.71), Eq. (3.76) reduces to

$$R = \frac{3}{5} \frac{(w^*)^2}{T}$$

As both w^* and T can be measured very accurately, this is currently the favorite way for the determination of the universal gas constant R [56, 57].

The curvature of the isobaric ideal gas heat capacity can be correlated very well with the Aly–Lee equation [58], which is based on the physical evaluation of the vibration part using statistical thermodynamics:

$$c_p^{id} = A + B \left(\frac{C/T}{\sinh(C/T)} \right)^2 + D \left(\frac{E/T}{\cosh(E/T)} \right)^2 \quad (3.77)$$

For the representation of high-precision data, the PPDS equation

$$\frac{c_p^{id}}{R} = B + (C - B)\, y^2 \left[1 + (y - 1)\left(D + Ey + Fy^2 + Gy^3 + Hy^4 \right) \right] \quad (3.78)$$

with

$$y = \frac{T}{A + T} \quad (3.79)$$

has proved to be useful, as it is flexible and its behavior at the limiting values is very well defined. For $T \to 0$, we get $y = 0$ and therefore $c_p^{id} = RB$. At $T \to \infty$, $y \to 1$, and $c_p^{id} = RC$. As c_p^{id} is a monotonically increasing function and taking Eqs. (3.68) and (3.71) into account, the constraint $C > B > 2.5$ is obligatory for any substance. If required, the limiting value at $T \to 0$ determined by the degrees of freedom for translation and rotation can be obtained by adding a corresponding artificial data point at a very low temperature. It should be mentioned that $c_p^{id} = RB$ and $c_p^{id} = RC$ are only the mathematical limits for $T \to 0$ and $T \to \infty$ but not necessarily the minimum and maximum values of the function, which is not in line with theory but usually acceptable for practical applications. For regression, the user can set the priorities himself whether the constraints are taken into account to get a better extrapolation behavior or left out to obtain a better accuracy in a limited temperature range. If the asymptotes are known, B and C can also be fixed. For Eq. (3.78), coefficients for a large number of substances are given in [28].

Acceptable results can often be obtained with the simpler equation:

$$c_p^{id} = A + BT + CT^2 + DT^3 \quad (3.80)$$

although its extrapolation capability is weak, especially to lower temperatures. For normal process simulations with a limited temperature range (200 … 500 K) it is usually sufficient, as well as for substances which decompose at relatively low temperatures anyway. Parameters for Eq. (3.80) are given in Appendix G.

It is useful to know the antiderivatives for the particular correlations for the calculation of enthalpies and entropies. They are listed in Appendix C.

For high-precision equations of state (see Section 2.5.2), multiparameter equations have been set up to obtain the best possible accuracy. One option is the extension of the Aly–Lee equation (3.77) to

$$c_p^{id} = A + B \left(\frac{C/T}{\sinh(C/T)} \right)^2 + D \left(\frac{E/T}{\cosh(E/T)} \right)^2 + F \left(\frac{G/T}{\sinh(G/T)} \right)^2$$
$$+ H \left(\frac{J/T}{\cosh(J/T)} \right)^2 \quad (3.81)$$

For Eq. (3.81), coefficients for 19 substances are given in [59]. Span [60] uses a combination of polynomial and Planck–Einstein terms

$$\frac{c_p^{id}}{R} = A_0 + \sum_{i=1}^{I_{Pol}} A_i T^{t_i} + \sum_{k=1}^{K_{PE}} B_k \left(\frac{\Theta_k}{T}\right)^2 \frac{\exp(\Theta_k/T)}{(\exp(\Theta_k/T) - 1)^2} \qquad (3.82)$$

where the A_i, t_i, B_k, and Θ_k are adjustable.

Specific isobaric heat capacities can be estimated with the Joback group contribution method [4]. The group contributions determine the coefficients of a polynomial of third degree:

$$\frac{c_p^{id}}{J/\text{mol K}} = \left(\sum \Delta_A - 37.93\right) + \left(\sum \Delta_B + 0.21\right) \cdot (T/K)$$
$$+ \left(\sum \Delta_C - 3.91 \cdot 10^{-4}\right) \cdot (T/K)^2$$
$$+ \left(\sum \Delta_D + 2.06 \cdot 10^{-7}\right) \cdot (T/K)^3 \qquad (3.83)$$

The particular coefficients for the groups are listed in Table 3.5. They become larger with increasing complexity of the molecule. If group contributions are missing, the method of Harrison and Seaton [61] can be applied, which is based on the contributions of the particular elements in a molecule.

Example 3.19

Estimate the ideal gas specific isobaric heat capacity of ethyl acetate (CH_3–COO–CH_2–CH_3) at $T = 298.15$ K using the Joback method. The molar mass of ethyl acetate is $M = 88.11$ g/mol.

Solution

The group increments for ethyl acetate are:
  2 · CH_3
  1 · CH_2
  1 · COO.

From the group contributions we obtain

$$\sum \Delta_A = 2 \cdot (19.5) + 1 \cdot (-0.909) + 1 \cdot (24.5)$$
$$= 62.591$$
$$\sum \Delta_B = 2 \cdot (-8.08 \cdot 10^{-3}) + 1 \cdot (9.5 \cdot 10^{-2}) + 1 \cdot (4.02 \cdot 10^{-2})$$
$$= 0.11904$$
$$\sum \Delta_C = 2 \cdot (1.53 \cdot 10^{-4}) + 1 \cdot (-5.44 \cdot 10^{-5}) + 1 \cdot (4.02 \cdot 10^{-5})$$
$$= 2.918 \cdot 10^{-4}$$
$$\sum \Delta_D = 2 \cdot (-9.67 \cdot 10^{-8}) + 1 \cdot (1.19 \cdot 10^{-8}) + 1 \cdot (-4.52 \cdot 10^{-8})$$
$$= -2.267 \cdot 10^{-7}$$

Table 3.5 Group contributions for estimating c_p^{id} with the Joback method.

Structural group	Δ_A	$10^2 \Delta_B$	$10^4 \Delta_C$	$10^8 \Delta_D$
$-CH_3$	19.500	−0.808	1.5300	−9.670
$>CH_2$	−0.909	9.500	−0.5440	1.190
$>CH-$	−23.000	20.400	−2.6500	12.000
$>C<$	−66.200	42.700	−6.4100	30.100
$=CH_2$	23.600	−3.810	1.7200	−10.300
$=CH-$	−8.000	10.500	−0.9630	3.560
$=C<$	−28.100	20.800	−3.0600	14.600
$=C=$	27.400	−5.570	1.0100	−5.020
$\equiv CH$	24.500	−2.710	1.1100	−6.780
$\equiv C-$	7.870	2.010	−0.0833	0.139
$-CH_2-$(ring)	−6.030	8.540	−0.0800	−1.800
$>CH-$ (ring)	−20.500	16.200	−1.6000	6.240
$>C<$ (ring)	−90.900	55.700	−9.0000	46.900
$=CH-$(ring)	−2.140	5.740	−0.0164	−1.590
$=C<$ (ring)	−8.250	10.100	−1.4200	6.780
$-F$	26.500	−9.130	1.9100	−10.300
$-Cl$	33.300	−9.630	1.8700	−9.960
$-Br$	28.600	−6.490	1.3600	−7.450
$-I$	32.100	−6.410	1.2600	−6.870
$-OH$ (alcohols)	25.700	−6.910	1.7700	−9.880
$-OH$ (phenols)	−2.810	11.100	−1.1600	4.940
$-O-$	25.500	−6.320	1.1100	−5.480
$-O-$ (ring)	12.200	−1.260	0.6030	−3.860
$>C=O$	6.450	6.700	−0.3570	0.286
$>C=O$(ring)	30.400	−8.290	2.3600	−13.100
$-CH=O$	30.900	−3.360	1.6000	−9.880
$-COOH$	24.100	4.270	0.8040	−6.870
$-COO-$	24.500	4.020	0.4020	−4.520
$=O$	6.820	1.960	0.1270	−1.780
$-NH_2$	26.900	−4.120	1.6400	−9.760
$>NH$	−1.210	7.620	.4860	1.050
$>NH$ (ring)	11.800	−2.300	1.0700	−6.280
$>N-$	−31.100	22.700	−3.2000	14.600
$-N-$				
$-N=$(ring)	8.830	−0.384	0.4350	−2.600
$=NH$	5.690	−0.412	1.2800	−8.880
$-CN$	36.500	−7.330	1.8400	−10.300
$-NO_2$	25.900	−0.374	1.2900	−8.880
$-SH$	35.300	−7.580	1.8500	−10.300
$-S-$	19.600	−0.561	0.4020	−2.760
$-S-$(ring)	16.700	0.481	0.2770	−2.110

The result for c_P^{id} is

$$\frac{c_P^{id}}{\text{J/mol K}} = (62.591 - 37.93) + (0.11904 + 0.21) \cdot (298.15)$$
$$+ (2.918 \cdot 10^{-4} - 3.91 \cdot 10^{-4}) \cdot (298.15)^2$$
$$+ (-2.267 \cdot 10^{-7} + 2.06 \cdot 10^{-7}) \cdot (298.15)^3 = 113.397$$

or $c_P^{id} = 1.287$ J/(g K), respectively. The value from the correlation in Appendix A is $c_P^{id} = 1.2897$ J/(g K).

As in the ideal gas state the molecules do not interfere with each other, there is an exact linear mixing rule for c_P^{id}:

$$c_{P,\text{mix}}^{id} = \sum_i x_i c_{P,i}^{id} \quad (3.84)$$

It depends on the unit used for c_P^{id} (J/(mol K) or J/(g K)) whether the mole fraction or the mass fraction is represented by x_i. Both forms are, of course, equivalent.

3.2.5
Liquid Heat Capacity

Analogously to the heat capacity of gases, the specific heat capacity of liquids determines the duty required for any temperature change of a liquid. As well, it has an influence on the heat transfer coefficient in heat exchangers.

At low temperatures, c_P^L is usually an almost linear function of temperature. A flat minimum often occurs (Figure 3.14). With increasing temperature, the slope becomes larger, and at the critical point it becomes infinity.

For a saturated liquid, as it often occurs in process engineering applications, the specific isobaric heat capacity does not make much sense, as a supply of heat does not yield a temperature increase but causes evaporation. Therefore, a specific heat capacity along the saturation line is used (c_σ), which is only a function of

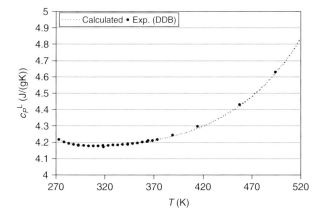

Figure 3.14 Liquid specific heat capacity of water as a function of temperature.

temperature. The relationship between the two heat capacities can be derived as follows:

$$c_\sigma = \lim_{dT \to 0} \frac{h^L(T+dT) - h^L(T)}{dT} = \lim_{dT \to 0} \frac{\left(\frac{\partial h^L}{\partial T}\right)_P dT + \left(\frac{\partial h^L}{\partial P}\right)_T dP^s}{dT} \quad (3.85)$$

with

$$dP^s = \frac{dP^s}{dT} dT \quad (3.86)$$

one obtains

$$c_\sigma = \lim_{dT \to 0} \frac{\left(\frac{\partial h^L}{\partial T}\right)_P dT + \left(\frac{\partial h^L}{\partial P}\right)_T \frac{dP^s}{dT} dT}{dT} = c_p^L + \left(\frac{\partial h^L}{\partial P}\right)_T \frac{dP^s}{dT} \quad (3.87)$$

For $(\partial h / \partial P)_T$, a more convenient expression is derived in Appendix C, A5 giving:

$$c_\sigma = c_p^L + \left[v - T\left(\frac{\partial v}{\partial T}\right)_P\right] \frac{dP^s}{dT} \quad (3.88)$$

The difference is negligible at low temperatures. As a rule of thumb, it can be neglected up to $T < 0.8\, T_c$ [62]. The difference can be estimated using the expression

$$c_\sigma = c_p^L - R \exp(20.1 T_r - 17.9) \quad (3.89)$$

Figure 3.15 illustrates the difference between c_p^L and c_σ and the quality of the estimation (3.89).

In the following, the difference between the two quantities is neglected. The specific heat capacity of liquids can be correlated with a polynomial:

$$c_p^L = A + BT + CT^2 + DT^3 + ET^4 + FT^5 \quad (3.90)$$

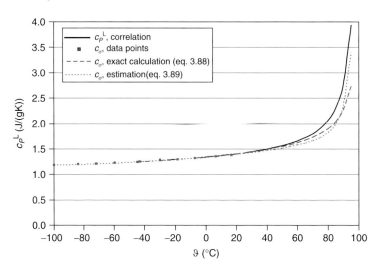

Figure 3.15 Difference between c_p^L and c_σ for R134a (1,1,1,2-tetrafluoroethane). (Data from [63].)

There are few substances where it is worth to fit all the coefficients. In most cases, a quadratic polynomial is sufficient. Due to the difficulties described above, data points in the critical area are often left out in the regression; nevertheless, average deviations of more than 1 ... 2% must often be accepted, which is caused by the experimental uncertainty of caloric measurements. For many components, heat capacity data for temperatures above the normal boiling point do not exist. In these cases, only a linear function is justified for the correlation, and the extrapolation to high temperatures becomes arbitrary. It can help to generate additional artificial data points with an estimation method. High-precision data can be correlated with the following PPDS equation

$$c_p^L = R\left(\frac{A}{\tau} + B + C\tau + D\tau^2 + E\tau^3 + F\tau^4\right) \quad (3.91)$$

with

$$\tau = 1 - \frac{T}{T_c} \quad (3.92)$$

Figure 3.16 illustrates the capabilities of Eqs. (3.90) and (3.91). The PPDS equation can reproduce the c_p^L of water over the entire temperature range with a remarkable accuracy, whereas the polynomial shows large deviations. However, it is fully acceptable if the values at high temperatures are excluded from the fit; it can be seen that the liquid heat capacity of water can be correlated with a polynomial up to 300 °C, which is sufficient for most applications. Coefficients for Eq. (3.91) can be found in [28].

For the estimation of c_p^L the method of Rowlinson/Bondi can be used [32], which is based on the ideal gas heat capacity and the corresponding states principle:

$$c_p^L = c_p^{id} + 1.45R + 0.45R(1 - T_r)^{-1} \\ + 0.25\omega R\left[17.11 + 25.2(1 - T_r)^{1/3}T_r^{-1} + 1.742(1 - T_r)^{-1}\right] \quad (3.93)$$

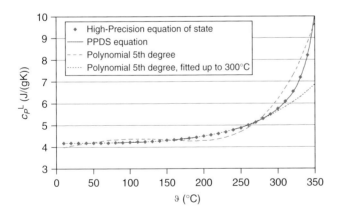

Figure 3.16 Comparison between polynomial and PPDS equation for the correlation of c_p^L of water. Reference values from high-precision equation of state [10].

One should be aware that liquid heat capacities of organic compounds are $c_p^L \approx 2\,\text{J/(g K)}$ at temperatures below the normal boiling point. If the estimated value is significantly different, one should be careful and check.

Example 3.20

Estimate the specific liquid heat capacity of methyl ethyl ketone at $T = 373.15$ K. Use $c_p^{id}(373.15\text{K}) = 1.671\,\text{J/(g K)}$, $T_c = 535.55$ K, $\omega = 0.323$, $M = 72.11\,\text{g/mol}$ [28]

Solution

With $c_p^{id}(373.15\text{ K}) = 1.671\,\text{J/(g K)} \cdot 72.11\,\text{g/mol} = 120.496\,\text{J/(mol K)}$ and

$$T_r = \frac{373.15}{535.55} = 0.697$$

one obtains

$$\frac{c_p^L}{\text{J/(mol K)}} = (120.496 + 1.45 \cdot 8.3143 + 0.45 \cdot 8.3143\,(0.303)^{-1}$$
$$+ 0.25 \cdot 0.323 \cdot 8.3143 \left[17.11 + 25.2\,(0.303)^{1/3}\,0.697^{-1}\right.$$
$$\left.+ 1.742\,(0.303)^{-1}\right] = 176.55$$
$$\Rightarrow c_p^L = 2.448\,\text{J/gK}$$

The value obtained from an adjusted correlation is 2.428 J/(g K).

It is worth mentioning that for liquids c_p^L and c_v^L can differ significantly. This might not be expected, as liquids are almost incompressible, and in many textbooks the incompressible fluid with $c_p^L = c_v^L$ is taken as a reference similar to the ideal gas [64]. For example, at $T = 300$ K, $P = 1$ bar benzene has a ratio $c_p^L/c_v^L = 1.39$ [10], and values in this range are more or less usual. The relationship between c_P and c_V can be expressed as

$$c_P = c_V + T\left(\frac{\partial P}{\partial T}\right)_v \left(\frac{\partial v}{\partial T}\right)_P \tag{3.94}$$

with

$$T\left(\frac{\partial P}{\partial T}\right)_v = P + \left(\frac{\partial u}{\partial v}\right)_T \tag{3.95}$$

which is valid both for liquids and gases. In Appendix C, a derivation of Eqs. (3.94) and (3.95) can be found.

Thus, the energy consumption of heating up a fluid at constant pressure is described by three independent parts:

$$c_P = c_V + P\left(\frac{\partial v}{\partial T}\right)_P + \left(\frac{\partial u}{\partial v}\right)_T \left(\frac{\partial v}{\partial T}\right)_P \tag{3.96}$$

While c_V describes the increase of the internal energy due to temperature elevation at constant volume, the second term denotes the work against the pressure field the fluid is exposed to. This term makes the difference between c_P and c_V for ideal gases, but it is negligible for liquids and solid substances, as the volume does not change very much with temperature. In the third term, the factor $(\partial u/\partial v)_T$ is huge for liquids. It describes the effort to overcome the attractive forces between the molecules when the volume is extended, and it is responsible for the differences between c_P and c_V. $(\partial u/\partial v)_T$ becomes zero for ideal gases according to their definition (no attractive forces between the molecules).

The linear mixing rule for c_P^L

$$c_{P,\text{mix}}^L = \sum_i x_i c_{P,i}^L \tag{3.97}$$

is usually a good approximation; however, it is not exact due to the neglection of the excess heat capacity (see Chapter 5). It depends again on the unit used for c_P^L (J/(mol K) or J/(g K)) whether mole fraction or mass fraction is represented by x_i.

3.2.6
Speed of Sound

The speed of sound plays an important role in the design of safety devices, as it determines the maximum possible flow rate of a fluid in case of a hazard. Furthermore, as a derivative property it has a considerable relevance for the adjustment of high-precision equations of state and for the indirect measurement of c_P^{id}, as explained in Section 3.2.4.

The general formula for the calculation of the speed of sound is [64]

$$(w^*)^2 = -v^2 \left(\frac{\partial P}{\partial v}\right)_S \tag{3.98}$$

For an ideal gas, Eq. (3.98) is equivalent to

$$w^{*,\text{id}} = \sqrt{\kappa R T} \tag{3.99}$$

using $\kappa = c_P^{\text{id}}/c_V^{\text{id}}$. For a real gas or liquid, the speed of sound can be calculated using a pressure-explicit equation of state (see Appendix C):

$$(w^*)^2 = -v^2 \left(\frac{\partial P}{\partial v}\right)_T \frac{c_P}{c_V} \tag{3.100}$$

Using the relationship (see Appendix C)

$$c_P = c_V + T \left(\frac{\partial P}{\partial T}\right)_v \left(\frac{\partial v}{\partial T}\right)_P \tag{3.101}$$

and Eq. (3.100), the speed of sound can as well be expressed as

$$(w^*)^2 = v^2 \left[\frac{T}{c_V} \left(\frac{\partial P}{\partial T}\right)_v^2 - \left(\frac{\partial P}{\partial v}\right)_T\right] \tag{3.102}$$

c_V can be determined by

$$c_V = c_V^{\text{id}} + T \int_\infty^v \left(\frac{\partial^2 P}{\partial T^2}\right)_v dv \qquad (3.103)$$

(see Appendix C).

Using these equations, the results for gases are usually sufficient. For liquids, a high-precision equation of state that describes the pressure-dependence of the liquid density is necessary. The expression for the speed of sound from high-precision equations of state formulated in terms of the *Helmholtz energy* (Eqs. (2.113) and (2.114)) is

$$\frac{(w^*)^2}{RT} = 1 + 2\delta\left(\frac{\partial \alpha^R}{\partial \delta}\right)_\tau + \delta^2\left(\frac{\partial^2 \alpha^R}{\partial \delta^2}\right)_\tau - \frac{\left[1 + \delta\left(\frac{\partial \alpha^R}{\partial \delta}\right)_\tau - \delta\tau\left(\frac{\partial^2 \alpha^R}{\partial \delta \partial \tau}\right)\right]^2}{\tau^2\left[\left(\frac{\partial^2 \alpha^{\text{id}}}{\partial \tau^2}\right)_\delta + \left(\frac{\partial^2 \alpha^R}{\partial \tau^2}\right)_\delta\right]} \qquad (3.104)$$

3.3
Correlation and Estimation of Transport Properties

Although transport properties are, strictly speaking, not part of chemical thermodynamics, the most common correlation and estimation methods are briefly introduced here, as they are needed for process simulation steps that are determined by mass transfer (e.g., absorption columns) and for the design of the equipment (e.g., heat transfer). It was decided to present the methods for calculating diffusion coefficients here as well, although they refer to mixtures and not to pure compounds.

3.3.1
Liquid Viscosity

The dynamic viscosity can be illustrated with the Couette flow (Figure 3.17). A fluid is located between two plates at the distance H. If a shearing force F is applied, a velocity gradient in the fluid is built up. The maximum velocity will occur at the point where the stress is applied, whereas the velocity is zero at the opposite side due to the wall adherence. For a Newtonian fluid, the velocity gradient is constant across the distance between the two plates. With A as the surface area of the upper plate, the shear stress τ is defined as

$$\tau = \frac{F}{A} \qquad (3.105)$$

and the dynamic viscosity η can be introduced via

$$\eta = \frac{\tau}{dw/dy} \qquad (3.106)$$

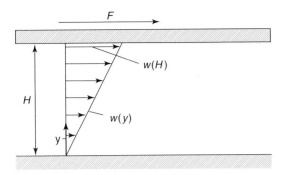

Figure 3.17 Explanation of the viscosity of a Newtonian fluid.

Sometimes also, the kinematic viscosity

$$\nu = \eta/\rho \tag{3.107}$$

is used.

The viscosity is a measure of internal friction in the fluid. A larger viscosity leads to more friction between two layers in the fluid and to a lower velocity gradient.

Liquid viscosity data are needed for the design of fluid transport and mixing processes (e.g., pumps, pressure drops in pipes or pipelines, and stirred vessels) and have a significant influence on the effectiveness of heat exchangers and diffusion processes. The required accuracy for viscosity calculations is lower than for thermodynamic properties, but the correct order of magnitude must be met.

The typical shape of the dynamic viscosity of a liquid as a function of the temperature is shown in Figure 3.18. Above the melting point, it decreases rapidly. The slope becomes more flat then but stays negative; the function is strictly

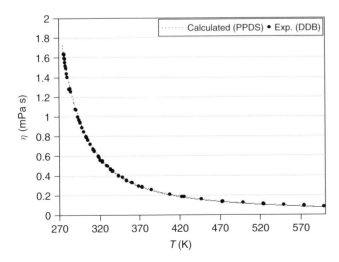

Figure 3.18 Dynamic viscosity of saturated liquid water as a function of temperature.

monotonic. A simple but useful correlation is

$$\ln \frac{\eta}{\eta_0} = A + \frac{B}{T} \tag{3.108}$$

with η_0 as an arbitrary viscosity unit to make the argument of the logarithm dimensionless. This equation can reproduce the curve qualitatively. To obtain a more accurate correlation, more terms must be added. An example is the extended Kirchhoff equation

$$\ln \frac{\eta}{\eta_0} = A + \frac{B}{T} + C \ln (T/K) + D (T/K)^E \tag{3.109}$$

There are some other equations available, where the measured viscosity data can be reproduced with high accuracy as a function of temperature. However, care should be taken as extrapolation of most of these equations is dangerous. It is recommended that any correlation for the liquid viscosity that is more complicated than Eq. (3.108) should be checked with a graphical representation.

Recently, a PPDS equation in the form

$$\frac{\eta}{\text{Pa s}} = E \cdot \exp \left[A \left(\frac{C - T}{T - D} \right)^{1/3} + B \left(\frac{C - T}{T - D} \right)^{4/3} \right] \tag{3.110}$$

has been applied. This equation seems to be the most accurate correlation for liquid viscosities. Furthermore, it seems to extrapolate quite well. When it is programmed, it must be taken care that the term in brackets $(C - T)/(T - D)$ sometimes turns out to be negative, so that it makes sense to write in these cases:

$$\left(\frac{C-T}{T-D} \right)^{1/3} = -\left(\frac{T-C}{T-D} \right)^{1/3} \quad \text{and} \quad \left(\frac{C-T}{T-D} \right)^{4/3} = -\left(\frac{T-C}{T-D} \right)^{1/3} \left(\frac{C-T}{T-D} \right)$$

In Figure 3.19, the deviation plots of the PPDS and the extended Kirchhoff equation for viscosity data of R134a (1,1,1,2-tetrafluoroethane) are compared. The superior flexibility of the PPDS equation is clearly demonstrated; especially the viscosities at low temperatures and in the vicinity of the critical point are much better reproduced. Coefficients for Eq. (3.110) can be found in [28].

In the past, a few methods for estimating the liquid viscosity have been proposed, like the group contribution method of Orrick/Erbar [32], Sastri and Rao [69, 70], or van Velzen et al. [71]. They can hardly be recommended, as their application is not easy and their results are often doubtful. For instance, the van Velzen method, often regarded as the most reliable one, produced a relative mean deviation of 92.8% in a benchmark with 670 components [72]. Although data points for the liquid viscosity often widely scatter and deviations can be easier tolerated than for the thermal properties, this value indicates that a lot of components, probably those which were not in the training set, are described poorly.

A new group contribution method developed by Rarey and Nannoolal [72] performs significantly better. Its relative mean deviation turned out to be 15.3% for 813 components. The method is defined by the following equation:

$$\ln \frac{\eta}{1.3 \text{ mPas}} = -B_\eta \frac{T - T_\eta}{T - \frac{T_\eta}{16}} \tag{3.111}$$

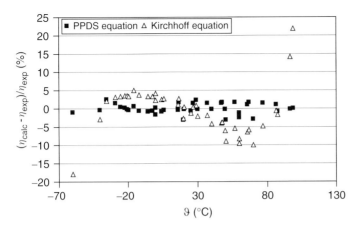

Figure 3.19 Deviation plots of the PPDS and the extended Kirchhoff equation for liquid viscosities of R134a. (Data from Okubo et al. [65], Oliveira and Wakeham [66], Assael et al. [67], and Lavrenchenko et al. [68].)

with

$$B_\eta = \frac{\sum_i v_i \Delta B_{\eta i} + \frac{1}{n_A} \sum_{i=1}^{m} \sum_{j=1}^{m} \frac{GIB_{ij}}{m-1}}{n_A^{2.5635} + 0.0685} + 3.7777 \quad (3.112)$$

and

$$\frac{T_\eta}{K} = 21.8444 \left(\frac{T_b}{K}\right)^{0.5} + \frac{\left(\sum_i v_i \Delta T_{\eta i} + \frac{2}{n_A} \sum_{i=1}^{m} \sum_{j=1}^{m} \frac{GIT_{ij}}{m-1}\right)^{0.9315}}{n_A^{0.6577} + 4.9259} - 231.1361 \quad (3.113)$$

n_A is the number of atoms in the molecule; hydrogen is not counted.

The group contributions for $\Delta B_{\eta i}$ and $\Delta T_{\eta i}$ can be taken from *www.ddbst.com*. T_η is the reference temperature where the dynamic viscosity of the substance equals $\eta = 1.3$ mPas. The terms with GIB and GIT are the group interaction contributions.

To each group, a priority number is assigned to avoid ambiguous group assignment. If two groups come into consideration, the one with the lower priority number should be used.

Example 3.21

Estimate the dynamic viscosity of liquid R123 (CF_3–$CHCl_2$) at $\vartheta_1 = 30\,°C$ and $\vartheta_2 = 50\,°C$ using the Rarey/Nannoolal method. The normal boiling point is $T_b = 300.97$ K.

The particular group contributions can be determined to be

Group ID	Frequency	ΔT_{η_i}	$\Sigma \Delta T_{\eta_i}$	ΔB_{η_i}	$\Sigma \Delta B_{\eta_i}$
21	3	35.2688	105.8064	0.0028323	0.0084969
26	2	313.1106	626.2212	0.0058228	0.0116456
7	2	103.4109	206.8218	0.0213473	0.0426946
121	1	241.8968	241.8968	−0.006142	−0.006142
124	1	−115.0418	−115.0418	−0.0259017	−0.0259017
			-------------		-------------
			1065.7044		0.0307934

With $n_A = 7$, one obtains

$$B_\eta = \frac{0.0307934}{7^{-2.5635} + 0.0685} + 3.7777 = 4.18655$$

and

$$\frac{T_\eta}{K} = 21.8444 \cdot 300.97^{0.5} + \frac{1065.7044^{0.9315}}{7^{0.6577} + 4.9259} - 231.1361 = 225.4035$$

giving

$$\ln \frac{\eta}{1.3 \text{ mPas}} = -4.18655 \cdot \frac{T - 225.4054\,K}{T - 14.0878\,K}$$

With this formula, the viscosity values $\eta(30\,°C) = 0.4216$ mPas and $\eta(50\,°C) = 0.3459$ mPas are obtained. The reference values [10] are $\eta_{\text{ref}}(30\,°C) = 0.403$ mPas and $\eta_{\text{ref}}(50\,°C) = 0.323$ mPas.

When estimating liquid viscosities, the estimation of the reference temperature T_η should only be used if no reliable data point is available. The estimation of B_η is generally more reliable than the estimation of T_η. If there is a reference data point, B_η should be estimated, while T_η can be adjusted with the data point using Eq. (3.111).

The dynamic viscosities of liquids increase with increasing pressure. According to Lucas [73], the effect can be taken into account by

$$\eta(T, P) = \eta(T, P^s(T)) \frac{1 + D(\Delta P_r / 2.118)^A}{1 + C\omega \Delta P_r} \tag{3.114}$$

with

$$\Delta P_r = \frac{P - P^s(T)}{P_c}$$

$$A = 0.9991 - \frac{4.674 \cdot 10^{-4}}{1.0523 T_r^{-0.03877} - 1.0513}$$

$$D = \frac{0.3257}{(1.0039 - T_r^{2.573})^{0.2906}} - 0.2086$$

$$C = -0.07921 + 2.1616 T_r - 13.404 T_r^2 + 44.1706 T_r^3 - 84.8291 T_r^4 + 96.1209 T_r^5 - 59.8127 T_r^6 + 15.6719 T_r^7 \tag{3.115}$$

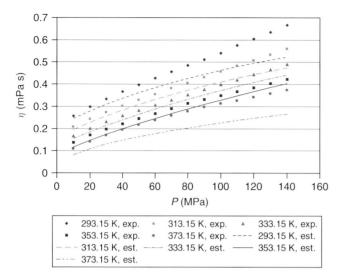

Figure 3.20 Dynamic viscosity of liquid R134a (1,1,1,2-tetrafluoroethane) as a function of pressure at various temperatures with the Lucas equation. (Data from Comunas et al. [74].)

Using Eq. (3.115), errors in the range of 10% can be expected. Figure 3.20 shows an example for R134a (1,1,1,2-tetrafluoroethane, CF_3-CH_2F), where the Lucas estimation reproduces the pressure dependence correctly up to $300 \cdots 500$ bar. The estimation is worse for pressures beyond this range. The isotherm at $100\,°C$ is pretty close to the critical temperature ($101.06\,°C$) and reproduced poorly.

For liquid mixtures, the dynamic viscosity is usually estimated with the mixing rule:

$$\ln \frac{\eta_{mix}}{\eta_0} = \sum_i x_i \ln \frac{\eta_i}{\eta_0} \qquad (3.116)$$

where x_i denotes for the mole fraction. This is one of the weakest areas in process simulation; using Eq. (3.116), more than the correct order of magnitude can hardly be obtained. With the help of an experimental value, the accuracy can be improved [6]. However, in most cases multicomponent mixtures are handled in process simulation, and experimental data are scarce. Thus, an improved correlation for binary mixtures does not necessarily yield significant improvements. Even a UNIFAC method has been developed [75], but due to its complexity its application is awkward. Example 3.22 illustrates the error caused by Eq. (3.116). In the Dortmund Data Bank, more than 13000 data sets are stored.

Example 3.22

Calculate the dynamic viscosity of a methanol–water mixture at $\vartheta = 40\,°C$ for $x_{methanol} = 0.5164$. $\eta_{water}(40\,°C) = 0.6796$ mPas, $\eta_{methanol}(40\,°C) = 0.4468$ mPas.

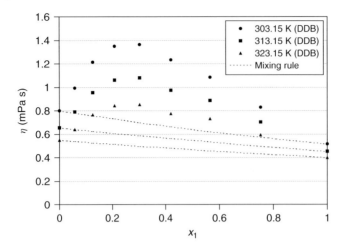

Figure 3.21 Viscosity of the system methanol (1)–water (2) at different temperatures.

Solution

$$\ln \frac{\eta}{\text{mPas}} = 0.5164 \ln 0.4468 + 0.4836 \ln 0.6796 = -0.6028 \Rightarrow \eta = 0.5473 \text{ mPas}$$

The actual value is $\eta_{\text{mix}} = 0.9345$ mPas. It is larger than both pure component values, therefore, Eq. (3.116) cannot reproduce it in principle. In Figure 3.21, the concentration dependence of the viscosity for the system methanol–water is shown.

3.3.2
Vapor Viscosity

According to the kinetic gas theory the viscosity of an ideal gas does not depend on density [44]. At low densities, there are few molecules available for momentum exchange. On the other hand, they can exchange more momentum per hit due to the larger mean free path. For ideal gases, both effects compensate, for real gases, the viscosity increases with density. The vapor viscosity is a strong function of the temperature, as the mean kinetic energy of the molecules increases with temperature so that they can exchange more momentum. For low pressures, the dynamic viscosity of ideal gases can be estimated with the Lucas equation [76]:

$$\frac{\eta^{\text{id}}}{10^{-7} \text{ Pa s}} = \frac{F_P^{\text{id}}}{\xi} \left[0.807 T_r^{0.618} - 0.357 \exp(-0.449 T_r) + 0.34 \exp(-4.058 T_r) + 0.018 \right] \tag{3.117}$$

Table 3.6 Dipole moment for a few substances [6] (1 Debye = 3.336 ·10^{-30} cm).

Substance	Formula	Dipole moment (Debye)
Hydrogen chloride	HCl	1.1
Carbon monoxide	CO	0.1
Carbon dioxide	CO_2	0.0
Water	H_2O	1.8
Ammonia	NH_3	1.5
Methane	CH_4	0.0
Ethanol	C_2H_5OH	1.7
Benzene	C_6H_6	0.0
Chlorobenzene	C_6H_5Cl	1.6
Aniline	$C_6H_5NH_2$	1.6

F_P^{id} is a correction factor for the influence of the polarity, which is characterized by the reduced dipole moment μ_r:

$$\mu_r = 52.46 \left(\frac{\mu}{\text{Debye}}\right)^2 \frac{P_c}{\text{bar}} \left(\frac{T_c}{\text{K}}\right)^{-2} \quad (3.118)$$

For some substances, the dipole moment itself can be taken from Table 3.6. The relationships for F_P^{id} are

$$F_P^{id} = 1 \qquad \text{for } 0 \leq \mu_r \leq 0.022$$
$$F_P^{id} = 1 + 30.55 \, (0.292 - z_c)^{1.72} \qquad \text{for } 0.022 \leq \mu_r \leq 0.075$$
$$F_P^{id} = 1 + 30.55 \, (0.292 - z_c)^{1.72} \, |0.96 + 0.1 \, (T_r - 0.7)| \quad \text{for } \mu_r \geq 0.075$$

(3.119)

ξ is called the *inverse reduced viscosity* and can be determined by

$$\xi = 0.176 \left(\frac{T_c}{\text{K}}\right)^{1/6} \left(\frac{M}{\text{g/mol}}\right)^{-1/2} \left(\frac{P_c}{\text{bar}}\right)^{-2/3} \quad (3.120)$$

The expected mean error of this method is considered to be low, so that measurements of this quantity are rarely carried out.

For process calculations, the dynamic viscosity of gases can be correlated with a simple polynomial

$$\eta^{id} = A + BT + CT^2 + DT^3 + ET^4 \quad (3.121)$$

or the DIPPR equation

$$\eta^{id} = \frac{AT^B}{1 + CT^{-1} + DT^{-2}} \quad (3.122)$$

which shows a more reliable extrapolation behavior.

Figure 3.22 shows a typical curvature of the vapor viscosity as a function of temperature.

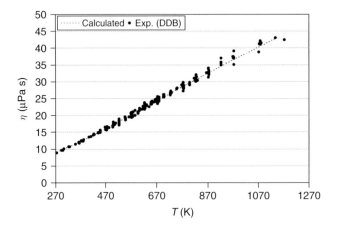

Figure 3.22 Dynamic viscosity of gaseous water as a function of temperature.

Example 3.23

Estimate the dynamic viscosity of ammonia at $T = 573.15$ K and $P = 0.1$ MPa.
The given values are:

$T_c = 405.5$ K
$P_c = 113.592$ bar
$z_c = 0.2553$
$\mu = 1.49$ Debye
$M = 17.031$ g/mol.

Solution

With

$$T_r = 573.15/405.5 = 1.4134$$
$$\mu_r = 52.46 \cdot 1.49^2 \cdot 113.592 \cdot 405.5^{-2} = 0.08046$$
$$\xi = 0.176 \, (405.5)^{1/6} \, (17.031)^{-1/2} \, (113.592)^{-2/3} = 0.004947$$
$$F_P^{id} = 1 + 30.55 \, (0.292 - 0.2553)^{1.72} \, |0.96 + 0.1 \, (1.4131 - 0.7)| = 1.1071$$

one obtains

$$\frac{\eta^{id}}{10^{-7} \text{ Pa s}} = \frac{1.1071}{0.004947} \left[0.807 \cdot 1.4134^{0.618} - 0.357 \exp(-0.449 \cdot 1.4134) \right.$$
$$\left. + 0.34 \exp(-4.058 \cdot 1.4134) + 0.018 \right]$$
$$= 185.6 \Rightarrow \eta^{id} = 18.56 \, \mu\text{Pas}$$

The literature value is $\eta^{id} = 20.1 \, \mu$Pas.

For $1 \leq T_r \leq 40$ and $0 \leq P_r \leq 100$, the pressure dependence of the vapor viscosity can be taken into account according to Lucas [76]:

$$\eta = \eta^{id} Y F_P \qquad (3.123)$$

with η^{id} from Eqs. (3.121) or (3.122), and Y via

$$Y = 1 + \frac{AP_r^E}{BP_r^F + \left(1 + CP_r^D\right)^{-1}} \qquad (3.124)$$

with

$$A = \frac{0.001245}{T_r} \exp\left(5.1726 T_r^{-0.3286}\right)$$

$$B = A(1.6553 T_r - 1.2723)$$

$$C = \frac{0.4489}{T_r} \exp\left(3.0578 T_r^{-37.7332}\right)$$

$$D = \frac{1.7368}{T_r} \exp\left(2.231 T_r^{-7.6351}\right)$$

$$E = 1.3088$$

$$F = 0.9425 \exp\left(-0.1853 T_r^{0.4489}\right) \qquad (3.125)$$

The factor F_P is calculated by

$$F_P = \frac{1 + \left(F_P^{id} - 1\right) Y^{-3}}{F_P^{id}} \qquad (3.126)$$

For $T_r < 1$ and $P < P^s(T_r)$, Lucas [76] suggests the function

$$\frac{\eta}{10^{-7}\text{ Pa s}} = Y \frac{F_P}{\xi} \qquad (3.127)$$

with

$$Y = 0.6 + 0.76 P_r^A + \left(6.99 P_r^B - 0.6\right)(1 - T_r)$$

$$A = 3.262 + 14.98 P_r^{5.508} \qquad (3.128)$$

$$B = 1.39 + 5.746 P_r$$

and the factor

$$F_P = \frac{1 + \left(F_P^{id} - 1\right)\left[\dfrac{Y}{\left(\eta^{id}/10^{-7}\text{ Pas}\right) \xi}\right]}{F_P^{id}} \qquad (3.129)$$

As a rule of thumb, the author estimates an accuracy of ±10%. For the quantum gases helium and hydrogen, a modified equation is given in [76]. Figure 3.23 gives an example of the performance of Eq. (3.127). The pressure dependence of the dynamic viscosity of gaseous n-pentane is quite significant for the 498.15 K isotherm due to its vicinity to the critical temperature. The estimation method of Lucas [76] works remarkably well.

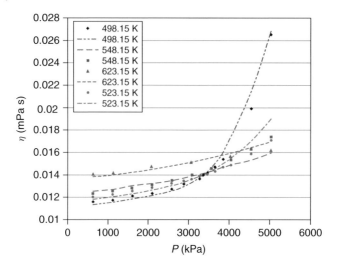

Figure 3.23 Dynamic viscosity of gaseous n-pentane as a function of pressure. (Data from [17].)

For the calculation of the dynamic viscosity of gaseous mixtures the Wilke mixing rule [78] can be applied:

$$\eta_{mix} = \sum_i \frac{y_i \eta_i}{\sum_j y_j F_{ij}} \tag{3.130}$$

with

$$F_{ij} = \frac{\left[1 + (\eta_i/\eta_j)^{1/2} (M_j/M_i)^{1/4}\right]^2}{\sqrt{8(1 + M_i/M_j)}} \tag{3.131}$$

where y_i is the mole fraction.

Figure 3.24 shows the nonlinearity of the mixing rule for the system nitrogen/n-heptane at $T = 344$ K.

For high pressure applications, one should apply Eqs. (3.123) or (3.127), using the mixing rules

$$T_{c,mix} = \sum_i y_i T_{ci} \tag{3.132}$$

$$P_{c,mix} = RT_{c,mix} \frac{\sum_i y_i z_{ci}}{\sum_i y_i v_{ci}} \tag{3.133}$$

$$M_{mix} = \sum_i y_i M_i \tag{3.134}$$

$$F_{P,mix}^{id} = \sum_i y_i F_{Pi}^{id} \tag{3.135}$$

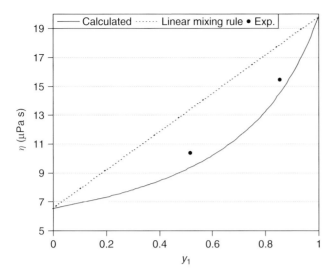

Figure 3.24 Dynamic viscosity as a function of concentration for the system nitrogen (1)/n-heptane (2) at $T = 344$ K. (Data from Carmichael *et al.* [77].)

3.3.3
Liquid Thermal Conductivity

Temperature gradients are the driving force for the heat transfer, where heat is transported from a region with high temperature to a region with low temperature. Heat transfer can occur within a phase or between phases. According to Fourier's law, the one-dimensional conductive heat transfer without phase change can be described by

$$\dot{Q} = -\lambda A \left(\frac{\partial T}{\partial x} \right) \tag{3.136}$$

where the negative sign refers to the opposite directions of heat flux and temperature gradient. A is the cross-flow area available for the heat transfer. The proportionality factor λ is the thermal conductivity, a material property describing the ability of a material to conduct heat.

In chemical processes, heat convection is much more important than heat conduction. Essentially, heat convection means that the heat transfer occurs between two phases, where at least one of them is a liquid or a gas. It is characterized by the fact that flows take place, where hot fluid flows into cold regions and vice versa. The convective heat transfer is therefore always related to flows, which can be generated by the temperature differences themselves (natural or free convection) or by a fluid flow engine, for example, a pump, a blower or a stirrer (forced convection).

Nevertheless, the thermal conductivity remains one of the key quantities for the description of heat transfer for heat convection as well. The usual approach for the calculation of convective heat transfer is

$$\dot{Q} = \alpha A \Delta T \tag{3.137}$$

where α is the heat transfer coefficient and ΔT is the temperature difference between the bulk of the two phases. For the calculation of α, the equations describing the convective heat transfer have the form

$$\text{Nu} = \frac{\alpha l}{\lambda} = F(\text{Re}, \text{Pr}) \tag{3.138}$$

for forced convection and

$$\text{Nu} = \frac{\alpha l}{\lambda} = f(\text{Gr}, \text{Pr}) \tag{3.139}$$

for natural convection. Nu is the so-called Nußelt number, which describes the ratio between convective heat transfer and heat transfer by thermal conduction. In both cases, l is the characteristic length, which has to be defined for every heat transfer arrangement in an appropriate way (e.g., inner tube diameter for heat transfer inside tubes). It can be seen that the heat transfer coefficient is proportional to the thermal conductivity as a first approximation. As the Prandtl number describing the influence of the physical properties

$$\text{Pr} = \frac{\eta c_P}{\lambda} \tag{3.140}$$

also depends on the thermal conductivity, the dependence is generally less distinctive. For example, for laminar, parallel flow along a flat plate, it can be derived [79] that

$$\text{Nu} = \frac{\alpha l}{\lambda} = 0.664 \cdot \text{Re}^{1/2} \text{Pr}^{1/3} \Rightarrow \alpha \sim \lambda^{2/3} \tag{3.141}$$

Nevertheless, for a reasonable design of heat transfer equipment, the knowledge of the thermal conductivity is indispensable.

Correlations valid over the whole range from the dilute gas to the dense liquid including the area in the vicinity of the critical point are available [80].

The thermal conductivity of liquids can usually be correlated with a polynomial of the fourth degree:

$$\lambda^L = A + BT + CT^2 + DT^3 + ET^4 \tag{3.142}$$

or the Jamieson [81] equation

$$\lambda^L = A\left(1 + B\tau^{1/3} + C\tau^{2/3} + D\tau\right) \tag{3.143}$$

where

$$\tau = 1 - \frac{T}{T_c} \tag{3.144}$$

Except for the critical region, the thermal conductivity is more or less a linear function of the temperature. Therefore, if only data for the low temperature region are available, only A and B are fitted in Eq. (3.143). For water, the thermal conductivity is much higher (0.6 ... 0.7 W/(Km)) than for organic components

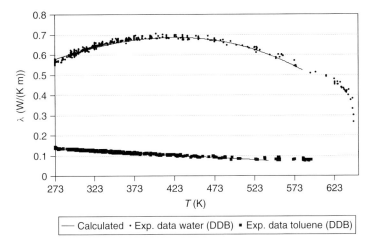

Figure 3.25 Thermal conductivity of water and toluene as a function of temperature.

(0.05 ... 0.2 W/(Km)) and shows a maximum at approx. 130 °C. Similarly, for some polyalcohols (ethylene glycol, glycerol) the thermal conductivity is higher than expected. The explanation for this phenomenon seems to be that the heat is transported by breaking and forming hydrogen bonds along the temperature gradient [82]. Figure 3.25 shows the thermal conductivity of water and toluene as a function of temperature. The pressure dependence is negligible for normal applications.

Thermal conductivities can be estimated with the Sato–Riedel equation

$$\frac{\lambda^L}{\text{W/K m}} = 1.111 \left(\frac{M}{\text{g/mol}}\right)^{-1/2} \frac{3 + 20(1 - T_r)^{2/3}}{3 + 20\left(1 - T_{b,r}\right)^{2/3}} \qquad (3.145)$$

Deviations of approx. 20% should be expected when Eq. (3.145) is used, which shows that it is worth to perform a data inquiry or to carry out a comparably inexpensive measurement, for instance with the hot-wire method [83], at least in critical cases.

Example 3.24

Estimate the thermal conductivity of liquid acetone at $T = 373.15$ K. The given data are

$T_c = 508.1$ K
$T_b = 329.23$ K
$M = 58.08$ g/mol.

Solution

$$\frac{\lambda^L}{\text{W/Km}} = 1.11\,(58.08)^{-1/2} \frac{3 + 20(1 - 373.15/508.1)^{2/3}}{3 + 20(1 - 329.23/508.1)^{2/3}} = 0.1265$$

Table 3.7 Values for Q used in Eq. (3.146).

T_r				P_r		
	1	5	10	50	100	200
0.8	0.036	0.038	0.038	0.038	0.038	0.038
0.7	0.018	0.025	0.027	0.031	0.032	0.032
0.6	0.015	0.02	0.022	0.024	0.025	0.025
0.5	0.012	0.0165	0.017	0.019	0.02	0.02

The experimental value is $\lambda = 0.1285 \, W/(Km)$.

The pressure dependence of the liquid thermal conductivity can usually be neglected. At very high pressures, it can be taken into account with the Missenard method [6]:

$$\frac{\lambda(P_r, T)}{\lambda(P_s, T)} = 1 + Q P_r^{0.7} \tag{3.146}$$

Values for Q are listed in Table 3.7. It should be emphasized that Eq. (3.146) is only able to evaluate a reasonable tendency for the pressure dependence, as Figure 3.26 shows.

The Li method is appropriate to be taken as a mixing rule [6]. Its formulation is

$$\lambda_{mix} = \sum_{i=1}^{n} \sum_{j=1}^{n} \frac{2\Phi_i \Phi_j}{\lambda_i^{-1} + \lambda_j^{-1}} \tag{3.147}$$

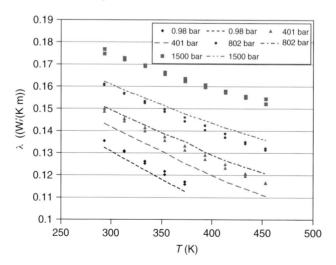

Figure 3.26 Pressure dependence of the liquid thermal conductivity of toluene and its estimation with the Missenard method. (Data from [17].)

with

$$\Phi_i = \frac{x_i v_i^L}{\sum_{j=1}^{n} x_j v_j^L} \qquad (3.148)$$

where x_i is the mole fraction.

Example 3.25

Determine the thermal conductivity of a liquid mixture consisting of benzene (1) and methyl formate (2) with $x_1 = 0.204$ at $T = 323$ K. Given data:

$\lambda_1 = 0.1376$ W/(Km)
$\lambda_2 = 0.1786$ W/(Km)
$M_1 = 78.114$ g/mol
$M_2 = 60.052$ g/mol
$\rho_1 = 847.2$ kg/m^3 => $v_1 = 9.22025 \cdot 10^{-5}$ m^3/mol
$\rho_2 = 930$ kg/m^3 => $v_2 = 3.4454 \cdot 10^{-5}$ m^3/mol.

Solution

$$\Phi_1 = \frac{0.204 \cdot 9.22025 \cdot 10^{-5}}{0.204 \cdot 9.22025 \cdot 10^{-5} + 0.796 \cdot 3.4454 \cdot 10^{-5}} = 0.4068$$

$$\Phi_2 = 1 - \Phi_1 = 0.5932$$

$$\frac{\lambda_{\text{mix}}}{\text{W/(Km)}} = \frac{2 \cdot 0.4068^2}{2 \cdot 0.1376^{-1}} + 2 \cdot \frac{2 \cdot 0.4068 \cdot 0.5932}{0.1376^{-1} + 0.1786^{-1}} + \frac{2 \cdot 0.5932^2}{2 \cdot 0.1786^{-1}} = 0.1606$$

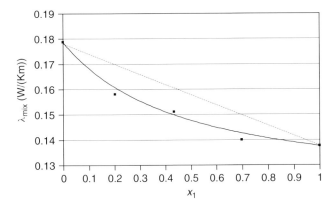

Figure 3.27 Liquid thermal conductivity as a function of concentration for the system benzene (1)/methyl formate (2) at $T = 323$ K. (Data from [84].) —— Li mixing rule --- linear mixing rule for comparison.

The experimental value is $\lambda = 0.158$ W/(Km) [84]. Figure 3.27 shows the curvature of the Li mixing rule.

3.3.4
Vapor Thermal Conductivity

Similar to the viscosity, the thermal conductivity of vapors and gases can be estimated from the kinetic gas theory. As well, its behavior is similar. It increases with temperature, and at moderate pressures (0.01 … 1 MPa) it is hardly a function of pressure. A fourth degree polynomial

$$\lambda^{id} = A + BT + CT^2 + DT^3 + ET^4 \tag{3.149}$$

or the PPDS equation

$$\lambda^{id} = \frac{\sqrt{T_r}}{A + \frac{B}{T_r} + \frac{C}{T_r^2} + \frac{D}{T_r^3}} \tag{3.150}$$

can be used for correlation. For temperature extrapolation, the latter equation has proved to be more reliable.

Figure 3.28 shows the thermal conductivity of gaseous water as a function of temperature as an example for a typical curvature.

For nonpolar components, the vapor thermal conductivity can be estimated by the Chung equation [85]

$$\lambda^{id} = \frac{3.75 \Psi \eta^{id} R}{M} \tag{3.151}$$

with

$$\Psi = 1 + \alpha \frac{0.215 + 0.28288\alpha - 1.061\beta + 0.26665\gamma}{0.6366 + \beta\gamma + 1.061\alpha\beta} \tag{3.152}$$

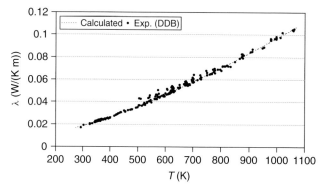

Figure 3.28 Thermal conductivity of gaseous water as a function of temperature.

and

$$\alpha = c_p^{id}/R - 2.5$$
$$\beta = 0.7862 - 0.7109\omega + 1.3168\omega^2 \qquad (3.153)$$
$$\gamma = 2 + 10.5 T_r^2$$

For polar components, the error will be larger than for nonpolar components. In these cases, the Roy–Thodos method [6] is recommended.

Example 3.26

Estimate the thermal conductivity of gaseous 2-methyl pentane at $T = 373.15$ K and $P = 0.1$ MPa using the Chung equation. The given data are

$M = 86.178$ g/mol
$\omega = 0.2801$
$T_c = 497.7$ K
$c_p^{id} = 2.007$ J/(gK)
$\eta = 8.18\,\mu$Pas.

Solution

With

$$\alpha = \frac{2.007 \cdot 86.178}{8.3143} - 2.5 = 18.303$$
$$\beta = 0.7862 - 0.7109 \cdot 0.2801 + 1.3168 \cdot 0.2801^2 = 0.6904$$
$$\gamma = 2 + 10.5 \cdot (373.15/497.7)^2 = 7.9023$$

one obtains

$$\Psi = 1 + 18.303 \frac{0.215 + 0.28288 \cdot 18.303 - 1.061 \cdot 0.6904 + 0.26665 \cdot 7.9023}{0.6366 + 0.6904 \cdot 7.9023 + 1.061 \cdot 18.303 \cdot 0.6904}$$
$$= 7.3519$$

and

$$\lambda^{id} = \frac{3.75 \cdot 7.3519 \cdot 8.18 \cdot 10^{-6} \cdot 8.3143}{86.178 \cdot 10^{-3}} \frac{W}{Km} = 0.0218 \text{ W/Km}$$

From data compilations, we find $\lambda^{id} = 0.0205$ W/(Km).

It should be mentioned that the pressure dependence of the thermal conductivity of gases is quite strange. For very small pressures ($P < 10^{-3}$ mbar), when the mean free path is large in comparison with the vessel dimensions, the thermal conductivity is proportional to the pressure and to the distance d between the walls of the vessel perpendicular to the heat flux direction:

$$\lambda = \frac{3}{8} P d \sqrt{\frac{3R}{MT}} \qquad (3.154)$$

At high pressures, it can be estimated according to Stiel and Thodos [86]:

$$\lambda = \lambda^{id} + 0.0122\Gamma^{-1}z_c^{-5}[\exp(0.535\rho/\rho_c) - 1] \quad \text{for } \rho/\rho_c < 0.5$$
$$\lambda = \lambda^{id} + 0.0114\Gamma^{-1}z_c^{-5}[\exp(0.67\rho/\rho_c) - 1.069] \quad \text{for } 0.5 < \rho/\rho_c < 2$$
$$\lambda = \lambda^{id} + 0.0026\Gamma^{-1}z_c^{-5}[\exp(1.155\rho/\rho_c) + 2.016] \quad \text{for } 2 < \rho/\rho_c < 2.8$$

(3.155)

Γ is given by

$$\frac{\Gamma}{(\text{W/Km})^{-1}} = 210 \left(\frac{T_c}{\text{K}}\right)^{1/6} \left(\frac{M}{\text{g/mol}}\right)^{1/2} \left(\frac{P_c}{\text{bar}}\right)^{-2/3} \quad (3.156)$$

The method is not considered to be accurate, an error between 10 … 20% has to be expected. For polar compounds, the method is not appropriate, like many other correlations which are listed in [6]. Fortunately, in process simulation applications neither Eq. (3.154) nor Eq. (3.155) are usually relevant. Figure 3.29 shows the pressure dependence of the thermal conductivity of methane and the performance of the Stiel–Thodos equation (3.155), which is in good qualitative agreement with the experimental data. For the calculation of the thermal conductivity of gaseous mixtures the mixing rule of Wassiljeva et al. [87] can be applied analogously to the Wilke mixing rule for the viscosity of gases with y_i as the mole fraction:

$$\lambda_{\text{mix}} = \sum_i \frac{y_i \lambda_i}{\sum_j y_j F_{ij}} \quad (3.157)$$

with

$$F_{ij} = \frac{\left[1 + (\eta_i/\eta_j)^{1/2} (M_j/M_i)^{1/4}\right]^2}{\sqrt{8(1 + M_i/M_j)}} \quad (3.158)$$

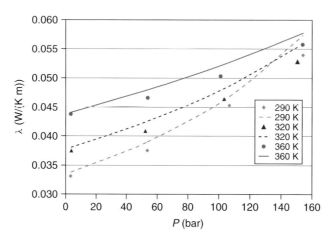

Figure 3.29 Experimental data and Stiel–Thodos estimation for the thermal conductivity of gaseous methane at high pressures for various temperatures. (Exp. data from Patek and Klomfar [83].)

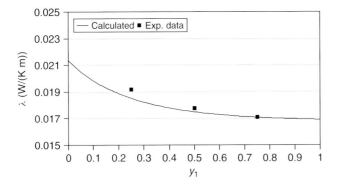

Figure 3.30 Vapor thermal conductivity as a function of concentration for the system benzene (1)/argon (2) at $T = 373.75$ K. (Exp. data from [88].)

Figure 3.30 shows that it is advantageous to use a nonlinear mixing rule like Eq. (3.157) for the thermal vapor conductivity.

3.3.5
Surface Tension

The surface tension indicates to which extent a liquid is prone to form drops. In process simulation, its role is comparatively small; it is used for hydrodynamic calculations of packings. Its most important application is the calculation of capillary heads, which are directly proportional to the surface tension [89].

The surface tension occurs in the boundary layer between vapor and liquid (Figure 3.31). At the interface, the attractive forces between the molecules act sidewise and toward the bulk liquid, whereas there are only little attractive forces from the vapor phase. Therefore, there is a resultant force which acts toward the bulk liquid, causing the surface layer to contract and form the smallest possible surface. In contrast, there is no resultant force for a molecule in the bulk of the liquid. The surface tension represents the energy required for creating additional surface between vapor and liquid; its unit is therefore J/m^2 or, respectively, N/m.

For pure liquids, the surface tension decreases with increasing temperature and becomes zero at the critical point (Figure 3.32). It is essentially determined by intermolecular forces, that is, by the difference of the attractive forces on the molecules in the surface between a gas and a liquid.

The surface tension can be correlated with the extended Watson equation

$$\sigma = A(1 - T_r)^{B + CT_r + DT_r^2 + ET_r^3} \tag{3.159}$$

where the coefficients C, D, and E can be set to 0 or fitted if necessary. If they are set to 0, $B = 1.22$ is often a good choice.

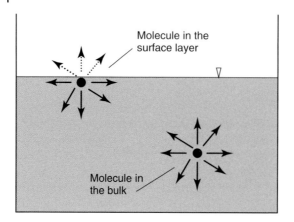

Figure 3.31 Forces on molecules in the bulk and in the surface layer.

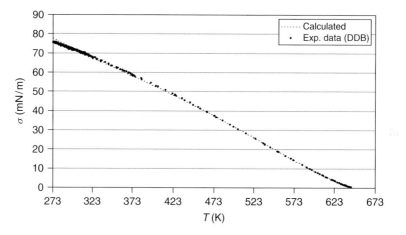

Figure 3.32 Surface tension of water as a function of temperature.

The surface tension can be estimated by the Brock–Bird–Miller equation, which is based on the corresponding states principle [90]:

$$\frac{\sigma}{\text{mN/m}} = \left(\frac{P_c}{\text{bar}}\right)^{2/3} \left(\frac{T_c}{\text{K}}\right)^{1/3} Q(1-T_r)^{11/9} \qquad (3.160)$$

with

$$Q = 0.1196 \left[1 + \frac{\frac{T_b}{T_c} \ln \frac{P_c}{1.01325 \text{bar}}}{1 - T_b/T_c}\right] - 0.279 \qquad (3.161)$$

For nonpolar substances, the error is usually below 5%.

Example 3.27

Estimate the surface tension of bromobenzene at $T = 323.15\text{ K}$ with the Brock–Bird–Miller equation. The given data are

$T_b = 429.05\text{ K}$
$T_c = 670.15\text{ K}$
$P_c = 4.52\text{ MPa}$.

Solution

With

$$Q = 0.1196\left[1 + \frac{\frac{429.05}{670.15}\ln\frac{45.2}{1.01325}}{1 - 429.05/670.15}\right] - 0.279 = 0.6489$$

one obtains

$$\sigma = 45.2^{2/3} \cdot 670.15^{1/3} \cdot 0.6489 \cdot (1 - 323.15/670.15)^{11/9}\text{ mN/m} = 32.23\text{ mN/m}$$

The experimental value is $\sigma = 33\text{ mN/m}$.

As a mixing rule, the equation

$$\frac{\sigma_{mix}}{\text{mN/m}} = \left(P^L_{mix}\frac{\rho^L_{mix}}{\text{mol/cm}^3} - P^V_{mix}\frac{\rho^V_{mix}}{\text{mol/cm}^3}\right)^4 \tag{3.162}$$

is usually applied. At low pressures, the vapor phase terms can be neglected. P is the parachor[6]:

$$P = \left(\frac{\rho}{\text{mol/cm}^3}\right)^{-1}\left(\frac{\sigma}{\text{mN/m}}\right)^{1/4} \tag{3.163}$$

which can be averaged by

$$P_{mix} = \sum_i\sum_j x_i x_j \frac{P_i + P_j}{2} \tag{3.164}$$

where x_i and x_j refer to the mole fractions of the particular components. For aqueous systems, Eq. (3.162) usually performs poorly, and special mixing rules should be applied [32].

Example 3.28

Estimate the surface tension of a mixture of nitromethane (1) and benzene (2) with a mole fraction $x_1 = 0.6$ at $T = 298\text{ K}$ using the mixing rule (3.162). The influence of the vapor in Eq. 3.162 should be neglected due to the low pressure. The given data are

---

6) Strictly, the parachor is not dimensionless as treated here. Its unit is $\text{cm}^3\text{ g}^{1/4}\text{ s}^{-1/2}\text{ mol}^{-1}$.

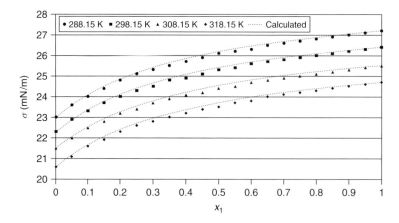

Figure 3.33 Surface tension of a liquid mixture of octanoic acid ethyl ester (1) and ethanol (2) at different temperatures. (Data from DDB [17].)

$M_1 = 61.04\,\text{g/mol}$
$M_2 = 78.114\,\text{g/mol}$
$\sigma_1 = 36.155\,\text{mN/m}$
$\sigma_2 = 28.23\,\text{mN/m}$
$\rho_1 = 1126.96\,\text{kg/m}^3 = 0.018463\,\text{mol/cm}^3$
$\rho_2 = 872.8\,\text{kg/m}^3 = 0.011173\,\text{mol/cm}^3$

Solution

One obtains

$$P_1 = 0.018463^{-1} \cdot 36.155^{1/4} = 132.815$$
$$P_2 = 0.011173^{-1} \cdot 28.23^{1/4} = 206.296$$

$$P_\text{mix} = 0.6^2 \cdot 132.815 + 2 \cdot 0.6 \cdot 0.4 \frac{132.815 + 206.296}{2} + 0.4^2 \cdot 206.296 = 162.207$$

$$\rho^L_\text{mix} = \left(\frac{x_1}{\rho_1} + \frac{x_2}{\rho_2}\right)^{-1} = \left(\frac{0.6}{0.018463} + \frac{0.4}{0.011173}\right)^{-1} \frac{\text{mol}}{\text{cm}^3} = 0.014642\,\text{mol/cm}^3$$

$$\sigma_\text{mix} = (162.207 \cdot 0.014642)^4\,\text{mN/m} = 31.8\,\text{mN/m}$$

which is well in line with the experimental value [32]. Figure 3.33 illustrates the influence of the concentration on the surface tension.

3.3.6
Diffusion Coefficients

Molecular diffusion is the dominating mass transfer mechanism in solids, polymers (e.g., membranes) and liquids where no flow takes place. For all calculations where mass transfer is decisive, binary diffusion coefficients are required. In process simulation, these so-called rate-based calculations are applied if the usual equilibrium approach cannot be regarded as valid.

Well-known examples are absorption columns, for example, the absorption of HCl in an exhaust air by water. Experience shows that quite large columns are required for this purpose, whereas the equilibrium calculation just requires one theoretical stage due to the electrolyte character of hydrogen chloride. Columns like that can only be successfully designed by application of a rate-based model (see also Section 14.5).

According to Fick's first law, the mass flux density can be set up as

$$\frac{\dot{n}_1}{F} = -\rho_{mix} D_{12} \left(\frac{dx}{dz}\right) \qquad (3.165)$$

where $\dot{n}_1$ is the mole flow of component 1 and z is the direction of the coordinate axis. The diffusion coefficient is symmetric, that is, $D_{12} = D_{21}$. It should be pointed out that Eq. 3.165 is a simplification; to be exact, activities, fugacities or chemical potentials should be used for the description of the driving force.[7] Typical orders of magnitude for diffusion coefficients are

Gases: $D = 5 \cdot 10^{-6} \ldots 10^{-5}$ m^2/s
Liquids: $D = 10^{-10} \ldots 10^{-9}$ m^2/s
Solids: $D = 10^{-14} \ldots 10^{-10}$ m^2/s.

One should note the large difference between diffusion coefficients in liquids and gases.

Diffusion coefficients are essentially determined by intermolecular forces.

Unlike to other quantities, they are rarely measured. Usually, one must rely on estimation methods. As the purpose of most of the mass transfer calculations is to obtain an impression of the effectivity of the mass transfer and the corresponding mass transfer models are approximations theirselves, this is normally acceptable.

For gases, the kinetic gas theory is able to derive relationships for the diffusion coefficient as well as it does for viscosity and thermal conductivity [44]. It depends both on temperature and pressure, but not on the composition. At pressures lower than 1 MPa, it is proportional to the reciprocal value of the pressure. At low pressures, it can be accurately estimated according to Fuller et al. [91]. The equation for the determination of vapor diffusion coefficients is

$$\frac{D_{12}}{\text{cm}^2/\text{s}} = \frac{0.00143 \left(\frac{T}{K}\right)^{1.75} \left[\left(\frac{M_1}{\text{g/mol}}\right)^{-1} + \left(\frac{M_2}{\text{g/mol}}\right)^{-1}\right]^{1/2}}{\frac{P}{\text{bar}} \sqrt{2} \left[\left(\sum \Delta_{V_1}\right)^{1/3} + \left(\sum \Delta_{V_2}\right)^{1/3}\right]^2} \qquad (3.166)$$

Δ_V is the so-called diffusion volume, which can be evaluated from the group contributions given in Table 3.8. The average error of the Fuller method is reported to be 4% [6].

---

7) Otherwise a miscibility gap e.g. between benzene and water would not be possible.

Table 3.8 Structural groups for the evaluation of the diffusion volume using the Fuller method.

Atomic and structural contributions			
C	15.9	Br	21.9
H	2.31	I	29.8
O	6.11	S	22.9
N	4.54	Aromatic ring	−18.3
F	14.7	Heterocyclic ring	−18.3
Cl	2.1		
Simple molecules			
He	2.67	CO	18.0
Ne	5.98	CO_2	26.9
Ar	16.2	N_2O	35.9
Kr	24.5	NH_3	20.7
Xe	32.7	H_2O	13.1
H_2	6.12	SF_6	71.3
D_2	6.84	Cl_2	38.4
N_2	18.5	Br_2	69.0
O_2	16.3	SO_2	41.8
Air	19.7		

Example 3.29

Determine the vapor diffusion coefficient of the binary mixture ammonia (NH_3) (1) and diethyl ether ($C_4H_{10}O$) (2) at $T = 288$ K and $P = 2$ bar using the Fuller method. The molar masses are $M_1 = 17.03$ g/mol and $M_2 = 74.12$ g/mol.

Solution

The diffusion volumes for the two substances are

$$\left(\sum \Delta_{v_1}\right) = 20.7$$

$$\left(\sum \Delta_{v_2}\right) = 4 \cdot 15.9 + 10 \cdot 2.31 + 6.11 = 92.81$$

After inserting into Eq. (3.166) one obtains

$$D_{12} = \frac{0.00143\,(288)^{1.75}\left[(17.03)^{-1} + (74.12)^{-1}\right]^{1/2}}{2\sqrt{2}\left[(20.7)^{1/3} + (92.81)^{1/3}\right]^2}\,\text{cm}^2/\text{s} = 0.0517\,\text{cm}^2/\text{s}$$

The literature value [92] is $D_{12} = 0.0505$ cm²/s.

The influence of pressure can be determined with the equation of Riazi and Whitson [93]:

$$\frac{(\rho D_{12})}{(\rho D_{12})^{\text{id}}} = 1.07 \left(\frac{\eta}{\eta^{\text{id}}}\right)^{B+CP/P_c} \tag{3.167}$$

with

$$B = -0.27 - 0.38\omega \tag{3.168}$$

and

$$C = -0.05 + 0.1\omega \quad (3.169)$$

As the diffusion coefficient always refers to mixtures, ω and P_c are supposed to be calculated by linear mixing rules:

$$\omega = y_1\omega_1 + y_2\omega_2$$
$$P_c = y_1 P_{c1} + y_2 P_{c2} \quad (3.170)$$

The disadvantage of this relationship is that the dynamic viscosity of the mixture at high pressure is necessary, which has to be estimated itself (Section 3.3.2). The expected errors of Eq. (3.167) are approx. 15%.

The diffusion coefficient of liquids is a complicated function of concentration, temperature, and pressure. A reliable calculation method does not exist. However, for most of the calculations used in process simulation it is sufficient to get the correct order of magnitude. For the liquid diffusion coefficient, it can be distinguished between the case of the ideal dilute solution and the general case of arbitrary concentrations.

In case of infinite dilution of solute 1 in solvent 2 a diffusion coefficient can be estimated in many ways [6]. They are all characterized by individual rules for single substances, limited application ranges, and input parameters that are hardly accessible. The Tyn/Calus equation is a good compromise between accuracy and applicability [6]:

$$\frac{D_{12}^\infty}{\text{cm}^2/\text{s}} = 8.93 \cdot 10^{-8} \left(\frac{v_2(T_b)}{\text{cm}^3/\text{mol}}\right)^{-1/3} \left(\frac{v_1(T_b)}{\text{cm}^3/\text{mol}}\right)^{1/6} \left(\frac{P_2}{P_1}\right)^{0.6} \frac{T}{K} \left(\frac{\eta_2}{\text{mPas}}\right)^{-1} \quad (3.171)$$

with P being the parachor, which can be estimated using Eq. (3.163).

The limitations and individual exception rules of the Tyn/Calus equation are as follows.

- The dynamic viscosity of the solvent should be less than 20–30 mPas.
- For water as solute, the parameters $v_1 = 37.4\,\text{cm}^3/\text{mol}$ and $P_1 = 105.2$ should be used.
- For organic acids as solutes the true values for v_1 and P_1 should be doubled, except the cases where water, methanol, or butanol are used as solvents.
- If the solvent is an alcohol and the solute is nonpolar, the values for v_1 and P_1 should be multiplied with a factor $8\eta_2/\text{mPas}$.

The expected error of the Tyn/Calus equation is estimated to be ~10%, but larger errors are not unusual.

The temperature dependence of Eq. (3.171) is only an approximation. The actual relationship is still not yet known, as well as the pressure dependence. The liquid diffusion coefficient decreases with increasing pressure, which is, however, only relevant for very high pressures.

The model of the ideal dilute solution is applicable, if the concentration of the solute does not exceed 5–10%. To apply the model to the general case of arbitrary concentrations, the Vignes correlation [94] can be used

$$D_{12} = (D_{12}^\infty)^{x_2} (D_{21}^\infty)^{x_1} \left(\frac{\partial \ln (x_1 \gamma_1)}{\partial \ln x_1}\right)_{T,P} \tag{3.172}$$

with γ as activity coefficient (see Section 4.8). Because of the Gibbs–Duhem equation (see Section 4.8) it does not matter which component is used for the differential quotient.

Example 3.30

Estimate the liquid diffusion coefficient of the mixture benzene (1) and toluene (2) at $T = 298$ K and $x_1 = 0.6$. The mixture is almost ideal, therefore, the activity coefficient is close to 1. The given values are

$v_1 = 95.84$ cm^3/mol
$v_2 = 118.3$ cm^3/mol
$\sigma_1 = 28.2$ mN/m
$\sigma_2 = 27.9$ mN/m
$\eta_1 = 0.601$ mPas
$\eta_2 = 0.553$ mPas.

Solution

From the specific volumes and the surface tensions, one obtains the parachors using Eq. (3.163):

$P_1 = 220.86$
$P_2 = 271.89$.

The diffusion coefficients of the ideal diluted solutions are

$$D_{12}^\infty = 8.93 \cdot 10^{-8} (118.3)^{-1/3} (95.84)^{1/6} \left(\frac{271.89}{220.86}\right)^{0.6} \cdot 298 \cdot (0.553)^{-1} \text{ cm}^2/\text{s}$$
$$= 2.38 \cdot 10^{-5} \text{ cm}^2/\text{s}$$
$$D_{21}^\infty = 8.93 \cdot 10^{-8} (95.84)^{-1/3} (118.3)^{1/6} \left(\frac{220.86}{271.89}\right)^{0.6} \cdot 298 \cdot (0.601)^{-1} \text{ cm}^2/\text{s}$$
$$= 1.89 \cdot 10^{-5} \text{ cm}^2/\text{s}$$

There is one experimental value available for D_{21}^∞, which is almost met by accident. For the actual concentrations, the Vignes correlation [94] yields

$$D_{12} = (2.38 \cdot 10^{-5})^{0.4} (1.89 \cdot 10^{-5})^{0.6} \text{ cm}^2/\text{s} = 2.07 \cdot 10^{-5} \text{ cm}^2/\text{s}$$

Additional Problems

P3.1 Regress a third order polynomial ($c_P = a + bT + cT^2 + dT^3$) to results of the equation for the heat capacity at constant pressure for graphite reported in "Butland, A.T.D. and Maddison, R.J. (1973). The specific heat of graphite: an evaluation of measurements. *J. Nucl. Mater.*, 49, 45–56":

$$c_P = \left(0.538657 + 9.11129 \cdot T - \frac{90.2725}{T} - \frac{43449.3}{T^2} \right.$$
$$\left. + \frac{1.50309 \cdot 10^7}{T^3} - \frac{1.43688 \cdot 10^9}{T^4}\right) \frac{\text{cal}}{\text{gK}}$$

P3.2 Calculate the enthalpy of vaporization of ethanol at 25 °C and 200 °C using the Watson equation. The enthalpy of vaporization at the normal boiling temperature of 351.41 K is 39.183 kJ/mol. The critical temperature of ethanol can be taken from Appendix A.

P3.3 Estimate the critical data T_c, P_c, and v_c of n-hexene using the Joback-method.

P3.4 Determine the enthalpy of formation as well as the Gibbs energy of formation for n-hexane and benzene at 25 °C and 1 atm (ideal gas) using the Joback group contribution method.

P3.5 Retrieve the vapor pressure data of naphthalene from the free DDBSP Explorer Version and export the values to MS-Excel. Regress the coefficients of the Wagner equation (Eq. (3.37)) to the experimental data using the Excel Solver Add-In. Use the Wagner equation to calculate the vapor pressure of the hypothetical subcooled liquid. From the vapor pressures of the hypothetical subcooled liquid and the sublimation pressure data found in the free DDBSP Explorer Version, estimate the melting point and heat of fusion for naphthalene. Compare the results to the values given in the DDB.

P3.6 Estimate the melting temperature of naphthalene using the Joback method and compare the result to the recommended value of 353.35 K. The estimation can either be performed by hand using the group contributions in Table 3.1 or with the help of the program Artist in the free DDBSP Explorer Version.

P3.7 Retrieve the saturated liquid density data for tetrahydrofuran in the free DDBSP Explorer Version and plot the data. In Artist, estimate the liquid density at the normal boiling point and compare the results to the data.

P3.8 In the free DDBSP Explorer Version, regress the liquid vapor pressure data for trichloromethane using the Antoine equation. Remove outliers that are not in agreement with the majority of the data. Check the manual to clarify the exact formulation of the Antoine equation. Convert the parameters so that the Antoine equation employs the natural logarithm and yields the vapor pressure in the unit kPa.

P3.9 Compare the liquid thermal conductivity data for ethylene glycol (1,2-ethanediol) and 1-butanol using the free DDBSP Explorer Version. Interpret the difference in the absolute value and the temperature dependence on a molecular basis.

P3.10 The boiling pressure of water at $\vartheta_1 = 90\,°C$ is $P_1^s = 0.702$ bar. At this temperature, the enthalpy of vaporization is 2282.5 J/g. Estimate the vapor pressure at $\vartheta_2 = 95\,°C$ using the Clausius-Clapeyron equation (Eq. (2.86)).

P3.11 Estimate the standard enthalpy of formation and the standard Gibbs energy of formation of ethylene glycol (CH_2OH-CH_2OH) in the ideal gas state using the Joback method. Can a similarly good result be expected as in case of ethyl acetate (Example 3.6)?

P3.12 An ideal gas is heated up from $T_1 = 300$ K to $T_2 = 350$ K. The specific duty is evaluated to be $q_{12} = 1040$ J/mol. Describe the chemical nature of the gas.

P3.13 Calculate the speed of sound in water vapor at $\vartheta = 100\,°C$, $P = 1$ bar. The vapor should be regarded as an ideal gas.

P3.14 For the refrigerant R134a, the following data are given:

$P^s(0\,°C) = 2.006$ bar $\qquad P^s(10\,°C) = 4.146$ bar
$\rho^V(0\,°C) = 14.428$ kg/m^3 $\qquad \rho^V(10\,°C) = 20.226$ kg/m^3
$\rho^L(0\,°C) = 1294.78$ kg/m^3 $\qquad \rho^L(10\,°C) = 1260.96$ kg/m^3

Estimate the enthalpy of vaporization at $\vartheta = 5\,°C$ using the Clausius-Clapeyron equation and interpolate the vapor pressure at $\vartheta = 5\,°C$ with the Hoffmann-Florin equation.

P3.15 At $P = 1.01325$ bar, the melting point of water is $\vartheta_m = 0\,°C$. How is the melting temperature affected, if the pressure is increased to 20 bar? Use the Clausius-Clapeyron equation for the solid-liquid equilibrium (Eq. (3.14)). The following values are given:

$\Delta h_m = 333$ J/g $\qquad \rho^L = 1000$ kg/m^3 $\qquad \rho^S = 917$ kg/m^3

P3.16 Estimate the whole set of physical properties (critical temperature, pressure, volume and compressibility, acentric factor, boiling and melting temperature, heat of fusion and heat of vaporization, standard enthalpy and standard Gibbs energy of formation, vapor pressure, liquid density, ideal gas heat capacity, liquid and vapor viscosity, liquid and vapor thermal conductivity, and surface tension; temperature-dependent properties at $\vartheta = 20\,°C$ of R134a (1,1,1,2-tetrafluoroethane, CH_2F-CF_3) without using any given information except its molar mass (102.03 g/mol) and its chemical structure. Compare the results with the values obtained with the data and equations given in Appendix A.

P3.17 Estimate the ideal gas heat capacity of methane at $T = 600$ K using the following information about its basic frequencies from spectroscopic data:

$\Theta_1 = \Theta_2 = \Theta_3 = 1876.6$ K

$\Theta_4 = \Theta_5 = 2186$ K

$\Theta_6 = 4190$ K

$\Theta_7 = \Theta_8 = \Theta_9 = 4343$ K

The characteristic temperature Θ is an abbreviation for $\Theta = h\nu_0/k$. Verify that the molecule has nine basic frequencies for the vibration.

P3.18 At approximately T = 380 K, the vapor pressure curves of water and n-hexane intersect. Calculate the difference in the vapor pressures, if both substances form drops with diameters of d = 2 nm.

P3.19 Estimate the dynamic viscosity of liquid 1,1,2-trichloroethane at T = 330 K using the Rarey/Nannoolal method.

P3.20 Estimate the vapor pressure of p-tolualdehyde at $\vartheta = 150\,°C$ using the Rarey/Moller method. The normal boiling point is $\vartheta_b = 206.3\,°C$. Use the Clausius-Clapeyron equation to determine the corresponding enthalpy of vaporization.

P3.21 The speed of sound in ammonia at T = 303.15 K, P = 0.01 bar is $w^* = 439.11$ m/s. Calculate the corresponding isobaric ideal gas heat capacity.

P3.22 Estimate the enthalpy of vaporization of benzene at the normal boiling point using the vapor pressure coefficients and the critical constants from Appendix A. For the vapor phase, use
a. the ideal gas law
b. the Soave-Redlich-Kwong equation of state.
Compare the results with the correlation given in Appendix A.

P3.23 Estimate the enthalpy of vaporization of ethanol at $\vartheta_1 = 25\,°C$ and $\vartheta_2 = 200\,°C$. Use Eq. (3.61) setting B = 0.38 and C = D = E = 0. Determine the parameter A from the reference value $\Delta h_v = 38\,470$ J/mol at T = 351.5 K.

References

1. Lydersen, A.L. (1955) Estimation of Critical Properties of Organic Compounds. Report 3, University of Wisconsin College of Engineering, Engineering Expirement Station, Madison, WI.
2. Ambrose, D., National Physical Laboratory (1978) *NPL Rep. Chem.*, **92**, corrected 1980.
3. Ambrose, D., National Physical Laboratory (1979) *NPL Rep. Chem.*, **98**.
4. Joback, K.G. and Reid, R.C. (1987) *Chem. Eng. Commun.*, **57**, 233–243.
5. Constantinou, L. and Gani, R. (1994) *AIChE J.*, **40**, 1697–1710.
6. Poling, B.E., Prausnitz, J.M., and O'Connell, J.P. (2001) *The Properties of Gases and Liquids*, McGraw-Hill, New York.
7. Nannoolal, Y., Rarey, J., and Ramjugernath, D. (2007) *Fluid Phase Equilib.*, **252**, 1–27.
8. Nannoolal, Y. (2006) Development and critical evaluation of group contribution methods for the estimation of critical properties, liquid vapor pressure and liquid viscosity of organic compounds. Thesis. University of Kwazulu-Natal.
9. (2005) *DIPPR Project 801*, Design Institute for Physical Property Data, AIChE.
10. Wagner, W. (2005) *FLUIDCAL, Software for the Calculation of Thermodynamic and Transport Properties of Several Fluids*, Ruhr-Universität Bochum.
11. Constantinou, L. and Gani, R. (1995) *Fluid Phase Equilib.*, **103**, 11–22.
12. Nannoolal, Y., Rarey, J., Ramjugernath, D., and Cordes, W. (2004) *Fluid Phase Equilib.*, **226**, 45–63.
13. Jakob, A. (1995) Thermodynamische Grundlagen der Kristallisation und ihre Anwendung in der Modellentwicklung. Thesis. University of Oldenburg.
14. Simon, F.E. and Glatzel, G. (1929) *Z. Anorg. Allg. Chem.*, **178**, 309–312.
15. Kechin, V.V. (1995) *J. Phys. Condens. Matter*, **7**, 531–535.
16. Hankinson, R.W. and Thomson, G.H. (1979) *AIChE J.*, **25**, 653–663.

17. Dortmund Data Bank www.ddbst.com.
18. Geier, S. October 17th, (2010) *W wie Wissen*, ARD.
19. Best, B. www.benbest.com/cryonics/pressure.html.
20. Wagner, W., Saul, A., and Pruß, A. (1994) *J. Phys., Chem. Ref. Data*, **23**, 515–527.
21. Domalski, E.S. and Hearing, E.D. (1993) *J. Phys. Chem. Ref. Data*, **22**, 805–1159.
22. Domalski, E.S. and Hearing, E.D. (1994) *J. Phys. Chem. Ref. Data*, **23**, 157.
23. Benson, S.W. (1968) *Thermochemical Kinetics*, Chapter 2, John Wiley & Sons, Inc., New York.
24. Benson, S.W., Cruickshank, F.R., Golden, D.M., Haugen, G.R., O'Neal, H.E., Rodgers, A.S., Shaw, R., and Walsh, R. (1969) *Chem. Rev.*, **69**, 279–324.
25. Anderson, T.F., Abrams, D.S., and Grens, E.A. (1978) *AIChE J.*, **24**, 20–29.
26. IK-CAPE (2002) Thermodynamics-Package for CAPE-Applications, www.dechema.de/dechema_media/Downloads/Informationssysteme/IK_CAPE_Equations.pdf.
27. PPDS, TUV NEL Ltd. (2007) www.ppds.co.uk, Glasgow.
28. Kleiber, M. and Joh, R. (2010) *VDI Heat Atlas*, Chapter D3.1, 2nd edn, Springer, Berlin, Heidelberg.
29. Wagner, W. (1973) *Cryogenics*, **13**, 470–482.
30. Moller, B. (2007) Development of an improved group contribution method for the prediction of vapor pressures of organic compounds. Master Thesis. University of KwaZulu-Natal.
31. McGarry, J. (1983) *Ind. Eng. Chem. Proc. Des. Dev.*, **22**, 313–322.
32. Reid, R.C., Prausnitz, J.M., and Poling, B.E. (1987) *The Properties of Gases and Liquids*, McGraw-Hill, New York.
33. Chase, J.D. (1987) *Ind. Eng. Chem. Res.*, **26**, 107–112.
34. Hoffmann, W. and Florin, F. (1943) *Z. VDI-Beih.* (2), 47–51.
35. Li, P., Ma, P.S., Yi, S.Z., Zhao, Z.G., and Cong, L.Z. (1994) *Fluid Phase Equilib.*, **101**, 101–119.
36. Riedel, L. (1957) *Kältetechnik*, **9**, 127–134.
37. Vetere, A. (1991) *Fluid Phase Equilib.*, **62**, 1–10.
38. Cox, E.R. (1923) *Ind. Eng. Chem.*, **15**, 592.
39. Schmid, B., Döker, M., and Gmehling, J. (2007) *Fluid Phase Equilib.*, **258**, 115–124.
40. Beyerlein, A.L., DesMarteau, D.D., Kul, I., and Zhao, G. (1998) *Fluid Phase Equilib.*, **150–151**, 287–296.
41. Kul, I., DesMarteau, D.D., and Beyerlein, A.L. (1999) *ASHRAE Trans.*, **106**, 351–357.
42. Moller, B., Rarey, J., and Ramjugernath, D. (2008) *J. Mol. Liq.*, **143**, 52–63.
43. Thomson, G.W. (1946) *Chem. Rev.*, **38**, 1.
44. Moore, W.J. and Hummel, D.O. (1986) *Physikalische Chemie*, Walter de Gruyter & Co., Berlin.
45. adsabs.harvard.edu/abs/2008AGUFM.A53J.08E
46. Widiatmo, J.V., Sato, H., and Watanabe, K. (1994) *J. Chem. Eng. Data*, **39**, 304–308.
47. Ihmels, E.C. (2002) *Experimentelle Bestimmung, Korrelation und Vorhersage von Dichten und Dampfdrücken*, Shaker-Verlag, Aachen.
48. Thomson, G.H., Brobst, K.R., and Hankinson, R.W. (1982) *AIChE J.*, **28**, 671.
49. Kleiber, M. and Joh, R. (2010) *VDI Heat Atlas*, Chapter D1, 2nd edn, Springer, Berlin.
50. Smith, B.D. and Srivastava, R. (1986) *The Thermodynamic Properties of Pure Compounds. Part A: Hydrocarbons and Alcohols*, Physical Data Series, Vol. 25, Elsevier, Amsterdam.
51. Smith, B.D. and Srivastava, R. (1986) *The Thermodynamic Properties of Pure Compounds. Part B: Halogenated Hydrocarbons and Ketons*, Physical Data Series, Vol. 25, Elsevier, Amsterdam.
52. Löffler, H.J. (1969) *Thermodynamik. Grundlagen und Anwendung auf reine Stoffe*, Springer, Berlin, Heidelberg.
53. Beekermann, W. and Kohler, F. (1995) *Int. J. Thermophys.*, **16**, 455–464.
54. Trusler, J.P.M and Zarari, M.P (1996) *J. Chem. Eng. Thermodyn.*, **28**, 329–335.

55. Lemmon, E., McLinden, M., and Wagner, W. (2009) *J. Chem. Eng. Data*, **54**, 3141–3180.
56. Trusler, J.P.M. (1991) *Physical Acoustics and Metrology of Fluids*, Institute of Physics, Bristol.
57. Moldover, M.R., Trusler, J.P.M., Edwards, T.J., Mehl, J.B., and Davies, R.S. (1988) *J. Res. Natl. Bur. Stand.*, **93**, 85–144.
58. Aly, F.A. and Lee, L.L. (1981) *Fluid Phase Equilib.*, **6**, 169–179.
59. Jaeschke, M. and Schley, P. (1995) *Int. J. Thermophys.*, **16**, 1381–1392.
60. Span, R. (2000) *Multiparameter Equations of State – A Accurate Source of Thermodynamic Property Data*, Springer, Berlin.
61. Harrison, B.K. and Seaton, W.H. (1988) *Ind. Eng. Chem. Res.*, **27**, 1536–1540.
62. Reid, R.C. and Sobel, J.E. (1965) *Ind. Eng. Chem. Fundam.*, **4**, 328–331.
63. Magee, J.W. (1992) *Int. J. Refrig.*, **15**, 372–380.
64. Baehr, H.D. (1992) *Thermodynamik*, 8 Auflage, Springer, Berlin.
65. Okubo, T., Hasuo, T., and Nagashima, A. (1992) *Int. J. Thermophys.*, **13**, 931–942.
66. Oliveira, C.M.B.P. and Wakeham, W.A. (1993) *Int. J. Thermophys.*, **14**, 33–44.
67. Assael, M.J., Dymond, J.H., and Polimatidou, S.K. (1994) *Int. J. Thermophys.*, **15**, 591.
68. Lavrenchenko, G.K., Ruvinskij, G.Y., Iljushenko, S.V., and Kanaev, V.V. (1992) *Int. J. Refrig.*, **15**, 386–392.
69. Sastri, S.R.S. and Rao, K.K. (1992) *Chem. Eng. J.*, **50**, 9.
70. Sastri, S.R.S. and Rao, K.K. (2000) *Fluid Phase Equilib.*, **175**, 311–323.
71. van Velzen, D., Lopes Cardozo, R., and Langenkamp, H. (1972) *Ind. Eng. Chem. Fundam.*, **11**, 20–25.
72. Nannoolal, Y., Rarey, J., and Ramjugernath, D. (2009) *Fluid Phase Equilib.*, **281**, 97–119.
73. Lucas, K. (1981) *Chem. Eng. Tech.*, **53**, 959–960.
74. Comunas, M.J.P., Baylaucq, A., Quinones-Cisneros, S.E., Zeberg-Mikkelsen, C.K., Boned, C., and Fernandez, J. (2003) *Fluid Phase Equilib.*, **210**, 21–32.
75. Gaston-Bonhomme, Y., Petrino, P., and Chevalier, J.L. (1994) *Chem. Eng. Sci.*, **49**, 1799–1806.
76. Lucas, K. and Luckas, M. (2002) *VDI-Wärmeatlas*, 9 Auflage, Springer-Berlin,
77. Carmichael, L.T. and Sage, B.H. (1966) *AIChE J.*, **12**, 559.
78. Wilke, C.R. (1950) *J. Chem. Phys.*, **18**, 517.
79. Gnielinski, V. (2010) Heat transfer in flow past a plane wall, *VDI Heat Atlas*, Chapter G4, 2nd edn, Springer, Berlin, Heidelberg.
80. Mathias, P.M., Parekh, V.S., and Miller, E.J. (2002) *Ind. Eng. Chem. Res.*, **41**, 989–999.
81. Jamieson, D.T. (1979) *J. Chem. Eng. Data*, **24**, 244.
82. Palmer, G. (1948) *Ind. Eng. Chem.*, **40**, 89–92.
83. Patek, J. and Klomfar, J. (2002) *Fluid Phase Equilib.*, **198**, 147–163.
84. Baroncini, C., Latini, G., and Pierpaoli, P. (1984) *Int. J. Thermophys.*, **5**, 387.
85. Chung, T.H., Ajlan, M., Lee, L.L., and Starling, K.E. (1988) *Ind. Eng. Chem. Res.*, **27**, 671.
86. Stiel, L.I. and Thodos, G. (1964) *AIChE J.*, **10**, 26.
87. Mason, E.A. and Saxena, S.C. (1958) *Phys. Fluids*, **1**, 361.
88. Bennett, L.A. and Vines, R.G. (1955) *J. Chem. Phys.*, **23**, 1587.
89. Falk, G. and Ruppel, W. (1976) *Energie und Entropie*, Springer, Berlin, Heidelberg, New York.
90. Brock, J.R. and Bird, R.B. (1955) *AIChE J.*, **1**, 174–177.
91. Fuller, E.N., Ensley, K., and Giddings, J.C. (1969) *J. Chem. Phys.*, **73**, 3679–3685.
92. Srivastava, B.N. and Srivastava, I.B. (1963) *J. Chem. Phys.*, **38**, 1183.
93. Riazi, M.R. and Whitson, C.H. (1993) *Ind. Eng. Chem. Res.*, **32**, 3081.
94. Vignes, A. (1966) *Ind. Eng. Chem. Fundam.* **5**, 189–199.

4
Properties of Mixtures

Up to now, we have considered different thermophysical properties for pure compounds as a function of temperature and pressure only. When dealing with mixtures, the influence of the composition has to be taken into account additionally. A mixture can be present in different states or phases, which may coexist in equilibrium. Usually, the coexisting phases do not have the same composition. There are different ways to describe the properties of a mixture:

- with the help of property changes upon mixing;
- by defining partial molar properties;
- by introducing excess properties.

For a better understanding, these options are explained here in a very general way, using an arbitrary state variable m:

$$m = f(T, P, z_1 \ldots z_n) \quad \text{or}$$
$$M = n_T m = f(T, P, n_1 \ldots n_n) \tag{4.1}$$

where m may be any state variable (e.g., h, u, s, g, c_v, c_P, or v) which is a function of the temperature T, the pressure P, and the molar composition z_i. m cannot be one of the intensive state variables T, P, or z_i. n_T represents the total mole number in the system, which is the sum of the mole numbers of the single species:

$$n_T = \sum_j n_j \tag{4.2}$$

The mole numbers n_i and the mole fractions z_i are related by the equation

$$z_i = \frac{n_i}{n_T} = \frac{n_i}{\sum_j n_j} \tag{4.3}$$

In this chapter, z_i is chosen as the symbol for the mole fraction representing not only the liquid state but the composition of a mixture in general. All the following relations are valid for the vapor phase as well.

As the expression $M = n_T m$ depends on the size of the considered system, M is an extensive state property in contrast to m.

Chemical Thermodynamics: for Process Simulation, First Edition.
Jürgen Gmehling, Bärbel Kolbe, Michael Kleiber, and Jürgen Rarey.
© 2012 Wiley-VCH Verlag GmbH & Co. KGaA. Published 2012 by Wiley-VCH Verlag GmbH & Co. KGaA.

4.1
Property Changes of Mixing

The difference between the real property of a mixture at a certain composition, temperature, and pressure and the mole fraction weighted average of the pure component properties at the same conditions (T, P) is called the *property change of mixing*. For an arbitrary property m one can write

$$\Delta m = m - \sum_j z_j m_j \tag{4.4}$$

or

$$\Delta M = n_T \Delta m = n_T m - \sum_j n_j m_j \tag{4.5}$$

For a better understanding, the property change of mixing is shown graphically in Figure 4.1. For this purpose, the state property m of a binary mixture is plotted versus z_1, the mole fraction of component 1. For $z_1 = 0$ and $z_1 = 1$ (the pure compounds) the state properties are identical with the pure component properties m_2 and m_1, or – in other words – the property change of mixing is 0 for pure components ($\Delta m = 0$).

The physical meaning of the property change of mixing can be illustrated by the example of the volume V – an extensive state property:

$$V(T, P, n_1, n_2 \ldots) = n_1 v_1(T, P) + n_2 v_2(T, P) + \ldots + n_T \Delta v(T, P, z_1, z_2 \ldots) \tag{4.6}$$

v_1, v_2... are the molar volumes of the pure components at the temperature T and the pressure P, Δv is the molar volume change of mixing at the same temperature and pressure. V is the total volume of the mixture.

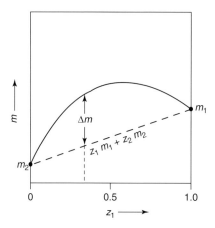

Figure 4.1 The property change of mixing for a binary system.

Example 4.1

$500\,cm^3$ ethanol (1) and $500\,cm^3$ water (2) are mixed at $25\,°C$. What is the volume of the mixture?

The molar volumes of the pure compounds at $25\,°C$ are

$$v_1 = 58.61\,cm^3/mol$$
$$v_2 = 18.06\,cm^3/mol$$

The volume change of mixing Δv for the above conditions is [1]

$$\Delta v(T, P, x_1) = -0.73\,cm^3/mol$$

Solution

The total number of moles follows from the given pure component volumes and molar volumes to

$$n_1 = (500/58.61)\,mol = 8.53\,mol$$
$$n_2 = (500/18.06)\,mol = 27.69\,mol$$

and therefore

$$n_T = n_1 + n_2 = 36.22\,mol.$$

After inserting this value, the volume of the mixture is calculated as

$$V = 500\,cm^3 + 500\,cm^3 + 36.22(-0.73)\,cm^3 = 973.56\,cm^3$$

Consequently, in this case the volume change of mixing is $\Delta V = -26.44\,cm^3$. The volume is smaller than the sum of the volumes of both pure components.

4.2 Partial Molar Properties

Forming the total differential of an extensive state variable leads to another option to describe the properties of a mixture:

$$d(n_T m) = \left(\frac{\partial (n_T m)}{\partial T}\right)_{P,n_i} dT + \left(\frac{\partial (n_T m)}{\partial P}\right)_{T,n_i} dP + \sum_i \left(\frac{\partial (n_T m)}{\partial n_i}\right)_{T,P,n_{j\neq i}} dn_i \quad (4.7)$$

The change in composition at constant temperature and pressure is described by the sum of differential quotients

$$\sum_i \left(\frac{\partial (n_T m)}{\partial n_i}\right)_{T,P,n_{j\neq i}} \quad (4.8)$$

The index n_j means that the number of moles of all components j is constant except the mole number of component i. The differential quotient for a single component is called the partial molar property:

$$\overline{m}_i \equiv \left(\frac{\partial(n_T m)}{\partial n_i}\right)_{T,P,n_{j\neq i}} \tag{4.9}$$

Partial molar properties are always determined at constant temperature, pressure, and concentrations except the one of the component regarded. When using partial molar properties, the total differential in Eq. (4.7) can be reformulated as

$$d(n_T m) = n_T \left(\frac{\partial m}{\partial T}\right)_{P,n_i} dT + n_T \left(\frac{\partial m}{\partial P}\right)_{T,n_i} dP + \sum_i \overline{m}_i dn_i \tag{4.10}$$

It can be shown by the Euler theorem (see Appendix C, B4) that

$$M = n_T m = \sum_i n_i \overline{m}_i \quad \text{or} \quad m = \sum_i z_i \overline{m}_i \tag{4.11}$$

The physical meaning of the partial molar property can be explained as follows. The sum of the mole-fraction-weighted partial molar properties gives the property of the mixture. The partial molar properties $\overline{m}_i$ are different from the state properties of the pure components m_i and must not be mixed up. They also depend on the composition.

The connection between the property changes of mixing and the partial molar properties can be derived by the combination of Eqs. (4.5) and (4.11):

$$\Delta m = \sum_i z_i (\overline{m}_i - m_i) \tag{4.12}$$

and

$$\Delta M = n_T \Delta m = \sum_i n_i (\overline{m}_i - m_i) \tag{4.13}$$

If the partial molar property $\overline{m}_i$ is identical with the respective pure component property for every single component, the property change of mixing Δm is zero.

For a binary mixture the characteristics of the partial molar properties can be illustrated graphically by plotting the state property m versus the mole fraction x_1. The total differential of the partial molar property gives an equation of the form

$$\overline{m}_i = \left(\frac{\partial(n_T m)}{\partial n_i}\right)_{T,P,n_{j\neq i}} = m\left(\frac{\partial n_T}{\partial n_i}\right)_{T,P,n_{j\neq i}} + n_T \left(\frac{\partial m}{\partial n_i}\right)_{T,P,n_{j\neq i}} \tag{4.14}$$

Using the relation

$$\left(\frac{\partial n_T}{\partial n_i}\right)_{T,P,n_{j\neq i}} = 1 \tag{4.15}$$

yields

$$\overline{m}_i = m + n_T \left(\frac{\partial m}{\partial n_i}\right)_{T,P,n_{j\neq i}} \tag{4.16}$$

For the binary case, replacement of the mole numbers n_i by the mole fractions $z_1 = n_1/n_T = n_1/(n_1 + n_2)$ gives

$$\left(\frac{dz_1}{dn_1}\right)_{n_2} = \frac{n_1 + n_2 - n_1}{(n_1 + n_2)^2} = \frac{n_2}{n_T^2} = \frac{z_2}{n_T} = \frac{1 - z_1}{n_T} \quad (4.17)$$

Thus, we get

$$\overline{m}_1 = m + (1 - z_1)\left(\frac{\partial m}{\partial z_1}\right)_{T,P} \quad (4.18)$$

and, as $dz_1 = -dz_2$:

$$\overline{m}_2 = m - z_1\left(\frac{\partial m}{\partial z_1}\right)_{T,P} \quad (4.19)$$

Graphically, the partial molar properties $\overline{m}_1$ and $\overline{m}_2$ can be read directly from a $m - z_1$-diagram as shown in Figure 4.2. At constant temperature and constant pressure, Eq. (4.19) is a straight line for m versus z_1:

$$m = \overline{m}_2 + z_1 \frac{dm}{dz_1} \quad (T, P = \text{constant}) \quad (4.20)$$

with the slope dm/dz_1 – or with other words – it is the tangent of the function $m = f(z_1)$ at a particular composition z_1. The axis intercept ($z_1 = 0$) of this tangent is the partial molar property $\overline{m}_2$. The same procedure is valid for the partial molar property $\overline{m}_1$ which is the axis intercept at ($z_1 = 1$) of the same tangent line.

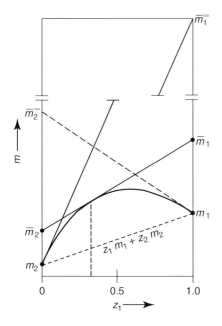

Figure 4.2 The partial molar properties for a binary mixture.

The tangents drawn at the end points ($z_1 = 0$) and ($z_1 = 1$) give the values for the partial molar properties at infinite dilution $\overline{m}_1^\infty$ and $\overline{m}_2^\infty$ and, of course, for the pure component properties $\overline{m}_2 = m_2$ and $\overline{m}_1 = m_1$.

Example 4.2

Calculate the partial molar properties in a liquid at infinite dilution using the Porter approach.

Solution

The so-called Porter equation is the simplest mathematical approach to describe the property change of mixing of a binary mixture as a function of the composition:

$$\Delta m = A x_1 x_2$$

This equation fulfills the constraint that the property change of mixing for the pure components is zero. The Porter approach is now used to calculate the partial molar properties at infinite dilution of component 1 in component 2 ($\overline{m}_1^\infty$) and vice versa ($\overline{m}_2^\infty$). The corresponding properties of state of the pure components are named m_1 and m_2:

$$\overline{m}_1 = m + (1 - x_1) \left(\frac{\partial m}{\partial x_1} \right)_{T,P} \tag{4.21}$$

$$\overline{m}_2 = m - x_1 \left(\frac{\partial m}{\partial x_1} \right)_{T,P} \tag{4.22}$$

$$m = \Delta m + x_1 m_1 + x_2 m_2$$

$$m = A x_1 x_2 + x_1 m_1 + x_2 m_2$$

$$\left(\frac{\partial m}{\partial x_1} \right)_{T,P} = A(1 - 2x_1) + m_1 - m_2$$

$$\overline{m}_1 = A x_1 x_2 + x_1 m_1 + x_2 m_2 + (1 - x_1)[A(1 - 2x_1) + m_1 - m_2]$$

At infinite dilution of component 1 in component 2 the mole fractions approach $x_1 \to 0$ and $x_2 \to 1$. Hence, it follows

$$\overline{m}_1^\infty = m_2 + A + m_1 - m_2 = m_1 + A$$

Analogously, $\overline{m}_2^\infty$ is derived as

$$\overline{m}_2^\infty = m_2 + A.$$

The results show that the Porter approach is only valid if the thermodynamic property shows a symmetrical profile.

4.3
Gibbs–Duhem Equation

The Gibbs–Duhem equation is one of the most important relationships of mixture thermodynamics. A general form of the Gibbs–Duhem equation can be derived in combination with the general definition of the partial molar properties. The Gibbs–Duhem equation allows for the development of so-called consistency tests, which are a necessary but not sufficient condition for the correctness of experimental data. The total differential of the function

$$n_T m = \sum_i n_i \overline{m}_i \tag{4.23}$$

is

$$d(n_T m) = \sum_i \overline{m}_i dn_i + \sum_i n_i d\overline{m}_i \tag{4.24}$$

At the same time Eq. (4.10) is valid

$$d(n_T m) = n_T \left(\frac{\partial m}{\partial T}\right)_{P,n_i} dT + n_T \left(\frac{\partial m}{\partial P}\right)_{P,n_i} dP + \sum_i \overline{m}_i dn_i \tag{4.10}$$

The combination of the two equations results in a general formulation of the Gibbs–Duhem equation

$$n_T \left(\frac{\partial m}{\partial T}\right)_{P,n_i} dT + n_T \left(\frac{\partial m}{\partial P}\right)_{T,n_i} dP - \sum_i n_i d\overline{m}_i = 0 \tag{4.25}$$

or, respectively,

$$\left(\frac{\partial m}{\partial T}\right)_{P,z_i} dT + \left(\frac{\partial m}{\partial P}\right)_{T,z_i} dP + \sum_i z_i d\overline{m}_i = 0 \tag{4.26}$$

At constant temperature and constant pressure, Eq. (4.26) reduces to

$$\sum_i z_i d\overline{m}_i = 0 \quad (T, P = \text{constant}) \tag{4.27}$$

This equation shows that the partial molar properties of the single components are not independent of each other, because the concentration dependences are correlated.

In thermodynamics, the Gibbs energy is a very important property, especially for the description of phase equilibria. To apply the Gibbs–Duhem equation to the Gibbs energy as a function of temperature, pressure, and composition $G = f(T, P, n_1, \ldots, n_n)$, its total differential is set up as

$$dG = \left(\frac{\partial G}{\partial T}\right)_{P,n_i} dT + \left(\frac{\partial G}{\partial P}\right)_{T,n_i} dP + \sum_i \left(\frac{\partial G}{\partial n_i}\right)_{T,P,n_{j\neq i}} \tag{4.28}$$

A comparison between this equation and the fundamental equation formulated for the Gibbs energy

$$dG = dH - TdS - SdT = VdP - SdT \tag{4.29}$$

gives

$$\left(\frac{\partial G}{\partial P}\right)_{T,n_i} = V \tag{4.30}$$

$$\left(\frac{\partial G}{\partial T}\right)_{P,n_i} = -S \tag{4.31}$$

and

$$\left(\frac{\partial G}{\partial n_i}\right)_{T,P,n_{j\neq i}} = \mu_i \tag{4.32}$$

μ_i is the chemical potential of component i. In a similar way (see Appendix C, B5), it can be derived that

$$\mu_i = \left(\frac{\partial G}{\partial n_i}\right)_{T,P,n_{j\neq i}} = \left(\frac{\partial U}{\partial n_i}\right)_{S,V,n_{j\neq i}} = \left(\frac{\partial H}{\partial n_i}\right)_{S,P,n_{j\neq i}} = \left(\frac{\partial A}{\partial n_i}\right)_{T,V,n_{j\neq i}}$$

$$= -T\left(\frac{\partial S}{\partial n_i}\right)_{u,V,n_{j\neq i}} \tag{4.33}$$

The chemical potential of component i is equal to the partial molar property of the Gibbs energy:

$$\mu_i = \left(\frac{\partial n_T g}{\partial n_i}\right)_{T,P,n_{j\neq i}} = \bar{g}_i \tag{4.34}$$

From Eq. (4.25) one obtains

$$SdT - VdP + \sum_i n_i d\mu_i = 0 \tag{4.35}$$

or, formulated with molar properties (divided by the total mole number n_T):

$$sdT - vdP + \sum_i z_i d\mu_i = 0 \tag{4.36}$$

This is a well known and common form of the Gibbs–Duhem equation. At constant temperature and constant pressure the Gibbs–Duhem equation reduces to

$$\sum_i z_i d\mu_i = 0 \quad (T, P = \text{constant}) \tag{4.37}$$

4.4
Ideal Mixture of Ideal Gases

It is convenient to start with the description of the properties of a real mixture with the ideal mixture as a first approximation. The deviation from ideal behavior is then taken into account by the addition of further terms. Therefore, the definition of an ideal mixture should be explained, before the excess properties are introduced. Starting point for the definition of an ideal mixture are the equations for a mixture of ideal gases. A mixture of ideal gases is characterized by the following behavior:

each species acts as if it alone occupies the entire volume V at temperature T. This assumption leads to the equation of state:

$$p_i^{id} = \frac{n_i RT}{V} \tag{4.38}$$

p_i is the partial pressure of component i in the mixture:

$$p_i \equiv y_i P \quad \text{or} \quad P \equiv \sum p_i \tag{4.39}$$

This relation is called *Dalton's law*. With the equation of state above, for example, the entropy of an ideal mixture can be calculated. According to Table 2.1, the total differential of the entropy for a pure ideal gas is

$$ds_i^{id} = \frac{c_{p_i}^{id}}{T} dT - \frac{R}{P} dP \tag{4.40}$$

Integration of this equation at constant temperature gives

$$s_i^{id}(T, P) = s_i^{id}(T, P^0) - R \ln \frac{P}{P^0} \tag{4.41}$$

The pressure P^0 is the pressure at an arbitrarily chosen reference state. As explained above, each species i in an ideal mixture behaves exactly like a pure compound which exerts the partial pressure p_i instead of the system pressure P. The partial molar entropy for a species in an ideal mixture, which is the entropy of a component in the mixture, results in

$$\bar{s}_i^{id}(T, P) = s_i^{id}(T, P^0) - R \ln \frac{p_i}{P^0}$$
$$= s_i^{id}(T, P^0) - R \ln \frac{P}{P^0} - R \ln y_i = s_i^{id}(T, P) - R \ln y_i \tag{4.42}$$

The entropy change of mixing results from Eqs. (4.12) and (4.42) to

$$\Delta s^{id} = \sum_i y_i \left(\bar{s}_i^{id} - s_i^{id} \right) \tag{4.43}$$

$$\Delta s^{id} = -R \sum_i y_i \ln y_i \tag{4.44}$$

As the mole fractions y_i vary between 0 and 1, the entropy change of mixing for an ideal mixture is always different from zero and positive.

In a similar way the Gibbs energy change of mixing may be derived. Integration of the corresponding equation in Table 2.1 gives

$$g_i^{id}(T, P) = g_i^{id}(T, P^0) + RT \ln \frac{P}{P^0} \tag{4.45}$$

When inserting again the partial pressure p_i instead of the total pressure P, the partial molar Gibbs energy of species i in an ideal mixture is derived as

$$\bar{g}_i^{id}(T, P) = g_i^{id}(T, P^0) + RT \ln \frac{p_i}{P^0} = g_i^{id}(T, P^0) + RT \ln \frac{P}{P^0} + RT \ln y_i$$
$$= g_i^{id}(T, P) + RT \ln y_i \tag{4.46}$$

The Gibbs energy change of mixing for an ideal mixture is then calculated as

$$\Delta g^{id} = \sum_i y_i \left(\bar{g}_i^{id} - g_i^{id}\right) = RT \sum_i y_i \ln y_i \qquad (4.47)$$

The Gibbs energy change of mixing for an ideal mixture does not equal zero as well, but different from the entropy change of mixing the Gibbs energy change of mixing is always negative.

The volume change of mixing of an ideal gas mixture Δv^{id} shows a different behavior. The pure component molar volume can directly be read from the ideal gas equation of state:

$$v_i^{id} = \frac{RT}{P} \qquad (4.48)$$

With the equation of state for the ideal gas mixture

$$p_i^{id} = \frac{n_i RT}{V} = y_i P = \frac{n_i}{\sum_j n_j} P \qquad (4.49)$$

the total volume of the mixture is calculated to be

$$V = \frac{RT}{P} \sum_i n_i \qquad (4.50)$$

and the partial molar volume results in

$$\bar{v}_i^{id} = \left(\frac{\partial V}{\partial n_i}\right)_{T,P,n_{j\neq i}} = \frac{RT}{P} \qquad (4.51)$$

As the partial molar volume and the molar volume of the pure species are identical, the volume change of mixing for an ideal gas is zero

$$\Delta v^{id} = 0 \qquad (4.52)$$

The same is true for the enthalpy change of mixing:

$$\Delta h^{id} = \Delta g^{id} + T\Delta s^{id} = RT \sum_i y_i \ln y_i - RT \sum_i y_i \ln y_i = 0 \qquad (4.53)$$

4.5
Ideal Mixture of Real Fluids

Analogously to ideal gases, an ideal mixture of real fluids can be defined, which is valid both for real gases and for liquids. This definition means that the properties of single species in a mixture differ from the ideal gas properties, but the mixing itself should not cause any additional real effect. This assumption facilitates the representation of different solution properties. The partial molar Gibbs energy can then be described analogously to Eq. (4.45) in the form

$$\bar{g}_i^{id} = g_i^{pure}(T, P) + RT \ln z_i \qquad (4.54)$$

Table 4.1 Property changes of ideal mixtures.

Δg^{id}	$RT \sum z_i \ln z_i$
Δs^{id}	$-R \sum z_i \ln z_i$
Δa^{id}	$RT \sum z_i \ln z_i$
Δu^{id}	0
Δh^{id}	0
Δv^{id}	0

where $g_i^{\text{pure}}(T, P)$ is the Gibbs energy of the pure real fluid. Now the superscript "id" means "ideal mixture of real fluids," in difference to the ideal mixture of ideal gases. Here, z_i is the mole fraction of component i valid for both the vapor and liquid state. The concept of the ideal mixture was introduced by Lewis. The concentration dependence of the Gibbs energy is the same for the ideal gas mixture as for the ideal mixture of real fluids. The property changes of mixing for the intensive state variables of an ideal mixture are therefore expressed by the same relations as the property changes of mixing of an ideal gas. These relations are summarized in Table 4.1.

As it can easily be seen, only Δs, Δg, and Δa are different from zero for an ideal mixture, which is caused by the entropy change of mixing.

In reality, the assumption of an ideal mixture is seldom fulfilled. Approximate ideal mixtures are formed by the molecules of similar size, geometry, and charge distribution, for example, mixtures of isotopes or isomers.

4.6
Excess Properties

As shown before, some property changes of mixing are different from zero even for an ideal mixture. Thus, the idea is obvious to define properties which describe the difference between a real and an ideal mixture. These properties are the so-called excess properties.

$$m^{\text{E}} = m - m^{\text{id}} \tag{4.55}$$

The excess property m^{E} represents the difference between the real state property m and the corresponding property of an ideal mixture m^{id}. In a similar way, the property change of mixing Δm can be expressed with the help of excess properties. The real property change of mixing is then composed of two parts:

- the ideal mixing part and
- the real mixing effect,

$$\Delta m = \Delta m^{\text{id}} + m^{\text{E}} \tag{4.56}$$

The real state property m itself is therefore composed of three parts:

1) the mole fraction weighted mean value of the pure component properties,

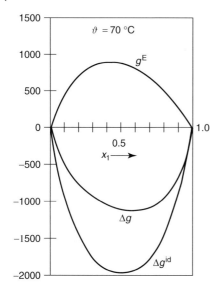

Figure 4.3 The Gibbs energy change of mixing, the ideal Gibbs energy change of mixing, and Gibbs excess energy for the system ethanol–water at 70 °C in the liquid phase.

2) the ideal property change of mixing, which is zero for some state properties, and
3) the excess property.

$$m = \sum_i x_i m_i + \Delta m^{id} + m^E \tag{4.57}$$

Figure 4.3 shows the different parts of the Gibbs energy change of mixing as calculated for the liquid mixture of ethanol and water at 70 °C. While the ideal Gibbs energy change of mixing is always negative (see Eq. (4.47)), the excess Gibbs energy may be positive or negative. At very low concentrations, the real behavior of the mixture converges more and more to the ideal mixture. The excess properties of pure substances are zero by definition. The general relations of thermodynamics are also true for the excess properties:

$$g^E = h^E - Ts^E = u^E + Pv^E - Ts^E \tag{4.58}$$

Partial molar excess properties can be defined analogously as

$$\overline{m}_i^E = \left(\frac{\partial(n_T m^E)}{\partial n_i}\right)_{T,P,n_{j \neq i}} \quad \text{and} \quad m^E = \sum_i x_i \overline{m}_i^E \tag{4.59}$$

The excess properties proved to be well-suited especially for the description of liquid mixtures.

4.7
Fugacity in Mixtures

As described in Section 2.4.3, the behavior of a real fluid can also be expressed by means of the fugacity, which is applicable to mixtures as well. The fugacity of a species in a mixture is defined in the same way as the fugacity of a pure component:

$$\bar{g}_i(T, P, x_i) = g_i^{\text{pure}}(T, P^0) + RT \ln \frac{f_i}{f_i^0} \tag{4.60}$$

$g_i^{\text{pure}}(T, P^0)$ is the Gibbs energy of the pure component at the system temperature and at an arbitrarily chosen reference pressure P^0. As for pure components, it is a limiting condition that the behavior of the real fluid is described by the ideal gas law, when the pressure approaches zero:

$$\lim_{P \to 0} \frac{f_i}{p_i} = \lim_{P \to 0} \frac{f_i}{y_i P} = 1 \tag{4.61}$$

The fugacity coefficient of a species in a mixture is also defined analogously to the definition of a pure component fugacity coefficient, where the partial pressure replaces the total pressure:

$$\varphi_i \equiv \frac{f_i}{y_i P} \quad \text{and therefore} \quad \lim_{P \to 0} \varphi_i = 1 \tag{4.62}$$

The fugacity coefficient in a mixture may be represented as a function of the variables temperature, pressure, or volume, corresponding to the equations for pure components in Section 2.4.3:

$$\ln \varphi_i = \frac{1}{RT} \int_0^P \left[\left(\frac{\partial V}{\partial n_i} \right)_{T,P,n_{j \neq i}} - \frac{RT}{P} \right] dP \tag{4.63}$$

or

$$\ln \varphi_i = \frac{1}{RT} \int_\infty^V \left[\frac{RT}{V} - \left(\frac{\partial P}{\partial n_i} \right)_{T,V,n_{j \neq i}} \right] dV - \ln z \tag{4.64}$$

The equations for mixtures additionally contain the mole numbers, that is, the composition as a variable. They are valid for both vapor and liquid. With the help of these equations, the fugacity coefficient can be determined from experimental $PvT\, x_i(y_i)$ data or it can be calculated using an equation of state.[1]

4.7.1
Fugacity of an Ideal Mixture

A relationship for the fugacity of species i in an ideal mixture is derived in a similar way as for the state properties.

---

1) A derivation of Eq. (4.64) can be found in Appendices C, D1.

The partial molar Gibbs energy of species i in an ideal mixture is

$$\bar{g}_i^{id}(T, P, y_i) = g_i^{pure}(T, P) + RT \ln y_i \qquad (4.65)$$

The Gibbs energy of the real pure component i is (see Eq. (2.60)):

$$g_i^{pure}(T, P) = g_i^{pure}(T, P^0) + RT \ln \frac{f_i^{pure}}{f_i^0} \qquad (4.66)$$

Inserting Eq. (4.66) into Eq. (4.65) one obtains

$$\bar{g}_i^{id}(T, P, y_i) = g_i^{pure}(T, P^0) + RT \ln \frac{y_i f_i^{pure}}{f_i^0} \qquad (4.67)$$

A comparison of Eqs. (4.60) and (4.67) shows that the fugacity of a component in an ideal mixture is calculated from the fugacity of the pure component multiplied by the mole fraction:

$$f_i^{id} = y_i f_i^{pure} \qquad (4.68)$$

This equation is also known as the *Lewis–Randall rule*.

4.7.2
Phase Equilibrium

Thermodynamic equilibrium between different phases exists if the following conditions[2] are fulfilled:

$$T^\alpha = T^\beta = \cdots = T^\varphi \qquad (4.69)$$
$$P^\alpha = P^\beta = \cdots = P^\varphi \qquad (4.70)$$
$$\mu_i^\alpha = \mu_i^\beta = \cdots = \mu_i^\varphi \qquad (4.71)$$

Besides temperature and pressure, the chemical potential of each species must be equal in all phases. It was shown in Section 4.3 that the chemical potential of a component is identical with the partial molar Gibbs energy.

$$\mu_i = \bar{g}_i \qquad (4.72)$$

On the other hand, the partial molar Gibbs energy may be expressed by the fugacity:

$$\bar{g}_i(T, P, z_i) = g_i^{pure}(T, P^0) + RT \ln \frac{f_i(T, P, z_i)}{f_i^0(T, P^0)} \qquad (4.73)$$

The properties of the pure component $g_i^{pure}(T, P^0)$ and f_i^0 are only functions of pressure and temperature; therefore, they show the same values in all phases.

---

[2] For the equilibrium on both sides of a membrane, Eq. (4.70) is not valid (Chapter 9).

Consequently, the conditions for phase equilibrium can be expressed by means of the fugacity as well:

$$\bar{g}_i^\alpha = \bar{g}_i^\beta = \cdots = \bar{g}_i^\varphi \tag{4.74}$$

$$f_i^\alpha = f_i^\beta = \cdots = f_i^\varphi \tag{4.75}$$

The last equation is a very important relationship for the thermodynamics of mixtures, and it is the starting point for practically all phase equilibrium calculations.

As it was done for the pure compounds (Section 2.4.3, Eq. (2.62)), it has to be defined that the fugacity becomes equal to the partial pressure at low pressures:

$$\lim_{P \to 0} \frac{f_i}{z_i P} = 1 \tag{4.76}$$

The reference state in Eq. (4.73) is not restricted to the pure substance. Analogously, it can be written as

$$\bar{g}_i(T, P, z_i) = \bar{g}_i^0\left(T, P^0, z_i^0\right) + RT \ln \frac{f_i(T, P, z_i)}{f_i^0(T, P^0, z_i^0)} \tag{4.77}$$

Considering two phases α and β, the equilibrium condition reads

$$\bar{g}_i^{0\alpha}\left(T, P^{0\alpha}, z_i^{0\alpha}\right) + RT \ln \frac{f_i^\alpha(T, P, z_i^\alpha)}{f_i^{0\alpha}(T, P^{0\alpha}, z_i^{0\alpha})} = \bar{g}_i^{0\beta}\left(T, P^{0\beta}, z_i^{0\beta}\right)$$

$$+ RT \ln \frac{f_i^\beta(T, P, z_i^\beta)}{f_i^{0\beta}(T, P^{0\beta}, z_i^{0\beta})} \tag{4.78}$$

Even in the case where the reference states $(P^{0\alpha}, z_i^{0\alpha})$ and $(P^{0\beta}, z_i^{0\beta})$ are not identical, Eq. (4.78) can be simplified using the relationship

$$\bar{g}_i^{0\alpha}\left(T, P^{0\alpha}, z_i^{0\alpha}\right) = \bar{g}_i^{0\beta}\left(T, P^{0\beta}, z_i^{0\beta}\right) + RT \ln \frac{f_i^{0\alpha}(T, P^{0\alpha}, z_i^{0\alpha})}{f_i^{0\beta}(T, P^{0\beta}, z_i^{0\beta})} \tag{4.79}$$

and the result is again

$$f_i^\alpha(T, P, z_i^\alpha) = f_i^\beta(T, P, z_i^\beta) \tag{4.80}$$

4.8
Activity and Activity Coefficient

For practical purposes – especially for the description of liquid mixtures – it is sometimes convenient to split the partial molar Gibbs energy into a term for the ideal mixture and a term for the excess Gibbs energy (see Section 4.6):

$$\bar{g}_i = \bar{g}_i^{id} + \bar{g}_i^E \tag{4.81}$$

$$\bar{g}_i = g_i^{pure}(T, P^0) + RT \ln x_i + RT \ln \frac{f_i}{x_i f_i^0} \tag{4.82}$$

For simplification, the auxiliary functions activity and activity coefficient are introduced:

$$a_i \equiv \frac{f_i}{f_i^0} \quad \text{and} \quad \gamma_i = \frac{a_i}{\zeta_i} \qquad (4.83)$$

Here, f_i^0 is the fugacity in an arbitrarily chosen reference state, whereas the temperature of the reference state must be equal to the temperature of the mixture. ζ_i represents an arbitrary measure of the concentration. Inserting Eq. (4.83) into Eq. (4.82), selecting the pure component at the system temperature T and a reference pressure P^0 as the standard state and then choosing the mole fraction in the liquid phase as the concentration measure leads to

$$\bar{g}_i = g_i^{\text{pure}}(T, P^0) + RT \ln x_i + RT \ln \gamma_i \qquad (4.84)$$

For an ideal mixture the excess part is equal to zero, since the activity coefficient is $\gamma_i = 1$. From this definition, the connection between the excess Gibbs energy and the activity coefficient can be seen clearly:

$$\left(\frac{\partial G^E}{\partial n_i}\right)_{T,P,n_{j\neq i}} = \left(\frac{\partial n_T g^E}{\partial n_i}\right)_{T,P,n_{j\neq i}} = \bar{g}_i^E = RT \ln \gamma_i \qquad (4.85)$$

or

$$g^E = \sum_i x_i \bar{g}_i^E = RT \sum_i x_i \ln \gamma_i \qquad (4.86)$$

The Gibbs–Duhem equation (Eq. (4.36)) may be expressed as well for the excess properties:

$$s^E dT - v^E dP + RT \sum_i x_i d \ln \gamma_i = 0 \qquad (4.87)$$

This equation is the starting point for the so-called consistency tests for experimental phase equilibrium data, as it will be shown in more detail in Chapter 5.

4.9
Application of Equations of State to Mixtures

The equations of state described in Section 2.5 for the description of thermodynamic properties of pure components can be applied to mixtures using appropriate mixing rules, which describe the concentration dependence of the parameters used. This will be explained in the following chapter.

For high-precision equations of state, there is currently no simple method available to apply them to mixtures where their advantages are maintained. For special mixtures such as natural gas, the development of a special equation has been considered to be useful [2]. For a few cases, the Lee–Kesler approach [3, 4],

$$z(T_r, P_r) = z_1(T_r, P_r) + \frac{\omega_{\text{mix}} - \omega_1}{\omega_2 - \omega_1}[z_2(T_r, P_r) - z_1(T_r, P_r)] \qquad (4.88)$$

has been successfully applied for binary mixtures; however, the procedure is based on fully empirical mixing rules, and the advantage in precision is usually lost. A generalized form of Eq. (4.88) is the Lee–Kesler–Plöcker equation [5], where the reference substance 1 is always methane and the reference substance 2 is always n-octane.

4.9.1
Virial Equation

Using statistical mechanics, exact mixing rules can be given for the Leiden form of the virial equation:

$$B = \sum_i \sum_j y_i y_j B_{ij} \tag{4.89}$$

$$C = \sum_i \sum_j \sum_k y_i y_j y_k C_{ijk} \tag{4.90}$$

where B and C are, respectively, the second and the third virial coefficients of the mixture. If the indices are identical, B and C refer to the virial coefficient of a particular pure substance. For different indices, B_{ij} and C_{ijk} are called *cross virial coefficients*.

The virial coefficients of the pure substances as well as the cross virial coefficients depend only on temperature, not on concentration. The cross second virial coefficients B_{ij} and B_{ji} are identical, as they represent the same interactions between molecule i and molecule j. Thus, the second virial coefficient of a binary mixture is given by

$$B = y_1^2 B_{11} + 2 y_1 y_2 B_{12} + y_2^2 B_{22} \tag{4.91}$$

For many applications it is useful to write Eq. (4.89) in a different way, using an excess virial coefficient:

$$B = \sum_i y_i B_{ii} + \frac{1}{2} \sum_i \sum_j y_i y_j \delta_{ij}, \tag{4.92}$$

with

$$\delta_{ij} \equiv 2 B_{ij} - B_{ii} - B_{jj} \tag{4.93}$$

In this representation, the second virial coefficient of the mixture is calculated as the sum of the arithmetic mean of the pure component values and an excess contribution δ_{ij}, which describes the deviations ($\delta_{ii} = \delta_{jj} = 0$) of the cross virial coefficients from the arithmetic mean. For the binary case, Eq. (4.92) reduces to

$$B = y_1 B_{11} + y_2 B_{22} + y_1 y_2 \delta_{12} \tag{4.94}$$

The excess virial coefficient can be positive or negative.

Cross second virial coefficients B_{ij} can often be found in the literature [6, 7], whereas hardly any data are available for third cross virial coefficients.

With the virial equation, the *PvT* behavior and the residual parts of the particular functions for gas mixtures can be calculated.

If the virial equation (Leiden form) is truncated after the second term, then one obtains according to Eq. (2.88)

$$z = 1 + \frac{B}{v}$$

or

$$P = \frac{RT}{v} + \frac{BRT}{v^2} \tag{4.95}$$

The fugacity coefficient from this equation is given by

$$\ln \varphi_i = \frac{2}{v} \sum_k y_k B_{ki} - \ln z. \tag{4.96}$$

For the Berlin form of the virial equation, this expression becomes

$$\ln \varphi_i = \left[2 \sum_k y_k B_{ki} - B \right] \frac{P}{RT} \tag{4.97}$$

The mathematical derivations of Eqs. (4.96) and (4.97) are given in Appendix C, D2 and D3.

4.9.2
Cubic Equations of State

For cubic equations of state, there are empirical mixing rules available for the coefficients a and b. They can be evaluated using the pure component properties a_{ii} and b_i of the components of the system, usually obtained from T_c and P_c. A number of empirical mixing rules for a and b have been suggested in the literature. The most popular one is the quadratic concentration dependence for the attractive parameter a:

$$a = \sum_i \sum_j z_i z_j a_{ij} \tag{4.98}$$

For the calculation of the cross coefficient a_{ij}, the geometric mean of the pure component parameters is corrected by a binary parameter k_{ij}:

$$a_{ij} = (a_{ii} a_{jj})^{0.5} (1 - k_{ij}) \tag{4.99}$$

Usually, the symmetric binary parameters k_{ij} (i.e., $k_{ij} = k_{ji}$) have small values in the range $-0.1 < k_{ij} < 0.1$. Nevertheless, they cannot be neglected, as they have a large influence on the results of phase equilibrium calculations. If the binary parameters are known, the phase equilibrium behavior of multicomponent mixtures can be described as well.

In most cases, the van der Waals co-volume b of a mixture is evaluated with a simple linear mixing rule:

$$b = \sum_i z_i b_i \tag{4.100}$$

Table 4.2 Residual parts of the particular thermodynamic quantities calculated with the Soave–Redlich–Kwong equation.

$g - g^{id}$	$-RT\ln\dfrac{P(v-b)}{RT} - \dfrac{a}{b}\ln\dfrac{v+b}{v} + Pv - RT$
$h - h^{id}$	$-\dfrac{1}{b}\left(a + \dfrac{XT}{2}\right)\ln\left(\dfrac{v+b}{v}\right) + Pv - RT$
$s - s^{id}$	$R\ln\dfrac{P(v-b)}{RT} - \dfrac{X}{2b}\ln\left(\dfrac{v+b}{v}\right)$

with

$$X = \frac{1}{\sqrt{T}} \sum_i \sum_j x_i x_j a_{ij} \left(\frac{m_i}{\sqrt{T_{c,i} \cdot a_i}} + \frac{m_j}{\sqrt{T_{c,j} \cdot a_j}} \right)$$

$$m_i = 0.480 + 1.574\omega_i - 0.176\omega_i^2$$

$$\alpha_i = \left[1 + m_i\left(1 - \sqrt{T_r}\right)\right]^2$$

These mixing rules Eqs. (4.98) and (4.100) are valid for both the vapor and the liquid phase; therefore, z_i was used as a mole fraction.

With the help of Eq. (4.64), the fugacity coefficients can be evaluated for any equation of state, for example, for the Soave–Redlich–Kwong equation [8]

$$\ln \varphi_k = \ln \frac{v}{v-b} - \frac{2\sum_i y_i a_{ik}}{RTb} \ln \frac{v+b}{v} + \frac{b_k}{v-b} - \ln \frac{Pv}{RT} \qquad (4.101)$$
$$+ \frac{ab_k}{RTb^2}\left(\ln \frac{v+b}{v} - \frac{b}{v+b}\right)$$

A derivation of Eq. (4.101) can be found in Appendix C, D4.[3]

For the evaluation of the fugacity coefficients and the residual parts, the specific volume of the phase being considered must be known under the given conditions. It can be evaluated by solving the cubic equation analytically or iteratively.

The residual parts of particular thermodynamic quantities calculated with the Soave–Redlich–Kwong equation, are listed in Table 4.2. For their derivations, see Appendix C, F1–F3.

It must be clear that classical mixing rules like Eq. (4.99) only enable the representation of nonpolar or slightly polar substances. In 1979, Huron and Vidal [9] were the first to develop a so-called g^E mixing rule which could be successfully applied to systems with polar components to make use of the advantages of both the g^E models and the equations of state. Instead of an arbitrarily defined mixing formula, g^E mixing rules use the information generated by an activity coefficient

---

3) The fugacity coefficient published in the original publication [8] is different, as another mixing rule for the parameter a (Eq. (4.99)) had been applied.

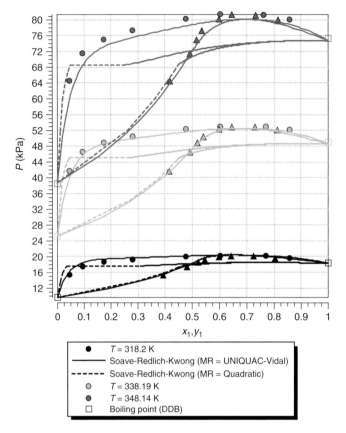

Figure 4.4 Vapor–liquid equilibrium of the system 2-propanol (1) + water (2). (- - - -) SRK + quadratic mixing rule, $k_{ij} = -0.1632$ (———) SRK + Huron–Vidal-g^E-mixing rule (UNIQUAC).

model. Thus, even difficult polar systems like 2-propanol/water can be described reliably, which is illustrated in Figure 4.4.

In both cases, the Soave–Redlich–Kwong equation of state has been used. The quadratic mixing rule with an adjusted k_{ij} interaction parameter fails completely, whereas the Huron–Vidal-g^E mixing rule succeeds, using UNIQUAC as the g^E model. Before focusing on the g^E mixing rules used in the PSRK- and VTPR equation of state, the Huron–Vidal g^E mixing rule is derived in the following section.

The principle of all g^E mixing rules is that an activity coefficient model and a cubic equation of state yield the same value for the excess Gibbs energy:

$$\frac{g^{E,EOS}}{RT} = \frac{g^{E,\gamma}}{RT} \quad (P = P^{\text{ref}}) \tag{4.102}$$

In contrast to the activity coefficient model, the g^E expression for the equation of state depends on the pressure. Therefore, a reference pressure P^{ref} must be defined, where Eq. (4.102) is valid. As a starting point for the derivation of a g^E mixing rule, the residual part of the Helmholtz energy is considered [10]:

$$a^R = a - a^{\text{id}} = \int_\infty^V -P dv - \int_\infty^V -\frac{RT}{v} dv \qquad (4.103)$$

For P, a pressure-explicit equation of state can be inserted, for example, the Peng–Robinson equation (Eq. (2.166)):

$$a^R = RT\left[\ln \frac{v}{v-b} - \frac{a}{2\sqrt{2}bRT} \ln \frac{v+b(1+\sqrt{2})}{v+b(1-\sqrt{2})}\right] \qquad (4.104)$$

A dimensionless inverse packing density u

$$u = v/b \qquad (4.105)$$

is introduced in Eq. (4.104):

$$a^R = RT\left[\ln \frac{u}{u-1} - \frac{a}{2\sqrt{2}bRT} \ln \frac{u+(1+\sqrt{2})}{u+(1-\sqrt{2})}\right] \qquad (4.106)$$

According to the relationship between excess property and residual part of the Helmholtz energy

$$a^E = a^R - \sum_i x_i a_i^R \qquad (4.107)$$

one obtains

$$\frac{a^E}{RT} = \ln \frac{u}{u-1} - \frac{a}{2\sqrt{2}bRT} \ln \frac{u+(1+\sqrt{2})}{u+(1-\sqrt{2})}$$
$$- \sum_i x_i \left[\ln \frac{u_i}{u_i-1} - \frac{a_{ii}}{2\sqrt{2}b_i RT} \ln \frac{u_i+(1+\sqrt{2})}{u_i+(1-\sqrt{2})}\right] \qquad (4.108)$$

Equation (4.108) can be simplified if the inverse packing density u is assumed to be constant both for the mixture and for the pure components:

$$u = u_i = \frac{v^L}{b} = \frac{v_i^L}{b_i} \qquad (4.109)$$

$$\frac{a^E}{RT} = -\frac{a}{2\sqrt{2}bRT} \ln \frac{u+(1+\sqrt{2})}{u+(1-\sqrt{2})} + \sum_i x_i \frac{a_{ii}}{2\sqrt{2}b_i RT} \ln \frac{u+(1+\sqrt{2})}{u+(1-\sqrt{2})} \qquad (4.110)$$

Equation (4.110) can be rearranged to

$$\frac{a}{bRT} = \sum_i x_i \frac{a_{ii}}{b_i RT} + \frac{a^E}{q_1 RT} \qquad (4.111)$$

with

$$q_1 = \frac{1}{2\sqrt{2}} \ln \frac{u + (1+\sqrt{2})}{u + (1-\sqrt{2})} \tag{4.112}$$

For the Soave–Redlich–Kwong equation of state, q_1 is equal to [11]:

$$q_1 = \ln \frac{u}{u+1} \tag{4.113}$$

In order to switch from a^E to g^E, we can make use of the relationship

$$g^E = a^E + P \cdot v^E \tag{4.114}$$

For $P \to \infty$, Huron and Vidal postulate that the liquid volume is equal to the co-volume b of the equation of state. Thus, $u = 1$, and the excess volume v^E reduces to zero.

$$v^L = b \quad \text{and} \quad v_i^L = b_i \quad (P^{\text{ref}} = \infty) \tag{4.115}$$

$$v^E = 0 \quad (P^{\text{ref}} = \infty) \tag{4.116}$$

Therefore, at infinite pressure $g^E = a^E$ (Eq. (4.114)), and the Huron–Vidal g^E mixing rule [9] can be written as

$$\frac{a}{bRT} = \sum_i x_i \frac{a_{ii}}{b_i RT} + \frac{g_\infty^E}{q_1 RT} \quad (P^{\text{ref}} = \infty) \tag{4.117}$$

In the literature, the abbreviations $\alpha = a/bRT$ and $\alpha_i = a_{ii}/b_i RT$ are often introduced into Eq. (4.117):

$$q_1 \left(\alpha - \sum_i x_i \alpha_i \right) = \frac{g_\infty^E}{RT} \quad (P^{\text{ref}} = \infty) \tag{4.118}$$

The parameter q_1 depends on the particular equation of state and the reference state. At infinite pressure ($u = 1$), one obtains for the Peng–Robinson equation

$$q_1 = -0.62323$$

The direct adjustment of the interaction parameters of the g^E model (usually Wilson, NRTL or UNIQUAC) to experimental data usually yields an accurate correlation of VLE data. Problems can arise if published interaction parameters are used (e.g., from the DECHEMA data series [12]). In most cases, these have been fitted to data at moderate pressures, whereas the Huron–Vidal g^E mixing rule has been derived for infinite pressure (see above), which can cause poor results.

In the recent years, the g^E mixing rules have been further optimized. The main target is to get a low reference pressure to obtain a maximum similarity to the activity coefficient models. The first step was the relationship of Mollerup [13] which connects the excess Helmholtz energy a^E and the excess Gibbs energy g_0^E at zero pressure:

$$\frac{a^E}{RT} = \frac{g_0^E}{RT} - \frac{Pv^E}{RT} + \sum_i x_i \ln \frac{b}{b_i} \tag{4.119}$$

which introduces the Flory–Huggins term $\sum_i x_i \ln(b/b_i)$. Setting $v^E = 0$, Michelsen [14] created the first-order Modified Huron–Vidal (MHV1) mixing rule at the reference pressure $P^{\text{ref}} = 0$:

$$q_1\left(\alpha - \sum_i x_i \alpha_i\right) = \frac{g_0^E}{RT} + \sum_i x_i \ln\frac{b}{b_i} \qquad (4.120)$$

For the Peng–Robinson equation, u is set constant, and Michelsen suggests

$$q_1 = -0.53$$

The assumption of a constant u-value has been confirmed by Ahlers [15], where the u parameters of 79 components are listed for both the Soave–Redlich–Kwong and the Peng–Robinson equation of state from specific liquid volumes at the normal boiling point.

Figure 4.5 shows the histograms of the distribution of u for both equations of state.

The u_i values of the Soave–Redlich–Kwong equation vary only slightly, with an average value of $u_i = 1.13$. After fitting to experimental VLE data, for the PSRK equation u was set to

$$u = 1.1$$

Thus, the g^E mixing rule for the PSRK equation is [10]

$$q_1\left(\alpha - \sum_i x_i \alpha_i\right) = \frac{g_0^E}{RT} + \sum_i x_i \ln\frac{b}{b_i} \qquad (4.121)$$

with $q_1 = -0.64663$ at $P^{\text{ref}} = 1$ atm

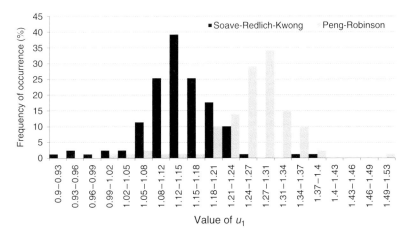

Figure 4.5 Histograms to justify the assumption of a constant u-value.

The fugacity coefficient of the PSRK equation is given by

$$\ln \varphi_i = \frac{b_i}{b}\left(\frac{Pv}{RT}-1\right) - \ln\frac{P(v-b)}{RT} - \left(\frac{1}{q_1}\ln\gamma_i + \frac{a_i}{RTb_i}\right.$$
$$\left. + \frac{1}{q_1}\left(\ln\frac{b}{b_i} + \frac{b_i}{b} - 1\right)\right)\ln\frac{v+b}{v} \quad (4.122)$$

with

A derivation of Eq. (4.122) is given in Appendix C, D5.

For the Peng–Robinson equation, u is determined in a similar way to be

$$u = 1.22489,$$

giving $q_1 = -0.53087$.

The volume translation used in the VTPR equation of state (see Eqs. (2.177–2.180)) has no influence on the results of the phase equilibrium calculations, it does not need to be considered in a g^E mixing rule [16]. Equation (2.180) is used as a mixing rule for the volume translation.

Several other g^E mixing rules have been developed, the most important ones are listed in Table 4.3.

Figure 4.6 shows the reproduction of g^E for the symmetric system (acetone–benzene) and for the asymmetric system (n-hexane–n-hexadecane) with the PSRK mixing rule and with the corresponding g^E model alone. In this case, the UNIFAC model (see Chapter 5) has been used. For the symmetric system, the experimental data are in good agreement with the results of the g^E model and the PSRK mixing rule, whereas for the asymmetric system the g^E model and the PSRK

Table 4.3 Overview on important g^E mixing rules.

Mixing rule	Reference state	Parameter
Huron and Vidal [9]: HV $q_1(\alpha - \sum_i x_i\alpha_i) = g_\infty^E/RT$	$P = \infty$	$q_1 = -0.693$ (SRK) $q_1 = -0.623$ (PR)
Michelsen [14]: MHV1 $q_1(\alpha - \sum_i x_i\alpha_i) = g_0^E/RT + \sum_i x_i\ln(b/b_i)$	$P = 0$	$q_1 = -0.593$ (SRK) $q_1 = -0.53$ (PR)
Holderbaum and Gmehling [17]: PSRK $q_1(\alpha - \sum_i x_i\alpha_i) = g_0^E/RT + \sum_i x_i\ln(b/b_i)$	$P = 1$ atm	$q_1 = -0.64663$ (SRK)
Dahl and Michelsen [18]: MHV2 $q_1(\alpha - \sum_i x_i\alpha_i) + q_2(\alpha^2 - \sum_i x_i\alpha_i^2) = g_0^E/RT + \sum_i x_i\ln(b/b_i)$	$P = 0$	$q_1 = -0.47832$ (SRK) $q_2 = -0.00469$ (SRK) $q_1 = -0.477$ (PR) $q_2 = -0.002$ (PR)
Boukouvalas et al. [16]: LCVM $\alpha = \lambda\alpha_{HV} + (1-\lambda)\alpha_{MHV1}$	$P = 0$	$q_{1,HV} = -0.623$ (PR) $q_{1,MHV1} = -0.53$ (PR) $\lambda = 0.36$

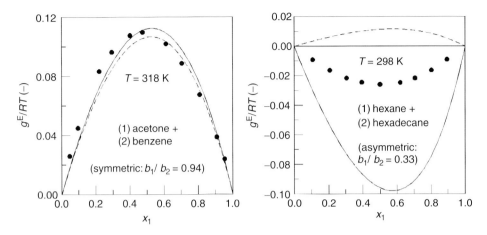

Figure 4.6 Calculation of the excess Gibbs energies for a symmetric and an asymmetric system, (- - -) PSRK, (———) UNIFAC, (•) experimental values.

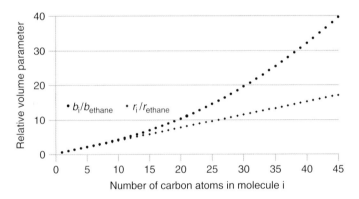

Figure 4.7 Dependence of the ratios b_i/b_{ethane} and r_i/r_{ethane} on the chain length of the alkane molecule i.

mixing rule differ significantly, neither of them meeting the experimental values. This can be considered as typical for the PSRK mixing rule [16, 19].

The reason for this behavior is that in asymmetric systems the difference between the combinatorial parts of the g^E model and the equation of state plays an important role [20].

To overcome this difficulty, it must be considered that there are two parameters which have almost the same physical meaning, as both the co-volume b and the relative van der Waals volume r represent the volume of the molecule, however, in a considerably different way, as illustrated in Figure 4.7.

The ratio of the co-volumes between an alkane molecule and the ethane molecule is depicted as a function of the number of carbon atoms in the alkane molecule, as well as the ratio of the corresponding relative van der Waals volumes. Up to n-octane, there is a good agreement, whereas for large molecules, that means for, asymmetric systems, large differences occur. A solution could be to replace the van der Waals volume in the g^E model with the co-volume of the equation of state. Starting with the PSRK-g^E mixing rule

$$\frac{a}{bRT} = \sum_i x_i \frac{a_{ii}}{b_i RT} + \frac{1}{q_1} \left(\frac{g_0^E}{RT} + \sum_i x_i \ln \frac{b}{b_i} \right) \qquad (4.123)$$

($q_1 = -0.64663$, $P^{\text{ref}} = 1$ atm)

the combinatorial part of the UNIFAC model (see Chapter 5) according to Eq. (4.124)

$$\frac{g^E_{\text{comb}}}{RT} = \sum_i x_i \ln V_i + 5 \sum_i x_i q_i \ln \left(\frac{F_i}{V_i} \right) \qquad (4.124)$$

depends on the ratio

$$V_i = \frac{r_i}{\sum_j x_j r_j} \qquad (4.125)$$

A similar ratio occurs in the Flory–Huggins term of the PSRK-g^E mixing rule:

$$\sum_i x_i \ln \frac{b}{b_i} = \sum_i x_i \ln V_i^* \qquad (4.126)$$

$$V_i^* = \frac{b_i}{\sum_j x_j b_j} \qquad (4.127)$$

Introducing Eqs. (4.124–4.127) into the mixing rule Eq. (4.123) yields

$$\frac{a}{bRT} = \sum_i x_i \frac{a_{ii}}{b_i RT} + \frac{1}{q_1} \left(\frac{g^E_{\text{res}}}{RT} + 5 \sum_i x_i q_i \ln \left(\frac{F_i}{V_i} \right) + \sum_i x_i \ln \frac{r_i}{\sum_j x_j r_j} \right.$$

$$\left. - \sum_i x_i \ln \frac{b_i}{\sum_j x_j b_j} \right) \qquad (4.128)$$

Substituting the r_i-parameters by the co-volumes b_i gives

$$\frac{a}{bRT} = \sum_i x_i \frac{a_{ii}}{b_i RT} + \frac{1}{q_1} \left(\frac{g^E_{\text{res}}}{RT} + 5 \sum_i x_i q_i \ln \left(\frac{F_i}{V_i} \right) \right)$$

($q_1 = -0.64663$, $P^{\text{ref}} = 1$ atm) \qquad (4.129)

The term $5 \sum x_i q_i \ln(F_i/V_i)$ is usually negligible. Thus, it can be written [21] as

$$\frac{a}{b} = \sum_i x_i \frac{a_{ii}}{b_i} + \frac{g^E_{\text{res}}}{q_1} \qquad (q_1 = -0.64663,\ P^{\text{ref}} = 1 \text{ atm}) \qquad (4.130)$$

which is similar to the Huron–Vidal mixing rule (Eq. (4.117)), however, P^{ref} is equal to 1 atm and not infinity, and only the residual part of the activity coefficient is used. The relative van der Waals volumes r_i are not used any more. The new g^E mixing rule (Eq. (4.130)) is called *VTPR-g^E mixing rule*, as it had been implemented in the VTPR equation of state [15]:

$$\frac{a}{b} = \sum_i x_i \frac{a_{ii}}{b_i} + \frac{g^E_{res}}{q_1} \quad (q_1 = -0.53087, P^{ref} = 1 \text{ atm}) \tag{4.131}$$

Introducing the dimensionless parameters $\alpha = a/bRT$ und $\alpha_i = a_{ii}/b_i RT$ one obtains

$$q_1 \left(\alpha - \sum_i x_i \alpha_i \right) = \frac{g^E_{res}}{RT} \quad (q_1 = -0.53087, P^{ref} = 1 \text{ atm}) \tag{4.132}$$

The partial molar quantity of the a parameter

$$\bar{a}_i = \left(\frac{\partial n_T a}{\partial n_i} \right)_{T,P,n_{j\neq i}} = b \left(\frac{RT \ln \gamma_i}{q_1} + \frac{a_{ii}}{b_i} \right)$$

$$+ \left(\frac{g^E_{res}}{q_1} + \sum_i x_i \frac{a_{ii}}{b_i} \right) (\bar{b}_i - b) \tag{4.133}$$

can be used to derive the fugacity coefficient, which is given by

$$\ln \varphi_i = \left(\frac{2 \sum_j z_j b_{ij}}{b} - 1 \right) \left(\frac{P(v+c)}{RT} - 1 \right) - \frac{Pc_i}{RT} - \ln \left(z + \frac{P(c-b)}{RT} \right)$$

$$- \frac{1}{2\sqrt{2}RT} \left(\frac{a_i}{b_i} + \frac{RT \ln \gamma_i}{q_1} \right) \ln \frac{v+c+\left(1+\sqrt{2}\right)b}{v+c+\left(1-\sqrt{2}\right)b} \tag{4.134}$$

For the b parameter, the mixing rule of Chen et al. [21]

$$b = \sum_i \sum_j x_i \cdot x_j \cdot b_{ij} \tag{4.135}$$

with

$$b_{ij}^{3/4} = 0.5 \left(b_i^{3/4} + b_j^{3/4} \right) \tag{4.136}$$

has been adopted for the VTPR equation of state instead of the linear mixing rule (Eq. (4.100)). Figure 4.8 illustrates the potential of Eq. (4.135) for the strongly asymmetric system ethane/hexadecane. The isotherms are well represented by VTPR, whereas the PSRK equation fails.

The caloric properties of the VTPR equation of state are given by

$$h(T,v,z) = \sum_i z_i h_i^{id}(T) + \frac{T \frac{\partial a}{\partial T} - a}{2\sqrt{2} b} \ln \frac{v+c+\left(1+\sqrt{2}\right)b}{v+c+\left(1-\sqrt{2}\right)b} + RT(z-1) \tag{4.137}$$

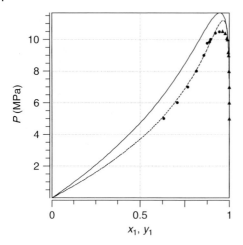

Figure 4.8 Representation of vapor–liquid equilibria of the system ethane (1) –n-hexadecane (2) at $T = 352.7$ K, --- VTPR, – PSRK. (Data from [22].)

and

$$s(T, v, z) = \sum_i z_i s_i^{id}(T, P) + R \sum_i z_i \ln z_i + R \ln \left(z - \frac{P(b-c)}{RT} \right)$$

$$+ \frac{\frac{\partial a}{\partial T}}{2\sqrt{2}b} \ln \frac{v + c + (1 + \sqrt{2})b}{v + c + (1 - \sqrt{2})b} \quad (4.138)$$

where

$$\frac{\partial a}{\partial T} = b \left(\sum_i z_i \frac{\partial a_i / \partial T}{b_i} + \frac{\partial g^E / \partial T}{q_1} \right) \quad (4.139)$$

$$\frac{\partial a_i}{\partial T} = 0.45724 \frac{R^2 T_{ci}^2}{P_{ci}} \frac{\partial \alpha_i}{\partial T} \quad (4.140)$$

$$\frac{\partial \alpha_i}{\partial T} = N_i \left[M_i - 1 - L_i M_i T_n^{N_i M_i} \right] \frac{T_n^{N_i M_i - N_i - 1}}{T_{ci}} \exp\left[L_i \left(1 - T_n^{N_i M_i} \right) \right] \quad (4.141)$$

where L_i, M_i and N_i are the parameters of the Twu-α-function.

Additional Problems

P4.1 The air inside a parking car has a relative humidity of $\varphi_1 = p_{H_2O}/P_{H_2O}^s = 0.6$ at $\vartheta_1 = 20°$C. At which temperature do the windows of the car get fogged, when the temperature decreases at night?
Consider the vapor phase to be ideal. The air shall be taken as a mixture of 80% nitrogen and 20% oxygen. The vapor pressure curve of water can be taken from Appendix A.

P4.2 A mixture of ethylene and propylene (T = 330 K, P = 23 bar, $y_{C_2H_4} = 0.2$) is cooled down in a heat exchanger. Which temperature has to be set, if the mixture should be $\Delta T = 5$ K above the dew point? The pressure drop of the heat exchanger is $\Delta P = 0.5$ bar. Use the PSRK equation of state. How is the result affected if hydrogen is added to the mixture so that the final concentration of hydrogen is $y_{H_2} = 0.06$?

P4.3 Calculate the compressibility factors z of the binary mixture propane (1)–methyl bromide (2) at T = 297 K and P = 5 bar for the concentrations $y_1 = 0.25$, $y_1 = 0.5$, and $y_1 = 0.75$ using the virial equation of state truncated after the second term. The second virial coefficients are

$$B_{11} = -394 \text{ cm}^3/\text{mol}$$
$$B_{22} = -567 \text{ cm}^3/\text{mol}$$
$$B_{12} = -411 \text{ cm}^3/\text{mol}$$

P4.4 Adopting the results of problem P4.3, calculate the fugacities of both components
a. in a mixture of ideal gases
b. in an ideal mixture of real gases
c. in a real mixture.

P4.5 Calculate the excess enthalpy of the system benzene (1)–cyclohexane (2) at $\vartheta = 25\,°C$ and $P = 1$ bar for a mole fraction of benzene of $x_1 = 0.5$ using the Soave-Redlich-Kwong equation of state. The required values can be taken from Appendix A. The binary interaction parameter is $k_{12} = 0.0246$.

P4.6 Acetone (5000 kg/h, $\vartheta_A = 40\,°C$) and chloroform (3000 kg/h, $\vartheta_C = 50\,°C$) are mixed at $P = 1$ bar. The mixing process is isobaric and adiabatic. The excess enthalpy can be calculated using the binary Wilson parameters given in Table 5.5. What is the temperature of the resulting stream?

P4.7 20 kg/h liquid ammonia ($-20\,°C$, 20 bar) and 100 kg/h liquid water ($20\,°C$, 20 bar) are mixed isobarically at 20 bar. The excess enthalpy is given by the formula:

$$h^E = \frac{A \cdot x_1 \cdot x_2}{\left(\frac{x_1}{B} + x_2 \cdot B\right)^2} \frac{J}{\text{mol}} \quad \text{with } A = -17\,235 \text{ and } B = 0.97645$$

Calculate the temperature of the mixture, assuming that $c_{P,H_2O} = 4.2$ J/g K and $c_{P,NH_3} = 4.8$ J/g K.

References

1. Grolier, J.-P.E. and Wilhelm, E. (1981) Fluid Phase Equilib., 6, 283.
2. Kunz, O., Klimeck, R., Wagner, W., and Jaeschke, M. (2007) The GERG-2004 Wide-Range Reference Equation of State for Natural Gases, GERG Technical Monograph TM15, Fortschr.-Ber. VDI, Reihe 6, Nr. 557, VDI-Verlag, Düsseldorf.
3. Arai, K., Inomata, H., and Saito, S. (1982) J. Chem. Eng. Data Jpn., 15 (1), 1–5.
4. Fontaine, J.M. (1989) Das Phasengleichgewicht Helium-Methan und die Beschreibung mit einer neuen

Zustandsgleichung. Thesis. TU Braunschweig.
5. Plöcker, U., Knapp, H., and Prausnitz, J. (1978) *Ind. Eng. Chem. Proc. Des. Dev.*, **17** (3), 324–332.
6. Warowny, W. and Stecki, J. (1979) *The Second Cross Virial Coefficients of Gaseous Mixtures*, PWN-Polish Scientific Publishers, Warsaw.
7. Dymond, J.H. and Smith, E.B. (1980) *The Virial Coefficients of Pure Gases & Mixtures*, Clarendon Press, Oxford.
8. Soave, G. (1972) *Chem. Eng. Sci.*, **27**, 1197–1203.
9. Huron, M.-J. and Vidal, J. (1979) *Fluid Phase Equilib.*, **3**, 255–271.
10. Fischer, K. and Gmehling, J. (1996) *Fluid Phase Equilib.*, **121**, 185.
11. Fischer, K. and Gmehling, J. (1995) *Fluid Phase Equilib.*, **112**, 1–22.
12. Gmehling, J., Onken, U., Arlt, W., Grenzheuser, P., Weidlich, U., Kolbe, B., and Rarey, J. (1977) *Vapor-Liquid Equilibrium Data Collection*, DECHEMA Chemistry Data Series, DECHEMA, Frankfurt.
13. Mollerup, J. (1986) *Fluid Phase Equilib.*, **25**, 323.
14. Michelsen, M.L. (1990) *Fluid Phase Equilib.*, **60**, 213.
15. Ahlers, J. (2003) Entwicklung einer universellen Gruppenbeitragszustandsgleichung. Thesis. Carl-von-Ossietzky-Universität, Oldenburg.
16. Boukouvalas, C., Spiliotis, N., Coutsikos, P., Tzouvaras, N., and Tassios, D. (1994) *Fluid Phase Equilib.*, **92**, 75.
17. Holderbaum, T. and Gmehling, J. (1991) *Fluid Phase Equilib.*, **79**, 251–265.
18. Dahl, S. and Michelsen, M. (1990) *AIChE J.*, **36**, 1829.
19. Kalospiros, N., Tzouvaras, N., Coutsikos, P., and Tassios, D.P. (1995) *AIChE J.*, **41**, 928.
20. Kontogeorgis, G.M. and Vlamos, P.M. (2000) *Chem. Eng. Sci.*, **55**, 2351.
21. Chen, J., Fischer, K., and Gmehling, J. (2002) *Fluid Phase Equilib.*, **200**, 411.
22. www.ddbst.com (accessed January 18, 2011).

5
Phase Equilibria in Fluid Systems

Conventional chemical plants can usually be divided into a preparation, reaction, and separation step (see Figure 5.1). Although the reactor can be considered as the heart or the central unit of the chemical plant, often 60–80% of the total costs are caused by the separation step, where the various thermal separation processes are applied to obtain the products with the desired purity, to recycle the unconverted reactants and to remove the undesired by-products. Because of the many advantages (energy used as separating agent, high-density differences between the two fluid phases (liquid, vapor)) in 90% of the cases distillation processes are applied in the chemical or the petrochemical industry, whereas in the pharmaceutical industry crystallization processes are far more important [1].

Different aspects have to be considered during the synthesis of separation processes. As the preliminary step the chemical engineer has to decide which separation processes should be used. Then he has to find out if separation problems occur. In the case of distillation these problems are typically azeotropic points, which do not allow separation by ordinary distillation. To understand distillation processes, the knowledge of residue curves and boundary lines is quite helpful. In the case of azeotropic points the engineer has to find an alternative way (e.g. separation at low or high pressure or by pressure swing distillation), or to select suitable solvents for the separation of the considered system (e.g. azeotropic or extractive distillation) or to choose a hybrid process (i.e., by combination of the distillation step with another separation process, for example, membrane separation, adsorption, etc.). Furthermore, the engineer has to design the equipment (e.g. to determine the number of theoretical stages needed or the height of the packing of the separation column) and in addition, he has to choose the optimum separation sequence. To treat the different aspects mentioned above, a reliable and detailed knowledge of the phase equilibrium behavior as a function of temperature, pressure, and composition for the multicomponent system, which has to be separated, is required.

The knowledge of the phase equilibrium behavior is not only important for the design of separation processes, but also for other applications, like the design of biphasic reactors, for example, gas–liquid reactors, the estimation of the fate of persistent chemicals in the environment, and so on.

In consequence, the typical question asked by the chemical engineer in the design phase is: "What is the composition and the pressure in phase β, when

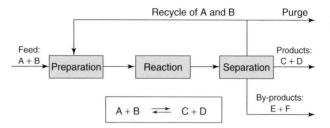

Figure 5.1 Simplified structure of a conventional chemical plant.

phase β is in equilibrium with phase α at given composition and temperature?" (see Figure 5.2). In most cases, multicomponent systems with nonpolar, polar, supercritical compounds, and electrolytes have to be considered. For example, for the system ethanol–water–sodium chloride–CO_2 several questions can be raised from Figure 5.2, such as:

- How strong does sodium chloride influence the solubility of CO_2 in the system ethanol–water?
- Does the system ethanol–water still show azeotropic behavior in the presence of sodium chloride?
- How is the solubility of sodium chloride in water influenced by the presence of ethanol and CO_2?
- Can the presence of sodium chloride cause a miscibility gap in the system ethanol–water?
- How strong is the pH-value influenced by the presence of CO_2?
- Which solvent can be applied to separate the azeotropic system ethanol–water by azeotropic or extractive distillation?
- Or can carbon dioxide directly be used for the separation of ethanol and water by supercritical extraction?

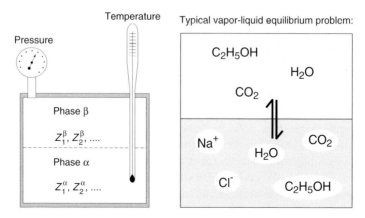

Figure 5.2 Equilibrium stage and typical separation problem.

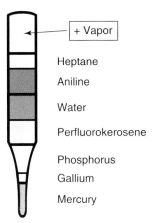

Figure 5.3 Seven liquid phases in equilibrium with the vapor phase [2].

However, the number of phases is not limited to two only. For example, in the case of hetero-azeotropic mixtures like butanol–water or ethanol–water–cyclohexane already two liquid phases exist besides the vapor phase. Hildebrand showed that in the system water–heptane–perfluorokerosene–aniline–phosphorus–gallium–mercury even seven liquid phases are in equilibrium with the vapor phase [2] (see Figure 5.3).

Depending on the state of the phases α and β vapor–liquid equilibria (VLE), liquid–liquid equilibria (LLE), solid–liquid equilibria (SLE), and so on, can be distinguished. In the case of VLE the phase equilibrium behavior is shown in Figure 5.4 as a Pxy-diagram for the binary system ethanol–water at 70 °C. For a given composition in the liquid phase the system pressure and the composition in

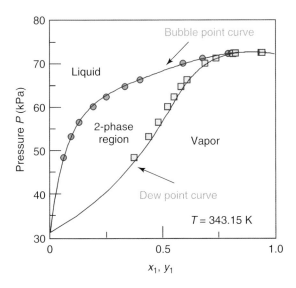

Figure 5.4 Pxy-diagram for the system ethanol (1)–water (2) at 70 °C [8].

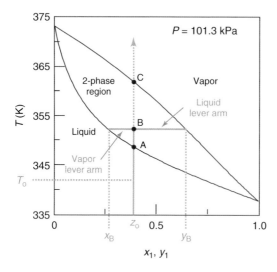

Figure 5.5 Illustration of the law of opposite lever arms on the basis of the binary system methanol (1)–water (2) at 101.3 kPa.

the vapor phase can be obtained from the diagram. Furthermore, it can be seen that at high ethanol concentrations the composition in the liquid phase is identical to the composition in the vapor phase, which makes a separation impossible by ordinary distillation.

In Pxy- and Txy-diagrams, the law of the opposite lever arms can be applied to determine the amount of vapor and liquid in the two-phase region. This is demonstrated in the Txy-diagram of the system methanol–water at 101.3 kPa (see Figure 5.5).

Consider a binary liquid mixture with the concentration z_0 and the temperature T_0. If the mixture is heated up, the bubble point line is reached in point A, and the first bubble is formed. When the mixture is further heated up, a further increase of temperature is obtained and more vapor is formed. At point B, the mixture consists of a liquid with the composition x_B and a vapor with the composition y_B. At point C, all liquid has been vaporized. Using n_T as the total number of moles, the mass balance yields for point B

$$n_T z_0 = \left(n^L + n^V\right) z_0 = n^L x_B + n^V y_B \tag{5.1}$$

which is equivalent to

$$\frac{n^L}{n^V} = \frac{y_B - z_0}{z_0 - x_B} \tag{5.2}$$

Therefore, the ratio between the amounts of vapor and liquid corresponds to the ratio of the lever arms located on the opposite side of the tie line.

From Figure 5.4 it can be seen that at 70 °C both compounds exist as liquids. Often the system temperature is above the critical temperature of one or more

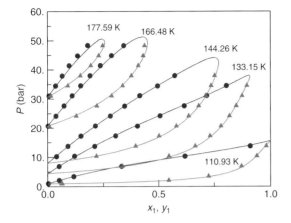

Figure 5.6 *Pxy*-diagram for the system nitrogen (1)–methane (2) at different temperatures.

components of the system considered. This is shown in Figure 5.6 for the binary system nitrogen–methane. Here the two phase regions do not cover the whole composition range for temperatures above the critical temperature of nitrogen ($T_c = 126.2$ K). Obviously now the binary system shows a critical point, where the length of the tie lines becomes zero.

But by applying the *Pxy*-diagram again the pressure and composition in the vapor phase for a given temperature and the corresponding composition in the liquid phase can be determined.

Often also the *K*-factors ($K_i = y_i/x_i$) are plotted as a function of the pressure, as shown in Figure 5.7 for the system nitrogen–methane. From this diagram the

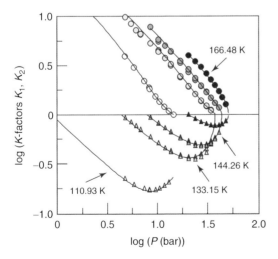

Figure 5.7 *K*-factors for the binary system nitrogen (1)–methane (2) as a function of pressure at different temperatures.

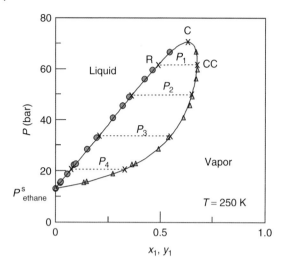

Figure 5.8 Isothermal Px-diagram for the binary system methane (1)–ethane (2) at 250 K.

K-factors for component 1 and 2 can be read directly for a given temperature and pressure. While for the low-boiler N_2 K-factors larger than one are obtained, K-factors smaller than one are observed for the high-boiler CH_4. At the same time the critical pressure of the mixture can be determined as the pressure where for a given isotherm both K-factors for the considered nonazeotropic system show a value of 1. This point can be found for all the isotherms above the critical temperature of nitrogen ($T_c = 126.2$ K).

With the help of an isothermal Pxy-diagram (see Figure 5.8) different phenomena which occur near the critical point, such as retrograde condensation or retrograde evaporation, can be explained.

For mole fractions lower than x_R the VLE behavior is similar to subcritical systems. Also in the range R–C a liquid and a vapor phase is obtained. But the vapor phase now is depleted of the low boiling component with increasing pressure from CC to C.

For compositions on the dew-point line in the range between C and CC a pressure decrease leads to the formation of a liquid phase. If the pressure is lowered further the amount of liquid phase will increase by condensation. At line P_1 the largest amount of liquid is found according to the law of the opposite lever arms. Below CC the system shows VLE behavior like subcritical systems again. This means vaporization instead of condensation is observed when the pressure is decreased, until the dew-point line is reached again and thus only vapor exists.

The lower limit of the region of retrograde condensation CC is often called *critical condensation point*. At CC the highest concentration of the low boiler in the vapor phase is obtained in equilibrium with the liquid phase. At this point the dew-point curve runs vertically and thus the slope for a given temperature is

$$\left(\frac{dP}{dy_1}\right)_{CC} = \infty \tag{5.3}$$

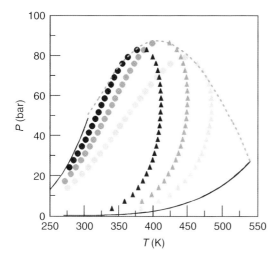

Figure 5.9 Experimental vapor pressures of ethane and heptane and experimental PT-data of the system ethane(1)–heptane (2) with different fixed compositions [3]—vapor pressure, ● liquid, ▲ vapor, ---- critical locus, $x_1 = y_1 = 0, 0.2658, 0.5871, 0.7709, 0.8871, 1$.

The upper limit of the retrograde region is the critical point C, which satisfies the following condition at constant temperature:

$$\left(\frac{dP}{dy_1}\right)_C = 0 \tag{5.4}$$

The phenomenon that a liquid is formed by lowering the pressure at constant temperature or, respectively, by increasing the temperature at constant pressure is called retrograde condensation.

Retrograde condensation plays an important role in technical applications, for example, in oil production, high-pressure pipelines, refrigeration processes and in natural gas reservoirs, where temperature and pressure are high enough to produce critical conditions.

The region in which vapor and liquid may coexist in a binary system is limited by the vapor pressure curves of the pure components and the critical line. In Figure 5.9 the vapor pressure curves of the pure compounds of the system ethane–heptane are shown together with the PT-curves of different fixed compositions of the liquid and the vapor phase. The intersections of the dew point and the bubble point curve for a given temperature and pressure mark the VLE for the chosen compositions in the liquid and the vapor phase. The critical points of a binary system can be found where a loop in Figure 5.9 is tangential to the envelope critical curve, also called *critical locus*.

The typical VLE behavior of a binary system above the critical temperature of one of the compounds looks like the behavior also shown in Figure 5.6

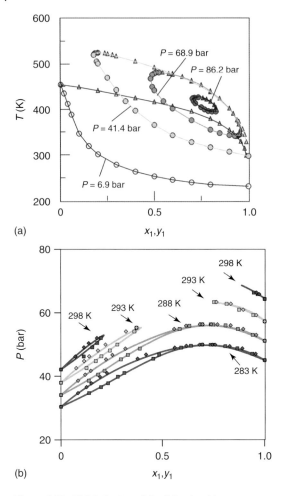

Figure 5.10 VLE behavior of the following binary systems near the critical point: (a) ethane (1)–heptane (2); (b) CO_2 (1)–ethane (2) experimental data taken from [3].

for the system nitrogen–methane. Other examples are CO_2–propane and argon–krypton.

In a few cases a different behavior is observed. In particular, this can happen if the system, for example, shows negative deviation from Raoult's law or a pressure maximum azeotrope. For the isobaric data of the system ethane–heptane and the isothermal data of the system CO_2–ethane this is shown in Figure 5.10. As can be seen for the system ethane–heptane, closed curves like islands appear at pressures of 68.9 and 86.2 bar. The reason is that at these pressures both components are supercritical (ethane: $P_c = 48.8$ bar, $T_c = 305.4$ K; heptane: $P_c = 27.3$ bar, $T_c = 540.3$ K) but the mixture is subcritical, which means coexisting liquid and vapor phase. For the system CO_2–ethane, the isotherms at 293 K and 298 K show

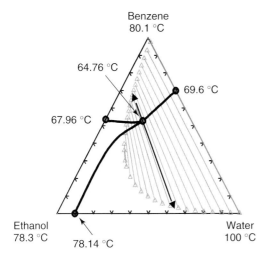

Figure 5.11 Ternary phase equilibrium diagram for the system ethanol–water–benzene at atmospheric pressure, ● – azeotropic points.

two critical points, one on the right and the other on the left-hand side, and thus no coexisting phases in the medium concentration range between the two critical points. This is caused by the fact that the system shows a pressure maximum azeotrope.

For multicomponent systems, the phase equilibrium behavior can become much more complicated. The phase equilibrium behavior of the ternary system ethanol–water–benzene at atmospheric pressure is shown in Figure 5.11. It can be seen that a ternary azeotrope exists besides three binary azeotropes. The binary system benzene–water shows a large miscibility gap, which results in a miscibility gap in the ternary system. In the diagram the binodal curve and a few tie lines are shown. The tie lines connect the two liquid phases in equilibrium. While the azeotropes ethanol–water and ethanol–benzene are homogeneous, the binary azeotrope benzene–water and the ternary azeotrope are heterogeneous azeotropes. The ternary azeotrope shows the lowest boiling point. This can be used to separate the azeotropic system ethanol–water by the so-called azeotropic distillation[1] (see Section 11.4). After condensation, the ternary azeotrope forms two liquid phases, a benzene and a water-rich phase. The compositions of the two liquid phases are marked in Figure 5.11 by the arrows. The occurrence of azeotropic behavior and the selection of suitable solvents for azeotropic distillation are discussed in more detail in Sections 5.6 and 11.4. In Figure 5.11 additionally the so-called boundary residual curves are shown. While in binary systems the azeotropic point cannot be crossed by ordinary distillation, boundary lines in ternary systems, and boundary

---

1) Nowadays, benzene is no more used because of its toxicity. In commercial plants, it has been widely replaced by cyclohexane.

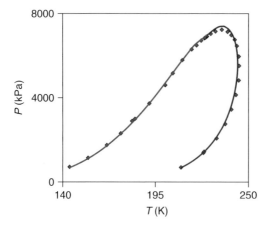

Figure 5.12 PT-diagram of a natural gas mixture consisting of 85.11 mol% methane, 10.07 mol% ethane, and 4.82 mol% propane [4].

surfaces in quaternary systems have the same consequences. How these residual curves are calculated is discussed in Section 11.3 in detail.

A natural gas is a typical multicomponent mixture; thus, a diagram like Figure 5.8 is not appropriate for illustration. Therefore, often PT-projections of phase diagrams, which are valid for a fixed overall concentration, are used (see Figure 5.12).

5.1
Thermodynamic Fundamentals

While a large number of phase equilibrium data are available for binary systems, much less data have been published for ternary systems and almost no data can be found for multicomponent systems. Of course the various phase equilibria for binary and multicomponent systems can be measured as a function of temperature or pressure and composition. Today highly sophisticated, reliable, and often computer-driven lab facilities are available to do so. Nevertheless, the measurement of the phase equilibrium behavior of multicomponent systems is very time consuming. For a ten component system, the number of experimental VLE data required is listed in Table 5.1, assuming the data are taken at constant pressure (e.g., atmospheric pressure) in 10 mol%-steps. For the pure components this means that only 10 normal boiling points have to be measured. Then there are 45 binary systems for which nine data points between 10 and 90 mol% are measured, so that in total 405 data points have to be measured. As can be seen in Table 5.1 there are additionally 120 ternary, 210 quaternary, 252 quinary, and many higher systems. For the 10-component system only one data point with 10 mol% of every component has to be experimentally determined. A total number of 92378

Table 5.1 Number of experimental data required for a ten-component system at a given pressure (e.g., atmospheric pressure), when the data are measured in 10 mol% steps.

Number of components	Number of systems	Data points/ system	Total number of data points/system	Total number of data points
1	10	1	10	10
2	45	9	405	415
3	120	36	4320	4735
4	210	84	17640	22375
5	252	126	31752	54127
6	210	126	26460	80587
7	120	84	10080	90667
8	45	36	1620	92287
9	10	9	90	92377
10	1	1	1	92378

data points results, which have to be measured. If 10 data points can be measured per working day, the measurements would last ~37 years [5].

Because of this time consuming effort reliable thermodynamic models are required, which allow the calculation of the phase equilibrium behavior of multicomponent systems using only a limited number of experimental data, for example, only binary data. From Table 5.1 it can be concluded that in this case only 42 days are required to measure all pure component and binary data of a ten-component system (in total 415 data points). Since a lot of binary VLE data can be found in the literature [3, 6], even less than 42 days of experimental work would be necessary.

Following Gibbs, phase equilibrium exists if the components show identical chemical potentials in the different phases α and β:

$$\mu_i^\alpha = \mu_i^\beta \tag{4.71}$$

The chemical potential is a thermodynamic quantity, which was first introduced by Gibbs. It is not an easily imaginable quantity. Later it was shown by Lewis (see Section 4.7.2) that the phase equilibrium condition given in Eq. (4.71) can be replaced by the following so-called isofugacity condition:

$$f_i^\alpha = f_i^\beta \tag{4.75}$$

At low pressures, except for strongly associating compounds the fugacities of the pure compounds are approximately identical to the vapor pressure or sublimation pressure depending on the state (liquid or solid). In case of mixtures, at low pressures the fugacity is nearly identical with the partial pressure of the compound considered.

For practical applications, Eqs. (4.71) and (4.75) are not very helpful, since the connection to the measurable quantities T, P and the composition in the liquid and vapor phase is missing to be able to calculate the required K-factors K_i or separation factors α_{ij} for the design of the different separation processes.[2] Therefore, auxiliary quantities such as activity coefficients γ_i and fugacity coefficients φ_i have been introduced.

The *fugacity coefficient* φ_i of component i can be defined as the ratio of the fugacity in the liquid phase L (vapor phase V) to the product of the mole fraction x_i (y_i) and system pressure P. In the vapor phase the product $y_i P$ can be substituted by the partial pressure p_i:

$$\varphi_i^L \equiv \frac{f_i^L}{x_i P} \tag{5.5}$$

$$\varphi_i^V \equiv \frac{f_i^V}{y_i P} = \frac{f_i^V}{p_i} \tag{5.6}$$

The *activity coefficient* γ_i is defined as follows:

$$\gamma_i \equiv \frac{f_i}{x_i f_i^0} \tag{5.7}$$

whereby the standard fugacity f_i^0 can be chosen arbitrarily.

Using the different definitions for the fugacities two different approaches can be derived for the description of phase equilibria. Starting from Eq. (4.75), the following relations for VLE are obtained [7]:

$$f_i^L = f_i^V \tag{5.8}$$

Approach A:

$$x_i \varphi_i^L = y_i \varphi_i^V \tag{5.9}$$

Approach B:

$$x_i \gamma_i f_i^0 = y_i \varphi_i^V P \tag{5.10}$$

In Approach A, the fugacity coefficients of the liquid φ_i^L and vapor phase φ_i^V are needed. They describe the deviation from ideal gas behavior and can be calculated with the help of equations of state, for example, cubic equations of state and reliable mixing rules. In Approach B, besides the activity coefficients γ_i a value for the standard fugacity f_i^0 is required. In the case of VLE usually the fugacity of the pure liquid at system temperature and system pressure is used as standard fugacity. For the calculation of the solubilities of supercritical compounds Henry constants are often applied as standard fugacity (see Section 5.7).

[2] In the case of distillation K_i is defined as the ratio of the vapor phase mole fraction to the liquid phase mole fraction ($K_i = y_i/x_i$), and the separation factor α_{ij} is the ratio of the K-factors ($\alpha_{ij} = K_i/K_j$).

Using Eqs. (5.8) and (5.6), the fugacity of the pure liquid at system temperature can directly be calculated, since the pressure is identical with the vapor pressure of the pure liquid which is in equilibrium with pure vapor:

$$f_i^0(T, P_i^s) = \varphi_i^L P_i^s = \varphi_i^V P_i^s \equiv \varphi_i^s P_i^s \tag{5.11}$$

where the fugacity coefficient in the liquid or vapor phase in the saturation state φ_i^L or φ_i^V can be replaced by the fugacity coefficient at saturation pressure φ_i^s.

To get the fugacity of the pure liquid not at the vapor pressure P_i^s, but at system pressure P, the compression or expansion of the pure liquid from the vapor pressure to the system pressure has to be taken into account. This can be done using Eq. (2.71):

$$\left(\frac{\partial \ln f_i^L}{\partial P}\right)_T = \frac{v_i^L}{RT} \tag{5.12}$$

With the assumption that the molar liquid volume v_i^L is constant in the pressure range covered, Eq. (5.13) is obtained for the standard fugacity at system temperature and system pressure, where the exponential term in Eq. (5.13) is called *Poynting factor* Poy$_i$:

$$f_i^0(T, P) = \varphi_i^s P_i^s \exp\frac{v_i^L(P - P_i^s)}{RT} = \varphi_i^s P_i^s \text{Poy}_i \tag{5.13}$$

Combining Eqs. (5.10) and (5.13) leads to the following relation for the description of VLE with the help of activity coefficients:

$$x_i \gamma_i \varphi_i^s P_i^s \text{Poy}_i = y_i \varphi_i^V P \tag{5.14}$$

Introducing the auxiliary quantity ϕ_i gives

$$x_i \gamma_i \phi_i P_i^s = y_i P \quad \text{with} \quad \phi_i = \frac{\varphi_i^s \text{Poy}_i}{\varphi_i^V} \tag{5.15}$$

If the pressure difference $P - P_i^s$ is not too large, the value of the Poynting factor is approximately 1. This is shown below for the system ethanol–water at 70 °C.

Example 5.1

Calculate the Poynting factor for ethanol and water at 70 °C for pressure differences of 1, 10 and 100 bar. At 70 °C, the following molar volumes can be used:
  ethanol: 61.81 cm³/mol, water 18.42 cm³/mol.

Solution

For a pressure difference of 1 bar (e.g. system pressure $P = 5$ bar, vapor pressure $P_i^s = 4$ bar) the following Poynting factors are obtained for ethanol and water:

$$\text{Poy}_{\text{ethanol}} = \exp\frac{0.06181 \cdot 1}{0.0831433 \cdot 343.15} = 1.0022$$

$$\text{Poy}_{\text{water}} = \exp\frac{0.01842 \cdot 1}{0.0831433 \cdot 343.15} = 1.0006$$

In the same way the following values are obtained for a pressure difference of 10 and 100 bar:

10 bar:	$Poy_{ethanol} = 1.022,$	$Poy_{water} = 1.006$
100 bar:	$Poy_{ethanol} = 1.242,$	$Poy_{water} = 1.067$

It can be seen that at typical pressure differences (e.g., $P - P_i^s < 1$ bar) in distillation processes the Poynting factors show values near unity. Because of the larger molar volume of ethanol, the deviation from unity is larger for ethanol than for water.

Besides the Poynting factor, the real vapor phase behavior has to be taken into account in Eqs. (5.14) and (5.15). This can be done with the help of equations of state. Since only the vapor phase nonideality has to be considered, simple equations of state, for example, the virial equation of state can be applied, which are only able to describe the PvT behavior of the vapor phase. For moderate pressures the use of second virial coefficients is sufficient. In the case of systems with strong associating compounds such as carboxylic acids or hydrogen fluoride this approach cannot be applied any more. In this case the deviation from ideal gas behavior caused by the strong interactions – comparable to chemical reactions – has to be taken into account by so-called chemical contributions (see Section 13.2).

Example 5.2

Calculate the fugacity coefficients φ_i^V, φ_i^s, and the auxiliary quantity ϕ_i for the system ethanol–water at 70 °C using the virial equation. At 70 °C the following second virial coefficients should be used for the system ethanol(1)–water(2):

$B_{11} = -1100 \, \text{cm}^3/\text{mol}$
$B_{12} = -850 \, \text{cm}^3/\text{mol}$
$B_{22} = -650 \, \text{cm}^3/\text{mol}$

The following liquid molar volumes can be used for the calculation of the Poynting factors: water 18.42 cm³/mol and ethanol: 61.81 cm³/mol.

Solution

The calculation procedure for the system ethanol (1)–water (2) is demonstrated for the data point $x_1 = 0.252$, $y_1 = 0.552$, and $P = 62.39$ kPa listed in Table 5.2.

Using Eq. (4.97) for the calculation of the fugacity coefficients in the vapor phase:

$$\ln \varphi_i^V = \left[2 \sum_j y_j B_{ij} - B \right] \frac{P}{RT} \tag{4.97}$$

Table 5.2 Vapor–liquid equilibrium data for the system ethanol (1)–water (2) at 70 °C [8].

x_1	y_1	P (kPa)
0.0	0.0	31.09[3]
0.062	0.374	48.33
0.095	0.439	53.2
0.131	0.482	56.53
0.194	0.524	60.12
0.252	0.552	62.39
0.334	0.583	64.73
0.401	0.611	66.34
0.593	0.691	70.11
0.680	0.739	71.23
0.793	0.816	72.35
0.810	0.826	72.41
0.943	0.941	72.59
0.947	0.945	72.59
1.0	1.0	72.3[3]

and for the pure compounds:

$$\ln \varphi_i^s = \frac{B_{ii} P_i^s}{RT} \tag{2.108}$$

where the second virial coefficient B of the mixture can be obtained using the following relation:

$$B = \sum_i \sum_j y_i y_j B_{ij} \tag{4.89}$$

Using Eq. (4.89) the following virial coefficient is obtained for the given vapor phase composition:

$$B = 0.552^2 \cdot (-1100) + 2 \cdot 0.552 \cdot 0.448 \cdot (-850) + 0.448^2 \cdot (-650)$$
$$= -886 \text{ cm}^3/\text{mol}$$

Using this value the fugacity coefficient of ethanol can be calculated directly using Eq. (4.97):

$$\ln \varphi_1^V = [2 (0.552 \cdot (-1100) + 0.448 \cdot (-850)) + 886] \frac{62.39}{8314.33 \cdot 343.15}$$
$$= -0.0238$$
$$\varphi_1^V = 0.9764$$

3) Unfortunately, in [8] the pure component vapor pressures were not measured. Therefore, these values were added by using the available constants for the Antoine equation (see Figure 5.30).

The fugacity coefficient of ethanol (1) in the saturation state is obtained as

$$\ln \varphi_1^s = \frac{-1100 \cdot 72.30}{8314.33 \cdot 343.15} = -0.02787$$

$$\varphi_1^s = 0.9725$$

In a similar way, the following values are obtained for water (2):

$$\varphi_2^V = 0.9862$$
$$\varphi_2^s = 0.9929$$

With the Poynting factors

$$\text{Poy}_{\text{ethanol}} = \exp\frac{0.06181 \cdot (0.6239 - 0.723)}{0.0831433 \cdot 343.15} = 0.9998$$

$$\text{Poy}_{\text{water}} = \exp\frac{0.01842 \cdot (0.6239 - 0.3109)}{0.0831433 \cdot 343.15} = 1.0002$$

the following ϕ_i values are obtained for this data point:

$$\phi_{\text{ethanol}} = \frac{0.9725 \cdot 0.9998}{0.9764} = 0.9958$$

$$\phi_{\text{water}} = \frac{0.9929 \cdot 1.0002}{0.9862} = 1.0070$$

For the whole composition range the ϕ_i-values are shown in Figure 5.13. It can be seen that in the whole composition range, the ϕ_i-values are between 0.98 and 1.01.

As shown in Example 5.2, for nonassociating compounds in contrast to strongly associating compounds such as carboxylic acids or HF the fugacity coefficients in the vapor phase φ_i^V and in the saturation state φ_i^s show very similar values at moderate pressures, so that ϕ_i-values around unity are obtained. This means that for nonassociating systems the following simplified relation can often be used to describe the VLE behavior:

$$x_i \gamma_i P_i^s \approx y_i P \tag{5.16}$$

Using the different approaches, the following relations are obtained to calculate the required K-factors K_i and relative volatilities (separation factors) α_{ij}:

$$K_i = \frac{y_i}{x_i} = \frac{\varphi_i^L}{\varphi_i^V} \qquad \alpha_{ij} = \frac{K_i}{K_j} = \frac{y_i/x_i}{y_j/x_j} = \frac{\varphi_i^L \varphi_j^V}{\varphi_i^V \varphi_j^L} \tag{5.17}$$

$$K_i = \frac{y_i}{x_i} \approx \frac{\gamma_i P_i^s}{P} \qquad \alpha_{ij} = \frac{K_i}{K_j} = \frac{y_i/x_i}{y_j/x_j} \approx \frac{\gamma_i P_i^s}{\gamma_j P_j^s} \tag{5.18}$$

As can be seen later, both approaches allow the calculation of the VLE behavior of multicomponent systems using binary data alone.

When the advantages and disadvantages of different approaches are compared, approach A (φ–φ approach) shows various important advantages over Approach B, for example, that the same auxiliary quantities are used to describe the real behavior

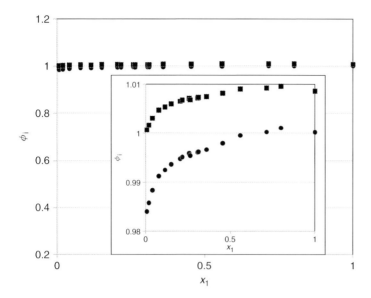

Figure 5.13 ϕ_i-values for the system ethanol(1)–water (2) at 70 °C.

in the liquid and vapor phase. No additional model is required to account for the real behavior of the vapor phase. Furthermore, no problem arises with supercritical compounds, since no standard fugacity (vapor pressure) is required. At the same time densities, enthalpies (including heats of vaporization), heat capacities, and so on, as a function of temperature, pressure, and composition can be calculated for both phases, which are required as additional information in Approach B – the so-called $\gamma-\varphi$-approach (see Section 5.2). The disadvantage is that for the calculation a computer is required. On the other hand, the strength of Approach B is its relative simplicity and the opportunity to have independent correlations for each quantity, which can be fitted as accurately as possible and as necessary.

5.2
Application of Activity Coefficient Models

The equation of state approach is very attractive for the calculation of VLE. But it requires an equation of state and reliable mixing rules, which are able to describe the PvT behavior not only of the vapor but also of the liquid phase with the required accuracy. In spite of the progress achieved in the last 20 years, up to now there is no universal equation of state and mixing rule which can be successfully applied to all kind of systems in a wide temperature and pressure range for pure compounds and mixtures.

For the calculation of VLE with Approach B often the simplified Eq. (5.16) is applied. Then besides the activity coefficients as a function of composition and temperature only the vapor pressures of the components are required for the calculation.

5 Phase Equilibria in Fluid Systems

Using Eq. (5.16) the required activity coefficients and the excess Gibbs energies can directly be derived from complete experimental VLE data. This is shown in Example 5.3 for the binary system ethanol–water measured at 70 °C.

Example 5.3

Calculate the activity coefficients and the excess Gibbs energies for the system ethanol (1)–water (2) at 70 °C as a function of composition using Eq. (5.16) and Table 5.2.

Solution

For the system ethanol (1)–water (2), the calculation of the activity coefficients and the excess Gibbs energy is demonstrated for a mole fraction $x_1 = 0.252$.

Using the simplified Eq. (5.16), the activity coefficients can be calculated by the following relation:

$$\gamma_i \approx \frac{y_i P}{x_i P_i^s}$$

For the selected composition the following activity coefficients are obtained for ethanol (1) and water (2):

$$\gamma_1 = \frac{0.552 \cdot 62.39}{0.252 \cdot 72.30} = 1.890$$

$$\gamma_2 = \frac{0.448 \cdot 62.39}{0.748 \cdot 31.09} = 1.202$$

With the help of these activity coefficients the excess Gibbs energy can be calculated using Eq. (4.86).

$$g^E = RT \left(x_1 \ln \gamma_1 + x_2 \ln \gamma_2 \right)$$

$$g^E = 8.31433 \cdot 343.15(0.252 \ln 1.890 + 0.748 \ln 1.202) = 850.2 \text{ J/mol}$$

$$\frac{g^E}{RT} = 0.252 \ln 1.890 + 0.748 \ln 1.202 = 0.298$$

For the other compositions the activity coefficients, the excess Gibbs energies and the dimensionless excess Gibbs energies (g^E/RT) are listed in Table 5.3. Furthermore the values are shown in the graphical form in Figure 5.14 together with the correlation results using the Wilson model (see Chapter 5.3).

Depending on the values of the activity coefficients γ_1 and γ_2 and the vapor pressures P_1^s and P_2^s, a very different VLE behavior is observed. In Figure 5.15, the vapor phase composition y_1, the activity coefficients $\ln \gamma_i$, the pressure P at isothermal conditions and the temperature T at isobaric conditions as a function of the mole fraction of component 1[4] in the liquid (vapor) phase are shown for binary

---

[4] In the case of binary VLE the low boiling substance is always designated as component 1.

Table 5.3 Experimental data [8] for the system ethanol (1)–water (2) at 70 °C and the derived activity coefficients and excess Gibbs energies.

x_1	y_1	P (kPa)	γ_1	γ_2	g^E (J/mol)	g^E/RT
0	0	31.09		1.000	0	0
0.062	0.374	48.33	4.032	1.037	345.0	0.1209
0.095	0.439	53.2	3.400	1.061	483.9	0.1696
0.131	0.482	56.53	2.877	1.084	594.6	0.2084
0.194	0.524	60.12	2.246	1.142	753.2	0.2640
0.252	0.552	62.39	1.890	1.202	850.2	0.2980
0.334	0.583	64.73	1.563	1.304	929.2	0.3257
0.401	0.611	66.34	1.398	1.386	940.9	0.3298
0.593	0.691	70.11	1.130	1.712	831.1	0.2913
0.68	0.739	71.23	1.071	1.869	703.3	0.2465
0.793	0.816	72.35	1.030	2.069	495.5	0.1737
0.81	0.826	72.41	1.021	2.133	459.3	0.1610
0.943	0.941	72.59	1.002	2.417	148.6	0.0521
0.947	0.945	72.59	1.002	2.423	138.9	0.0487
1	1	72.3	1.000		0	0

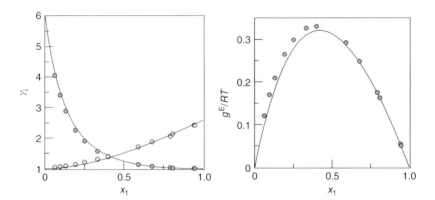

Figure 5.14 Concentration dependence of the activity coefficients and of the dimensionless excess Gibbs energy for the system ethanol (1)–water (2) at 70 °C [8] —— Wilson model.

systems with very different real behavior. In the two diagrams on the right-hand side the pressure and temperature are not only given as a function of the liquid phase (continuous boiling point line) but also as a function of the vapor phase composition (dashed dew-point line).

While the first system benzene–toluene shows nearly ideal behavior ($\gamma_i \approx 1$), the activity coefficients for the next three systems steadily increase (positive deviation from Raoult's law). The influence of the activity coefficients can particularly be

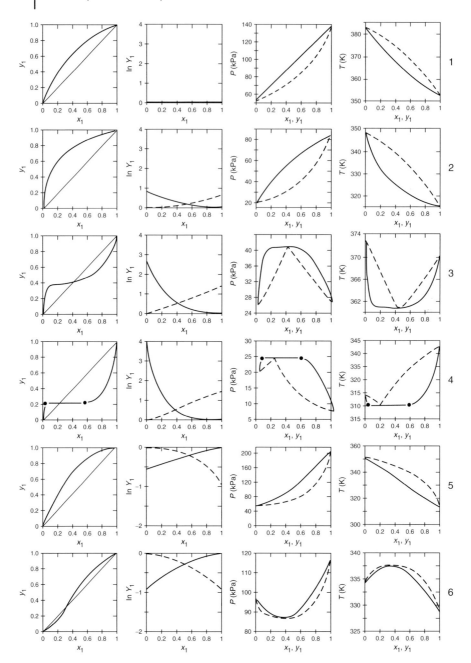

Figure 5.15 Different types of vapor–liquid equilibrium diagrams for the following binary systems: (1) benzene (1)–toluene (2); (2) methanol (1)–water (2); (3) 1-propanol (1)–water (2); (4) 1-butanol (1)–water (2); (5) dichloromethane (1)–2-butanone (2); (6) acetone (1)–chloroform (2).

recognized from the pressure as a function of the liquid phase mole fraction x_1 at a given temperature. While a straight line for the pressure is obtained in the case of the nearly ideal system benzene–toluene following Raoult's law, higher pressures than the values obtained using Raoult's law are observed for the system methanol–water ($\gamma_i > 1$). With increasing activity coefficients as in the case of the system 1-propanol–water ($\gamma_i \gg 1$) the pressure even shows a maximum. At the same time a minimum of the temperature is observed in the isobaric case. At the pressure maximum (respectively minimum of the temperature) the boiling point line and the dew-point line meet. This means that the composition in the liquid and the vapor phase becomes identical and in the y–x-diagram an intersection of the 45° line is observed. These points are called *azeotropic points*. Systems with an azeotropic point cannot be separated by ordinary distillation.

When the values of the activity coefficients further increase, two liquid phases can occur, as in the case of the system 1-butanol–water. If the two liquid phase region (shown by the horizontal line) intersect the 45° line in the y–x-diagram, a so-called heterogeneous azeotropic point occurs. In the case of heterogeneous azeotropic points the condensation of the vapor leads to the formation of two liquid phases. In the system 1-butanol–water a butanol-rich and a water-rich phase is formed. The pressure (temperature) and the vapor phase composition show constant values for the binary system in the whole heterogeneous region.

Besides the large number of systems with positive deviation from Raoult's law ($\gamma_i > 1$), sometimes systems with negative deviation from Raoult's law ($\gamma_i < 1$), are observed. In Figure 5.15, the systems dichloromethane–2-butanone and acetone–chloroform were chosen as examples. Because of the strong hydrogen bonding effects between the two compounds, associates with low volatility are formed in these systems. This results in the fact that the pressure above the liquid mixture is lower than the pressure assuming ideal behavior ($\gamma_i = 1$). Depending on the vapor pressures azeotropic behavior can also occur in systems with negative deviations from Raoult's law, as in the system acetone–chloroform. However, in contrast to systems with positive deviations from Raoult's law, in these systems azeotropic points with a pressure minimum (temperature maximum) are formed. Binary systems with negative deviation from Raoult's law ($\gamma_i < 1$) cannot show two liquid phases. The occurrence and disappearance of binary azeotropes are discussed in more detail in Section 5.6.

5.3
Calculation of Vapor–Liquid Equilibria Using g^E-Models

As discussed in Section 5.1, besides the vapor pressure of the pure compounds an activity coefficient model is required, which allows the calculation of the VLE behavior using only binary experimental data. Using Eq. (4.85) an analytical expression for the activity coefficients can be derived if an expression for the excess Gibbs energy is available. By definition, the expression for the excess Gibbs energy must obey the following boundary condition:

$$g^E \to 0 \text{ for } x_i \to 1$$

For the binary case the excess Gibbs energy g^E shows a value of 0 for $x_1 = 1$ and $x_2 = 1$. The simplest expression which obeys the boundary conditions is the Porter equation [9]:

$$\frac{g^E}{RT} = A x_1 x_2 \tag{5.19}$$

In this equation, A is a parameter which can be fitted to experimental data. Using Eq. (4.85) an analytical expression for the activity coefficients can be derived directly from Porter's expression [9]. For the derivation it is advisable to replace the mole fractions by the mole numbers:

$$\frac{(n_1 + n_2) g^E}{RT} = \frac{G^E}{RT} = \frac{A n_1 n_2}{n_1 + n_2}$$

$$\ln \gamma_1 = \left(\frac{\partial G^E / RT}{\partial n_1}\right)_{T,P,n_2} = \frac{A n_2 (n_1 + n_2) - A n_1 n_2}{(n_1 + n_2)^2}$$

$$\ln \gamma_1 = A x_2^2$$

In the same way, the following expression is obtained for component 2:

$$\ln \gamma_2 = A x_1^2$$

Porter's equation can be applied if g^E shows a symmetric curvature, this means an extreme value at equimolar composition ($x_1 = 0.5$). This behavior is only observed for chemically similar compounds of similar size.

For the description of g^E/RT for all other binary systems a more flexible expression is required. The simplest way is the introduction of further adjustable parameters, as in the Redlich–Kister expansion [10]:

$$\frac{g^E}{RT} = x_1 x_2 \left[A + B(x_1 - x_2) + C(x_1 - x_2)^2 + \cdots\right] \tag{5.20}$$

With the help of the flexible Redlich–Kister expansion all kinds of concentration dependencies of g^E for binary systems can be described. The contribution of the different parameters to the value of the excess Gibbs energy is shown in Figure 5.16. However, both the Porter and the Redlich–Kister model can only be used for binary systems. Furthermore, the correct temperature dependence of the activity coefficients cannot be described using temperature-independent parameters.

In practice, g^E-models are required which allow the calculation of the real behavior of multicomponent systems in the whole composition and a wide temperature range using binary data alone. The largest part of the VLE data (88.5%) has been published for binary systems. Only 10.3% of the VLE data published are for ternary and approx. 1% for quaternary systems [3]. This means there is nearly no chance to find the required experimental VLE data for quaternary and higher systems.

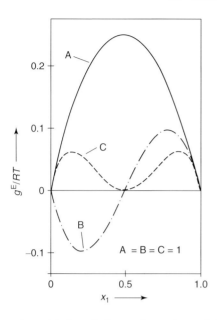

Figure 5.16 Contribution of the different parameters of the Redlich–Kister expansion to the value of the excess Gibbs energy.

A sophisticated thermodynamic model should take into account the various contributions to the excess Gibbs energy to be able to describe not only the concentration, but also the temperature and pressure dependence correctly:

$$g^E = h^E - Ts^E = u^E + Pv^E - Ts^E \tag{5.21}$$

Most of the excess properties are available experimentally. While the g^E-values can be obtained from VLE measurements (see Example 5.3), the excess enthalpies h^E are obtained from calorimetric and the excess volumes v^E from density measurements. When the excess properties mentioned before are known, other excess properties, for example the excess entropy s^E, can directly be derived, as shown in the next example.

Example 5.4

Construct a diagram with the thermodynamic excess properties g^E, h^E, and $-Ts^E$ for the system ethanol (1)–water (2) from the VLE data of Mertl [8] (Tables 5.2 and 5.3) and the excess enthalpies [11] in Table 5.4.

Solution

The g^E-values can directly be calculated from the activity coefficients derived from the experimental VLE-data. The values for g^E/RT were already listed in Table 5.3. At a concentration of $x_1 = 0.252$ the following value for the excess Gibbs energy is obtained:

$$g^E = 8.31433 \cdot 343.15 \cdot 0.298 = 850.2 \text{ J/mol}.$$

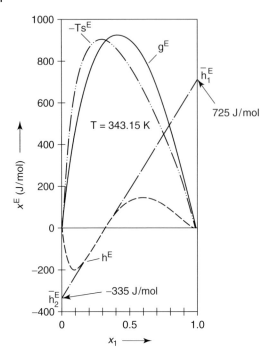

Figure 5.17 Excess Gibbs energy, excess enthalpy, and $-Ts^E$ for the system ethanol (1)–water (2) at 70 °C.

The value for $-Ts^E$ can be calculated from the difference $g^E - h^E$:

$$-Ts^E = g^E - h^E$$

For the whole composition range the results are shown in Figure 5.17.

When the excess properties are known, the values of the activity coefficients can be extrapolated to other conditions. Applying the van't Hoff equation (see Appendix C), the following temperature and pressure dependences of the excess Gibbs energy are obtained:

$$\left(\frac{\partial (g^E/T)}{\partial T}\right)_{P,x} = -\frac{h^E}{T^2} \tag{5.22}$$

respectively:

$$\left(\frac{\partial (g^E/T)}{\partial (1/T)}\right)_{P,x} = h^E \tag{5.23}$$

and

$$\left(\frac{\partial g^E}{\partial P}\right)_{T,x} = v^E \tag{5.24}$$

5.3 Calculation of Vapor–Liquid Equilibria Using g^E-Models

The temperature dependence of the molar excess enthalpy can be expressed by the molar excess heat capacity c_P^E.

$$\left(\frac{\partial h^E}{\partial T}\right)_{P,x} = c_P^E \tag{5.25}$$

With the help of these expressions, the temperature and pressure dependence of the activity coefficients can be derived directly:

$$\left(\frac{\partial \ln \gamma_i}{\partial (1/T)}\right)_{P,x} = \frac{\bar{h}_i^E}{R} \tag{5.26}$$

$$\left(\frac{\partial \ln \gamma_i}{\partial P}\right)_{T,x} = \frac{\bar{v}_i^E}{RT} \tag{5.27}$$

The partial molar excess properties vary with composition. They can be derived directly from the curvature of the excess enthalpies h^E or excess volumes v^E as a function of the mole fraction. How the partial molar properties can be determined by the tangent line for the excess enthalpy and a composition of $x_1 = 0.252$ is shown in Figure 5.17. Using the partial molar excess values $\bar{h}_i^E$ and $\bar{v}_i^E$ the activity coefficient at a different temperature or pressure can be determined. But it has to be considered that these partial molar excess properties do not only depend on composition but also on temperature and pressure.

While the pressure influence on the activity coefficient can usually be neglected in the case of VLE, the temperature dependence should be considered. This is shown in Example 5.5.

Example 5.5

Estimate the activity coefficients for $x_1 = 0.252$ at $50\,°C$ for the system ethanol (1)–water (2) using the activity coefficients given in Table 5.3 and the experimental excess enthalpy data from Table 5.4. Simplifying, it should be assumed that the excess enthalpy h^E is constant in the temperature range considered.

Activity coefficients at $70\,°C$:

$$\gamma_1 = 1.890$$
$$\gamma_2 = 1.202$$

Solution

For a mole fraction of $x_1 = 0.252$, the following values for the partial molar excess enthalpy can be read from Figure 5.17:

$$\bar{h}_1^E = 725\,\text{J/mol}$$
$$\bar{h}_2^E = -335\,\text{J/mol}$$

Table 5.4 Excess enthalpy data [11] for the system ethanol (1)–water (2) at 70 °C.

x_1	h^E (J/mol)	x_1	h^E (J/mol)
0.0303	−108.7	0.3962	61.6
0.0596	−173.7	0.4502	101.3
0.0896	−200.1	0.4980	129.7
0.1238	−194.0	0.5802	151.3
0.1239	−196.4	0.5889	153.3
0.1697	−160.9	0.6976	135.8
0.1905	−149.9	0.7439	115.0
0.2402	−92.2	0.8022	84.0
0.3021	−24.8	0.8457	62.0
0.3514	22.8	0.8957	39.3

While for the partial molar excess enthalpy of ethanol (1) a positive value is obtained, a negative value is obtained for water (2) for this composition. Following Eq. (5.26) one obtains with the help of these values

$$\ln \gamma_i(T_2) = \ln \gamma_i(T_1) + \frac{\overline{h}_i^E}{R}\left(\frac{1}{T_2} - \frac{1}{T_1}\right),$$

the values for the activity coefficients at 50 °C can be calculated:

$$\ln \gamma_1(323.15\,\text{K}) = \ln 1.890 + \frac{725}{8.31433}\left(\frac{1}{323.15} - \frac{1}{343.15}\right)$$

$$\gamma_1(323.15\,\text{K}) = 1.920$$

$$\ln \gamma_2(323.15\,\text{K}) = \ln 1.202 - \frac{335}{8.31433}\left(\frac{1}{323.15} - \frac{1}{343.15}\right)$$

$$\gamma_2(323.15\,\text{K}) = 1.193.$$

It can be recognized that different temperature dependencies are observed for the two compounds involved, caused by the different sign of the partial molar excess enthalpies. While the activity coefficient for ethanol decreases, the activity coefficient for water increases with increasing temperature in the temperature range covered. But as can be seen from Figure 5.17 the temperature dependence of the partial molar excess enthalpies strongly depends on composition. For example, for compositions $x_1 < 0.1$ negative partial molar excess enthalpies for ethanol would result.

Example 5.6

Calculate the activity coefficient at infinite dilution of ethanol (1) in n-decane (2) at 353.15 and 433.15 K from the γ_1^∞ value at 338.65 K [3]:

$$\gamma_1^\infty = 15.9$$

assuming that the value of the partial molar excess enthalpy of ethanol $\overline{h}_1^{E,\infty} = 19000$ J/mol [3] is constant in the temperature range covered.

Solution

Using Eq. (5.26), the following activity coefficients result:

$$\ln \gamma_1^\infty (353.15 \text{ K}) = \ln 15.9 + \frac{19000}{8.31433}\left(\frac{1}{353.15} - \frac{1}{338.65}\right)$$
$$= 2.7663 - 0.2771 = 2.4892$$
$$\gamma_1^\infty (353.15 \text{ K}) = 12.05$$

$$\ln \gamma_1^\infty (433.15 \text{K}) = \ln 15.9 + \frac{19000}{8.31433}\left(\frac{1}{433.15} - \frac{1}{338.65}\right)$$
$$= 2.7663 - 1.4722 = 1.2941$$
$$\gamma_1^\infty (433.15 \text{ K}) = 3.65$$

It can be seen that the activity coefficient at infinite dilution of ethanol in *n*-decane decreases by a factor greater than 4 when the temperature is increased from 338 to 433 K.

From these results it can be concluded that the temperature dependence cannot be neglected. While positive values of the partial molar excess enthalpies lead to a decrease of the activity coefficients with increasing temperature, negative values of the partial molar excess enthalpies lead to an increase of the activity coefficients with increasing temperature. The variation of the molar excess enthalpy with composition and temperature is often very complex. In the system ethanol–water around 70 °C even the sign changes with composition, as shown in Figure 5.18.

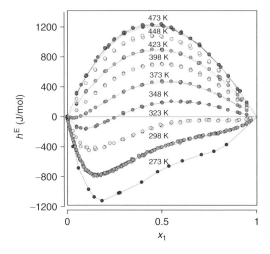

Figure 5.18 Selected excess enthalpy data at different temperatures for the system ethanol–water [3].

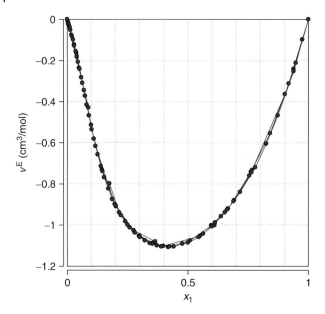

Figure 5.19 Excess volumes of the system ethanol (1)–water (2) at 20 °C [3].

Example 5.7

Estimate the difference between the sum of the volumes of the pure compounds and the volume of the resulting binary mixture of 0.4 mol ethanol and 0.6 mol water at 20 °C. Additionally please check the influence of a pressure difference of 2 bar on the activity coefficient qualitatively.

In the table below the densities at 20 °C and the molar masses are given [3]. (see also Example 4.1)

Compound	Density (g/cm³)	Molar mass (g/mol)
Ethanol	0.7893	46.069
Water	0.9982	18.015

The excess volumes for the system ethanol (1)–water (2) at 20 °C are shown in Figure 5.19.

Solution

To determine the volume of the pure compounds first the masses have to be calculated:

$$m_{\text{ethanol}} = 0.4 \cdot 46.069 = 18.43 \text{ g} \quad V_{\text{ethanol}} = \frac{18.43}{0.7893} = 23.35 \text{ cm}^3$$

$$m_{\text{H}_2\text{O}} = 0.6 \cdot 18.015 = 10.81 \text{ g} \quad V_{\text{H}_2\text{O}} = \frac{10.81}{0.9982} = 10.83 \text{ cm}^3$$

This means that the total volume of the pure compounds under ideal conditions is 34.18 cm^3.

For the calculation of the correct mixture volume the excess volume has to be known. From Figure 5.19 an excess volume of -1.10 cm^3/mol for a composition of $x_1 = 0.4$ can be read. Using this value the volume of the mixture can be calculated using Eq. (4.57):

$$v = \sum x_i v_i + v^E$$
$$v = 34.18 - 1.10 = 33.08 \text{ cm}^3/\text{mol}$$

This means that the volume of the mixture is approx. 3.2% lower than the volume of the pure compounds, this means starting from 100 cm^3 only 96.8 cm^3 remain.

For $x_1 = 0.4$ for both components a partial molar excess volume of -1.1 cm^3/mol (intersection at $x_1 = 0$ and $x_1 = 1$ of the slope at $x_1 = 0.4$) is obtained. Using this value the following change of the activity coefficients is obtained, when the pressure is increased by 2 bar:

$$\ln \frac{\gamma_i(P_2)}{\gamma_i(P_1)} = \frac{\bar{v}_i^E}{RT}(P_2 - P_1) = \frac{-0.0011 \cdot 2}{0.0831433 \cdot 293.15} = -9.026 \cdot 10^{-5}$$

$$\frac{\gamma_i(P_2)}{\gamma_i(P_1)} = 0.9999$$

Because of the negative sign of the partial molar excess volume the activity coefficient decrease with increasing pressure. But it can be seen that in contrast to the temperature influence caused by h^E, the pressure influence on the activity coefficients is negligible for typical pressure differences observed for VLE. But for large pressure differences the effect has to be taken into account. This is demonstrated in Section 5.8 for LLE.

The excess properties h^E and v^E do not only depend on temperature but also on pressure. This is shown in Figure 5.20 for the excess volumes of the system ethanol–water at 298 K. While for an equimolar mixture approximately a value of -1 cm^3/mol is observed at low pressures, the excess volume decreases to values smaller than -0.3 cm^3/mol at pressures above 2000 bar.

The temperature dependence of the excess Gibbs energy and the activity coefficients can be derived from a g^E–h^E-diagram (Figure 5.21). Depending on the sign of the excess properties g^E and h^E four quadrants are obtained.[5]

---

5) However, following Eq. 5.26 not the excess enthalpy but the partial molar excess enthalpy is the determining property to describe the temperature dependence of the activity coefficients. Depending on the curvature of h^E as a function of composition for positive (negative) values of h^E negative (positive) partial molar excess enthalpies can be obtained, for example, if an S-shaped curvature occurs as shown in Figure 5.18. Therefore, the following statements are only valid conditionally. But in most cases the sign of the partial molar and the molar excess enthalpy are identical (exception for S-shaped curves).

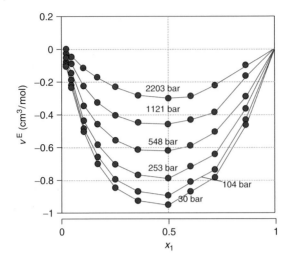

Figure 5.20 Excess volume of the system ethanol (1)–water (2) at 323 K as a function of pressure [3, 12].

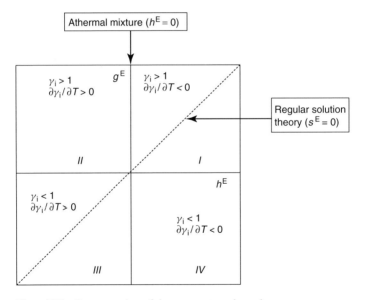

Figure 5.21 Representation of the temperature dependence of the activity coefficients in a g^E–h^E-diagram.

For positive deviations from Raoult's law ($g^E > 0$, $\gamma_i > 1$), depending on the sign of the excess enthalpy two cases can be distinguished. In the case of endothermic behavior ($h^E > 0$), the excess Gibbs energy and therewith, the values of the activity coefficient decrease with increasing temperature.

Most of the binary systems belong to quadrant I, this means they show positive deviations from Raoult's law and endothermic h^E behavior. This means that these systems become more ideal with increasing temperature. In the case of exothermal behavior (quadrant II), the activity coefficient will increase with temperature; this means stronger deviation from Raoult's law is obtained with increasing temperature. Only a few systems belong to quadrant II ($g^E > 0, h^E < 0$).[6] Just these systems, for example, alcohol–water-, alkyl amine–water systems, and so on, are of great technical importance. Systems with negative deviation from Raoult's law ($g^E < 0$ (quadrants III and IV)) are significantly more rare. In these systems the interaction energies between different components are stronger than those between the pure components, as for example, in the system acetone–chloroform. Because of the strong hydrogen bondings complexes are formed, which are less volatile. The systems which belong to quadrant III become more ideal, since for the excess Gibbs energy less negative values are obtained with increasing temperature. For systems in quadrant IV stronger negative deviation from Raoult's law are observed with increasing temperature. In Figure 5.21 also the lines for the so-called athermal mixture ($h^E = 0$, i.e., $g^E = -Ts^E$) and the regular solution ($s^E = 0$, i. e., $g^E = h^E$) are shown.

It would be desirable to apply analytical expressions for the activity coefficient, which are not only able to describe the concentration dependence, but also the temperature dependence correctly. Presently, there is no approach completely fulfilling this task. But the newer approaches, as for example, the Wilson [13], NRTL (nonrandom two liquid theory) [14], and UNIQUAC (universal quasi-chemical theory) equation [15] allow for an improved description of the real behavior of multicomponent systems from the information of the binary systems. These approaches are based on the concept of local composition, introduced by Wilson [13]. This concept assumes that the local composition is different from the overall composition because of the interacting forces. For this approach, different boundary cases can be distinguished:

- Because of the very similar interacting forces, the local composition is identical with the macroscopic composition (random mixture, almost ideal behavior $\gamma_i \approx 1$).
- Two liquid phases are formed, since molecule 2 has no tendency to locate near molecule 1 and vice versa (strong positive deviation from Raoult's law, $\gamma_i \gg 1$).
- The interacting forces between the different molecules are much larger than those between the same molecules, so that complexes are formed (negative deviation from Raoult's law, $\gamma_i < 1$).

The different equations are represented in detail in the literature. To derive a reliable g^E- resp. activity coefficient model using Eq. (4.85) the different excess properties (h^E, s^E, v^E) should be taken into account. Flory [16] and Huggins [17, 18] independently derived an expression for g^E starting from the excess entropy

---

6) Sometimes only in a limited concentration range (see Figure 5.18).

of athermal polymer solutions that means $h^E = 0$ using the lattice theory. In these mixtures with molecules very different in size, volume fractions ϕ_i instead of mole fractions are used. In binary systems the volume fraction can be calculated by the following expressions using the molar volumes v_i:

$$\phi_1 = \frac{x_1 v_1}{x_1 v_1 + x_2 v_2} \qquad \phi_2 = \frac{x_2 v_2}{x_1 v_1 + x_2 v_2} \qquad (5.28)$$

Using the expression for the excess entropy:

$$s^E = -R\left(x_1 \ln \frac{\phi_1}{x_1} + x_2 \ln \frac{\phi_2}{x_2}\right) \qquad (5.29)$$

an expression for the excess Gibbs energy can be derived,

$$g^E = -Ts^E = RT\left(x_1 \ln \frac{\phi_1}{x_1} + x_2 \ln \frac{\phi_2}{x_2}\right) \qquad (5.30)$$

which can be used to derive an expression for the activity coefficients γ_i with the help of Eq. (4.85) for an athermal solution ($h^E = 0$, see Figure 5.21):

$$\ln \gamma_i = \ln \frac{\phi_i}{x_i} + 1 - \frac{\phi_i}{x_i} \qquad (5.31)$$

With the help of this expression it can be shown that strong negative deviations from Raoult's law result for systems with compounds very different in size. From Eq. (5.31) it can easily be understood why the removal of the remaining monomers from polymer solutions is much more difficult than expected.

Example 5.8

Calculate the activity coefficient of the monomer in a polymer using the athermal Flory–Huggins equation. For the calculation the following volumes should be used:

$$v_1 = 70 \text{ cm}^3/\text{mol} \quad v_2 = 70000 \text{ cm}^3/\text{mol}$$

Solution

The calculation is performed for a mole fraction of $x_1 = 0.2$. For this composition the following volume fractions are obtained:

$$\phi_1 = \frac{0.2 \cdot 70}{0.2 \cdot 70 + 0.8 \cdot 70000} = 2.499 \cdot 10^{-4} \qquad \phi_2 = 0.99975$$

Using these values the activity coefficient γ_1 can be calculated directly:

$$\ln \gamma_1 = \ln \frac{2.499 \cdot 10^{-4}}{0.2} + 1 - \frac{2.499 \cdot 10^{-4}}{0.2} = -5.686$$

$$\gamma_1 = 3.39 \cdot 10^{-3}$$

For the whole composition range the activity coefficient of the monomer (1) as a function of the weight fraction of the polymer is shown in Figure 5.22. It can be

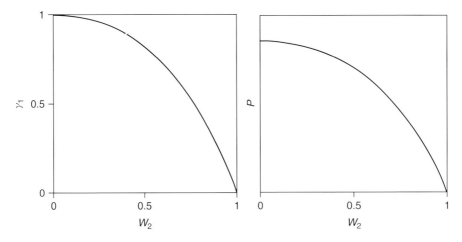

Figure 5.22 Calculated activity coefficients of the monomer and resulting system pressures as a function of the weight fraction of the polymer for a polymer solution using the Flory–Huggins equation.

seen that the volatility (activity coefficient) of the monomer drastically decreases with decreasing composition of the monomer.

To be able to account for contributions caused by the excess enthalpy h^E, a simple one parameter term was added to the Flory–Huggins equation:

$$\frac{g^E}{RT} = x_1 \ln \frac{\phi_1}{x_1} + x_2 \ln \frac{\phi_2}{x_2} + \phi_1 \phi_2 \chi \qquad (5.32)$$

Starting from Eq. (5.32) the following expression is obtained for the activity coefficient of the monomer:

$$\ln \gamma_1 = \ln \frac{\phi_1}{x_1} + 1 - \frac{\phi_1}{x_1} + \chi \phi_2^2 \qquad (5.33)$$

Wilson started from a similar equation as Flory and Huggins for the derivation of his equation for arbitrary mixtures apart from polymers [13]. However, instead of the true volume fractions Wilson used the so-called local volume fraction ξ_i in the expression for the excess Gibbs energy:

$$\frac{g^E}{RT} = \sum_i x_i \ln \frac{\xi_i}{x_i} \qquad (5.34)$$

Local mole fractions were introduced by Wilson to define the local volume fraction, where the deviation from the macroscopic concentration is taken into account with the help of interaction energies between the different compounds using Boltzmann factors. With the introduction of the auxiliary quantities Λ_{ij} the equations for g^E and γ_i can be derived.[7] The great advantage of the Wilson equation is that only binary

---

7) The detailed derivation of the expression for the activity coefficients of the Wilson equation starting from Eq. (5.34) is given in Appendix C, E1.

Example 5.9

Compare the experimental vapor phase mole fractions published by Hiaki et al. [19] for the system acetone (1)–chloroform (2)–methanol (3) at 1 atm with the calculated ones using the binary Wilson parameters $\Delta \lambda_{ij}$ listed in Table 5.5 assuming ideal vapor phase behavior.

For the calculation, the following molar volumes:

$$v_1 = 74.04 \text{ cm}^3 \text{ mol}^{-1}; \ v_2 = 80.67 \text{ cm}^3 \text{ mol}^{-1}; \ v_3 = 40.73 \text{ cm}^3 \text{ mol}^{-1}$$

and constants for the Antoine equation ($\log P_i^s (\text{mm Hg}) = A - B/(\vartheta(^\circ C) + C)$) should be used to calculate the vapor pressures.

Component	A	B	C
Acetone	7.1327	1219.97	230.653
Chloroform	6.95465	1170.97	226.232
Methanol	8.08097	1582.27	239.7

Solution

The calculation should be performed for the following composition at $T = 331.42$ K:

$$x_1 = 0.229, x_2 = 0.175, x_3 = 0.596, y_1 = 0.250, y_2 = 0.211, y_3 = 0.539.$$

Using the Wilson interaction parameters listed in Table 5.5 at 331.42 K the following values for $\Delta \lambda_{ij}$ are obtained:

$$\Delta \lambda_{12} = 375.2835 - 3.78434 \cdot 331.42 + 0.0079107 \cdot 331.42^2 = -10.02 \text{ K}$$

Table 5.5 Wilson interaction parameters (K) for the ternary system acetone (1)–chloroform (2)–methanol (3) (definition of a_{ij}, b_{ij}, c_{ij}, see Eq. (5.35)).

i	j	a_{ij} (K)	a_{ji} (K)	b_{ij}	b_{ji}	c_{ij} (K^{-1})	c_{ji} (K^{-1})
1	2	375.2835	−1722.58	−3.78434	6.405502	7.91073E−03	−7.47788E−03
1	3	31.1208	747.217	−0.67704	−0.256645	8.68371E−04	−1.24796E−03
2	3	−1140.79	3596.17	2.59359	−6.2234	3.10E−05	3.00E−05

In the same way one obtains

$$\Delta\lambda_{21} = -421.03 \text{ K}, \Delta\lambda_{13} = -97.88 \text{ K}, \Delta\lambda_{31} = 525.08 \text{ K},$$
$$\Delta\lambda_{23} = -277.82 \text{ K}, \Delta\lambda_{32} = 1536.91 \text{ K}.$$

In the next step the Wilson parameters Λ_{ij} used in Table 5.6 can be determined:

$$\Lambda_{ij} = \frac{v_j}{v_i} \cdot \exp\left[-\frac{\Delta\lambda_{ij}}{T}\right]$$

$$\Lambda_{12} = \frac{80.67}{74.04} \cdot \exp\left[\frac{10.02}{331.42}\right] = 1.1230$$

In the same way the other Wilson parameters are obtained:

$$\Lambda_{21} = 3.2695$$
$$\Lambda_{13} = 0.7391, \Lambda_{31} = 0.3728$$
$$\Lambda_{23} = 1.1675, \Lambda_{32} = 0.01918$$

With the help of these parameters Λ_{ij} the required activity coefficients can be calculated. For γ_1 one obtains

$$\ln \gamma_1 = -\ln(x_1\Lambda_{11} + x_2\Lambda_{12} + x_3\Lambda_{13}) + 1 - \frac{x_1\Lambda_{11}}{x_1\Lambda_{11} + x_2\Lambda_{12} + x_3\Lambda_{13}}$$
$$- \frac{x_2\Lambda_{21}}{x_1\Lambda_{21} + x_2\Lambda_{22} + x_3\Lambda_{23}} - \frac{x_3\Lambda_{31}}{x_1\Lambda_{31} + x_2\Lambda_{32} + x_3\Lambda_{33}}$$

For the selected composition of $x_1 = 0.229$, $x_2 = 0.175$, and $x_3 = 0.596$ the following activity coefficient is obtained:

$$\ln \gamma_1 = -\ln(0.229 + 0.175 \cdot 1.123 + 0.596 \cdot 0.7391)$$
$$+ 1 - \frac{0.229}{0.229 + 0.175 \cdot 1.123 + 0.596 \cdot 0.7391}$$
$$- \frac{0.175 \cdot 3.2695}{0.229 \cdot 3.2695 + 0.175 + 0.596 \cdot 1.1675}$$
$$- \frac{0.596 \cdot 0.3728}{0.229 \cdot 0.3728 + 0.175 \cdot 0.01918 + 0.596} = 0.2016$$
$$\gamma_1 = 1.223$$

Similarly the other activity coefficients are calculated as

$$\gamma_2 = 1.101, \gamma_3 = 1.205$$

For the vapor pressures one obtains at the measured temperature of 331.42 K:

$$P_i^s = 10^{A_i - \frac{B_i}{\vartheta + C_i}}$$

$$P_1^s = 10^{7.1327 - \frac{1219.97}{58.27 + 230.653}} = 813.25 \text{ mm Hg}, P_2^s = 689.91 \text{ mm Hg},$$
$$P_3^s = 589.94 \text{ mm Hg}$$

Then the partial pressures and the total pressure can be calculated:

$$p_i = x_i \cdot \gamma_i \cdot P_i^s$$

$$P = 0.229 \cdot 1.223 \cdot 813.25 + 0.175 \cdot 1.101 \cdot 689.91 + 0.596 \cdot 1.205 \cdot 589.94$$
$$P = 227.76 + 132.93 + 423.68 = 784.37 \text{ mm Hg}$$

The vapor phase composition is obtained from the ratio of the partial and the total pressure:

$$y_1 = \frac{p_1}{P} = \frac{227.76}{784.37} = 0.2904, \quad y_2 = 0.1694, \quad y_3 = 0.5402$$

Since the calculated pressure is greater than the constant experimental pressure of 760 mm Hg the calculated temperature has to be decreased in the next step until the experimental and the calculated pressure are identical. For the liquid composition considered this is fulfilled at a temperature of 330.60 K, where nearly the same values are obtained for the vapor phase mole fraction.

In the same way the vapor phase mole fractions can be calculated for all other data published by Hiaki et al. [19]. The experimental and calculated values are shown in Figure 5.23.

It can be seen that nearly perfect agreement between the experimental and calculated vapor phase mole fractions is obtained. Furthermore, the complex topology and the ternary saddle point are predicted correctly, as shown in Figure 5.24.

This means that the Wilson equation based on the local composition concept allows the prediction of the VLE behavior of multicomponent systems from binary data.

Later, further g^E-models based on the local composition concept were published, such as the NRTL [14] and the UNIQUAC [15] equation, which also allow the prediction of the activity coefficients of multicomponent systems using only binary parameters. In the case of the UNIQUAC equation the activity coefficient is calculated by a combinatorial and a residual part. While the temperature-independent combinatorial part takes into account the size and the shape of the molecule, the interactions between the different compounds are considered by the residual part. In contrast to the Wilson equation the NRTL und UNIQUAC equation can also be used for the calculation of LLE.

The analytical expressions of the activity coefficients for binary and multicomponent systems for the three g^E-models are given in Table 5.6. While for the Wilson and the UNIQUAC model two binary interaction parameters ($\Delta\lambda_{12}, \Delta\lambda_{21}$ resp. $\Delta u_{12}, \Delta u_{21}$) are used, in the case of the NRTL equation besides the two binary interaction parameters ($\Delta g_{12}, \Delta g_{21}$) additionally a nonrandomness factor α_{12} is required for a binary system, which is often not fitted but set to a defined value. For the Wilson equation additionally molar volumes and for the UNIQUAC equation relative van der Waals volumes and surface areas are required. These values are easily available.

Table 5.6 Important expressions for the excess Gibbs energy and the derived activity coefficients.

Model	Parameters	Expressions for the activity coefficients
Wilson [13]		$\dfrac{g^E}{RT} = \sum_i x_i \ln \dfrac{\xi_i}{x_i}$ or $\dfrac{g^E}{RT} = -\sum_i x_i \ln \sum_j x_j \Lambda_{ij}$
	$\Delta \lambda_{12}{}^a$	$\ln \gamma_1 = -\ln(x_1 + \Lambda_{12} x_2) + x_2 \left(\dfrac{\Lambda_{12}}{x_1 + \Lambda_{12} x_2} - \dfrac{\Lambda_{21}}{\Lambda_{21} x_1 + x_2} \right)$
	$\Delta \lambda_{21}$	$\ln \gamma_2 = -\ln(x_2 + \Lambda_{21} x_1) - x_1 \left(\dfrac{\Lambda_{12}}{x_1 + \Lambda_{12} x_2} - \dfrac{\Lambda_{21}}{\Lambda_{21} x_1 + x_2} \right)$
	$\Delta \lambda_{ij}$	$\ln \gamma_i = -\ln \left(\sum_j x_j \Lambda_{ij} \right) + 1 - \sum_k \dfrac{x_k \Lambda_{ki}}{\sum_j x_j \Lambda_{kj}}$
		$\ln \gamma_1^\infty = 1 - \ln \Lambda_{12} - \Lambda_{21}$
		$\ln \gamma_2^\infty = 1 - \ln \Lambda_{21} - \Lambda_{12}$
NRTL [14]		$\dfrac{g^E}{RT} = \sum_i x_i \dfrac{\sum_j \tau_{ji} G_{ji} x_j}{\sum_j G_{ji} x_j}$
	$\Delta g_{12}{}^b$	$\ln \gamma_1 = x_2^2 \left[\tau_{21} \left(\dfrac{G_{21}}{x_1 + x_2 G_{21}} \right)^2 + \dfrac{\tau_{12} G_{12}}{(x_2 + x_1 G_{12})^2} \right]$
	Δg_{21}	
	α_{12}	$\ln \gamma_2 = x_1^2 \left[\tau_{12} \left(\dfrac{G_{12}}{x_2 + x_1 G_{12}} \right)^2 + \dfrac{\tau_{21} G_{21}}{(x_1 + x_2 G_{21})^2} \right]$
	$\Delta g_{ij}, \alpha_{ij}$	$\ln \gamma_i = \dfrac{\sum_j \tau_{ji} G_{ji} x_j}{\sum_k G_{ki} x_k} + \sum_j \dfrac{x_j G_{ij}}{\sum_k G_{kj} x_k} \left(\tau_{ij} - \dfrac{\sum_n x_n \tau_{nj} G_{nj}}{\sum_k G_{kj} x_k} \right)$
UNIQUAC [15]		$g^E = g^{E,C} + g^{E,R}$
		$\dfrac{g^{E,C}}{RT} = \sum_i x_i \ln \dfrac{\phi_i}{x_i} + \dfrac{z}{2} \sum_i q_i x_i \ln \dfrac{\theta_i}{\phi_i}$
		$\dfrac{g^{E,R}}{RT} = -\sum_i q_i x_i \ln \left(\sum_j \theta_j \tau_{ji} \right)$
	$\Delta u_{12}{}^c$	$\ln \gamma_1 = \ln \gamma_1^C + \ln \gamma_1^R$
	Δu_{21}	$\ln \gamma_1^C = 1 - V_1 + \ln V_1 - 5 q_1 \left(1 - \dfrac{V_1}{F_1} + \ln \dfrac{V_1}{F_1} \right)$
		$\ln \gamma_1^R = -q_1 \ln \dfrac{q_1 x_1 + q_2 x_2 \tau_{21}}{q_1 x_1 + q_2 x_2}$
		$\quad + q_1 q_2 x_2 \left[\dfrac{\tau_{21}}{q_1 x_1 + q_2 x_2 \tau_{21}} - \dfrac{\tau_{12}}{q_1 x_1 \tau_{12} + q_2 x_2} \right]$
		$\ln \gamma_2 = \ln \gamma_2^C + \ln \gamma_2^R$
		$\ln \gamma_2^C = 1 - V_2 + \ln V_2 - 5 q_2 \left(1 - \dfrac{V_2}{F_2} + \ln \dfrac{V_2}{F_2} \right)$
		$\ln \gamma_2^R = -q_2 \ln \dfrac{q_1 x_1 \tau_{12} + q_2 x_2}{q_1 x_1 + q_2 x_2}$
		$\quad + q_1 q_2 x_1 \left[\dfrac{\tau_{12}}{q_1 x_1 \tau_{12} + q_2 x_2} - \dfrac{\tau_{21}}{q_1 x_1 + q_2 x_2 \tau_{21}} \right] $

(continued overleaf)

Table 5.6 (continued)

Model	Parameters	Expressions for the activity coefficients
	Δu_{ij}	$\ln \gamma_i = \ln \gamma_i^C + \ln \gamma_i^R$
		$\ln \gamma_i^C = 1 - V_i + \ln V_i - 5q_i \left(1 - \frac{V_i}{F_i} + \ln \frac{V_i}{F_i}\right)$
		$\ln \gamma_i^R = q_i \left(1 - \ln \frac{\sum_j q_j x_j \tau_{ji}}{\sum_j q_j x_j} - \sum_j \frac{q_j x_j \tau_{ij}}{\sum_k q_k x_k \tau_{kj}}\right)$

$^a \Lambda_{ij} = \frac{V_j}{V_i} \exp(-\Delta\lambda_{ij}/T)$, $\Lambda_{ii} = 1$;
V_i molar volume of component i[8];
$\Delta\lambda_{ij}$ interaction parameter between component i and j (K).

$^b \tau_{ij} = \Delta g_{ij}/T$, $\tau_{ii} = 0$; $G_{ij} = \exp(-\alpha_{ij}\tau_{ij})$, $G_{ii} = 1$;
Δg_{ij} interaction parameter between component i and j (K);
α_{ij} nonrandomness parameter: $\alpha_{ij} = \alpha_{ji}$.

$^c \gamma_i^C$ combinatorial part of the activity coefficient of component i;
γ_i^R residual part of the activity coefficient of component i;
$\tau_{ij} = \exp(-\Delta u_{ij}/T)$, $\tau_{ii} = 1$.
Δu_{ij} interaction parameter between component i and j (K)
r_i relative van der Waals volume of component i
q_i relative van der Waals surface area of component i
$V_i = \frac{r_i}{\sum_j r_j x_j}$ volume fraction/mole fraction of component i
$F_i = \frac{q_i}{\sum_j q_j x_j}$ surface area fraction/mole fraction of component i

---

While for the interaction parameters $\Delta\lambda_{ij}, \Delta g_{ij}, \Delta u_{ij}$ the unit K can be used, often for the published interaction parameters, for example, given in [6] the unit of a molar energy can be found. That is the case when in the denominator of the exponential term RT instead of T is used. The unit then depends on the choice of the unit of the general gas constant R (J/mol K, cal/mol K, etc.). When a large temperature range is covered, temperature-dependent parameters have to be used to describe the temperature dependence of the activity coefficients with the required accuracy, this means, following the Gibbs–Helmholtz relation the excess enthalpies resp. partial molar excess enthalpies in the temperature range covered. In this textbook the following temperature dependence of the binary interaction parameters is used:

$$\Delta\lambda_{ij}(T) = \Delta g_{ij}(T) = \Delta u_{ij}(T) = a_{ij} + b_{ij}T + c_{ij}T^2 \tag{5.35}$$

The application of the g^E-models is also explained in Appendix C, E2.

---

8) In practice usually constant, this means temperature-independent molar volumes V_i are used.

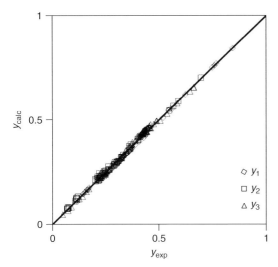

Figure 5.23 Experimental [3, 19] and calculated vapor phase mole fractions for the system acetone (1)–chloroform (2)–methanol (3) at atmospheric pressure.

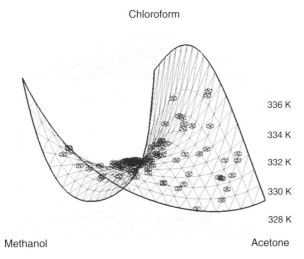

Figure 5.24 Experimental Tx-data [3, 19] and calculated Tx behavior of the ternary system acetone–chloroform–methanol at 1 atm using binary Wilson parameters.

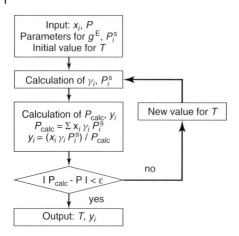

Figure 5.25 Flow diagram for the calculation of isobaric VLE data assuming ideal behavior in the vapor phase.

5.4
Fitting of g^E-Model Parameters

The quality of the design of a distillation column by solving the MESH equations[9] mainly depends on the accuracy of the K-factors (separation factors) [1]. Using one of the g^E-models given in Table 5.6 these values can be calculated for the system to be separated, if the binary parameters are available. However, for the proper design binary parameters have to be used which describe the K-factors resp. separation factors α_{ij} of the system to be separated over the entire composition and temperature range considered reliably.

The separation factors mainly depend on composition and temperature. The correct composition dependence is described with the help of activity coefficients. Following the Clausius–Clapeyron equation presented in Section 2.4.4 the temperature dependence is mainly influenced by the slope of the vapor pressure curves (enthalpy of vaporization) of the components involved. But also the activity coefficients are temperature-dependent following the Gibbs–Helmholtz equation (Eq. (5.26)). This means that besides a correct description of the composition dependence of the activity coefficients also an accurate description of their temperature dependence is required. For distillation processes at moderate pressures, the pressure effect on the activity coefficients (see Example 5.7) can be neglected. To take into account the real vapor phase behavior, equations of state, for example, the virial equation, cubic equations of state, such as the Redlich–Kwong, Soave–Redlich–Kwong (SRK), Peng–Robinson (PR), the association model, and so on, can be applied.

Assuming ideal vapor phase behavior in phase equilibrium calculations, besides carefully chosen binary g^E-model parameters only reliable vapor pressures are needed. The simple calculation procedure for the isobaric case is shown in Figure 5.25. In the isobaric case initial values for the temperature are required. During the calculation the temperature has to be changed in a way that the difference

---

9) MESH equations: these are the resulting balance equations for the ideal stage concept for the material balance (M), equilibrium conditions (E), summation conditions (S), and the heat balance (H).

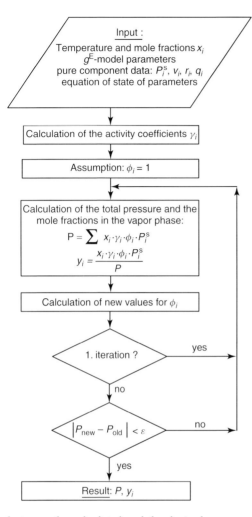

Figure 5.26 Flow diagram for the calculation of isothermal VLE taking into account the nonideal behavior of the vapor phase.

between the calculated and the desired pressure is smaller than a chosen value ε. The calculation for the isothermal case is still simpler, since the temperature does not have to be guessed and then adjusted by iteration. The calculation procedure shown in Figure 5.25 can also be applied for multicomponent systems.

If the real vapor phase behavior and the Poynting factor have to be taken into account, the procedure is a little more complicated. The procedure is shown in Figure 5.26. As input, information about the molar volumes and the real vapor phase behavior is required additionally, for example, the parameters of the equation of state chosen, the parameters for the association constants, and so on. With the help of this information, the Poynting factors and the fugacity coefficients for the mixture and the pure compounds are calculated.

A prerequisite for the correct description of the real behavior of multicomponent systems is a reliable description of the binary subsystems with the help of the fitted binary g^E-model parameters.

For fitting the binary interaction parameters nonlinear regression methods are applied, which allow adjusting the parameters in such a way that a minimum deviation of an arbitrary chosen objective function F is obtained. For this job, for example, the Simplex–Nelder–Mead method [21] can be applied successfully. The Simplex–Nelder–Mead method in contrast to many other methods [22] is a simple search routine, which does not need the first and the second derivate of the objective function with respect to the different variables. This has the great advantage that computational problems, such as "underflow" or "overflow" with the arbitrarily chosen initial parameters can be avoided.

As objective function for fitting the required g^E-model parameters different types of objective functions and experimental or derived properties X, for example, vapor phase mole fraction, pressure, temperature, K-factor K_i, separation factor α_{12}, and so on, can be selected, where either the relative or the absolute deviation of the experimental and correlated values (pressure, temperature, vapor phase composition, etc.) can be minimized,

$$F = \frac{1}{n} \sum \sum (X_{\text{calc},i,j} - X_{\exp,i,j})^2 \stackrel{!}{=} \text{Min}$$

$$F = \frac{1}{n} \sum \sum |X_{\text{calc},i,j} - X_{\exp,i,j}| \stackrel{!}{=} \text{Min}$$

$$F = \frac{1}{n} \sum \sum \left[\frac{X_{\text{calc},i,j} - X_{\exp,i,j}}{X_{\exp,i,j}}\right]^2 \stackrel{!}{=} \text{Min}$$

$$F = \frac{1}{n} \sum \sum \left|\frac{X_{\text{calc},i,j} - X_{\exp,i,j}}{X_{\exp,i,j}}\right| \stackrel{!}{=} \text{Min}$$

In the case of complete data, this means VLE data, where P, T, x_i, y_i is given, also the deviation between the experimental and predicted activity coefficients or excess Gibbs energies can be used to fit the required binary parameters. Furthermore the parameters can be determined by a simultaneous fit to different properties to cover properly the composition and temperature dependence of the activity coefficients. For example, the deviation of the derived activity coefficients can be minimized together with the deviations of the activity coefficients at infinite dilution, excess enthalpies, and so on. Accurate activity coefficients at infinite dilution measured with sophisticated experimental techniques are of special importance, since they deliver the only reliable information about the real behavior in the dilute range [23],[10] for example, at the top or the bottom of a distillation column. Excess enthalpies measured using flow calorimetry are important too, since they provide the most reliable information about the temperature dependence of the activity

10) With the required care it is quite simple to measure reliable activity coefficients at infinite dilution of a low boiling substance in a high boiling compound, for example, with the help of the dilutor technique, gas–liquid chromatography, ebulliometry, Rayleigh distillation, and so on. It is much more difficult to measure these values for high boiling components in low boiling compounds, for example, water in ethylene oxide, NMP in benzene, etc. But these values are of special importance for the proper design of distillation columns. In the case of positive deviation from Raoult's law the greatest separation effort is required for the removal of the last traces of the high boiling compounds at the top of the column.

coefficients via the Gibbs–Helmholtz equation (see Eq. (5.26)), and thus for the separation factors and K-factors.

In the past in different papers e.g. [70] the maximum likelihood method was recommended. In this method the experimental errors of all measured quantities are taken into account in the objective function. But later it was found out that this procedure did not improve the results for ternary, quaternary, and higher systems using the fitted binary parameters.

The impact of inaccurate g^E-model parameters can be very serious. The parameters have a major influence on the investment and operating costs (number of stages, reflux ratio). The influence of the g^E-model parameters on the results is especially large if the separation factor is close to unity. Poor parameters can either lead to the calculation of nonexisting azeotropes in zeotropic systems (see Section 11.1) or the calculation of zeotropic behavior in azeotropic systems. Poor parameters can also lead to a miscibility gap which does not exist.[11] In the case of positive deviation from Raoult's law a separation problem often occurs at the top of the column, where the high boiler has to be removed, since at the top of a distillation column the most unfortunate separation factors are obtained.

Starting from Eq. (5.18), the following separation factors at the top and the bottom of a distillation column (low boiler: component 1) are obtained:

top of the column ($x_1 \rightarrow 1$):

$$\alpha_{12}^\infty = \frac{P_1^s}{\gamma_2^\infty P_2^s}$$

bottom of the column ($x_2 \rightarrow 1$):

$$\alpha_{12}^\infty = \frac{\gamma_1^\infty P_1^s}{P_2^s}$$

From these equations it can easily be seen that for positive deviation from Raoult's law ($\gamma_i > 1$) the smallest separation factors and therewith the greatest separation problems occur at the top of the column. To determine the separation factor at the top of the column, one divides by a number larger than unity (γ_2^∞), while in the bottom of the column one multiplies with a number larger than unity (γ_1^∞). While, for example, for the system acetone–water separation factors a little above unity are obtained at the top of the column at atmospheric pressure, separation factors greater than 40 are observed at the bottom of the column (see Chapter 11). In the

---

11) Process simulators often contain extensive data banks with pure component and mixture parameters, for example, default g^E-model parameters. This allows for generating the required input very fast. But the user should use these data and parameters with care. Even the simulator companies mention that these default values should not directly be used for process simulation. The user should ask the company expert for phase equilibrium thermodynamics to check the pure component data and mixture parameters carefully. In Figures 5.31 and 5.34 it is shown what can happen when the default values provided by the process simulator companies are used. An additional example for bad default values is given in Appendix F.

In Section 11.1 it is shown how the thermophysical properties should be checked prior to process simulation.

Table 5.7 Type of VLE data published.

Type of VLE data	Measured values				Percentage of the published VLE data
	x_i	y_i	P	T	
Isothermal complete	✓	✓	✓	constant	18.33
Isobaric complete	✓	✓	constant	✓	28.37
Isothermal Px data	✓	–	✓	constant	29.66
Isobaric Tx data	✓	–	constant	✓	9.11
Isothermal xy data	✓	✓	–	constant	3.12
Isobaric xy data	✓	✓	constant	–	2.03
Isothermal yP data	–	✓	✓	constant	0.54
Isobaric yT data	–	✓	constant	✓	0.16
Isoplethic PT data	constant	–	✓	✓	6.84
Complete data	✓	✓	✓	✓	1.85

case of negative deviations from Raoult's law the separation problem usually occurs at the bottom of the column.

Published VLE data are often of questionable quality. For fitting reliable g^E-model parameters accurate experimental VLE data should be used. An overview about the different types of VLE data published together with the proportion of such data from all published VLE data is given in Table 5.7.

In most cases the measurements are performed at isothermal or isobaric conditions. Occasionally measurements are also performed at constant composition. Sometimes none of the properties is kept constant. In less than 50% of the cases all values (x_i, y_i, T, P) are measured. The reason is that any three of the four values (x_i, y_i, T, P) are sufficient to derive the fourth quantity. Because of the greater experimental effort required, seldom dew-point data (T, P, y_i) are measured. But these data are of special importance to determine reliable separation factors for high boiling compounds (e.g., water) in low boiling compounds (ethylene oxide) at the top of the column, which at the end mainly determine the number of stages of a distillation column (see Chapter 11).

As can be seen from Table 5.7, the measurement of complete isobaric data is very popular. The reason is that a great number of chemical engineers prefer isobaric data, since distillation columns run at nearly isobaric conditions. But the measurement of isobaric data shows several disadvantages compared to isothermal data. This was already discussed in detail by Van Ness [24]. For example, the temperature dependence of the vapor pressure P_i^s has to be taken into account. At the same time the temperature and composition dependence of the g^E-model has to be regarded. Therefore Van Ness [24] comes to the following conclusion:

In the early unsophisticated days of chemical engineering VLE data were taken at constant pressure for direct application in the design of distillation columns, which were treated as though they operated at uniform pressure. There is no longer excuse for taking isobaric data, but regrettably the practice persists. Rigorous thermodynamic treatment of isobaric data presents problems that do not arise with isothermal data. Their origin is the need to take into account not only the composition dependence of the excess Gibbs energy but also its temperature dependence.

Since the measurement of temperature and pressure is more accurate than concentration measurements, Van Ness recommended the measurement of Px-data at isothermal conditions. Indeed, today mainly isothermal Px-data are measured. In the cell of the static equipments the precise liquid composition is usually achieved by injection of the degassed liquids with the help of precise piston pumps. The change of the feed composition by evaporation can easily be taken into account, when the volume of the cell and the pressure is known. Depending on the vapor volume the change of the feed composition is smaller than 0.1 mol% at moderate pressures. By this method a much more precise determination of the liquid composition is achieved than by analytical measurements. The measurement of the pressure and the temperature can be realized very precisely.

5.4.1
Check of VLE Data for Thermodynamic Consistency

Not in all cases, the quality of the published data is sufficient. The quality of complete data (P, T, x_i, y_i) can be checked with the help of thermodynamic consistency tests. A large number of consistency tests have been developed. Most often the so-called area test is applied. The derivation of the required equations for the area test is started from the following equation (see Section 4.3):

$$\frac{dG^E}{RT} = \left(\frac{\partial G^E/RT}{\partial P}\right)_{T,n_i} dP + \left(\frac{\partial G^E/RT}{\partial T}\right)_{P,n_i} dT + \sum \left(\frac{\partial G^E/RT}{\partial n_i}\right)_{T,P,n_j \neq n_i} dn_i$$

(5.36)

By substitution

$$\left(\frac{\partial G^E}{\partial P}\right)_{T,n_i} = V^E$$

(5.37)

$$\left(\frac{\partial G^E/T}{\partial T}\right)_{P,n_i} = -\frac{H^E}{T^2}$$

(5.38)

$$\left(\frac{\partial G^E/RT}{\partial n_i}\right)_{T,P,n_j \neq n_i} = \ln \gamma_i$$

(5.39)

and applying of molar properties, the following relation is obtained for a binary system ($dx_1 = -dx_2$).

$$\frac{dg^E}{RT} = \frac{v^E}{RT}dP - \frac{h^E}{RT^2}dT + \ln\frac{\gamma_1}{\gamma_2}dx_1 \tag{5.40}$$

After integration from $x_1 = 0$ to $x_1 = 1$ an expression is obtained, which can be applied for the graphical examination of complete VLE data[12] for thermodynamic consistency.

$$\int_{x_1=0}^{x_1=1} \frac{dg^E}{RT} = \int_{x_1=0}^{x_1=1} \ln\frac{\gamma_1}{\gamma_2}dx_1 - \int_{x_1=0}^{x_1=1} \frac{h^E}{RT^2}dT + \int_{x_1=0}^{x_1=1} \frac{v^E}{RT}dP = 0 \tag{5.41}$$

In the case of isothermal or isobaric data one term in the equation above can be cancelled. Since the pressure dependence can usually be neglected, in the case of isothermal VLE data the following simple relation can be used for checking the thermodynamic consistency of VLE data:

$$\int_{x_1=0}^{x_1=1} \ln\frac{\gamma_1}{\gamma_2}dx_1 = 0 \tag{5.42}$$

The consistency test (Redlich–Kister test) is performed by plotting the logarithmic value of the ratio of the activity coefficients as a function of the mole fraction x_1. If the VLE data are thermodynamically consistent the area above and below the x-axis should be equal.

In the case of isobaric data the excess enthalpy part has to be taken into account. This can be done if the excess enthalpies for the system investigated are known. Since the excess enthalpies are usually not known, in the area test the contribution is taken into account empirically using the quantity J as suggested by Redlich and Kister [25]:

$$J = 150\frac{|\Delta T_{max}|}{T_{min}}(\%) \tag{5.43}$$

The value of J strongly depends on the temperature difference ΔT_{max}. In the case of zeotropic systems this is the difference of the boiling points.

In the DECHEMA Chemistry Data Series [6] a deviation of $D < 10\%$ is allowed to pass the thermodynamic consistency test successfully,

$$D = \frac{|A - B|}{A + B}100(\%) \tag{5.44}$$

where A is the area above the x-axis and B is the area below the x-axis.

12) As well isothermal VLE data, where only the liquid and vapor phase mole fractions, this means K-factors are measured (e.g., by headspace gas-chromatography) can be checked for thermodynamic consistency with the help of the area test, since the system pressure cancels out when the ratio of the activity coefficients is calculated.

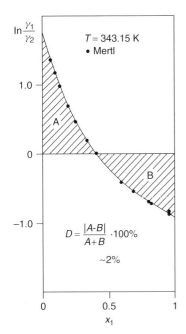

Figure 5.27 Check of isothermal complete VLE data of the system ethanol (1)-water (2) [8] for thermodynamic consistency with the help of the area test.

While in the isothermal case a deviation of $D < 10\%$ is accepted, in the isobaric case a larger area deviation is allowed to take into account the contribution of the excess enthalpy by Eq. (5.43):

$$D - J < 10\%$$

Example 5.10

Check the isothermal VLE data of the system ethanol (1)–water (2) measured by Mertl [8] at 70 °C (see Table 5.3) for thermodynamic consistency using the area test.

Solution

For the judgment of the quality of the VLE data the logarithmic values of the ratio of the activity coefficients $\ln \gamma_1/\gamma_2$ have to be plotted against the mole fraction of ethanol. The required activity coefficients are given in Table 5.3. For example, for a mole fraction of $x_1 = 0.252$ the following ratio is obtained:

$$\ln \frac{\gamma_1}{\gamma_2} = \ln \frac{1.890}{1.202} = 0.453$$

For the whole composition range the values are shown in Figure 5.27. It can be seen that the areas above and below the x-axis are nearly identical ($\approx 2\%$ deviation). This means that the data published by Mertl [8] can be considered as thermodynamically consistent.

Another option to check complete VLE data for thermodynamic consistency was developed by Van Ness et al. [26] resp. Fredenslund et al. [27]. In this consistency

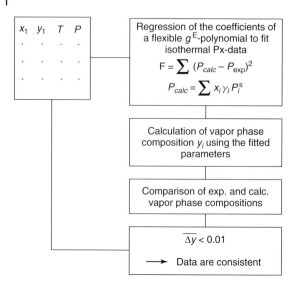

Figure 5.28 Flow diagram of the consistency test of Van Ness et al. [26] and Fredenslund et al. [27].

test only a part of the redundant phase equilibrium data, usually the TPx-data, are used to fit the g^E-model parameters of a flexible g^E-model, such as the Redlich–Kister expansion to minimize, for example, the deviation in pressure. In the next step the fitted parameters are used to calculate the data, which were not used for fitting the parameters, this means the corresponding vapor phase compositions. When a mean deviation between the experimental and calculated vapor phase mole fraction of <0.01 is obtained, the VLE data are considered as thermodynamically consistent. A flow diagram of this consistency test is shown in Figure 5.28.

Often isothermal Px-data are measured (see Table 5.7). They cannot be checked for thermodynamic consistency. But if the data can be described very accurately with the help of a consistent g^E-model, these VLE data can also be considered as thermodynamically consistent. The same is true for other incomplete VLE data listed in Table 5.7.

With the help of the g^E-model only the deviations from Raoult's law should be described. The correct deviations from Raoult's law can only be obtained, if the exact values of the pure component vapor pressures are used during the fitting procedure. This is shown below for fitting the NRTL parameters for the nearly ideal but nevertheless azeotropic system 2-propanol–*tert*-butanol simultaneously to two isothermal Px-data sets measured at 40 °C. The very different results of the fitting procedure are shown in Figure 5.29. Obviously, the two data sets show a systematic small difference in the pressure measurement. While a correct description of the Px-data and the azeotropic VLE behavior is obtained, if the vapor pressure data of the authors are used for fitting the parameters. Total disagreement is observed if the VLE calculation is performed using the vapor pressures calculated by Antoine

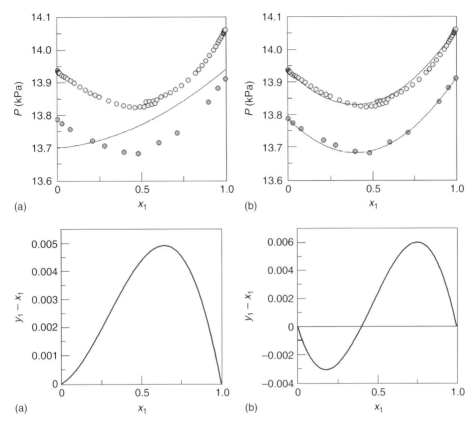

Figure 5.29 Results for the binary system 2-propanol (1) –tert-butanol (2) at 40 °C. (a) Using Antoine constants for both compounds ($\Delta g_{12} = -37.984$ cal/mol, $\Delta g_{21} = 29.94$ cal/mol, $\alpha_{12} = 0.3642$), (b) using the pure component vapor pressures of the authors ($\Delta g_{12} = -54.254$ cal/mol, $\Delta g_{21} = 26.41$ cal/mol, $\alpha_{12} = 0.3680$) for fitting the NRTL-parameters.

constants from literature. From Figure 5.29 it can directly be seen that of course for the objective function F:

$$F = \sum_i \left(\frac{P_{\text{exp},i} - P_{\text{calc},i}}{P_{\text{exp},i}} \right)^2$$

a much lower value is obtained, if the pure component vapor pressures given by the authors are used.

For a large number of binary systems the required binary g^E-model parameters for the Wilson, NRTL, and UNIQUAC equation and the results of the consistency tests can be found in the VLE Data Collection of the DECHEMA Chemistry Data Series published by Gmehling et al. [6]. One example page is shown in Figure 5.30. It shows the VLE data for the system ethanol and water at 70 °C published by Mertl [8]. On every page of this data compilation the reader will find the system, the reference, the Antoine constants with the range of validity, the experimental

226 | 5 Phase Equilibria in Fluid Systems

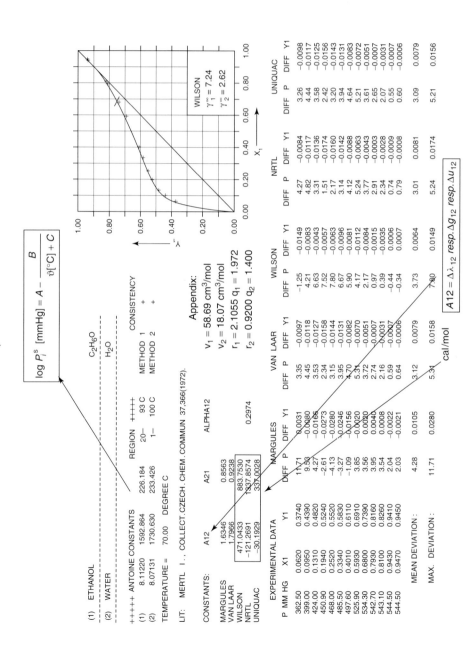

Figure 5.30 Example page of the VLE Data Collection [6].

VLE data, the results of two thermodynamic consistency tests, and the parameters of different g^E-models, such as the Wilson, NRTL, and UNIQUAC equation. Additionally, the parameters of the Margules [28] and van Laar [29] equation are listed.[13] Furthermore, the calculated results for the different models are given. For the model which shows the lowest mean deviation in vapor phase mole fraction the results are additionally shown in graphical form together with the experimental data and the calculated activity coefficients at infinite dilution. In the appendix of the data compilation the reader will find the additionally required pure component data, such as the molar volumes for the Wilson equation, the relative van der Waals properties for the UNIQUAC equation, and the parameters of the dimerization constants for carboxylic acids. Usually, the Antoine parameter A is adjusted to A' to start from the vapor pressure data given by the authors, and to use the g^E-model parameters only to describe the deviation from Raoult's law.[14] Since in this data compilation only VLE data up to 5000 mm Hg are presented, ideal vapor phase behavior is assumed when fitting the parameters. For systems with carboxylic acids the association model is used to describe the deviation from ideal vapor phase behavior.

In practice almost exclusively VLE data are used to fit the required parameters. Since a distillation column works nearly at constant pressure, most chemical engineers prefer thermodynamically consistent isobaric VLE-data in contrast to isothermal VLE-data to fit the model parameters. But that can cause problems, in particular if the boiling points of the two compounds considered are very different [24], as for example, for the binary system ethanol–n-decane. The result of the Wilson equation after fitting temperature-independent binary parameters only to reliable isobaric data at 1 atm is shown in Figure 5.31 for the system ethanol–n-decane, where the sum of the relative deviations of the activity coefficients was used as objective function.

From the results shown in Figure 5.31 it can be seen that already for VLE poor results are obtained. In particular large deviations are obtained at low ethanol concentrations. This is not only true for the Txy behavior, but also for the activity coefficients, although the activity coefficients were used to fit the Wilson parameters. The reason for the observed large deviations is that with temperature-independent parameters the observed temperature dependence cannot be described correctly. This conclusion can also be drawn when looking at the calculated excess enthalpies shown in Figure 5.31. Reliable g^E-model parameters should be able to describe the excess enthalpies following the Gibbs–Helmholtz equation. Apparently, excess enthalpies are obtained which strongly deviate from the experimental values [3], in particular at 90 and 140 °C. Of course wrong h^E-values will mean an incorrect temperature dependence of the activity coefficients. For the activity coefficients of ethanol at infinite dilution this is shown in Figure 5.31.

13) Both models (Margules, van Laar) are hardly used for process simulation today.
14) Unfortunately, Mertl [8] has not given the pure component vapor pressures. But with Antoine constants used reliable vapor pressures are obtained as can be seen from Figure 5.4

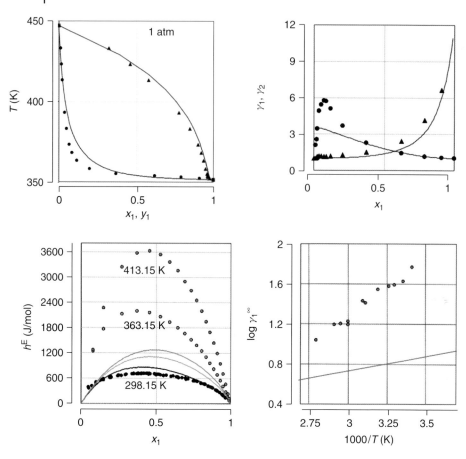

Figure 5.31 Result of the fit of temperature-independent Wilson parameters to consistent isobaric VLE data at 1 atm of the system ethanol (1)–n-decane (2) and calculated results for the excess enthalpies and activity coefficients at infinite dilution for the same system using the fitted parameters —— Wilson (Wilson parameters: $\Delta\lambda_{12} = 1284.12$ cal/mol, $\Delta\lambda_{21} = 1172.85$ cal/mol, $v_1 = 58.68$ cm^3/mol, $v_2 = 195.92$ cm^3/mol) • experimental [3].

In the case of isobaric data the temperature will change with composition. In particular at low ethanol concentrations the temperature alters drastically. It can be seen that the temperature change between $x_1 = 0$ to $x_1 = 0.1$ is nearly 100 K. Since the system ethanol (1)–n-decane (2) shows strong endothermic behavior with large positive values of the partial molar excess enthalpies for ethanol ($\overline{h}_1^{E\infty} \approx 19000$ J/mol, see Figures 5.31 and 5.32), which results in a decrease of the activity coefficient of ethanol with increasing temperature (see Example 5.6) following the Gibbs–Helmholtz equation. This leads to a maximum value of γ_1 at a mole fraction of approx. $x_1 = 0.1$ for the isobaric data as shown in Figure 5.31. It can easily be understood that this curvature can not be fitted correctly using temperature-independent parameters.

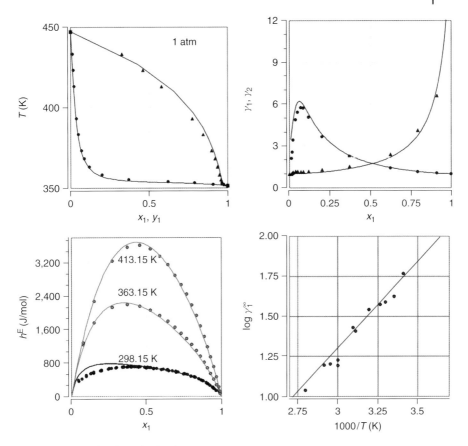

Figure 5.32 Calculated results for the VLE at 1 atm, activity coefficients γ_i, excess enthalpies as a function of composition and activity coefficients at infinite dilution as $f(T)$ for the system ethanol(1)–n-decane(2) using temperature-dependent Wilson parameters fitted simultaneously to VLE, h^E and γ^∞ data ● ▲ – experimental [3].

To obtain the correct values at infinite dilution and the correct temperature dependence resp. excess enthalpies, besides VLE data further reliable thermodynamic information should be taken into account for fitting temperature-dependent g^E-model parameters. Temperature-dependent Wilson parameters fitted simultaneously to VLE, excess enthalpies and activity coefficients at infinite dilution of the system ethanol–n-decane are given in Table 5.8. The results for VLE, activity coefficients as a function of composition and at infinite dilution and excess enthalpies obtained using these parameters are shown in Figure 5.32 together with the experimental values. It can be seen that with the temperature-dependent binary Wilson parameters (recommended values) not only the VLE behavior, but also the activity coefficients and the excess enthalpies as a function of composition and temperature are described correctly.

Table 5.8 Temperature-dependent Wilson parameters for the system ethanol (1)–n-decane (2): $\Delta\lambda_{ij}$ (cal/mol) $= a_{ij} + b_{ij}T + c_{ij}T^2$.

	a_{ij}	b_{ij}	$c_{ij} \cdot 10^4$
$\Delta\lambda_{12}$	4841.1	−7.9999	2.7050
$\Delta\lambda_{21}$	1276.8	−2.1230	5.4421

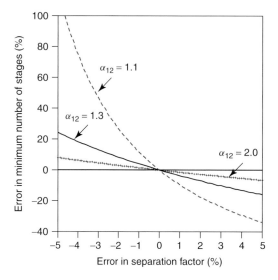

Figure 5.33 Influence of the error of the separation factor on the minimum number of theoretical stages calculated using the Fenske equation.

Already small deviations between the experimental and calculated separation factors can lead to a very different number of stages required for a given separation problem. This is especially true if the separation factor is not far away from unity. For three different separation factors the influence of an error on the minimum number of theoretical stages calculated by the Fenske equation [30] is shown in Figure 5.33,

$$N_{\text{th,min}} = \frac{\log \frac{(x_1/x_2)_d}{(x_1/x_2)_b}}{\log \alpha_{12}}$$

where d is the distillate and b the bottom product. From Figure 5.33 it can be recognized that for a separation factor of 1.1 an error of −4.5%, the minimum number of theoretical stages is nearly doubled. In the case of a separation factor of 2.0 the minimum number of theoretical stages is increased by less than 7%.

5.4.2
Recommended g^E-Model Parameters

As discussed in the chapters above reliable model parameters are most important. While mainly VLE data are used in the chemical industry, it is recommended to use all kinds of reliable data (phase equilibrium data (VLE, γ^∞, azeotropic data, SLE of eutectic systems, etc.), excess enthalpies) for fitting simultaneously g^E-model parameters, which often have to be temperature dependent. To account only for the deviations from Raoult's law, it is recommended to use the pure component vapor pressures measured by the authors for every data set. This can be done by multiplying the vapor pressure with a correction factor, for the Antoine equation, this corresponds to changing the parameter A to A'. Sometimes a large number of experimental data are available. Then of course the data used should be distributed equally over the whole temperature (pressure) range. Since often a lot of VLE data at atmospheric pressure are reported, perhaps some of the data have to be removed or at least a lower weighting factor for the numerous data should be used. The same is true for excess enthalpies. Most authors have measured excess enthalpies around room temperature. For fitting temperature-dependent model parameters the whole temperature range should be covered. While consistent VLE data (azeotropic data) provide the information about the composition

Table 5.9 Recommended NRTL interaction parameters (cal/mol) for different binary systems.

a_{12} (cal/mol)	a_{21} (cal/mol)	b_{12} (cal/(mol K))	b_{21} (cal/(mol K))	c_{12} (cal/(mol K^2))	c_{21} (cal/(mol K^2))	α_{12}
Acetone (1)–cyclohexane (2)						
1423.8	2880.0	−2.9548	−10.136	0.0008073	0.011832	0.4212
Acetone (1)–benzene (2)						
−16.064	110.25	0.37896	0.6426	−0.0004859	−0.0004819	0.6991
Benzene (1)–cyclohexane (2)						
1403.1	8.4201	−7.5455	1.8632	0.01316	−0.006295	0.3
Acetone (1)–water (2)						
833.97	−3146.5	0.83106	17.457	−0.0037816	−0.013622	0.5466
Ethanol (1)–1,4–dioxane (2)						
412.58	1334.9	0.41917	−3.7148	–	–	0.3548
Acetone (1)–chloroform (2)						
5563.9	−12824.0	−43.308	83.358	0.061888	−0.11637	0.0850
Acetone (1)–methanol (2)						
477.04	327.46	0.03815	−0.73193	−0.002931	0.0018051	0.7
Chloroform (1)–methanol (2)						
5378.5	−1471.6	6.37397	−13.015	−0.032724	0.032599	0.055

dependence of the activity coefficients, excess enthalpies deliver the most important information about the temperature dependence. Accurate activity coefficients at infinite dilution deliver the only reliable information about the dilute composition range. In the case of simple eutectic systems, also SLE data can be used. SLE data of eutectic systems deliver supporting data at low temperature, while excess enthalpies at high temperature can be used as supporting data at high temperature, to fit reliable temperature-dependent parameters for the temperature range covered. For fitting the parameters simultaneously to all kind of data, weighting factors w_i are used, so that the objective function looks like

$$F = w_{\text{VLE}} \sum_{n_{\text{VLE}}} \Delta \text{VLE} + w_{\gamma^\infty} \sum_{n_{\gamma^\infty}} \Delta \gamma^\infty + w_{h^E} \sum_{n_{h^E}} \Delta h^E$$

$$+ w_{\text{LLE}} \sum_{n_{\text{LLE}}} \Delta \text{LLE} + w_{\text{SLE}} \sum_{n_{\text{SLE}}} \Delta \text{SLE} + w_{\text{AZD}} \sum_{n_{\text{AZD}}} \Delta \text{AZD} \quad (5.45)$$

For several binary systems recommended model parameters for the g^E-models Wilson, NRTL, and UNIQUAC are given in Tables 5.9–5.11. Typical results for the system acetone–water using the NRTL model are shown in Figure 5.34. It can be

Table 5.10 Recommended Wilson interaction parameters (cal/mol) for different binary systems.

a_{12} (cal/mol)	a_{21} (cal/mol)	b_{12} (cal/(mol K))	b_{21} (cal/(mol K))	c_{12} (cal/(mol K^2))	c_{21} (cal/(mol K^2))
Acetone (1)–cyclohexane (2)					
3109.2	1670.7	−10.622	−4.5189	0.013757	0.0012266
Acetone (1)–benzene (2)					
−113.72	201.96	2.5292	−1.5516	−0.0035364	0.0026447
Benzene (1)–cyclohexane (2)					
1558.96	−203.7	−8.2383	3.2140	0.012856	−0.0075245
Acetone (1)–water (2)					
−1305.7	1054.9	1.4341	3.4522	0.009414	−0.005907
Ethanol (1)–1,4-dioxane (2)					
1404.7	136.80	−3.3856	0.82783	–	–
Acetone (1)–chloroform (2)					
375.28	−1722.6	−3.7843	6.4055	0.0079107	−0.0074779
Acetone (1)–methanol (2)					
−60.756	863.79	−0.06114	−1.0533	–	–
Chloroform (1)–methanol (2)					
−1140.8	3596.2	2.5936	−6.2234	0.000031	0.000030
Ethanol (1)–n-decane (2)					
4841.1	1276.8	−7.9999	−2.123	0.0002705	0.0005442

Table 5.11 Recommended UNIQUAC interaction parameters (cal/mol) for different binary systems.

a_{12} (cal/mol)	a_{21} (cal/mol)	b_{12} (cal/(mol K))	b_{21} (cal/(mol K))	c_{12} (cal/(mol K^2))	c_{21} (cal/(mol K^2))
Acetone (1)–cyclohexane (2)					
259.18	560.82	−1.0167	0.041374	–	–
Acetone (1)–benzene (2)					
−75.46	120.20	−0.10062	0.44835	−0.0008052	0.0004704
Benzene (1)–cyclohexane (2)					
566.78	−116.17	−3.8155	2.2017	0.006297	−0.004641
Acetone (1)–water (2)					
2619.0	33.80	−6.3149	−4.6102	0.0008817	0.012937
Ethanol (1)–1,4-dioxane (2)					
−27.083	762.43	0.47646	−1.9128	–	–
Acetone (1)–chloroform (2)					
101.70	−853.91	−4.4866	6.9067	0.010999	−0.013211
Acetone (1)–methanol (2)					
324.48	59.98	1.2457	−0.8408	−0.003207	0.0013662
Chloroform (1)–methanol (2)					
3561.3	−416.85	−9.4697	−0.33196	0.0073773	0.002515

seen that nearly perfect agreement between experimental and correlated VLE data, activity coefficients at infinite dilution as $f(T)$, azeotropic composition as $f(T)$ and excess enthalpies as $f(T)$ is obtained. In Figure 5.35 the results for default values given in a process simulator are shown. The difference in quality can easily be recognized. For all properties much better results are obtained using the recommended NRTL parameters. This is especially true for the excess enthalpies and as consequence for the temperature dependence of the activity coefficient at infinite dilution. But reliable activity coefficients at infinite dilution are of particular importance for the design of distillation columns, where at the top (bottom) the last traces of high (low) boiler have to be removed. As described the separation factors at infinite dilution mainly influence the number of theoretical stages required for a distillation column. The procedure for fitting recommended temperature-dependent g^E-model parameters is described in more detail by Rarey–Nies et al. [31] and Tochigi et al. [32].

For some binary systems the use of temperature-dependent parameters is essential, since with temperature-independent parameters excess enthalpies above certain values cannot be described anymore with the chosen g^E-model (Novák [20]) using temperature-independent parameters. Problems can also arise if systems, such as alkane–alcohol systems, show strong deviations from Raoult's law, this means large activity coefficients at infinite dilution, but no miscibility gap. Typical

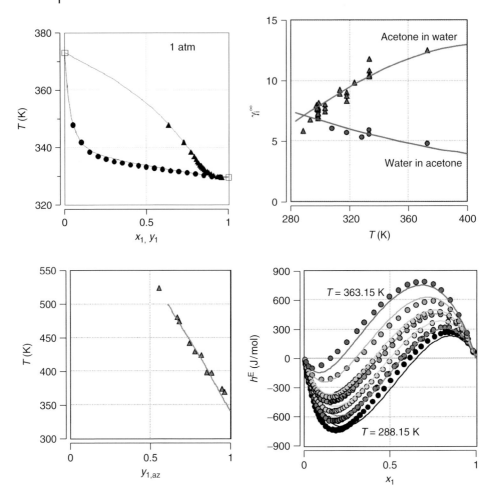

Figure 5.34 Results for acetone (1)–water (2) using recommended temperature dependent NRTL parameters fitted simultaneously to consistent VLE, h^E, and γ^∞ data.

examples are ethanol–decane or cyclohexane–1-propanol. When the activity coefficients at infinite dilution are described correctly with the NRTL or the UNIQUAC model a miscibility gap is calculated. As long as homogeneous behavior is described with these models, too low activity coefficients at infinite dilution result (Novák [20]). Only with the Wilson equation the correct activity coefficients at infinite dilution and homogeneous behavior can be described as shown before.

In Section 11.1 the importance of reliable g^E-model parameters for the synthesis and design of extractive distillation processes is demonstrated for the separation of cyclohexane from benzene using NMP as entrainer. Furthermore for the system acetone–water it is shown how default values can lead to poor separation factors or even not existing azeotropic points at the top of the column ($x_{\text{acetone}} \rightarrow 1$).

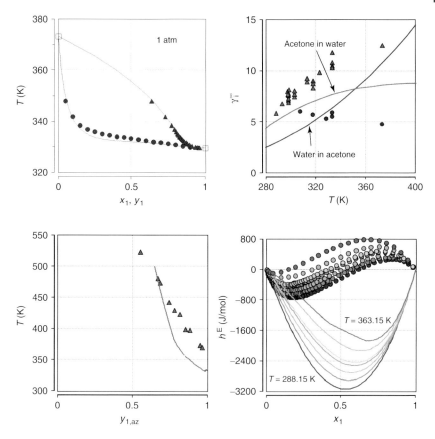

Figure 5.35 Results for acetone (1)–water (2) using the NRTL parameters provided by the process simulator.

5.5
Calculation of Vapor–Liquid Equilibria Using Equations of State

As mentioned before, Approach A (also called φ–φ approach) compared to Approach B (also called γ–φ approach) has the great advantage that supercritical compounds can be handled easily and that besides the phase equilibrium behavior various other properties such as densities, enthalpies including enthalpies of vaporization, heat capacities and a large number of other important thermodynamic properties can be calculated via residual functions for the pure compounds and their mixtures. For the calculation besides the critical data and the acentric factor for the equation of state and reliable mixing rules, only the ideal gas heat capacities of the pure compounds as a function of temperature are additionally required. A perfect equation of state with perfect mixing rules would provide perfect results. This is the reason why after the development of the van der Waals equation of state in 1873 an enormous number of different equations of state have been suggested.

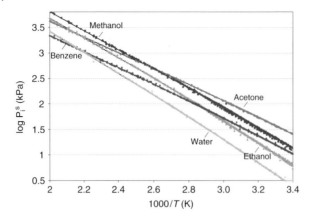

Figure 5.36 Experimental [3] and calculated vapor pressures for selected solvents using the PR equation of state and the Twu-α-function.

In principle, for the calculation of VLE any equation of state can be used which is able to describe the PvT behavior of the vapor and the liquid phase, for example, cubic equations of state, further developments of the virial equation, or Helmholtz equations of state. Most popular in chemical industry are further developments of the cubic van der Waals equation of state. Great improvements were obtained by modification of the attractive part, by introducing the temperature dependence of the attractive parameter with the help of a so-called α-function and the development of improved mixing rules, the so-called g^E-mixing rules, which allow the applicability to asymmetric systems and systems with polar compounds.

Exemplarily a few typical results of cubic equations of state used in practice are shown below. A prerequisite for the reliable description of VLE data of binary and multicomponent systems is the reliable description of the pure component vapor pressures. With the introduction of an α-function for the description of the temperature dependence of the attractive parameter $a(T)$ and the usage of the acentric factor ω as third parameter the results for pure component vapor pressures were significantly improved. In Figure 5.36 the experimental and calculated vapor pressures for five solvents are shown, where the PR equation of state with the Twu α-function was used. It can be seen that nearly perfect agreement is obtained in the wide temperature range covered. Even the slopes are described reliably. This means that following the Clausius–Clapeyron equation also the enthalpies of vaporization are described correctly. From the slopes it can be concluded that the enthalpies of vaporization increase from benzene to the alcohols, and then to water. This leads to the fact that in binary systems, for example, acetone–methanol, or ethanol–benzene the low boiler at low temperature can become the high boiler at higher temperatures.

In Figure 5.37 the experimental and calculated enthalpies of vaporization using the SRK and the volume translated PR equation of state for 11 different compounds in a wide temperature range up to the critical temperature are shown. It can be

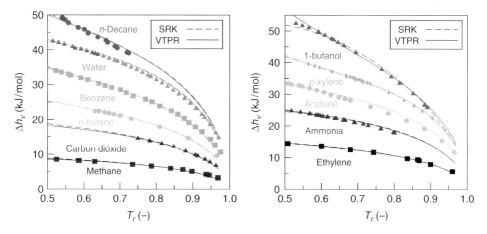

Figure 5.37 Experimental [3] and calculated enthalpies of vaporization using the Soave–Redlich–Kwong and the volume translated PR equation of state.

seen that both equations of state provide excellent agreement with the experimental findings. That is not surprising, since with the reliably calculated vapor pressures and volumes of the vapor and liquid phase, reliable enthalpies of vaporization should be obtained following the Clausius–Clapeyron equation (see Eq. (2.86)).

Example 5.11

Calculate the liquid density of cyclohexane at the normal boiling point ($T_b = 353.85$ K, $P = 1$ atm) with the help of the PR equation of state.
Pure component properties:

Component	M (g mol^{-1})	T_c (K)	P_c (bar)	ω
Cyclohexane	84.16	553.8	40.8	0.213

Solution

For the calculation of the liquid density at the normal boiling point for cyclohexane first the parameters a and b of the PR equation of state have to be determined from the critical data and the acentric factor using Eqs. (2.167)–(2.169).

$$a_c = 0.45724 \frac{R^2 T_c^2}{P_c} = 0.45724 \frac{0.08314^2 \cdot 553.8^2}{40.8}$$

$$= 23.758 \left(\text{dm}^3\right)^2 \text{bar/mol}^2$$

$$b = 0.0778 \frac{RT_c}{P_c} = 0.0778 \frac{0.08314 \cdot 553.8}{40.8} = 0.087798 \text{ dm}^3$$

$$T_r = \frac{T}{T_c} = \frac{353.85}{553.8} = 0.63895$$

$$\alpha(T) = \left[1 + \left(0.37464 + 1.54226\omega - 0.26992\omega^2\right)\left(1 - T_r^{0.5}\right)\right]^2$$

$$\alpha(T) = \left[1 + \left(0.37464 + 1.54226 \cdot 0.213 - 0.26992 \cdot 0.213^2\right)\right.$$
$$\left. \cdot \left(1 - 0.63895^{0.5}\right)\right]^2 = 1.2965$$

$$a(T) = a_c \cdot \alpha(T) = 23.758 \cdot 1.2965 = 30.8022 \, (\text{dm}^3)^2 \, \text{bar/mol}^2$$

In the next step, the molar liquid volume has to be determined for which the right-hand side of the PR equation of state gives a value of 1 atm. This can be done iteratively or by solving the cubic equation.

$$P = \frac{RT}{v - b} - \frac{a(T)}{v(v + b) + b(v - b)}$$

For the given conditions a molar volume of $v = 0.11068 \, \text{dm}^3 \, \text{mol}^{-1}$ is obtained.

$$P = \frac{0.08314 \cdot 353.85}{0.11068 - 0.087798}$$
$$- \frac{30.8022}{0.11068 \cdot (0.11068 + 0.087798) + 0.087798 \cdot (0.11068 - 0.087798)}$$

$$P = 1285.687 - 1284.681 = 1.006 \, \text{bar} \approx 1 \, \text{atm}$$

In the next step with the help of the molar liquid volume and the molar mass the liquid density at the normal boiling point can be calculated:

$$\rho = \frac{84.16}{0.11068} = 760.38 \, \text{g dm}^{-3}$$

Experimentally, a density of $719 \, \text{g dm}^{-3}$ was determined for cyclohexane at the normal boiling point [3]. This means that the calculated value using the PR equation of state is ~6% too high. In Figure 5.38 the calculated liquid densities of cyclohexane are shown together with the experimental liquid densities for a wide temperature range ($T_r = 0.5$–0.8). At the same time the experimental and liquid densities for five more solvents are shown in this diagram. It can be seen that with the exception of water the calculated liquid densities using the PR equation of state are too high. The largest density deviations are obtained for the very polar compound water. It seems that the difference between the experimental and calculated densities is nearly constant. Using the SRK equation of state, even larger deviations between the experimental and calculated liquid densities are obtained. But a reliable description of the pure component densities is a prerequisite for the calculation of reliable mixture densities for multicomponent systems. Peneloux et al. [33] showed that the results for the liquid densities can be improved by introducing a translation parameter c (see Section 2.5.5).

In Figure 5.39 for the temperature range $T_r = 0.5$–0.8 experimental and liquid densities for the same solvents as in Figure 5.38 are shown. But for the calculation

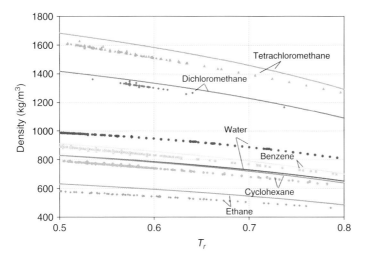

Figure 5.38 Experimental and calculated liquid densities using the PR equation of state for six different solvents in the temperature range $T_r = 0.5-0.8$.

now the volume translated PR equation of state has been used, where the translation parameter was adjusted to the experimental liquid density at $T_r = 0.7$ (see Eq. (2.178)). This ensures that with the volume translation perfect results are obtained at $T_r = 0.7$. Finally, not only for $T_r = 0.7$, but also for other temperatures improved results are obtained. From Figure 5.39 it can be seen that with the

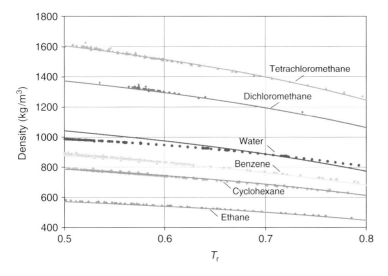

Figure 5.39 Experimental and calculated liquid densities using the volume translated PR equation of state for six different solvents in the temperature range $T_r = 0.5-0.8$.

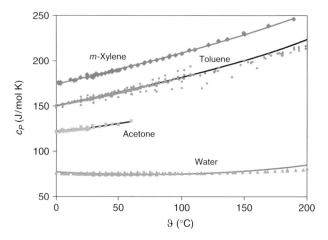

Figure 5.40 Experimental [3] and calculated liquid heat capacities using VTPR.

exception of the strong polar compound water very good agreement is obtained between the experimental and calculated densities in the whole temperature range covered. For 44 compounds investigated in the temperature range ($T_r = 0.5$–0.8) a relative mean deviation smaller than 2% was obtained. This mean relative deviation obtained for VTPR is much smaller than the deviations obtained for the PR (mean deviation approx. 6%) or the SRK equation of state (mean deviation approx. 12%) for the same 44 compounds.

Equations of state not only allow to calculate densities, enthalpies of vaporization, but also other thermodynamic properties, such as heat capacities, enthalpies, entropies, internal energy, Gibbs energy, Helmholtz energy, and other important properties, for example, Joule–Thomson coefficients, and so on. In Figure 5.40 experimental and calculated liquid heat capacities using the VTPR equation of state for five different solvents in the temperature range 0–200 °C are shown. As can be seen the agreement between experimental and calculated data is within approx. 2%. The calculated results of course depend on the quality of the heat capacities of the ideal gas, the parameters of the α-function, and further parameters. In [34] and Chapter 2 it was shown that the results can still be improved when other thermodynamic data are used in addition for fitting the parameters of the α-function.

5.5.1
Fitting of Binary Parameters of Cubic Equations of State

To describe the behavior of mixtures (enthalpies of vaporization, densities, heat capacities, phase equilibria, etc.) using equations of state, binary parameters are required. The different mixing rules suggested were already discussed in Section 4.9.2. While empirical mixing rules, for example, quadratic mixing rules could only be applied for nonpolar systems, the range of applicability of equations of state with modern g^E-mixing rules was extended to polar systems, for example, systems with water, alcohols, ketones, and so on.

It was shown in Section 4.9.2 that in the quadratic mixing rules a binary parameter k_{12} is required to describe the behavior of the binary system. For fitting the binary parameter usually VLE data are used. With the help of all the required binary parameters k_{ij} (in the case of a ternary system: k_{12}, k_{13}, k_{23}) the ternary or multicomponent system can then be calculated.

In Figure 5.41 for the binary system n-butane–CO_2 the experimental results are shown together with the calculated results for $k_{12} = 0$ and for the fitted binary parameter $k_{12} = 0.1392$. It can be seen that the agreement is highly improved when going from $k_{12} = 0$ to $k_{12} = 0.1392$. Furthermore, it is remarkable that k_{12} seems to be temperature-independent over a wide temperature range. It is clear that starting from a poor description of the binary system as in the case of $k_{12} = 0$, there is no chance to obtain good results for a ternary or a multicomponent system.

For a long time the empirical quadratic mixing rules were used in gas-processing or petrochemistry. But poor results were obtained for systems with polar compounds. This is exemplarily shown for the systems acetone–water and isopropanol–water in Figures 5.42 and 5.43. It can be seen from the diagrams on the left-hand side that unsatisfactory results are obtained, if the binary parameter k_{12} of the empirical mixing rules is fitted to these systems. Huron and Vidal [35] carefully investigated the advantages of g^E-models and equations of state and developed the so-called g^E-mixing rules. Now, in g^E-mixing rules the parameters of a g^E-model, for example, of the Wilson, NRTL, or UNIQUAC model are fitted to calculate the attractive parameter of the chosen cubic equation of state. For the two systems mentioned the results are shown on the right-hand side of Figures 5.42 and 5.43. The improvements obtained are significant. The application of the new mixing rules now allowed using equations of state also for process simulation in

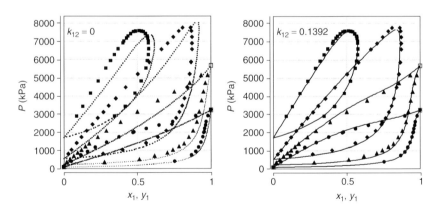

Figure 5.41 VLE results for the system n-butane(1)–CO_2(2) using the binary parameter $k_{12} = 0$ and an adjusted binary parameter (● – 270 K; ▲ – 292.6 K; ◆ – 325.01 K; ■ – 377.6 K) [3].

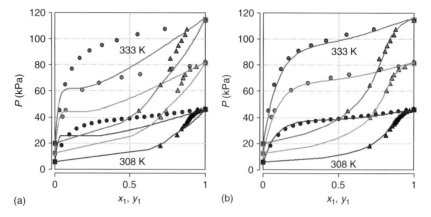

Figure 5.42 Experimental and calculated VLE data for the system acetone (1)–water (2) using the PR equation of state with classical mixing rules ($k_{12} = -0.2428$) (a) and the Soave–Redlich–Kwong equation of state with g^E-mixing rules (NRTL, $\Delta g_{12} = 257.9$ cal/mol, $\Delta g_{21} = 1069$ cal/mol, $\alpha_{12} = 0.2$) (b) at 308, 323 and 333 K.

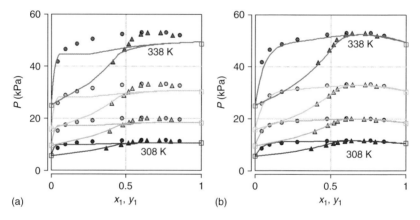

Figure 5.43 Experimental and calculated VLE data for the system isopropanol (1)–water (2) using the PR equation of state with classical mixing rules ($k_{12} = -0.168$) (a) and the Soave–Redlich–Kwong equation of state with g^E-mixing rules (NRTL, $\Delta g_{12} = -339.0$ cal/mol, $\Delta g_{21} = 1914$ cal/mol, $\alpha_{12} = 0.2$) (b) at 308, 318, 328, and 338 K.

chemical industry. Using the binary parameters derived from VLE data equations of state directly allow the calculation of all other mixture properties.

Instead of the binary parameter k_{12} in the case of g^E-mixing rules, the parameters of the Wilson, NRTL, or UNIQUAC equation are fitted. Depending on the strength of the temperature dependence either constant or temperature-dependent parameters have to be fitted. With the help of temperature-dependent parameters

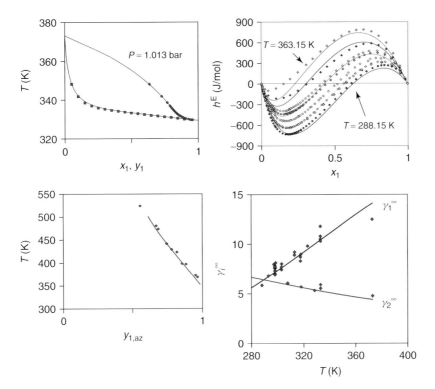

Figure 5.44 Experimental and correlated VLE, h^E, azeotropic, and γ^∞ data using the VTPR equation of state with g^E-mixing rules (UNIQUAC with temperature-dependent parameters); UNIQUAC parameters: a(acetone, H_2O) = 523.84 K, a(H_2O, acetone) = −937.99 K, b(acetone, H_2O) = −1.3221, b(H_2O, acetone) = 4.4338, c(acetone, H_2O) = 0.00125 (K^{-1}), c(H_2O, acetone) = 0.00043 (K^{-1}).

even the temperature dependence of the excess enthalpies can be described with the required accuracy. For the system acetone–water the results are shown in Figure 5.44. It can be seen that besides the VLE behavior, the excess enthalpies, activity coefficients at infinite dilution and the azeotropic data as a function of temperature can be described with the required accuracy.

With the help of the binary parameters k_{12} or g^E-model parameters now the phase equilibrium behavior, densities, enthalpies, Joule–Thomson coefficients, and so on, for binary, ternary and multicomponent systems can be calculated. For the calculation of the VLE behavior the procedure is demonstrated in the following example for the binary system nitrogen–methane using classical mixing rules. The same procedure can be applied to calculate the VLE behavior of multicomponent systems and with g^E-mixing rules as well.

Example 5.12

With the help of the SRK equation of state the system pressure and vapor phase composition for the binary system nitrogen (1)–methane (2) for a liquid mole fraction of nitrogen $x_1 = 0.2152$ at 144.26 K should be calculated. The required pure component data and the binary parameter are given below:

Component	T_c (K)	P_c (bar)	ω
N_2	126.15	33.94	0.045
CH_4	190.63	46.17	0.010

Binary parameter: $k_{12} = 0.0267$ [36].

Solution

The calculation has to be performed iteratively. The objective of the iterative procedure is to find the pressure and vapor phase composition for which the following equilibrium condition is fulfilled:

$$x_i \varphi_i^L = y_i \varphi_i^V \qquad (5.9)$$

for both components. To start with the calculation, first of all estimated values for the vapor phase composition and system pressure are required. In this case, a vapor phase mole fraction $y_1 = 0.6$ and a pressure $P = 20$ bar were chosen.

During the iterative procedure these values have to be changed until the equilibrium condition (5.9) is fulfilled. A flow diagram for this procedure is shown in Figure 5.45. For the considered example the first step of this procedure is described in detail below.

First the pure component parameters for both compounds have to be calculated with the help of Eqs. (2.162–2.164) at the given temperature (144.26 K) using the critical data P_c, T_c, and the acentric factor ω:

$$a_{11}(T) = 0.42748 \frac{0.08314^2 \cdot 126.15^2}{33.94} \alpha_1(T)$$

$$= 1.3856 \cdot \alpha_1(T)(dm^3)^2 \text{ bar I mol}^2$$

$$T_{r1} = \frac{144.25}{126.15} = 1.1436$$

$$\alpha_1(T) = \left[1 + \left(0.48 + 1.574 \cdot 0.045 - 0.176 \cdot 0.045^2\right)\left(1 - 1.1436^{0.5}\right)\right]^2$$

$$\alpha_1(T) = 0.9251$$

$$a_{11}(T) = 1.2818 \left(dm^3\right)^2 \text{ bar/mol}^2$$

$$b_1 = 0.08664 \frac{0.08314 \cdot 126.15}{33.94} = 0.02677 \text{ dm}^3$$

$$a_{22}(T) = 0.42748 \frac{0.08314^2 \cdot 190.63^2}{46.17} \alpha_2(T) = 2.3259\, \alpha_2(T)\,(dm^3)^2 \text{ bar/mol}^2$$

5.5 Calculation of Vapor–Liquid Equilibria Using Equations of State

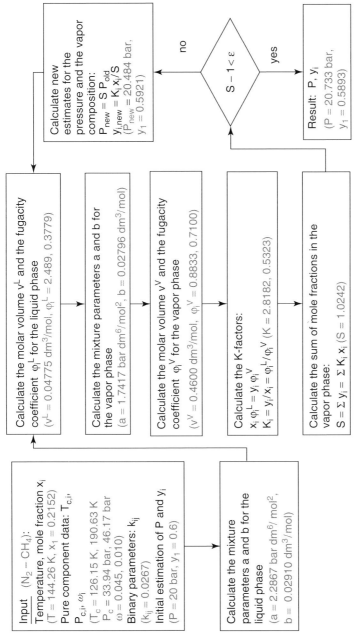

Figure 5.45 Flow diagram for the calculation of isothermal vapor–liquid equilibria using the SRK equation of state.

$$T_{r2} = \frac{144.26}{190.63} = 0.7568$$

$$\alpha_2(T) = \left[1 + (0.48 + 1.574 \cdot 0.01 - 0.176 \cdot 0.01^2)(1 - 0.7568^{0.5})\right]^2$$

$$\alpha_2(T) = 1.1331$$

$$a_{22}(T) = 2.6356 (\text{dm}^3)^2 \text{ bar/mol}^2$$

$$b_2 = 0.08664 \frac{0.08314 \cdot 190.63}{46.17} = 0.02974 \text{ dm}^3/\text{mol}$$

In the next step, the cross parameter a_{12} and the mixture parameters a and b for the liquid phase with the help of the mixing rules (Eqs. (4.98–4.100)) are calculated.

$$a_{12}(T) = a_{21}(T) = (1.2818 \cdot 2.6356)^{0.5}(1 - 0.0267)$$
$$= 1.7889 \,(\text{dm}^3)^2 \text{ bar/mol}^2$$

$$a(T) = 0.2152^2 \cdot 1.2818 + 2 \cdot 0.2152 \cdot 0.7848 \cdot 1.7889 + 0.7848^2 \cdot 2.6356$$
$$= 2.2869 (\text{dm}^3)^2 \text{ bar/mol}^2$$

$$b = 0.2152 \cdot 0.02677 + 0.7848 \cdot 0.02974 = 0.0291 \text{ dm}^3/\text{mol}$$

With the help of these parameters, the molar volume of the liquid has to be determined in a way that the given pressure $P = 20$ bar is obtained with the SRK equation of state.

This calculation can be performed iteratively or by solving the cubic equation of state. For the given pressure of 20 bar a molar liquid volume of $0.0477563 \text{ dm}^3/\text{mol}$ is obtained:

$$P = \frac{RT}{v-b} - \frac{a(T)}{v(v+b)} = \frac{0.08314 \cdot 144.26}{0.04776 - 0.0291}$$
$$- \frac{2.2869}{0.04776(0.04776 + 0.0291)} = 20 \text{ bar}$$

With the help of the calculated molar volume and the various pure component and mixture parameters of the liquid phase both fugacity coefficients for the liquid phase φ_i^L can be calculated using Eq. (4.101). For nitrogen (1) the following value is obtained:

$$\ln \varphi_1^L = \ln \frac{0.04776}{0.04776 - 0.0291}$$
$$- \frac{2(0.2152 \cdot 1.2818 + 0.7848 \cdot 1.7889)}{0.08314 \cdot 144.26 \cdot 0.0291} \ln \frac{0.04776 + 0.0291}{0.04776}$$
$$+ \frac{0.02677}{0.04776 - 0.0291} - \ln \frac{20 \cdot 0.04776}{0.08314 \cdot 144.26}$$
$$+ \frac{2.2869 \cdot 0.02677}{0.08314 \cdot 144.26 \cdot 0.0291^2}$$
$$\cdot \left(\ln \frac{0.04776 + 0.0291}{0.04776} - \frac{0.0291}{0.04776 + 0.0291} \right)$$

$\varphi_1^L = 2.4892$

In the same way, the fugacity coefficient for methane (2) is obtained:

$\varphi_2^L = 0.3779$

In the next step, the mixture parameters for the vapor phase ($y_1 = 0.6$) have to be determined:

$$a(T) = 0.6^2 \cdot 1.2818 + 2 \cdot 0.6 \cdot 0.4 \cdot 1.7889 + 0.4^2 \cdot 2.6356$$
$$= 1.7418 \, (\text{dm}^3)^2 \, \text{bar/mol}^2$$

$$b = 0.6 \cdot 0.02677 + 0.4 \cdot 0.02974 = 0.2796 \, \text{dm}^3/\text{mol}$$

With the help of these parameters a molar volume of $0.4601 \, \text{dm}^3/\text{mol}$ for the vapor phase is obtained for the pressure of 20 bar using the SRK equation of state. With this volume and the parameters for the vapor phase the following fugacity coefficients are obtained:

$\varphi_1^V = 0.8833$

$\varphi_2^V = 0.7100$

With the help of the fugacity coefficients obtained first the K-factors ($K_i = y_i/x_i$) for the two components can be calculated using Eq. (5.17):

$$K_1 = \frac{y_1}{x_1} = \frac{\varphi_1^L}{\varphi_1^V} = \frac{2.4892}{0.8833} = 2.8181$$

$$K_2 = \frac{y_2}{x_2} = \frac{\varphi_2^L}{\varphi_2^V} = \frac{0.3779}{0.7100} = 0.5323$$

Then it can be checked whether the equilibrium condition is fulfilled. In equilibrium the sum of the mole fractions in the vapor phase should be equal to 1.

$$S = \sum x_i K_i = 0.2152 \cdot 2.8181 + 0.7848 \cdot 0.5323 = 0.6065 + 0.4177 = 1.0242$$

It can be seen that with the estimated vapor phase composition and pressure the equilibrium conditions are not fulfilled. This means that new values have to be estimated for the vapor phase composition, for example, by normalizing the vapor phase mole fractions:

$$y_1 = \frac{0.6065}{1.0242} = 0.5922$$

$$y_2 = \frac{0.4177}{1.0242} = 0.4078$$

Furthermore, a new pressure is estimated using the K-factor method:

$$P_{\text{new}} = P_{\text{old}} \cdot S = 20.484 \, \text{bar}$$

The iteration can be stopped when the sum of the calculated mole fractions

$$S = \sum y_i = \sum x_i K_i,$$

only deviates by a small value for example, $\varepsilon = 10^{-5}$ from the desired value of 1.

After a few iterations the stop criterion $\varepsilon < 10^{-5}$ is fulfilled. This means that the correct equilibrium composition and pressure are obtained.

x_1	$y_{1,exp}$	$y_{1,calc}$	P_{exp} (bar)	P_{calc} (bar)
0.2152	0.5804	0.5893	20.684	20.733

At these conditions the following fugacity coefficients are obtained for the two compounds in the different phases:

φ_1^L	φ_2^L	φ_1^V	φ_2^V
2.4106	0.3655	0.8803	0.6984

In Figure 5.6, the calculated results using the SRK equation of state are shown together with the experimental data for different temperatures and the whole composition range for the system nitrogen (1)–methane (2).

The whole procedure is given in the form of a flow diagram in Figure 5.45. The same procedure shown for the binary system nitrogen–methane can be applied for multicomponent systems. For the calculation besides the critical data T_c, P_c, and the acentric factors ω_i of the compounds involved only the binary parameters k_{ij} for the quadratic mixing rule or the g^E-model parameters in the case of g^E-mixing rules are required.

For the quaternary system $N_2-CO_2-H_2S$–methanol calculated with the help of the SRK equation of state using binary parameters k_{ij} respectively g^E-mixing rule parameters the calculated and experimental results are shown in Figure 5.46. From the results it can be concluded that in this case only slightly improved results are obtained using g^E-model parameters.

5.6
Conditions for the Occurrence of Azeotropic Behavior

At the azeotropic point, the mole fractions of all components in the liquid phase are identical with the mole fractions in the vapor phase for homogeneous systems.

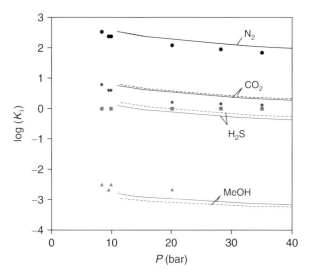

Figure 5.46 Experimental [3] and predicted K-factors for the system $N_2-CO_2-H_2S$-methanol at $-15\,^\circ C$: ----- SRK + quadratic mixing rule; –– SRK + g^E-mixing rule.

This leads to the fact that all K-factors and all relative volatilities show a value of 1 at the azeotropic point and that the system cannot be separated by ordinary distillation. A reliable knowledge of all azeotropic points for the system to be separated is of essential importance for the synthesis and design of separation processes.

For a binary system, the following relations are valid for homogeneous systems at the azeotropic point using the simplified Eq. (5.18) of Approach B:

$$\alpha_{12} = \frac{K_1}{K_2} = \frac{\gamma_1/x_1}{\gamma_2/x_2} = \frac{\gamma_1 P_1^s}{\gamma_2 P_2^s} = 1 \rightarrow \frac{\gamma_2}{\gamma_1} = \frac{P_1^s}{P_2^s} \text{ or } \frac{\gamma_1}{\gamma_2} = \frac{P_2^s}{P_1^s} \quad (5.46)$$

Using an equation of state (Approach A) the following relation is obtained for the azeotropic point:

$$\alpha_{12} = \frac{K_1}{K_2} = \frac{\gamma_1/x_1}{\gamma_2/x_2} = \frac{\varphi_1^L \varphi_2^V}{\varphi_1^V \varphi_2^L} = 1 \rightarrow \frac{\varphi_1^L}{\varphi_1^V} = \frac{\varphi_2^L}{\varphi_2^V} \quad (5.47)$$

It can be seen that starting from Eq. (5.46) azeotropic behavior always occurs if for a given composition the ratio of the pure component vapor pressures P_1^s/P_2^s is identical to the ratio of the activity coefficients γ_2/γ_1. The typical curvature of the γ_2/γ_1-ratio in logarithmic form for an azeotropic system with positive and negative deviation from Raoult's law at constant temperature is shown in Figure 5.47a,b, respectively. The azeotropic composition can directly be obtained from the intersection of the straight line for the vapor pressure ratio and the curve for the ratio of the activity coefficients γ_2/γ_1.

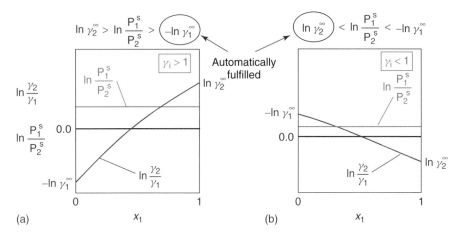

Figure 5.47 Examination of the azeotropic behavior of binary homogeneous systems at constant temperature (component 1 = low boiling compound).[15]

From Eq. (5.46), it can be seen that azeotropic behavior can easily occur in a binary system if the vapor pressures of the two components are very similar, since in this case already very small deviations from Raoult's law are sufficient to fulfill the equation and to create an azeotropic point either with positive or negative deviation from Raoult's law. If the vapor pressures are identical (e.g., at the Bancroft point), the binary system shows the azeotropic behavior.

From Figure 5.47, it can be concluded that the occurrence of azeotropic points can be calculated if besides the activity coefficients at infinite dilution the ratio of the vapor pressures is known. Azeotropic behavior occurs if the following condition for positive resp. negative deviation from Raoult's law is fulfilled (see Figure 5.47):

$$\ln \gamma_2^\infty > \ln \frac{P_1^s}{P_2^s} \qquad -\ln \gamma_1^\infty > \ln \frac{P_1^s}{P_2^s} \qquad (5.48)$$

positive resp. negative deviation from Raoult's law

Since approx. 90% of the systems show positive deviation from Raoult's law, in most cases pressure maximum (temperature minimum) azeotropes are observed.

The activity coefficients at infinite dilution and the vapor pressures depend on temperature following the Gibbs–Helmholtz (Eq. 5.26) and the Clausius–Clapeyron equation (Eq. 3.64) [7], respectively. The result of the temperature dependences is that azeotropic behavior can occur or disappear with increasing or decreasing temperature (pressure). To understand if azeotropic

---

15) When a strange composition dependence of the activity coefficients exists also two azeotropes with pressure minimum and pressure maximum can be found in a binary system, for example, the system benzene–hexafluorobenzene shows this behavior.

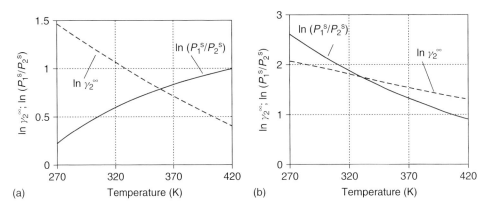

Figure 5.48 Temperature dependence of ln γ_2^∞ (- -) calculated with the help of the NRTL equation and the ratio of the vapor pressures ln P_1^s/P_2^s (—) for the system ethanol (1)–1,4-dioxane (2) (a) and acetone (1)–water (2) (b).

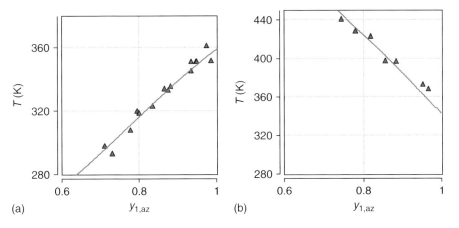

Figure 5.49 Azeotropic behavior of the system ethanol (1)–1,4-dioxane (2) (a) and acetone (1)–water (2) (b). ▲ experimental [37]; — calculated using NRTL.

behavior occurs or disappears, only the knowledge of the temperature dependence of γ_2^∞ (for positive deviations from Raoult's law) and the ratio of the vapor pressures is required. For the systems ethanol–1,4-dioxane and acetone–water the experimental azeotropic data are shown in Figures 5.48 and 5.49 together with the calculated results using the NRTL equation with the parameters given in Table 5.9. While for the first system the azeotropic behavior disappears at higher temperature, the opposite is true for the second system acetone–water, where azeotropic behavior occurs at temperatures above 70 °C. As can be seen from Figure 5.49 the occurrence and disappearance of the azeotropic behavior for both systems is described reliably with the NRTL model.

Example 5.13

Determine the azeotropic points of the system acetone (1)–methanol (2) at 50, 100, and 150 °C by the ratio of the activity coefficients and vapor pressures using the Wilson equation with the interaction parameters given in the table below.

For the calculation, the following molar volumes:
$v_1 = 74.04 \text{ cm}^3 \text{ mol}^{-1}$; $v_2 = 40.73 \text{ cm}^3 \text{ mol}^{-1}$ should be used.

Parameters of the Antoine equation: $\log P_i^s(\text{mm Hg}) = A - \dfrac{B}{\vartheta(°C) + C}$

Compound	A	B	C
Acetone	7.1327	1219.97	230.653
Methanol	8.08097	1582.27	239.7

Wilson parameters

i	j	a_{ij} (cal mol^{-1})	b_{ij} (cal mol^{-1} K^{-1})
1	2	−60.756	−0.06114
2	1	863.79	−1.0533

Solution

Exemplarily for the calculation of the ratio of the activity coefficients the following composition at a temperature of 100 °C is used: $x_1 = 0.2$ and $x_2 = 0.8$.

First the Wilson interaction parameters $\Delta \lambda_{ij}$ at 100 °C have to be calculated:

$$\Delta \lambda_{ij} = a_{ij} + b_{ij} \cdot T$$

$$\Delta \lambda_{12} = -60.756 - 0.06114 \cdot 373.15 = -83.57 \text{ cal mol}^{-1}$$

$$\Delta \lambda_{21} = 863.79 - 1.0533 \cdot 373.15 = 470.75 \text{ cal mol}^{-1}$$

Then the Wilson parameters Λ_{ij} can be determined:

$$\Lambda_{ij} = \frac{v_j}{v_i} \cdot \exp\left[-\frac{\Delta \lambda_{ij}}{RT}\right]$$

$$\Lambda_{12} = \frac{40.73}{74.04} \cdot \exp\left[-\frac{-83.57}{1.98721 \cdot 373.15}\right] = 0.6157$$

$$\Lambda_{21} = \frac{74.04}{40.73} \cdot \exp\left[-\frac{470.75}{1.98721 \cdot 373.15}\right] = 0.9635$$

With the help of these parameters the required activity coefficients can be calculated. Exemplarily the calculation is shown for the activity coefficient of acetone (1):

$$\ln \gamma_1 = -\ln(x_1 + x_2 \Lambda_{12}) + x_2 \left(\frac{\Lambda_{12}}{x_1 + x_2 \Lambda_{12}} - \frac{\Lambda_{21}}{x_1 \Lambda_{21} + x_2}\right)$$

$$\ln \gamma_1 = -\ln(0.2 + 0.8 \cdot 0.6157)$$
$$+ 0.8 \cdot \left(\frac{0.6157}{0.2 + 0.8 \cdot 0.6157} - \frac{0.9635}{0.2 \cdot 0.9635 + 0.8} \right) = 0.3021$$
$$\gamma_1 = 1.353$$

Similarly the following value is obtained for γ_2:

$$\gamma_2 = 1.024$$

With the help of these activity coefficients the ratio can be estimated:

$$\frac{\gamma_1}{\gamma_2} = \frac{1.353}{1.024} = 1.3213$$

Using the Antoine constants given above one obtains for the ratio of the vapor pressures P_2^s/P_1^s at 100 °C:

$$\frac{P_2^s}{P_1^s} = \frac{2649.26}{2774.1} = 0.955$$

For the whole composition range and the different temperatures selected the results are shown in Figure 5.50. From the diagram it can be seen that a strong temperature (pressure) dependence of the azeotropic composition is observed (50 °C: $x_{1,az} \approx 0.8$; 150 °C: $x_{1,az} \approx 0.2$). This is mainly caused by the different enthalpies of vaporization of the two compounds considered. While acetone shows an enthalpy of vaporization of 29.4 kJ/mol at 50 °C, the value for methanol is

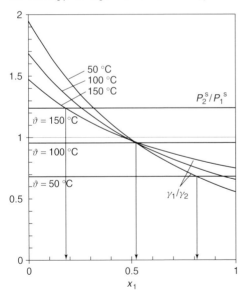

Figure 5.50 Ratio of the activity coefficients γ_1/γ_2 and ratio of the vapor pressures P_2^s/P_1^s of the system acetone (1)–methanol (2) for the determination of the azeotropic points at 50, 100, and 150 °C.

36.1 kJ/mol. The strong pressure dependence observed can be applied in practice for the separation of azeotropic systems by pressure swing distillation, this means using two distillation columns running at different pressures. Since the enthalpies of vaporization of water are higher than for organic compounds, pressure swing distillation is mostly used for the separation of water from organic compounds, as for example, the separation of tetrahydrofuran–water, acetonitrile–water, and so on.

The temperature dependence of the separation factor (see Eq. (5.18)) and of the azeotropic composition of binary systems depends on the type of azeotrope (pressure maximum, pressure minimum), the temperature dependence of the vapor pressures, and the composition and temperature dependence of the activity coefficients. These dependencies can be described with the help of the heats of vaporization and partial molar excess enthalpies following the Clausius–Clapeyron respectively the Gibbs–Helmholtz equation [38] (derivation see Appendix C, B9):

$$\left(\frac{\partial y_1}{\partial T}\right)_{az} = \frac{(y_1 y_2)_{az}\left(\Delta h_{v1} - \Delta h_{v2} + \overline{h}_2^E - \overline{h}_1^E\right)}{RT^2\left[1 - (\partial y_1/\partial x_1)_{az}\right]} \qquad (5.49)$$

where the expression $(\partial y_1/\partial x_1)_{az}$ shows values < 1 for systems with positive deviations from Raoult's law and values > 1 for systems with negative deviations from Raoult's law. In most cases the difference of the enthalpies of vaporization is larger than the difference of the partial molar excess enthalpies.

Example 5.14

Calculate the temperature dependence of the azeotropic composition ($y_{1,az} = 0.9$) of the system ethanol (1)–water (2) at 70 °C using Eq. (5.49). At 70 °C the compounds show the following enthalpies of vaporization:

$\Delta h_{v,\text{ethanol}}$: 39800 J/mol
$\Delta h_{v,\text{water}}$: 42000 J/mol

Solution

Besides the enthalpies of vaporization additionally the difference of the partial molar excess enthalpies and the slope $\partial y_1/\partial x_1$ at the azeotropic point at 70 °C is required. This information can be derived from Figures 5.17 and 5.30. For the difference of the partial molar excess enthalpies $\overline{h}_2^E - \overline{h}_1^E$ approximately a value of 500 J/mol and for the slope a value of 0.9 is obtained.

Using these values the temperature dependence can be calculated:

$$\left(\frac{\partial y_1}{\partial T}\right)_{az} = \frac{0.9 \cdot 0.1 \cdot (39800 - 42000 + 500)}{8.31433 \cdot 343.15^2 (1 - 0.9)} = -1.56 \cdot 10^{-3} \text{ K}^{-1}$$

This means that the mole fraction of ethanol in the azeotrope will decrease with increasing temperature and that the azeotropic behavior should disappear at a lower temperature. This is in agreement with the experimental findings. From the enthalpies used for the calculation, it can be recognized that this is mainly caused by the higher enthalpy of vaporization of water compared to ethanol.

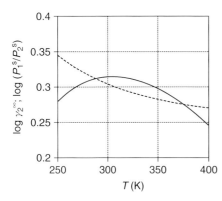

Figure 5.51 γ_2^∞ (—) and the ratio of the vapor pressures P_1^s/P_2^s in logarithmic form (- - -) as a function of temperature for the system acetone (1)–carbon tetrachloride (2).

Depending on the sign of the partial molar excess enthalpies as a function of temperature, the activity coefficient γ_2^∞ can show a maximum or a minimum in the considered temperature range following the Gibbs–Helmholtz relation. This means that the condition for azeotropic points can be fulfilled either only in a small temperature range or at low and again at high temperature. For the maximum case the condition is fulfilled for the system acetone–carbon tetrachloride. The reason is that the sign of the partial molar excess enthalpy changes because of the S-shaped heat of mixing behavior of this system. This results in a maximum for γ_2^∞, so that the condition for azeotropic behavior is fulfilled only in a limited temperature range. Figure 5.51 shows the curvature of γ_2^∞ and the ratio of the vapor pressures in logarithmic form as a function of temperature. The experimental and predicted excess enthalpies and the experimental and predicted azeotropic data using modified UNIFAC are shown in Figures 5.52 and 5.53. It can be seen that modified UNIFAC (see Section 5.9.3.1) is able to predict the occurrence and disappearance of the azeotropic behavior.

Instead of g^E-models also equations of state can be used for the determination of azeotropic behavior of binary or multicomponent systems. In Figure 5.54 the experimental and predicted azeotropic points using the group contribution equation of state VTPR (see Section 5.9.4) for the system ethane–CO_2 up to pressures of 80 bar are shown.

As mentioned before, azeotropic behavior always occurs if the compounds to be separated have identical vapor pressures (Bancroft point). Since the slope of the vapor pressure curve, following the Clausius–Clapeyron equation, depends on the value of the enthalpy of vaporization, a low boiler may become the high boiler with rising temperature, if the enthalpy of vaporization is smaller than the one for the second compound. This is shown for ethanol and benzene in Figure 5.55. While benzene is the low boiler at low temperatures, the opposite becomes true at higher temperatures, since the molar enthalpy of vaporization of the polar component ethanol is larger than the molar enthalpy of vaporization of benzene (see Appendix A).

Even the system water–deuterated water shows a Bancroft point, since deuterated water with a higher normal boiling point (101.4 °C instead of 100 °C) shows a larger enthalpy of vaporization than water. For example, at 25 °C the enthalpy

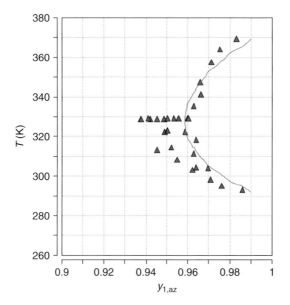

Figure 5.52 Experimental and predicted azeotropic composition of the system acetone (1)–carbon tetrachloride (2) as a function of temperature ▲ experimental [3] — predicted using modified UNIFAC.

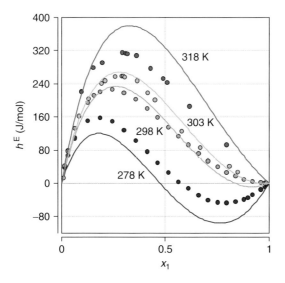

Figure 5.53 Experimental and predicted excess enthalpies of the system acetone (1)–carbon tetrachloride (2) as a function of temperature ● experimental data at 5, 25, 30, 45 °C [3] — predicted using modified UNIFAC.

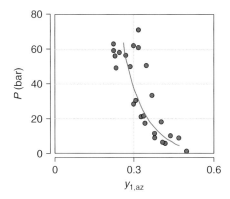

Figure 5.54 Experimental and predicted azeotropic composition of the system ethane (1)–CO_2 (2) as a function of pressure. ● experimental [3], —— predicted using VTPR (see Section 5.9.4).

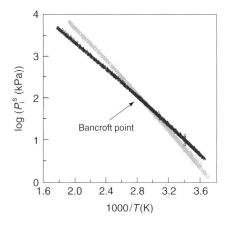

Figure 5.55 Vapor pressure of ethanol and benzene as a function of the inverse temperature experimental data [3] ● ethanol ▲ benzene.

of vaporization for water is 43.9 kJ/mol and for D_2O 45.35 kJ/mol. The result is that at temperatures around 220 °C the vapor pressures of water and deuterated water become identical following the Clausius–Clapeyron equation, so that even the system water–deuterated water shows azeotropic behavior in a limited temperature range (493–495 K) near the Bancroft point, as shown in Figure 5.56.

Azeotropic behavior is not limited to binary systems only. Also ternary and quaternary azeotropic points are observed. For the determination of the azeotropic points in ternary and quaternary systems, thermodynamic models (g^E-models, equations of state, group contribution methods) can again be applied [40]. Azeotropic points in homogeneous systems can be found with the help of nonlinear regression methods. At the azeotropic point all separation factors α_{ij} show a value of 1 in the case of homogeneous systems. This means that the following condition has to be fulfilled:

$$F = \sum_{i}^{n} \sum_{j>i}^{n} |\alpha_{ij} - 1| = 0 \tag{5.50}$$

For the azeotropic system acetone–chloroform–methanol, the three possible separation factor curves with a value of 1 are shown in Figure 5.57. The intersection

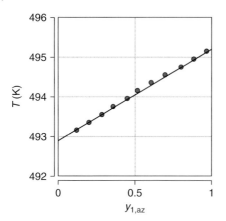

Figure 5.56 Experimental azeotropic points of water (1)–deuterated water (2) [3, 39].

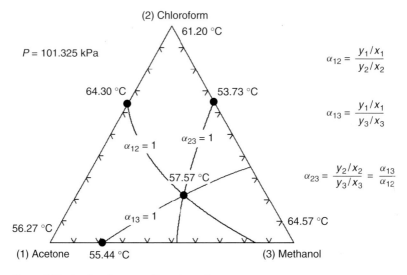

Figure 5.57 Predicted contour lines ($\alpha_{ij} = 1$) using modified UNIFAC for the ternary system acetone (1)–chloroform (2)–methanol (3) at atmospheric pressure).

of two of these curves, for example, $\alpha_{12} = 1$ and $\alpha_{13} = 1$ leads to an azeotropic point, since at the intersection point the criterion for the separation factor α_{23} is automatically fulfilled (see Figure 5.57).

In the case of heterogeneous azeotropic mixtures a different calculation procedure has to be applied. For the ternary system water–ethanol–benzene this is shown in Figure 5.58. Heterogeneous azeotropic behavior occurs if a pressure maximum can be found along the binodal curve. The required pressures can be calculated starting from the composition of the heterogeneous binary system up to the critical point. In doing so, one can start from the composition in the organic or the aqueous phase. The result is shown on Figure 5.58b. It can be seen that for both procedures a pressure maximum occurs, this means heterogeneous azeotropic

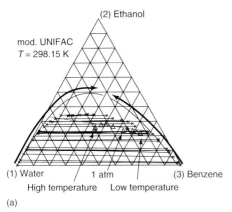 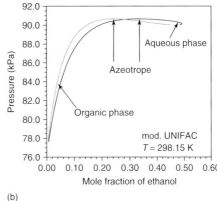

Figure 5.58 Experimental and calculated LLE behavior using modified UNIFAC for the system water–ethanol–benzene at 298.15 K (a) pressure as a function of composition of the benzene rich phase resp. water-rich phase (b).

behavior is obtained for this system. In Figure 5.58, the temperature dependence of the azeotropic composition is also shown. It can be seen that the concentration of water in the azeotrope increases with increasing temperature (pressure). The main reason for the observed temperature dependence is that the vapor pressure of water increases faster than the vapor pressures of ethanol and benzene, because of the larger enthalpy of vaporization.

Azeotropic behavior is also obtained for quaternary systems. But fortunately, azeotropic points in quinary or higher systems do not exist, since with increasing number of components it becomes more and more unlikely that for one composition all the separation factors become exactly unity.

For a large number of systems experimental (a)zeotropic information can be found in a comprehensive data compilation [37]. The knowledge of the azeotropic points is of special importance during the synthesis of separation processes and the selection of suitable solvents for azeotropic distillation.

5.7
Solubility of Gases in Liquids

The objective of absorption processes is the separation of gas mixtures or the removal of undesired compounds from gas mixtures. For the selection of the optimal solvent or solvent mixture (absorbent) and the design of absorption processes a reliable knowledge of the gas solubility as a function of temperature and pressure is of special importance.[16]

16) The reliable knowledge of gas solubilities is also required for the design of gas–liquid reactors.

Industrially important absorption processes are for example the removal of sour gases (CO_2, H_2S) from natural gas or synthesis gas, the removal of carbon dioxide in chemical plants such as ethylene oxide plants, the removal of SO_2 from flue gas, or the absorption of CO_2 in power plants (carbon capture and storage (CCS)), and so on. One has to distinguish physical and chemical absorption

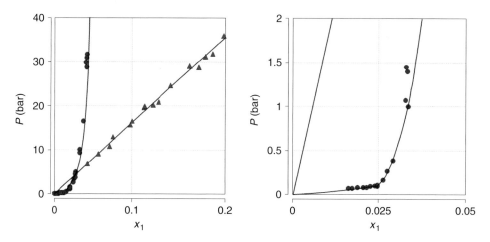

Figure 5.59 Experimental solubilities [3] of CO_2 in methanol (▲) and aqueous monoethanolamine solution (30 mass%) (●) at $T = 313.15$ K.

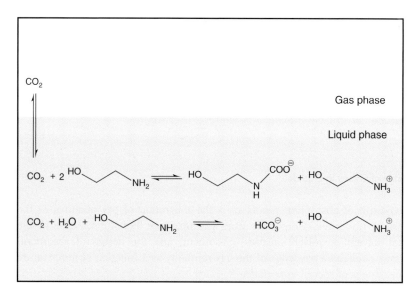

Figure 5.60 Chemical reactions which have to be considered besides the gas solubility for the absorption of CO_2 in aqueous monoethanolamine solutions.

processes. For physical absorption processes, only the knowledge of the phase equilibrium behavior is required. For chemical absorption processes, all chemical equilibria have to be taken into account in addition to the phase equilibria. Often, reaction kinetics and mass transfer has to be regarded as well. In Figure 5.59 the solubility of carbon dioxide in methanol (physical absorption) and aqueous monoethanolamine solution (chemical absorption) is shown for a temperature of 313.15 K.[17] The reactions which have to be considered for the absorption of CO_2 in monoethanolamine are shown in Figure 5.60. From Figure 5.59 it can be recognized that chemical absorption shows great advantages compared to physical absorption, when the partial pressure of the gas to be absorbed is low, as in the case of CCS-processes. Physical absorption shows advantages at high partial pressures.

For the calculation of gas solubilities for physical absorption processes both approaches discussed in Section 5.1 (Eqs. (5.9) and (5.10)) for VLE calculations can be applied.

$$x_i \varphi_i^L = y_i \varphi_i^V \tag{5.9}$$

$$x_i \gamma_i f_i^0 = y_i \varphi_i^V P \tag{5.10}$$

While there is no difference for the calculation of gas solubilities in comparison to VLE in the case of the equation of state (approach A), for approach B, there is the problem that the standard state (pure liquid at system temperature and system pressure) used for VLE calculations cannot be used anymore, since supercritical compounds are not existent as liquid. This means an alternative standard fugacity is required for the γ–φ-approach.

5.7.1
Calculation of Gas Solubilities Using Henry Constants

An alternative is the usage of the Henry constant H_{ij} as standard fugacity f_i^0. Using the Henry constant as standard fugacity, the following expression is obtained for the calculation of gas solubilities in a binary system:

$$x_1 \gamma_1^* H_{12} = y_1 \varphi_1 P \tag{5.51}$$

1 = solute, 2 = solvent

The *Henry constant* is defined as

$$H_{12} = \lim_{\substack{x_1 \to 0 \\ x_2 \to 1 \\ P \to P_2^s}} \frac{f_1}{x_1} \tag{5.52}$$

and is exactly valid only for the case where the partial pressure of the gas is equal to zero, this means when the total pressure is identical to the vapor pressure of the solvent.

Since in the case of gas solubilities normally only a small concentration range is covered in the liquid phase, simple expressions for the description of the

---

17) A temperature of 313.15 K was chosen to be able to compare the solubilities. In practice methanol is used as absorbent (Rectisol process) at much lower temperatures.

concentration dependence of the activity coefficient γ_1^* like for example, Eq. (5.53) (so-called Porter equation) can be applied:

$$\ln \gamma_1^* = A \left(x_2^2 - 1\right) \tag{5.53}$$

The simple expressions used have to satisfy the asymmetric convention ($\gamma_1^* = 1$ for $x_1 \to 0$, $x_2 \to 1$). However, one should have in mind that equations of state have significant advantages if γ_1^* is significantly different from unity. The application of the Henry constant should be restricted to low solute concentrations $x_1 < 0.03$, so that Eq. (5.51) reduces to the usually applied form:

$$x_1 H_{12} = y_1 \varphi_1 P \tag{5.54}$$

At low pressures the fugacity coefficient also shows values close to unity, so that the following simple expression (so-called Henry's law) is obtained:

$$x_1 H_{12} = y_1 P = p_1 \tag{5.55}$$

A comparison of Eq. (5.55) with Eq. (5.16) at infinite dilution of component 1 shows that in the subcritical region the Henry constant corresponds to

$$H_{12} \approx \gamma_1^\infty \cdot P_1^s \tag{5.56}$$

At low pressures the following statements can be derived from Henry's law:

- the gas solubility is proportional to the partial pressure;
- the gas solubility is proportional to the reciprocal value of the Henry constant;
- the temperature dependence of the gas solubility is only determined by the temperature dependence of the Henry constant.

In comparison to the standard fugacity "pure liquid at system temperature and system pressure" used for VLE calculations, there is the great disadvantage of the Henry constant that it is not a pure component property, but has to be derived from experimental gas solubility data.

The value of the Henry constant can be very different. It strongly depends on the properties of the gas (T_c, P_c) and strength of the interactions with the solvent. In Table 5.12 Henry constants for various gases in water are listed for a temperature of 25 °C. It can be seen that these values differ by orders of magnitude. While for the light gases (He, Ar, H_2, N_2, O_2, CO, CH_4, SF_6) Henry constants greater than 40000 bar are observed, values around 1000 bar are found for CO_2, H_2S, C_2H_2 in water, where it is surprising that the Henry constant for the relatively large compound SF_6 is even greater than for helium. In Table 5.13, Henry constants for six gases in four solvents are given for a temperature of 25 °C. The values show that also the interactions between the gas and the solvent play an important role. For all the gases the values are significantly different between the polar solvent methanol and the nonpolar solvent n-heptane, caused by the different intermolecular forces between the compounds. Looking at the Henry constant of the sour gases (CO_2, H_2S) and methane in methanol, it seems that methanol is a highly selective absorbent for the removal of sour gases from natural gas. This effect is realized in the so-called Rectisol process [41]. Furthermore, Henry constants show strong,

Table 5.12 Henry constants of various gases in water at 25 °C [3].

Gas	H_{ij} (bar)
He	144000
Ar	40000
H_2	71000
N_2	83500
O_2	44200
H_2S	580
CO	58000
CO_2	1660
CH_4	40200
C_2H_2	1350
C_2H_4	11700
C_2H_6	30400
SF_6	236000

Table 5.13 Henry constants (bar) of various gases in different organic liquids at 25 °C [3].

	H_2	N_2	O_2	H_2S	CO_2	CH_4
Methanol	6100	3900	2200	33.5	145	1180
Acetone	3400	1850	1200	14.5	50	545
Benzene	3850	2300	1260	19.0	105	490
Heptane	1450	760	500	23.4	78	210

nonlinear temperature dependence. In Figures 5.61 and 5.62, Henry constants for various systems are shown as a function of temperature. For the three systems helium, nitrogen, and oxygen in water shown in Figure 5.61 even a maximum of the Henry constant is observed. This means that the gas solubility for a given partial pressure can increase as well as decrease with increasing temperature depending on the temperature range considered. In Figure 5.62 it is shown that for the systems with hydrogen the Henry constant decreases with increasing temperature, while the opposite behavior is observed for methane in methanol and carbon dioxide in toluene.

In process simulators, the temperature dependence of the Henry constants is often described by the following expression:

$$\ln \frac{H_{12}(T)}{\text{bar}} = A_{12} + \frac{B_{12}}{T} + C_{12} \ln \frac{T}{K} + D_{12}T + \frac{E_{12}}{T^2} \quad (5.57)$$

Usually the data situation does not justify adjusting all parameters. In most cases only two of them (A, B or A, D) can be fitted. The value of the constant A depends on the chosen unit for the pressure.

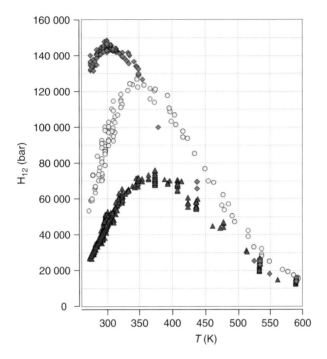

Figure 5.61 Henry constants of He, N_2, and O_2 in water as a function of temperature experimental [3] ◆ helium, ● nitrogen, ▲ oxygen.

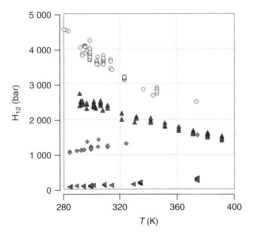

Figure 5.62 Henry constants of various gases in organic solvents as a function of temperature. Experimental data from [3]. ● hydrogen in benzene, ▲ hydrogen in cyclohexane, ◆ methane in methanol, ◄ carbon dioxide in toluene.

Example 5.15

Determine the Henry constant for carbon dioxide (1) in water (2) with the help of the following phase equilibrium data at 50 °C:

$x_1 \cdot 10^3$	p_1 (kPa)
0.342	101.33
0.683	202.65
1.354	405.3
2.02	607.95
2.66	810.6
3.3	1013.3
3.93	1216
4.55	1419
5.15	1621
5.75	1824

$P^s_{H_2O}$ = 12.3 kPa at 50 °C, virial coefficients: $B_{11} = -102$ cm^3/mol, $B_{12} = -198$ cm^3/mol, $B_{22} = -812$ cm^3/mol.

Solution

For the calculation of the fugacity coefficients the total pressure has to be determined first. Since the partial pressure of water is approximately identical with the vapor pressure at 50 °C (12.3 kPa), the total pressure can directly be calculated, for example, for $p_1 = 405.3$ kPa (data point 3):

$$P = 405.3 + 12.3 = 417.6 \text{ kPa}$$

With these values the vapor phase composition is obtained:

$$y_1 = \frac{405.3}{417.6} = 0.9705$$

With this information, the second virial coefficient B and the fugacity coefficient for carbon dioxide can be calculated using Eq. (4.89):

$$B = 0.9705^2 (-102) + 2 \cdot 0.9705 \cdot 0.0295 (-198)$$
$$+ 0.0295^2 (-812) = -108 \text{ cm}^3/\text{mol}$$

With Eq. (4.87) follows for the fugacity coefficient:

$$\ln \varphi_1 = \{2 [0.9705 (-102) + 0.0295 (-198)] + 108\} \frac{417.6}{8314.33 \cdot 323.15}$$

$$\varphi_1 = 0.9843$$

In the next step the fugacity f_1 and the ratio f_1/x_1 can be calculated:

$$f_1 = y_1 \varphi_1 P = \varphi_1 p_1 = 0.9843 \cdot 405.3 = 398.94 \text{ kPa}$$
$$f_1/x_1 = 398.94/0.001354 = 294600 \text{ kPa} = 2946 \text{ bar}$$

Table 5.14 Experimental gas solubility data, fugacities, fugacity coefficients and the ratio f_1/x_1 for the system CO_2–water at 50 °C.

$x_1 \cdot 10^3$	p_1 (kPa)	φ_1	f_1 (kPa)	f_1/x_1 (bar)
0.342	101.33	0.9960	100.92	2951
0.683	202.65	0.9920	201.01	2943
1.354	405.3	0.9843	398.94	2946
2.02	607.95	0.9768	593.83	2940
2.66	810.6	0.9693	785.7	2954
3.3	1013.3	0.9618	974.63	2953
3.93	1216	0.9545	1160.6	2953
4.55	1419	0.9471	1344.0	2954
5.15	1621	0.9399	1523.6	2958
5.75	1824	0.9327	1701.2	2959

For the other data points the fugacity coefficients φ_1, fugacities f_1 and ratios f_1/x_1 are given in Table 5.14.

From a diagram (see Figure 5.63a) where the ratio f_1/x_1 is plotted against the liquid mole fraction of carbon dioxide, the Henry constant can be determined at the mole fraction $x_1 = 0$. Besides the ratio f_1/x_1, additionally the ratio p_1/x_1 is shown in Figure 5.63. While the ratio f_1/x_1 stays nearly constant, the values for the ratio p_1/x_1 are distinctly different already at low partial pressures. But the extrapolation to $x_1 = 0$ ($p_1 \to 0$) leads to the same value for the Henry constant ($H_{1,2} \approx 2950$ bar).

Another option to determine the Henry constant is the plot of the fugacity f_1 over the mole fraction x_1. That is shown in Figure 5.63b. In this diagram the Henry

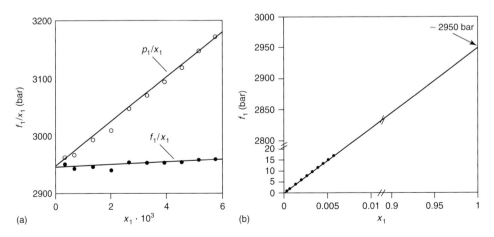

Figure 5.63 f_1/x_1 (p_1/x_1) resp. f_1 as a function of x_1 for the system CO_2(1)–water (2) at 50 °C.

constant is obtained from the straight line through the experimental data at the intersection at $x_1 = 1$. Both procedures lead to the same value.

Henry's law is valid exactly only for $P = P_2^s$. However, the pressure dependence of the Henry constant is relatively low, but it can be taken into account by

$$\left(\frac{\partial \ln H_{12}}{\partial P}\right)_T = \frac{\bar{v}_1^\infty}{RT} \qquad (5.58)$$

where $\bar{v}_1^\infty$ [18] is the partial molar volume of the dissolved gas (1) in the solvent (2) at infinite dilution. Assuming that the partial molar volume is constant, the Henry constant at the pressure P can directly be calculated using Eq. (5.59), which is known as *Krichevsky–Kasarnovsky equation*.

$$\ln H_{12}(P) = \ln H_{12}(P_2^s) + \frac{\bar{v}_1^\infty (P - P_2^s)}{RT} \qquad (5.59)$$

For mixed solvents, an empirical logarithmic mixing rule

$$\ln \frac{H_{i,\text{mix}}}{\text{bar}} = \frac{\sum_j x_j \ln \frac{H_{ij}}{\text{bar}}}{\sum_j x_j} \qquad (5.60)$$

can be applied, where the summation is only carried out for solvents for which the Henry constant is known.

This mixing rule makes sense only in cases where the Henry constants for the gases in the highly concentrated compounds of the solvent are known.

Example 5.16

Calculate the Henry constant of CO_2 in a liquid mixture of methanol, water and trioxane at 25 °C. The concentrations and the Henry constants of CO_2 in the pure solvents are given in the following table:

	x	$H_{CO_2,j}$ (bar) at $\vartheta = 25\,°C$
Methanol	0.39	145
Water	0.6	1660
Trioxane	0.01	unknown

Solution

Using Eq. (5.60), one has to take into account that only the concentrations of methanol and water are counted, as the Henry constant of CO_2 in trioxane is not

---

[18] The partial molar volumes at infinite dilution can be obtained from the observed volume change during absorption.

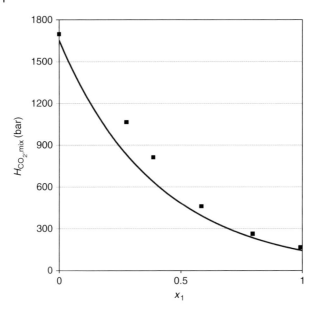

Figure 5.64 Henry constant of CO_2 in the system methanol (1)–water (2) at 25 °C — correlation; ■ experimental data from [42].

known:

$$\ln\frac{H_{CO_2,\text{mix}}}{\text{bar}} = \frac{\sum_j x_j \ln \frac{H_{CO_2,j}}{\text{bar}}}{\sum_j x_j} = \frac{0.39 \cdot \ln 145 + 0.6 \cdot \ln 1660}{0.39 + 0.6}$$

$$= 6.4545 \Rightarrow H_{CO_2,\text{mix}} = 635 \text{ bar}$$

There is an experimental value for $x_{\text{methanol}} = 0.3885$ and $x_{\text{water}} = 0.6115$ at $\vartheta = 25\,°C$, giving $H_{CO_2,\text{mix}} = 749$ bar [42]. Clearly, the simple mixing rule can by far not be taken as exact; however, at least the correct order of magnitude is met, which is usually sufficient in process simulation. Figure 5.64 shows the relationship of the Henry constant H_{CO_2} at $\vartheta = 25\,°C$ as a function of the methanol concentration. It can be seen that in this case the mixing rule works in a qualitatively correct way.

When dissolving a gas in the liquid phase the enthalpy of the gas changes similarly to the enthalpy change of vaporization. The enthalpy of solution at infinite dilution $\Delta h_i^{\infty L}$ – the enthalpy difference between the gaseous and the dissolved solute – can be expressed with the help of the Henry constant as derived below.

Starting from the phase equilibrium condition

$$\mu_i^V = \mu_i^L \tag{5.61}$$

and

$$\mu_i^V = \mu_i^{0V} + RT \ln \frac{f_i^V}{f_i^{0V}} \tag{5.62}$$

where the standard fugacity is based on the pure component, and

$$\mu_i^L = \mu_i^{\infty L} + RT \ln(x_i \gamma_i^*) \tag{5.63}$$

where the standard fugacity is based on the state of infinite dilution, one obtains

$$\frac{\mu_i^{0V} - \mu_i^{\infty L}}{RT} = \ln \frac{x_i f_i^{0V}}{f_i^V} \tag{5.64}$$

as γ_i^* becomes unity at infinite dilution. Considering

$$H_{i,\text{mix}} = \lim_{x_1 \to 0} \frac{f_i^L}{x_i} \quad \text{and} \quad f_i^L = f_i^V$$

one obtains

$$\frac{\mu_i^{0V} - \mu_i^{\infty L}}{RT} = -\ln \frac{H_{i,\text{mix}}}{f_i^{0V}} \tag{5.65}$$

Using the van't Hoff equation (see Appendix C, A7), differentiation of both sides with respect to temperature yields

$$-\frac{h_i^{0V} - h_i^{\infty L}}{RT^2} = -\frac{d \ln \frac{H_{i,\text{mix}}}{f_i^{0V}}}{dT} = -\frac{f_i^{0V}}{H_{i,\text{mix}} f_i^{0V}} \frac{1}{dT} \frac{dH_{i,\text{mix}}}{dT} = -\frac{1}{H_{i,\text{mix}}} \frac{dH_{i,\text{mix}}}{dT} \tag{5.66}$$

and therefore

$$\Delta h_i^{\infty L} = h_i^{0V} - h_i^{\infty L} = \frac{RT^2}{H_{i,\text{mix}}} \frac{dH_{i,\text{mix}}}{dT} \tag{5.67}$$

The application of Henry's law is recommended especially for systems with a single solvent, such as the solubility of nitrogen in water or, as shown below, the solubility of methane in benzene. In multicomponent mixtures with one or more

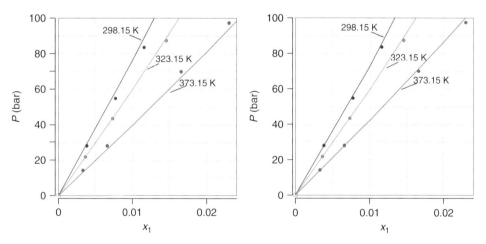

Figure 5.65 Experimental and calculated Px-data using the SRK equation of state with quadratic mixing rules for the system nitrogen (1)–NMP (2) at different temperatures. (a) $k_{12} = 0.3403$, (b) $k_{12} = -0.07938 + 0.001297\,T$.

supercritical compounds the use of Henry constants must be examined carefully. In this case the use of an equation of state should be preferred.

5.7.2
Calculation of Gas Solubilities Using Equations of State

As mentioned before, the great advantage of the equation of state approach is that for the calculation of VLE no standard fugacity (vapor pressure, Henry constant) is required. This means that there is no difference in the calculation procedure for VLE and gas solubilities. In Figure 5.65 typical results are shown for the system nitrogen–NMP (N-Methylpyrrolidone) at different temperatures. For the correlation the SRK equation of state with quadratic mixing rules was used. It can be seen that the results can be slightly improved when a binary parameter k_{12} with a linear temperature dependence is used for the temperature range covered.

Example 5.17

Calculate the Henry constant for methane in benzene at 60 °C with the help of the SRK equation of state ($k_{12} = 0.08$ [36]).
Pure component properties

Component	T_c (K)	P_c (bar)	ω
Methane	190.63	46.17	0.010
Benzene	562.6	49.24	0.212

Solution

The calculation can be carried out in the same way as shown in Example 5.12 for the system nitrogen–methane. For the calculation only initial values for the pressure and vapor phase mole fraction are required. With the calculated fugacity coefficients new values for the pressure and vapor phase mole fractions can be calculated. This iterative procedure is stopped at a given convergence criterion. For a mole fraction of $x_1 = 0.01$ in the liquid phase in equilibrium the following values are obtained:

$y_1 = 0.9014$
$P = 5.666$ bar

$\varphi_1^L = 89.70 \quad \varphi_2^L = 0.0915$
$\varphi_1^V = 0.9951 \quad \varphi_2^V = 0.9193$

Using these values the Henry constant can be calculated:

$$H_{12} = \frac{f_1}{x_1} = \frac{y_1 \varphi_1^V P}{x_1} = \frac{0.9014 \cdot 0.9951 \cdot 5.666}{0.01} = 508.2 \text{ bar}$$

or

$$H_{12} = \frac{f_1}{x_1} = \frac{x_1 \varphi_1^L P}{x_1} = \varphi_1^L P = 89.70 \cdot 5.666 = 508.2 \text{ bar}$$

In fact the Henry constant should be calculated for $x_1 \to 0$. But when the interaction energies are not too strong as in this case, the Henry constant can be used up to a few mole-%.

5.7.3
Prediction of Gas Solubilities

If no experimental data are available gas solubilities can be predicted today with the help of group contribution equations of state, such as Predictive Soave–Redlich–Kwong (PSRK) [43] or VTPR [44]. These models are introduced in Sections 5.9.4 and 5.9.5.

Up to the 1970s, methods based on the regular solution theory and the fugacity of a hypothetical liquid were suggested for the prediction of gas solubilities. This procedure can lead to reasonable results as long as only nonpolar components are regarded. According to the method of Prausnitz and Shair [45], the reduced standard fugacity of the solute (hypothetical liquid) is described by the following expression:

$$\ln \frac{f_1^0 (1.013 \text{ bar})}{P_{c1}} = 7.81 - \frac{8.06}{T_{r1}} - 2.94 \ln T_{r1} \tag{5.68}$$

which is valid in the temperature range $0.7 < T_{r1} < 2.5$. Figure 5.66 shows the temperature dependence of the reduced standard fugacity of the hypothetical liquid.

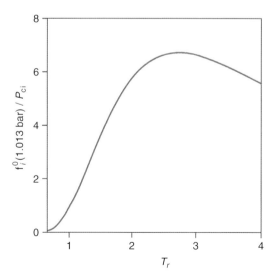

Figure 5.66 Temperature dependence of the reduced standard fugacity of the solute (hypothetical liquid).

5 Phase Equilibria in Fluid Systems

Table 5.15 Hypothetical liquid molar volumes and solubility parameters at $\vartheta = 25\,°C$.

Gas	v_i (10^{-6} m^3/mol)	δ_i (J/m^3)$^{0.5}$	T_c (K)	P_c (bar)
N$_2$	32.4	5279	126.2	33.9
CO	32.1	6405	132.9	35
O$_2$	33	8185	154.6	50.4
Ar	57.1	10906	150.8	48.7
CH$_4$	52	11622	190.4	46
CO$_2$	55	12277	304.1	73.8
Kr	65	13095	209.4	55
C$_2$H$_4$	65	13505	282.4	50.4
C$_2$H$_6$	70	13505	305.4	48.8
Rn	70	18068	377	62.8
Cl$_2$	74	17802	416.9	79.8

According to the regular solution theory, the corresponding activity coefficient at infinite dilution can be expressed as

$$\ln \gamma_1^\infty = \frac{v_1}{RT}(\delta_1 - \delta_2)^2 \tag{5.69}$$

with the solubility parameter

$$\delta_i(T) = \left(\frac{\Delta h_{vi}(T) - RT}{v_i}\right)^{0.5} \tag{5.70}$$

For supercritical gases no liquid phase and thus no values for v_i and Δh_{vi} exist. In Table 5.15 hypothetical values for the molar liquid volume and the solubility parameter for some well-known light gases at $\vartheta = 25°C$ are listed. As nothing better is available, these values are also applied at other temperatures as well. The Henry constant can finally be calculated using Eq. (5.56), where instead of the vapor pressure the fugacity of the hypothetical liquid is used.

Example 5.18

Estimate the Henry constant of methane (1) in benzene (2) at $\vartheta = 60\,°C$ using the method of Prausnitz and Shair.

Solution

The values for the solute from Table 5.15 are

$$v_1 = 52 \cdot 10^{-6}\ \text{m}^3/\text{mol}$$
$$\delta_1 = 11622\ (\text{J/m}^3)^{0.5}$$

$$T_{c1} = 190.4\ \text{K}$$
$$P_{c1} = 46.0\ \text{bar}$$

For benzene, the solubility parameter can be determined using the following information:

$$M = 78.114 \text{ g/mol}$$

$$\Delta h_{v2}(60°C) = 408.7 \text{ J/g}$$

$$\rho_2(60°C) = 837.9 \text{ kg/m}^3 \Rightarrow v_2 = 93.22 \cdot 10^{-6} \text{ m}^3/\text{mol}$$

leading to

$$\delta_2 = \left(\frac{408.7 \cdot 78.114 - 8.31433 \cdot 333.15}{93.22 \cdot 10^{-6}}\right)^{0.5} = 17685 \text{ (J/m}^3)^{0.5}$$

Thus, the activity coefficient of the solute at infinite dilution is calculated via

$$\ln \gamma_1^\infty = \frac{v_1}{RT}(\delta_1 - \delta_2)^2 = \frac{52 \cdot 10^{-6}}{8.3143 \cdot 333.15}(11622 - 17685)^2$$

$$= 0.6901 \Rightarrow \gamma_1^\infty = 1.994$$

The standard fugacity can be determined with Eq. (5.68) for $T_{r_1} = 333.15/190.4 = 1.75$:

$$f_1^0 = 46.0 \exp\left(7.81 - \frac{8.06}{1.75} - 2.94 \ln 1.75\right) = 218.7 \text{ bar}$$

The result for the Henry constant at $\vartheta = 60°C$ is

$$H_{12}(60°C) = 218.7 \text{ bar} \cdot 1.994 = 436.1 \text{ bar}$$

The experimental value is approx. 513 bar [46].

5.8
Liquid–Liquid Equilibria

In Section 5.2 it was shown that strongly real behavior leads to the formation of two liquid phases with different compositions. The concentration differences of the compounds in the different phases can be used, for example, for the separation by extraction. As in the case of other phase equilibria, the fugacities in the different liquid phases are identical in the case of LLE:

$$f_i' = f_i'' \quad i = 1, 2, \ldots, n \tag{5.71}$$

As shown before, the fugacities can either be described using activity coefficients or fugacity coefficients. Using activity coefficients the following relation is obtained:

$$(x_i \gamma_i f_i^0)' = (x_i \gamma_i f_i^0)'' \tag{5.72}$$

Since the standard fugacity f_i^0 is the same for the two liquid phases, the following simple equation results from Eq. (5.72):

$$x_i' \gamma_i' = x_i'' \gamma_i'' \tag{5.73}$$

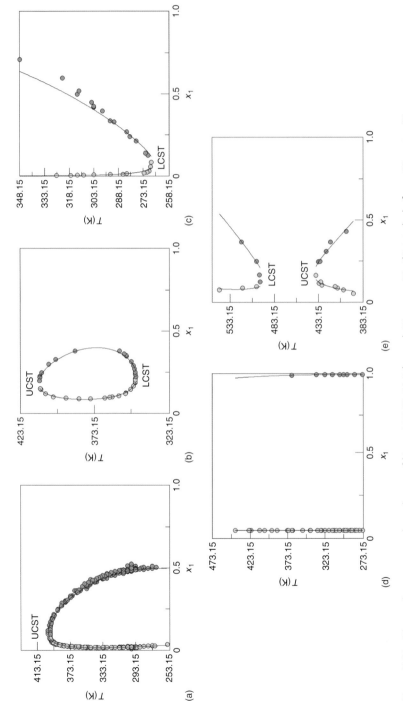

Figure 5.67 Observed temperature dependences of binary LLE [3] (a) 1-butanol (1)–water (2), (b) tetrahydrofuran (1)–water (2), (c) dipropylamine (1)–water (2), (d) n-hexane (1)–water (2), (e) benzene (1)–sulfur (2).

The product $x_i \gamma_i$ is also called *activity* a_i (see section 4.8). This means that the so-called isoactivity criterion has to be fulfilled in the case of LLE.

Using fugacity coefficients, a similar relation results:

$$\left(x_i \varphi_i^L\right)' = \left(x_i \varphi_i^L\right)'' \tag{5.74}$$

Two liquid phases always occur in the case of strong positive deviation from Raoult's law. The LLE behavior as a function of temperature only depends on the temperature dependence of the activity coefficients. The possible temperature dependencies for binary systems at constant pressure[19] are shown in Figure 5.67 in the form of the temperature-concentration-diagrams, the so-called binodal curves.

In most cases the mutual solubility rises with increasing temperature until the system becomes homogeneous above the upper critical solution temperature (UCST). This behavior is shown in Figure 5.67a. The other cases shown in Figure 5.67 occur more rarely than this behavior. In case (c) the mutual solubility increases with decreasing temperature until the two-phase region completely disappears below the lower critical solution temperature (LCST). Sometimes, even both critical solution temperatures occur (case (b)). Finally, there are systems with a miscibility gap over the entire temperature range. Cases (a), (c), and (d) can be regarded as special cases of (b), as in many cases the binodal line is interrupted by the melting curve, the boiling curve, or both. A very complex behavior is found for sulfur with aromatic compounds, for example, benzene–sulfur. For this system the LLE behavior disappears at the UCST. But at higher temperatures again a miscibility gap occurs.

For the ternary case, the most frequently observed curve shapes are shown in the form of triangular diagrams in Figure 5.68. Like in binary systems, the two-phase region is limited by the binodal curve. The two liquid phases in equilibrium are connected by so-called tie lines. From the tie line end points the distribution coefficient K_i between the two phases ' and " can be calculated.

$$K_i = \frac{x_i''}{x_i'} \tag{5.75}$$

In so-called closed systems (case (a)), which are observed for about 75% of the systems, only one binary pair shows a miscibility gap. For these systems, a critical point C arises, where both liquid phases show the same concentration. Case (b) presents a system where two binary pairs show partial miscibility (open system). This behavior occurs in about 20% of all cases. Besides these most important cases, however, there are a large number of other possibilities [47]. For example, there are systems where all binary subsystems are homogeneous, but a miscibility gap (island) is found in the ternary system (see Figure 5.76). Additionally, there is the chance that three liquid phases are formed.

---

19) For not too large pressure differences, for example a few bar, the pressure influence can usually be neglected for condensed phases. But as shown in Sections 5.8.2 and 8.1.4, high pressures can have a significant influence on the LLE and SLE behavior.

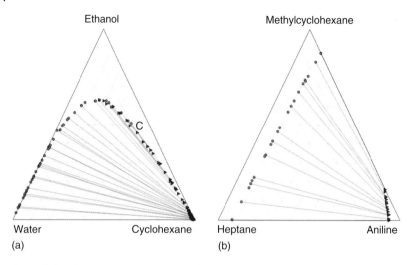

Figure 5.68 The most important types of ternary LLE [3] at a temperature of 25 °C.

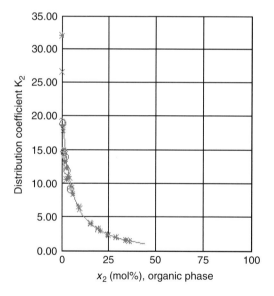

Figure 5.69 Experimental [3] and calculated partition coefficients of ethanol in the system water(1)–ethanol(2)–cyclohexane(3) at 298.15 K using UNIQUAC.

The distribution coefficients are not constant. They strongly depend on the concentration. For the system water–ethanol–cyclohexane the distribution coefficients for ethanol are shown in Figure 5.69. It can be seen that the largest distribution coefficients are obtained at infinite dilution. These values at infinite dilution are called *Nernst distribution coefficients*.

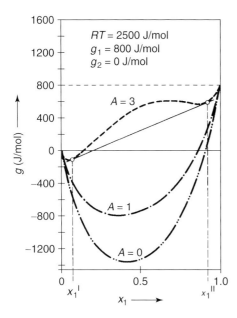

Figure 5.70 Concentration dependence of the molar Gibbs energy for systems with different strong real behavior A = coefficient of the Porter equation (Eq. (5.19)).

Equation (5.73) forms the basis for the calculation of LLE. As can be learned from this equation, the concentration and the temperature dependence of LLE is described via the activity coefficients. However, the occurrence of two liquid phases and critical solution temperatures cannot be understood alone by this equation (isoactivity criterion).

Looking at the concentration dependence of the Gibbs energy, it is easier to understand the formation of two liquid phases. For an ideal binary system, the molar Gibbs energy at a given composition (see Chapter 4) can be calculated by

$$g_{ideal} = x_1 g_1 + x_2 g_2 + RT(x_1 \ln x_1 + x_2 \ln x_2) \tag{5.76}$$

Since the mole fractions are always smaller than 1, the last term is negative and zero for $x_1 = 1$ and $x_2 = 1$. Therefore, the molar Gibbs energy as a function of composition shows a minimum (see Figure 5.70 with A = 0). However, in case of a real system the excess Gibbs energy has to be added:

$$g_{real} = g_{ideal} + g^E \tag{5.77}$$

$$g_{real} = g_{ideal} + RT(x_1 \ln \gamma_1 + x_2 \ln \gamma_2) \tag{5.78}$$

With increasing positive deviation from Raoult's law the positive contribution of the excess Gibbs energy is enlarged and, consequently, the molar Gibbs energy at the composition considered. This is shown in Figure 5.70. For example, it is assumed that the contribution of the excess Gibbs energy can be taken into account by Porter's approach ($g^E/RT = A x_1 x_2$). The resulting curve shape of the molar

Gibbs energy for different values of the parameter A is shown in Figure 5.70. With increasing g^E-values, this means increasing values of the parameter A, the molar Gibbs energy becomes larger. Because of the equilibrium criterion (minimum of Gibbs energy), in the case of strongly real behavior (i.e., in Figure 5.70 for $A = 3$) two liquid phases with the concentration x_1' and x_1'' are formed. At these concentrations the molar Gibbs energy of the two phases shows a lower value than a homogeneous mixture. For the Gibbs energy for a mole fraction of $x_1 = 0.5$, an approximate value of 540 J/mol is obtained for the homogeneous composition, while for the heterogeneous composition an approximate value of 250 J/mol results. In Figure 5.70 the values of the molar Gibbs energy for the composition of the coexisting liquid phases ' and '' are shown by the tangent and for the homogeneous composition by the dashed line.

According to this, the formation of two liquid phases can only occur if the curve shape of the Gibbs energy as a function of the composition shows an inflexion point, that is, there must be a region where the following condition is valid:

$$\left(\frac{\partial^2 g}{\partial x^2}\right)_{T,P} < 0 \tag{5.79}$$

The same criterion can be applied for the Gibbs energy of mixing:

$$\left(\frac{\partial^2 \Delta g}{\partial x^2}\right)_{T,P} < 0 \tag{5.80}$$

In turn, the different g^E-models can be used to describe the contribution of the excess Gibbs energy or activity coefficients. An exception is the Wilson model. No miscibility gap can be represented by this equation because the Wilson equation describes a monotone behavior of the composition for each parameter combination, that is, $(\partial^2 \Delta g/\partial x^2 > 0)$ (see Appendix C, E3).

It is more complicated to calculate LLE in multicomponent systems accurately than to describe vapor-liquid or solid-liquid equilibria. The reason is that in the case of LLE the activity coefficients have to describe not only the concentration dependence but also the temperature dependence correctly, whereas in the case of the other phase equilibria (VLE, SLE) the activity coefficients primarily have to describe the deviation from ideal behavior (Raoult's law resp. ideal solid solubility), and the temperature dependence is mainly described by the standard fugacities (vapor pressure resp. melting temperature and heat of fusion).

This is the main reason why up to now no reliable prediction of the LLE behavior is possible. Even the calculation of the LLE behavior of ternary systems using binary parameters can lead to poor results for the distribution coefficients and the binodal curve. Fortunately, it is quite easy to measure LLE data of ternary and higher systems up to atmospheric pressure.

Example 5.19

Calculate the miscibility gap for the system n-butanol (1)–water (2) at 50 °C and additionally the corresponding pressure and vapor phase composition using the

UNIQUAC equation with the help of the UNIQUAC parameters fitted to VLE data [6]:

$$\Delta u_{12} = 129.7 \text{ cal/mol}, \quad \Delta u_{21} = 489.6 \text{ cal/mol}$$

Pure component properties:

Component	r_i	q_i	P_i^s (kPa)
n-Butanol	3.4543	3.052	4.61
Water	0.92	1.4	12.36

Solution

For the calculation of the miscibility gap the procedure shown in Figure 5.73 can be applied. However, for a binary system the miscibility gap can also be determined graphically. For the graphical approach the activity coefficients are evaluated for different concentrations. For illustration, the activity coefficient γ_1 is calculated for a mole fraction $x_1 = 0.05$. Using the UNIQUAC equation, the values for τ_{12} and τ_{21} are determined (see Table 5.6):

$$\tau_{12} = \exp\frac{-\Delta u_{12}}{RT} = \exp\frac{-129.7}{1.98721 \cdot 323.15} = 0.8171 \qquad \tau_{21} = 0.4685$$

Furthermore for the calculation of the combinatorial part the values for V_1 and F_1 are required:

$$V_1 = \frac{r_1}{r_1 x_1 + r_2 x_2} = \frac{3.4543}{3.4543 \cdot 0.05 + 0.92 \cdot 0.95} = 3.3$$

$$F_1 = \frac{q_1}{q_1 x_1 + q_2 x_2} = \frac{3.052}{3.052 \cdot 0.05 + 1.4 \cdot 0.95} = 2.0585$$

Then the combinatorial part of the activity coefficient can be calculated

$$\ln \gamma_1^C = 1 - V_1 + \ln V_1 - 5q_1 \left(1 - \frac{V_1}{F_1} + \ln \frac{V_1}{F_1}\right)$$

$$\ln \gamma_1^C = 1 - 3.3 + \ln(3.3) - 5 \cdot 3.052 \left(1 - \frac{3.3}{2.0585} + \ln \frac{3.3}{2.0585}\right) = 0.8956$$

In the next step the contribution of the residual part is determined:

$$\ln \gamma_1^R = -q_1 \ln \frac{q_1 x_1 + q_2 x_2 \tau_{21}}{q_1 x_1 + q_2 x_2} + q_1 q_2 x_2 \left[\frac{\tau_{21}}{q_1 x_1 + q_2 x_2 \tau_{21}} - \frac{\tau_{12}}{q_1 x_1 \tau_{12} + q_2 x_2}\right]$$

$$\ln \gamma_1^R = -3.052 \cdot \ln \frac{3.052 \cdot 0.05 + 1.4 \cdot 0.95 \cdot 0.4685}{3.052 \cdot 0.05 + 1.4 \cdot 0.95}+ 3.052 \cdot 1.4 \cdot 0.95$$

$$\cdot \left[\frac{0.4685}{3.052 \cdot 0.05 + 1.4 \cdot 0.95 \cdot 0.4685} - \frac{0.8171}{3.052 \cdot 0.05 \cdot 0.8171 + 1.4 \cdot 0.95} \right]$$

$$\ln \gamma_1^R = -3.052 \ln \frac{0.7757}{1.4826} + 4.0592 \left(\frac{0.4685}{0.7757} - \frac{0.8171}{1.45469} \right)$$

$$= 1.977 + 0.1716 = 2.1486$$

$$\ln \gamma_1 = \ln \gamma_1^R + \ln \gamma_1^C = 2.1486 + 0.8956 = 3.0442$$

$$\gamma_1 = 20.99$$

In the same way, the activity coefficient of component 2 and the activities for component 1 and 2 can be calculated:

$$x_1 = 0.05 \quad \gamma_1 = 20.99 \quad a_1 = 1.0495 \quad x_2 = 0.95 \quad \gamma_2 = 1.028 \quad a_2 = 0.9766$$

For other mole fractions x_1 the following activities are calculated:

x_1	0.005	0.01	0.0015	0.02	0.05	0.1	0.2	0.4	0.6
a_1	0.2824	0.4972	0.6598	0.9801	1.0495	0.9605	0.7253	0.6141	0.6802
a_2	0.9953	0.9913	0.9878	0.9848	0.9762	0.9842	1.0332	1.0951	0.9766

In the case of a miscibility gap, an intersection is obtained when the activity of component 2 is plotted against the activity of component 1 for different mole fractions x_1 as in the example considered (see Figure 5.71). At the intersection the

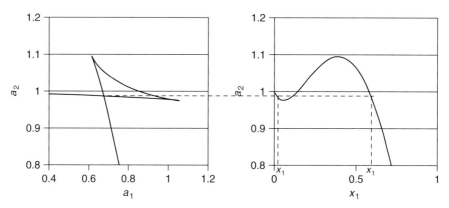

Figure 5.71 Graphical determination of LLE for binary systems exemplary shown for the system butanol (1)–water (2) at 50 °C.

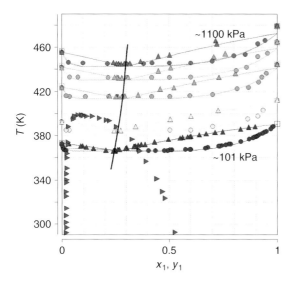

Figure 5.72 Experimental and calculated VLE and azeotropic data using the UNIQUAC parameters given in Example 5.19 together with experimental LLE data [3] —— azeotropic composition.

equilibrium condition (Eq. (5.73)) is fulfilled. This means that the same activities are obtained for two different compositions. As can be seen, the intersection is obtained at approx. $x_1 = 0.015$ and $x_1 = 0.59$, when the activity of component 2 is plotted against the mole fraction x_1 (Figure 5.72).

For the system n-butanol (1)–water (2) at 50 °C the following values are obtained in equilibrium:

$x_1' = 0.01556$ $\gamma_1' = 43.40$ $a_1' = 0.6753$ $x_1'' = 0.5906$ $\gamma_1'' = 1.143$ $a_1'' = 0.6751$
$x_2' = 0.9844$ $\gamma_2' = 1.003$ $a_2' = 0.9873$ $x_2'' = 0.4094$ $\gamma_2'' = 2.412$ $a_2'' = 0.9875$

Using these values obtained for phase ' or phase " directly the corresponding pressure and mole fraction in the vapor phase can be calculated.

$$P = x_1 \gamma_1 P_1^s + x_2 \gamma_2 P_2^s$$
$$P = 0.01556 \cdot 43.4 \cdot 4.61 + 0.9844 \cdot 1.003 \cdot 12.36)$$

$$= 3.113 + 12.204 = 15.317 \text{ kPa}$$
$$y_1 = \frac{p_1}{P} = \frac{3.113}{15.317} = 0.2032$$

Because the activities in phase ' and " are identical, the same results are obtained starting from the composition in phase ", this means for $x_1 = 0.5906$.

For the entire composition range, the VLLE results are shown in Figure 5.72, together with the calculated and experimental data for a few isobaric VLE data, the calculated azeotropic composition as $f(T)$ and experimental LLE data.[20] It can be recognized that at atmospheric pressure the system n-butanol(1)–water(2) shows a heterogeneous azeotrope with a mole fraction of approximately $y_{1,az} = 0.25$ and a temperature of 366 K. At other pressures, the azeotropic composition will change. The change of the azeotropic composition depends not only on the temperature dependence of the vapor pressures, but also on the temperature dependence of the activity coefficients. In Figure 5.72 the typical temperature dependence is shown in the form of isobaric Txy-diagrams. While at atmospheric pressure a heterogeneous azeotropic point occurs, homogeneous azeotropic behavior is observed at higher pressures (temperatures). The temperature dependence of the azeotropic behavior is discussed in detail in Section 5.6.

While the calculation of binary LLE can be performed graphically, the calculation of LLE for ternary and higher systems has to be performed iteratively. One possible procedure for a multicomponent system is shown in Figure 5.73 in the form of a flow diagram. The method takes into account the isoactivity conditions (Eq. (5.73)) and the material balance.

Starting from the mole numbers n_i (initial feed stream to the equilibrium stage) with a composition in the two phase region, mole numbers n'_i (composition) are estimated for the liquid phase. From the difference $n_i - n'_i$, the mole numbers n''_i (composition) in the second liquid phase can be calculated. Then the activity coefficients of the components in the two liquid phases are determined. In the next step it is checked if the isoactivity condition is fulfilled. Of course, after the first step the isoactivity condition will not be fulfilled. Therefore, the estimated mole numbers n'_i have to be changed in the right way. Using the K-factor method, the following equation is obtained for the variation of the mole numbers starting from Eq. (5.73) and the material balance:

$$x'_i \gamma'_i = x''_i \gamma''_i \quad \text{resp.} \quad \frac{n'_i}{n'_T} \gamma'_i = \frac{n''_i}{n''_T} \gamma''_i$$

with

$$n'_T = \Sigma n'_i \quad \text{and} \quad n''_T = \Sigma n''_i$$

one obtains

$$n''_i = n'_i \frac{n''_T \gamma'_i}{n'_T \gamma''_i}$$

Taking into account the material balance $n''_i = n_i - n'_i$

$$n_i - n'_i = n'_i \frac{n''_T \gamma'_i}{n'_T \gamma''_i} \rightarrow n_i = n'_i + n'_i \frac{n''_T \gamma'_i}{n'_T \gamma''_i} = n'_i \cdot \left[1 + \frac{n''_T \gamma'_i}{n'_T \gamma''_i}\right]$$

new mole numbers in phase ' can be calculated using the following relation:

20) A comparison of the calculated LLE in Example 5.19 at 50 °C with the experimental findings shows a disagreement. While 59 mol% butanol were determined for the butanol rich phase, experimentally less than 50 mol% was found. The reason is that parameters fitted to VLE data do not describe the LLE behavior correctly.

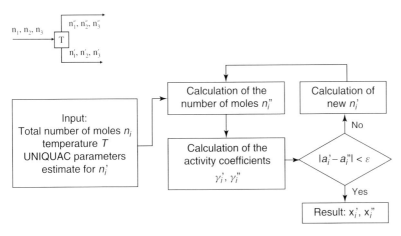

Figure 5.73 Flow diagram for the calculation of LLE using the K-factor method.

$$n'_{i,\text{new}} = \frac{n_i}{1 + \gamma'_i n''_T / \gamma''_i n'_T}$$

The iteration is stopped when the activities in the different phases are identical within a small value ε. If the feed composition is outside the two-phase region, the solution will lead to the trivial solution, where the composition of the two liquid phases is identical.

Example 5.20

Calculate the LLE composition for the system water (1)–ethanol (2)–benzene (3) for the following feed stream: $n_1 = 1$ mol, $n_2 = 0.3$ mol, and $n_3 = 1$ mol at 25 °C with the help of the UNIQUAC model.

UNIQUAC parameters $\Delta u_{ij}(K)$ fitted to LLE data at 25 °C:

Compound	1	2	3
1	0	526.02	309.64
2	−318.06	0	−91.532
3	1325.1	302.57	0

Relative van der Waals properties:

Compound	r_i	q_i
Water	0.9200	1.400
Ethanol	2.1055	1.972
Benzene	3.1878	2.400

Solution

To start the calculation, first the mole numbers in phase ' have to be estimated. Let us assume that $n'_1 = 0.8$ mol, $n'_2 = 0.1$ mol, and $n'_3 = 0.2$ mol are in phase ', for the given feed $n''_1 = 0.2$ mol, $n''_2 = 0.2$ mol, and $n''_3 = 0.8$ mol remain for phase ''. Then for these phases in the next step the activity coefficients can be calculated with the help of the UNIQUAC equation.

UNIQUAC parameters τ_{ij}

Compound	1	2	3
1	1	0.1713	0.35397
2	2.9060	1	1.3593
3	0.01174	0.36247	1

The following composition is obtained for phase ': $x'_1 = 0.8/1.1 = 0.7273$, $x'_2 = 0.0909$, $x'_3 = 0.1818$.

For this composition the activity coefficients have to be calculated. That is exemplarily shown for component 1:

$$V_1 = \frac{r_1}{r_1 x_1 + r_2 x_2 + r_3 x_3}$$

$$= \frac{0.92}{0.92 \cdot 0.7273 + 2.1055 \cdot 0.0909 + 3.1878 \cdot 0.1818} = 0.6389$$

$$F_1 = \frac{q_1}{q_1 x_1 + q_2 x_2 + q_3 x_3} = \frac{1.4}{1.4 \cdot 0.7273 + 1.972 \cdot 0.0909 + 2.4 \cdot 0.1818}$$

$$= 0.8569$$

$$\ln \gamma_1^C = 1 - V_1 + \ln V_1 - 5 q_1 \left(1 - \frac{V_1}{F_1} + \ln \frac{V_1}{F_1}\right)$$

$$\ln \gamma_1^C = 1 - 0.6389 + \ln(0.6389) - 5 \cdot 1.4 \left(1 - \frac{0.6389}{0.8569} + \ln \frac{0.6389}{0.8569}\right)$$

$$= 0.1873$$

$$\ln \gamma_1^R = q_1 \left(1 - \ln \frac{q_1 x_1 + q_2 x_2 \tau_{21} + q_3 x_3 \tau_{31}}{q_1 x_1 + q_2 x_2 + q_3 x_3} - \frac{q_1 x_1}{q_1 x_1 + q_2 x_2 \tau_{21} + q_3 x_3 \tau_{31}}\right.$$

$$\left. - \frac{q_2 x_2 \tau_{12}}{q_1 x_1 \tau_{12} + q_2 x_3 \tau_{32}} - \frac{q_3 x_3 \tau_{13}}{q_1 x_1 \tau_{13} + q_2 x_2 \tau_{23} + q_3 x_3}\right)$$

$$\ln \gamma_1^R = 1.4 \cdot \left(1 - \ln \frac{1.4 \cdot 0.7273 + 1.972 \cdot 0.0909 \cdot 2.906 + 2.4 \cdot 0.1818 \cdot 0.01174}{1.4 \cdot 0.7273 + 1.972 \cdot 0.0909 + 2.4 \cdot 0.1818}\right.$$

$$-\frac{1.4 \cdot 0.7273}{1.4 \cdot 0.7273 + 1.972 \cdot 0.0909 \cdot 2.906 + 2.4 \cdot 0.1818 \cdot 0.01174}$$

$$-\frac{1.972 \cdot 0.0909 \cdot 0.1713}{1.4 \cdot 0.7273 \cdot 0.1713 + 1.972 \cdot 0.0909 + 2.4 \cdot 0.1818 \cdot 0.36247}$$

$$\left.-\frac{2.4 \cdot 0.1818 \cdot 0.35397}{1.4 \cdot 0.7273 \cdot 0.35397 + 1.972 \cdot 0.0909 \cdot 1.3593 + 2.4 \cdot 0.1818}\right)$$

$$= 0.2640$$

$$\ln \gamma_1 = \ln \gamma_1^C + \ln \gamma_1^R = 0.1873 + 0.2640 = 0.4513$$

$$\gamma_1 = 1.570$$

The activity coefficients for all components in both phases are given in the following table:

Phase '			Phase "		
n_i	x_i	γ_i	n_i	x_i	γ_i
0.8	0.7273	1.570	0.2	0.1667	8.856
0.1	0.0909	0.2948	0.2	0.1667	0.860
0.2	0.1818	18.11	0.8	0.6667	1.425

Using these data improved mole numbers are calculated for phase ' with the help of the K-factor method. Then the mole numbers in phase ", the compositions and the activity coefficients are calculated again. For n_1' one obtains

$$n_1' = \frac{1}{1 + \frac{1.57 \cdot 1.2}{8.856 \cdot 1.1}} = 0.8379$$

and for all other values:

Phase '			Phase "		
n_i	x_i	γ_i	n_i	x_i	γ_i
0.8379	0.7458	1.181	0.1621	0.1378	20.99
0.2183	0.1943	0.7311	0.0817	0.0694	0.5809
0.0673	0.0600	36.77	0.9327	0.7928	1.258

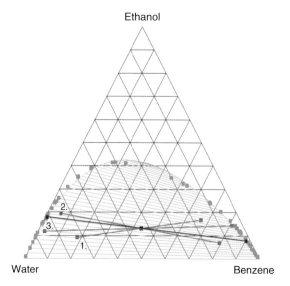

Figure 5.74 The first three steps of the K-factor method together with the experimental and calculated LLE behavior of the ternary system water–ethanol–benzene at 25 °C using UNIQUAC.

After a few steps, convergence – this means LLE – is obtained, since the changes of the calculated mole numbers are below a small value ε. The final values are:

Phase '			Phase "		
n_i	x_i	γ_i	n_i	x_i	γ_i
0.9799	0.8112	1.053	0.0201	0.0184	46.35
0.2153	0.1782	1.006	0.0847	0.0776	2.310
0.0128	0.0106	88.49	0.9872	0.9040	1.039

The results of the first three steps and the final LLE results for the system water–ethanol–benzene at 25 °C are shown in Figure 5.74. It can be seen that after three steps the equilibrium composition is nearly reached, also for poor initial estimates.

5.8.1
Temperature Dependence of Ternary LLE

The temperature dependence of LLE of ternary systems can be very different, as shown for binary systems in Figure 5.67. In most cases, the miscibility gap

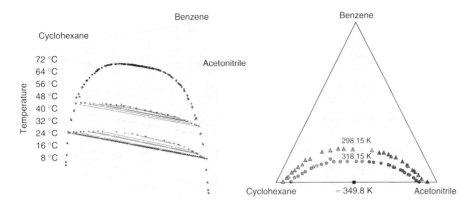

Figure 5.75 Qualitative progress of the temperature dependence of ternary liquid–liquid equilibria.

becomes smaller with increasing temperature. For a closed system this behavior is shown in Figure 5.75. In the case presented, the mutual solubility increases with increasing temperature, this means, the range of concentration where two liquid phases coexist decreases more and more until the heterogeneous region disappears above the UCST of the binary system AB.

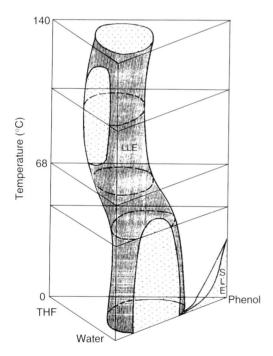

Figure 5.76 LLE behavior of the ternary system tetrahydrofuran–water–phenol as a function of temperature.

But as mentioned above, the temperature dependence can be much more complex. For the system tetrahydrofuran–water–phenol this is shown in Figure 5.76. At temperatures below 66 °C, only the system phenol–water shows a miscibility gap, whereas the other two binaries are homogeneous in this temperature range. The binary miscibility gap extends into the ternary area. Above the UCST of the system phenol–water, the formation of an island curve is observed, where all binary systems are homogeneous while the ternary system is heterogeneous. At approx. 72 °C, the LCST of the system tetrahydrofuran–water is reached. The binary system shows a miscibility gap up to a temperature of about 137 °C. Above the UCST of the system tetrahydrofuran–water again an island curve is formed in the ternary system. Up to now this complex LLE behavior can not be described with the help of a g^E-model, even with linear or quadratic temperature-dependent model parameters. Since phenol has a melting point at approx. 41 °C, in Figure 5.76 additionally the SLE behavior for the system phenol–water is shown.

5.8.2
Pressure Dependence of LLE

Although it was mentioned at the beginning of Section 5.8 that pressure differences of a few bar only have a negligible influence on the LLE behavior, in practice often higher pressures are realized. Already a slight volume compression of a liquid can lead to very high pressures. In centrifugal extractors often higher pressures are observed. The influence of the pressure on the activity coefficients (LLE) can be taken into account if the excess volumes are known. The influence can directly be calculated using Eq. (5.27). The activity coefficients will decrease with increasing pressure in the case of negative partial molar excess volumes, as shown in Example 5.7. This means that the miscibility gap becomes smaller with

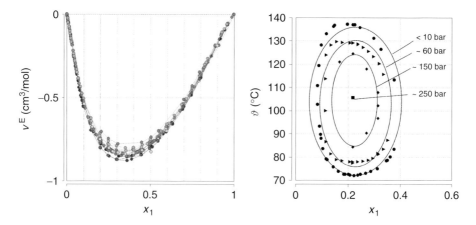

Figure 5.77 Excess volumes and LLE behavior of the system tetrahydrofuran (1)–water (2) [3] as a function of pressure.

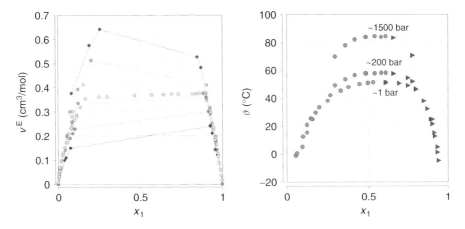

Figure 5.78 Excess volumes and LLE behavior of the system methanol (1)–n-heptane (2) [3] as a function of pressure.

increasing pressure. The opposite is true in the case of positive partial molar excess volumes. The influence of the sign of the excess volumes on the LLE behavior is exemplarily shown in Figures 5.77 and 5.78 for the systems tetrahydrofuran–water and methanol–heptane.

For the system tetrahydrofuran–water, negative excess volumes are observed. This results in the fact that the system becomes homogeneous at pressures around 250 bar.

In contrast, the miscibility gap becomes larger with increasing pressure for the system methanol–heptane (see Figure 5.78) because of the positive partial molar excess volumes for this system.

5.9
Predictive Models

Both approaches (g^E-models ($\gamma-\varphi$ approach), equations of state ($\varphi-\varphi$ approach)) allow the calculation of multicomponent systems using binary information alone. However, often the required experimental binary data are missing.

Assuming that 1000 compounds are of technical interest, phase equilibrium information for about 500000 binary systems are required to fit the required binary parameters to describe all possible binary and multicomponent systems. Although more than 64500 VLE data sets for nonelectrolyte systems have been published up to now, VLE data are available for only 10300 binary systems, since for a few systems a large number of data sets were published, for example, for the systems ethanol–water, ammonia–water, water–carbon dioxide, methanol–water, methane–nitrogen more than 150 data sets are available. This means that only for ~2% of the required systems at least one VLE data set is available. If only consistent VLE data are accepted or if more than one VLE data set is desired, the percentage even decreases to ∼1.2%. If also information about the dilute range and the temperature dependence in the form of γ^∞ and h^E should be used to fit the required model parameters, the percentage of the available systems is less than 0.2%, although approximately 62500 γ^∞-values and 21000 h^E-data sets have been published, which are stored in the Dortmund Data Bank [3].

Since the assumption of ideal behavior can lead to very erroneous results and measurements are very time consuming, reliable predictive models with a large range of applicability would be desirable.

Because of the importance of distillation processes, first it was the objective to develop models only for the prediction of VLE. The first predictive model with a wide range of applicability was developed by Hildebrand and Scatchard [48]. The so-called regular solution theory is based on considerations of van Laar, who was a student of van der Waals and used the van der Waals equation of state to derive an expression for the excess Gibbs energy [49]. Since the two parameters a and b of the van der Waals equation of state can be obtained from critical data, it should be possible to calculate the required activity coefficients using critical data. However, the results were strongly dependent on the mixing rules applied.

5.9.1
Regular Solution Theory

Hildebrand and Scatchard [48] showed that better results are obtained, if instead of the van der Waals constants a and b molar volumes v_i and so-called solubility parameters δ_i are used instead. For binary systems the following relations are obtained for the activity coefficients:

$$\ln \gamma_1 = \frac{v_1 \Phi_2^2 (\delta_1 - \delta_2)^2}{RT}$$
$$\ln \gamma_2 = \frac{v_2 \Phi_1^2 (\delta_1 - \delta_2)^2}{RT} \tag{5.81}$$

Φ_i volume fraction of component $i = (x_i v_i)/\Sigma x_j v_j$
δ_i solubility parameter of component i.

The solubility parameter δ_i can be calculated using values for the enthalpy of vaporization and the molar volume v_i at 298.15 K:

$$\delta_i = \left(\frac{\Delta h_{vi} - RT}{v_i}\right)^{0.5} \tag{5.82}$$

Δh_{vi} molar heat of vaporization of component i (cal/mol).

For regular solutions the solubility parameters δ_i and molar volumes v_i can be assumed to be constant for a larger temperature range. For a few compounds the parameters are given in Table 5.16.

Table 5.16 Molar volumes and solubility parameters for selected compounds.

Compound	v_i (cm³/mol)	δ_i (cal/cm³)$^{0.5}$
Carbon tetrachloride	97	8.6
Carbon disulfide	61	10.0
n-Pentane	116	7.1
Benzene	89	9.2
Cyclohexane	109	8.2
Hexene-1	126	7.3
n-Hexane	132	7.3
Toluene	107	8.9
n-Heptane	148	7.4
n-Octane	164	7.5

The regular solution theory is not limited to binary systems. It can directly be applied for the calculation of activity coefficients in multicomponent systems:

$$\ln \gamma_i = \frac{v_i}{RT} (\delta_i - \bar{\delta})^2 \tag{5.83}$$

where the mean solubility parameter:

$$\bar{\delta} = \sum_i \Phi_i \delta_i = \frac{\sum_i x_i v_i \delta_i}{\sum_i x_i v_i} \tag{5.84}$$

can be obtained by summation over all compounds.

However, the regular solution theory can only be applied for nonpolar systems and systems with positive deviations from Raoult's law.

Example 5.21

Estimate the activity coefficients at infinite dilution for the system benzene (1)–cyclohexane (2) at 353.15 K.

Solution

With the values given in Table 5.16 the values at infinite dilution ($\Phi_2 = 1, \Phi_1 = 1$) can directly be estimated:

$$\ln \gamma_1^\infty = \frac{89 (9.2 - 8.2)^2}{1.98721 \cdot 353.15} = 0.1268 \quad \ln \gamma_2^\infty = \frac{109 (9.2 - 8.2)^2}{1.98721 \cdot 353.15} = 0.1553$$

$$\gamma_1^\infty = 1.135 \qquad \qquad \gamma_2^\infty = 1.168$$

Experimentally, higher values were measured ($\gamma_1^\infty \approx 1.35, \gamma_2^\infty \approx 1.44$) [3].

5.9.2
Group Contribution Methods

Group contribution methods do not show these weaknesses discussed for the regular solution theory.

In group contribution methods it is assumed that the mixture does not consist of molecules but of functional groups. In Figure 5.79 this is shown for the systems ethanol–n-hexane. Ethanol can be subdivided in a methyl-, methylene- and alcohol-group and n-hexane in two methyl- and four methylene-groups. It can be shown that the required activity coefficients can be calculated if only the interaction parameters between the functional groups are known. For example, if the group interaction parameters between the alkane and the alcohol group are known, not only the activity coefficients (VLE behavior) of the system ethanol–n-hexane, but also for all other alkane–alcohol or alcohol–alcohol systems can be predicted. The great advantage of group contribution methods is that the number of functional groups is much smaller than the number of possible molecules.

The required equation of the solution of groups concept can be derived from the excess Gibbs energy of the groups in the mixture and the excess Gibbs energy in the pure compound.

For the pure compound i built up by functional groups one can derive the following expression for the molar (g^E) and total Gibbs energy (g^E):

$$\frac{g^{E(i)}}{RT} = \sum_k X_k^{(i)} \ln \Gamma_k^{(i)}$$

$$\frac{G^{E(i)}}{RT} = n^{(i)} \sum_k \nu_k^{(i)} \ln \Gamma_k^{(i)} \qquad \sum_i \frac{G^{E(i)}}{RT} = \sum_i \sum_k n^{(i)} \nu_k^{(i)} \ln \Gamma_k^{(i)}$$

For a mixture built up by functional groups one can write

$$\frac{g^{E(m)}}{RT} = \sum_k X_k \ln \Gamma_k$$

$$\frac{G^{E(m)}}{RT} = \sum_i \left(n^{(i)} \nu^i \right) \sum_k X_k \ln \Gamma_k \qquad \frac{G^{E(m)}}{RT} = \sum_i \sum_k n^{(i)} \nu_k^i \ln \Gamma_k$$

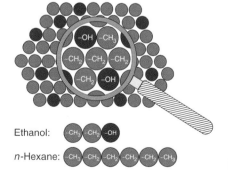

Figure 5.79 Group contribution concept.

From the difference of the excess Gibbs energies for the groups in the mixture and in the pure compound (standard state) one can derive an expression for the required activity coefficient γ_i:

$$\frac{G^E}{RT} = \sum_i n^{(i)} \ln \gamma_i = \frac{G^{E(m)}}{RT} - \sum_i \frac{G^{E(i)}}{RT}$$

$$= \sum_i \sum_k n^{(i)} v_k^{(i)} \ln \Gamma_k - \sum_i \sum_k n^{(i)} v_k^{(i)} \ln \Gamma_k^{(i)}$$

$$= \sum_i n^{(i)} \left[\sum_k v_k^{(i)} \left(\ln \Gamma_k - \ln \Gamma_k^{(i)} \right) \right]$$

which leads to the equation for the solution of groups concept:

$$\ln \gamma_i = \sum_k v_k^{(i)} \left(\ln \Gamma_k - \ln \Gamma_k^{(i)} \right)$$

with

$\Gamma_k^{(i)}$	group activity coefficient of group k in pure component i
Γ_k	group activity coefficient of group k in the mixture
$v_k^{(i)}$	number of groups k in component i
$v^{(i)}$	number of groups in component i

$$v^{(i)} = \sum_i v_k^{(i)} = v_1^{(i)} + v_2^{(i)} + \ldots + v_m^{(i)}$$

$n^{(i)}$	number of moles of component i
$n_T = \sum_i n^{(i)}$	total number of moles in the mixture
$n^{(i)} v^{(i)}$	total number of moles of groups of component i
$\sum_i n^{(i)} v^{(i)}$	total number of moles of groups in the mixture
$X_k^{(i)} = \frac{v_k^{(i)}}{v^{(i)}}$	group mole fraction of group k in compound i
$X_k = \frac{\sum_i n^{(i)} v_k^{(i)}}{\sum_i n^{(i)} v^{(i)}}$	group mole fraction of group k in the mixture

5.9.3
UNIFAC Method

The first group contribution method for the prediction of VLE (activity coefficients) was the so-called analytical solution of groups (ASOG) method [50, 51], developed within Shell. The ASOG method uses the Wilson model to describe the concentration dependence of the group activity coefficients required in the solution of groups concept.

In 1975, the UNIFAC group contribution method was published by Fredenslund et al. [27, 52, 53]. Like the ASOG method, the UNIFAC method is based on the solution of groups concept. But in the UNIFAC method, the activity coefficients are calculated from a combinatorial and a residual part exactly like in the UNIQUAC model (see Eq. (5.85)). While the temperature-independent combinatorial part takes into account the size and form of the molecules, that is, the entropic contribution, the residual part considers the enthalpic interactions:

$$\ln \gamma_i = \ln \gamma_i^C + \ln \gamma_i^R \tag{5.85}$$

The combinatorial part $\ln \gamma_i^C$ can be calculated using the following equation, which is identical to the UNIQUAC model:

$$\ln \gamma_i^C = 1 - V_i + \ln V_i - 5q_i\left(1 - \frac{V_i}{F_i} + \ln \frac{V_i}{F_i}\right) \tag{5.86}$$

where V_i (volume/mole fraction ratio) and F_i (surface area/mole fraction ratio) can be calculated for a given composition using the relative van der Waals volumes r_i and van der Waals surface areas q_i of the molecules:

$$V_i = \frac{r_i}{\sum_j r_j x_j} \tag{5.87}$$

$$F_i = \frac{q_i}{\sum_j q_j x_j} \tag{5.88}$$

For the UNIFAC group contribution method the relative van der Waals properties r_i and q_i can be obtained using the relative van der Waals group volumes R_k and relative van der Waals group surface areas Q_k, which can be derived from x-ray data. Tabulated values for R_k and Q_k can be found by Hansen et al. [53]. They can also be derived from the tabulated van der Waals properties published by Bondi [54]. For selected groups the R_k and Q_k values are given in the Appendix H:

$$r_i = \sum_k v_k^{(i)} R_k \tag{5.89}$$

$$q_i = \sum_k v_k^{(i)} Q_k \tag{5.90}$$

where $v_k^{(i)}$ is the number of functional groups of type k in compound i.

The temperature-dependent residual part $\ln \gamma_i^R$ takes into account the interactions between the different compounds. In group contribution methods, this part is calculated via the solution of groups concept using group activity coefficients Γ_k and $\Gamma_k^{(i)}$:

$$\ln \gamma_i^R = \sum_k v_k^{(i)} \left(\ln \Gamma_k - \ln \Gamma_k^{(i)}\right) \tag{5.91}$$

Γ_k and $\Gamma_k^{(i)}$ are the group activity coefficients for group k in the mixture, respectively, for the pure compound i. For the description of the concentration dependence of the group activity coefficients the UNIQUAC equation is used:

$$\ln \Gamma_k = Q_k \left[1 - \ln\left(\sum_m \Theta_m \Psi_{mk}\right) - \sum_m \frac{\Theta_m \Psi_{km}}{\sum_n \Theta_n \Psi_{nm}}\right] \tag{5.92}$$

The surface area fractions Θ_m and the group mole fractions X_m of group m can be calculated using the following relations:

$$\Theta_m = \frac{Q_m X_m}{\sum_n Q_n X_n} \tag{5.93}$$

$$X_m = \frac{\sum_j \nu_m^{(j)} x_j}{\sum_j \sum_n \nu_n^{(j)} x_j} \tag{5.94}$$

The parameter Ψ_{nm} contains the group interaction parameter a_{nm} between the functional groups n and m, for example, between alkanes and ketones:

$$\Psi_{nm} = \exp\left(-\frac{a_{nm}}{T}\right) \tag{5.95}$$

These functional groups are called *main groups*. They often consist of more than one subgroup. For example, in the case of alkanes one has to distinguish between CH_3-, CH_2-, $CH-$, and C-groups. The different alkane subgroups all have different values for the van der Waals properties. The same is true for the ketone group, where one has to distinguish between the CH_3CO-, CH_2CO-, and $CHCO-$group. In the UNIFAC method, for every main group combination two temperature-independent group interaction parameters (a_{nm}, a_{mn}) are required, which were fitted almost exclusively to consistent experimental vapor–liquid equilibrium data stored in the Dortmund Data Bank [3]. Since the interactions are defined per area, depending on the subgroup different strong interactions are calculated for e.g. CH_3-CH_2CO and CH_2-CH_2CO pairs. By definition, the group interaction parameters between identical main groups (a_{nn}, a_{mm}) are equal to 0. This means that the parameters Ψ_{nn} and Ψ_{mm} become unity. The van der Waals properties and the published group interaction parameters can be found in the internet (see Appendix H).

Example 5.22

Calculate the VLE of the system n-hexane (1)–2-butanone (2) at 60 °C for a mole fraction of $x_1 = 0.5$ with the help of the UNIFAC method assuming ideal behavior of the vapor phase.

Vapor pressures and structural information:

Component	P_i^s at 60 °C (kPa)	CH_3	CH_2	CH_3CO
n-Hexane	75.85	2	4	–
2-Butanone	51.90	1	1	1

van der Waals properties:

Group	R_k	Q_k
CH_3	0.9011	0.848
CH_2	0.6744	0.540
CH_3CO	1.6724	1.488

Group interaction parameters a_{nm} between the main group alkanes (CH_2) and ketones (CH_2CO):

a_{nm} (K)	CH_2	CH_2CO
CH_2	0.0	476.4
CH_2CO	26.76	0.0

Solution

First of all the van der Waals properties of the two compounds can be calculated with the help of the van der Waals properties of the groups:

$$r_1 = 2 \cdot 0.9011 + 4 \cdot 0.6744 = 4.4998$$
$$q_1 = 2 \cdot 0.848 + 4 \cdot 0.54 = 3.856$$
$$r_2 = 1 \cdot 0.9011 + 1 \cdot 0.6744 + 1 \cdot 1.6724 = 3.2479$$
$$q_2 = 1 \cdot 0.848 + 1 \cdot 0.54 + 1 \cdot 1.488 = 2.876$$

Using these van der Waals properties for $x_1 = 0.5$ the following values are obtained for V_i and F_i:

$$V_1 = \frac{4.4998}{0.5\,(4.4998 + 3.2479)} = 1.1616$$

$$V_2 = \frac{3.2479}{0.5\,(4.4998 + 3.2479)} = 0.8384$$

$$F_1 = \frac{3.856}{0.5\,(3.856 + 2.876)} = 1.1456$$

$$F_2 = \frac{2.876}{0.5\,(3.856 + 2.876)} = 0.8544$$

With the help of these values the combinatorial part can be calculated. For n-hexane (1)

$$\ln \gamma_1^C = 1 - 1.1616 + \ln 1.1616 - 5 \cdot 3.856 \left(1 - \frac{1.1616}{1.1456} + \ln \frac{1.1616}{1.1456}\right)$$
$$= -0.00994$$

For 2-butanone (2) the following value is obtained:

$$\ln \gamma_2^C = -0.001210$$

For the calculation of the group activity coefficients in the mixture first of all the parameters Ψ_{nm}, the group mole fractions and surface area fractions have to be determined. For the parameters Ψ_{nm} the following values are obtained:

$$\Psi_{CH_3,CH_3CO} = \Psi_{CH_2,CH_3CO} = \exp\frac{-476.4}{333.15} = 0.2393$$

$$\Psi_{CH_3CO,CH_3} = \Psi_{CH_3CO,CH_2} = \exp\frac{-26.76}{333.15} = 0.9228$$

$$\Psi_{CH_3,CH_3} = \Psi_{CH_2,CH_2} = \Psi_{CH_3CO,CH_3CO} = \Psi_{CH_2,CH_3} = \Psi_{CH_3,CH_2} = 1$$

The following group mole fractions and surface area fractions are obtained for the considered binary system at $x_1 = 0.5$:

$$X_{CH_3} = \frac{(2+1)\,0.5}{(6+3)\,0.5} = 0.3333$$

$$X_{CH_2} = \frac{(4+1)\,0.5}{(6+3)\,0.5} = 0.5556$$

$$X_{CH_3CO} = \frac{0.5}{(6+3)\,0.5} = 0.1111$$

$$\Theta_{CH_3} = \frac{0.848 \cdot 0.3333}{0.848 \cdot 0.3333 + 0.54 \cdot 0.5556 + 1.488 \cdot 0.1111} = 0.3779$$

$$\Theta_{CH_2} = 0.4011$$

$$\Theta_{CH_3CO} = 0.2210$$

Now all values are available to calculate the group activity coefficients in the binary system:

$$\ln \Gamma_{CH_3} = 0.848 \bigg[1 - \ln(0.3779 + 0.4011 + 0.221 \cdot 0.9228)$$

$$- \frac{0.3779 + 0.4011}{0.3779 + 0.4011 + 0.221 \cdot 0.9228}$$

$$- \frac{0.221 \cdot 0.2393}{(0.3779 + 0.4011)\,0.2393 + 0.221} \bigg]$$

$$\ln \Gamma_{CH_3} = 0.080458$$

$$\ln \Gamma_{CH_2} = 0.051235$$

$$\ln \Gamma_{CH_3CO} = 0.92872$$

For the pure compounds the following group mole fractions and surface area fractions are obtained:
For n-hexane (1):

$$X^{(1)}_{CH_3} = 0.3333 \qquad X^{(1)}_{CH_2} = 0.6667$$

$$\Theta_{CH_3} = \frac{0.848 \cdot 0.3333}{0.848 \cdot 0.3333 + 0.54 \cdot 0.6667} = 0.4398$$

$$\Theta_{CH_2} = 0.5602$$

For 2-butanone (2):

$$X^{(1)}_{CH_3} = 0.3333 \quad X^{(1)}_{CH_2} = 0.3333 \quad X^{(1)}_{CH_3CO} = 0.3333$$

$$\Theta_{CH_3} = \frac{0.848 \cdot 0.3333}{0.848 \cdot 0.3333 + 0.54 \cdot 0.3333 + 1.488 \cdot 0.3333} = 0.2949$$

$$\Theta_{CH_2} = 0.1878$$

$$\Theta_{CH_3CO} = 0.5173$$

With these values the group activity coefficients in the pure compounds can be calculated. For pure n-hexane (1) one obtains

$$\ln \Gamma^{(1)}_{CH_3} = 0.0 \qquad \ln \Gamma^{(1)}_{CH_2} = 0.0$$

and for 2-butanone (2)

$$\ln \Gamma^{(2)}_{CH_3} = 0.848 \left[1 - \ln(0.2949 + 0.1878 + 0.5173 \cdot 0.9228) \right.$$

$$- \frac{0.2949 + 0.1878}{0.2949 + 0.1878 + 0.5173 \cdot 0.9228}$$

$$\left. - \frac{0.5173 \cdot 0.2393}{(0.2949 + 0.1878) 0.2393 + 0.5173} \right]$$

$$\ln \Gamma^{(2)}_{CH_3} = 0.29038$$

$$\ln \Gamma^{(2)}_{CH_2} = 0.18491$$

$$\ln \Gamma^{(2)}_{CH_3CO} = 0.262$$

Herewith all values are available to calculate the residual part of the activity coefficients following the solution of groups concept and finally to calculate the required activity coefficients:

$$\ln \gamma^R_1 = 2(0.80458 - 0) + 4(0.051235 - 0) = 0.365856$$

$$\ln \gamma_1 = \ln \gamma^R_1 + \ln \gamma^C_1 = 0.365856 - 0.00994 = 0.35592$$

$$\gamma_1 = 1.4275$$

$$\ln \gamma^R_2 = (0.080458 - 0.29038) + (0.051235 - 0.18491)$$

$$+ (0.92872 - 0.262) = 0.32312$$

$$\ln \gamma_2 = \ln \gamma^R_2 + \ln \gamma^C_2 = 0.32312 - 0.01210 = 0.31102$$

$$\gamma_2 = 1.3648$$

Assuming ideal vapor phase behavior the knowledge of the activity coefficients allows calculating the partial pressures, total pressure, and the vapor phase mole fraction:

$$P = p_1 + p_2 = 0.5 \cdot 1.4275 \cdot 75.85 + 0.5 \cdot 1.3648 \cdot 51.9 = 89.55 \text{ kPa}$$

$$y_1 = \frac{p_1}{P} = \frac{0.5 \cdot 1.4275 \cdot 75.85}{89.55} = 0.6045$$

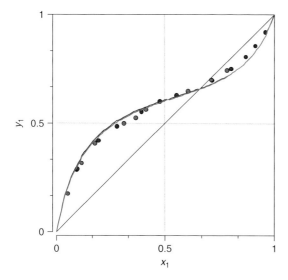

Figure 5.80 Experimental and predicted y–x-data for the system n-hexane (1)-2-butanone (2) at 60 °C.

The results for the whole composition range are shown in Figure 5.80. It can be seen that good agreement between experimental and predicted VLE is observed. In Figure 5.81 it is shown that good results are obtained not only for the system n-hexane–2-butanone, but also for all the other alkane–ketone systems. It can be seen that even the azeotropic points are predicted accurately. It is worth mentioning that for all systems shown the same two group interaction parameters were used, which describe the interaction between the alkane and ketone group.

Because of the reliable results obtained for VLE and the large range of applicability, the method was directly integrated into the different process simulators. However, in spite of the reliable results for VLE, UNIFAC also shows a few weaknesses, for example, unsatisfying results are obtained:

- for the activity coefficients at infinite dilution,
- for the excess enthalpies, this means the temperature dependence of the activity coefficients following the Gibbs–Helmholtz relation, and
- for strongly asymmetric systems, this means for compounds very different in size.

For the system 2-butanone–n-hexane the predicted results of the excess enthalpy using UNIFAC are shown in Figure 5.82 together with the experimental data. It can be seen that the predicted excess enthalpies are not in agreement with the experimental values. This means that an extrapolation to high or low temperatures will produce incorrect results. The same is true for all other alkane–ketone systems, as shown in Figure 5.86.

All these weaknesses are not surprising, since with the VLE data used to fit the required temperature-independent group interaction parameters no information

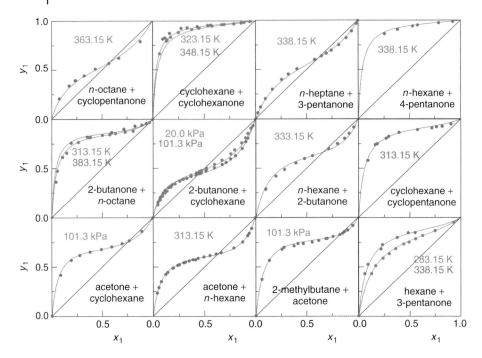

Figure 5.81 Experimental [3] and predicted VLE data for alkane–ketone systems using UNIFAC.

about the temperature dependence (excess enthalpies), very asymmetric systems and the very dilute region is used, since VLE data are usually only measured between 5 and 95 mol% for symmetric or slightly asymmetric systems. An extrapolation to infinite dilution can be very dangerous. However, activity coefficients at infinite dilution measured with special techniques (gas stripping or dilutor technique, ebulliometry, gas–liquid chromatography) provide the required information for the dilute composition range. At the same time systems investigated by gas-liquid chromatography are very asymmetric, since the compounds involved (stationary phase, solutes) show very different volatility. VLE data measured at different temperatures (pressures) deliver an idea about the temperature dependence, but measurements are time consuming. The most accurate information about the temperature dependence is obtained from excess enthalpies measured by isothermal flow calorimetry.

5.9.3.1 Modified UNIFAC (Dortmund)

To reduce the weaknesses of UNIFAC, the modified UNIFAC method was developed [55]. The main differences compared to original UNIFAC are:

- an empirically modified combinatorial part was introduced to improve the results for asymmetric systems;
- temperature-dependent group interaction parameters are used;

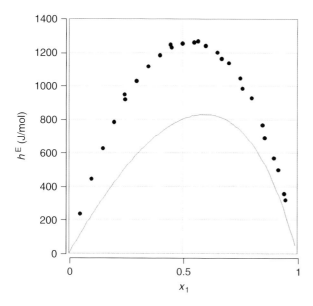

Figure 5.82 Experimental [3] and predicted excess enthalpies using UNIFAC of the system 2-butanone (1)−n-hexane (2) at 25 °C.

- additional main groups, for example, for cyclic alkanes, formic acid, and so on, were added.

For fitting the temperature-dependent group interaction parameters of modified UNIFAC, in contrast to original UNIFAC, besides VLE data the following data are also used:

- activity coefficients at infinite dilution,
- excess enthalpy data,
- excess heat capacity data,
- LLE data,
- SLE data of simple eutectic systems, and
- azeotropic data.

The various thermodynamic properties deliver different important information for fitting reliable temperature-dependent parameters. The contributions can be summarized as follows:

- VLE (azeotropic data) provide the information about the activity coefficients for a wide composition range (5–95 mol%);
- the required data for the dilute range are delivered by the activity coefficients at infinite dilution;
- at the same time, γ^∞-values measured by gas–liquid chromatography provide reliable information about the real behavior of asymmetric systems.

- excess enthalpies (excess heat capacities) deliver the required information about the temperature dependence;
- h^E-values at high temperature (often at 140 °C) together with SLE data of simple eutectic systems at low temperature are important supporting data for fitting reliable temperature-dependent group interaction parameters.

For fitting the parameters simultaneously to the different types of data, weighting factors are used for the different contributions to the objective function:

$$F = w_{VLE} \sum \Delta VLE + w_{\gamma^\infty} \sum \Delta \gamma^\infty + w_{h^E} \sum \Delta h^E + w_{c_p^E} \sum \Delta c_p^E$$
$$+ w_{LLE} \sum \Delta LLE + w_{SLE} \sum \Delta SLE + w_{AZD} \sum \Delta AZD \quad (5.96)$$

The modifications of modified UNIFAC compared to original UNIFAC are summarized below. The combinatorial part is calculated using the following slightly modified empirical equation:

$$\ln \gamma_i^C = 1 - V_i' + \ln V_i' - 5q_i \left(1 - \frac{V_i}{F_i} + \ln \frac{V_i}{F_i}\right) \quad (5.97)$$

for which besides V_i the following volume/mole fraction ratio V_i' is used:

$$V_i' = \frac{r_i^{3/4}}{\sum_j r_j^{3/4} x_j} \quad (5.98)$$

To describe the temperature dependence, linear or quadratic temperature-dependent parameters were introduced in Eq. (5.95):

$$\Psi_{nm} = \exp\left(-\frac{a_{nm} + b_{nm}T + c_{nm}T^2}{T}\right) \quad (5.99)$$

While linear temperature-dependent group interaction parameters are already required to describe the VLE behavior and excess enthalpies simultaneously, quadratic temperature-dependent parameters are used when the system shows a strong temperature dependence of the excess enthalpies.

Most important for the application of group contribution methods for the synthesis and design of separation processes is a comprehensive and reliable parameter matrix with reliable parameters. The present status of modified UNIFAC is shown in Figure 5.84. Today parameters are available for 91 main groups. In the recent years new main groups were introduced for the different types of amides, isocyanates, epoxides, anhydrides, peroxides, carbonates, various sulfur compounds, and so on. In the last year the range of applicability was even extended to systems with ionic liquids [56].

Because of the importance of modified UNIFAC for process development the range of applicability is continuously extended by filling the gaps in the parameter table and revising some of the existing parameters with the help of systematically measured data and by using new experimental data published and stored in the Dortmund Data Bank [3]. For fitting temperature-dependent parameters, in particular excess enthalpy data covering a wide temperature range are desirable. These data can be measured using for example isothermal flow

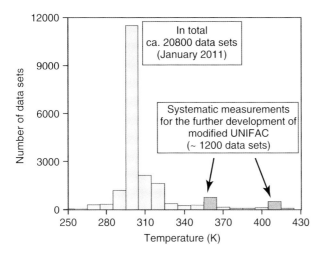

Figure 5.83 Available excess enthalpy data as a function of temperature.

calorimetry. But unfortunately most of the published data were measured near room temperature (see Figure 5.83). To complete the data base and in particular to get the required supporting data at high temperature for fitting the temperature-dependent group interaction parameters of modified UNIFAC, nearly 1200 data sets were systematically measured by isothermal flow calorimetry. Additionally, a large number of VLE data, SLE of eutectic systems, and activity coefficients at infinite dilution were measured systematically in our laboratory.

Since 1996, further extension (i.e., revision of the existing parameters, filling of gaps in the parameter matrix or the introduction of new main groups) was carried out within the UNIFAC consortium. The current status of the complete parameter matrix is always available via internet [57]. A great part of the modified UNIFAC parameters was published by Gmehling *et al.* [58]. The published van der Waals properties and modified UNIFAC group interaction parameters are given in the internet (see Appendix I). But a great part of the group interaction parameters were revised using a larger database to fit the parameters. The revised and the new fitted parameters are only available for the sponsors of the company consortium [57].

Modified UNIFAC is an ideal thermodynamic model for process development. With the help of this predictive model easily various process alternatives can be compared, suitable solvents for separation processes like azeotropic distillation, extractive distillation, extraction can be selected, the influence of solvents on chemical equilibrium conversion can be predicted, and so on.

Modified UNIFAC can also be applied to provide artificial data for fitting the missing binary parameters of the parameter matrix of a g^E-model. But if the key components of a separation step are considered, for the final design an experimental examination of the results is recommended.

The progress achieved when going from UNIFAC to modified UNIFAC can be recognized from a comparison of the results for 2200 consistent binary VLE data sets. Using the UNIQUAC equation for the correlation of the 2200 VLE data sets

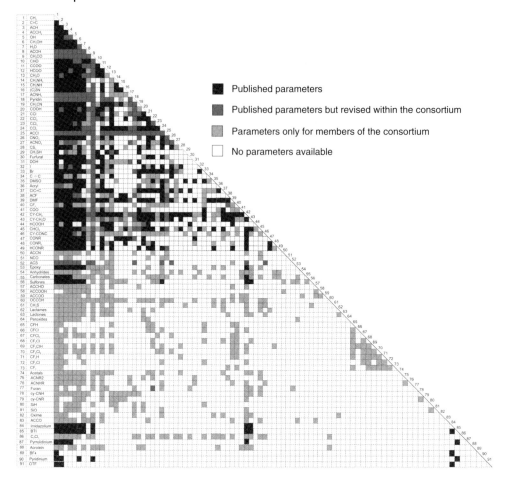

Figure 5.84 Present status of the modified UNIFAC method.

a mean absolute deviation of 0.0058 for the vapor phase mole fraction is obtained. While for original UNIFAC method a mean deviation of 0.0141 results, a mean deviation of 0.0088 of the vapor phase mole fractions is achieved with modified UNIFAC (see Figure 5.85). This means that the deviation with UNIFAC compared to a correlation of the VLE data using the UNIQUAC model was improved by nearly a factor of 3 from 0.0083 (0.0141-0.0058) to 0.0030 (0.0088-0.0058). As can be seen from Figure 5.85, similar improvements are also obtained for the predicted temperatures and pressures.

Not only the results for VLE, but also for SLE, LLE, excess enthalpies, excess heat capacities, activity coefficients at infinite dilution were distinctly improved when going from UNIFAC to modified UNIFAC. For excess enthalpies this is shown in Figure 5.86. It can be seen that in all cases the predicted results are in good agreement with the experimental findings in the case of modified UNIFAC, while

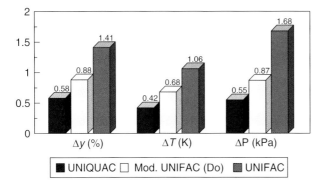

Figure 5.85 Mean absolute deviation in vapor phase mole fraction, temperature, and pressure for the correlation, respectively, prediction of 2200 binary consistent VLE data.

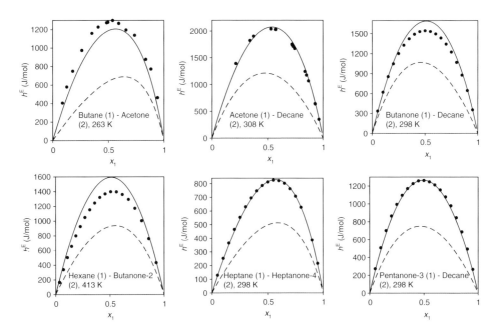

Figure 5.86 Experimental [3] and predicted excess enthalpies for different binary alkane – ketone systems: – – – original UNIFAC; —— modified UNIFAC.

the results of original UNIFAC show strong deviations as already discussed in Section 5.9.3.

Typical results for VLE, excess enthalpies, SLE, activity coefficients at infinite dilution, excess heat capacities, and azeotropic data for systems of alkanes with ketones are shown in Figures 5.87 and 5.88. While in Figure 5.87 results are

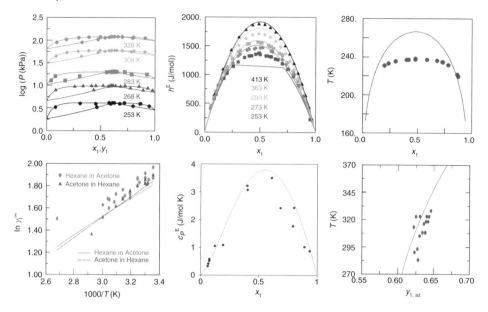

Figure 5.87 Experimental and predicted results for the system acetone (1) – hexane (2) —— modified UNIFAC; ▲ ● ■ experimental [3].

presented for the system acetone–n-hexane, in Figure 5.88 a comparison of the predicted and experimental results for VLE, h^E, SLE, azeotropic data, LLE and γ^∞ for different ketones with various alkanes is presented. In the case of SLE, additionally the curvature assuming ideal behavior is shown by the dashed lines. The improvements obtained when taking into account the real behavior is obvious. Of course, the same group interaction parameters are applied for all the predictions. As can be seen in all cases, good agreement is obtained for the different phase equilibria and excess properties, although a wide temperature range (-100 to $160\,°C$) is covered. The correct description of the temperature dependence is achieved by the reliable prediction of the excess enthalpies in the temperature range covered.

In the meantime the range of applicability of modified UNIFAC was even extended to systems with ionic liquids [56]. In Figure 5.89 the experimental and predicted activity coefficients at infinite dilution of various n-alkanes in different alkyl-methyl-imidazolium bistrifluoromethylsulfonylimides are shown as a function of temperature. It can be seen that not only the temperature dependence, but also the dependence of the activity coefficients from the number of C-atoms of the alkanes and the alkyl rests is properly described.

Besides the prediction of phase equilibria the group contribution methods UNIFAC or modified UNIFAC can be applied for other applications of great practical interest, for example, the calculation of octanol–water partition coefficients

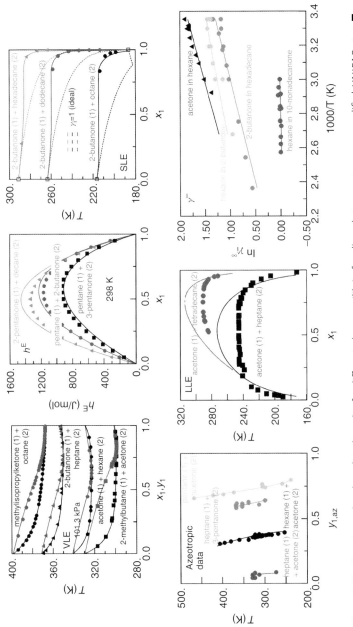

Figure 5.88 Experimental and predicted results for different phase equilibria for alkane-ketone systems — modified UNIFAC; ▲ ● ■ experimental [3].

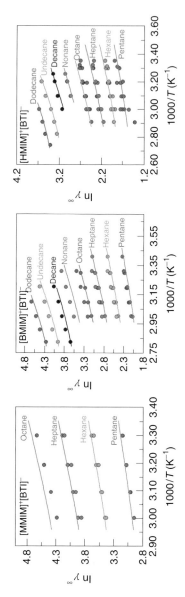

Figure 5.89 Experimental [3] and predicted activity coefficients at infinite dilution of various n-alkanes in three alkyl (methyl, butyl, hexyl)-methyl-imidazolium bis-trifluoromethylsulfonylimides as a function of temperature.

of persistent chemicals [59] to decide about their fate in the environment, or the flash points of flammable liquid mixtures [60].

5.9.3.2 Weaknesses of the Group Contribution Methods UNIFAC and Modified UNIFAC

As shown before the group contribution method modified UNIFAC is a powerful and reliable predictive g^E-model. It was continuously further developed in the last 20 years so that the method provides reliable results and a large range of applicability. But in spite of the great advantages compared to original UNIFAC (better results for excess enthalpies, activity coefficients at infinite dilution, asymmetric systems) it shows the typical weaknesses of a group contribution approach. Hence, for example

- Isomer effects cannot be predicted. This means the same activity coefficients are obtained, for example, for o-/m-/p-xylene or phenanthrene/anthracene with the different solvents. But at least in the case of VLE or SLE calculation this is not a great problem, since the required standard fugacities, that is, vapor pressure, melting point, and heat of fusion are of much greater importance than small differences of the activity coefficients. Similar problems are also observed for other predictive models, for example, the quantum chemical approach.
- Unreliable results are obtained for group contribution methods in the case, if a large number of functional groups have to be taken into account, as in the case of pharmaceuticals or when the molecule shows groups such as, –C(Cl)(F)(Br) as for example, in refrigerants. But also in these cases similar problems are observed for other approaches, for example, the quantum-chemical methods.
- Furthermore, poor results are obtained for the solubilities and activity coefficients at infinite dilution of alkanes or naphthenes in water. This was accepted by the developers of modified UNIFAC to achieve reliable VLE results, for example, for alcohol/water systems. The reason was that starting from experimental γ^∞-values of approx. 250000 for n-hexane in water at room temperature it was not possible to fit alcohol–water parameters which deliver γ^∞-values for hexanol in water of 800 and at the same time describe the azeotropic composition of ethanol and higher alcohols with water properly and obtain homogeneous behavior for alcohol–water systems up to C_3-alcohols and heterogeneous behavior starting from C_4-alcohols. To allow for a prediction of hydrocarbon solubilities in water an empirical relation was developed [61, 62], which allows the estimation of the solubilities of hydrocarbons in water and of water in hydrocarbons (see below).
- For the system tert-butanol–water a miscibility gap is predicted, although tertiary butanol in contrast to 1-butanol, 2-butanol, and isobutanol forms a homogeneous mixture with water.

As mentioned before, unsatisfying results of modified UNIFAC are obtained for the activity coefficients at infinite dilution and the solubilities of hydrocarbons in water. Typical results are given in Table 5.17. From the listed solubilities it can be

Table 5.17 Experimental [3] and predicted solubilities of n-hexane and cyclohexane in water at 25 °C using modified UNIFAC.

Hydrocarbon	Solubility in water x_{exp}	Solubility in water x_{calc}
n-Hexane	$2.5 \cdot 10^{-6}$	$1.5 \cdot 10^{-4}$
Cyclohexane	$1.3 \cdot 10^{-5}$	$1.7 \cdot 10^{-3}$

Table 5.18 Parameters for the empirical estimation of hydrocarbon solubilities in water.

Hydrocarbon	A	B	C
Alkanes [62]	1.104	0.0042	-2.817
Naphthenes	1.3326	0.006427	-3.676
Alkenes	1.523	0.00603	-3.0418

seen that the solubilities and therewith the activity coefficients at infinite dilution are approximately a factor 100 off.

To obtain satisfying results for the solubility of alkanes the following empirical relation was suggested by Banerjee [61] for the temperature range 273–373 K:

$$\log \frac{c_{\text{hydrocarbon in water}}}{\text{mol l}^{-1}} = A \cdot \log\left(\frac{55.56}{\gamma^{\infty}_{\text{hydrocarbon in water at 298.15K*}}}\right) + B \cdot T + C$$

*predicted using modified UNIFAC (5.100)

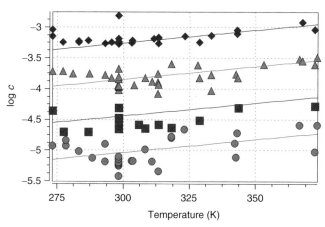

Figure 5.90 Experimental [3] and predicted solubilities c (mol/l) of alkanes in water as a function of temperature using Eq. (5.100) [62]: ♦ n-pentane; ▲ n-hexane; ■ n-heptane; ● n-octane.

Table 5.19 Parameters for the empirical estimation of water solubilities in hydrocarbons.

Hydrocarbon	A	B	C
Alkanes	2.3171	0.01796	−3.672
Naphthenes	0.1806	0.01532	−7.172
Alkenes	−1.1104	0.011	−8.332

For the different hydrocarbons the parameters A, B, and C were fitted to solubility data stored in the Dortmund Data Bank and are given in Table 5.18. For the calculation of the hydrocarbon solubilities only the activity coefficient of the hydrocarbon in water at 25 °C predicted using modified UNIFAC is required.

Typical results for different alkanes are shown in Figure 5.90. It can be seen that the predicted results are in good agreement with the experimental findings.

For the calculation of the solubility of water in alkanes a similar equation can be applied. Also in this case the required parameters A, B, and C were fitted to solubility data stored in the Dortmund Data Bank:

$$\log x_{\text{water in hydrocarbon}} = A \cdot \log \left(\frac{1}{\gamma^\infty_{\text{water in hydrocarbon at 298.15 K}^*}} \right) + B \cdot T + C$$

*predicted using modified UNIFAC (5.101)

The parameters A, B and C for Eq. (5.101) are given in Table 5.19

Example 5.23

Calculate the solubilities of n-hexane in water and water in n-hexane at 298.15 K with the help of the empirical relations given above.

Solution

First the activity coefficients at infinite dilution of n-hexane in water and water in n-hexane at 25 °C have to be calculated. Using the modified UNIFAC parameters given in Appendix I an activity coefficient at infinite dilution of 6618 for n-hexane in water and a value of 135.9 for water in n-hexane is obtained. With these values directly the solubilities can be calculated:

$$\log \frac{c_{n\text{-hexane in water}}}{\text{mol l}^{-1}} = 1.104 \cdot \log \left(\frac{55.56}{6618} \right) + 0.0042 \cdot 298.15 - 2.817$$

$$= -3.8566$$

$$c_{n\text{-hexane in water}} = 1.39 \cdot 10^{-4} \text{ mol/l}$$

$$x_{n\text{-hexane in water}} \approx 2.53 \cdot 10^{-6}$$

$$\log x_{\text{water in }n\text{-hexane}} = 2.3171 \cdot \log\left(\frac{1}{135.9}\right) + 0.01796 \cdot 298.15 - 3.672$$

$$= -3.260$$

$$x_{\text{water in }n\text{-hexane}} = 0.000549$$

Experimentally n-hexane solubilities between $9.44 \cdot 10^{-5}$ and $1.55 \cdot 10^{-4}$ mol/l are reported (see Figure 5.90). As well the water solubility is in very good agreement with the experimental values.

5.9.4
Predictive Soave–Redlich–Kwong (PSRK) Equation of State

As can be recognized from the results shown before, modified UNIFAC is a very powerful predictive model for the development and design of chemical processes, in particular separation processes. However, modified UNIFAC is a g^E-model. This means that it cannot handle supercritical compounds. For supercritical compounds either Henry constants have to be introduced or Approach A has to be used. In the latter case, an equation of state is required, which is able to describe the PvT behavior of both the vapor (gas) and the liquid phase.

As mentioned in Section 2.5 the first equation of state which was able to describe the PvT behavior of the liquid and the vapor phase was developed by van der Waals. With only two parameters a and b, the van der Waals equation of state is able to describe the different observed phenomena, such as condensation, evaporation, the two phase region and the critical behavior. But the calculated densities, vapor

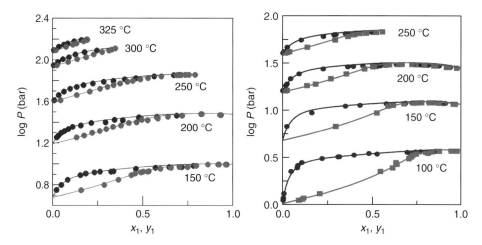

Figure 5.91 Experimental and predicted VLE data using PSRK for the systems ethanol (1)–water (2) (left hand side) and acetone (1)–water (2) (right hand side) at subcritical and supercritical conditions.

pressures, etc. using the van der Waals equation of state were only in qualitative agreement with the experimental findings.

Therefore today improved cubic equations of state like the SRK [63] or the PR equation [64] are used. But up to 1979 the application of the equations of state approach was limited to nonpolar or slightly polar compounds in particular because of the empirical quadratic mixing rules used. Huron and Vidal [35] combined the advantages of g^E-models and equations of state by introducing more sophisticated so-called g^E-mixing rules (see Section 4.9.2). With the application of original UNIFAC for the prediction of the required g^E-values in the mixing rule predictive group contribution equations of state were developed [43]. While in the approach of Huron and Vidal infinite pressure is taken as reference state, in the group contribution equation of state PSRK (predictive SRK) atmospheric pressure is used. The great advantage of this approach is that in the PSRK method the already available UNIFAC parameters can directly be used. But now the UNIFAC parameters can be applied at supercritical conditions. For the systems

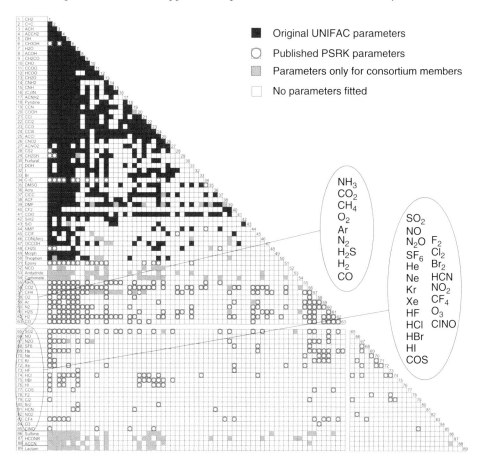

Figure 5.92 Current parameter matrix of the group contribution equation of state PSRK.

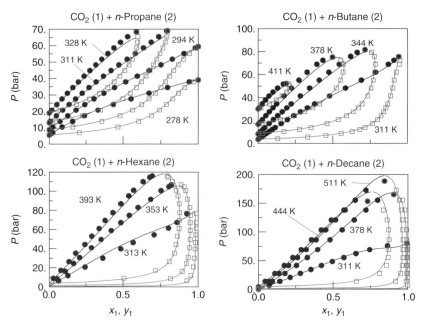

Figure 5.93 Experimental and predicted VLE data using PSRK for various CO_2–alkane systems at subcritical and supercritical conditions.

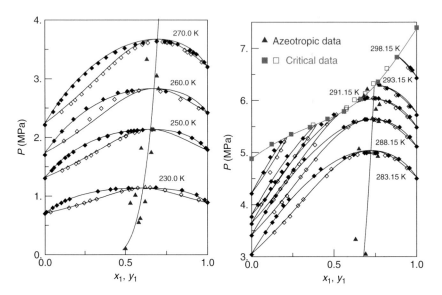

Figure 5.94 Experimental and predicted VLE, azeotropic and critical data of the system CO_2 (1)–ethane (2) using PSRK as a function of temperature.

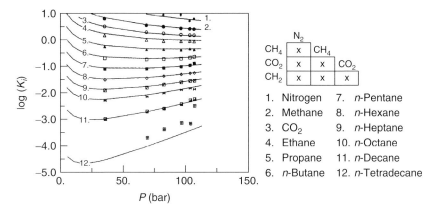

Figure 5.95 Experimental [65] and predicted K-factors for a 12-component system using PSRK at a temperature of 322 K as a function of pressure and the required parameter matrix for the group contribution equation of state PSRK.

ethanol–water and acetone–water this is shown in Figure 5.91. It can be seen that the results are very satisfying.

To use PSRK for process synthesis and design a large matrix with reliable parameters is desirable. The possibility to handle systems at supercritical conditions all of a sudden allowed including also gases like CO_2, CH_4, H_2S, H_2, and so on, as new functional groups in the parameter matrix. In total, 30 different gases were added as new main groups. The required parameters for the gases were fitted to VLE data of low boiling substances and gas solubilities stored in the Dortmund Data Bank [3]. The current PSRK parameter matrix is given in Figure 5.92.

Typical VLE results for different CO_2–alkane systems are shown in Figures 5.93 and 5.94. While in Figure 5.93 only VLE data for four different CO_2-alkane (propane, butane, hexane, decane) are shown, for the system ethane[21]–CO_2 additionally the experimental and predicted azeotropic and critical data are shown. As can be seen, excellent results are obtained for all systems considered. This means that the group contribution concept can also be applied for the gases included in the PSRK matrix.

Predicted results using PSRK for a 12 component system at 322 K are shown in Figure 5.95 in the form of the K-factors ($K_i = y_i/x_i$) as a function of pressure. Using classical mixing rules 66 binary parameters would be required. In the case of a group contribution equation of state the number of required parameters in this case goes down to 6, since all alkanes are described with the same group interaction parameters. This is a great advantage of group contribution equations of state in comparison to the typical equation of state approach, in particular for processes such as the gas-to-liquid process, where a large number of alkanes, alkenes, alcohols besides a few gases have to be handled.

---

21) In PSRK, ethane is built up by two methyl groups.

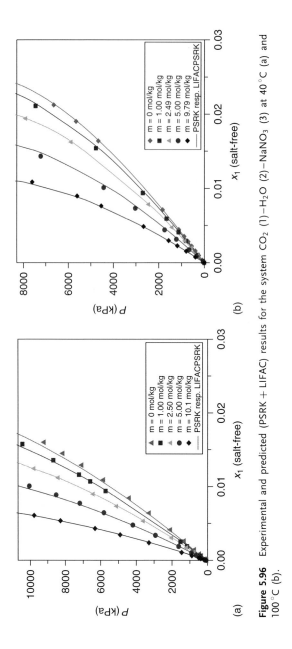

Figure 5.96 Experimental and predicted (PSRK + LIFAC) results for the system CO_2 (1)–H_2O (2)–$NaNO_3$ (3) at 40 °C (a) and 100 °C (b).

PSRK was even extended to systems with strong electrolytes by using the electrolyte model LIFAC [66] instead of the original UNIFAC method for calculating the excess Gibbs energy. The LIFAC method takes into account the middle range and long range interactions of the electrolytes by the Debye–Hückel and a modified *Pitzer* term (see Section 7.3.5). The PSRK model in combination with the LIFAC model allows the prediction of salting in and salting out effect of strong electrolytes on VLE and gas solubilities. In Figure 5.96 the influence of sodium nitrate on the solubility of carbon dioxide in water for different salt concentrations at 40 °C and 100 °C is shown. As can be seen not only the salt effect but also the temperature dependence for this ternary system is described with the required accuracy.

5.9.5
VTPR Group Contribution Equation of State

The PSRK model [43] provides reliable predictions of VLE and gas solubilities. Therefore, PSRK was implemented in most process simulators and is well accepted as a predictive thermodynamic model for the synthesis and design of the different processes in chemical, gas processing, and petroleum industry. But also PSRK shows all the weaknesses of UNIFAC and the SRK equation of state. Since the SRK equation of state is used in the group contribution equation of state PSRK, poor results are obtained for liquid densities of the pure compounds and the mixtures. Furthermore, poor results are obtained for activity coefficients at infinite dilution, heats of mixing and very asymmetric systems because of the use of original UNIFAC. Ahlers and Gmehling [44] developed a generalized group contribution

Table 5.20 Main differences between the new group contribution equation of state VTPR and the PSRK model.

Module	PSRK	VTPR
Equation of state	Soave–Redlich–Kwong	Volume-translated Peng–Robinson
α-Function	Generalized Mathias–Copeman	Generalized Twu
Mixing rule for the parameter a	$\dfrac{a}{bRT} = \sum_i x_i \dfrac{a_{ii}}{b_i RT} + \dfrac{1}{A}\left(\dfrac{g^E}{RT} + \sum_i x_i \ln\dfrac{b}{b_i}\right)$ $A = -0.64663$	$\dfrac{a}{b} = \sum_i x_i \dfrac{a_{ii}}{b_i} + \dfrac{g^{E,R}}{A}$ $A = -0.53087$
Mixing rule for the parameter b	$b = \Sigma x_i b_i$	$b_{ij}^{3/4} = \left(b_{ii}^{3/4} + b_{jj}^{3/4}\right)/2$ $b = \sum_i \sum_j x_i x_j b_{ij}$
g^E information	(a) original UNIFAC (b) temp-depend. PSRK parameters	Temp-depend. VTPR parameters
Database	VLE, GLE	VLE, GLE, h^E, SLE, γ^∞

equation of state called *VTPR*, where most of the weaknesses of PSRK were removed (see also Sections 2.5.5 and 4.9.2). The main differences between PSRK and VTPR are summarized in Table 5.20.

A better description of liquid densities is achieved, by using the volume translated PR (Peneloux *et al.* [33]) instead of the SRK equation of state, which is used in the PSRK model. Based on the ideas of Chen *et al.* [67] an improved g^E-mixing-rule is used. The prediction of asymmetric systems is improved by using a quadratic *b* mixing-rule with a modified combination rule [67]. The improvements obtained when going from the group contribution equation of state PSRK to VTPR can be recognized from the predicted results using these models for symmetric alkane–alkane systems shown in Figure 5.97 and asymmetric alkane–alkane

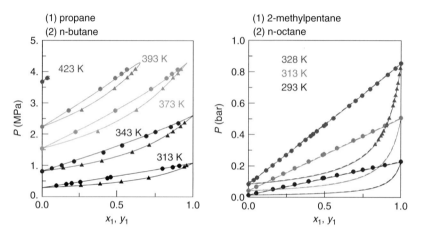

Figure 5.97 Experimental and predicted VLE data for symmetric alkane–alkane systems ---- PSRK —— VTPR.

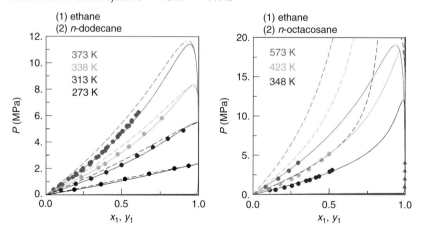

Figure 5.98 Experimental and predicted VLE data for asymmetric alkane–alkane systems ---- PSRK —— VTPR.

systems shown in Figure 5.98. For the prediction of alkane–alkane systems no interaction parameters are required for both models. This means that the results mainly depend on the mixing rules used. As can be seen from the Pxy-diagrams much better results are predicted using VTPR instead of PSRK in the case of the asymmetric systems ethane–dodecane and ethane–octacosane, while nearly the same results are obtained for the symmetric systems propane–butane and 2-methylpentane–n-octane.

In the case of the group contribution equation of state VTPR, instead of temperature-independent group interaction parameters from original UNIFAC, temperature-dependent group interaction parameters as in modified UNIFAC are used. As for modified UNIFAC, the required temperature-dependent group interaction parameters of VTPR are fitted simultaneously to a comprehensive data base. Besides VLE data for systems with sub and supercritical compounds, gas

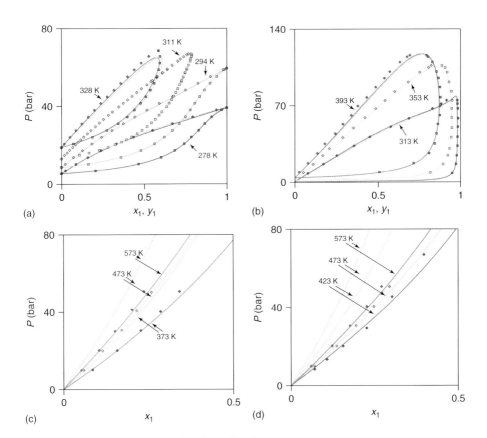

Figure 5.99 Experimental and predicted VLE data for different CO_2–alkane systems (a) $CO_2(1)$ – propane(2) (b) $CO_2(1)$ – n-hexane(2) (c) $CO_2(1)$ – n-eicosane(2) (d) $CO_2(1)$ – n-octacosane(2) ··· PSRK —— VTPR group contribution equation of state.

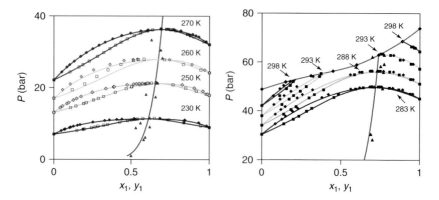

Figure 5.100 Experimental and predicted VLE data, azeotropic points and critical data for the system CO_2 (1)–ethane (2) using VTPR.

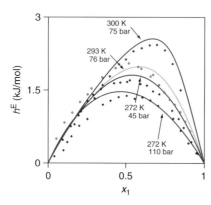

Figure 5.101 Experimental and predicted excess enthalpy data for the system CO_2(1)–ethane (2).

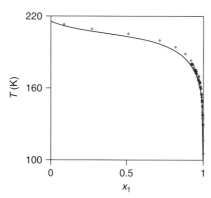

Figure 5.102 Experimental and predicted SLE data for the system ethane (1)–CO_2 (2) ● experimental [3, 68] —— group contribution equation of state VTPR.

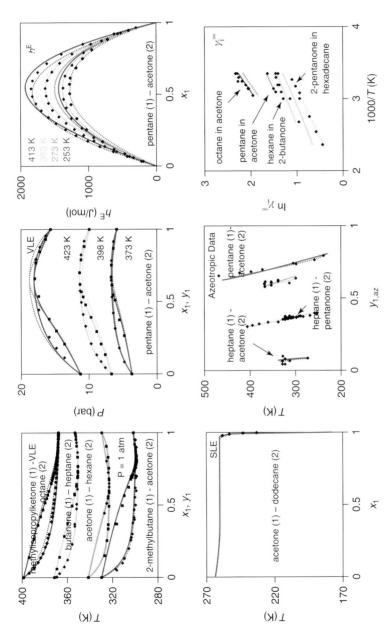

Figure 5.103 Experimental and predicted phase equilibrium data and excess enthalpies for alkanes with ketones predicted using modified UNIFAC respectively the group contribution equation of state VTPR; ··· modified UNIFAC, —— group contribution equation of state VTPR.

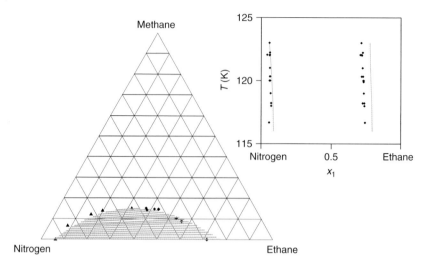

Figure 5.104 Experimental and predicted LLE data using VTPR for the ternary system nitrogen–ethane–methane at 122 K and the binary system ethane–nitrogen [3].

solubilities, SLE of eutectic systems, activity coefficients at infinite dilution and excess enthalpies covering a large temperature and pressure range are used.

The results obtained for the different pure component properties and various phase equilibria of the new group contribution equation of state are very promising [44]. In Figure 5.99 the predicted VLE results for different alkane–CO_2-systems using PSRK and the group contribution equation of state VTPR are presented. While both methods show similar results for the slightly asymmetric system CO_2–propane and CO_2–n-hexane, again much better results are achieved for the strongly asymmetric system CO_2–eicosane and CO_2–octacosane with the group contribution equation of state VTPR because of the improved mixing rules.

Using the same parameters, VLE, azeotropic data, critical data and excess enthalpies for the system CO_2–ethane were predicted. A comparison of the predicted and experimental results is shown in Figures 5.100 and 5.101. It can be seen that as in the case of PSRK (see Figure 5.94) excellent agreement between the predicted results and the experimental findings is obtained. In Figure 5.101 the experimental and predicted excess enthalpies for the systems CO_2–ethane using the group contribution equation of state VTPR are shown. It can be seen that nearly a perfect description of the VLE, azeotropic, and the critical line is obtained. Furthermore, not only the temperature, but also the pressure dependence of the excess enthalpies is described correctly with the group contribution equation of state VTPR. Perhaps it has to be mentioned again that for all the predictions (VLE, azeotropic data, critical line, h^E) shown in Figures 5.99–5.101 the same parameters were used to describe the interactions between CO_2 and alkanes.

Using the same group interaction parameters, other phase equilibria can be predicted as well. The predicted SLE behavior of the binary system ethane–CO_2

5.9 Predictive Models

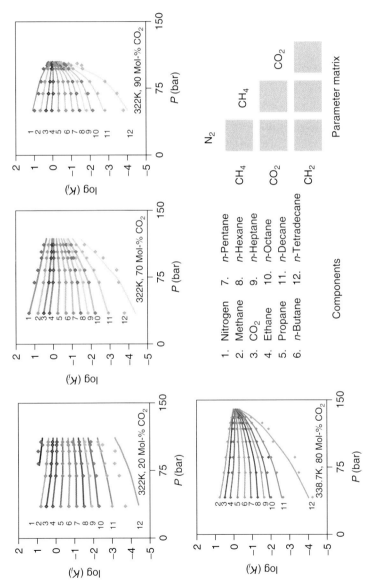

Figure 5.105 Experimental [65] and predicted K-factors for a 12 component system.

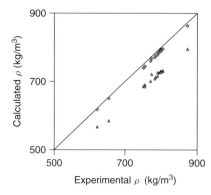

Figure 5.106 Experimental and calculated liquid densities using PSRK and VTPR for the quaternary system pentane–hexane–benzene–cyclohexane at 298.15 K: ▲ PSRK; ♦ VTPR.

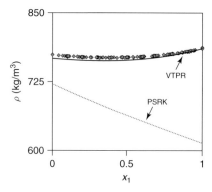

Figure 5.107 Experimental and predicted densities for the binary system acetone (1)–cyclohexane (2) at 298.15 K: ---- PSRK; — VTPR.

using VTPR is shown in Figure 5.102 together with the experimental data. It can be seen that the parameters fitted to a comprehensive data base can be successfully applied also at very low temperatures.

The next example (see Figure 5.103) shows the results of the group contribution equation of state VTPR in comparison to the results of modified UNIFAC for different VLE, excess enthalpies, SLE, azeotropic data, activity coefficients at infinite dilution for various alkane–ketone systems. It can be seen that with the group contribution equation of state VTPR similarly good results are obtained for the different phase equilibria and excess enthalpies as obtained with modified UNIFAC. But, besides the prediction of the different phase equilibria of subcritical compounds, the method can directly be applied for systems with supercritical compounds, for example, it can directly be applied for the calculation of gas solubilities. At the same time various other thermophysical properties (densities, enthalpies, for example, enthalpies of vaporization, heat capacities, Joule–Thomson coefficients, etc.) for pure compounds and mixtures for the liquid or gas phase can be predicted for the given condition (temperature, pressure, composition). The main disadvantage is that the available parameter matrix of the group contribution equation of state is still limited. But work is in progress to extend it.

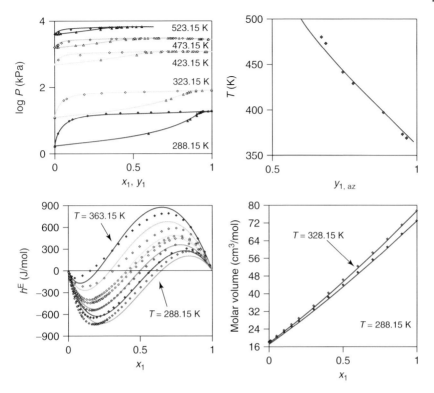

Figure 5.108 Correlation results of the equation of state VTPR for the system acetone(1)–water(2) using temperature dependent UNIQUAC parameters $a_{12} = 472.46$ K, $a_{21} = -585.54$ K, $b_{12} = -0.40712$, $b_{21} = 2.5101$, $c_{12} = 3.607 \cdot 10^{-4}$ K^{-1}, $c_{21} = 1.595 \cdot 10^{-5}$ K^{-1}.

At the same time the group contribution equation of state VTPR in contrast to modified UNIFAC can be applied for the prediction of phase equilibria including compounds not covered by modified UNIFAC, for example, the various gases. The predicted LLE results for the ternary system nitrogen–CO_2–methane at 122 K and the binary system nitrogen–CO_2 as a function of temperature are shown in Figure 5.104 together with the experimental data.

A group contribution equation of state shows in particular great advantages compared to the usual equation of state approach in the case of multicomponent mixtures, when the multicomponent mixture consists of gases and various alkanes, alcohols, alkenes, and so on. The reason is that the same parameters can be used for all alkanes, alcohols, alkenes, so that the size of the parameter matrix is small in comparison to the typical equation of state approach. The results of VTPR for a 12 component system consisting of nitrogen–methane–CO_2–alkanes are shown in Figure 5.105. As can be seen, excellent results are obtained with the six required parameters (66 binary parameters would be required for the classical equation of state approach).

As mentioned already several times, using equations of state besides phase equilibria also other properties, such as densities, heat capacities, enthalpies, Joule–Thomson coefficients, and so on, for pure compounds and mixtures can be calculated. In Figure 5.106 the improvements in liquid densities for a four component system are shown, when instead of the PSRK the VTPR group contribution equation of state is used. The main reason for the improvements using VTPR comes from the fact, that the pure component densities are already much better described by the VTPR equation of state. In Figure 5.107 the predicted densities of the PSRK and the VTPR group contribution equation of state are shown together with the experimental densities. The improvement when going from PSRK to VTPR is significant.

VTPR can also be applied to correlate experimental data. Then instead of the group contribution method a g^E-model, for example, the UNIQUAC, Wilson, or NRTL equation can be applied. This is of great interest when reliable experimental data are available. For the system acetone–water the correlation results are shown in Figure 5.108. It can be seen that nearly perfect results are obtained for VLE, excess enthalpies and the azeotropic composition. Similarly good results can be obtained with the help of a g^E-model (see Figure 5.34). But as can be seen from the VLE data, now the model can be applied at supercritical conditions. At the same time other properties, for example, densities, and so on, can be calculated.

An overview about the development of group contribution methods and group contribution equations of state for the prediction of phase equilibria and other thermophysical properties can be found in [69].

Additional Problems

P5.1 Calculate the pressure and the vapor phase mole fraction for the system ethanol (1)–water (2) at 70 °C with the help of the different g^E-models (Wilson, NRTL, UNIQUAC) for an ethanol mole fraction of $x_1 = 0.252$ using the interaction parameters, auxiliary parameters, and Antoine constants given in Figure 5.30 and assuming ideal vapor phase behavior. Besides total and partial pressures and vapor phase composition, calculate also K-factors and separation factors. Repeat the calculation using the quantities defined in Eq. (5.15) $\phi_1 = 0.9958$ and $\phi_2 = 1.0070$.

P5.2 Regress the binary interaction parameters of the UNIQUAC model to the isobaric VLE data of the system ethanol (1) - water (2) measured by Kojima et al. at 1 atm and listed below. As objective function, use:
  a. relative quadratic deviation in the activity coefficients
  b. quadratic deviation in boiling temperatures
  c. relative quadratic deviation in vapor phase compositions
  d. relative deviation in separation factors.

Adjust the vapor pressure curves using a constant factor to exactly match the author's pure component vapor pressures.

x_1	y_1	T (K)	x_1	y_1	T (K)
0.0000	0.0000	373.15	0.5500	0.6765	352.57
0.0500	0.3372	363.15	0.6000	0.6986	352.28
0.1000	0.4521	359.08	0.6500	0.7250	352.00
0.1500	0.5056	357.12	0.7000	0.7550	351.75
0.2000	0.5359	356.05	0.7500	0.7840	351.57
0.2500	0.5589	355.29	0.8000	0.8167	351.45
0.3000	0.5794	354.67	0.8500	0.8591	351.37
0.3500	0.5987	354.14	0.9000	0.8959	351.35
0.4000	0.6177	353.67	0.9500	0.9474	351.39
0.4500	0.6371	353.25	1.0000	1.0000	351.48
0.5000	0.6558	352.90	–	–	–

Reference: Kojima, K., Tochigi, K., Seki, H., Watase, K., and Kagaku, Kogaku (1968) **32**, 149–153.

P5.3 Compare the experimental data for the system ethanol–water measured at 70 °C (see Figure 5.30 resp. Table 5.2) with the results of the group contribution method modified UNIFAC and the group contribution equation of state VTPR.

P5.4 Calculate the Pxy-diagram at 70 °C for the system ethanol(1)–benzene(2) assuming ideal vapor phase behavior using the Wilson equation. The binary Wilson parameters Λ_{12} and Λ_{21} should be derived from the activity coefficients at infinite dilution (see Table 5.6). Experimentally the following activity coefficients at infinite dilution were determined at this temperature:

$$\gamma_1^\infty = 7.44 \quad \gamma_2^\infty = 4.75$$

P5.5 Determine the azeotropic composition of the following homogeneous binary systems
a. acetone–water
b. ethanol–1,4-dioxane
c. acetone–methanol.
at 50, 100, and 150 °C using the group contribution method modified UNIFAC.

P5.6 In the manual of a home glass distillery (s. Figure P5.1) the following recommendation is given: "After some time liquid will drip out of the cooler. You are kindly requested to collect the first small quantity and not to use it, as first a methanol enrichment takes place." Does this recommendation make sense? The purpose of the glass distillery is to enrich ethanol. Consider the wine to be distilled as a mixture of ethanol (10 wt%), methanol (200 wt ppm), and water. The one stage distillation takes place at atmospheric pressure.

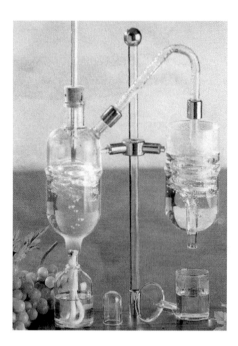

Figure P5.1 Home glass distillery

Calculate the percentages of methanol and ethanol removed from 200 g feed, when 10 g of the distillate is withdrawn. For the calculation the modified UNIFAC method should be applied. The constants for the Antoine equation for ethanol and water can directly be taken from Figure 5.30. For methanol the vapor pressure constants and the molar mass are given in Appendix A. For the calculation ideal vapor phase behavior should be assumed.

P5.7   Calculate the VLE behavior, h^E data, azeotropic data, and activity coefficients at infinite dilution for the system n-pentane–acetone at 373 K, 398 K, and 423 K using modified UNIFAC. The results are shown graphically in Figure 5.103.

The vapor pressure constants are given in Appendix A. Experimental data can be downloaded from the textbook material page on *www.ddbst.com*. For the calculation by modified UNIFAC ideal vapor phase behavior should be assumed.

P5.8   Using the free Explorer Version of DDB/DDBSP, search for mixture data for the system acetone–n-hexane.

   a. Plot the experimental pressure as function of liquid and vapor phase composition together with the predictions using UNIFAC, modified UNIFAC, and PSRK for the data sets at 318 K and 338 K.

b. How large are the differences in the azeotropic composition as shown in the plot of separation factor vs. composition?

c. Plot the experimental heats of mixing data as function of liquid phase composition together with the predictions of UNIFAC, modified UNIFAC, and PSRK for the data sets at 243 K, 253 K, and 298 K. Interpret the linear part in some of the calculated heat of mixing curves.

d. Plot the experimental LLE data together with the results of UNIFAC and modified UNIFAC. What led to the improved results in case of modified UNIFAC?

P5.9 Using the free Explorer Version of DDB/DDBSP, search for mixture data for the systems CO_2–n-hexane and CO_2–hexadecane. Plot the experimental high pressure VLE data (HPV) together with the predictions using PSRK. Compare the results to those of VTPR (Figure 5.99d) and examine the results for SLE in the binary mixture CO_2–n-hexane.

P5.10 Calculate the activity coefficients in the system methanol (1)–toluene (2) from the data measured by Ocon et al. [71] at atmospheric pressure assuming ideal vapor phase behavior. Try to fit the untypical behavior of the activity coefficients of methanol as function of composition using temperature independent g^E-model parameters (Wilson, NRTL, UNIQUAC). Explain why the activity coefficients of methanol show a maximum at high toluene concentration.

The vapor pressure constants are given in Appendix A. Experimental data as well as molar volumes, r and q values can be downloaded from the textbook page on *www.ddbst.com*. For the calculation, ideal vapor phase behavior should be assumed.

P5.11 Predict the Henry constants of methane, carbon dioxide, and hydrogen sulfide in methanol in the temperature range −50 to 200 °C with the help of the group contribution methods PSRK and VTPR.

Compare the predicted Henry constants with experimental values from the textbook page on *www.ddbst.com*.

P5.12 Predict the solubility of methane, carbon dioxide, and hydrogen sulfide in methanol at a temperature of −30 °C for partial pressures of 5 bar, 10 bar, and 20 bar using the PSRK and VTPR group contribution equations of state. Compare the results with the solubilities obtained using Henry's law and the Henry constants predicted in problem P5.11.

P5.13 In the free DDBSP Explorer Version, search for data for all subsystems of the system methanol–methane–carbon dioxide.

a. Compare the available gas solubility data with the results of the PSRK method via the data prediction option in DDBSP.

b. Plot the available high pressure VLE data (HPV) for the system methanol–carbon dioxide together with the predicted curve using the PSRK method. Examine and familiarize yourself with the different graphical representations.

c. Regress the dataset 2256 using the Soave-Redlich-Kwong equation of state with the quadratic mixing rule and a g^E mixing rule with activity coefficient calculation via the UNIQUAC model. Explain the differences.

P5.14 In the free DDBSP Explorer Version, search for all mixture data for the system benzene–water. Calculate the solubility of benzene in water from the experimental activity coefficients at infinite dilution and compare the results to the experimental LLE data.

P5.15 Examine with the help of the regular solution theory, UNIFAC and modified UNIFAC if the binary systems benzene–cyclohexane and benzene–n-hexane show an azeotropic point at 80 °C. In case of the regular solution theory, calculate the solubility parameter from the saturated liquid density and the heat of vaporization using Eq. (5.70). All required data are given in Appendices A, H, and I.

References

1. Baerns, M., Behr, A., Brehm, A., Gmehling, J., Hofmann, H., Onken, U., and Renken, A. (2006) *Technische Chemie*, Wiley-VCH Verlag GmbH, Weinheim.
2. Hildebrand, J.H. (1949) *J. Phys. Coll. Chem.*, **53**, 944–947.
3. Dortmund Data Bank *www.ddbst.com*.
4. Parikh, J.S., Bukacek, R.F., Graham, L., and Leipziger, S. (1984) *J. Chem. Eng. Data*, **29**, 301–303.
5. Novak, J.P., Matous, J., and Pick, J. (1987) *Liquid – Liquid Equilibria*, Elsevier, Amsterdam.
6. Gmehling, J., Onken, U. *et al.* (1977) *Vapor–Liquid Equilibrium Data Collection*, DECHEMA Chemistry Data Series, 37 parts, Vol. I, DECHEMA, Frankfurt.
7. Gmehling, J. and Kolbe, B. (1992) *Thermodynamik*, Wiley-VCH Verlag GmbH, Weinheim.
8. Mertl, L. (1972) *Collect. Czech. Chem. Commun.*, **37**, 366–374.
9. Porter, A.W. (1921) *Trans. Faraday Soc.*, **16**, 336–345.
10. Redlich, O., Kister, A.T., and Turnquist, C.E. (1952) *Chem. Eng. Prog. Symp. Ser.*, **48** (2), 49–61.
11. Larkin, J.A. (1975) *J. Chem. Thermodyn.*, **7**, 137.
12. Chaudry, M.S. and Lamb, J.A. (1987) *J. Chem. Eng. Data*, **32**, 431–434.
13. Wilson, G.M. (1964) *J. Am. Chem. Soc.*, **86**, 127.
14. Renon, H. and Prausnitz, J.M. (1968) *AIChE J.*, **14**, 135.
15. Abrams, D. and Prausnitz, J.M. (1975) *AIChE J.*, **21**, 116.
16. Flory, P.J. (1941) *J. Chem. Phys.*, **9**, 660.
17. Huggins, M.L. (1941) *J. Phys. Chem.*, **9**, 440.
18. Huggins, M.L. (1942) *Ann. Acad. N. Y. Acad. Sci.*, **43**, 1.
19. Hiaki, T., Kurihara, K., and Kojima, K. (1994) *J. Chem. Eng. Data*, **39**, 714–719.
20. Voňka, P., Novák, J.P., Suška, J., and Pick, J. (1975) *Chem. Eng. Commun.*, **2**, 51.
21. Nelder, J.A. and Mead, R. (1965) *Comput. J.*, **7**, 308.
22. Hoffmann, U. and Hofmann, H. (1971) *Einführung in die Optimierung*, Wiley-VCH Verlag GmbH, Weinheim.
23. Gmehling, J., Menke, J. *et al.* (1986) *Activity Coefficients at Infinite Dilution*, DECHEMA Chemistry Data Series, 6 parts, Vol. IX, DECHEMA, Frankfurt.
24. Van Ness, H.C. (1995) *Pure Appl. Chem.*, **67**, 859–872.
25. (a) Redlich, O. and Kister, A.T. (1948) *Ind. Eng. Chem.*, **40**, 345; (b) Herington, E.F.H. (1947) *Nature*, **160**, 610; (c) Herington, E.F.H. (1951) *J. Inst. Petrol.*, **37**, 457.
26. Van Ness, H.C., Byer, S., and Gibbs, R.E. (1973) *AIChE J.*, **19**, 238.
27. Fredenslund, Aa., Gmehling, J., and Rasmussen, P. (1977) *Vapor–Liquid*

Equilibria Using UNIFAC, Elsevier, Amsterdam.
28. Margules, M. (1895) Akad, S.-B. *Wiss. Wien, Math. Naturwiss. Kl. II*, **104**, 1234.
29. van Laar, J.J. (1910) *Z. Phys. Chem.*, **72**, 723.
30. Fenske, M.R. (1931) *Ind. Eng. Chem.*, **24**, 482.
31. Rarey-Nies, J.R., Tiltmann, D., and Gmehling, J. (1989) *Chem. Ing. Tech.*, **61**, 407–410.
32. Tochigi, K., Rarey, J., and Gmehling, J. (2009) *J. Chem. Eng. Jpn.*, **42**, 376–380.
33. Peneloux, A., Rauzy, E., and Freze, R. (1982) *Fluid Phase Equilib.*, **8**, 7–23.
34. Diedrichs, A., Rarey, J., and Gmehling, J. (2006) *Fluid Phase Equilib.*, **248**, 56–69.
35. Huron, M.-J. and Vidal, J. (1979) *Fluid Phase Equilib.*, **3**, 255.
36. Knapp, H., Döring, R., Oellrich, L., Plöcker, U., and Prausnitz, J.M. (1982) *Vapor–Liquid Equilibria for Mixtures of Low Boiling Substances*, DECHEMA Chemistry Data Series, Vol. **VI**, DECHEMA, Frankfurt.
37. Gmehling, J., Menke, J., Krafczyk, J., and Fischer, K. (2004) *Azeotropic Data*, 3 parts, Wiley-VCH Verlag GmbH, Weinheim.
38. Novak, J.P., Matous, J., and Vonka, P. (1991) *Collect. Czech. Chem. Commun.*, **56**, 745–749.
39. Zieborak, K. (1966) *Z. Phys. Chem.*, **231**, 248–258.
40. Gmehling, J. and Möllmann, C. (1998) *Ind. Eng. Chem. Res.*, **37**, 3112–3123.
41. Hochgesand, G. (1970) *Ind. Eng. Chem.*, **62**, 37–43.
42. Sada, E., Kito, S., and Ito, Y. (1976) 170th Meeting ACS, Chicago, August 27–28, 1975, pp. 374–380.
43. (a) Holderbaum, T. and Gmehling, J. (1991) *Fluid Phase Equilib.*, **70**, 251–265; (b) Horstmann, S., Fischer, K., and Gmehling, J. (2000) *Fluid Phase Equilib.*, **167**, 173–186.
44. (a) Ahlers, J. and Gmehling, J. (2001) *Fluid Phase Equilib.*, **191**, 177–188; (b) Ahlers, J. and Gmehling, J. (2002) *Ind. Eng. Chem. Res.*, **41**, 3489–3498; (c) Ahlers, J. and Gmehling, J. (2002) *Ind. Eng. Chem. Res.*, **41**, 5890–5899.
45. Prausnitz, J.M. and Shair, F.H. (1961) *AIChE J.*, **7**, 682.
46. Horiuti, J. (1931) *Sci. Pap. Inst. Phys. Chem. Res. (Jpn.)*, **17**, 125–256.
47. Sørensen, J.M., Arlt, W., Macedo, E., and Rasmussen, P. (1979) *Liquid–Liquid Equilibrium Data Collection*, DECHEMA Chemistry Data Series, 4 parts, DECHEMA, Frankfurt.
48. (a) Scatchard, G. (1931) *Chem. Rev.*, **8**, 321–333; (b) Hildebrand, J. and Wood, S.E. (1933) *J. Chem. Phys.* **1**, 817–822.
49. Prausnitz, J.M. and Gmehling, J. (1980) *Thermische Verfahrenstechnik – Phasengleichgewichte*, Krausskopff Verlag, Mainz.
50. Derr, E.L. and Deal, C.H. (1969) *Inst. Chem. Eng. Symp. Ser.*, **32**, 40–51.
51. Kojima, K. and Tochigi, K. (1979) *Prediction of Vapor–Liquid Equilibria by the ASOG Method*, Kodansha-Elsevier, Tokyo.
52. Fredenslund, Aa., Jones, R.L., and Prausnitz, J.M. (1975) *AIChE J.*, **21**, 1086–1099.
53. Hansen, K.H., Schiller, M., Fredenslund, Aa., Gmehling, J., and Rasmussen, P. (1991) *Ind. Eng. Chem. Res.*, **30**, 2352–2355.
54. Bondi, A. (1967) *Chem. Rev.*, **67**, 565.
55. Weidlich, U. and Gmehling, J. (1987) *Ind. Eng. Chem. Res.*, **26**, 1372–1381.
56. Nebig, S. and Gmehling, J. (2011) *Fluid Phase Equilib.*, **302**, 220–225.
57. UNIFAC Consortium www.unifac.org.
58. Gmehling, J., Li, J., and Schiller, M. (1993) *Ind. Eng. Chem. Res.*, **32**, 178–193.
59. Wienke, G. and Gmehling, J. (1998) *Toxicol. Environ. Chem.*, **65**, 57–86. (Erratum: (1998) **67**, 275.)
60. Gmehling, J. and Rasmussen, P. (1982) *Ind. Eng. Chem. Fundam.*, **21**, 186.
61. Banerjee, S. (1985) *Environ. Sci. Technol.*, **19**, 369–370.
62. Jakob, A., Grensemann, H., Lohmann, J., and Gmehling, J. (2006) *Ind. Eng. Chem. Res.*, **45**, 7924–7933.
63. Soave, G. (1972) *Chem. Eng. Sci.*, **27**, 1197.
64. Peng, D.Y. and Robinson, D.B. (1976) *Ind. Eng. Chem. Fundam.*, **15**, 59–64.

65. Turek, E.A., Metcalfe, R.S., Yarborough, L., and Robinson, R.L. (1984) *Soc. Petrol Eng. J.*, **24**, 308–324.
66. Yan, W., Topphoff, M., Rose, C., and Gmehling, J. (1999) *Fluid Phase Equilib.*, **162**, 97–113.
67. Chen, J., Fischer, K., and Gmehling, J. (2002) *Fluid Phase Equilib.*, **200**, 411–429.
68. Jensen, R.H. and Kurata, F. (1971) *AIChE J.*, **17**, 357–364.
69. Gmehling, J. (2009) *J. Chem. Thermodyn.*, **41**, 731–747.
70. Anderson, T.F., Abrams, D.A., Grens, E.A., (1978) *AIChE J.*, **24**(1), 20–29.
71. Ocon, J., Tojo, G., Espada, L., (1969) *Anal. Quim.*, **65** 641–648.

6
Caloric Properties

6.1
Caloric Equations of State

The caloric properties are the internal energy u, the enthalpy h, the entropy s, the Helmholtz energy a, and the Gibbs energy g. They can be calculated if an equation of state, the specific isobaric heat capacity of the ideal gas c_p^{id}, and an appropriate mixing rule are available. As standalone values, the caloric properties are not valid. As explained in Chapter 2, a reference point must always be defined. In process simulations, the standard enthalpy of formation and the standard Gibbs energy of formation in the ideal gas state at $\vartheta = 25\,°C$ and $P = 101\,325$ Pa are used in most cases. Thus, the caloric properties defined in this way can be used for describing chemical reactions as well.

In the following sections, expressions for the caloric properties are derived both for pressure-explicit and for volume-explicit equations of state, which have to be valid for the regarded temperature and pressure range.

6.1.1
Internal Energy and Enthalpy

The specific internal energy u and the specific enthalpy h are related by

$$h = u + Pv \qquad (6.1)$$

As the term Pv can easily be calculated both by pressure- and by volume-explicit equations of state, they can be treated together. First, the starting points can be fixed as

$$h_0 = \Delta h_f^0 \qquad (6.2)$$

and

$$u_0 = \Delta h_f^0 - RT_0 \qquad (6.3)$$

with $T_0 = 298.15$ K, $P_0 = 1.01325$ bar in the ideal gas state. The ideal gas state at T_0, P_0 can be hypothetical, most substances are in the liquid state at these conditions. As the enthalpy of an ideal gas does not depend on the pressure, one can write

$$h_0(T_0, P_0) = h(T_0, P \to 0) \quad \text{and} \quad u_0(T_0, P_0) = u(T_0, P \to 0) \qquad (6.4)$$

Chemical Thermodynamics: for Process Simulation, First Edition.
Jürgen Gmehling, Bärbel Kolbe, Michael Kleiber, and Jürgen Rarey.
© 2012 Wiley-VCH Verlag GmbH & Co. KGaA. Published 2012 by Wiley-VCH Verlag GmbH & Co. KGaA.

The total differential of the specific internal energy u is

$$du = \left(\frac{\partial u}{\partial T}\right)_v dT + \left(\frac{\partial u}{\partial v}\right)_T dv = c_v dT + \left(\frac{\partial u}{\partial v}\right)_T dv \tag{6.5}$$

With the fundamental equation

$$du = Tds - Pdv \tag{6.6}$$

and the total differential of the specific entropy

$$ds = \left(\frac{\partial s}{\partial T}\right)_v dT + \left(\frac{\partial s}{\partial v}\right)_T dv \tag{6.7}$$

one obtains

$$du = T\left(\frac{\partial s}{\partial T}\right)_v dT + \left[T\left(\frac{\partial s}{\partial v}\right)_T - P\right] dv \tag{6.8}$$

The comparison of the coefficients in Eqs. (6.5) and (6.8) yields

$$\left(\frac{\partial u}{\partial T}\right)_v = T\left(\frac{\partial s}{\partial T}\right)_v \Rightarrow \left(\frac{\partial s}{\partial T}\right)_v = \frac{1}{T}\left(\frac{\partial u}{\partial T}\right)_v \tag{6.9}$$

$$\left(\frac{\partial u}{\partial v}\right)_T = T\left(\frac{\partial s}{\partial v}\right)_T - P \Rightarrow \left(\frac{\partial s}{\partial v}\right)_T = \frac{(\partial u/\partial v)_T + P}{T} \tag{6.10}$$

The mixed quadratic derivatives of the specific entropy

$$\frac{\partial^2 s}{\partial T \partial v} = \frac{1}{T}\frac{\partial^2 u}{\partial T \partial v} \tag{6.11}$$

and

$$\frac{\partial^2 s}{\partial v \partial T} = -\frac{1}{T^2}\left[\left(\frac{\partial u}{\partial v}\right)_T + P\right] + \frac{1}{T}\left[\frac{\partial^2 u}{\partial v \partial T} + \left(\frac{\partial P}{\partial T}\right)_v\right] \tag{6.12}$$

must be identical. Thus, one gets

$$-\frac{1}{T^2}\left[\left(\frac{\partial u}{\partial v}\right)_T + P\right] + \frac{1}{T}\left(\frac{\partial P}{\partial T}\right)_v = 0 \tag{6.13}$$

and therefore

$$\left(\frac{\partial u}{\partial v}\right)_T = T\left(\frac{\partial P}{\partial T}\right)_v - P \tag{6.14}$$

Combining Eqs. (6.5) and (6.14) one obtains

$$du = c_v dT + \left[T\left(\frac{\partial P}{\partial T}\right)_v - P\right] dv \tag{6.15}$$

Considering that there are ideal gas conditions at $v \to \infty$, the result is

$$u(T, v) = u_0 + \int_{T_0}^{T} c_v^{id} dT + \int_{\infty}^{v}\left[T\left(\frac{\partial P}{\partial T}\right)_v - P\right] dv \tag{6.16}$$

Subsequently, the specific enthalpy can be calculated by

$$h(T, v) = h_0 + \int_{T_0}^{T} c_v^{id} dT + \int_{\infty}^{v} \left[T \left(\frac{\partial P}{\partial T} \right)_v - P \right] dv + Pv - RT_0 \qquad (6.17)$$

For a volume-explicit equation of state, we can analogously derive an expression for the specific enthalpy:

The total differential of the specific enthalpy is

$$dh = \left(\frac{\partial h}{\partial T} \right)_P dT + \left(\frac{\partial h}{\partial P} \right)_T dP = c_P dT + \left(\frac{\partial h}{\partial P} \right)_T dP \qquad (6.18)$$

With the fundamental equation

$$dh = Tds + vdP \qquad (6.19)$$

and the total differential of the specific entropy

$$ds = \left(\frac{\partial s}{\partial T} \right)_P dT + \left(\frac{\partial s}{\partial P} \right)_T dP \qquad (6.20)$$

one obtains

$$dh = T \left(\frac{\partial s}{\partial T} \right)_P dT + \left[T \left(\frac{\partial s}{\partial P} \right)_T + v \right] dP \qquad (6.21)$$

The comparison of the coefficients in Eqs. (6.18) and (6.21) yields

$$\left(\frac{\partial h}{\partial T} \right)_P = T \left(\frac{\partial s}{\partial T} \right)_P \Rightarrow \left(\frac{\partial s}{\partial T} \right)_P = \frac{1}{T} \left(\frac{\partial h}{\partial T} \right)_P \qquad (6.22)$$

$$\left(\frac{\partial h}{\partial P} \right)_T = T \left(\frac{\partial s}{\partial P} \right)_T + v \Rightarrow \left(\frac{\partial s}{\partial P} \right)_T = \frac{(\partial h/\partial P)_T - v}{T} \qquad (6.23)$$

The mixed quadratic derivatives of the specific entropy

$$\frac{\partial s}{\partial T \partial P} = \frac{1}{T} \frac{\partial^2 h}{\partial T \partial P} \qquad (6.24)$$

and

$$\frac{\partial s}{\partial P \partial T} = -\frac{1}{T^2} \left[\left(\frac{\partial h}{\partial P} \right)_T - v \right] + \frac{1}{T} \left[\frac{\partial^2 h}{\partial P \partial T} - \left(\frac{\partial v}{\partial T} \right)_P \right] \qquad (6.25)$$

must be identical. Thus, one gets

$$\frac{1}{T^2} \left[\left(\frac{\partial h}{\partial P} \right)_T - v \right] + \frac{1}{T} \left(\frac{\partial v}{\partial T} \right)_P = 0 \qquad (6.26)$$

and therefore

$$\left(\frac{\partial h}{\partial P} \right)_T = v - T \left(\frac{\partial v}{\partial T} \right)_P \qquad (6.27)$$

Combining Eqs. (6.18) and (6.27), one gets

$$dh = c_P dT + \left[v - T \left(\frac{\partial v}{\partial T} \right)_P \right] dP \qquad (6.28)$$

Considering that there are ideal gas conditions at $P^+ \to 0$, one obtains

$$h(T, P) = h_0 + \int_{T_0}^{T} c_p^{id} dT + \int_0^P \left[v - T \left(\frac{\partial v}{\partial T} \right)_P \right] dP \qquad (6.29)$$

Subsequently, the specific internal energy can be calculated by

$$u(T, P) = u_0 + \int_{T_0}^{T} c_p^{id} dT + \int_0^P \left[v - T \left(\frac{\partial v}{\partial T} \right)_P \right] dP - Pv + RT_0 \qquad (6.30)$$

To obtain the specific internal energy and enthalpy for liquid mixtures, the excess enthalpy has to be considered:

$$h_{mix} = \sum_i x_i h_i + h^E(T, x) \qquad (6.31)$$

$$u_{mix} = \sum_i x_i u_i + h^E(T, x) - Pv^E(T, x) \qquad (6.32)$$

For vapor mixtures, the transfer from the pure components to the mixture is performed in the ideal gas state (see Section 6.2) so that no excess properties occur.

6.1.2
Entropy

Using the various expressions for the partial differentials of the specific entropy in Eqs. (6.10), (6.14), (6.23), and (6.27), one obtains for the partial differentials of the specific entropy

$$\left(\frac{\partial s}{\partial v} \right)_T = \left(\frac{\partial P}{\partial T} \right)_v \qquad (6.33)$$

and

$$\left(\frac{\partial s}{\partial P} \right)_T = -\left(\frac{\partial v}{\partial T} \right)_P \qquad (6.34)$$

For the residual part of the specific entropy, it can be written

$$(s - s^{id})(T, v) = (s - s^{id})(T, v \to \infty) + \int_\infty^v \left[\left(\frac{\partial s}{\partial v} \right)_T - \left(\frac{\partial s^{id}}{\partial v} \right)_T \right] dv \qquad (6.35)$$

For $v \to \infty$, $s = s^{id}$, and the first term vanishes. With Eq. (6.33), we get

$$\left(\frac{\partial s^{id}}{\partial v} \right)_T = \frac{R}{v} \qquad (6.36)$$

and therefore

$$s(T, v) = s_0(T_0, v_0) + \int_{T_0}^{T} \frac{c_v^{id}}{T} dT + R \ln \frac{v}{v_0} + \int_\infty^v \left[\left(\frac{\partial P}{\partial T} \right)_v - \frac{R}{v} \right] dv \qquad (6.37)$$

With T and P as coordinates, it can analogously be derived as

$$(s - s^{id})(T, P) = (s - s^{id})(T, P \to 0) + \int_0^P \left[\left(\frac{\partial s}{\partial P} \right)_T - \left(\frac{\partial s^{id}}{\partial P} \right)_T \right] dP \qquad (6.38)$$

For $P \to 0$, $s = s^{id}$, and the first term vanishes again. With Eq. (6.34), we get

$$\left(\frac{\partial s^{id}}{\partial P}\right)_T = -\frac{R}{P} \qquad (6.39)$$

and therefore

$$s(T, P) = s_0(T_0, P_0) + \int_{T_0}^{T} \frac{c_P^{id}}{T} dT - R \ln \frac{P}{P_0} + \int_0^P \left[\frac{R}{P} - \left(\frac{\partial v}{\partial T}\right)_P\right] dP \qquad (6.40)$$

In Eqs. (6.37) and (6.40), T_0 and P_0 or, respectively, T_0 and v_0 refer to an arbitrary reference point for the specific entropy. In process simulation, the standard entropy is taken as the reference entropy so that chemical reactions can be described in a consistent way without further conversions. It is defined at standard conditions ($T_0 = 298.15$ K, $P_0 = 101325$ Pa) in the ideal gas state, even if it is fictive. From the data that are usually available, it can be calculated by

$$s_{abs} = \frac{\Delta h_f^0 - \Delta g_f^0}{T_0} \qquad (6.41)$$

For mixtures, the entropy of mixing and the excess entropy have to be regarded. With

$$\Delta s^{id} = -R \sum_i x_i \ln x_i \qquad (6.42)$$

and

$$s^E = \frac{h^E - g^E}{T} \qquad (6.43)$$

one obtains

$$s_{mix} = \sum_i x_i s_i + s^E + \Delta s^{id} \qquad (6.44)$$

6.1.3
Helmholtz Energy and Gibbs Energy

Helmholtz and Gibbs energy are derived according to their definition

$$a = u - Ts \qquad (6.45)$$

and

$$g = h - Ts \qquad (6.46)$$

Using Eqs. (6.30) and (6.40), one can write

$$a(T, P) = u_0 + \int_{T_0}^{T} c_P^{id} dT + \int_0^P \left[v - T\left(\frac{\partial v}{\partial T}\right)_P\right] dP - Pv + RT_0$$

$$- Ts_0(T_0, P_0) - T \int_{T_0}^{T} \frac{c_P^{id}}{T} dT + RT \ln \frac{P}{P_0} - T \int_0^P \left[\frac{R}{P} - \left(\frac{\partial v}{\partial T}\right)_P\right] dP$$

$$= u_0 + \int_{T_0}^{T} c_P^{id} dT + \int_{0}^{P} \left[v - \frac{RT}{P} \right] dP - Pv + RT_0$$

$$- Ts_0(T_0, P_0) - T \int_{T_0}^{T} \frac{c_P^{id}}{T} dT + RT \ln \frac{P}{P_0} \qquad (6.47)$$

Analogously, we obtain

$$g(T, P) = h_0 + \int_{T_0}^{T} c_P^{id} dT + \int_{0}^{P} \left[v - T \left(\frac{\partial v}{\partial T} \right)_P \right] dP$$

$$- Ts_0(T_0, P_0) - T \int_{T_0}^{T} \frac{c_P^{id}}{T} dT + RT \ln \frac{P}{P_0} - T \int_{0}^{P} \left[\frac{R}{P} - \left(\frac{\partial v}{\partial T} \right)_P \right] dP$$

$$= h_0 + \int_{T_0}^{T} c_P^{id} dT + \int_{0}^{P} \left[v - \frac{RT}{P} \right] dP$$

$$- Ts_0(T_0, P_0) - T \int_{T_0}^{T} \frac{c_P^{id}}{T} dT + RT \ln \frac{P}{P_0} \qquad (6.48)$$

$$a(T, v) = u_0 + \int_{T_0}^{T} c_v^{id} dT + \int_{\infty}^{v} \left[T \left(\frac{\partial P}{\partial T} \right)_v - P \right] dv$$

$$- Ts_0(T_0, v_0) - T \int_{T_0}^{T} \frac{c_v^{id}}{T} dT - RT \ln \frac{v}{v_0} - T \int_{\infty}^{v} \left[\left(\frac{\partial P}{\partial T} \right)_v - \frac{R}{v} \right] dv$$

$$= u_0 + \int_{T_0}^{T} c_v^{id} dT + \int_{0}^{v} \left[\frac{RT}{v} - P \right] dv - Ts_0(T_0, v_0) - T \int_{T_0}^{T} \frac{c_v^{id}}{T} dT$$

$$- RT \ln \frac{v}{v_0} \qquad (6.49)$$

and

$$g(T, v) = h_0 + \int_{T_0}^{T} c_v^{id} dT + \int_{\infty}^{v} \left[T \left(\frac{\partial P}{\partial T} \right)_v - P \right] dv + Pv - RT_0$$

$$- Ts_0(T_0, v_0) - T \int_{T_0}^{T} \frac{c_v^{id}}{T} dT - RT \ln \frac{v}{v_0} - T \int_{\infty}^{v} \left[\left(\frac{\partial P}{\partial T} \right)_v - \frac{R}{v} \right] dv$$

$$= h_0 + \int_{T_0}^{T} c_v^{id} dT + \int_{\infty}^{v} \left[\frac{RT}{v} - P \right] dv + Pv - RT_0$$

$$- Ts_0(T_0, v_0) - T \int_{T_0}^{T} \frac{c_v^{id}}{T} dT - RT \ln \frac{v}{v_0} \qquad (6.50)$$

For mixtures, we can derive

$$g_{mix} = \sum_{i} x_i g_i + g^E(T, x) - T \Delta s^{id} \qquad (6.51)$$

and

$$a_{mix} = \sum_i x_i a_i + g^E(T,x) - Pv^E(T,x) - T\Delta s^{id} \qquad (6.52)$$

6.2 Enthalpy Description in Process Simulation Programs

For the calculation of the enthalpy of gases, calculations with the ideal gas heat capacity are a good approximation for low pressures. For high pressures, the influence of the intermolecular forces on the enthalpy has to be taken into account, usually by applying a cubic equation of state. In most cases, these forces are attractive, so that additional energy is necessary to move the molecules away from each other, that is, to lower the density. If this energy is not added, the substance cools down when it is expanded. This effect has technical applications in the liquefaction of gases by adiabatic throttling.

The enthalpy difference at a given temperature from the ideal gas state at $P \to 0$ and a state at an arbitrary pressure in the gas or liquid phase can be calculated with an equation of state. For nonassociating substances, the cubic equations are often used in process simulations. The corresponding expressions for the pure component enthalpy departure for the most often applied equations are:

- **Peng–Robinson**

$$(h - h^{id})(T,P) = h(T,P) - h(T, P\to 0)$$
$$= RT(z-1) - \frac{1}{\sqrt{8}b}\left(a - T\frac{da}{dT}\right)\ln\left(\frac{v+(1+\sqrt{2})b}{v+(1-\sqrt{2})b}\right) \qquad (6.53)$$

with

$$\frac{da}{dT} = -0.45724\frac{R^2 T_c^2}{P_c}\left[1 + m\left(1 - \sqrt{\frac{T}{T_c}}\right)\right]\frac{m}{\sqrt{TT_c}}$$

$$m = 0.37464 + 1.54226\omega - 0.26992\omega^2$$

- **Soave–Redlich–Kwong**

$$(h - h^{id})(T,P) = h(T,P) - h(T, P\to 0)$$
$$= RT(z-1) - \frac{1}{b}\left(a - T\frac{\partial a}{\partial T}\right)\ln\left(\frac{v+b}{v}\right) \qquad (6.54)$$

with

$$\frac{da}{dT} = -0.42748\frac{R^2 T_c^2}{P_c}\left[1 + m\left(1 - \sqrt{\frac{T}{T_c}}\right)\right]\frac{m}{\sqrt{TT_c}}$$

$$m = 0.48 + 1.574\omega - 0.176\omega^2$$

For the specific volume, both the liquid and the vapor can be inserted, depending on the phase that is regarded.

In process simulation, it is necessary that the enthalpy can be continuously described in the liquid as well as in the vapor phase. With this purpose, the problem arises that the particular quantities that contribute to the enthalpy are not independent from each other. For any calculation strategy (route), there will be one quantity left which can be calculated using the others. The three most widely used routes are discussed below.

6.2.1
Route A: Vapor as Starting Phase

It is the most common approach to use the vapor as the starting phase when the $\gamma - \varphi$-approach is used. It works as follows.

For all components (index i), the reference point is set at ideal gas standard conditions ($T_0 = 298.15$ K, $P \to 0$, $h_0 = \Delta h_f^0$):

$$h_i^{id}(T_0, P \to 0) = \Delta h_{f,i}^0 \tag{6.55}$$

Therefore, the enthalpy description is consistent with respect to chemical reactions.

To calculate the enthalpy of a vapor, first the ideal gas enthalpy is calculated at system temperature T at $P \to 0$ using c_p^{id}:

$$h_i^{id}(T, P \to 0) = \Delta h_{f,i}^0 + \int_{T_0}^{T} c_{Pi}^{id} dT \tag{6.56}$$

At this point, the transition from the pure components to the mixture takes place:

$$h_i^{id}(T, P \to 0, y_i) = \sum_i y_i h_i^{id}(T, P \to 0) \tag{6.57}$$

Because of the ideal gas state, there is no excess term.

To obtain the enthalpy for a real vapor phase, the residual part $(h - h^{id})$ must be added:

$$h^v(T, P, y_i) = h^{id}(T, P \to 0, y_i) + (h - h^{id})(T, P, y_i) \tag{6.58}$$

If the enthalpy of a liquid has to be calculated (Figure 6.1), the enthalpy of the saturated vapor has to be evaluated for each component by applying $P = P_i^s$ to Eq. (6.58). Then, the enthalpy of vaporization has to be subtracted:

$$h_i^L(T) = \Delta h_{f,i}^0 + \int_{T_0}^{T} c_{Pi}^{id} dT + (h - h^{id})_i(T, P_i^s) - \Delta h_{vi}(T) \tag{6.59}$$

With this route, any liquid of a pure component is considered to be in the saturation state at the given temperature. Enthalpy changes due to further compression are neglected, which is a good approximation at least at low pressures. The enthalpy of a liquid mixture can then be evaluated by

$$h^L(T, x_i) = \sum_i x_i h_i^L(T) + h^E(T, x_i) \tag{6.60}$$

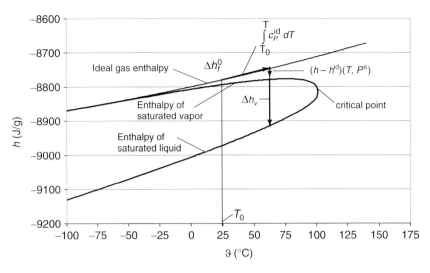

Figure 6.1 Calculation of the liquid enthalpy of 1,1,1,2-tetrafluoroethane (R134a) by route A.

which has been discussed in Chapter 5 as well. If the binary parameters have not been adjusted to h^E values as well, h^E should be neglected.

Unfortunately, this route causes large errors in the determination of the specific heat capacity of the liquid, which is defined as the slope of h^L. Figure 6.2 shows that even for a well-known pure substance like water the errors can be larger than 10% (old fit in Figure 6.2), which cannot be accepted and which often causes severe problems, for example, in the design of heat exchangers for liquids.

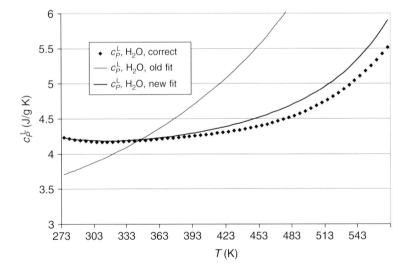

Figure 6.2 Results for c_P^L of water as a function of temperature using route A.

6 Caloric Properties

The reason for this behavior is the poor representation of the slope of the enthalpy of vaporization.

A procedure has been developed [1] to fit the coefficients for the correlation of the enthalpy of vaporization (e.g., Eqs. (3.61) and (3.62)) to c_p^L data as well, using the temperature derivation of Eq. (6.59). It is well known that functions which give approximately the same values can differ significantly in their derivatives, especially if the slopes are flat. At temperatures far below the critical point the enthalpy of vaporization can be regarded as such a function. Figure 6.3 shows two fits for the enthalpy of vaporization of water. In the temperature range from 0–200 °C, they seem to differ by an almost constant value, in fact, the difference is never more than 1% but not constant. The surprising result for the derivatives is shown in Figure 6.4. There are significant differences. In both cases, the dashed line represents the best fit to the high precision data for the enthalpy of vaporization of water. For the full line, the coefficients for the correlation of Δh_v have been fitted

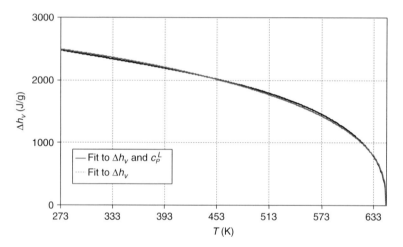

Figure 6.3 Two fits of the enthalpy of vaporization of water.

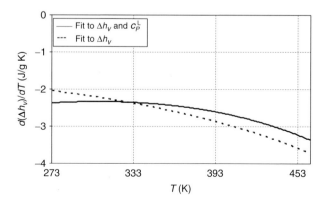

Figure 6.4 Slopes of the two fits.

using the objective function

$$F = w_1 \sum_i \left(\frac{\Delta h_{vi,calc} - \Delta h_{vi,exp}}{\Delta h_{vi,exp}} \right)^2 + w_2 \sum_i \left(\frac{c^L_{Pi,calc} - c^L_{Pi,exp}}{c^L_{Pi,exp}} \right)^2 \quad (6.61)$$

between 0 and 200 °C. The values for $(h-h^{id})$ have been calculated with the PR equation of state, not in order to be exact but to represent the conditions that occur in process simulation runs. The weighting factors w_1 and w_2 have been chosen by a trial-and-error procedure in a way that for both Δh_v and $c_P{}^L$ the results are satisfactory. Figure 6.2 shows that the results for $c_P{}^L$ have improved significantly in the range between 0 and 200 °C, which was used for the adjustment (new fit). The extrapolation is considered to be acceptable, taking into account that due to the increasing deviations of $(h-h^{id})$ high precision results cannot be expected. In general, the deterioration in the correlation for the enthalpy of vaporization is acceptable. For most substances, it can still be represented within 1%. It should be noted that in most cases all the parameters in Eqs. (3.61) or (3.62) can be used for the combined adjustment to liquid heat capacity and enthalpy of vaporization, as a sufficient amount of data is usually available if two quantities are used. For a fit only to the enthalpy of vaporization, it is often recommended to truncate Eq. 3.61 after the first summand.

The main advantage of this method is that no additional correlations and routes have to be implemented in the simulator. Only the parameters for the heat of vaporization are affected. Moreover, the procedure has become a consistency test. If the data for vapor pressure, $c_P{}^{id}$, $c_P{}^L$, and Δh_v are correct, the equation of state used gives reasonable values for $(h-h^{id})$, and no extrapolations of any correlation take place during the parameter adjustment, it should be possible to represent both the liquid heat capacity and the enthalpy of vaporization. If the procedure does not work at once for a substance, all the input data should again be checked carefully. In most of the cases, ambiguous data sources or raw data based on bad estimations can be detected. After correction, the problems often disappear.

For associating substances (see Section 13.2), the combined fitting procedure is therefore modified in the following way: the association constants and the coefficients for the enthalpy of vaporization (Eqs. (3.61) and (3.62)) are simultaneously fitted to vapor density data, specific isobaric vapor heat capacities, liquid heat capacities, and enthalpies of vaporization, giving the objective function

$$F = w_1 \sum_i \left(\frac{\Delta h_{vi,calc} - \Delta h_{vi,exp}}{\Delta h_{vi,exp}} \right)^2 + w_2 \sum_i \left(\frac{c^L_{Pi,calc} - c^L_{Pi,exp}}{c^L_{Pi,exp}} \right)^2$$
$$+ w_3 \sum_i \left(\frac{c^V_{Pi,calc} - c^V_{Pi,exp}}{c^V_{Pi,exp}} \right)^2 + w_4 \sum_i \left(\frac{v^V_{i,calc} - v^V_{i,exp}}{v^V_{i,exp}} \right)^2 \quad (6.62)$$

A lot of effort is required for implementing this procedure, as there are more weighting factors that have to be evaluated in a trial-and-error procedure for each substance individually, taking care that the plots of all the four quantities are reasonable.

For acetic acid, five parameters for the enthalpy of vaporization and four parameters describing the association (two each for dimerization and tetramerization, see Section 13.2) had to be fitted simultaneously. Good results have been obtained. While the reproduction of the vapor density and the vapor heat capacity did not change significantly, large progress was made for c_P^L and Δh_v (Figures 6.5 and 6.6). Up to the normal boiling point, where the assumption of an ideal gas mixture of the particular associates is valid, c_P^L is reproduced very well.

6.2.2
Route B: Liquid as Starting Phase

To overcome the difficulties with c_P^L, process simulators usually offer the option to start with a reference state $h_{\text{ref},i}(T_{\text{ref},i})$ in the liquid phase. The enthalpy of the liquid phase is evaluated for each component by integration of the liquid heat capacity with respect to temperature:

$$h_i^L(T) = \int_{T_{\text{ref},i}}^{T} c_{p_i}^L dT + h_{\text{ref},i}(T_{\text{ref},i}) \tag{6.63}$$

Analogously to Eq. (6.60), the enthalpy of a liquid mixture can be evaluated by taking the excess enthalpy into account. Again, the liquid enthalpy refers to a saturated liquid, enthalpy changes due to further compression are neglected.

For the calculation of a vapor enthalpy, the liquid heat capacity for each component is integrated from the reference temperature to a transition temperature T_{trans}, which can be chosen arbitrarily. A good choice is the normal boiling point,

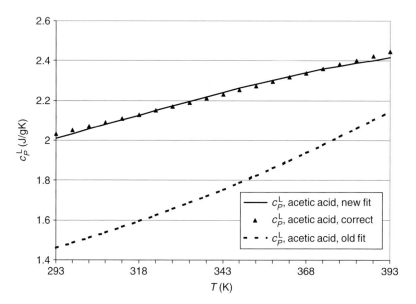

Figure 6.5 Improved calculation for c_P^L of acetic acid.

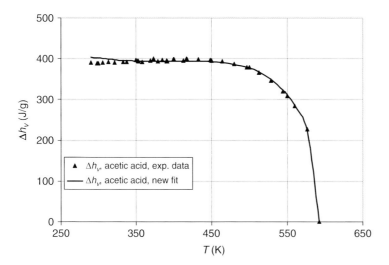

Figure 6.6 Acceptable representation of the enthalpy of vaporization of acetic acid after the simultaneous fit.

as it exists for almost every substance:

$$h_i^L(T_{\text{trans},i}) = \int_{T_{\text{ref},i}}^{T_{\text{trans},i}} c_{P_i}^L dT + h_{\text{ref},i}(T_{\text{ref},i}) \tag{6.64}$$

Only at this transition temperature, the correlation for the enthalpy of vaporization is applied to calculate the enthalpy of saturated vapor at the transition temperature. Then, $(h-h^{\text{id}})_i$ is subtracted to obtain the enthalpy of the ideal gas at transition temperature:

$$h_i^{\text{id}}(T_{\text{trans},i}) = h_i^L(T_{\text{trans},i}) + \Delta h_{vi}(T_{\text{trans},i}) - (h - h^{\text{id}})(T_{\text{trans},i}, P_i^s) \tag{6.65}$$

To move to the system temperature, the ideal gas heat capacity is integrated:

$$h_i^{\text{id}}(T, P \rightarrow 0) = h_i^{\text{id}}(T_{\text{trans},i}, P \rightarrow 0) + \int_{T_{\text{trans},i}}^{T} c_{P_i}^{\text{id}} dT \tag{6.66}$$

The transition from the pure components to the mixture is achieved according to Eq. (6.57). Finally, $(h-h^{\text{id}})$ at system temperature and pressure is added:

$$h^V(T, P) = h^{\text{id}}(T, P \rightarrow 0) + (h - h^{\text{id}})(T, P, y_i) \tag{6.67}$$

$h_{\text{ref},i}(T_{\text{ref},i})$ has to be chosen in a way that at standard conditions (25 °C, ideal gas) the standard enthalpy of formation Δh_f^0 is yielded to maintain the consistency with respect to chemical reactions. A good choice for T_{ref} is the melting point. The method is illustrated in Figure 6.7.

Certainly, the correlated liquid heat capacity is reproduced with this method. In the low-pressure region, Route B works quite well. Although the correlation for Δh_v is used only at T_{trans}, it is quite accurately reproduced. At high pressures, Δh_v becomes inaccurate, and it does not vanish at the critical point. Often, the

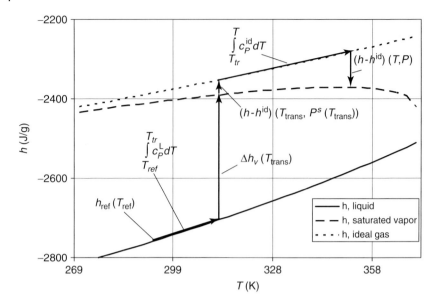

Figure 6.7 Calculation of the enthalpy of saturated vapor using route B.

correlation for c_P^L is based only on values below the normal boiling point, and a linear function is sufficient for the correlation. The extrapolation of such a correlation to higher temperatures yields results that are significantly too low, which is quite annoying, as users believe that the choice of Route B guarantees the correct c_P^L.

Moreover, the effort to define all the arbitrary reference constants (T_{ref}, h_{ref}, T_{trans} for each component treated this way), is quite high. In commercial simulators, it is not as widely accepted as Route A and often restricted to certain applications.

6.2.3
Route C: Equation of State

As discussed in Chapter 2, all the information about the nonideality can be derived from an equation of state. Similarly to Route A, the ideal gas at standard conditions is taken as the reference state:

$$h_i^{\text{id}}(T, P \to 0) = \Delta h_{f,i}^0 + \int_{T_0}^{T} c_{P_i}^{\text{id}} dT \qquad (6.68)$$

The enthalpy of the ideal gas mixture is calculated using Eq. (6.57). To obtain the enthalpy for a real phase, $(h - h^{\text{id}})$ must be added, which is calculated by an equation of state valid both for the liquid and for the vapor phase:

$$h^v(T, P, y_i) = h^{\text{id}}(T, P \to 0, y_i) + (h - h^{\text{id}})(T, P, y_i) \qquad (6.69)$$

For the Soave–Redlich–Kwong and the Peng-Robinson equation, expressions for $(h - h^{\text{id}})$ are given in Eq. (6.53) and Eq. (6.54), respectively. The relationship for the Soave–Redlich–Kwong equation is derived in Appendix C, F1. Of course, the user has to take care that the equation of state used is capable enough for this evaluation. With this approach, the enthalpy of a liquid does not necessarily refer to saturation conditions. Enthalpy changes due to further compression are taken into account, as the enthalpy always refers to a temperature T and a pressure P. However, for most equations of state like the cubic equations one can hardly expect that the effect will be calculated quantitatively correct. For a pure component, the Clausius–Clapeyron equation is inherently integrated in the calculation procedure, as both the phase equilibrium condition $\varphi^L = \varphi^V$ and the Clausius–Clapeyron equation are directly equivalent to $g^L = g^V$ (see Section 2.4.4). For the cubic equations of state, the reproduction of the enthalpy of vaporization works quite well for nonpolar components. For polar components, the accuracy is usually not sufficient, unless an individual α-function is used. With a good α-function to obtain a correct vapor pressure representation, it can be expected that Δh_v will be quite well reproduced even for polar components.

For $c_P{}^L$, the same problem arises as for Route A. Analogously, it can be overcome by fitting the α-function both to vapor pressure and $c_P{}^L$ data. If a generalized equation of state like the Peng–Robinson or the Soave–Redlich–Kwong equation of state is used, there is no way to overcome this difficulty.

For the handling of enthalpies in a process simulation program, the change of a model between two blocks is often critical. This problem has much to do with the enthalpy description. Between the two blocks, the simulation program hands over the values for P and h to describe the state of the stream. According to the particular models used in the two blocks, the stream is assigned with two different temperatures that may differ significantly.

Example 6.1

A stream consisting of n-hexane ($T_1 = 450$ K, saturated liquid) is coming out of a block which uses the Peng–Robinson equation for the vapor phase. It is transferred to another block using an activity coefficient model with an ideal gas phase at $P = 1$ bar, where it is evaporated. Which error will be produced due to the model change?

Solution

According to the Peng–Robinson equation of state, the specific enthalpy of the stream leaving the first block is determined to be $h_1 = -163.3$ kJ/mol at $P_1 = 12.226$ bar. Using an activity coefficient model with an ideal gas phase, the coordinates for $P_2 = P_1$ and $h_2 = h_1$ refer to the liquid state at $T_2 = 430.96$ K. Using $c_P{}^L$ from Appendix A, the corresponding enthalpy difference

$$\Delta h \approx c_P^l \Delta T = 3.1 \frac{J}{gK} \cdot 19K = 58.9 \frac{J}{g} \tag{6.70}$$

will be missing in the energy balance.

Therefore, care must be taken when the thermodynamic model is changed in a flowsheet. It is recommended that a dummy block is introduced between two blocks operating with different models. Another option is to carry out a model change where both models yield at least similar results, maybe in a block operating at low pressure. In general, one should stick to the most comprehensive model in a flowsheet; nevertheless there are cases (e.g., association, electrolytes occurring only in parts of the flowsheet) where a model change cannot be avoided.

In process engineering applications, users are often forced to enter an enthalpy of vaporization referring to a liquid process stream, which is usually a mixture. However, an enthalpy of vaporization is characterized by an evaporation procedure where temperature and pressure remain the same, and where liquid and vapor concentrations are identical. This is the case only for pure substances. In principle, an enthalpy of vaporization could also be applied to azeotropic mixtures, but azeotropic concentrations also vary with temperature so that such a value would only make sense at one certain point.

In general, for mixtures it must be distinguished between an integral and a differential evaporation. An integral evaporation means that the whole mixture will be transferred from the bubble point to dew point. In this case, during the evaporation the temperature of the remaining liquid will rise until the boiling temperature of the most heavy boiling component is reached. It is possible to assign a number and to declare it as an enthalpy of vaporization, but it would not refer to a certain temperature. The differential evaporation means that only the very first bubble of the liquid is vaporized. For this type, it is possible to assign a pressure and a temperature, but its interpretation is much more complex, as the following example shows.

Example 6.2

A vessel containing a liquid mixture of ethylene glycol (EG) ($C_2H_6O_2$, 90 wt%) and water (10 wt%) is exposed to a fire. The safety valve shall actuate at $P = 1$ bar. The heat transfer from the fire to the vessel can be characterized by the empirical standard equation [2]

$$\frac{\dot{Q}}{\text{kW}} = 43.2 \left(\frac{A}{\text{m}^2}\right)^{0.82}$$

The surface exposed amounts to $A = 40 \, \text{m}^2$, corresponding to a heat input of 890 kW. From a phase equilibrium calculation, the boiling point of the system is calculated to be $T = 410.6$ K. At this temperature, the enthalpies of vaporization of the two components are

$$\Delta h_{v,\text{H}_2\text{O}}(T) = 2153 \, \text{J/g (water)}$$

$$\Delta h_{v,\text{EG}}(T) = 944 \, \text{J/g (ethylene glycol)}$$

Calculate an enthalpy of vaporization which enables the user to calculate the mass flow at the very first actuation.

Solution

- **Option 1**
  For simplicity, the enthalpy of vaporization is obtained from averaging the pure component values in the liquid phase:
  $$\Delta h_{v,\mathrm{mix}} = x_{H_2O}\Delta h_{v,H_2O}(T) + x_{EG}\Delta h_{v,EG}(T)$$
  $$= 0.1 \cdot 2153\ \mathrm{J/g} + 0.9 \cdot 944\ \mathrm{J/g} = 1065\ \mathrm{J/g}$$

  The advantage is that no further calculation step is necessary. However, as shown in the following, the liquid concentrations are only applicable if the vapor concentration is identical. For this wide boiling mixture, this is by far not the case, as shown below. Option 1 is a bad choice.

- **Option 2**
  In the vapor phase, the concentration of the first bubble can be evaluated by calculating a flash block in a simulation program. It turns out that
  $$y_{H_2O} = 0.749$$
  $$y_{EG} = 0.251$$

  As it makes sense to assign the enthalpy of vaporization to the amounts of substances which have been vaporized, the averaging procedure can be done in the vapor phase:
  $$\Delta h_{v,\mathrm{mix}} = y_{H_2O}\Delta h_{v,H_2O}(T) + y_{EG}\Delta h_{v,EG}(T)$$
  $$= 0.749 \cdot 2153\ \mathrm{J/g} + 0.251 \cdot 944\ \mathrm{J/g} = 1850\ \mathrm{J/g}$$

  This value is much more plausible; however, as can be seen in the following, there is still room for improvement.

- **Option 3**
  A differential flash is performed with the simulator program, defined by the set pressure of the safety valve and a small vapor fraction of 0.001, representing the first bubble in a differential flash.
  Dividing the duty by the generated mass flow, we get a direct relationship between the heat input and the vapor flow. In this case, the result is
  $$\Delta h_{v,\mathrm{mix}} = \frac{Q}{\Delta m_{v,\mathrm{incr}}} = \frac{103189\ \mathrm{J}}{0.04385\ \mathrm{kg}} = 2353\ \mathrm{J/g}$$

  This value for the enthalpy of vaporization obtained with Option 3 is considerably higher than the one obtained with Option 2, as the flash calculation can take the temperature rise of the liquid into account. In this case, where a wide boiling mixture is regarded, even this incremental evaporation with the 0.001 vapor fraction causes a temperature rise in the liquid by 0.06 K, which has a considerable influence on the calculated heat of vaporization. For illustration, the mass balance is listed in Table 6.1: In fact, this example is an extreme one due to the wide boiling range. Usually, Option 2 should yield a better approximation and could be used for a rough check. Nevertheless, as there is no way to avoid the flash

Table 6.1 Mass balance for the incremental flash in Example 6.2.

		Feed	Vapor	Liquid
Mass in vessel	kg			
H_2O		10	0.033	9.967
$C_2H_6O_2$		90	0.011	89.989
Mass fraction				
H_2O		0.1	0.7490	0.0997
$C_2H_6O_2$		0.9	0.2510	0.9003
Total mole inventory	kmol	2.005	0.002	2.003
Total mass inventory	kg	100	0.044	99.956
Total volume	m^3	0.0999	0.0685	0.0999
Temperature (K)	K	410.62	410.68	410.68
Pressure	Pa	100 000	100 000	100 000
Vapor fraction		0.0	1.0	0.0
Liquid fraction		1.0	0.0	1.0
Enthalpy	J/kmol	−3.9E+08	−2.51E+08	−3.96E+08
Enthalpy	J/kg	−7.93E+06	−1.14E+07	−7.93E+06
Enthalpy	J	−7.93E+08	−5.03E+05	−7.93E+08

calculation to get the vapor concentrations, the more accurate Option 3 is even easier to apply.

The mass flow through the safety valve can then be determined to be

$$\dot{m} = \frac{\dot{Q}}{\Delta h_{v,mix}} = \frac{890 \text{ kW}}{2353 \text{ J/g}} = 0.378 \text{ kg/s} = 1362 \text{ kg/h}$$

A more sophisticated approach for the calculation of mass flows in pressure relief cases is presented in Section 14.4.

Example 6.3

In a heat exchanger, liquid water (10 000 kg/h) at $P = 6$ bar is heated up from $\vartheta_1 = 100\,°C$ to $\vartheta_2 = 150\,°C$. Calculate the necessary duty of the heat exchanger using the following four options.

1) Use a high-precision equation of state.
2) Use Route B (liquid as starting phase). c_p^L can be taken from Appendix A.
3) Use Route A (vapor as starting phase). Take the Peng–Robinson equation for the vapor phase calculations. c_p^{id} and Δh_v can be taken from Appendix A.
4) Use Route A (vapor as starting phase). Take the Peng–Robinson equation for the vapor phase calculations. c_p^{id} can be taken from Appendix A. Δhv should be fitted simultaneously to c_p^L data as well.

Interpret the various results.

6.2 Enthalpy Description in Process Simulation Programs | 351

Solution

1) From the high-precision equation of state, the following enthalpies are obtained:

$$h_1^L(\vartheta_1, P) = 419.451 \text{ J/g} \quad h_2^L(\vartheta_2, P) = 632.256 \text{ J/g}$$

$$\dot{Q} = \dot{m}(h_2 - h_1) = 10000 \, \frac{\text{kg}}{\text{h}} (632.256 - 419.451) \, \frac{\text{J}}{\text{g}}$$

$$= 2128050 \, \frac{\text{kJ}}{\text{h}} = 591.1 \text{ kW}$$

As the values for the enthalpies correspond to the Steam Table [3], these values can be taken as reference values.

2) The expression for the specific enthalpy of a liquid in Route B is (Eq. (6.63))

$$h_i^L(T) = \int_{T_{ref,i}}^T c_{pi}^L dT + h_{ref,i}(T_{ref,i})$$

The enthalpy difference between the two states becomes

$$h_2^L - h_1^L = \int_{T_1}^{T_2} c_p^L dT = -T_c \int_{\tau_1}^{\tau_2} c_p^L d\tau$$

with $\tau = 1 - T/T_c$. The antiderivative of the c_p^L function from Appendix A is

$$\int c_p^L d\tau = R \left(A \ln \tau + B\tau + \frac{C}{2}\tau^2 + \frac{D}{3}\tau^3 + \frac{E}{4}\tau^4 + \frac{F}{5}\tau^5 \right) + \text{const.}$$

giving

$$h_2^L - h_1^L = -RT_c \left(A \ln \frac{\tau_2}{\tau_1} + B(\tau_2 - \tau_1) + \frac{C}{2} (\tau_2^2 - \tau_1^2) + \frac{D}{3} (\tau_2^3 - \tau_1^3) \right.$$
$$\left. + \frac{E}{4} (\tau_2^4 - \tau_1^4) + \frac{F}{5} (\tau_2^5 - \tau_1^5) \right)$$

With $T_c = 647.096$ K, $\tau_1 = 1 - 373.15/647.096 = 0.42335$ and $\tau_2 = 1 - 423.15/647.096 = 0.34608$ and the parameters from Appendix A one gets

$$h_2^L - h_1^L = -\frac{8.3143}{18.015} \frac{\text{J}}{\text{mol K}} \frac{\text{mol}}{\text{g}} 647.096 \text{ K}.$$

$$\left\{ 0.25598 \ln \frac{0.34608}{0.42335} + 12.54595(0.34608 - 0.42335) \right.$$

$$+ \frac{-31.40896}{2} (0.34608^2 - 0.42335^2)$$

$$+ \frac{97.7665}{3} (0.34608^3 - 0.42335^3)$$

$$+ \frac{-145.4236}{4} (0.34608^4 - 0.42335^4)$$

$$\left. + \frac{87.0185}{5} (0.34608^5 - 0.42335^5) \right\}$$

$$= -298.648 \frac{J}{g}(-0.05159 - 0.96943 + 0.93369 - 1.12186 + 0.64628 - 0.15027)$$

$$= 212.99 \frac{J}{g}$$

$$\dot{Q} = \dot{m}(h_2 - h_1) = 10000 \frac{kg}{h} 212.98 \frac{J}{g} = 2129\,799 \frac{kJ}{h} = 591.61 \text{ kW}$$

This result close to the number obtained for Option 1 is not surprising, as Route B is especially adapted for the calculation of enthalpies in the liquid phase. The c_P^L equation is fitted to the Steam Table; therefore, in this example an almost exact result should be obtained.

3) Using Route A, the enthalpies are calculated via

$$h^L(T) = \Delta h_f^0 + \int_{T_0}^{T} c_p^{id} dT + (h - h^{id})(T, P^s) - \Delta h_v(T) \tag{6.59}$$

With Δh_f^0 and the functions for c_p^{id}, P^s, and Δh_v from Appendix A and the antiderivative for c_p^{id} from Appendix C, F4, one gets

$$\Delta h_f^0 = -241820 \frac{J}{mol} = -241820 \frac{J}{18.015g} = -13423.3 \frac{J}{g}$$

$$\int_{298.15K}^{423.15K} c_p^{id} dT = 235.96 \frac{J}{g} \qquad \int_{298.15K}^{373.15K} c_p^{id} dT = 140.88 \frac{J}{g}$$

$$P^s(150\,°C) = 4.7608 \text{ bar} \qquad P^s(100\,°C) = 1.0138 \text{ bar}$$

$$\Delta h_v(150\,°C) = 2115.2 \frac{J}{g} \qquad \Delta h_v(100\,°C) = 2256.56 \frac{J}{g}$$

The residual part of the enthalpy of the saturated vapor can be determined with the Peng–Robinson equation of state using Eq. (6.53):

$$(h - h^{id})(T, P) = h(T, P) - h(T, P \to 0) = RT(z - 1)$$

$$-\frac{1}{\sqrt{8}b}\left(a - T\frac{\partial a}{\partial T}\right) \ln\left(\frac{v + (1+\sqrt{2})b}{v + (1-\sqrt{2})b}\right)$$

with

$$\frac{da}{dT} = -0.45724 \frac{R^2 T_c^2}{P_c}\left[1 + m\left(1 - \sqrt{\frac{T}{T_c}}\right)\right]\frac{m}{\sqrt{TT_c}}$$

$$m = 0.37464 + 1.54226\omega - 0.26992\omega^2$$

Getting

$$v^V(150\,°C, 4.7608 \text{ bar}) = 7.1714 \cdot 10^{-3} \text{ m}^3/\text{mol}$$
$$v^V(100\,°C, 1.0138 \text{ bar}) = 3.0336 \cdot 10^{-2} \text{ m}^3/\text{mol}$$

$$z^V(150\,°C, 4.7608 \text{ bar}) = 0.9704 \quad z^V(100\,°C, 1.0138 \text{ bar}) = 0.9913$$

one obtains

$$(h^v - h^{id})(150\,°C, 4.7608\,\text{bar}) = -15.901\,\frac{J}{g}$$

$$(h^v - h^{id})(100\,°C, 1.0138\,\text{bar}) = -3.985\,\frac{J}{g}$$

The results for the enthalpies are

$$h^L(150\,°C) = -13423.3\,\frac{J}{g} + 235.96\,\frac{J}{g} - 15.901\,\frac{J}{g} - 2115.2\,\frac{J}{g}$$

$$= -15318.4\,\frac{J}{g}$$

$$h^L(100\,°C) = -13423.3\,\frac{J}{g} + 140.88\,\frac{J}{g} - 3.985\,\frac{J}{g} - 2256.56\,\frac{J}{g}$$

$$= -15543.0\,\frac{J}{g}$$

and for the heat duty

$$\dot{Q} = \dot{m}(h_2 - h_1) = 10000\,\frac{\text{kg}}{\text{h}}(-15318.4 + 15543.0)\,\frac{J}{g} = 2245650\,\frac{\text{kJ}}{\text{h}}$$

$$= 623.8\,\text{kW}$$

The deviation from the reference value in Option 1 is ~5.5%. This is not acceptable for the design of a heat exchanger. Errors like this are typical for the calculation of liquid enthalpies with Route A, although the parameters for c_p^{id} and Δh_v given in Appendix A give accurate results.

4) The reason for the significant deviation of Option 3 is the poor representation of c_p^L with Route A because of a limited quality of the Peng–Robinson equation in vapor enthalpy calculations and possibly a poor reproduction of the slope of the enthalpy of vaporization as a function of temperature. Over a limited temperature range, this can be corrected by an alternative adjustment of the coefficients in the correlation of the enthalpy of vaporization (Eq. (3.62)). With the new coefficients

$$A = 6.83738 \quad B = 13.36976 \quad C = -13.82304 \quad D = 3.7741 \quad E = 1.525$$

it is possible to represent both c_p^L and the enthalpy of vaporization with an average accuracy of less than 1% in the temperature range between 0 and 250 °C.

The new values for the enthalpy of vaporization are

$$\Delta h_v\,(150\,°C) = 2109.0\,\frac{J}{g} \quad \Delta h_v\,(100\,°C) = 2241.5\,\frac{J}{g}$$

The enthalpies become

$$h^L(150\,°C) = -13423.3\,\frac{J}{g} + 235.96\,\frac{J}{g} - 15.901\,\frac{J}{g} - 2109.0\,\frac{J}{g} = -15312.2\,\frac{J}{g}$$

$$h^L(100\,°C) = -13423.3\,\frac{J}{g} + 140.88\,\frac{J}{g} - 3.985\,\frac{J}{g} - 2241.5\,\frac{J}{g} = -15527.9\,\frac{J}{g}$$

and the heat duty is calculated to be

$$\dot{Q} = \dot{m}(h_2 - h_1) = 10000 \,\frac{\text{kg}}{\text{h}} (-15312.2 + 15527.9) \,\frac{\text{J}}{\text{g}} = 2157\,000 \,\frac{\text{kJ}}{\text{h}}$$
$$= 599.2 \text{ kW}$$

The error in the duty has reduced to ~1.4%, which could be accepted in a design calculation. The alternative coefficients for the enthalpy of vaporization can be regarded as a compromise; the accuracy in the representation of the enthalpy of vaporization has decreased but is still acceptable. However, in the documentation of a simulation run it should be clearly pointed out that these coefficients are a compromise for Route A, which can represent both the specific liquid heat capacity and the enthalpy of vaporization in the temperature range $0 \ldots 250\,°C$, where the Peng–Robinson equation of state is used for calculating the residual part of the enthalpy.

6.3
Caloric Properties in Chemical Reactions

The option of using the standard enthalpy of formation as the reference point in process simulators has the considerable advantage that no further conversions between the reference points of the particular components have to be done. The energy balance of a block involving a chemical reaction can simply be performed by balancing between inlet and outlet of a block. This is illustrated in the following two examples. Chemical reactions are mainly covered in Chapter 12.

Example 6.4
Liquid n-butanol ($C_4H_{10}O$, 100 kg/h, $\vartheta = 50\,°C$, $P = 1$ bar) is combusted with 2000 kg/h air (80 mass% nitrogen, 20 mass% oxygen, $\vartheta = 50\,°C$, $P = 1$ bar) in a steam generator. The flue gas outlet conditions are $\vartheta = 200\,°C$, $P = 1$ bar. Calculate the duty of the combustor and the amount of saturated steam generated at $P = 5$ bar, if the feed water is fed to the boiler as a saturated liquid.

For the vapor phase, the ideal gas equation of state can generally be applied.

Solution

The reaction which takes place in the combustor

$$C_4H_{10}O + 6O_2 \longrightarrow 4CO_2 + 5H_2O$$

This means that 6 mol O_2 per mole of n-butanol are consumed. The molar masses involved are

$M_{\text{butanol}} = 74.123$ g/mol
$M_{O_2} = 31.999$ g/mol
$M_{CO_2} = 44.009$ g/mol

$M_{H_2O} = 18.015$ g/mol.

The mole flow of n-butanol is

$$\dot{n}_{\text{butanol}} = \frac{\dot{m}_{\text{butanol}}}{M_{\text{butanol}}} = \frac{100 \text{ kg/h}}{74.123 \text{ g/mol}} = 1.3491 \text{ kmol/h}$$

The oxygen demand is therefore

$$\dot{m}_{O_2,\text{consumption}} = \dot{n}_{\text{butanol}} \cdot 6 \cdot M_{O_2} = 1.3491 \text{ kmol/h} \cdot 6 \cdot 31.999 \text{ g/mol}$$
$$= 259.019 \text{ kg/h}$$

Thus, the flue gas flow consists of

$$\dot{m}_{O_2} = \dot{m}_{O_2,\text{air}} - \dot{m}_{O_2,\text{consumption}} = 2000 \text{ kg/h} \cdot 0.2 - 259.019 \text{ kg/h}$$
$$= 140.981 \text{ kg/h}$$

$$\dot{m}_{N_2} = \dot{m}_{N_2,\text{air}} = 2000 \text{ kg/h} \cdot 0.8 = 1600 \text{ kg/h}$$
$$\dot{m}_{CO_2} = \dot{n}_{\text{butanol}} \cdot 4 \cdot M_{CO_2} = 1.3491 \text{ kmol/h} \cdot 4 \cdot 44.009 \text{ g/mol} = 237.49 \text{ kg/h}$$
$$\dot{m}_{H_2O} = \dot{n}_{\text{butanol}} \cdot 5 \cdot M_{H_2O} = 1.3491 \text{ kmol/h} \cdot 5 \cdot 18.015 \text{ g/mol} = 121.52 \text{ kg/h}$$

corresponding to the mass fractions

$w_{O_2} = 0.0671$
$w_{N_2} = 0.7619$
$w_{CO_2} = 0.1131$
$w_{H_2O} = 0.0579$.

Figure 6.8 illustrates the inlet and outlet streams of the combustor. All enthalpies are calculated via Route A.

- n-butanol:

$$h^L(T) = \Delta h_f^0 + \int_{T_0}^{T} c_p^{id} dT + (h - h^{id})(T, P^s) - \Delta h_v(T) \tag{6.71}$$

Using the data given in Appendix A and the antiderivatives for c_p^{id} in Appendix C, F4 one gets

$$\Delta h_f^0 = -274600 \text{ J/mol} \cdot \frac{1 \text{ mol}}{74.123 \text{ g}} = -3704.65 \text{ J/g}$$

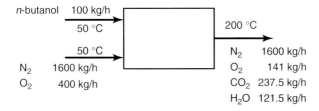

Figure 6.8 Inlet and outlet streams of the combustor.

$$\int_{25\,°C}^{50\,°C} c_p^{id}\,dT = 37.63\text{ J/g}$$

$$h - h^{id} = 0 \text{ (ideal gas)}$$

$$\Delta h_v\,(50\,°C) = 681.24\text{ J/g}$$

$$\Rightarrow h_{butanol}^L = -3704.65\text{ J/g} + 37.63\text{ J/g} + 0 - 681.24\text{ J/g} = -4348.26\text{ J/g}$$

- Air:
  According to Eq. (6.57) one can write

  $$h_{air}^{id}(50\,°C) = w_{O_2}h_{O_2}^{id}(50\,°C) + w_{N_2}h_{N_2}^{id}(50\,°C)$$

  Using the data given in Appendix A and the antiderivatives for c_p^{id} in Appendix C, F4 one gets

  $$h_{O_2}^{id}(T) = \Delta h_{f,O_2}^0 + \int_{25\,°C}^{50\,°C} c_p^{id}\,dT = 0 + 22.99\text{ J/g} = 22.99\text{ J/g}$$

  $$h_{N_2}^{id}(T) = \Delta h_{f,N_2}^0 + \int_{25\,°C}^{50\,°C} c_p^{id}\,dT = 0 + 26.0\text{ J/g} = 26.0\text{ J/g}$$

  $$\Rightarrow h_{air}^{id}(50\,°C) = 0.2 \cdot 0.22.99\text{ J/g} + 0.8 \cdot 26.0\text{ J/g} = 25.40\text{ J/g}$$

- Flue gas:
  According to Eq. (6.57) one can write

  $$h_{fluegas}^{id}(200\,°C) = w_{O_2}h_{O_2}^{id}(200\,°C) + w_{N_2}h_{N_2}^{id}(200\,°C) + w_{CO_2}h_{CO_2}^{id}(200\,°C)$$
  $$+ w_{H_2O}h_{H_2O}^{id}(200\,°C)$$

  Using the data given in Appendix A and the antiderivatives for c_p^{id} in Appendix C, F4 one gets

  $$h_{O_2}^{id}(T) = \Delta h_{f,O_2}^0 + \int_{25\,°C}^{200\,°C} c_p^{id}\,dT = 0 + 164.13\text{ J/g} = 164.13\text{ J/g}$$

  $$h_{N_2}^{id}(T) = \Delta h_{f,N_2}^0 + \int_{25\,°C}^{200\,°C} c_p^{id}\,dT = 0 + 182.67\text{ J/g} = 182.67\text{ J/g}$$

  $$h_{CO_2}^{id}(T) = \Delta h_{f,CO_2}^0 + \int_{25\,°C}^{200\,°C} c_p^{id}\,dT = -393510\text{ J/mol} \cdot \frac{1\text{ mol}}{44.009\text{ g}}$$
  $$+ 161.77\text{ J/g} = -8779.81\text{ J/g}$$

  $$h_{H_2O}^{id}(T) = \Delta h_{f,H_2O}^0 + \int_{25\,°C}^{200\,°C} c_p^{id}\,dT = -241814\text{ J/mol} \cdot \frac{1\text{ mol}}{18.015\text{ g}}$$
  $$+ 332.27\text{ J/g} = -13090.66\text{ J/g}$$

$$\Rightarrow h_{\text{fluegas}}^{\text{id}}(200\,°C) = 0.0671 \cdot 164.13\,\text{J/g} + 0.7619 \cdot 182.67\,\text{J/g}$$
$$+ 0.1131 \cdot (-8779.81)\,\text{J/g} + 0.0579 \cdot (-13090.66)\,\text{J/g}$$
$$= -1600.76\,\text{J/g}$$

The duty of the steam generator can be calculated to be

$$\dot{Q} = 2100\,\text{kg/h} \cdot (-1600.76)\,\text{J/g} - 100\,\text{kg/h} \cdot (-4348.26)\,\text{J/g}$$
$$- 2000\,\text{kg/h} \cdot 25.40\,\text{J/g}$$
$$= -2977\,570\,\text{kJ/h}$$
$$= -827.1\,\text{kW}$$

As the enthalpy of vaporization of water at $P = 5\,\text{bar}$ is $\Delta h_v = 2109\,\text{J/g}$, this corresponds to an amount of steam of

$$\dot{m} = \frac{827.1\,\text{kW}}{2109\,\text{J/g}} = \frac{827100\,\text{J/s}}{2109\,\text{J/g}} = 392.18\,\text{g/s} = 1.412\,\text{t/h}$$

Similarly to the enthalpy, the Gibbs energy can be calculated using the appropriate routes and the standard Gibbs energy of formation as the reference point. Again, the standard Gibbs energy of formation of the elements is zero. For normal compounds, the standard Gibbs energy of formation can be derived by combining the standard enthalpy of formation and the standard entropies in a reaction which forms the compounds starting from the elements. This is demonstrated in Example 6.5.

Example 6.5

Evaluate the standard Gibbs energy of formation of HCl by evaluation of its formation reaction from the elements at $T_0 = 298.15\,\text{K}$, $P_0 = 101\,325\,\text{Pa}$. An ideal gas phase can be assumed.

The standard entropies are [4]

- $s_{\text{standard}}(\text{HCl}) = 186.902\,\text{J/(mol K)}$
- $s_{\text{standard}}(\text{Cl}_2) = 223.081\,\text{J/(mol K)}$
- $s_{\text{standard}}(\text{H}_2) = 130.68\,\text{J/(mol K)}$.

The standard enthalpies of formation can be taken from Appendix A.

Solution

The formation reaction of HCl is

$$\tfrac{1}{2}\,\text{H}_2 + \tfrac{1}{2}\,\text{Cl}_2 \longrightarrow \text{HCl}$$

The Gibbs energy of formation can be calculated as

$$\Delta g_f^0 = h_{f,\text{HCl}} - \tfrac{1}{2} h_{f,\text{H}_2} - \tfrac{1}{2} h_{f,\text{Cl}_2} - T_0 \left(s_{\text{standard},\text{HCl}} - \tfrac{1}{2} s_{\text{standard},\text{H}_2} - \tfrac{1}{2} s_{\text{standard},\text{Cl}_2} \right)$$
$$= \Delta h_{f,\text{HCl}}^0 - T_0 \left(s_{\text{standard},\text{HCl}} - \tfrac{1}{2} s_{\text{standard},\text{H}_2} - \tfrac{1}{2} s_{\text{standard},\text{Cl}_2} \right)$$

$$= -92310 \, \frac{\text{J}}{\text{mol}} - 298.15 \, \text{K} \cdot \left(186.902 - \frac{1}{2} \cdot 130.68 - \frac{1}{2} \cdot 223.081\right) \frac{\text{J}}{\text{mol K}}$$

$$= -95298 \, \frac{\text{J}}{\text{mol}}$$

The value tabulated in Appendix A is $\Delta g_f^0 = -95300$ J/mol.

The following example shall illustrate the procedure of using the standard Gibbs energy of formation in a process simulator; again, it should be pointed out that chemical reactions are explained in more detail in Chapter 12.

Example 6.6

The Deacon reaction

$$2\text{HCl} + 0.5 \text{O}_2 \rightleftharpoons \text{H}_2\text{O} + \text{Cl}_2$$

plays an important role in exhaust air treatment. Thermal combustion of exhaust air takes place in the temperature range from 800 to 1200 °C, where the chlorine in chlorinated components is completely converted to HCl. However, due to the equilibrium of the Deacon reaction elementary chlorine might be formed, which can hardly be absorbed by water in the flue gas scrubber. Calculate the Gibbs energies of the particular components at $\vartheta_1 = 1000\,°\text{C}$ and $\vartheta_2 = 70\,°\text{C}$. The total pressure shall be $P = 1.01325$ bar in both cases. An ideal gas phase should be assumed. The concentrations in the flue gas are $y_{\text{H}_2\text{O}} = 0.025$, $y_{\text{O}_2} = 0.1$, $y_{\text{CO}_2} = 0.05$, $y_{\text{N}_2} = 0.8245$, and $y_{\text{HCl}} = 5 \cdot 10^{-4}$. What are the consequences of the results?

Solution

In Chapter 12, it will be explained that the difference of the Gibbs energies of the pure components at the total pressure P is important for the calculation of the chemical equilibrium. For pure components with an ideal gas phase at $P = P_0 = 1.01325$ bar, g^{id} can be calculated according to Eq. (6.48)

$$g^{\text{id}}(T,P) = h_0 + \int_{T_0}^{T} c_P^{\text{id}} dT - Ts_0(T_0,P_0) - T\int_{T_0}^{T} \frac{c_P^{\text{id}}}{T} dT$$

$$= \Delta h_f^0 - T_0 s_0 + \int_{T_0}^{T} c_P^{\text{id}} dT - (T - T_0)s_0(T_0,P_0) - T\int_{T_0}^{T} \frac{c_P^{\text{id}}}{T} dT$$

$$= \Delta g_f^0 + \int_{T_0}^{T} c_P^{\text{id}} dT - (T - T_0)s_0(T_0,P_0) - T\int_{T_0}^{T} \frac{c_P^{\text{id}}}{T} dT$$

$$= \Delta g_f^0 + \int_{T_0}^{T} c_P^{\text{id}} dT - (T - T_0)\frac{\Delta h_f^0 - \Delta g_f^0}{T_0} - T\int_{T_0}^{T} \frac{c_P^{\text{id}}}{T} dT$$

$$= \frac{T}{T_0}\Delta g_f^0 + \left(1 - \frac{T}{T_0}\right)\Delta h_f^0 + \int_{T_0}^{T} c_P^{\text{id}} dT - T\int_{T_0}^{T} \frac{c_P^{\text{id}}}{T} dT$$

Using the coefficients from Appendix A and the antiderivatives for c_p^{id} from Appendix C, F4, we obtain at $\vartheta = 70\,°C$:

$$\frac{T}{T_0} = \frac{343.15\,\text{K}}{298.15\,\text{K}} = 1.1509 \quad 1 - \frac{T}{T_0} = 1 - \frac{343.15\,\text{K}}{298.15\,\text{K}} = -0.1509$$

$$g_{HCl}^{id}(70\,°C) = -1.1509 \cdot 95300\,\frac{J}{\text{mol}} - 0.1509 \cdot (-92310)\,\frac{J}{\text{mol}} + 1312\,\frac{J}{\text{mol}}$$
$$- 343.15\,\text{K} \cdot 4.097\,\frac{J}{\text{mol K}}$$
$$= -95845\,\frac{J}{\text{mol}}$$

$$g_{O_2}^{id}(70\,°C) = 0 + 0 + 1327\,\frac{J}{\text{mol}} - 343.15\,\text{K} \cdot 4.145\,\frac{J}{\text{mol K}} = -95.36\,\frac{J}{\text{mol}}$$

$$g_{H_2O}^{id}(70\,°C) = -1.1509 \cdot 228590\,\frac{J}{\text{mol}} - 0.1509 \cdot (-241814)\,\frac{J}{\text{mol}} + 1519\,\frac{J}{\text{mol}}$$
$$- 343.15\,\text{K} \cdot 4.745\,\frac{J}{\text{mol K}}$$
$$= -226704\,\frac{J}{\text{mol}}$$

$$g_{Cl_2}^{id}(70\,°C) = 0 + 0 + 1544\,\frac{J}{\text{mol}} - 343.15\,\text{K} \cdot 4.822\,\frac{J}{\text{mol K}} = -110.67\,\frac{J}{\text{mol}}$$

$$g_{CO_2}^{id}(70\,°C) = -1.1509 \cdot 394370\,\frac{J}{\text{mol}} - 0.1509 \cdot (-393510)\,\frac{J}{\text{mol}} + 1716\,\frac{J}{\text{mol}}$$
$$- 343.15\,\text{K} \cdot 5.356\,\frac{J}{\text{mol K}}$$
$$= -394622\,\frac{J}{\text{mol}}$$

$$g_{N_2}^{id}(70\,°C) = 0 + 0 + 1311\,\frac{J}{\text{mol}} - 343.15\,\text{K} \cdot 4.096\,\frac{J}{\text{mol K}} = -94.54\,\frac{J}{\text{mol}}$$

At $\vartheta = 1000\,°C$ one obtains

$$\frac{T}{T_0} = \frac{1273.15\,\text{K}}{298.15\,\text{K}} = 4.2702 \quad 1 - \frac{T}{T_0} = 1 - \frac{1273.15\,\text{K}}{298.15\,\text{K}} = -3.2702$$

$$g_{HCl}^{id}(1000\,°C) = -4.2702 \cdot 95300\,\frac{J}{\text{mol}} - 3.2702 \cdot (-92310)\,\frac{J}{\text{mol}} + 29902\,\frac{J}{\text{mol}}$$
$$- 1273.15\,\text{K} \cdot 43.831\,\frac{J}{\text{mol K}}$$
$$= -130979\,\frac{J}{\text{mol}}$$

$$g_{O_2}^{id}(1000\,°C) = 0 + 32378\,\frac{J}{\text{mol}} - 1273.15\,\text{K} \cdot 46.979\,\frac{J}{\text{mol K}} = -27433\,\frac{J}{\text{mol}}$$

$$g_{H_2O}^{id}(1000\,°C) = -4.2702 \cdot 228590\,\frac{J}{\text{mol}} - 3.2702 \cdot (-241814)\,\frac{J}{\text{mol}}$$
$$+ 37743\,\frac{J}{\text{mol}} - 1273.15\,\text{K} \cdot 54.277\,\frac{J}{\text{mol K}}$$
$$= -216705\,\frac{J}{\text{mol}}$$

$$g^{id}_{Cl_2}(1000\,°C) = 0 + 35667\frac{J}{mol} - 1273.15\,K \cdot 52.603\frac{J}{mol\,K} = -31305\frac{J}{mol}$$

The Gibbs energy of the reaction can be calculated to be at 70 °C:

$$\Delta g^{id}_R = g^{id}_{H_2O} + g^{id}_{Cl_2} - 2g^{id}_{HCl} - 0.5g^{id}_{O_2}$$

$$= (-226704 - 110.67 + 2 \cdot 95845 + 0.5 \cdot 95.36)\frac{J}{mol} = -35077\frac{J}{mol}$$

and at 1000 °C:

$$\Delta g^{id}_R = g^{id}_{H_2O} + g^{id}_{Cl_2} - 2g^{id}_{HCl} - 0.5g^{id}_{O_2}$$

$$= (-216705 - 31305 + 2 \cdot 130979 + 0.5 \cdot 27433)\frac{J}{mol} = 27665\frac{J}{mol}$$

It will be explained in Chapter 12 that the chemical equilibrium of this reaction is characterized by the equilibrium constant

$$K = \frac{y_{H_2O}y_{Cl_2}}{y^2_{HCl}y^{0.5}_{O_2}} = \exp\left(-\frac{\Delta g^{id}_R}{RT}\right)$$

The equilibrium constants are calculated to be

$$K(70\,°C) = \exp\left(-\frac{-35077\,J/mol}{8.3143\frac{J}{mol\,K} \cdot 343.15\,K}\right) = 218503$$

$$K(1000\,°C) = \exp\left(-\frac{27665\,J/mol}{8.3143\frac{J}{mol\,K} \cdot 1273.15\,K}\right) = 0.0733$$

Consider a typical flue gas with the initial concentrations $y_{H_2O} = 0.025$, $y_{O_2} = 0.1$, and $y_{HCl} = 5 \cdot 10^{-4}$. It is a good approximation that at $\vartheta = 1000\,°C$ these three concentrations remain unaffected by the reaction. The equilibrium chlorine concentration is

$$y_{Cl_2, 1000\,°C} = K(1000\,°C) \cdot \frac{y^2_{HCl}y^{0.5}_{O_2}}{y_{H_2O}} \approx 0.0733 \cdot \left(\frac{(5 \cdot 10^{-4})^2 \cdot 0.1^{0.5}}{0.025}\right) = 2.3 \cdot 10^{-7}$$

meaning that only 0.092% of the HCl is converted to Cl_2 in the flue gas. At $\vartheta = 70\,°C$, the situation is completely different. Setting

$$y_{Cl_2} \approx (5 \cdot 10^{-4} - y_{HCl})/2$$

according to the stoichiometry, one gets

$$2.5 \cdot 10^{-4} - 0.5y_{HCl} = K(70\,°C) \cdot \frac{y^2_{HCl}y^{0.5}_{O_2}}{y_{H_2O}} \approx 218503 \cdot \frac{y^2_{HCl} \cdot 0.1^{0.5}}{0.025}$$

$$= 2.76 \cdot 10^6 y^2_{HCl}$$

with the solution $y_{HCl} = 9.43 \cdot 10^{-6}$, corresponding to $y_{Cl_2} = 2.45 \cdot 10^{-4}$ (245 vol ppm), which means that the chlorine concentration has increased by three orders of magnitude. This would not be acceptable for release to atmosphere. Fortunately, the reaction is very slow at low temperatures so that elementary chlorine, which can hardly be removed from the flue gas by the typical caustic wash, is not extensively formed [5].

6.4
The G-Minimization Technique

For complicated equilibrium calculations, the G-minimization technique is a useful option for the evaluation of phase equilibria, especially if both phase and reaction equilibria are involved. In contrast to the equilibrium conditions, the minimum of G is not only a necessary but a sufficient equilibrium condition. As well, for complicated equilibria it is often the only way to keep the overview. The only knowledge that must be available is the functional relationship for the Gibbs energy g and a clear concept for the minimum evaluation task. The following example shall illustrate the method.

Example 6.7

Solve the equilibrium calculations for the Deacon reaction in Example 6.6 at $\vartheta = 70\,°C$ using the G-minimization technique.

Solution

Writing n_T for the total number of moles in the flue gas and Δn for the number of chlorine moles formed in the reaction, the number of moles of the particular components in the flue gas before and after reaction is shown in Table 6.2.

The Gibbs energy of the system before the reaction takes place is

$$G = n_{HCl}\, g_{HCl} + n_{O_2}\, g_{O_2} + n_{H_2O}\, g_{H_2O} + n_{Cl_2}\, g_{Cl_2} + n_{CO_2}\, g_{CO_2}$$
$$+ n_{N_2}\, g_{N_2} - n_T T \Delta s^{id}$$

After reaction, the expression for the Gibbs energy changes according to Table 6.2.

Table 6.2 Overview on the numbers of moles for the particular components.

Before reaction	After reaction
$n_{H_2O} = n_T \cdot y_{H_2O} = 0.025 n_{tot}$	$n_{H_2O} = 0.025 n_T + \Delta n$
$n_{Cl_2} = n_T \cdot y_{Cl_2} = 0$	$n_{Cl_2} = \Delta n$
$n_{O_2} = n_T \cdot y_{O_2} = 0.1 n_T$	$n_{O_2} = 0.1 n_T - 0.5 \Delta n$
$n_{HCl} = n_T \cdot y_{HCl} = 5 \cdot 10^{-4} n_T$	$n_{HCl} = 5 \cdot 10^{-4} n_T - 2\Delta n$
$n_{CO_2} = n_T \cdot y_{CO_2} = 0.05 n_T$	$n_{CO_2} = 0.05 n_T$
$n_{N_2} = n_T \cdot y_{N_2} = 0.8245 n_T$	$n_{N_2} = 0.8245 n_T$
Total number of moles:	
n_T	$n_T - 0.5 \Delta n$

Using the expressions for the particular numbers of moles, the Gibbs energy after reaction can be expressed as

$$G = (5 \cdot 10^{-4} n_T - 2\Delta n) g_{HCl} + (0.1 n_T - 0.5\Delta n) g_{O_2} + (0.025 n_T + \Delta n) g_{H_2O}$$
$$+ \Delta n g_{Cl_2} + 0.05 n_T g_{CO_2} + 0.8245 n_T g_{N_2} - (n_T - 0.5\Delta n) \cdot T\Delta S^{id}$$

After division by n_T and applying the abbreviation $X = \Delta n/n_T$, one obtains

$$\frac{G}{n_T} = (5 \cdot 10^{-4} - 2X)g_{HCl} + (0.1 - 0.5X) g_{O_2} + (0.025 + X)g_{H_2O} + X g_{Cl_2}$$
$$+ 0.05 g_{CO_2} + 0.8245 g_{N_2} - (1 - 0.5X) \cdot T\Delta S^{id}$$

The expression for Δs^{id} is

$$\Delta s^{id} = -R \sum_i y_i \ln y_i$$

$$= -R \left[\frac{5 \cdot 10^{-4} - 2X}{1 - 0.5X} \ln \frac{5 \cdot 10^{-4} - 2X}{1 - 0.5X} + \frac{0.1 - 0.5X}{1 - 0.5X} \ln \frac{0.1 - 0.5X}{1 - 0.5X} \right]$$
$$- R \left[\frac{0.025 + X}{1 - 0.5X} \ln \frac{0.025 + X}{1 - 0.5X} + \frac{X}{1 - 0.5X} \ln \frac{X}{1 - 0.5X} \right]$$
$$- R \left[\frac{0.05}{1 - 0.5X} \ln \frac{0.05}{1 - 0.5X} + \frac{0.8245}{1 - 0.5X} \ln \frac{0.8245}{1 - 0.5X} \right]$$

Inserting the specific Gibbs energies calculated in Example 6.6 gives

$$\frac{G/n_T}{J/mol} = (5 \cdot 10^{-4} - 2X)(-95845) + (0.1 - 0.5X)(-95.36) + (0.025 + X)$$
$$\cdot (-226704) + X(-110.67) + 0.05 \cdot (-394622) + 0.8245 \cdot (-94.54)$$
$$+ 8.3143 \cdot 343.15 \cdot \left[(5 \cdot 10^{-4} - 2X) \ln \frac{5 \cdot 10^{-4} - 2X}{1 - 0.5X} \right.$$
$$\left. + (0.1 - 0.5X) \ln \frac{0.1 - 0.5X}{1 - 0.5X} \right]$$
$$+ 8.3143 \cdot 343.15 \cdot \left[(0.025 + X) \ln \frac{0.025 + X}{1 - 0.5X} + X \ln \frac{X}{1 - 0.5X} \right]$$
$$+ 8.3143 \cdot 343.15 \cdot \left[0.05 \ln \frac{0.05}{1 - 0.5X} + 0.8245 \ln \frac{0.8245}{1 - 0.5X} \right]$$

where g has to be minimized by varying X. As $n_{HCl} \geq 0$, it can be derived from the expression in Table 6.2 that $0 \leq X \leq 2.5 \cdot 10^{-4}$. A plot of the function (see Figure 6.9) indicates that there is a flat minimum in the range $2 \cdot 10^{-4} < X < 2.5 \cdot 10^{-4}$. The criterion for the evaluation of the minimum is that the first derivative of the function becomes 0. In this case, it is sufficient to calculate the derivative numerically with a small step size $\Delta X = 10^{-8}$:

$$g'(X) \approx \frac{g(X + \Delta X) - g(X - \Delta X)}{2\Delta X}$$

The zero root of the function can be found by an application of the Regula Falsi, taking care of the possible range of values. The algorithm [6] is defined by

$$X_{i+2} = X_{i+1} - \frac{g'(X_{i+1})}{g'(X_{i+1}) - g'(X_i)} (X_{i+1} - X_i)$$

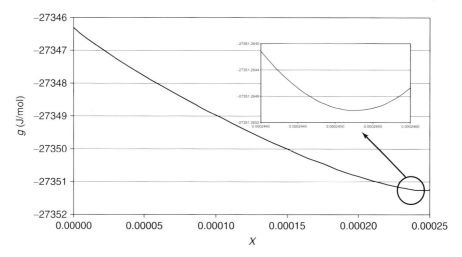

Figure 6.9 Plot and detail magnification of the specific Gibbs energy in Example 6.7 with respect to the extent of reaction X.

Table 6.3 Iteration history in Example 6.7.

Iteration step	X	g(X) (J/mol)	g'(X) (J/mol)
1	0.00024100	−27351.25626023	−3715.91704607
2	0.00024800	−27351.25920358	4949.05198139
3	0.00024400	−27351.26412045	−1364.79993671
4	0.00024487	−27351.26492186	−466.83853725
5	0.00024532	−27351.26501482	61.04382919
6	0.00024526	−27351.26501634	−2.50511221
7	0.00024527	−27351.26501634	−0.01309672
8	0.00024527	−27351.26501634	0.00000000

The iteration history is listed in Table 6.3, giving the solution $X = \Delta n/n_T = 2.452655 \cdot 10^{-4}$. Therefore, the concentrations after reaction can be determined according to Table 6.2: The results differ only slightly from the values obtained in Example 6.5, due to the numerical compromises made in both calculations.

$y_{H_2O} = n_{H_2O}/(n_T - 0.5\Delta n) = (0.025 + 2.452655 \cdot 10^{-4})/(1 - 0.5 \cdot 2.452655 \cdot 10^{-4}) = 0.02525$
$y_{Cl_2} = n_{Cl_2}/(n_T - 0.5\Delta n) = 2.452655 \cdot 10^{-4}/(1 - 0.5 \cdot 2.452655 \cdot 10^{-4}) = 2.453 \cdot 10^{-4}$
$y_{O_2} = n_{O_2}/(n_T - 0.5\Delta n) = (0.1 - 0.5 \cdot 2.452655 \cdot 10^{-4})/(1 - 0.5 \cdot 2.452655 \cdot 10^{-4}) = 0.09989$
$y_{HCl} = n_{HCl}/(n_T - 0.5\Delta n) = (5 \cdot 10^{-4} - 2 \cdot 2.452655 \cdot 10^{-4})/(1 - 0.5 \cdot 2.452655 \cdot 10^{-4})$
$\quad = 9.47 \cdot 10^{-6}$
$y_{CO_2} = n_{CO_2}/(n_T - 0.5\Delta n) = 0.05/(1 - 0.5 \cdot 2.452655 \cdot 10^{-4}) = 0.05$
$y_{N_2} = n_{N_2}/(n_T - 0.5\Delta n) = 0.8245/(1 - 0.5 \cdot 2.452655 \cdot 10^{-4}) = 0.8246$

Additional Problems

P6.1 A process simulation software reports the enthalpy of a saturated liquid stream of 53 kg/h of cyclohexane at 45 °C as −2.301 Gcal/h. Estimate this value using the data given in Appendices A and B and compare the results.

P6.2 How much heat per mol of water must be removed to reach a final temperature of T = 1000 K when the oxyhydrogen gas reaction

$$H_2 + 0.5\, O_2 \rightleftharpoons H_2O$$

is performed? Use the ideal gas equation of state and the physical properties from Appendix A. The starting temperature is $T_{start} = 400$ K. Assume stoichiometric composition and total conversion.

P6.3 Calculate the enthalpy difference when pure liquid water at P = 1 bar is heated up from $\vartheta_1 = 20\,°C$ to $\vartheta_2 = 70\,°C$ using both Route A and Route B (see Section 6.2). Which is the better option? Use the Peng-Robinson equation of state for the vapor phase.

P6.4 Calculate the enthalpy of vaporization of liquid propylene at $\vartheta_1 = -100\,°C$, $\vartheta_2 = 0\,°C$, $\vartheta_3 = 50\,°C$, $\vartheta_4 = 70\,°C$, and $\vartheta_5 = 90\,°C$ using Route B (see Section 6.2). Use the Peng-Robinson equation of state for the vapor phase. Up to which temperatures can Route B be regarded as a useful option?

References

1. Kleiber, M. (2003) *Ind. Eng. Chem. Res.*, **42**, 2007–2014.
2. ANSI/API (2007) 521. *Pressure Relieving and Depressurizing Systems*, 5th edn, ANSI/API.
3. Wagner, W. and Kretzschmar, H.J. (2008) *International Steam Tables – Properties of Water and Steam Based on the Industrial Formulation of the IAPWS-IF97*, Springer, Berlin.
4. Baehr, H.D. and Kabelac, S. (2006) *Thermodynamik*, 13th edn, Springer, Berlin.
5. Domschke, T. (1996) *Chem. Ing. Tech.*, **68**, 575–579.
6. Press, W.H., Teukolsky, S.A., Vetterling, W.T. and Flannery, B.P. (1992) *Numerical Recipes in Fortran*, 2nd edition, Cambridge Univ. Press, Cambridge.
7. Sofyan, Y., Ghajar, A.J., and Gasem, K.A.M. (2003) *Ind. Eng. Chem. Res.*, **42**, 3786–3801.

7
Electrolyte Solutions

7.1
Introduction

An electrolyte is a substance that conducts electric current as a result of its dissociation into positively and negatively charged ions in solutions or melts. Ions with a positive charge are called *cations*, ions with negative charge *anions*, respectively. The most typical electrolytes are acids, bases, and salts dissolved in a solvent, very often in water. But molten electrolytes and solid electrolytes in the absence of any solvent are also possible, for example, molten sodium chloride or solid silver iodide.

The charge of an ion is an integer multiple of the elementary charge

$$e = 1.602189 \cdot 10^{-19} \text{ C} \tag{7.1}$$

The product of the elementary charge e and the Avogadro number N_A is the Faraday constant

$$F = eN_A = 96484.56 \text{ C/mol} \tag{7.2}$$

The total charge of an ion is given by the number z. Examples are

H_3O^+	$z = 1$
Cl^-	$z = -1$
Ca^{2+}	$z = 2$
SO_4^{2-}	$z = -2$

In a macroscopic solution, the total number of charges becomes zero, since the solution is always neutral, otherwise there would be an electric current:

$$q = \sum n_i z_i = 0 \tag{7.3}$$

For electrolyte solutions, the particular ions are formed by dissociation reactions like

$$NaCl \rightleftharpoons Na^+ + Cl^-$$

$$CaSO_4 \rightleftharpoons Ca^{2+} + SO_4^{2-}$$

Chemical Thermodynamics: for Process Simulation, First Edition.
Jürgen Gmehling, Bärbel Kolbe, Michael Kleiber, and Jürgen Rarey.
© 2012 Wiley-VCH Verlag GmbH & Co. KGaA. Published 2012 by Wiley-VCH Verlag GmbH & Co. KGaA.

$$H_3PO_4 + H_2O \rightleftarrows H_3O^+ + H_2PO_4^-$$

$$H_2PO_4^- + H_2O \rightleftarrows H_3O^+ + HPO_4^{2-}$$

$$HPO_4^{2-} + H_2O \rightleftarrows H_3O^+ + PO_4^{3-}$$

$$NH_3 + H_2O \rightleftarrows NH_4^+ + OH^-$$

The H^+ ion cannot exist as a pure proton, it is always attached to a water molecule (H_3O^+).

It can be distinguished between strong and weak electrolytes. While strong electrolytes like HCl, H_2SO_4, HNO_3, NaCl, or NaOH dissociate almost completely, weak electrolytes do so only to a small extent. Sometimes, their electrolyte character can be neglected at all. Examples are formic acid (HCOOH), acetic acid (CH_3COOH), HF, H_2S, SO_2, NH_3, or CO_2, which react to some extent with water to dissociated products. The degree of dissociation α is defined as

$$\alpha = \frac{n_{dissociated}}{n_{total}} \tag{7.4}$$

For electrolyte solutions the molality is often used as a concentration scale instead of the mole fraction. *Molality* is defined as the number of moles of a solute in 1 kg solvent (solv). The relationship between molality and mole fraction is

$$m_i = \frac{n_i}{n_{solv} M_{solv}} = \frac{x_i}{x_{solv} M_{solv}} \tag{7.5}$$

and for electrolyte mixtures

$$x_i = \frac{n_i}{\sum_{j \neq solv} n_j + n_{solv}} = \frac{m_i}{\sum_{j \neq solv} m_j + \frac{1}{M_{solv}}} \tag{7.6}$$

respectively. The solvent may be a single component as well as a mixture of components. The unit of the molality is mol/kg solvent (solvent mixture).[1] For simplicity, most of the following derivations are written down for a single component solvent, but they may be easily extended to a solvent mixture.

Example 7.1

Calculate the mole fraction of a 1-molal aqueous solution of NaCl. $M_{H_2O} = 18.015$ g/mol.

---

1) Another concentration scale which is often found in the literature is molarity. The molarity is defined as the number of moles of a solute in 1 l of solvent. This scale has the disadvantage that additionally the density of the solvent as a function of temperature (and pressure) must be taken into account. Therefore, only the molality scale is used in the following text.

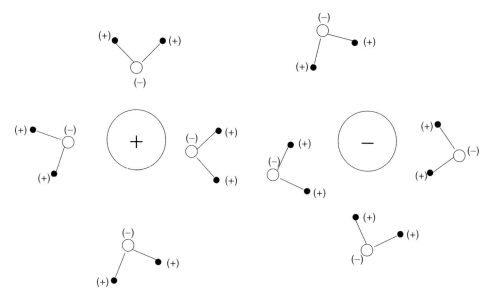

Figure 7.1 Schematic structure of an aqueous electrolyte solution.

Solution

Applying Eq. (7.6), one obtains

$$x_i = \frac{m_i}{\sum_{j \neq \text{solv}} m_j + \frac{1}{M_{\text{solv}}}} = \frac{1\,\frac{\text{mol}}{\text{kg}}}{1\,\frac{\text{mol}}{\text{kg}} + \frac{1}{18.015\,\frac{\text{g}}{\text{mol}}}} = \frac{1\,\frac{\text{mol}}{\text{kg}}}{1\,\frac{\text{mol}}{\text{kg}} + \frac{1000\,\text{mol}}{18.015\,\text{kg}}} = 0.0177^{[2]}$$

The molecular structure of an electrolyte solution is significantly determined by the electrostatic interactions between the charged ions (Coulomb–Coulomb interactions) and by the long-range interactions of the charged ions and the dipole moments of the solvent (Coulomb–dipole interactions). Figure 7.1 illustrates the schematic distribution of water as a strongly polar solvent around a cation and an anion.

The oxygen atom in the water molecule has a negative partial charge due to its high electronegativity. Therefore, the water molecule in the vicinity of an ion is arranged in a way that the oxygen atom is directed toward the positively charged cations. Vice versa, the hydrogen atoms in the water molecule are partially positively charged and oriented toward the anions. Around the ions, a shell of solvent molecules (solvation sphere) is formed, which accompanies the ion even when it moves.

---

2) In many textbooks, the factor 1000, coming from the simple unit conversion 1000 g = 1 kg, is already integrated into Eq. (7.6). However, this might lead to confusion as in this case quantities with different units (mol/kg and mol/g) would have to be added in the denominator without unit conversion.

Table 7.1 Heats of solution at infinite dilution in water for a few electrolytes at $T = 298.15$ K.

	NaCl (S)	NaOH (S)	HCl (G)	SO$_3$ (G)
$\Delta h_{sol}^\infty/(\text{kJ/mol})$	3.87	−44.51	−74.85	−227.72

[a] Data from Luckas and Krissmann [1].

The corresponding procedure is called *solvation*. It is an exothermic process. On the other hand, the dissociation of the electrolyte is an endothermic process, because the ionic lattice has to be destroyed, which is related to a need of energy. The overall heat of solution is the sum of both contributions. It is usually dominated by the solvation and therefore remains exothermic. However, there are many exceptions. Table 7.1 gives a few heats of solution to illustrate the order of magnitude.

For the behavior of the electrolyte solution, the relative dielectric constant ε of the solvent is an important parameter. According to Coulomb's law, the force between two ions i and j is

$$F_{ij} = \frac{1}{4\pi\varepsilon_0\varepsilon} \frac{q_i q_j}{r_{ij}^2} \tag{7.7}$$

with the vacuum dielectric constant $\varepsilon_0 = 8.854188 \cdot 10^{-12}\ \text{C}^2\ \text{N}^{-1}\ \text{m}^{-2}$. The forces between the ions are strongly reduced if the relative dielectric constant ε of the

Table 7.2 Relative dielectric constants of selected solvents at $T = 298.15$ K.

Solvent	ε
Water	78.54
Nitromethane	38.0
Nitrobenzene	34.82
Methanol	32.63
Ethanol	24.35
1-Propanol	20.33
2-Propanol	19.4
1-Butanol	17.43
2-Butanol	16.7
Acetone	20.56
Acetic acid	6.25
Diethyl ether	4.33
Benzene	2.27
Cyclohexane	2.02

[a] Data from Luckas and Krissmann [1].

Table 7.3 Coefficients for the calculation of the dielectric constant (Eq. (7.8)).

Component	A	B	C	D	E	T_{min} (K)	T_{max} (K)
Water	289.8229	−1.1480	$1.7843 \cdot 10^{-3}$	$-1.053 \cdot 10^{-6}$	0	273	643
Methanol	301.6681	−2.3343	$7.9275 \cdot 10^{-3}$	$-1.2858 \cdot 10^{-5}$	$7.964 \cdot 10^{-9}$	163	511
Ethanol	191.9472	−1.3540	$3.8877 \cdot 10^{-3}$	$-4.1286 \cdot 10^{-6}$	0	163	353

solvent is high. For water, the relative dielectric constant is extremely high, causing extremely good conditions for the dissociation of the electrolyte, as the attractive Coulomb forces between the ions are reduced by almost two orders of magnitude by the solvation shells. Table 7.2 gives values for the relative dielectric constants for a few selected solvents.

It must be emphasized that the dielectric constant is usually strongly temperature-dependent. Simple polynomials can be used for its correlation over wide temperature ranges. In Table 7.3, the parameters for the equation

$$\varepsilon = A + B\frac{T}{K} + C\left(\frac{T}{K}\right)^2 + D\left(\frac{T}{K}\right)^3 + E\left(\frac{T}{K}\right)^4 \tag{7.8}$$

are given for the three most important solvents water, methanol, and ethanol (A. Mohs, private communication).

7.2
Thermodynamics of Electrolyte Solutions

For the calculation of the chemical potentials of the particular components in an electrolyte solution, it is necessary to define an appropriate standard and reference state. It is convenient to define a standard state where the values of both the concentration and the activity coefficient and hence the activity are unity. For the solvent, there is no problem to use the pure component at the required pressure and temperature as standard state as shown in Chapter 4:

$$\mu_i = \mu_i^0 + RT \ln (x_i \gamma_i) \tag{7.9}$$

For an electrolyte component, this standard state does not make sense, since the properties of electrolyte components such as HCl or NaCl in the pure state are totally different from the properties of an electrolyte solution. Instead, it is useful to define a hypothetical ideal solution at unit concentration as standard state:

$$\mu_i = \mu_i^{*\xi}(T, P) + RT \ln (\xi_i \gamma_i^*) \tag{7.10}$$

ξ_i is an arbitrarily chosen concentration scale. If the concentration scale is molality, Eq. (7.10) is written as

$$\mu_i = \mu_i^{*m}(T, P) + RT \ln \left(\frac{m_i}{m_i^0} \gamma_i^{*m}\right) \tag{7.11}$$

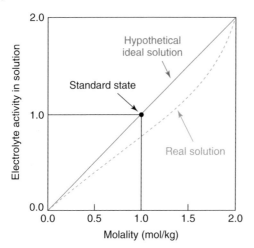

Figure 7.2 Illustration of the standard state of an electrolyte solution.

where m_i^0 is the unit molality (1 mol/kg). For convenience, it is a convention to omit the writing of m_i^0 bearing in mind that m_i^0 provides for the right dimensions in Eq. (7.11). $\mu_i^{*m}(T, P)$ is the chemical potential of an hypothetical ideal solution at a concentration of 1 mol/kg and γ_i^{*m} the molality activity coefficient:

$$\mu_i = \mu_i^{*m}(T, P) + RT \ln \left(m_i \gamma_i^{*m}\right) \tag{7.12}$$

Figure 7.2 gives a brief illustration of this definition. It shows that the real solution approaches the behavior of the hypothetical ideal solution in the dilute concentration range. Therefore, the asymmetric convention for the activity coefficient is applied in this chapter, similar as in Chapter 5 for the description of gas solubilities using Henry constants. At infinite dilution of the solute in the solvent (zero molality) the activity coefficient is one.

Analogously, the mole fraction can be used as the concentration measure, again with the asymmetric convention for the activity coefficient:

$$\mu_i = \mu_i^*(T, P) + RT \ln \left(x_i \gamma_i^*\right) \tag{7.13}$$

A relationship between the standard chemical potentials formulated for mole fractions and molalities, respectively, can be derived from

$$\mu_i = \mu_i^*(T, P) + RT \ln \left(x_i \gamma_i^*\right) = \mu_i^{*m}(T, P) + RT \ln \left(m_i \gamma_i^{*m}\right) \tag{7.14}$$

At infinite dilution of the solute the following relations are valid:

$$\lim_{x_{solv} \to 1} \frac{x_i}{m_i} = M_{solv} \quad \lim_{x_{solv} \to 1} \gamma_i^* = 1 \quad \lim_{x_{solv} \to 1} \gamma_i^{*m} = 1 \tag{7.15}$$

Introducing these relations into Eq. (7.14) gives

$$\mu_i^*(T, P) = \mu_i^{*m}(T, P) - RT \ln \frac{x_i \gamma_i^*}{m_i \gamma_i^{*m}} = \mu_i^{*m}(T, P) - RT \ln \frac{M_{solv}}{\text{kg/mol}} \tag{7.16}$$

7.2 Thermodynamics of Electrolyte Solutions

The relationship between the activity coefficients for the different concentration scales follows immediately from combining Eqs. (7.6), (7.14), and (7.16):

$$\gamma_i^* = \gamma_i^{*m}(1 + M_{\text{solv}} m_i) \tag{7.17}$$

In practical applications, it is advantageous not to set up the chemical potential for the dissociated ions but for the entire electrolyte component itself.

Taking a completely dissociated electrolyte, each molecule consisting of ν_c cations and ν_a anions, the chemical potential of the entire electrolyte $\mu_\pm$ can be written as

$$\mu_\pm = \left(\frac{\partial G}{\partial n_\pm}\right)_{T,P,n_{\text{solv}}} = \left(\frac{\partial G}{\partial n_c}\right)\left(\frac{\partial n_c}{\partial n_\pm}\right) + \left(\frac{\partial G}{\partial n_a}\right)\left(\frac{\partial n_a}{\partial n_\pm}\right) \tag{7.18}$$

and with

$$n_\pm = n_c/\nu_c + n_a/\nu_a \tag{7.19}$$

one gets

$$\mu_\pm = \nu_c \mu_c + \nu_a \mu_a \tag{7.20}$$

After introduction of the activity coefficients, Eq. (7.20) can be expressed as

$$\begin{aligned}\mu_\pm &= \nu_c \mu_c^{*m} + \nu_c RT \ln\left(\frac{m_c}{m^*}\gamma_c^{*m}\right) + \nu_a \mu_a^{*m} + \nu_a RT \ln\left(\frac{m_a}{m^*}\gamma_a^{*m}\right) \\ &= \nu_c \mu_c^{*m} + \nu_a \mu_a^{*m} + RT \ln\left[\left(\frac{m_c}{m^*}\right)^{\nu_c}\left(\frac{m_a}{m^*}\right)^{\nu_a}\gamma_c^{*m\nu_c}\gamma_a^{*m\nu_a}\right]\end{aligned} \tag{7.21}$$

By using the definitions

$$\nu_\pm = \nu_c + \nu_a \tag{7.22}$$

for the number of ions in the electrolyte,

$$m_\pm = m_c^{\nu_c/\nu_\pm} m_a^{\nu_a/\nu_\pm} \tag{7.23}$$

for the mean ionic molality,

$$\gamma_\pm^* = \gamma_c^{\nu_c/\nu_\pm} \gamma_a^{\nu_a/\nu_\pm} \tag{7.24}$$

for the mean ionic activity coefficient, and

$$a_\pm^* = a_c^{\nu_c/\nu_\pm} a_a^{\nu_a/\nu_\pm} \tag{7.25}$$

and thus at molality scale

$$a_\pm^{*m} = m_\pm \gamma_\pm^{*m} \tag{7.26}$$

for the mean ionic activity, we obtain the mean chemical potential of the electrolyte:

$$\mu_\pm = \nu_c \mu_c^{*m} + \nu_a \mu_a^{*m} + \nu_\pm RT \ln a_\pm^{*m} \tag{7.27}$$

or

$$\mu_\pm = \mu_\pm^{*m} + \nu_\pm RT \ln a_\pm^{*m} \tag{7.28}$$

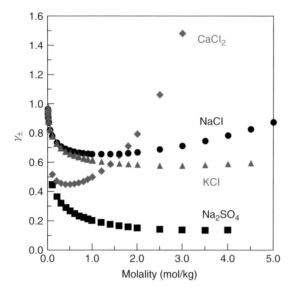

Figure 7.3 Experimental mean ionic activity coefficients for different salts in the aqueous solution.

Some standard potential values for electrolyte components can be taken from Appendix D. Both dissociated (Eq. (7.27)) and nondissociated electrolytes (Eq. (7.28)) can be treated by using the standard Gibbs energy:

$$\mu_i^{*m} = \Delta g_{fi}^0(*, m) \tag{7.29}$$

from the parameter tables, where the expression in brackets denotes for the kind of reference state. For the mean ionic activity coefficient $\gamma_\pm$, Figure 7.3 shows some experimental values for different salts.

The activity of the solvent can be derived from the Gibbs–Duhem equation:

$$x_{\text{solv}} d \ln a_{\text{solv}} + \sum_{i \neq \text{solv}} x_i d \ln a_i^{*m} = 0 \tag{7.30}$$

The result is [1]:

$$\ln a_{\text{solv}} = -M_{\text{solv}} \left[\sum_{i \neq \text{solv}} m_i + \sum_{j \neq \text{solv}} \int_0^{m_j} \sum_{i \leq j} m_i \frac{\partial \ln \gamma_i^{*m}}{\partial m_j} dm_j \right] \tag{7.31}$$

Equation (7.31) is derived in Appendix C, B8.

If only fully dissociated electrolytes are dissolved in the solvent, Eq. (7.31) reduces to

$$\ln a_{\text{solv}} = -M_{\text{solv}} \left[\sum_{i \neq \text{solv}} \nu_{i\pm} m_{i,\text{tot}} + \sum_{j \neq \text{solv}} \int_0^{m_{j,\text{tot}}} \sum_{i \leq j} \nu_{i\pm} m_{i,\text{tot}} \frac{\partial \ln \gamma_{i\pm}^{*m}}{\partial m_{j,\text{tot}}} dm_{j,\text{tot}} \right] \tag{7.32}$$

where i is now the index for the electrolytes instead for the ions. $m_{i,\text{tot}}$ denotes the molality of an electrolyte which totally dissociates in the solvent. If there is only one electrolyte, we get a further simplification to

$$\ln a_{\text{solv}} = -M_{\text{solv}} \nu_{i\pm} \left[m_{\text{tot}} + \int_0^{m_{\text{tot}}} m_{\text{tot}} \frac{\partial \ln \gamma_{i\pm}^{*m}}{\partial m_{\text{tot}}} dm_{\text{tot}} \right] \qquad (7.33)$$

Often, the so-called osmotic coefficient is used. It denotes for the ratio between the osmotic pressures of a real and an ideal solution:

$$\Phi = \frac{\Pi_{\text{real}}}{\Pi_{\text{ideal}}} \qquad (7.34)$$

It can be derived as follows [2].

Consider an electrolyte solution and a pure solvent being separated by a semipermeable membrane which can be passed only by the solvent. In this case, the solvent will pass through the membrane to dilute the electrolyte solution until the chemical equilibrium has been reached. In the solution, a considerable pressure can be built up. The equilibrium condition is

$$f_{1\text{solv}}(P_1, x_{\text{solv}} = 1) = f_{2\text{solv}}(P_2, x_{\text{solv}} = x_2) \qquad (7.35)$$

this means,

$$(\varphi_{\text{solv}}^s P_{\text{solv}}^s \text{Poy}_{\text{solv}})_1 = (a_{\text{solv}} \varphi_{\text{solv}}^s P_{\text{solv}}^s \text{Poy}_{\text{solv}})_2 \qquad (7.36)$$

giving

$$\exp\left(\frac{v_{\text{solv}}(P_1 - P_{\text{solv}}^s)}{RT}\right) = a_{\text{solv}} \exp\left(\frac{v_{\text{solv}}(P_2 - P_{\text{solv}}^s)}{RT}\right) \qquad (7.37)$$

leading to the expression

$$a_{\text{solv}} = \exp\left(-\frac{v_{\text{solv}}}{RT}(P_2 - P_1)\right) \qquad (7.38)$$

Thus, the osmotic pressure of the real solution is given by

$$\Pi_{\text{real}} = P_2 - P_1 = -\frac{RT \ln a_{\text{solv}}}{v_{\text{solv}}} \qquad (7.39)$$

For the ideal solution, Eq. (7.39) turns into

$$\Pi_{\text{ideal}} = P_2 - P_1 = -\frac{RT \ln x_{\text{solv}}}{v_{\text{solv}}} = \frac{RT}{v_{\text{solv}}} \ln\left(1 - \sum_{i \neq \text{solv}} x_i\right) \qquad (7.40)$$

For dilute solutions, $\ln\left(1 - \sum_{i \neq \text{solv}} x_i\right) \approx -\sum_{i \neq \text{solv}} x_i$ and $x_i \approx m_i M_{\text{solv}}$, the result is

$$\Pi_{\text{ideal}} = \frac{RT M_{\text{solv}}}{v_{\text{solv}}} \sum_{j \neq \text{solv}} m_j \qquad (7.41)$$

Therefore, the osmotic coefficient defined by Eq. (7.34) can be expressed by

$$\Phi = \frac{\Pi_{real}}{\Pi_{ideal}} = \frac{\ln a_{solv}}{M_{solv} \sum_{j \neq solv} m_j} \qquad (7.42)$$

which is usually taken for practical applications.

All necessary enthalpy expressions for electrolyte solutions can be rigorously derived using the Gibbs–Helmholtz equation:

$$h_i = -T^2 \frac{\partial (\mu_i/T)}{\partial T} \qquad (7.43)$$

which had been introduced in Section 2.4.3. Luckas and Krissmann [1] give the following relationships:

$$h_i^L(T, P, x_i) = h_i^{L,pure}(T, P^0) - RT^2 \left[\frac{\partial \ln \gamma_i}{\partial T} + \frac{\partial}{\partial T} \int_{p^0}^{P} \frac{v_i^L}{RT} dP \right] \qquad (7.44)$$

applicable for the solvents and

$$h_i^L(T, P, x_i) = h_i^{*m}(T_0, P_0) + \int_{T_0}^{T} c_{Pi}^{*m} dT - RT^2 \left(\frac{\partial \ln \gamma_i^{*m}}{\partial T} + \frac{\partial}{\partial T} \int_{P_0}^{P} \frac{v_i^*}{RT} dP \right) \qquad (7.45)$$

applicable for the ions, where values for h_i^{*m} and c_{Pi}^{*m} can be taken from Appendix D.

It must be clear that reliable results can only be expected if sufficiently accurate expressions for the temperature dependence of the activity coefficients are available.

It should be clearly pointed out that the c_{Pi}^{*m} do not have the typical character of heat capacities. In fact, they are just conversion terms to account for the temperature dependence of the standard enthalpies and standard Gibbs energies. They can even have negative values (see Appendix D).

7.3
Activity Coefficient Models for Electrolyte Solutions

7.3.1
Debye–Hückel Limiting Law

For the development of activity coefficient models for electrolyte solutions, the theory of Debye and Hückel is usually the starting point. It can be regarded as an exact equation to describe the behavior of an electrolyte system at infinite dilution.

For the derivation of the Debye–Hückel limiting law, the following assumptions are made [2]:

1) Only the electrostatic forces between the ions are regarded. All the other forces are negligible.
2) The electrostatic interaction energies are small in comparison to the thermal energies.
3) The ions are regarded as punctual charges with a spherical field.

4) The dielectric constant of the solution is equal to the one of the solvent.
5) The electrolyte is completely dissociated.
6) The distribution of the ions around a center ion is governed by Boltzmann's law due to the electric potential:

$$\frac{c_i(r)}{c_i^{(0)}} = \exp\left(\frac{-z_i\, e\, \varphi^{el}(r)}{kT}\right) \tag{7.46}$$

where $c_i(r)$ is the volume concentration of ionic species in a volume element at distance r from the center and $c_i^{(0)}$ is the same property when all the ions are uniformly distributed. $\varphi^{el}(r)$ is the electric potential at distance r from the center.

A detailed derivation of the Debye–Hückel theory is given in [3, 4]. After introducing the molal ionic strength

$$I = \frac{1}{2}\sum_i m_i z_i^2 \tag{7.47}$$

an approximation for the result is

$$\log \gamma_i^m = A_m z_i^2 I^{1/2} \tag{7.48}$$

with

$$A_m(T) = 1.8248 \cdot 10^6 \, \frac{\sqrt{\frac{\rho_{solv}}{g/cm^3}}}{\left(\frac{T}{K}\varepsilon_r\right)^{1.5}} \sqrt{\frac{kg}{mol}} \tag{7.49}$$

where A_m is characteristic for the solvent.

An expression for the mean activity coefficient of a salt can be derived by substitution and taking into account electroneutrality:

$$\log \gamma_\pm^{*m} = -A_m |z_+ z_-| I^{1/2} \tag{7.50}$$

which is the first approximation for $\gamma_\pm$ of the Debye–Hückel theory in the very dilute region. With the Debye–Hückel limiting law, the activity of the solvent can be determined to be [1]

$$\ln a_{solv} = M_{solv}\left(-\sum_{i\neq solv} v_{i\pm} m_{i,tot} + \frac{2\ln 10}{3} A_m I^{3/2}\right) \tag{7.51}$$

Due to the long-range Coulomb forces, even very dilute electrolyte solutions do not behave like ideal solutions. For water, at $\vartheta = 25\,°C$ and $P = 1$ atm, one obtains

$$A_m = 0.5108\sqrt{kg/mol},\ \varepsilon_r = 78.41,\ \rho_{solv} = 997\ kg/m^3$$

and

$$\log \gamma_\pm^{*m}\, (25\,°C, \text{aqueous}) = -0.51|z_+ z_-| I^{1/2} \tag{7.52}$$

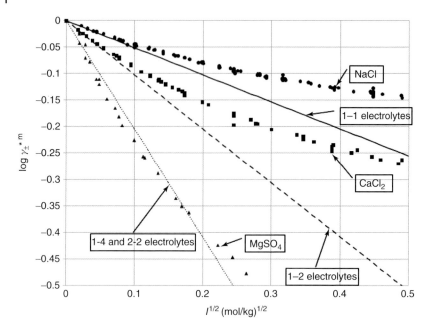

Figure 7.4 Mean logarithmic ionic activity coefficient for different aqueous systems with strong electrolytes as a function of the ionic strength.

Figure 7.4 gives an impression of the agreement of the Debye–Hückel limiting law with experimental data for a 1–1 electrolyte (NaCl), a 1–2 electrolyte (CaCl$_2$), and a 2–2 electrolyte (MgSO$_4$). In each case, it can be seen that for low concentrations the Debye–Hückel equation is a good approximation. However, at higher concentrations, ion–ion repulsion, dispersive forces, and ion–solvent solvation cause stronger deviations, so that the Debye–Hückel limiting law has to be extended. The Debye–Hückel approach is considered to be valid up to an ionic strength of $I < 0.005$ mol/kg.

7.3.2
Bromley Extension

With increasing electrolyte concentration, the short-range interactions become more and more dominating. Therefore, in activity coefficient models the Debye–Hückel term, which describes the long-range interactions, has to be extended by a term describing the short-range interactions. A well-known empirical extension of the Debye–Hückel theory is the Bromley equation [5]:

$$\log \gamma_\pm^{*m} = \frac{-A_m |z_+ z_-| I^{1/2}}{1 + I^{1/2}} + \frac{(0.06 + 0.6\beta)|z_+ z_-| I}{\left(1 + \dfrac{1.5 I}{|z_+ z_-|}\right)^2} + \beta I \tag{7.53}$$

where β is a temperature-dependent adjustable parameter. Values are given in [5]. The Bromley equation is valid up to an ionic strength $I < 6$ mol/kg.

7.3.3
Pitzer Model

One of the most widely used activity coefficient models has been proposed by Pitzer in 1973 [6–10]. In principle, it is a series expansion of the Gibbs energy, analogous to the virial equation of state; however, unlike to that, it is not directly justified by statistical mechanics. The expression is

$$\frac{G^E}{m_{solv} RT} = f(I) + \sum_i \sum_j \lambda_{ij}(I) m_i m_j + \sum_i \sum_j \sum_k \mu_{ijk} m_i m_j m_k \tag{7.54}$$

The first term $f(I)$ represents a modified Debye–Hückel term for the long-range interactions:

$$f(I) = -A_\phi(T) \frac{4I}{b} \ln\left(1 + b\sqrt{I}\right) \tag{7.55}$$

The second and the third terms take into account the short-range interactions, where binary and ternary parameters have to be adjusted as a function of the ionic strength. The Gibbs energy refers to the mass of the solvent.

For the particular functions, several options have been discussed [6, 7]. In the finally favored form for practical applications, the ternary interactions are neglected. The activity coefficients are expressed as

$$\ln \gamma_i^{*m} = \frac{z_i^2}{2} f_\gamma + 2 \sum_i m_j B_{ij} + \sum_j \left(\sum_k m_k |z_k| \right) m_j C_{ij}$$
$$+ \frac{z_i^2}{2} \sum_j \sum_k m_j m_k B'_{jk} + \frac{|z_i|}{2} \sum_j \sum_k m_j m_k C_{jk} \tag{7.56}$$

with

$$f_\gamma = -2A_\phi \left[\frac{\sqrt{I}}{1 + b\sqrt{I}} \right] + \frac{2}{b} \ln\left(1 + b\sqrt{I}\right) \tag{7.57}$$

$$B_{ij} = \beta_{ij}^{(0)} + \frac{2\beta_{ij}^{(1)}}{\alpha_1^2 I} \left\{ 1 - (1 + \alpha_1 \sqrt{I}) \exp\left(-\alpha_1 \sqrt{I}\right) \right\}$$
$$+ \frac{2\beta_{ij}^{(2)}}{\alpha_2^2 I} \left\{ 1 - (1 + \alpha_2 \sqrt{I}) \exp\left(-\alpha_2 \sqrt{I}\right) \right\} \tag{7.58}$$

$$B'_{ij} = \frac{2\beta_{ij}^{(1)}}{\alpha_1^2 I^2} \left\{ -1 + (1 + \alpha_1 \sqrt{I} + 0.5\alpha_1^2 I) \exp\left(-\alpha_1 \sqrt{I}\right) \right\}$$
$$+ \frac{2\beta_{ij}^{(2)}}{\alpha_2^2 I} \left\{ 1 - (1 + \alpha_2 \sqrt{I} + 0.5\alpha_2^2 I) \exp\left(-\alpha_2 \sqrt{I}\right) \right\} \tag{7.59}$$

$$C_{ij} = \frac{C_{ij}^{\phi}}{2\sqrt{|z_i z_j|}} \tag{7.60}$$

and

$$A_{\phi} = -61.44534 \exp\left(\frac{T/K - 273.15}{273.15}\right) + 2.864468\left[\exp\left(\frac{T/K - 273.15}{273.15}\right)\right]^2$$
$$+ 183.5379 \ln\frac{T/K}{273.15} - 0.6820223(T/K - 273.15)$$
$$+ 0.0007875695[(T/K)^2 - 273.15^2] + 58.95788\frac{273.15}{T/K} \tag{7.61}$$

The adjustable parameters are β_{ij}^0, $\beta_{ij}^{(1)}$, and C_{ij}^{ϕ}. For the other parameters, the standard values are set to

$$b = 1.2$$
$$\alpha_1 = 2$$
$$\beta_{ij}^{(2)} = 0$$

For representing the special behavior of 2–2 electrolytes ($z_+ = |z_-| = 2$), the standard values are set to

$$b = 1.2$$
$$\alpha_1 = 1.4$$
$$\alpha_2 = 12$$

In this case, $\beta_{ij}^{(2)}$ is used as an additional adjustable parameter. The matrices B_{ij} and C_{ij} are symmetric, that is, $B_{ij} = B_{ji}$, $C_{ij} = C_{ji}$, and $B_{ii} = C_{ii} = 0$.

The activity coefficient of the solvent can be expressed as

$$\frac{1}{M_{\text{solv}}} \ln \gamma_{\text{solv}} = f(I) - If_\gamma - \sum_i \sum_j m_i m_j (B_{ij} + B'_{ji} I) - \sum_i \sum_j \left(\sum_k m_k |z_k|\right) m_i m_j C_{ij} \tag{7.62}$$

Collections of useful Pitzer parameters can be taken from Rosenblatt [11], Zemaitis et al. [12], and Pitzer [13]. Temperature-dependent interaction parameters should be used if a wide temperature range must be covered, as it is common practice for nonelectrolyte systems as well. It is important especially for strong acids and bases. The Pitzer equation is usually valid up to molalities of 6 mol/kg.

7.3.4
Electrolyte-NRTL Model by Chen

Both the long-range and the short-range interactions have to be taken into account for the calculation of the phase equilibrium behavior in multicomponent electrolyte mixtures. While the long-range interactions can be accounted for by the Debye–Hückel term, the short-range interactions can be described by the local composition models Wilson, NRTL, and UNIQUAC/UNIFAC. The presumption that the local concentrations around a selected molecule, which are determined by short-range interaction forces, can be significantly different from the global concentration is valid as well for electrolyte solutions.

However, the influence of long-range interactions has to be taken into account by a term describing the Debye–Hückel theory. For this term, in the general case the density and the dielectric constant of the mixed solvent have to be determined (see Eq. (7.49)). As the reference state of the electrolyte components refers to the infinitely diluted solution in pure water, the Debye–Hückel term must be corrected by the so-called Born term, which takes into account the difference between the dielectric constants of water and the solvent mixture [14]:

$$\Delta_{\text{Born}} g^E = \frac{e^2 N_A}{8\pi \varepsilon_0} \left(\frac{1}{\varepsilon_{\text{solv}}} - \frac{1}{\varepsilon_{H_2O}} \right) \sum_i x_i \frac{z_i^2}{r_i} \qquad (7.63)$$

For the ionic radius r_i, the default value $r_i = 3 \cdot 10^{-10}$ m has been recommended. Thus, the Gibbs energy of a system described by a local composition model can be expressed as

$$g^E = g^E_{\text{DH}} + g^E_{\text{LC}} + \Delta_{\text{Born}} g^E \qquad (7.64)$$

The most widely used local composition model is the NRTL electrolyte model from Chen [15, 16], which is implemented in most of the commercial process simulators. Its great advantage is that it merges with the NRTL equation if electrolytes are missing. The binary parameters are valid for both models. Therefore, no model change is necessary in a flow-sheet if electrolytes are removed by a unit operation. On the other hand, the availability of the Chen model is a reason to prefer the NRTL equation to Wilson and UNIQUAC for nonelectrolyte components, since in large processes one can never exclude process steps with electrolytes, even if it is not planned at the beginning of the project. In this case, no further changes on the binary parameter matrix would be necessary.

In a liquid containing electrolyte components, there are three types of cells that have to be distinguished (see Figure 7.5). If there is a neutral molecule in the center of the cell, it can be surrounded by cations, anions, and other molecules. For cells with a cation or an anion in the center, it is assumed that because of the strong repulsive forces between equally charged ions, no ion with the same charge can be found in the direct neighborhood of the central ion.

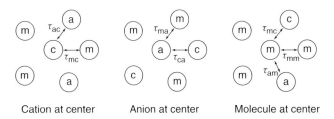

Cation at center   Anion at center   Molecule at center

Figure 7.5 Three types of cells according to the assumptions of like-ion repulsion and electroneutrality.

For the local concentrations Θ_{ij} (concentration of species i in direct neighborhood to the central species j), the following equations can be set up:

- **Cation at center:**

$$\Theta_{mc} + \Theta_{ac} = 1 \tag{7.65}$$

- **Anion at center:**

$$\Theta_{ma} + \Theta_{ca} = 1 \tag{7.66}$$

- **Molecule at center:**

$$\Theta_{cm} + \Theta_{am} + \Theta_{mm} = 1 \tag{7.67}$$

Additionally, for the cells with a molecule at center, Chen *et al.* [15] postulates local electroneutrality, that is,

$$\Theta_{am}|z_a| = \Theta_{cm} z_c \tag{7.68}$$

where z_a and z_c are the charges of the ions. Analogously to the NRTL equation [17], the local concentrations are related to the overall mole fractions via

$$\frac{\Theta_{ji}}{\Theta_{ii}} = \frac{x_j C_j}{x_i C_i} G_{ji} \tag{7.69}$$

$$\frac{\Theta_{ji}}{\Theta_{ki}} = \frac{x_j C_j}{x_k C_k} G_{ji,ki} \tag{7.70}$$

where $C_i = |z_i|$ for ions, $C_i = 1$ for molecules, and

$$G_{ij} = \exp(-\alpha_{ij} \tau_{ij}) \tag{7.71}$$

For molecule–molecule interactions, the *interaction parameters* τ_{ij} are defined as a function of temperature as for the conventional NRTL equation:

$$\tau_{ij} = A_{ij} + \frac{B_{ij}}{T} + F_{ij} \ln (T/K) + G_{ij} T \tag{7.72}$$

The interaction parameters between molecules and electrolytes are called *pair parameters*. They are described with a slightly different temperature dependence:

$$\begin{aligned} \tau_{m,ca} &= C_{m,ca} + \frac{D_{m,ca}}{T} + E_{m,ca} \left(\frac{298.15 \text{ K} - T}{T} + \ln \frac{T}{298.15 \text{ K}} \right) \\ \tau_{ca,m} &= C_{ca,m} + \frac{D_{ca,m}}{T} + E_{ca,m} \left(\frac{298.15 \text{ K} - T}{T} + \ln \frac{T}{298.15 \text{ K}} \right) \end{aligned} \tag{7.73}$$

The same Eq. (7.73) holds for electrolyte–electrolyte interactions, if they share either one common cation or one common anion:

$$\begin{aligned} \tau_{ca,c'a} &= C_{ca,c'a} + \frac{D_{ca,c'a}}{T} + E_{ca,c'a} \left(\frac{298.15 \text{ K} - T}{T} + \ln \frac{T}{298.15 \text{ K}} \right) \\ \tau_{ca,ca'} &= C_{ca,ca'} + \frac{D_{ca,ca'}}{T} + E_{ca,ca'} \left(\frac{298.15 \text{ K} - T}{T} + \ln \frac{T}{298.15 \text{ K}} \right) \end{aligned} \tag{7.74}$$

It must be emphasized that the pair parameters are not specific for an ion but for an electrolyte.

The NRTL nonrandomness factors are usually set to a uniform value of $\alpha_{ij} = \alpha_{ji,ki} = 0.2$. Due to the condition of electroneutrality (Eq. (7.68)), the number of adjustable parameters in the Chen model is reduced as

$$\tau_{cm} = \tau_{am} = \tau_{ca,m} \tag{7.75}$$

$$\tau_{mc,ac} = \tau_{ma,ca} = \tau_{m,ca} \tag{7.76}$$

$$\alpha_{cm} = \alpha_{am} = \alpha_{ca,m} \tag{7.77}$$

$$\alpha_{mc,ac} = \alpha_{ma,ca} = \alpha_{m,ca} \tag{7.78}$$

The final working equations are

$$\ln \gamma_i^* = \ln \gamma_{i,DH}^* + \Delta_{\text{Born}} \ln \gamma_i^* + \ln \gamma_{i,NRTL}^* \tag{7.79}$$

For the Debye–Hückel term, a modification from Pitzer [18] is used:

$$\ln \gamma_{i,DH}^* = \frac{-A_\phi}{\sqrt{\frac{M_{\text{solv}}}{\text{kg/mol}}}} \left[\left(\frac{2z_i^2}{14.9} \right) \ln \left(1 + 14.9\sqrt{I_x}\right) + \frac{z_i^2 \sqrt{I_x} - 2I_x^{1.5}}{1 + 14.9\sqrt{I_x}} \right] \tag{7.80}$$

For A_ϕ, Eq. (7.61) is used.

The Born term is

$$\Delta_{\text{Born}} \ln \gamma_i^* = \frac{z_i^2 e^2 N_A}{8\pi \varepsilon_0 R T r_i} \left(\frac{1}{\varepsilon_{\text{solv}}} - \frac{1}{\varepsilon_{H_2O}} \right) \tag{7.81}$$

where r_i is the Born radius, if no better values are available, $r_i = 3 \cdot 10^{-10}$ m is an appropriate choice.

The ionic strength based on the mole fractions is defined as

$$I_x = \frac{1}{2} \sum_i x_i z_i^2 \tag{7.82}$$

The mixing rule of the dielectric constant is

$$\varepsilon_{\text{solv}} = \sum_i w_i \varepsilon_i \tag{7.83}$$

The dielectric constants of the pure solvents are expressed as temperature functions:

$$\varepsilon_i(T) = A_i + B_i \left(\frac{1}{T} - \frac{1}{298.15 \text{ K}} \right) \tag{7.84}$$

The NRTL term is different for molecules and ions. Using the abbreviation

$$X_i = x_i C_i \tag{7.85}$$

the activity coefficients are

- for molecules:

$$\ln \gamma_{m,\text{NRTL}} = \frac{\sum_j X_j G_{jm} \tau_{jm}}{\sum_k X_k G_{km}} + \sum_{m'} \frac{X_{m'} G_{mm'}}{\sum_k X_k G_{km'}} \left(\tau_{mm'} - \frac{\sum_k X_k G_{km'} \tau_{km'}}{\sum_k X_k G_{km'}} \right)$$

$$+ \sum_c \sum_{a'} \frac{X_{a'}}{\sum_{a''} X_{a''}} \frac{X_c G_{mc,a'c}}{\sum_{k \neq c} X_k G_{kc,a'c}} \left(\tau_{mc,a'c} - \frac{\sum_{k \neq c} X_k G_{kc,a'c} \tau_{kc,a'c}}{\sum_{k \neq c} X_k G_{kc,a'c}} \right)$$

$$+ \sum_a \sum_{c'} \frac{X_{c'}}{\sum_{c''} X_{c''}} \frac{X_a G_{ma,c'a}}{\sum_{k \neq a} X_k G_{ka,c'a}} \left(\tau_{ma,c'a} - \frac{\sum_{k \neq a} X_k G_{ka,c'a} \tau_{ka,c'a}}{\sum_{k \neq a} X_k G_{ka,c'a}} \right) \quad (7.86)$$

- for cations:

$$\frac{1}{z_c} \ln \gamma_{c,\text{NRTL}} = \sum_{a'} \frac{X_{a'}}{\sum_{a''} X_{a''}} \frac{\sum_{k \neq c} X_k G_{kc,a'c} \tau_{kc,a'c}}{\sum_{k \neq c} X_k G_{kc,a'c}}$$

$$+ \sum_m \frac{X_m G_{cm}}{\sum_k X_k G_{km}} \left(\tau_{cm} - \frac{\sum_k X_k G_{km} \tau_{km}}{\sum_k X_k G_{km}} \right)$$

$$+ \sum_a \sum_{c'} \frac{X_{c'}}{\sum_{c''} X_{c''}} \frac{X_a G_{ca,c'a}}{\sum_{k \neq a} X_k G_{ka,c'a}} \left(\tau_{ca,c'a} - \frac{\sum_{k \neq a} X_k G_{ka,c'a} \tau_{ka,c'a}}{\sum_{k \neq a} X_k G_{ka,c'a}} \right)$$

$$(7.87)$$

- for anions:

$$\frac{1}{|z_a|} \ln \gamma_{a,\text{NRTL}} = \sum_{c'} \frac{X_{c'}}{\sum_{c''} X_{c''}} \frac{\sum_{k \neq a} X_k G_{ka,c'a} \tau_{ka,c'a}}{\sum_{k \neq a} X_k G_{ka,c'a}}$$

$$+ \sum_m \frac{X_m G_{am}}{\sum_k X_k G_{km}} \left(\tau_{am} - \frac{\sum_k X_k G_{km} \tau_{km}}{\sum_k X_k G_{km}} \right)$$

$$+ \sum_c \sum_{a'} \frac{X_{a'}}{\sum_{a''} X_{a''}} \frac{X_c G_{ac,a'c}}{\sum_{k \neq c} X_k G_{kc,a'c}} \left(\tau_{ac,a'c} - \frac{\sum_{k \neq c} X_k G_{kc,a'c} \tau_{kc,a'c}}{\sum_{k \neq c} X_k G_{kc,a'c}} \right)$$

$$(7.88)$$

The following mixing rules are used:

$$\alpha_{cm} = \frac{\sum_a X_a \alpha_{ca,m}}{\sum_{a'} X_{a'}} \quad \alpha_{am} = \frac{\sum_c X_c \alpha_{ca,m}}{\sum_{c'} X_{c'}} \quad (7.89)$$

$$G_{cm} = \frac{\sum_a X_a G_{ca,m}}{\sum_{a'} X_{a'}} \qquad G_{am} = \frac{\sum_c X_c G_{ca,m}}{\sum_{c'} X_{c'}} \qquad (7.90)$$

According to Chen and Evans [16], the interaction parameters are related by

$$\tau_{ma,ca} = \tau_{am} - \tau_{ca,m} + \tau_{m,ca} \qquad (7.91)$$

$$\tau_{mc,ac} = \tau_{cm} - \tau_{ca,m} + \tau_{m,ca} \qquad (7.92)$$

The normalization to the state of ideal dilution can be achieved by

$$\ln \gamma_i^* = \ln \gamma_i - \ln \gamma_i^\infty \qquad (7.93)$$

Analytical expressions for the activity coefficient at infinite dilution have been derived [12] for pure solvents (index W):

- for molecules: :

$$\ln \gamma_{m,W}^\infty = \tau_{Wm} + G_{mW}\tau_{mW} \qquad (7.94)$$

- for cations: :

$$\frac{1}{z_c} \ln \gamma_{c,W}^\infty = \sum_{a'} \frac{X_{a'}}{\sum_{a''} X_{a''}} \tau_{Wc,a'c} + G_{cW}\tau_{cW} \qquad (7.95)$$

- for anions: :

$$\frac{1}{|z_a|} \ln \gamma_{a,W}^\infty = \sum_{c'} \frac{X_{c'}}{\sum_{c''} X_{c''}} \tau_{Wa,c'a} + G_{aW}\tau_{aW} \qquad (7.96)$$

Due to its availability for a long time, the widely developed parameter matrix in the commercial process simulators and its compatibility with the usual NRTL equation, the NRTL electrolyte model from Chen [15, 16] has become the most widely used electrolyte model. Nevertheless, it gives considerable errors for multicomponent systems, that is, if two or more electrolyte components or a solvent mixture is considered. In these cases, the results of the NRTL electrolyte model can only be taken as a qualitative estimation.

Example 7.2

Calculate the activity coefficients of the particular components in a 30 wt% aqueous solution of caustic soda at $\vartheta = 100\,°C$ with the Chen NRTL electrolyte model. As well, calculate the equilibrium pressure. An ideal vapor phase and complete dissociation should be assumed. The following values are given:

P^s (H_2O, 100 °C) = 1.0138 bar
ρ^l (H_2O, 100 °C) = 957 kg/m^3 = 0.957 g/cm^3
ε_{H_2O} (100 °C, H_2O) = 55.3 [19]
M_{H_2O} = 18.015 g/mol, M_{Na^+} = 22.99 g/mol, M_{OH^-} = 17.01 g/mol.

7 Electrolyte Solutions

Solution

According to Eq. (7.79), the activity coefficient consists of the parts:
$$\ln \gamma_i^* = \ln \gamma_{i,\text{DH}}^* + \Delta_{\text{Born}} \ln \gamma_i^* + \ln \gamma_{i,\text{NRTL}}^*$$

The calculation is started with the Debye–Hückel term

$$\ln \gamma_{i,\text{DH}}^* = \frac{-A_\phi}{\sqrt{\dfrac{M_{\text{solv}}}{\text{kg/mol}}}} \left[\left(\frac{2z_i^2}{14.9}\right) \ln\left(1 + 14.9\sqrt{I_x}\right) + \frac{z_i^2 \sqrt{I_x} - 2I_x^{1.5}}{1 + 14.9\sqrt{I_x}} \right] \quad (7.97)$$

using Eq. (7.61)

$$A_\phi = -61.44534 \exp\left(\frac{373.15 - 273.15}{273.15}\right)$$
$$+ 2.864468 \left[\exp\left(\frac{373.15 - 273.15}{273.15}\right)\right]^2$$
$$+ 183.5379 \ln \frac{373.15}{273.15} - 0.6820223 \cdot (373.15 - 273.15)$$
$$+ 0.0007875695 \cdot [373.15^2 - 273.15^2] + 58.95788 \frac{273.15}{373.15} = 0.45945$$

Switching from weight fractions to mole fractions gives

$$W_{\text{Na}^+} = 0.3 \cdot \frac{22.99}{22.99 + 17.01} = 0.1724$$
$$W_{\text{OH}^-} = 0.3 - 0.1724 = 0.1276$$

$$x_{\text{Na}^+} = x_{\text{OH}^-} = \frac{\dfrac{0.1724}{22.99}}{\dfrac{0.1724}{22.99} + \dfrac{0.1276}{17.01} + \dfrac{0.7}{18.015}} = 0.1393$$

$$x_{\text{H}_2\text{O}} = 1 - 2 \cdot 0.1393 = 0.7215$$

and therefore

$$I_x = \frac{1}{2}\left(0.1393 \cdot (1)^2 + 0.1393 \cdot (-1)^2\right) = 0.1393$$

The Debye–Hückel contributions are

$$\ln \gamma_{\text{Na}^+,\text{DH}}^* = \ln \gamma_{\text{OH}^-,\text{DH}}^* = \frac{-0.45945}{\sqrt{0.018015}} \left[\left(\frac{2 \cdot 1^2}{14.9}\right) \ln\left(1 + 14.9\sqrt{0.1393}\right)\right.$$
$$\left. + \frac{1^2 \cdot \sqrt{0.1393} - 2 \cdot 0.1393^{1.5}}{1 + 14.9\sqrt{0.1393}}\right] = -1.00479$$

$$\ln \gamma_{\text{H}_2\text{O},\text{DH}}^* = \frac{-0.45945}{\sqrt{0.018015}} \left[\left(\frac{2 \cdot 0^2}{14.9}\right) \ln\left(1 + 14.9\sqrt{0.1393}\right)\right.$$
$$\left. + \frac{0^2 \cdot \sqrt{0.1393} - 2 \cdot 0.1393^{1.5}}{1 + 14.9\sqrt{0.1393}}\right] = 0.05424$$

7.3 Activity Coefficient Models for Electrolyte Solutions

The Born terms are

$$\Delta_{Born} \ln \gamma^*_{Na^+} = \Delta_{Born} \ln \gamma^*_{OH^-} = 0$$

as water is a pure solvent.

For the calculation of the NRTL contributions, the following terms are necessary[3]:

$$C_{H_2O} = C_{Na^+} = C_{OH^-} = 1$$

$$\tau_{H_2O,NaOH} = C_{H_2O,NaOH} + \frac{D_{H_2O,NaOH}}{T} + E_{H_2O,NaOH}\left(\frac{298.15\ K - T}{T} + \ln\frac{T}{298.15\ K}\right)$$

$$= 6.738 + \frac{1420.242\ K}{373.15\ K} + 3.0139\left(\frac{298.15\ K - 373.15\ K}{373.15\ K} + \ln\frac{373.15\ K}{298.15\ K}\right) = 10.6146$$

$$\tau_{NaOH,H_2O} = C_{NaOH,H_2O} + \frac{D_{NaOH,H_2O}}{T} + E_{NaOH,H_2O}\left(\frac{298.15\ K - T}{T} + \ln\frac{T}{298.15\ K}\right)$$

$$= -3.771 + \frac{-471.82\ K}{373.15\ K} + 2.1366\left(\frac{298.15\ K - 373.15\ K}{373.15\ K} + \ln\frac{373.15\ K}{298.15\ K}\right) = -4.9854$$

$$\alpha_{NaOH,H_2O} = \alpha_{H_2O,NaOH} = 0.2$$

$$G_{H_2O,NaOH} = \exp(-0.2 \cdot \tau_{H_2O,NaOH}) = 0.1197$$

$$G_{NaOH,H_2O} = \exp(-0.2 \cdot \tau_{NaOH,H_2O}) = 2.7104$$

$$\tau_{ma,ca} = \tau_{am} - \tau_{ca,m} + \tau_{m,ca} = \tau_{ca,m} - \tau_{ca,m} + \tau_{m,ca} = \tau_{m,ca} = \tau_{H_2O,NaOH}$$

$$\tau_{mc,ac} = \tau_{cm} - \tau_{ca,m} + \tau_{m,ca} = \tau_{ca,m} - \tau_{ca,m} + \tau_{m,ca} = \tau_{m,ca} = \tau_{H_2O,NaOH}$$

The activity coefficients can be calculated in the following way:

- for water

$$\sum_j X_j G_{jm} \tau_{jm} = X_c G_{cm}\tau_{cm} + X_a G_{am}\tau_{am} + X_m G_{mm}\tau_{mm}$$

$$= 0.1393 \cdot 2.7105 \cdot (-4.9857) + 0.1393 \cdot 2.7105 \cdot (-4.9857)$$
$$+ 0 = -3.764$$

$$\sum_k X_k G_{km} = X_c G_{cm} + X_a G_{am} + X_m G_{mm}$$

$$= 0.1393 \cdot 2.7105 + 0.1393 \cdot 2.7105 + 0.7215 \cdot 1 = 1.4764$$

$$\sum_{k \neq c} X_k G_{kc,ac}\tau_{kc,ac} = X_a G_{ac,ac}\tau_{ac,ac} + X_m G_{mc,ac}\tau_{mc,ac}$$

$$= 0.1393 \cdot 1 \cdot 0 + 0.7215 \cdot 0.1197 \cdot 10.6146 = 0.9165$$

$$\sum_{k \neq a} X_k G_{ka,ca}\tau_{ka,ca} = X_c G_{ca,ca}\tau_{ca,ca} + X_m G_{ma,ca}\tau_{ma,ca}$$

$$= 0.1393 \cdot 1 \cdot 0 + 0.7215 \cdot 0.1197 \cdot 10.6146 = 0.9165$$

---

3) The parameters have been taken from the ASPEN Plus Parameter Databank.

$$\sum_{k \neq a} X_k G_{ka,ca} = X_c G_{ca,ca} + X_m G_{ma,ca}$$

$$= 0.1393 \cdot 1 + 0.7215 \cdot 0.1197 = 0.2256$$

$$\sum_{k \neq c} X_k G_{kc,ac} = X_a G_{ac,ac} + X_m G_{mc,ac}$$

$$= 0.1393 \cdot 1 + 0.7215 \cdot 0.1197 = 0.2256$$

Applying Eq. (7.86) yields

$$\ln \gamma_{H_2O,NRTL} = \frac{-3.764}{1.4764} + \frac{0.7215 \cdot 1}{1.4764}\left(0 - \frac{-3.764}{1.4764}\right)$$
$$+ \frac{0.1393}{0.1393}\frac{0.1393 \cdot 0.1197}{0.2256}\left(10.6146 - \frac{0.9165}{0.2256}\right)$$
$$+ \frac{0.1393}{0.1393}\frac{0.1393 \cdot 0.1197}{0.256}\left(10.6146 - \frac{0.9165}{0.2256}\right)$$
$$= -2.5494 + 1.24578 + 0.48406 + 0.48406 = -0.3355$$
$$\Rightarrow \gamma_{H_2O,NRTL} = 0.715$$

For water as the solvent, the activity coefficient at infinite dilution of water in itself is unity. Thus, Eq. (7.93) yields

$$\ln \gamma^*_{i,DH} = \ln \gamma_{i,DH} - \ln \gamma_i^\infty = \ln \gamma_{i,DH} \qquad (7.98)$$

Together with the Debye–Hückel contribution, one obtains

$$\ln \gamma_{H_2O} = \ln \gamma_{H_2O,DH} + \ln \gamma_{H_2O,NRTL}$$
$$= 0.0542 - 0.3355 = -0.2813 \Rightarrow \gamma_{H_2O} = 0.7548$$

- for the Na^+ cation (c) (Eq. (7.87)):

$$\frac{1}{1}\ln \gamma_{c,NRTL} = \frac{0.1393}{0.1393}\frac{0.9165}{0.2256} + \frac{0.7215 \cdot 2.7105}{1.4764}\left(-4.9857 - \frac{-3.764}{1.4764}\right)$$
$$+ \frac{0.1393}{0.1393}\frac{0.1393 \cdot 1}{0.2256}\left(0 - \frac{0.9165}{0.2256}\right) = -1.6721$$

$$\ln \gamma^*_{c,NRTL} = \ln \gamma_{c,NRTL} - \ln \gamma^\infty_{c,NRTL} = \ln \gamma_{c,NRTL} - (\tau_{mc,ac} + G_{cm}\tau_{cm})$$
$$= -1.6721 - (10.6146 + 2.7105 \cdot (-4.9857)) = 1.227$$

$$\ln \gamma^*_c = \ln \gamma^\infty_{c,DH} + \ln \gamma^*_{c,NRTL} = -1.00479 + 1.227 = 0.2222 \Rightarrow \gamma^*_c = 1.2488$$

- for the OH^- anion (a) (Eq. (7.88)):

$$\frac{1}{1}\ln \gamma_{a,NRTL} = \frac{0.1393}{0.1393}\frac{0.9165}{0.2256} + \frac{0.7215 \cdot 2.7105}{1.4764}\left(-4.9857 - \frac{-3.764}{1.4764}\right)$$
$$+ \frac{0.1393}{0.1393}\frac{0.1393 \cdot 1}{0.2256}\left(0 - \frac{0.9165}{0.2256}\right) = -1.6721$$

$$\ln \gamma^*_{a,\text{NRTL}} = \ln \gamma_{a,\text{NRTL}} - \ln \gamma^\infty_{a,\text{NRTL}} = \ln \gamma_{a,\text{NRTL}} - (\tau_{ma,ca} + G_{am}\tau_{am})$$
$$= -1.6721 - (10.6146 + 2.7105 \cdot (-4.9857)) = 1.227$$

$$\ln \gamma^*_a = \ln \gamma^*_{a,\text{DH}} + \ln \gamma^*_{a,\text{NRTL}} = -1.00479 + 1.227 = 0.2222 \Rightarrow \gamma^*_a = 1.2488$$

The equilibrium pressure can be evaluated by

$$P = \gamma_{H_2O,\text{NRTL}} \cdot x_{H_2O} \cdot P^S_{H_2O} = 0.7548 \cdot 0.7215 \cdot 1.0138 \text{ bar} = 0.5521 \text{ bar}$$

In the literature, the equilibrium pressure is reported to be $P = 0.551$ bar [20].

7.3.5
LIQUAC Model

Based on the UNIQUAC equation, an electrolyte model has been developed by Li et al. [21], called *LIQUAC*.

In the LIQUAC model, the excess Gibbs energy is calculated by three contributions taking into account the long-range (LR), the middle-range (MR), and the short-range (SR) interactions:

$$g^E = g^E_{LR} + g^E_{MR} + g^E_{SR} \qquad (7.99)$$

The LR term represents the interactions caused by the Coulomb electrostatic forces, expressed by a modified Debye–Hückel term. The physical validity of the term is limited to the very dilute region. The purpose of this term is just to provide the true limiting law at infinite dilution.

The activity coefficient caused by the LR interactions for the ions and the solvent is calculated by the following equations in the LIQUAC model:

$$\ln \gamma^{LR}_{\text{ion}} = \frac{-z_j^2 A I^{1/2}}{1 + bI^{1/2}} \qquad (7.100)$$

$$\ln \gamma^{LR}_{\text{solv}} = \left[\frac{2AM_{\text{solv}}\rho}{b^3 \rho_{\text{solv}}}\right]\left[1 + bI^{0.5} - \frac{1}{1 + bI^{0.5}} - 2\ln\left(1 + bI^{0.5}\right)\right] \qquad (7.101)$$

with the Debye–Hückel parameters A and b:

$$A = \frac{1.327757 \cdot 10^5 \left(\dfrac{\rho}{\text{kg/m}^3}\right)^{0.5}}{\left(\varepsilon \dfrac{T}{K}\right)^{1.5}} \qquad (7.102)$$

$$b = \frac{6.359696 \left(\dfrac{\rho}{\text{kg/m}^3}\right)^{0.5}}{\left(\varepsilon \dfrac{T}{K}\right)^{1.5}} \qquad (7.103)$$

For a binary mixture, ε can be calculated using Osters rule:

$$\varepsilon = \varepsilon_1 + \left[\frac{(\varepsilon_2 - 1)(2\varepsilon_2 + 1)}{2\varepsilon_2} - (\varepsilon_1 - 1)\right]\varphi_2 \qquad (7.104)$$

where the index 1 is water and 2 the other component.

For a multicomponent mixture, ε can be estimated by

$$\varepsilon = \sum_{solv} \varphi_{solv} \varepsilon_{solv} \qquad (7.105)$$

φ_i is the volume fraction of the solvent i, defined by

$$\varphi_i = \frac{V_i}{\sum_{solv} V_j} \qquad (7.106)$$

The ionic strength is given by Eq. (7.47):

$$I = 0.5 \sum_i \frac{m_i}{\text{mol/kg}} z_i^2$$

The MR term is the contribution of the indirect effects of the charge interactions, such as the charge–dipole interactions and charge-induced dipole interactions, to the excess Gibbs energy. In the LIQUAC model, the MR term is given by

$$\frac{G_{MR}^E}{RT} = m_{solv} \sum_i \sum_j B_{ij} m_i m_j \qquad (7.107)$$

where B_{ij} is the interaction coefficient between species i and j (ion or molecule), similar to the virial coefficient representing the indirect effects.

In the LIQUAC model B_{ij} represents all indirect effects caused by the charges, where the ionic strength dependence is described by the following simple relation:

$$B_{ij}(I) = b_{ij} + c_{ij} \exp(a_1 \cdot I^{0.5} + a_2 \cdot I) \qquad (7.108)$$

where b_{ij} and c_{ij} are the adjustable MR parameters between species i and j ($b_{ij} = b_{ji}, c_{ij} = c_{ji}$), a_1 and a_2 are empirical constants determined with the help of a few experimental data for electrolyte systems. The best results were obtained with the following equations:

$$B_{ion,ion} = b_{ion,ion} + c_{ion,ion} \exp(-I^{0.5} + 0.13 \cdot I) \qquad (7.109)$$

$$B_{ion,solv} = b_{ion,solv} + c_{ion,solv} \exp(-1.2 \cdot I^{0.5} + 0.13 \cdot I) \qquad (7.110)$$

This means the ion–ion interactions are different to the ion–solvent interactions.

In the LIQUAC model the interactions between like-charged ions are neglected in the MR part, so that Eq. (7.107) can be simplified to

$$\frac{G_{MR}^E}{RT} = m_{solv} \left[\sum_{solv}\sum_{ion} B_{solv,ion}(I) m_{solv} m_{ion} + \sum_c \sum_a B_{ca}(I) m_c m_a\right] \qquad (7.111)$$

where c covers all cations and a covers all anions.

7.3 Activity Coefficient Models for Electrolyte Solutions

By differentiating Eq. (7.111) with respect to the number of moles of solvent or ions, respectively, one obtains the following equations for the calculation of the MR part of the activity coefficient for the particular species:

$$\ln \gamma_{\text{solv}}^{\text{MR}} = \sum_{\text{ion}} B_{\text{solv,ion}}(I) m_{\text{ion}}$$

$$- \frac{M_{\text{solv}}}{\overline{M}} \sum_{\text{solv}} \sum_{\text{ion}} \left[B_{\text{solv,ion}}(I) + I \cdot B'_{\text{solv,ion}}(I) \right] x'_{\text{solv}} m_{\text{ion}}$$

$$- M_{\text{solv}} \sum_{c} \sum_{a} \left[B_{ca}(I) + I \cdot B'_{ca}(I) \right] m_c m_a \quad (7.112)$$

$$\ln \gamma_j^{\text{MR}} = \frac{\sum_{\text{solv}} B_{j,\text{solv}}(I) x_{\text{solv}}}{\overline{M}} + \left[\frac{z_j^2}{2\overline{M}} \right] \sum_{\text{solv}} \sum_{\text{ion}} B'_{\text{solv,ion}}(I) x'_{\text{solv}} m_{\text{ion}}$$

$$+ \sum_{a} B_{j,a}(I) m_a + \left[\frac{z_j^2}{2} \right] \sum_{c} \sum_{a} B'_{ca}(I) m_c m_a - \frac{B_{j,\text{solv}}(I)}{M_{\text{solv}}} \quad (7.113)$$

where $B'_{ij}(I) = dB_{ij}(I)/dI$, and x'_i is the salt-free mole fraction. $\overline{M}$ is the molecular weight of the solvent mixture calculated by

$$\overline{M} = \sum x'_{\text{solv}} M_{\text{solv}} \quad (7.114)$$

A subset of the LIQUAC interaction parameters required for the MR term is given in Table 7.4.

Table 7.4 Selected interaction parameters for the LIQUAC model [21].

Component i	Component j	a_{ij} (K)	a_{ji} (K)	b_{ij}	c_{ij}
H_2O	Na^+	219.4	−299.8	−7.432	1.576
H_2O	K^+	537.5	−452.3	−7.475	1.577
H_2O	Ca^{2+}	1137.0	−759.9	−19.84	3.149
H_2O	Cl^-	22.93	159.1	7.387	−1.576
H_2O	Br^-	22.41	282.5	5.952	−1.576
H_2O	NO_3^-	−17.99	347.6	7.462	−1.576
H_2O	SO_4^{2-}	−364.0	789.4	14.28	−3.151
Methanol	Na^+	495.5	601.3	5.702	1.198
Methanol	K^+	555.3	641.1	5.555	1.452
Methanol	Cl^-	−378.2	−231.6	−5.62	−1.006
Ethanol	Na^+	−172.2	−304.8	3.811	1.435
Ethanol	Cl^-	1519.0	−164.9	−3.62	−1.206
Acetone	K^+	−741.7	152.2	2.799	1.649
Na^+	Cl^-	89.17	−483.7	0.1925	0.1165
Na^+	SO_4^{2-}	269.3	−919.1	0.06311	0.3924
K^+	Cl^-	−417.2	244.7	0.09109	−0.1161
Ca^{2+}	Cl^-	339.0	0.0	0.4088	−0.2575

For the SR part in the LIQUAC model, the UNIQUAC equation is used (see Chapter 5), where the relative van der Waals surface areas and volumes of the ions were fixed to one. This means that the activity coefficients are split into a combinatorial (C) and a residual part (R):

$$\ln \gamma_i^{SR} = \ln \gamma_i^C + \ln \gamma_i^R \quad (7.115)$$

For the calculation of the temperature-independent combinatorial part only the van der Waals surface areas and volumes are required.

$$\ln \gamma_i^C = 1 - V_i + \ln V_i - 5 q_i \cdot \left[1 - \frac{V_i}{F_i} + \ln\left(\frac{V_i}{F_i}\right) \right] \quad (7.116)$$

For the calculation of the residual part, besides the surface area the interaction parameters are necessary:

$$\ln \gamma_i^R = q_i \left(1 - \ln \frac{\sum_j q_j x_j \tau_{ji}}{\sum_k q_k x_k} - \sum_j \frac{q_j x_j \tau_{ij}}{\sum_k q_k x_k \tau_{kj}} \right) \quad (7.117)$$

with

$$V_i = \frac{r_i}{\sum_k r_k x_k} \qquad F_i = \frac{q_i}{\sum_k q_k x_k} \qquad \tau_{ij} = \exp\left(-\frac{a_{ij}}{T}\right) \quad (7.118)$$

Finally, the activity coefficient of the solvent is calculated by

$$\ln \gamma_{solv} = \ln \gamma_{solv}^{LR} + \ln \gamma_{solv}^{MR} + \ln \gamma_{solv}^{SR} \quad (7.119)$$

where the activity coefficient is defined on the mole fraction scale.

For the ions, the SR term of the activity coefficient has to be normalized to the infinite dilution reference state using the relation:

$$\ln \gamma_{ion}^{SR} = \left(\ln \gamma_{ion}^C + \ln \gamma_{ion}^R \right) - \left(\ln \gamma_{ion}^{C,\infty} + \ln \gamma_{ion}^{R,\infty} \right) \quad (7.120)$$

For the combinatorial and the residual part of the UNIQUAC equation one obtains the following expressions for the values at infinite dilution:

$$\ln \gamma_{ion}^{C,\infty} = 1 - \frac{r_{ion}}{r_{solv}} + \ln\left(\frac{r_{ion}}{r_{solv}}\right) - 5 q_{ion} \left[1 - \frac{r_{ion} q_{solv}}{r_{solv} q_{ion}} + \ln\left(\frac{r_{ion} q_{solv}}{r_{solv} q_{ion}}\right) \right] \quad (7.121)$$

$$\ln \gamma_{ion}^{R,\infty} = q_{ion} \left(1 - \ln \tau_{solv,ion} - \tau_{ion,solv} \right) \quad (7.122)$$

For the calculation of the activity coefficient of an ion j at the chosen standard state the equation

$$\ln \gamma_j = \ln \gamma_j^{LR} + \ln \gamma_j^{MR} + \ln \gamma_j^{SR} - \ln\left(\frac{M_{solv}}{M} + M_{solv} \sum_{ion} m_{ion} \right) \quad (7.123)$$

is used, where molalities are used as the concentration measure and the hypothetical ideal solution of unit molality is the standard state for the ions.

7.3 Activity Coefficient Models for Electrolyte Solutions

Table 7.5 UNIQUAC parameters (K) for selected solvents (Li et al. [21]).

				m			
n	Water	Methanol	Ethanol	1-Propanol	2-Propanol	1-Butanol	Acetone
Water	0.0	289.6	162.4	130.0	83.6	−23.46	−45.90
Methanol	−181.0	0.0	−101.7	−13.20	39.7	40.39	−54.2
Ethanol	−14.5	130.2	0.0	−69.6	−176.5	162.3	–
1-Propanol	106.2	57.1	126.7	0.0	289.4	–	–
2-Propanol	110.5	−26.7	241.9	−200.9	0.0	–	–
1-Butanol	308.8	121.5	16.39	–	–	0.0	–
Acetone	337.9	223.8	–	–	–	–	0.0

Table 7.6 Relative van der Waals volumes (r_{solv}) and van der Waals surface areas (q_{solv}) for selected solvents (the relative van der Waals properties of the ions r_{ion} and q_{ion} were set to a value of 1).

Solvent	r_{solv}	q_{solv}
Water	0.92	1.4
Methanol	1.431	1.432
Ethanol	2.106	1.972
1-Propanol	2.780	2.512
2-Propanol	2.779	2.508
1-Butanol	3.454	3.052
Acetone	2.574	2.336

A part of the published LIQUAC parameters [21] and van der Waals properties are given in Tables 7.5 and 7.6.

Example 7.3

Predict the mean activity coefficient of Na_2SO_4 in a 1.5 molal aqueous solution at 298.15 K using the LIQUAC model. For the calculation the following properties should be used:

$$\rho_{H_2O} = 994.99 \text{ kg/m}^3 \quad \varepsilon_{r,H_2O} = 78.248 \quad M = 18.015 \text{ g/mol}$$

The other required parameters can be found in Tables 7.4–7.6.

Solution

In the first step, the LR contributions for the ions to the activity coefficients should be calculated using Eq. (7.100). For this purpose, the Debye–Hückel parameters A

and b and the ionic strength are required:

$$z_{Na^+} = 1 \quad z_{SO_4^{2-}} = -2 \quad m_{Na^+} = 3 \text{ mol/kg} \quad m_{SO_4^{2-}} = 1.5 \text{ mol/kg}$$

$$A = 1.327757 \cdot 10^5 \frac{994.99^{0.5}}{(78.248 \cdot 298.15)^{1.5}} = 1.1754$$

$$b = 6.359696 \frac{(994.99)^{0.5}}{(78.248 \cdot 298.15)^{0.5}} = 1.3134$$

$$I = 0.5 \left(1^2 \cdot 3 + 2^2 \cdot 1.5\right) = 4.5$$

For the Na^+ ions one obtains

$$\ln \gamma_{Na^+}^{LR} = -\frac{1.1754 \cdot 1^2 \cdot 4.5^{0.5}}{1 + 1.3134 \cdot 4.5^{0.5}} = -0.6585$$

In the same way the value for SO_4^{2-} can be calculated.

$$\ln \gamma_{SO_4^{2-}}^{LR} = -\frac{1.1754 \cdot (-2)^2 \cdot 4.5^{0.5}}{1 + 1.3134 \cdot 4.5^{0.5}} = -2.63415$$

In the next step, the MR term for the ions should be calculated using Eq. (7.113):

$$B_{Na^+,SO_4^{2-}} = b_{ion,ion} + c_{ion,ion} \exp\left(-I^{0.5} + 0.13 I\right)$$
$$= 0.06311 + 0.3924 \cdot \exp\left(-4.5^{0.5} + 0.13 \cdot 4.5\right) = 0.14754$$

$$B'(I) = \frac{dB(I)}{dI} = \left(\frac{-0.5}{I^{0.5}} + 0.13\right) c_{ion,ion} \exp\left(-I^{0.5} + 0.13 I\right)$$

For the MR term of Na^+ one obtains

$$\ln \gamma_{Na^+}^{MR} = \frac{1^2}{2 \cdot 0.018015} \left(1 \cdot 3 \cdot (-0.0339) + 1 \cdot 1.5 \cdot 0.0678\right) + 1.5 \cdot 0.14754$$
$$+ \frac{1^2}{2} \left(1.5 \cdot 3 \cdot (-0.00892)\right) = 0.2004$$

For SO_4^{2-} the following value is obtained:

$$\ln \gamma_{SO_4^{2-}}^{MR} = 0.3587$$

In the last step, the SR term is calculated using the UNIQUAC equation. The required van der Waals properties and interaction parameters are summarized in the table below.

| | | | | a_{nm} (K) | |
n	r	q	Na^+	SO_4^{2-}	H_2O
Na^+	1	1	0	269.3	−299.8
SO_4^{2-}	1	1	−919.1	0	798.4
H_2O	0.92	1.4	219.4	−364	0

7.3 Activity Coefficient Models for Electrolyte Solutions

The required parameters τ_{nm} can be calculated directly for the considered temperature of 298.15 K using the interaction parameters a_{nm}.

		τ_{nm} at 298.15 K	
n	Na$^+$	SO$_4^{2-}$	H$_2$O
Na$^+$	1	0.40526	2.7334
SO$_4^{2-}$	21.817	1	0.07082
H$_2$O	0.4791	3.3901	1

For the given composition the following surface fractions, volume/mole fractions and surface area/mole fractions are obtained for the given surface areas, volumes, and composition:

	x_i	V_i	F_i
Na$^+$	0.0500	1.0799	0.7299
SO$_4^{2-}$	0.0250	1.0799	0.7299
H$_2$O	0.9250	0.9935	1.0219

which allows for the determination of both the combinatorial and the residual contribution to obtain the activity coefficients for the mixture and at infinite dilution:

$$\ln \gamma_{Na^+}^C = 1 - 1.0799 + \ln(1.0799)$$
$$- 5 \cdot 1 \cdot \left[1 - \frac{1.0799}{0.7299} + \ln\left(\frac{1.0799}{0.7299}\right)\right] = 0.43593$$

$$\ln \gamma_{Na^+}^R = 1 \cdot \left[1 - \ln \frac{1 \cdot 0.5 \cdot 1 + 1 \cdot 0.025 \cdot 21.817 + 1.4 \cdot 0.925 \cdot 0.4791}{1 \cdot 0.05 + 1 \cdot 0.025 + 1.4 \cdot 0.925}\right.$$
$$- \left(\frac{1 \cdot 0.05 \cdot 1}{1 \cdot 0.05 \cdot 1 + 1 \cdot 0.025 \cdot 21.817 + 1.4 \cdot 0.925 \cdot 0.4791}\right.$$
$$+ \frac{1 \cdot 0.025 \cdot 0.40526}{1 \cdot 0.05 \cdot 0.40526 + 1 \cdot 0.025 \cdot 1 + 1.4 \cdot 0.925 \cdot 3.3901}$$
$$\left.\left.+ \frac{1.4 \cdot 0.925 \cdot 2.7334}{1 \cdot 0.05 \cdot 2.7334 + 1 \cdot 0.025 \cdot 0.07082 + 1.4 \cdot 0.925 \cdot 1}\right)\right]$$
$$= -1.3934$$

$$\ln \gamma_{Na^+}^{C,\infty,SR} = 1 - \frac{1}{0.92} + \ln\left(\frac{1}{0.92}\right)$$
$$- 5 \cdot 1 \cdot \left[1 - \frac{1 \cdot 1.4}{0.92 \cdot 1} + \ln\left(\frac{1 \cdot 1.4}{0.92 \cdot 1}\right)\right] = 0.50585$$

$$\ln \gamma_{Na^+}^{R,\infty,SR} = 1 \cdot (1 - \ln(0.4791) - 2.7334) = -0.99755$$

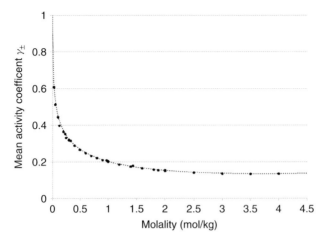

Figure 7.6 Experimental and predicted mean ion activity coefficients of Na$_2$SO$_4$ using LIQUAC at 298.15 K: • experimental [22].

$$\ln \gamma_{Na^+}^{*,SR} = 0.43593 - 1.3934 - 0.50585 + 0.99755 = -0.46577$$

Still, the correction term has to be taken into account to convert the activity coefficient to the molality scale:

$$\ln \gamma_{Na^+}^{*} = -0.6585 + 0.2004 - 0.46577$$
$$- \ln \left(\frac{0.018015}{0.018015} + 0.018015 \cdot 4.5 \right) = -1.002$$

$$\gamma_{Na^+}^{*} = 0.3671$$

In the same way the activity coefficient for the ion SO$_4^{2-}$ can be calculated:

$$\gamma_{SO_4^{2-}}^{*} = 0.03789$$

so that all information is available to determine the mean activity coefficient for Na$_2$SO$_4$:

$$\gamma_{\pm Na_2SO_4} = \sqrt[\nu]{\gamma_+^{\nu_+} \gamma_-^{\nu_-}} = \sqrt[3]{0.3671^2 \cdot 0.03789^1} = 0.1722$$

This value is in good agreement with the experimental value, as can be seen from Figure 7.6, where the available mean activity coefficients of Na$_2$SO$_4$ are shown as a function of the salt concentration together with the predicted values using the LIQUAC model at a temperature of 298.15 K.

Example 7.4

For the results shown in Example 7.3, the osmotic coefficient of water should be calculated for the same composition and temperature.

Solution

For the calculation of the osmotic coefficient, the different parts (LR, MR, and SR) have to be calculated with the help of the LIQUAC equation using the properties and parameters before.

For the different parts one obtains

$$\ln \gamma_{H_2O}^{SR} = 0.02023$$
$$\ln \gamma_{H_2O}^{MR} = -0.00856$$
$$\ln \gamma_{H_2O}^{LR} = 0.01606$$

Using these values directly, the activity coefficient and the osmotic coefficient for water can be calculated using Eq. (7.119):

$$\ln \gamma_{H_2O} = 0.02023 - 0.00856 + 0.01606 = 0.02769$$
$$\gamma_{H_2O} = 1.0281$$

Using Eq. (7.42) for the osmotic coefficient ϕ, a value of

$$\Phi = -\frac{\ln a_{H_2O}}{M_{H_2O} \sum_{j \neq H_2O} m_j} = -\frac{\ln(0.925 \cdot 1.0281)}{18.015 \frac{g}{mol}(1.5 + 3)\frac{mol}{kg}} = 0.6198$$

This value is in good agreement with the experimental values, as can be seen from Figure 7.7, where experimental osmotic coefficients for water are shown as a function of the Na_2SO_4-concentration for a temperature of 298.15 K. The corresponding group contribution model based on the UNIFAC method is called *LIFAC* and has been proposed by Yan *et al.* [23]. The current version of both models (mod. LIQUAC and mod. LIFAC, respectively) have been described by Kiepe *et al.* [24]. Its particular strength is the accurate calculation of mean activity coefficients of strong electrolytes up to high concentrations.

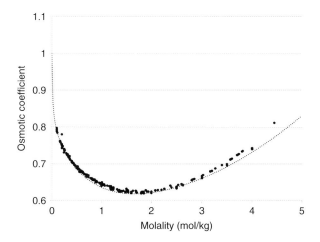

Figure 7.7 Osmotic coefficient of water as a function of the Na_2SO_4-concentration at 298 K: ● experimental [22].

7.3.6
MSA Model

During the last decade, the statistical thermodynamics of electrolytes have been continuously developed. The MSA theory (Mean Spherical Approximation) can yield analytical expressions for parameters which have a certain physical meaning (e.g., ionic diameter). To maintain the advantages of the NRTL electrolyte model and overcome its difficulties, the MSA theory has been successfully combined with the NRTL equation by Kunz and his co-workers [25].

The Born term is no more necessary, as the reference state for the ions is not the pure aqueous solution but the infinite dilution of ions in the solvent mixture. Thus, the MSA-NRTL has only two terms:

$$\ln \gamma_i^* = \ln \gamma_i^{*,\text{NRTL}} + \ln \gamma_i^{*,\text{MSA}} \qquad (7.124)$$

The exact calculation equations are given in [25], where it has also been proved that the Gibbs–Duhem equation is fulfilled. As well, NRTL parameters have been fitted up to molalities of 30 mol/kg for a number of systems. Together with the ionic diameters, they are listed in [25]. Osmotic and mean ionic activity coefficients could be reproduced in an excellent way for a number of systems. Furthermore, the parameters fitted to binary systems have been successfully applied to ternary systems, that is, one salt in a binary solvent mixture, which always causes problems with the Electrolyte NRTL model [25].

7.4
Dissociation Equilibria

In principle, the dissociation of electrolytes obeys to the law of mass actions as described in Chapter 12 for conventional chemical reactions. For evaluation, the Gibbs energies of formation of the compounds involved are needed. For some of the most important electrolyte compounds, they are listed in Appendix D.

Nowadays, in modern process simulation programs the possible electrolyte reactions and the equations for equilibrium calculation are generated automatically. The user only has to check the reactions in order to simplify the chemistry as much as possible, that is, to eliminate equilibrium reactions with extremely high or low equilibrium constants, as in these cases the equilibrium will be completely on one side of the reaction. Considering these reactions as equilibrium reactions will often yield to a drastically increased calculation time and has often a bad influence on the convergence.

The following section will outline the important foundations of the particular kinds of dissociation equilibria to ensure an appropriate interpretation of process simulation results with electrolytes.

The most profound electrolyte reaction is the dissociation of water itself, although it can usually be neglected in process simulations. Nevertheless, it is necessary for an understanding of the pH value, which is an interesting quantity for chemical considerations and for material selection.

For simplicity, the water dissociation is formulated as

$$H_2O(L) \rightleftarrows H^+ + OH^-$$

In this consideration, it is neglected that the proton (H^+) cannot exist in an aqueous solution. It is always attached to a water molecule, giving an H_3O^+ complex.

The chemical equilibrium can be written as

$$K_{H_2O}(T) = \exp\left[-\frac{\mu_{H^+}^{*m} + \mu_{OH^-}^{*m} - \mu_{L,H_2O}}{RT}\right] = \frac{a_{H^+}^{*m} a_{OH^-}^{*m}}{a_{L,H_2O}} \quad (7.125)$$

with

$$\mu_{L,H_2O} = \Delta g_{f,H_2O}^0 = -237\,129 \text{ J/mol}$$
$$\mu_{OH^-}^{*m} = \Delta g_{f,OH^-}^0 = -157\,244 \text{ J/mol}$$
$$\mu_{H^+}^{*m} = \Delta g_{f,H^+}^0 = 0$$

The dissociation constant for the molar case can be determined to be

$$K_{H_2O}(T_0) = \exp\left(\frac{-\Delta g_R^0}{RT}\right) = \exp\frac{-79885}{8.31433 \cdot 298.15}$$
$$= \frac{a_{H^+}^{*m} a_{OH^-}^{*m}}{a_{L,H_2O}} = 1.0105 \cdot 10^{-14}$$

at $T_0 = 298.15$ K. Obviously, only a low amount of the water is dissociated. Thus, the water can be taken as a pure component with $a_{L,H_2O} = 1$, and the ions can be regarded to be ideally diluted with $\gamma_{H^+}^{*m} = \gamma_{OH^-}^{*m} = 1$, giving at $T = T_0$

$$m_{H^+} m_{OH^-} = 1.011 \cdot 10^{-14} \frac{\text{mol}^2}{\text{kg}^2} \quad (7.126)$$

and, as

$$m_{H^+} = m_{OH^-} \quad (7.127)$$

$$m_{H^+} = m_{OH^-} = 1.005 \cdot 10^{-7} \frac{\text{mol}}{\text{kg}} \quad (7.128)$$

Since the dissociation is endothermic, K_{H_2O} increases with temperature, leading to higher dissociation rates.

The pH value is defined as follows:

$$\text{pH} = -\log a_{H^+}^{*m} \quad (7.129)$$

At 298.15 K, pH $= 7$ refers to pure water. If acids are added, additional H^+ ions are introduced, lowering the concentration of OH^- according to Eq. (7.126). Thus, the pH will decrease. If alkaline substances are added, the OH^- ions introduced lower the H^+ concentration and therefore increase the pH value.

Commercial simulators evaluate the pH value by calculating the concentration of H^+ ions from solving the equilibrium equations of the electrolyte

Example 7.5

Explain the removal of a $CaSO_4$ layer in a heat exchanger by stepwise adding Na_2CO_3 and HCl described in [26]. The influence of the crystal water (giving $CaSO_4 \cdot 2\,H_2O$) can be neglected.

Solution

The solubility products of $CaSO_4$ and $CaCO_3$ at $\vartheta = 25\,°C$ are [27]

$$K_{sp}(CaCO_3) = 10^{-8}\,(mol/kg)^2$$
$$K_{sp}(CaSO_4) = 6.1 \cdot 10^{-5}\,(mol/kg)^2$$

At the beginning, the aqueous solution is saturated with $CaSO_4$ according to the solubility product. Adding Na_2CO_3, the solubility product of $CaCO_3$ is exceeded almost immediately. A layer of solid $CaCO_3$ will be formed. Concerning $CaSO_4$, the solution is no more saturated, as most of the Ca^{2+} ions have been removed from the aqueous solution. Therefore, additional $CaSO_4$ will be dissolved, and the cycle starts again. At the end, almost all of the Ca^{2+} ions remain in the solid $CaCO_3$ layer according to its solubility product, whereas the sulfate ions are completely dissolved in the aqueous solution. There is no problem to remove the $CaCO_3$ layer from the heat exchanger surface. After addition of hydrochloric acid, the Ca^{2+} ions are dissolved according to

$$CaCO_3(S) + 2H^+ + 2Cl^- \longrightarrow Ca^{2+} + H_2O + CO_2(G) + 2Cl^-$$

Since the gaseous CO_2 is removed from the solution by degassing, this reaction is not an equilibrium limited reaction.

7.5
Influence of Salts on the Vapor-Liquid Equilibrium Behavior

The solubility of organic components in water can often be considerably influenced by adding electrolytes. For example, if 1 mol sodium chloride is added to 1 kg of water, the solubility of benzene is reduced by approx. 30% and that of nitrogen by approx. 25% [28]. This reduction of solubility is called *"salting out."* As well, a "salting in" effect exists. For instance, the solubility of methane in water can be increased by adding alkyl ammonium salts [28].

Figure 7.8 schematically illustrates the salting out effect. The addition of an electrolyte means in this case that molecules of solvent 1 are attracted by the electrolyte. Therefore, the number of solvent 1 molecules available for dissolving solvent 2 molecules decreases. Solvent 1 molecules condense from the vapor phase, giving an increase in the liquid concentration of solvent 1 and an increase

7.5 Influence of Salts on the Vapor-Liquid Equilibrium Behavior

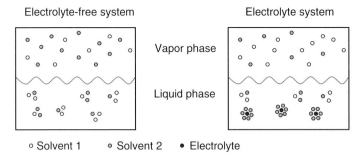

Figure 7.8 Influence of salts on the vapor–liquid equilibrium behavior.

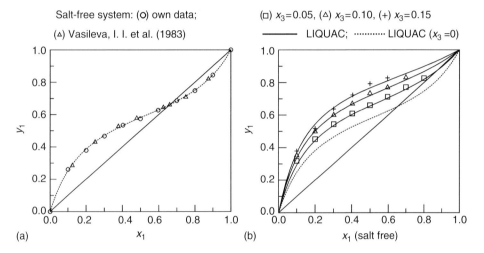

Figure 7.9 Vapor–liquid equilibrium diagram for the ternary system methyl acetate (1)–methanol (2)–LiNO$_3$ (3) at 101.33 kPa. (Experimental data from Vasileva et al. [30] and Topphoff et al. [31].)

in the vapor concentration of solvent 2. In other words, solvent 2 has been salted out.

Figure 7.9 shows a possible technical application of the salting out effect. At ambient pressure, methyl acetate–methanol forms a narrow boiling system with an azeotrope that is difficult to separate (Figure 7.9a). By adding approx. 10 mol% lithium nitrate, the azeotrope vanishes due to the salting out effect of the methyl acetate (Figure 7.9b), and it can be separated from methanol by distillation. The drawback of this process is that in the next step the methanol must be separated from the salt, which is easy with respect to thermodynamics but requires solids processing. Nevertheless, similar process steps are discussed in the literature [29].

7.6
Complex Electrolyte Systems

Thermodynamic calculations of solutions containing several substances that form electrolyte equilibria require a very careful approach. In contrast to conventional vapor–liquid equilibrium (VLE) calculations, the knowledge about the subsystems does not necessarily lead to a satisfactory representation. Moreover, even new components can be formed which do not occur in the subsystems.

A well-known example is the determination of the simultaneous solubility of carbon dioxide and ammonia in salt-containing waste water systems. As one sour gas (CO_2) and one alkaline gas (NH_3) are involved, they can react to nonvolatile ions, which considerably increase the solubility of these gases. Figure 7.10 gives a listing of the reactions occurring in the system CO_2–NH_3–H_2O. The electrolyte character of the mixture is mainly determined by the formation of the carbamate ion NH_2COO^-, which could not be predicted by conventional electrolyte dissociation reactions.

Similar reactions take place in the systems CO_2–MEA (monoethanolamine)–H_2O and CO_2–MDEA (methyldiethanolamine)–H_2O, which are technically important in the field of CO_2 removal from flue gases, giving a carbamate ion for the MEA reaction

$$2NH_2CH_2CH_2OH + CO_2 \rightleftharpoons NH_3^+CH_2CH_2OH + COO^-NHCH_2CH_2OH$$

and a normal dissociation reaction of the carbon dioxide in the MDEA reaction:

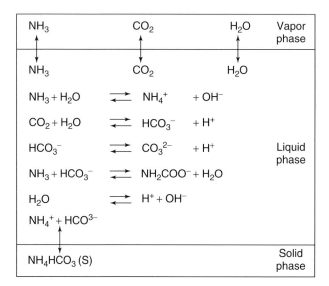

Figure 7.10 Phase and reaction equilibria in the system CO_2–NH_3–H_2O.

$$CH_3N(CH_2CH_2OH)_2 + H_2O + CO_2 \rightleftharpoons CH_3NH^+(CH_2CH_2OH)_2 + HCO_3^-$$

Details are explained in [33–36].
More information about the parameterization of complex electrolyte equilibria can be taken from [36–42].

Additional Problems

P7.1 Calculate the mole fractions of all true species in a 0.1 molal aqueous solution of sodium thiosulfate assuming complete dissociation. Is this a valid assumption?

P7.2 Calculate the mean ionic molality (Eq. (7.23)) and the molal ionic strength (Eq. (7.47)) of a
- 1 molal aqueous solution of hydrogen chloride
- 0.01 molal ethanolic solution of sulfuric acid

assuming total dissociation of the electrolyte compound.

P7.3 Search the DDB (free Explorer Version) for osmotic coefficients for the system water–sodium chloride. From the data at 333 K, calculate the activity coefficient of the solvent water as function of the molality of the salt.

P7.4 Search the DDB (free Explorer Version) for vapor-liquid equilibrium data for the system methanol–sodium chloride. Calculate the activity coefficient of methanol as function of salt concentration.

P7.5 Search the DDB (free Explorer Version) for salt solubility data of sodium chloride in the system potassium chloride–water. Plot and interpret the experimental data.

P7.6 Calculate the saturation pressure of an aqueous NaOH solution at $\vartheta = 100\,°C$ at NaOH concentrations of 10, 30, and 50 wt%. Use the Electrolyte NRTL model. The necessary parameters are given in Example 7.2.

P7.7 Calculate the saturation pressure of an aqueous H_2SO_4 solution at $\vartheta = 100\,°C$ at H_2SO_4 concentrations of 10, 30, and 50 wt%. Use the Electrolyte NRTL model. The required parameters can be downloaded from the textbook page on www.ddbst.com.

P7.8 Calculate the liquid heat capacity of a 20 wt% NaCl solution at $\vartheta = 80\,°C$ using the Electrolyte NRTL model. The required parameters can be downloaded from the textbook page on www.ddbst.com.

P7.9 Calculate the saturation pressure of a 20 wt% ammonia solution using the NRTL equation and the Electrolyte NRTL model at $\vartheta = 20\,°C$. An ideal vapor phase can be assumed. The pK_b value of ammonia at 25 °C is 4.75. Further required parameters can be downloaded from the textbook page on www.ddbst.com. Do you think that the use of the electrolyte model is necessary in this case? How is the situation affected if significant amounts of CO_2 are added to the system?

References

1. Luckas, M. and Krissmann, J. (2001) *Thermodynamik der Elektrolytlösungen*, Springer, Berlin, Heidelberg.
2. Polka, H.M. (1993) Experimentelle Bestimmung und Berechnung von Dampf-Flüssig-Gleichgewichten für Systeme mit starken Elektrolyten. Thesis. University of Oldenburg, Germany.
3. Maurer, G. (2004) Electrolyte solutions, in *Thermodynamics for Environment* (eds P. Giericz and S. Malanowski), Information Processing Centre, Warszawa.
4. Moore, W.J. and Hummel, D.O. (1986) *Physikalische Chemie*, Walter de Gruyter, Berlin, New York.
5. Bromley, L.A. (1973) *AIChE J.*, **19**, 313–320.
6. Pitzer, K.S. (1973) *J. Phys. Chem.*, **77**, 268–277.
7. Pitzer, K.S. and Mayorga, G. (1973) *J. Phys. Chem.*, **77**, 2300–2308.
8. Pitzer, K.S. (1977) *Acc. Chem. Res.*, **10**, 371.
9. Pitzer, K.S. and Kim, J.J. (1974) *J. Am. Chem. Soc.*, **78**, 5701–5706.
10. Pitzer, K.S. (1979) *Activity Coefficients in Electrolyte Solutions*, Chapter 7, vol. 1, CRC Press, Boca Raton.
11. Rosenblatt, G.M. (1981) *AIChE J.*, **27**, 619–626.
12. Zemaitis, J.F., Clark, D.M., Rafal, M., and Scrivner, N.C. (1986) *Handbook of Aqueous Electrolyte Thermodynamics*, AIChE, New York.
13. Pitzer, K.S. (1991) *Activity Coefficients in Electrolyte Solutions*, CRC Press, Boca Raton.
14. Austgen, D.M., Rochelle, G.T., Peng, X., and Chen, C.C. (1989) *Ind. Eng. Chem. Res.*, **28**, 1060–1073.
15. Chen, C.C., Britt, H.I., Boston, J.F., and Evans, L.B. (1982) *AIChE J.*, **28**, 588–596.
16. Chen, C.C. and Evans, L.B. (1986) *AIChE J.*, **32**, 444–454.
17. Renon, H. and Prausnitz, J.M. (1968) *AIChE J.*, **14**, 135–145.
18. Pitzer, K.S. (1980) *J. Am. Chem. Soc.*, **102**, 2902–2906.
19. en.wikipedia.org/wiki/Relative_permittivity. January 2011.
20. Krey, J. (1972) *Z. Phys. Chem. Neue Folge*, **81**, 252–273.
21. Li, J., Polka, H.M., and Gmehling, J. (1994) *Fluid Phase Equilib.*, **94**, 89–114.
22. Dortmund Data Bank www.ddbst.com. January 2011.
23. Yan, W.D., Topphoff, M., Rose, C., and Gmehling, J. (1997) *Fluid Phase Equilib.*, **162**, 97–113.
24. Kiepe, J., Noll, O., and Gmehling, J. (2006) *Ind. Eng. Chem. Res.*, **45**, 2361–2373.
25. Papaiconomou, N., Simonin, J.P., Bernard, O., and Kunz, W. (2002) *Phys. Chem. Chem. Phys.*, **4**, 4435–4443.
26. Neumaier, R. (1971) *Handbuch neuzeitlicher Pumpenanlagen*. Lederle oHG, 3rd edition.
27. Rauscher, K., Voigt, J., Wilke, I., and Wilke, K.T. (1982) *Chemische Tabellen und Rechentafeln für die analytische Praxis*, 7th edn, Verlag Harri Deutsch, Frankfurt.
28. Maurer, G. (1983) *Fluid Phase Equilib.*, **13**, 269–296.
29. Ligero, E.L. and Ravagnani, T.M.K. (2003) *Chem. Eng. Proc.*, **42**, 543–552.
30. Vasileva, I.I., Marinicheva, A.N., and Susarerv, M.P. (1983) *Vestn. Leningr. Univ. Fiz. Khim.*, 113–114.
31. Topphoff, M., Kiepe, J., and Gmehling, J. (2001) *J. Chem. Eng. Data*, **46**, 1333–1337.
32. Lemoine, B., Li, Y.G., Cadours, R., Bouallou, C., and Richon, D. (2000) *Fluid Phase Equilib.*, **172**, 261–277.
33. Rolker, J. and Arlt, W. (2006) *Chem. Ing. Tech.*, **78**, 416–424.
34. Rudolph, J. (2006) *Chem. Ing. Tech.*, **78**, 381–388.
35. Bishnoi, S. and Rochelle, G.T. (2002) *Ind. Eng. Chem. Res.*, **41**, 604–612.
36. Jödecke, M. (2004) Experimentelle und theoretische Untersuchungen zur Löslichkeit von Kohlendioxid in wässrigen, salzhaltigen Lösungen mit organischen Komponenten. Thesis. Shaker Verlag.
37. Jödecke, M., Xia, J., Perez-Salado Kamps, A., and Maurer, G. (2003) *Chem. Ing. Tech.*, **75**, 376–379.

38. Bieling, V., Rumpf, B., Strepp, F., and Maurer, G. (1989) *Fluid Phase Equilib.*, **53**, 251–259.
39. Göppert, U. and Maurer, G. (1988) *Fluid Phase Equilib.*, **41**, 153–185.
40. Lichtfers, U. and Rumpf, B. (2000) *Chem. Ing. Tech.*, **72**, 1526–1530.
41. Müller, G., Bender, E., and Maurer, G. (1988) *Ber. Bunsenges. Phys. Chem.*, **92**, 148–160.
42. Xia, J., Rumpf, B., and Maurer, G. (1999) *Fluid Phase Equilib.*, **165**, 99–119.

8
Solid–Liquid Equilibria

Besides distillation, absorption and extraction, crystallization processes are often applied in separation sequences. This is particularly the case if the components have a very low vapor pressure (e.g., salts, active pharmaceutical ingredients (APIs)) or decompose at higher temperature. Furthermore, crystallization has advantages compared to distillation if the separation factor shows values near unity and cannot be influenced by an entrainer as in the case of isomers, for example, m-xylene/p-xylene.

In process technology it can be distinguished between melt and solution crystallization. An example for melt crystallization is the separation of p-xylene from a xylene blend for the production of polyethylene terephthalate (PET). Solution crystallization can, for example, be applied for the purification of APIs with the help of a suitable solvent or solvent mixture or the crystallization of salts from the aqueous solution.

For the synthesis and design of crystallization processes, the reliable knowledge of the solid–liquid equilibrium (SLE) behavior, respectively, the solubility of solids in a solvent or solvent mixture is of special importance. Additionally, reliable kinetic information about metastable zone nucleation and crystal growth is required. Information about the SLE behavior, respectively, the driving forces is also required to avoid blocking of, for example, in pipelines or other process units.

The SLE behavior is more complex than for other phase equilibria, for example, vapor–liquid or liquid–liquid equilibria (LLE), since it may be necessary to take into account limited miscibility of the solid solutions, polymorphism, and solvate formation of the solid state, the formation of more or less stable intermolecular compounds in the solid state or the occurrence of LLE.

This means different types of SLE have to be distinguished. The type of SLE mainly depends on the mutual solubility of the components in the solid phase. There are systems with no, partial and total miscibility in the solid phase. The formation of solid solutions mainly depends on how well the lattice structure of compound 2 fits to the lattice structure of compound 1. Thus, complete solid solubility only occurs when the species have the same crystal structure and size.

There are two ways that solid solutions are formed. In a substitutional solid solution, compound 2 occupies the lattice sites of compound 1, as long as the crystal can accommodate compound 2 without altering its basic structure. For example, the system molybdenum–tungsten shows total miscibility in the solid phase. The reason is that both metals have a body center cubic crystal lattice with nearly the same distance to the nearest neighbors 2.72 resp. 2.73 Å. In general, a binary pair must have these characteristics to be completely miscible. Binary pairs with these characteristics are relatively rare and more likely in alloys (noble gases) than in mixtures of organic compounds.

As well, the real behavior in the liquid and solid phase has an influence on the SLE behavior. In some cases one or more solid compounds are formed. At the same time miscibility gaps can appear. Furthermore, many substances occur in more than one crystalline form. This phenomenon is called *polymorphism* for compounds and allotropy in the case of elements. If the transition curve lies within the region of stable states the phenomenon is called *enantiotropy*. In the metastable region it is called *monotropy*.

The different types of SLE, even the complex ones, are discussed in detail in [1, 2]. Some of the most important SLE diagrams are shown in Figure 8.1 in the form of Tx-diagrams, since condensed phases are not very sensitive to pressure.[1] In Figure 8.1a the SLE behavior of the simple eutectic system benzene–naphthalene is shown. This is the widest spread SLE behavior in the case of organic compounds. In such simple eutectic systems the solid phase does not form mixed crystals. In an eutectic system the eutectic temperature is lower than the melting points of the participating components. Simple eutectic systems with total immiscibility in the solid phase are of special importance in separation technology, since for these systems one theoretical stage is sufficient to obtain the pure compounds.

In metallurgy low melting eutectic mixtures are often used for soldering. The eutectic 4-component mixture with 50 wt% Bi, 10 wt% Cd, 26.7 wt% Pb, 13.3 wt% Sn, known as *Wood's metal* shows a very low eutectic temperature (~70 °C).

Besides eutectic systems also other SLE are known, where solid solutions are formed. The system anthracene–phenanthrene with nearly ideal mixture behavior of the liquid and solid phase shows total miscibility in the solid phase (see Figure 8.1b). With increasing real behavior in the participating phases the curvatures of the liquidus[2] and solidus line[3] become different. If the deviation from ideal behavior in the liquid or solid phase becomes strong enough, even extreme temperatures are obtained, for example, a temperature minimum is observed for the system naphthylamine–naphthalene shown in Figure 8.1c. A temperature minimum is also found for the system argon–methane (see Figure 8.1d). But different to the system naphthylamine–naphthalene the system argon–methane shows only partial miscibility in the solid phase. Besides systems with temperature minimum, also systems with a temperature maximum are observed. For example, the enantiomeric system D-carvoxime-L-carvoxime shows total miscibility in the

---

1) This is true for pressures up to a few bar. But higher pressures will lead to different melting points and eutectic compositions (see Section 3.1.4), what can be used to overcome weaknesses of crystallization processes by pressure swing crystallization.

2) The liquidus line is the locus of temperature in the Tx-diagram above which only the liquid is stable.

3) The solidus line is the locus of temperature in the Tx-diagram below which only the solid is stable.

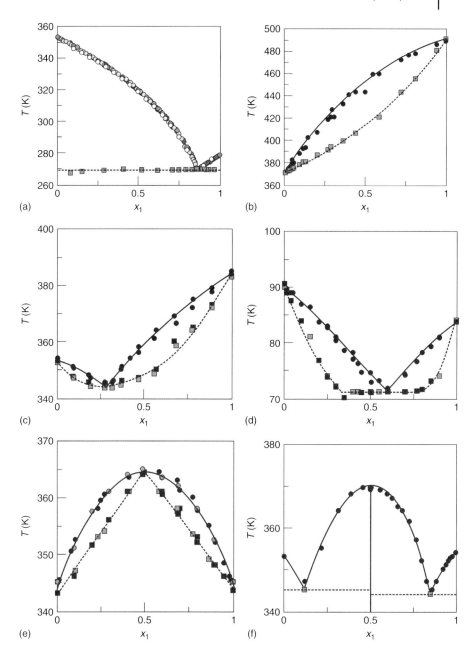

Figure 8.1 Selected types of SLE. (a) Benzene (1)–naphthalene (2); (b) anthracene (1)–phenanthrene (2); (c) 2-aminonaphthalene (1)–naphthalene (2); (d) argon (1)–methane (2); (e) D-carvoxime (1)–L-carvoxime (2); (f) naphthalene (1)–2,4,6-trinitrotoluene (2): ● liquidus line; ■ solidus line.

solid phase but a temperature maximum, which is caused by less positive deviations from Raoult's law in the solid than in the liquid phase (see Figure 8.1e). Sometimes also intermolecular compound formation in the solid state with a congruent melting point are formed. For the system naphthalene-2,4,6-trinitrotoluene the SLE behavior is shown in Figure 8.1f. The formation of intermolecular compounds in the solid state can be applied for the separation of systems by adductive crystallization. One example of industrial interest is the separation of *p*-xylene from *m*-xylene with the help of carbon tetrachloride, which forms a 1:1-complex with *p*-xylene. As well, the majority (>90% [3]) of the enantiomers form a racemic compound.

8.1
Thermodynamic Relations for the Calculation of Solid–Liquid Equilibria

As for all other phase equilibria, the calculation of SLE starts from the isofugacity condition (Eq. (8.1)):

$$f_i^L = f_i^S \tag{8.1}$$

where L is the index for the liquid and S the index for the solid phase.

The fugacity can be described with the help of activity coefficients and standard fugacities for the liquid phase L and the solid phase S.

$$x_i^L \gamma_i^L f_i^{0L} = x_i^S \gamma_i^S f_i^{0S} \quad \text{or} \quad \frac{x_i^L \gamma_i^L}{x_i^S \gamma_i^S} = \frac{f_i^{0S}}{f_i^{0L}} \tag{8.2}$$

In the case of simple eutectic systems (Figure 8.1a, complete immiscibility in the solid state) the solid will crystallize in pure form, this means, f_i^S is identically with f_i^{0S}:

$$x_i^L \gamma_i^L f_i^{0L} = f_i^{0S} \quad \text{or} \quad x_i^L \gamma_i^L = \frac{f_i^{0S}}{f_i^{0L}} \tag{8.3}$$

To be able to calculate the solubility in the liquid and solid phase, only the activity coefficients γ_i^L and γ_i^S and the ratio of the standard fugacities (f_i^{0S}/f_i^{0L}) of the hypothetical liquid and of the solid at system temperature is required. Since the standard fugacities f_i^{0L} and f_i^{0S} are nearly identical with the vapor pressure of the subcooled liquid and the sublimation pressure, the solubility at a given temperature can directly be calculated from the vapor pressure of the hypothetical (subcooled) liquid[4] and the sublimation pressure of the solid. For example, the vapor pressures and the sublimation pressures of naphthalene are shown in Figure 8.2 as a function of temperature. As can be seen from Figure 8.2, the ratio of the sublimation pressure below the melting temperature as shown in Figure 8.2.

---

4) The vapor pressure of the hypothetical liquid can be obtained by extrapolation of the vapor pressure curve to temperatures

8.1 Thermodynamic Relations for the Calculation of Solid–Liquid Equilibria

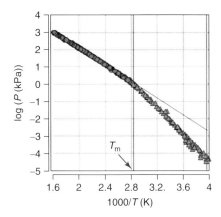

Figure 8.2 Vapor pressures and sublimation pressures of naphthalene – hypothetical liquid.

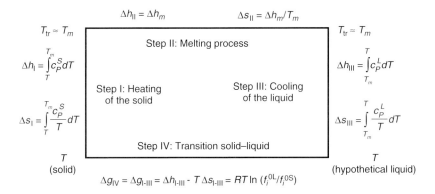

Figure 8.3 Thermodynamic cycle for the derivation of an expression for the ratio of standard fugacities f_i^{0L}/f_i^{0S}.

to the vapor pressure of the hypothetical liquid is equal to 1 at the melting point and becomes smaller than 1 at lower temperatures.

The required ratio of the standard fugacities (vapor pressures) can be obtained in a very convenient way by the thermodynamic cycle shown in Figure 8.3.

From fundamental thermodynamics, the ratio of the standard fugacities can directly be expressed by the change of the Gibbs energy (step IV):

$$\Delta g_{IV} = RT \ln \frac{f_i^{0L}}{f_i^{0S}} \tag{8.4}$$

The change of the Gibbs energy Δg_{IV} can also be obtained by the enthalpy and entropy changes of the steps I–III:

$$\Delta g_{IV} = \Delta g_{I-III} = \Delta h_{I-III} - T\Delta s_{I-III} \tag{8.5}$$

With the simplification that the melting temperature T_m is nearly identical with the triple point temperature T_{tr} and that the heat of fusion Δh_m is approximately identical with the enthalpy change of the solid–liquid phase transition at the triple point, the following expressions are obtained for the enthalpy and entropy change (steps I–III) for component i:

$$\Delta h_{\text{I-III}} = \int_T^{T_{m,i}} c_{P,i}^S \, dT + \Delta h_{m,i} + \int_{T_{m,i}}^T c_{P,i}^L \, dT = \Delta h_{m,i} + \int_{T_{m,i}}^T \Delta c_{P,i} \, dT$$

$$\Delta s_{\text{I-III}} = \int_T^{T_{m,i}} \frac{c_{P,i}^S}{T} \, dT + \frac{\Delta h_{m,i}}{T_{m,i}} + \int_{T_{m,i}}^T \frac{c_{P,i}^L}{T} \, dT = \frac{\Delta h_{m,i}}{T_{m,i}} + \int_{T_{m,i}}^T \frac{\Delta c_{P,i}}{T} \, dT$$

(8.6)

where $\Delta c_{P,i} = c_{P,i}^L - c_{P,i}^S$

Assuming that the heat capacity difference is constant, combining Eqs. (8.3–8.6) provides the general relation for the calculation of SLE:

$$-\ln \frac{x_i^L \gamma_i^L}{x_i^S \gamma_i^S} = \ln \frac{f_i^{0L}}{f_i^{0S}} = \frac{\Delta h_{m,i}}{RT} \left(1 - \frac{T}{T_{m,i}}\right) - \frac{\Delta c_{P,i}(T_{m,i} - T)}{RT} + \frac{\Delta c_{P,i}}{R} \ln \frac{T_m}{T}$$

(8.7)

In the case of simple eutectic systems, Eq. (8.7) simplifies to

$$-\ln x_i^L \gamma_i^L = \ln \frac{f_i^{0L}}{f_i^{0S}} = \frac{\Delta h_{m,i}}{RT} \left(1 - \frac{T}{T_{m,i}}\right) - \frac{\Delta c_{P,i}(T_{m,i} - T)}{RT} + \frac{\Delta c_{P,i}}{R} \ln \frac{T_m}{T}$$

(8.8)

Not far away from the melting point, the contributions of the heat capacity differences can be neglected, since the two contributions have the tendency to cancel out because of the different signs. After neglecting these terms, a very simple relation is obtained, where besides the activity coefficients γ_i only the enthalpy of fusion $\Delta h_{m,i}$ and the melting temperature $T_{m,i}$ are required for the calculation of SLE.

$$\ln \frac{x_i^L \gamma_i^L}{x_i^S \gamma_i^S} = -\frac{\Delta h_{m,i}}{RT} \left(1 - \frac{T}{T_{m,i}}\right)$$

(8.9)

In the case of eutectic systems one obtains

$$\ln x_i^L \gamma_i^L = -\frac{\Delta h_{m,i}}{RT} \left(1 - \frac{T}{T_{m,i}}\right)$$

(8.10)

8.1.1
Solid–Liquid Equilibria of Simple Eutectic Systems

Following Eq. (8.10), in the case of simple eutectic systems the solubility of a solid compound in the liquid phase x_i^L can be calculated for a given temperature, if besides the enthalpy of fusion and the melting temperature information about the real behavior, that is, the activity coefficient, is available. Following Eq. (8.10), even similar compounds, as for example, anthracene and phenanthrene, can show very different solubilities in the various solvents because of the different pure component properties Δh_m and T_m. The solubility of a compound will decrease with decreasing temperature, increasing melting temperature, increasing enthalpy of fusion and increasing positive deviation from Raoult's law.

If phase transitions in the solid phase occur additionally between the considered temperature and the melting temperature, these enthalpy and entropy effects have to be considered in the thermodynamic cycle (Figure 8.3). Neglecting the terms with the heat capacity differences, the following equation can be derived for simple eutectic systems:

$$\ln x_i^L \gamma_i^L = -\frac{\Delta h_{m,i}}{RT}\left(1 - \frac{T}{T_{m,i}}\right) - \sum \frac{\Delta h_{trs,i}}{RT}\left(1 - \frac{T}{T_{trs,i}}\right) \quad (8.11)$$

In this case, for the calculation of the solid solubilities x_i^L additional information about the transition temperature T_{trs} and enthalpy of transition Δh_{trs} is required.

Example 8.1

Calculate the solubility of anthracene and phenanthrene in benzene at 20 °C. For the calculation it should be assumed that the binary systems show ideal mixture behavior ($\gamma_i = 1$).

anthracene   phenanthrene

Pure component properties:

Compound	T_m (K)	Δh_m (J/mol)
Anthracene	489.6	28860
Phenanthrene	369.4	18640

Solution

- Anthracene:

$$\ln x_i = -\frac{28860}{8.31433 \cdot 293.15}\left(1 - \frac{293.15}{489.6}\right) = -4.7511$$

$$x_i = 0.0086$$

- Phenanthrene:

$$\ln x_i = -\frac{18640}{8.31433 \cdot 293.15}\left(1 - \frac{293.15}{369.4}\right) = -1.5785$$

$$x_i = 0.2063$$

From the results it can be recognized that the solubilities of the chemically very similar compounds anthracene and phenanthrene differ by more than one order of magnitude. The much lower solubility of anthracene is thereby caused by the higher melting temperature and larger enthalpy of fusion (lower standard fugacity f_i^{0S}).

Example 8.2

Calculate the SLE behavior of benzene (2) with the different xylene isomers (1) assuming ideal behavior of the liquid phase.

Component	T_m (K)	Δh_m (J/mol)
Benzene	278.68	9952
o-xylene	247.65	13604
m-xylene	225.35	11544
p-xylene	286.45	16790

Solution

The calculation of the SLE behavior is demonstrated for the system benzene (2)-p-xylene (1) at a temperature of $-10\,°C$.

- Solubility of benzene (2):

$$\ln x_2 = -\frac{9952}{8.31433 \cdot 263.15}\left(1 - \frac{263.15}{278.68}\right) = -0.2535$$

$$x_2 = 0.7761$$

- Solubility of p-xylene (1):

$$\ln x_1 = -\frac{16\,790}{8.31433 \cdot 263.15}\left(1 - \frac{263.15}{286.45}\right) = -0.6242$$

$$x_1 = 0.5357$$

In the same way, the solubilities for other temperatures and the other xylene isomers can be calculated. The results for all three systems are shown in Figure 8.4 together with the experimental results. It can be seen that not only the solubilities but also the eutectic points are in good agreement with the experimental findings. This means that the assumption of ideal behavior is reasonable for the systems considered.

In the case of nonideal systems, the real behavior has to be taken into account using activity coefficients obtained from g^E-models, e.g. group contribution methods like modified UNIFAC. The required activity coefficients can of course also be calculated using an equation of state or group contribution equation of state.

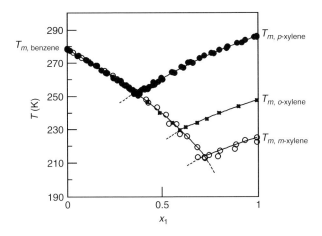

Figure 8.4 Experimental [4] and calculated SLE of benzene (2) with the different xylene isomers (1): —— liquidus line; - - - - unstable region.

Example 8.3

Calculate the solubility of benzene(2) in ethanol(1) at 260 K. The activity coefficients in the liquid phase should be taken into account by the UNIQUAC equation. The UNIQUAC parameters and melting data are:

$\Delta u_{12} = -43.0$ K	$r_1 = 2.1055$	$q_1 = 1.972$
$\Delta u_{21} = 384.09$ K	$r_2 = 3.1878$	$q_2 = 2.400$
$\Delta h_{m,2} = 9952$ J/mol	$T_{m,2} = 278.68$ K	

Solution

Using Eq. (8.10), the activity in the liquid phase at 260 K can be calculated:

$$\ln x_2 \gamma_2 = -\frac{9952}{8.31433 \cdot 260}\left(1 - \frac{260}{278.68}\right) = -0.3086$$

$$x_2 \gamma_2 = 0.7345$$

In the case of an ideal system ($\gamma_2 = 1$), a solubility of $x_2 = 0.7345$ would be obtained. For the determination of the correct benzene solubility at 260 K, the mole fraction x_2 has to be determined for which the product $x_2\,\gamma_2$ results in an activity of $a_2 = 0.7345$. The calculation must be performed iteratively, since the activity coefficient γ_2 depends on composition. For a temperature of 260 K the following activity coefficient and solubility is obtained after a few iterations:

$$\gamma_2 = 3.0176 \quad x_2 = 0.2434 \quad x_2\gamma_2 = 0.7345$$

The calculated SLE of the system ethanol–benzene in the whole composition range is shown in Figure 8.5, together with the experimental data. It can be recognized

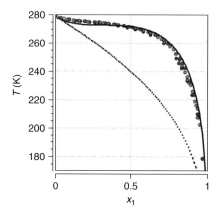

Figure 8.5 Experimental and calculated SLE behavior of the system ethanol (1)–benzene (2): ● experimental [4]; ···· ideal; —— UNIQUAC.

that the calculated results are in good agreement with the experimental findings. Furthermore, it becomes clear that the assumption of ideal behavior would lead to large errors.

Example 8.4

Construct the SLE diagram for the ternary system o-xylene–m-xylene–p-xylene assuming eutectic behavior. Since isomer compounds are considered, ideal behavior in the liquid phase can be assumed.

Compound	T_m (K)	Δh_m (J/mol)
o-xylene	247.65	13604
m-xylene	225.35	11544
p-xylene	286.45	16790

Solution

The ideal solubility is exemplarily calculated for p-xylene at a temperature of 260 K:

$$\ln x_3 = -\frac{16790}{8.31433 \cdot 260}\left(1 - \frac{260}{286.45}\right) = -0.7172$$

$$x_3 = 0.4881$$

In the case of ideal behavior this value is independent on the amounts of m-xylene and o-xylene. The results for the whole composition range are shown in Figure 8.6. From the diagram it can be seen that the lowest temperature (around 210 K) is obtained for the ternary eutectic point. It can also be seen that in a large

8.1 Thermodynamic Relations for the Calculation of Solid–Liquid Equilibria | 415

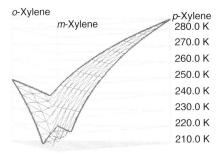

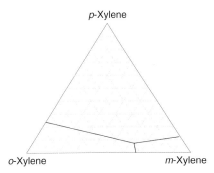

Figure 8.6 Predicted SLE diagram and eutectic lines for the ternary system o-xylene–m-xylene–p-xylene assuming ideal behavior.

composition range pure p-xylene crystallizes, which can be used in practice to separate p-xylene from a xylene blend.

Example 8.5

Calculate the SLE diagram of the eutectic system CCl_4(1)–n-octane(2) with the help of the modified UNIFAC model and the Wilson model. Please take into account that CCl_4 shows a phase transition at 225.35 K.

Pure component properties:

Component	T_m (K)	Δh_m (J/mol)	T_{trs} (K)	Δh_{trs} (J/mol)
CCl_4	250.77	2528.4	225.35	4560
n-octane	216.35	20638	–	–

Molar volumes and Wilson interaction parameters $\Delta\lambda_{ij}$:

Component	Molar volume (cm³/mol)	Interaction parameters (cal/mol)
(1) CCl_4	97.09	$\Delta\lambda_{12} = 33.81$
(2) n-octane	163.54	$\Delta\lambda_{21} = 166.1$

Solution

While for the calculation of the CCl$_4$ composition below the transition temperature T_{trs} Eq. (8.11) has to be applied, above this temperature Eq. (8.10) can be used.

In the first step, the solubility of carbon tetrachloride at 240 K is calculated:

$$\ln x_1\gamma_1 = -\frac{2528.4}{8.31433 \cdot 240}\left(1 - \frac{240}{250.77}\right) = -0.0544$$

$$x_1\gamma_1 = 0.9471$$

Assuming ideal behavior, the solubility would be calculated to be $x_1 = 0.9471$. Using the Wilson equation by an iterative procedure a value of $x_1 = 0.9454$ ($\gamma_1 = 1.0017$) is obtained. This means at the high concentration only a small influence of the real behavior is observed.

In the next step, one solubility point below the transition temperature at 215 K is calculated. Starting from Eq. (8.11):

$$\ln x_1\gamma_1 = -\frac{2528.4}{8.31433 \cdot 215}\left(1 - \frac{215}{250.77}\right) - \frac{4560}{8.31433 \cdot 215}\left(1 - \frac{215}{225.35}\right)$$
$$= -0.3189$$

$$x_1\gamma_1 = 0.7269$$

Taking into account the real behavior, a mole fraction of $x_1 = 0.6995$ ($\gamma_1 = 1.0391$) is obtained iteratively. For the whole composition range the SLE diagram is shown in Figure 8.7. It can be seen that nearly perfect agreement is achieved for modified UNIFAC. Even the inflection point at the transition temperature is described correctly. For the Wilson equation, deviations can be recognized at low temperatures. The reason is that contrary to modified UNIFAC the Wilson parameters are not adjusted to excess enthalpies, which leads to a poor temperature extrapolation.

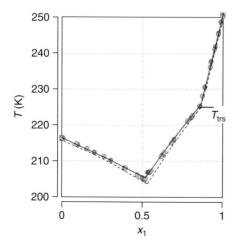

Figure 8.7 Experimental and predicted SLE data for the system carbon tetrachloride (1)–n-octane (2): ● experimental; – – – – Wilson; —— mod. UNIFAC.

Example 8.6

Calculate the ideal solubility of naphthalene (3) in a mixture of ethanol (1)–2,2,4-trimethylpentane (2) at 298.15 K as a function of the ethanol concentration. Compare the results with the results obtained taking into account the real mixture behavior using modified UNIFAC. ($T_{m,\text{naphthalene}} = 353.35$ K, $\Delta h_{m,\text{naphthalene}} = 19110$ J/mol)

Solution

In this example the solubility of naphthalene shall be calculated for a composition $x_1 = 0.5$ on the solute free basis. Using Eq. (8.10), one obtains

$$\ln x_3 \gamma_3 = \frac{-19110}{8.31433 \cdot 298.15}\left(1 - \frac{298.15}{353.35}\right) = -1.204$$

$$x_3 \gamma_3 = 0.2999$$

Using modified UNIFAC, iteratively the mole fraction of naphthalene has to be determined, for which the product $x_3 \gamma_3$ provides a value of 0.2999. This is the case for $x_3 = 0.099974$ ($x_1 = x_2 = 0.450013$). For this composition an activity coefficient of $\gamma_3 = 2.9998$ is calculated. The solubilities of naphthalene for the whole composition range of the solvent mixture are shown in Figure 8.8, together with the experimental values. From Figure 8.8 it can be seen that the predicted results using modified UNIFAC are in good agreement with the experimental findings. The solubility increase from ethanol to 2,2,4-trimethylpentane is correctly described, whereas the calculated ideal solubilities are a factor 3–4 off.

8.1.1.1 Freezing Point Depression

As shown before, small amounts of a solute (2) dissolved in a solvent (1) will lower the freezing point. An expression for the freezing point depression can directly be derived starting from Eq. (8.10) for simple eutectic systems, where in the case of diluted solutes ideal behavior can be assumed.

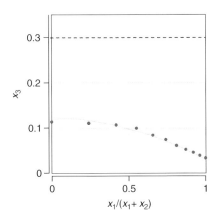

Figure 8.8 Experimental and predicted solubility of naphthalene (3) in ethanol (1) –2,2,4-trimethylpentane (2) at 298.15 K: ● experimental; —— modified UNIFAC; - - - - ideal.

Starting from the equilibrium condition (8.10)

$$\ln x_1 = -\frac{\Delta h_{m,1}}{RT}\left(1 - \frac{T}{T_{m,1}}\right) = -\frac{\Delta h_{m,1}}{R}\left(\frac{T_{m,1} - T}{TT_{m,1}}\right) \qquad (8.10)$$

and with $x_1 = 1 - x_2$ using the simplification for small values of x_2:

$$\lim_{x_2 \to 0} \ln(1 - x_2) = -x_2$$

one obtains

$$x_2 = \frac{\Delta h_{m,1}}{RT}\left(\frac{T_{m,1} - T}{T_{m,1}}\right) = \frac{\Delta h_{m,1}\Delta T}{RTT_{m,1}}$$

For small freezing point depressions ΔT ($T \approx T_{m,1}$) one can write

$$\Delta T \approx \frac{RT_{m,1}^2}{\Delta h_{m,1}} x_2$$

Replacing the mole fraction x_2 by the molality m_2:

$$x_2 = \frac{m_2}{m_T}, \quad m_T \approx m_1 \approx \frac{1}{M_1}, \quad x_2 \approx m_2 M_1$$

leads to the following equation for the melting point depression ΔT:

$$\Delta T = \frac{RT_{m,1}^2 m_2 M_1}{\Delta h_{m,1}} = K_{cry} m_2 \qquad (8.12)$$

where the cryoscopic constant K_{cry} can be calculated with the help of the molar mass, the melting temperature and the heat of fusion of the solvent (1):

$$K_{cry} = \frac{RT_{m,1}^2 M_1}{\Delta h_{m,1}} \qquad (8.13)$$

Following Eq. (8.12) the freezing point depression is proportional to the cryoscopic constant and the concentration of the solute (m_2 (mol/kg)) in the molality scale. Since the cryoscopic constant K_{cry} of the solvent depends on the molar mass, the melting temperature T_m and the enthalpy of fusion, the various liquids show different values. From Eq. (8.12) it can be seen that the measurement of the freezing point depression allows the determination of the molar mass of unknown compounds, if the melting temperature and the enthalpy of fusion is known.

Example 8.7

Calculate the cryoscopic constant of water and naphthalene using the following pure component data.

Component	T_m (K)	Δh_m (J/mol)
Water	273.15	5996.1
Naphthalene	353.35	19110

Solution

$$K_{cry,water} = \frac{8.31433 \cdot 273.15^2 \cdot 18.015}{5996.1} \frac{\text{J K}^2\text{mol g}}{\text{mol K J mol}} = 1.864 \frac{\text{K kg}}{\text{mol}}$$

$$K_{cry,naphthalene} = \frac{8.31433 \cdot 353.35^2 \cdot 128.173}{19110} = 6.963 \frac{\text{K kg}}{\text{mol}}$$

Similar values can be found in [5].

8.1.2
Solid–Liquid Equilibria of Systems with Solid Solutions

8.1.2.1 Ideal Systems

For systems with mutual miscibility in the solid phase – so-called solid solutions – Eq. (8.9) has to be used instead Eq. (8.10). In the case of ideal systems ($\gamma_i = 1$) the concentrations x_i^L and x_i^S of the liquidus and solidus line in the whole concentration range can directly be determined from the melting temperatures $T_{m,i}$ and the enthalpies of fusion $\Delta h_{m,i}$ of the compounds involved. For compounds 1 and 2 of a binary system the following expressions are obtained:

$$\ln x_1^L = \ln x_1^S - \frac{\Delta h_{m,1}}{RT}\left(1 - \frac{T}{T_{m,1}}\right) \tag{8.14}$$

$$\ln\left(1 - x_1^L\right) = \ln\left(1 - x_1^S\right) - \frac{\Delta h_{m,2}}{RT}\left(1 - \frac{T}{T_{m,2}}\right) \tag{8.15}$$

By substitution one obtains for x_1^S:

$$x_1^S = \frac{1 - \exp\left[-\frac{\Delta h_{m,2}}{RT}\left(1 - \frac{T}{T_{m,2}}\right)\right]}{\exp\left[-\frac{\Delta h_{m,1}}{RT}\left(1 - \frac{T}{T_{m,1}}\right)\right] - \exp\left[-\frac{\Delta h_{m,2}}{RT}\left(1 - \frac{T}{T_{m,2}}\right)\right]} \tag{8.16}$$

which can be used to calculate the SLE behavior for ideal systems with total solubility in the solid phase.

Example 8.8

Calculate the SLE data of the system anthracene (1)–phenanthrene (2), assuming that both the liquid and the solid phase behave ideally.

	Pure component data	
Component	T_m (K)	Δh_m (J/mol)
Anthracene	489.6	28860
Phenanthrene	369.4	18640

Solution

In this example the calculation is performed for a temperature of 430 K. Using Eq. (8.16) one obtains

$$x_1^S = \frac{1 - \exp\left[-\frac{18640}{8.31433 \cdot 430}\left(1 - \frac{430}{369.4}\right)\right]}{\exp\left[-\frac{28860}{8.31433 \cdot 430}\left(1 - \frac{430}{489.6}\right)\right] - \exp\left[-\frac{18640}{8.31433 \cdot 430}\left(1 - \frac{430}{369.4}\right)\right]}$$

$$= 0.6836$$

Using this value for x_1^S directly the mole fraction in the liquid phase x_1^L can be calculated:

$$\ln x_1^L = \ln 0.6836 - \frac{28860}{8.31433 \cdot 430}\left(1 - \frac{430}{489.6}\right) = -1.363$$

$$x_1^L = 0.2559$$

For the whole composition range the calculated SLE behavior is shown in Figure 8.9. It can be seen that the experimental and calculated data assuming ideal behavior are in qualitative agreement. Taking into account the deviations from ideal behavior in the liquid and solid phase improved results for both the liquidus and the solidus line are obtained. This is also shown in Figure 8.9, where the Wilson parameters for the liquid and solid phase were fitted to the SLE data (see also Section 8.1.2.2).

8.1.2.2 Solid–Liquid Equilibria for Nonideal Systems

Following Eq. (8.9) in the case of nonideal systems the knowledge about the real behavior (e.g., activity coefficients γ_i) both in the liquid and in the solid phase is required as additional information for the calculation of SLE. The deviation from ideal behavior can be taken into account using for example, g^E-models or equations of state. Starting from Eq. (8.9) for nonideal systems, the following expression is

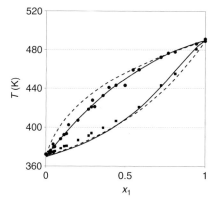

Figure 8.9 Experimental and calculated SLE behavior for the system anthracene (1)–phenanthrene (2): - - - - ideal; —— Wilson ($\Delta\lambda_{12}^L = -273.6$ K, $\Delta\lambda_{21}^L = -322.7$ K $\Delta\lambda_{12}^S = -647.3$ K, $\Delta\lambda_{21}^S = 417.5$ K) $v_1^L = v_1^S = 138.92$ cm³/mol, $v_2^L = v_2^S = 125.03$ cm³/mol.

obtained for the solidus line:

$$x_1^S = \frac{1 - \frac{\gamma_2^S}{\gamma_2^L}\exp\left[-\frac{\Delta h_{m,2}}{RT}\left(1 - \frac{T}{T_{m,2}}\right)\right]}{\frac{\gamma_1^S}{\gamma_1^L}\exp\left[-\frac{\Delta h_{m,1}}{RT}\left(1 - \frac{T}{T_{m,1}}\right)\right] - \frac{\gamma_2^S}{\gamma_2^L}\exp\left[-\frac{\Delta h_{m,2}}{RT}\left(1 - \frac{T}{T_{m,2}}\right)\right]} \quad (8.17)$$

In a similar way, an expression for the liquidus line can be derived:

$$x_1^L = \frac{1 - \frac{\gamma_2^S}{\gamma_2^L}\exp\left[-\frac{\Delta h_{m,2}}{RT}\left(1 - \frac{T}{T_{m,2}}\right)\right]}{\frac{\gamma_1^S}{\gamma_1^L}\exp\left[-\frac{\Delta h_{m,1}}{RT}\left(1 - \frac{T}{T_{m,1}}\right)\right] - \frac{\gamma_2^S}{\gamma_2^L}\exp\left[-\frac{\Delta h_{m,2}}{RT}\left(1 - \frac{T}{T_{m,2}}\right)\right]}$$

$$\cdot \left(\frac{\gamma_1^S}{\gamma_1^L}\exp\left[-\frac{\Delta h_{m,1}}{RT}\left(1 - \frac{T}{T_{m,1}}\right)\right]\right) \quad (8.18)$$

To describe the nonideal behavior, the various g^E-models introduced in Chapter 5.3 can be applied. The parameters can be fitted in a way that a minimum deviation to the experimental SLE data is obtained. In the case of strong deviation from ideal behavior, sometimes an extreme temperature (maximum or minimum temperature) is observed as shown in Figure 8.1c. In both cases, x_i^L is equal to x_i^S at the extremum. Starting from Eq. (8.9) the following simplified equation is obtained for the minimum or maximum temperature T_{extreme}:

$$\ln \frac{\gamma_i^L}{\gamma_i^S} = -\frac{\Delta h_{m,i}}{R}\left(\frac{1}{T_{\text{extreme}}} - \frac{1}{T_{m,i}}\right) \quad (8.19)$$

From this equation it becomes clear that a temperature maximum occurs when the activity coefficient in the liquid phase is larger than in the solid phase. For a temperature minimum the opposite is true.

With the assumption that the simple Porter equation

$$\ln \gamma_1^L = A(x_2^L)^2 \quad \ln \gamma_2^L = A(x_1^L)^2$$
$$\ln \gamma_1^S = B(x_2^S)^2 \quad \ln \gamma_2^S = B(x_1^S)^2$$

is used to describe the deviations from ideal behavior, the SLE diagram can be calculated from the extreme temperature and its associated composition. At the extreme temperature T_{extreme} Eq. (8.19) can be written in the following form for components 1 and 2 (with $x_i^L = x_i^S$):

$$(A - B)\left(x_2^L\right)^2 = \ln \frac{\gamma_1^L}{\gamma_1^S} = -\frac{\Delta h_{m,1}}{R}\left(\frac{1}{T_{\text{extreme}}} - \frac{1}{T_{m,1}}\right) \quad (8.20)$$

$$(A - B)\left(x_1^L\right)^2 = \ln \frac{\gamma_2^L}{\gamma_2^S} = -\frac{\Delta h_{m,2}}{R}\left(\frac{1}{T_{\text{extreme}}} - \frac{1}{T_{m,2}}\right) \quad (8.21)$$

If the extreme temperature and the composition are known, it is possible to obtain the difference of the Porter parameters $A - B$. The result of this approach is shown in Example 8.9 for a system with temperature maximum and in Example 8.10 for a system with temperature minimum.

Example 8.9

Calculate the SLE diagram for the binary system D-carvoxime (1)–L-carvoxime (2) with a congruent melting point at a composition $x_1 = 0.5$, $T_{max} = 365.15$ K using the Porter expression and the following pure component data:

Component	Δh_m (J/mol)	T_m (K)
D-carvoxime	17008	343.15
L-carvoxime	17008	343.15

L-carvoxime   D-carvoxime

Solution

$$(A - B) = -\frac{17008}{(0.5)^2 \cdot 8.31433} \left(\frac{1}{365.15} - \frac{1}{343.15} \right) = 1.313$$

But unfortunately only the difference of the Porter parameters can be calculated. However, from the difference $(A > B)$ it can be concluded that the activity coefficients in the liquid phase are larger than in the solid phase. Since ideal behavior in the liquid phase can be assumed for a mixture of enantiomers (i.e., Porter parameter $A = 0$), the results for a value of $B = -1.313$ are shown in Figure 8.10. For a given value of x_i^L, the temperature and the value for x_i^S can be calculated iteratively, so that Eqs. (8.20) and (8.21) are fulfilled. It can be seen that for the system D-carvoxime–L-carvoxime quite good agreement with the experimental findings [4] is obtained.

Example 8.10

Calculate the SLE diagram for the binary system diphenylacetylene (1) –N,N-diphenylhydrazine (2) with a minimum melting point at a composition $x_1 = 0.826$, $T_{min} = 323.5$ K using the Porter expression and the following pure component data:

Component	Δh_m (J/mol)	T_m (K)
Diphenylacetylene	21387	335.65
N,N-diphenylhydrazine	17633	404.15

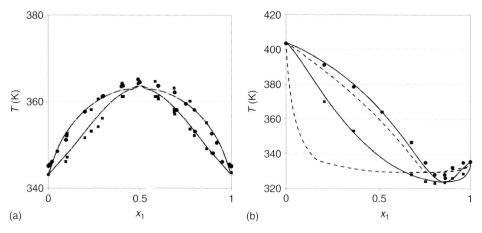

Figure 8.10 Experimental and calculated SLE behavior using the Porter model ($B = -1.313$) for the system D-carvoxime (1)-L-carvoxime (2) (a), respectively, the Porter ($B = 1.917$) and the Wilson model for the system diphenylacetylene (1)-N,N-diphenylhydrazine (2) with "rough" Wilson parameters (b) ($\Delta\lambda_{12}{}^L = -760$ K, $\Delta\lambda_{21}{}^L = 100$ K, $\Delta\lambda_{12}{}^S = -160$ K, $\Delta\lambda_{21}{}^S = -370$ K), $v_1{}^L = v_1{}^S = 180.03$ cm³/mol, $v_2{}^L = v_2{}^S = 159.10$ cm³/mol: - - - Porter; —— Wilson.

Solution

Again Eqs. (8.20) and (8.21) can be used to calculate the differences $A - B$. Since the pure component parameters are not identical as in Example 8.9, different values are obtained for the differences $(A - B)$.[5]

$$(A - B) = -\frac{21387}{(0.174)^2 \cdot 8.31433}\left(\frac{1}{323.5} - \frac{1}{335.65}\right) = -9.507$$

$$(A - B) = -\frac{17633}{(0.826)^2 \cdot 8.31433}\left(\frac{1}{323.5} - \frac{1}{404.15}\right) = -1.917$$

As in Example 8.9, only the differences of the parameters are obtained. But for a system with a temperature minimum the activity coefficients in the solid phase are larger than in the liquid phase. The experimental and calculated data assuming ideal behavior in the liquid phase ($A = 0$) are shown in Figure 8.10 for a value $B = 1.917$. It can be seen that the results using the simple Porter equation are not satisfying. Therefore, Wilson parameters were directly fitted to the experimental data. It can be seen that much better results are obtained, if the nonideal behavior in the liquid and solid phase is taken into account.

---

5) With Porters equation only symmetric systems can be described reliably.

8.1.3
Solid–Liquid Equilibria with Intermolecular Compound Formation in the Solid State

In the case that an intermolecular compound is formed in the solid state

$$A_n B_m(S) \rightleftharpoons nA + mB$$

the chemical equilibrium of the compound formation has to be taken into account.

The SLE behavior of the system naphthalene-2,4,6-trinitrotoluene with intermolecular compound formation is shown in Figure 8.1f. The formation of an intermolecular compound can be applied for the separation of systems by adductive crystallization. Most enantiomers form an intermolecular compound (racemic compound).

At the melting point of the intermolecular compound $A_n B_m$ formed, one can write (see Section 12.2):

$$\Delta g_R^0 = -RT\ln K(T_m) = -RT\ln \frac{x_{A_0}^n x_{B_0}^m \gamma_{A_0}^n \gamma_{B_0}^m}{x_{A_n B_m} \gamma_{A_n B_m}} \tag{8.22}$$

Since the activity of the precipitated intermolecular compound $A_n B_m$ is equal to unity one can replace the equilibrium constant by the solubility product K_{sp}:

$$\Delta g_R^0 = -RT\ln K_{sp}(T_m) = -RT\ln x_{A_0}^n x_{B_0}^m \gamma_{A_0}^n \gamma_{B_0}^m \tag{8.23}$$

where the composition x_{i0} of the compounds A and B in the liquid phase can be described by the stoichiometric coefficients n and m:

$$x_{A_0} = \frac{n}{n+m} \qquad x_{B_0} = \frac{m}{n+m} \tag{8.24}$$

so that Eq. (8.23) can be written in the following form:

$$K(T_m) = \left(\frac{n}{n+m}\right)^n \left(\frac{m}{n+m}\right)^m \gamma_{A_0}^n \gamma_{B_0}^m \tag{8.25}$$

To calculate the equilibrium constant $K(T)$ or $K_{sp}(T)$ at the temperature T, the van't Hoff equation (see Eq. (12.21)) can be used:

$$\frac{d\ln K}{dT} = \frac{\Delta h_R^0}{RT^2}$$

$$\ln K(T) = \ln K(T_m) - \frac{\Delta h_R^0}{R}\left(\frac{1}{T} - \frac{1}{T_m}\right)$$

which can also be described by

$$\ln x_A^n x_B^m \gamma_A^n \gamma_B^m = \ln x_{A_0}^n x_{B_0}^m \gamma_{A_0}^n \gamma_{B_0}^m - \frac{\Delta h_R^0}{R}\left(\frac{1}{T} - \frac{1}{T_m}\right) \tag{8.26}$$

8.1 Thermodynamic Relations for the Calculation of Solid–Liquid Equilibria

Assuming ideal behavior, the following simplified relation is obtained

$$\ln x_A^n x_B^m = \ln x_{A_0}^n x_{B_0}^m - \frac{\Delta h_R^0}{R}\left(\frac{1}{T} - \frac{1}{T_m}\right)$$

$$= \ln\left[\left(\frac{n}{m+n}\right)^n \left(\frac{m}{n+m}\right)^m\right] - \frac{\Delta h_R^0}{R}\left(\frac{1}{T} - \frac{1}{T_m}\right) \quad (8.27)$$

or

$$\ln x_A^n x_B^m = n\ln n + m\ln m - (m+n)\ln(m+n) - \frac{\Delta h_R^0}{R}\left(\frac{1}{T} - \frac{1}{T_m}\right) \quad (8.28)$$

The standard enthalpy of reaction Δh_R^0 is identical to the enthalpy of fusion Δh_m, because the compound $A_n B_m$ dissociates completely into A and B, when melting.

Unfortunately, the required properties T_m and Δh_m of the compound formed are often not known. But the melting temperature of the adduct $A_n B_m$ can be obtained from experimental SLE data. The heat of fusion can be measured calorimetrically. For a large number of compounds these data are stored in [4].

For the estimation of the heat of fusion Δh_m of the compound formed several methods have been suggested in the literature, for example:

- by using weighted average values of the entropies of mixing of the compounds involved,
- by determining this value from experimental SLE data assuming ideal behavior,
- as additional parameter, when fitting the g^E-model parameters to experimental SLE data,

Example 8.11

Calculate the SLE behavior of the binary system CCl_4 (A)–p-xylene (B) which forms an 1:1 adduct AB. Perform the calculation assuming ideal behavior using the following experimental melting temperatures and enthalpies of fusion [4]:

$T_{m,A} = 250.77$ K $\quad \Delta h_{m,A} = 2528.4$ J/mol
$T_{m,B} = 286.45$ K $\quad \Delta h_{m,B} = 16790$ J/mol
$T_{m,AB} = 269.15$ K $\quad \Delta h_{m,AB} = 27698$ J/mol

Solution

The calculation of the SLE diagram should be performed using the experimental data. For the solubility of the intermolecular compound, the calculation is performed exemplarily for a temperature of 260 K.[6]

$$\ln(x_A \cdot (1 - x_A)) = -2\ln 2 - \frac{27698.1}{8.31433 \cdot 260}\left(1 - \frac{260}{269.15}\right)$$
$$= -1.386 - 0.4356 = -1.8216$$

---

6) Instead of the temperature also a value for x_A ($x_B = 1 - x_A$) can be chosen, for which then directly the temperature can be calculated.

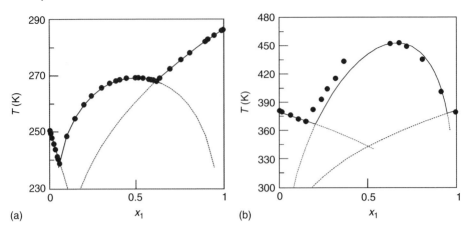

Figure 8.11 Experimental (●) and calculated SLE behavior of the systems: p-xylene (1)–CCl$_4$ (2) (a) acridine (1)–1,3-dihydroxybenzene (2) (b).

After solving the quadratic equation

$$x_A - x_A^2 = e^{-1.8216} = 0.1618$$
$$x_A^2 - x_A + 0.1618 = 0$$
$$x_A = 0.5 \pm \sqrt{0.25 - 0.1618}$$

one obtains two values for the mole fraction x_A:

$$x_{A,1} = 0.7970$$
$$x_{A,2} = 0.2030$$

The right-hand and left-hand branch of the SLE diagram, that is, the solubilities of A and B can then be calculated as shown in the examples before for simple eutectic systems. For the whole composition range, the results are shown in Figure 8.11. It can be seen that a good agreement with the experimental data is observed.

In Figure 8.11, additionally the results for the system acridine–1,3-dihydroxybenzene, which forms a 2:1 adduct are shown.[7] In this case also qualitative agreement is obtained.

---

7) For the calculation the following experimental values were used: acridine: $T_m = 380.95$ K, $\Delta h_m = 19\,687$ J/mol,

1,3-dihydroxybenzene: $T_m = 379.15$ K, $\Delta h_m = 21003$ J/mol, adduct A_2B: $T_m = 452.65$ K, $\Delta h_m = 23956$ J/mol.

8.1.4
Pressure Dependence of Solid–Liquid Equilibria

Although condensed phases are not sensitive to pressure, high pressures will cause a different melting point (see Section 3.1.4). While the melting temperatures of most compounds increase with increasing pressure, the opposite is true for water because of the negative P–T-slope of the SLE line at low and moderate pressures. The pressure dependence can be described with the help of the Clausius–Clapeyron or the Simon–Glatzel equation [6] (see Section 3.1.4). Besides the change of the melting temperature, high pressures will also influence the activity coefficients (see Section 5.3). Depending on the influence of the pressure on the melting points and activity coefficients this can lead to a change of the eutectic composition and therefore to an option to overcome the weaknesses of crystallization. By pressure swing crystallization, both compounds can then be obtained with high purity in the case of eutectic systems. As an example the pressure influence on the SLE behavior for the system lauric acid – myristic acid is shown in Figure 8.12.

8.2
Salt Solubility

The solubility of salts in water can be calculated in a similar way to that of the chemical equilibrium of reversible reactions (see Chapter 12). Let us consider the following reversible "chemical reaction" of a salt without and with crystal water:

$$M_{\nu_+}X_{\nu_-} \rightleftarrows \nu_+ M^+_{aq} + \nu_- X^-_{aq}$$

$$M_{\nu_+}X_{\nu_-} \cdot n\,H_2O \rightleftarrows \nu_+ M^+_{aq} + \nu_- X^-_{aq} + n\,H_2O$$

For the first reaction the following equilibrium constant (solubility product) is obtained in the saturation state [7], since the precipitating salt shows an activity of

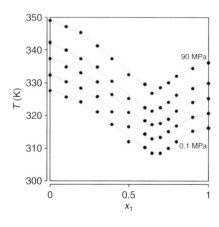

Figure 8.12 Solid–liquid equilibria of the system lauric acid (1)–myristic acid (2) at different pressures (0.1, 20, 40, 60, and 90 MPa): ● experimental [4].

one:
$$K_{sp} = a_{M^+}^{v_+} a_{X^-}^{v_-} = (m_\pm \gamma_\pm)^v \qquad (8.29)$$

with:
$$v = v_+ + v_-$$

In the case of salts which form hydrates, additionally the water activity has to be taken into account in the solubility product:

$$K_{sp} = a_{M^+}^{v_+} a_{X^-}^{v_-} a_{H_2O}{}^n = (m_\pm \gamma_\pm)^v a_{H_2O}{}^n \qquad (8.30)$$

The mean activity coefficients $\gamma_\pm$ and molalities $m_\pm$ can be calculated in the following way using the number of cations v_+ and anions v_-:

$$\gamma_\pm = \left(\gamma_+^{v_+} \gamma_-^{v_-}\right)^{1/v} \qquad m_\pm = \left(m_+^{v_+} m_-^{v_-}\right)^{1/v} \qquad (8.31)$$

The solubility product K_{sp} can be calculated for a given temperature, as in the case of a typical chemical reaction, using the tabulated standard thermodynamic properties of formation for the Gibbs energy and enthalpy and the molar heat capacity of the salt and the ions in aqueous solution (aq).

$$\Delta g_R^0(T) = \sum_i v_i \Delta g_{B,i}^0(T) = RT \ln K_{sp}(T) \qquad (8.32)$$

The temperature dependence of the solubility product can be determined with the help of the van't Hoff equation:

$$\frac{d \ln K_{sp}}{dT} = \frac{\Delta h_R^0}{RT^2} \qquad (8.33)$$

If the temperature dependence of the standard enthalpy of reaction can be neglected in the temperature range covered, Eq. (8.33) can be integrated directly:

$$\ln K_{sp}(T) = \ln K_{sp}(T_0) - \frac{\Delta h_R^0}{R}\left(\frac{1}{T} - \frac{1}{T_0}\right) \qquad (8.34)$$

If the temperature dependence of the standard enthalpy of reaction cannot be neglected, it can be taken into account using the molar heat capacities as in the case of typical chemical reactions (Kirchhoff's law).

For selected salts and ions in water the thermodynamic standard properties are listed in Table 8.1. To be able to determine the salt solubility from the solubility product, only an electrolyte model, such as Pitzer, Electrolyte NRTL, LIQUAC [8], or LIFAC [9] for the calculation of the mean activity coefficients $\gamma_\pm$, and in the case of hydrated salts additionally the activity of water is required (see Chapter 7).

Example 8.12

Calculate the ideal solubility ($\gamma_\pm = 1$) of NaCl, KCl, and NH_4Cl in water at 25 °C with the help of the standard thermodynamic properties listed in Table 8.1. What are the solubilities when the deviation from ideal behavior is taken into account using the LIQUAC model.

8.2 Salt Solubility

Table 8.1 Thermodynamic standard properties for selected salts and ions in the aqueous solution.

Salt	Molar heat capacity (J mol^{-1} K^{-1})				$\Delta h^0_{f,298\,K}$ (kJ/mol)	$\Delta g^0_{f,298\,K}$ (kJ/mol)
	298.15 K	300 K	400 K	500 K		
NaCl	50.503	50.544	52.374	53.907	−411.12	−384.024
KCl	51.713	51.71	52.483	54.209	−431.30	−409.34
NH$_4$Cl	84.1	–	–	–	−314.00	−203.60
Na$_2$SO$_4$	128.151	128.487	145.101	–	−1387.816	−1269.848
Na$_2$SO$_4 \cdot$10 H$_2$O	−246.60	–	–	–	−4327.26	−3646.85
Na$^+$ (aq)	46.4	–	–	–	−240.12	−261.905
K$^+$ (aq)	21.8	–	–	–	−252.38	−283.27
NH$_4^+$ (aq)	79.9	–	–	–	−132.51	−79.31
Cl$^-$ (aq)	−136.4	–	–	–	−167.159	−131.228
SO$_4^-$ (aq)	−293.0	–	–	–	−909.27	−744.53

Solution

For the calculation of the solubility product at 25 °C only the Gibbs energies of formation Δg_f^0 at 25 °C of the compounds involved are required. With the values given in Table 8.1 for sodium chloride, the following solubility product is obtained:

$$\Delta g_R^0 = \sum_i v_i \Delta g_{f,i}^0 = \Delta g_{f,Na^+(aq)}^0 + \Delta g_{f,Cl^-(aq)}^0 - \Delta g_{f,NaCl}^0$$

$$\Delta g_R^0 = -261.905 - 131.228 + 384.024 = -9.109 \, \text{kJ/mol}$$

$$K_{sp} = e^{-\frac{\Delta g_R^0}{RT}} = e^{\frac{9109}{8.31433 \cdot 298.15}} = 39.432$$

Using this value, directly the ideal solubility of NaCl at 25 °C can be calculated from Eq. (8.29):

$$m_{NaCl} = \sqrt{39.432} = 6.277 \, \text{mol/kg}$$

In the same way, the ideal solubility of KCl and NH$_4$Cl at 25 °C can be calculated using the standard thermodynamic properties listed in Table 8.1:

$\Delta g_{R,KCl}^0 = -5.158 \, \text{kJ/mol}$ $\quad K_{KCl} = 8.01 \quad$ $m_{KCl} = 2.83 \, \text{mol/kg}$

$\Delta g_{R,NH_4Cl}^0 = -6.938 \, \text{kJ/mol}$ $\quad K_{NH_4Cl} = 16.42 \quad$ $m_{NH_4Cl} = 4.05 \, \text{mol/kg}$

In Figure 8.13 the ideal solubilities of the three salts are shown as a function of temperature together with the experimental data and the results of the LIQUAC model. It can be seen that the predicted solubilities of NaCl at 25 °C in contrast to the calculated solubilities of KCl and NH$_4$Cl assuming ideal behavior ($\gamma_\pm = 1$) are nearly identical with the experimental values. This is caused by the fact that accidentally the mean activity coefficient of NaCl for the calculated solubility of about 6 mol/kg is approx. 1 (see Figure 8.14).

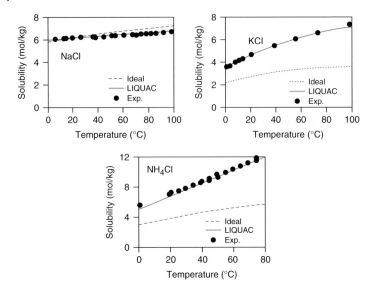

Figure 8.13 Experimental, ideal and predicted salt solubilities using LIQUAC as a function of temperature.

Using LIQUAC in all cases good agreement between the experimental and the predicted solubilities is obtained as can be seen from Figure 8.13. The reason for the good agreement is that the mean activity coefficients of the different salts are predicted correctly, as shown exemplarily for a temperature of 25 °C in Figure 8.14.

With the help of an electrolyte model it can also be calculated whether the salt crystallizes with or without crystal water at a given temperature and how the presence of other salts influences the solubility. In Figure 8.15, the predicted solubilities of Na_2SO_4 and $Na_2SO_4 \cdot 10\ H_2O$ using LIQUAC are shown together with the experimental data. As can be recognized, the results using LIQUAC with

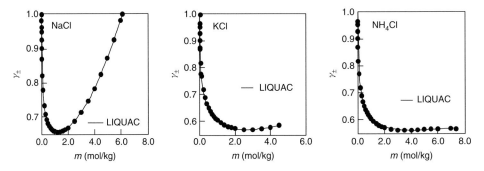

Figure 8.14 Mean activity coefficients of NaCl, KCl, and NH_4Cl in water at 25 °C.

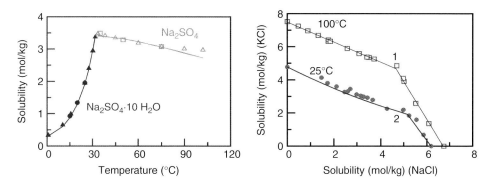

Figure 8.15 Experimental and predicted solubilities of sodium sulfate (left hand side) resp. mutual solubilities of NaCl and KCl in water (right hand side) in the temperature range from 0 to 100 °C, ——— LIQUAC.

the standard thermodynamic properties given in Table 8.1 are in good agreement with the experimental findings. Furthermore, in Figure 8.15 the experimental and predicted mutual solubilities of NaCl and KCl at 25 and 100 °C using LIQUAC are shown. As can be seen also in this case, the calculated mutual solubilities are in good agreement with the experimental data. The solubilities of this system are of special importance for the industrial production of KCl from sylvinite (mixture of potassium chloride (sylvine) and sodium chloride (halite)).

The procedure for the calculation of salt solubilities in aqueous solutions can be extended to organic solvents or solvent mixtures, starting from the condition that the fugacity (chemical potential) of a precipitated salt in phase equilibrium is identical in water, an organic solvent or the aqueous solution (see Figure 8.16). This means that the already available standard thermodynamic properties in the aqueous phase given in Table 8.1 can be used to determine the salt solubility in organic solvents or solvent mixtures [10].

For potassium chloride in a water–methanol mixture, the predicted results using LIQUAC are shown together with the experimental values in Figure 8.17. As can be seen from the diagram, good agreement with the experimental solubilities is obtained.

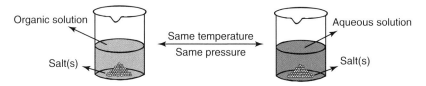

Figure 8.16 Starting point for the derivation of the required equations for the calculation of salt solubilities in organic solvents starting from the tabulated standard thermodynamic properties of the ions in an aqueous solution.

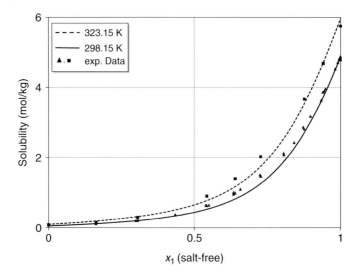

Figure 8.17 Experimental and predicted solubility (salt-free basis) of KCl in the system water (1)–methanol (2)
——— - - - LIQUAC.

8.3
Solubility of Solids in Supercritical Fluids

For the design of extraction processes of mostly high boiling liquid or solid compounds with the help of supercritical fluids, for example, the extraction of caffeine from coffee beans using carbon dioxide the phase equilibrium behavior as a function, of the pressure without or in the presence of co-solvents is required. As in the case of all other phase equilibria, the isofugacity condition has to be fulfilled:

$$f_i^L = f_i^V \text{ resp. } f_i^S = f_i^V \tag{8.35}$$

This means that – not only for the calculation of the solubilities of liquids but also of solids in supercritical fluids – equations of state can be applied. For the calculation of liquids in supercritical fluids the equation derived for vapor–liquid equilibria (VLE) or gas solubilities can be applied directly:

$$x_i \varphi_i^L = y_i \varphi_i^V \tag{5.9}$$

The fugacity of the pure solid compound 2 can be described by the sublimation pressure, the fugacity coefficient in the saturation state and the Poynting factor, so that the following phase equilibrium relation is obtained for the calculation of the solubility of the solid 2 in the supercritical fluid:

$$\varphi_2^s P_2^{subl} \text{Poy}_2 = y_2 \varphi_2^V P \tag{8.36}$$

Since the sublimation pressure of the solid is low, the fugacity coefficient in the saturation state shows values which are nearly unity, so that finally the following

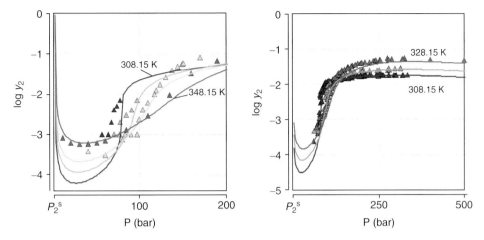

Figure 8.18 Experimental and calculated solubilities for the systems $CO_2(1)$–octanol (2) (left hand side) and $CO_2(1)$–naphthalene (2) (right-hand side) at different temperatures as a function of pressure: ▲ experimental [4]; —— SRK, $k_{12} = 0.110$ (a), $T =$ 308.15, 318.15, 328.15, 348.15 K; —— SRK, $k_{12} = 0.103$ (b), $T = 308.15$ K, 318.15 K, 328.15 K, Antoine constants (sublimation pressure) $\log P_i^s$ (mm Hg) $= A - B/(\vartheta + C)$. ($A = 10.0368, B = 2893.34, C = 235.373, v_2^S = 111.94$ cm^3/mol).

expression is obtained for the solubility of the solid compound 2 in the supercritical fluid.

$$y_2 = \frac{P_2^{\text{subl}} \text{Poy}_2}{\varphi_2^V P} \tag{8.37}$$

For the calculation of the solubility y_2 reliable sublimation pressures are required. For the calculation of the Poynting factor, additionally the molar volume of the solid compound is required. Then the equation of state only has to describe the fugacity coefficient in the gas phase ϕ_2^V as a function of pressure.

The typical progression of the solubility y_2 as a function of pressure is shown in Figure 8.18 for different temperatures for the systems CO_2–octanol and CO_2–naphthalene. While octanol is a liquid, naphthalene is a solid in the temperature range covered. For both cases also the results of the Soave–Redlich–Kwong equation with quadratic mixing rules to calculate the fugacity coefficient are shown together with the experimental values. While in the case of octanol Eq. (5.9) is used, Eq. (8.37) is used in the case of naphthalene.

For the pure compound 2, this means at the vapor (sublimation) pressure P_2^s the mole fraction y_2 is equal to unity. With the addition of CO_2 the vapor phase concentration of component 2 decreases. Since at low pressures the fugacity coefficient and the Poynting factor are approximately unity, the mole fraction y_2 at moderate pressure becomes approximately:

$$y_2 \approx \frac{P_2^{\text{subl}}}{P}$$

At higher pressures the Poynting factor and in particular the fugacity coefficient ϕ_2^V become important, so that the solubility y_2 as a function of pressure has the typical curvature shown in Figure 8.18 together with the experimental data and the correlated results using the Soave–Redlich–Kwong (SRK) equation of state).

Additional Problems

P8.1 Calculate the solubility of the isomeric compounds anthracene and phenanthrene in benzene and ethanol at 25 °C assuming ideal behavior and using the group contribution method modified UNIFAC.

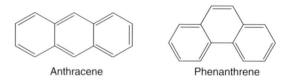

Anthracene Phenanthrene

Pure component properties:

	T_m (K)	Δh_m (kJ/mol)
Anthracene	489.7	28.860
Phenanthrene	369.5	18.644

Compare the calculated results with the solubilities measured at 25 °C:

	$x_{anthracene}$	$x_{phenanthrene}$
Benzene	0.008	0.21
Ethanol	0.00044	0.013

P8.2 Calculate the cryoscopic constants of benzene and camphor using the following pure component data:

Component	T_m (K)	Δh_m (J/mol)	M (g/mol)
Benzene	278.68	9944	78.113
Camphor	451.75	4397	152.236

P8.3 Calculate the freezing point depression of naphthalene for the case that 0.2 g of benzene is added to 50 g naphthalene. ($K_{cry,naphthalene}$ = 6.96 K kg/mol, M = 128.173 g/mol).

P8.4 Calculate the molar mass of an unknown component which leads to a freezing point depression of 0.223 K of naphthalene, when 0.25 g of the unknown compound is added to 100 g naphthalene. ($K_{cry,naphthalene}$ = 6.96 K kg/mol).

P8.5 Calculate the solubility of m-xylene (1) in p-xylene (2) and p-xylene (1) in m-xylene (2) at a temperature of 225 K.
Pure component properties:

Compound	T_m (K)	Δh_m (J/mol)
m-Xylene	225.35	11544
p-Xylene	286.45	16790

P8.6 Calculate the SLE diagram for the binary eutectic systems 1,2,3-trichlorobenzene–n-decane and 1,2,4-trichlorobenzene–n-tetradecane assuming
a. ideal behavior
b. taking into account the real behavior using modified UNIFAC.
Compare the results with the published data that can be downloaded from the textbook page on *www.ddbst.com*.
Pure component properties:

Compound	T_m (K)	Δh_m (J/mol)
1,2,3-Trichlorobenzene	326.85	17400
1,2,4-Trichlorobenzene	290.22	18599
n-Decane	243.45	28757
n-Tetradecane	279.02	45062

P8.7 Calculate the SLE behavior of the system p-toluidine–phenol with 1 : 1 compound formation assuming ideal behavior, where the melting point and enthalpy of fusion of the 1 : 1 compound should be adjusted to get a reliable description of the SLE behavior. Published data can be downloaded from the textbook page on *www.ddbst.com*.

Pure component properties:

Compound	T_m (K)	Δh_m (J/mol)
p-Toluidine	316.90	18899
Phenol	314.05	11281

P8.8 Calculate the solubility of solid carbon dioxide in propane with the help of the group contribution equations of state PSRK and VTPR assuming simple eutectic behavior. Compare the results with the results assuming ideal behavior and the experimental data that can be downloaded from the textbook page on www.ddbst.com. All required parameters can be found in Appendix A.

P8.9 Estimate the enthalpy of fusion of benzene (1) by regressing the following solubility data of benzene in toluene assuming ideal ($\gamma_1 = 1$) and simple eutectic behavior.

x_1	T (K)	x_1	T (K)
1.000	278.65	0.826	267.35
0.929	272.95	0.687	256.85

P8.10 Determine the solid-liquid equilibrium temperature of the ideal ternary system m-xylene (1)–o-xylene (2)–p-xylene (3) for a composition of $x_1 = 0.1$ and $x_2 = 0.1$ with the help of the melting temperatures and enthalpies of fusion given in Example 8.4 in the textbook. Which component will crystallize?

P8.11 Is it, at least theoretically possible that a eutectic binary mixture with liquid-liquid immiscibility shows a eutectic composition inside the miscibility gap?

P8.12 In the free DDBSP Explorer Version, search for solid-liquid equilibrium data for the mixture 2-propanol–benzene. Regress the two datasets simultaneously using the Wilson, NRTL, and UNIQUAC model. Check the performance of the three models. Compare the data to the results of the predictive methods UNIFAC, modified UNIFAC, and PSRK. Examine the different graphical representations.

P8.13 For the calculation of the solubility of sucrose in water Peres and Macedo [11] proposed the following physical property parameters:

Component	T_m (K)	Δh_m (J/mol)	Δc_P (J/mol K)
Sucrose	459.15	46187	316.1153−1.1547 (T-298.15 K)

The solubility of sucrose in water at 25 °C is 2.0741 g per g water corresponding to a sucrose mole fraction of 0.09842. Calculate the activity coefficient of sucrose in water at 25 °C at this concentration with and without taking into account the heat capacity difference between subcooled liquid and solid sucrose.

P8.14 The vapor pressures of liquid anthracene and phenanthrene can be described by the Antoine equation using the Antoine parameters given below. Calculate the vapor-liquid-solid equilibrium (VLSE) along the solid-liquid saturation curve assuming ideal mixture behavior in the liquid phase. Melting points and heats of fusion of both components are given in example P8.1 above. Compare the vapor-liquid separation factors to those of an isothermal VLE data set at 220 °C (calculate assuming ideal liquid mixture behavior).

Component	A	B	C
Anthracene	7.28683	2429.04	209.567
Phenanthrene	7.43238	2456.35	204.303

Antoine equation: $\log P_i^s (\text{mm Hg}) = A - B/(\vartheta(°C) + C)$

References

1. Haase, R. and Schönert, H. (1969) *Solid–Liquid Equilibrium, The International Encyclopedia of Physical Chemistry and Chemical Physics*, Pergamon Press, Oxford.
2. Hansen, M. and Anderko, K. (1958) *Constitution of Binary Alloys*, Metallurgy and Metallurgical Engineering Series, 2nd edn, McGraw-Hill, New York.
3. Jaques, J. and Collet, A. (1994) *Enantiomers, Racemates, and Resolutions*, Krieger Publishing Company, Malabar, FL.
4. Dortmund Data Bank, www.ddbst.com. January 2011.
5. Brdicka, R. (1973) *Grundlagen der Physikalischen Chemie*, VEB Deutscher Verlag der Wissenschaften, Berlin.
6. Simon, F.E. and Glatzel, G. (1929) *Z. Anorg. Allg. Chem.*, **178**, 309–312.
7. Li, J., Lin, Y., and Gmehling, J. (2005) *Ind. Eng. Chem. Res.*, **44**, 1602–1609.
8. Li, J., Polka, H.M., and Gmehling, J. (1994) *Fluid Phase Equilib.*, **94**, 89–114.
9. Yan, W., Topphoff, M., Rose, C., and Gmehling, J. (1999) *Fluid Phase Equilib.*, **162**, 97–113.
10. Huang, J., Li, J., and Gmehling, J. (2009) *Fluid Phase Equilib.*, **275**, 8–20.
11. Peres, A.M. and Macedo, E.A. (1996) *Fluid Phase Equilib.*, **123**, 71–95.

9
Membrane Processes

9.1
Osmosis

In many technical applications, liquids with different concentrations are separated by so-called semipermeable membranes, this means membranes which are permeable only for the solvent but not for the dissolved species. Phase equilibrium between these two liquids can only be achieved by diffusion of the solvent through this membrane. This happens in a way that the solvent is transported from the solution with the lower solute concentration to the solution with the higher concentration. This phenomenon is called *osmosis*.

Completely semipermeable membranes do hardly exist. Most of the membranes used in practical applications can generally be passed by substances with a low molar mass, whereas large molecules are rejected. In most of the cases, the solvent is water, an example for a membrane is cellophane.

The equilibrium between the solutions is specified by equal temperatures and fugacities in the two phases, whereas the pressures will be different as shown below.

Moreover, the transport of the solvent through the membrane can be decreased by increasing the pressure of the solution with the high concentration. The equilibrium pressure, where the flow through the membrane vanishes, is called *osmotic pressure*. If the pressure is further increased, the direction of the flow changes and the solvent is transported to the solution with the lower concentration. This process is called *reverse osmosis*. It is especially useful when it is necessary to separate ionic materials from aqueous solutions. Many ionic species can be removed with more than 90% efficiency in a single stage [1]. Of course, the use of multiple stages can even increase the separation. A well-known example is the desalination of sea water for the production of potable water. In contrast to the usual evaporation processes, reverse osmosis processes show a significantly lower energy consumption.

Osmosis, osmotic equilibrium, and reverse osmosis are illustrated in Figure 9.1.

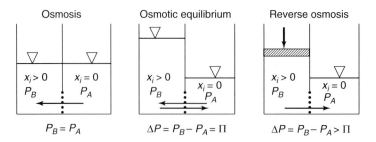

Figure 9.1 Osmosis, osmotic equilibrium, and reverse osmosis.

The osmotic pressure in the equilibrium state can be determined by the equilibrium condition

$$f_j^A = f_j^B \tag{9.1}$$

where the index j denotes the solvent which can penetrate through the membrane. Evaluation of the fugacities gives ($x_j = \gamma_j = 1$ on side A)

$$\varphi_j^s P_j^s \exp\frac{v_j^L(P_A - P_j^s)}{RT} = x_j \gamma_j \varphi_j^s P_j^s \exp\frac{v_j^L(P_B - P_j^s)}{RT} \tag{9.2}$$

and, subsequently,

$$\Pi = P_B - P_A = -\frac{RT}{v_j^L}\ln(x_j \gamma_j) \tag{9.3}$$

For a dilute solution, the approximations

$$\gamma_j \approx 1$$

$$\ln x_j = \ln\left(1 - \sum_i x_i\right) \approx -\sum_i x_i$$

$$\sum_i c_i = \frac{\sum_i n_i}{V} \approx \frac{\sum_i x_i}{v_j^L}$$

are valid, and we obtain

$$\Pi = RT\sum_i c_i \tag{9.4}$$

where x_i is the mole fraction and c_i is the volume concentration of solute i.

Example 9.1

Calculate the osmotic pressure of

1) 5 wt% aqueous glucose ($C_6H_{12}O_6$) solution
2) 5 wt% aqueous sodium chloride (NaCl) solution

at $\vartheta = 25\,°C$. The molar masses are as follows:

$$M_{glucose} = 180.16\,g/mol$$

$$M_{NaCl} = 58.44\,g/mol; \quad M_{Na^+} = 22.99\,g/mol; \quad M_{Cl^-} = 35.453\,g/mol$$

$$M_{H_2O} = 18.015\,g/mol$$

For simplicity, the liquid density should be assumed to be $1000\,kg/m^3$. The approximation (Eq. (9.4)) shall still be valid.

Solution

1) The mole fraction of the glucose solution is

$$x_{glucose} = \frac{\frac{0.05}{180.16\,g/mol}}{\frac{0.05}{180.16\,g/mol} + \frac{0.95}{18.015\,g/mol}} = 0.00524$$

For the mean molar mass, we get

$$M_{av} = 0.00524 \cdot 180.16\,g/mol + (1 - 0.00524) \cdot 18.015\,g/mol$$
$$= 18.864\,g/mol$$

giving a volume concentration of

$$c_{glucose} = \frac{x_{glucose}\,\rho^L}{M_{av}} = \frac{0.00524 \cdot 1000\,kg/m^3}{18.864\,kg/kmol} = 0.278\,\frac{kmol}{m^3}$$

The osmotic pressure can then be calculated to be

$$\Pi = RT c_{glucose} = \frac{8.314\,J \cdot 298.15\,K \cdot 278\,mol}{mol\,K\,m^3}$$
$$= 6.89 \cdot 10^5\,Pa = 6.89\,bar$$

2) In an aqueous solution, the sodium chloride will completely dissociate into Na^+ and Cl^- ions. Therefore, the mole number of dissolved species in 1 g of solution is

$$n_{diss} = \frac{0.05\,g/g}{58.44\,g/mol} \cdot 2 = 0.001711\,\frac{mol}{g}$$

The mole number of the solvent in 1 g of solution is

$$n_{H_2O} = \frac{0.95\,g/g}{18.015\,g/mol} = 0.0527\,\frac{mol}{g}$$

Thus, the mole fraction of the dissolved species is

$$x_{diss} = \frac{n_{diss}}{n_{diss} + n_{H_2O}} = \frac{0.001711}{0.001711 + 0.0527} = 0.03145$$

Considering the molar masses of the ions, one obtains for the mean molar mass:

$$M_{av} = \frac{1}{2} \cdot 0.03145 \cdot 22.99 \text{ g/mol} + \frac{1}{2} \cdot 0.03145 \cdot 35.453 \text{ g/mol}$$
$$+ (1 - 0.03145) \cdot 18.015 \text{ g/mol}$$
$$= 18.367 \text{ g/mol}$$

giving a volume concentration of

$$c_{diss} = \frac{x_{diss}\, \rho_L}{M_{av}} = \frac{0.03145 \cdot 1000 \text{ kg/m}^3}{18.367 \text{ kg/kmol}} = 1.712 \frac{\text{kmol}}{\text{m}^3}$$

The osmotic pressure can then be calculated to be

$$\Pi = RT c_{diss} = \frac{8.314 \text{ J} \cdot 298.15 \text{ K} \cdot 1712 \text{ mol}}{\text{mol K m}^3}$$
$$= 42.45 \cdot 10^5 \text{ Pa} = 42.45 \text{ bar}$$

Due to the dissociation, the osmotic pressures of electrolytes tend to be considerably higher than for molecular substances.

If the osmotic pressure is measured, it can be used for the determination of the molar masses of complex substances by reversing the calculation in Example 9.1.

Osmotic equilibria are important especially in biological systems [2]. For example, the cell membranes can be regarded as semipermeable. Inside the cell, the concentration of dissolved substances is considerably higher than outside. Due to the osmotic pressure, water tends to be transported into the cell. To avoid bursting of the cell, there are rigid cell walls with a great mechanical stability, which can absorb the osmotic pressure. Antibiotics like penicillin destroy the cell wall, and water can flow into the cell due to the osmotic pressure, leading to the explosion of the cell due to the osmotic pressure.

As well, many things from everyday life are based on osmosis [3].

The conservation of food by salting or sugaring leads to the dewatering of the food by osmosis; the consequence is that microorganisms can no more multiply due to the low water concentration.

If vegetables are being cooked, it is advantageous to add salt to the water, as vegetables themselves contain salt. This way, the osmotic transport of water into the vegetables is prevented, which would lead to a significant loss of taste.

If lettuce is exposed to its dressing, it loses its stability by the loss of water, which is transported to the dressing by means of osmosis.

Finally, cherries often burst after a rain, as osmosis causes the water to diffuse into the fruit, which leads to an osmotic pressure inside the fruit, exceeding the stability of the skin of the cherry fruit.

On cruise liners, several hundred tons of potable water per day are produced by reverse osmosis of sea water.

9.2 Pervaporation

The importance of membrane processes in separation technology is steadily increasing. Membranes separate a feed mixture into a permeate, that is, the part of a stream which passes the membrane, and a retentate, which does not pass the membrane (Figure 9.2). In thermal process engineering, membrane separations are often an energy-saving alternative for the separation of azeotropes, as the separation effect is almost independent from the vapor–liquid equilibrium which is used in distillation. The two components separate because of their different ability to pass the membrane, which is illustrated in Figure 9.2.

One of the important membrane processes is pervaporation, that is, the partial vaporization through a membrane. The mass transfer of a component through a membrane to the permeate side depends mainly on its affinity to the membrane material. General description of mass transfer in pervaporation consists of the following five steps [4]:

1) transfer of a component from the feed bulk to the membrane surface,
2) sorption of the component inside the membrane,
3) diffusion of the component across the membrane,
4) desorption of the component as vapor on the permeate side, and
5) transfer of the component from the membrane surface to the permeate bulk.

To describe the mass transport through the membrane, it can be assumed that the flux of component i through the membrane is proportional to its driving force, which is the difference in fugacity on either side of the membrane, and inversely proportional to its thickness. The flux J_i can be described as

$$J_i = P_i \frac{f_{i,f} - f_{i,P}}{l} \tag{9.5}$$

where P_i is the permeability, f is the fugacity, and l is the membrane thickness.

In a pervaporation process, the feed to the membrane module is kept at a pressure high enough to maintain it in the liquid phase. The other side of the membrane is kept at a pressure below the dew point of the permeate so that it leaves the module as vapor, meaning that the components are evaporating while passing through the membrane. Assuming an ideal vapor phase, the fugacities for modeling a pervaporation process can be expressed by partial pressures on the permeate side and by saturation pressures of each component at the feed side:

$$J_i = P_i^* (\gamma_i x_i P_i^S - y_i P_P) \tag{9.6}$$

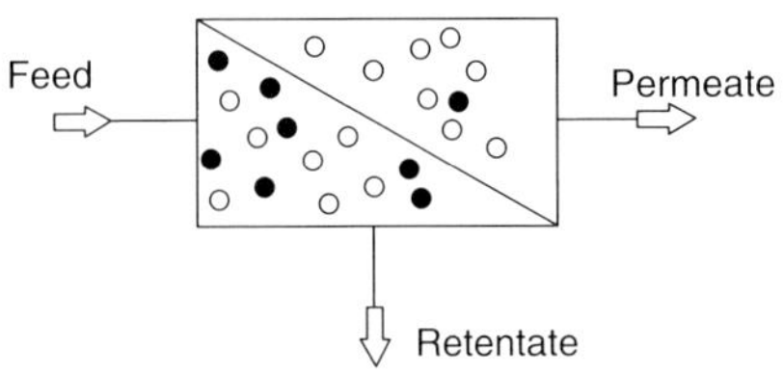

Figure 9.2 Principle of a membrane separation process.

where y_i is the mole fraction of component i in the permeate and P_p is the permeate pressure. $P_i^* = P_i/l$ is called the *apparent permeability constant* and characterizes the properties of the membrane. Thermodynamics determine the driving forces for the membrane separation, whereas an appropriate expression for the permeability is still an important target of current research [4].

Additional Problems

P9.1 The osmotic pressure of a salt solution in water at 25 °C has been measured as 7.2 atm against pure water. Estimate the freezing and boiling temperature of the solution from the data given in Appendix A.

P9.2 An aqueous solution contains 6 g/l of a polymer. At 300 K, the osmotic pressure is 30 mbar. Estimate the average molar mass of the polymer, if ideal behavior is assumed. Is this average the mass average or number average molar mass (see Section 10.1)?

P9.3 In an aqueous sugar solution, the freezing point of water at atmospheric pressure is lowered by 0.1 K. Estimate the osmotic pressure of the solution at 20 °C. Use the cryoscopic constant of water given in Example 8.7.

P9.4 The osmotic pressure of an aqueous glycerol solution is $\Pi = 12$ bar. Estimate the glycerol concentration of the solution. The structural formula of glycerol is $CH_2OH-CHOH-CH_2OH$.

References

1. Smith, R. (2005) *Chemical Process – Design and Integration*, John Wiley & Sons, Ltd, Chichester.
2. Morris, J.G. (1976) *Physikalische Chemie für Biologen*, Verlag Chemie, Weinheim.
3. Wikipedia The Free Encyclopedia, www.wikipedia.org, January 2011.
4. Bozek, E. (2010) Chemical reaction – pervaporation hybrid process and its application to transesterification of methyl acetate with n-butanol. Thesis. University of Oldenburg, Germany.

10
Polymer Thermodynamics*

10.1
Introduction

A polymer is a large molecule (macromolecule) composed of repeating structural units typically connected by covalent chemical bonds. Due to the extraordinary range of properties of polymeric materials, polymers play an essential and ubiquitous role in everyday life from plastics and elastomers to natural biopolymers such as DNA and proteins that are essential for life. A simple example is polyethylene, whose repeating unit is based on ethylene monomers. Synthetic polymer materials like polyamides (e.g., nylon and perlon), polyethylene, fluoropolymers (e.g., teflon), and silicones have formed the basis for the polymer industry. Most commercially important polymers today are entirely synthetic and produced in high volume. A world-scale polyethylene plant produces 300 000 tons per year or more. Synthetic polymers today are applied in almost all industries and areas of life. They are widely used as adhesives and lubricants, or structural material components for products ranging from toys to aircraft. They have been employed in a variety of biomedical applications ranging from implantable devices to controlled drug delivery. Polymers like poly(methyl methacrylate) are used as photoresist materials in semiconductor manufacturing and as dielectrics with a small dielectric constant for use in high-performance microprocessors. Recently, polymers have also been employed as flexible substrates in the development of organic light-emitting diodes for electronic displays. An increasing number of high-performance materials are polymer blends, made by mixing two or more ingredients.

The production and processing of polymers are also influenced by the presence of phase separation and segregation, which may be either necessary or highly undesirable. For example, segregation of highly viscous phases during a polymerization process may lead to catastrophic consequences like plugging lines or overheating reactors. In order to manufacture polymers having tailor-made properties, polymer fractionation can be established utilizing the liquid–liquid phase split [1, 2]. Knowledge of thermodynamic data of polymer containing systems is a necessity

*By Sabine Enders, TU Berlin

for industrial and laboratory processes. Such data serve as essential tools for understanding the physical behavior of polymer systems, for studying intermolecular interactions, and for gaining insights into the molecular nature of mixtures.

Because of their size, macromolecules are not conveniently described in terms of stoichiometry alone. The structure of simple macromolecules, such as linear homopolymers, may be described in terms of the individual monomer subunit and total molar-mass distribution. Complicated bio-macromolecules, synthetic dendrimers, or hyperbranched polymers require multifaceted structural description like degree of branching.

The most important difference between a mixture made from two low-molar-mass substances and a mixture made from a polymer and a low-molar-mass component is the difference in the size of the molecules. This difference in size leads to extremely unsymmetrical thermodynamic functions. For this reason, the polymer is divided into virtual segments and all thermodynamic quantities will be considered as segment-molar instead of molar. The segment number r_i is defined by

$$r_i = \frac{L_i}{L_{standard}} \tag{10.1}$$

where L_i is a suitable physical property (i.e., molar volume, molar mass, degree of polymerization, or van der Waals volume) and $L_{standard}$ is the physical property of an arbitrarily chosen standard segment (i.e., solvent or monomer unit). The segment molar quantities are given by

$$\bar{\bar{z}} = \frac{z}{\sum_{i=1}^{n} r_i x_i} \tag{10.2}$$

where z is an arbitrary molar quantity, x_i is the mole fraction of the component i in the mixture, and n is the number of components in the considered mixture. All molar quantities in this chapter are segment molar quantities; therefore, the double prime is omitted in the following text.

In practice, all polymers contain many usually homologous components, and hence polymers can be characterized by a molar-mass distribution. Depending on the polymerization mechanism, different analytical functions are available from the literature [1, 2]. In the case of radically polymerized products, the Schulz–Flory distribution $W(r)$ can be derived from the reaction mechanism [1]:

$$W(r) = \frac{k^k}{\Gamma(k) r_N} \left(\frac{r}{r_N}\right)^k \exp\left(-k \frac{r}{r_N}\right) \tag{10.3}$$

where k is a measure of the reciprocal nonuniformity, r is the segment number, r_N is the number-average segment number, and $\Gamma(k)$ is the Gamma function. The function $W(r)$ describes the distribution of the segment numbers of all polymer molecules present in the system, or respectively the molar mass. The influence of k on the distribution function at constant number average segment number is demonstrated in Figure 10.1, where the integral distribution function is given by

$$F(r) = \int_0^r W(r) dr \tag{10.4}$$

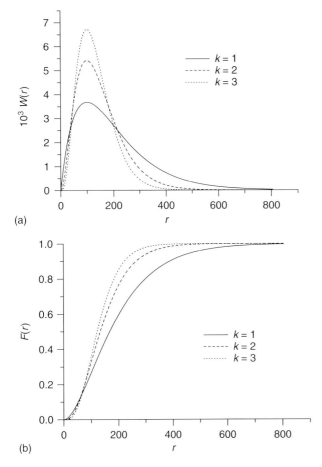

Figure 10.1 Differential (a) and integral (b) Schulz–Flory distribution function with $r_N = 100$ and different k values.

Continuous distribution functions (i.e., Eq. (10.3)) can be characterized by their moments, which are defined by

$$r^{(n)} = \int_0^\infty W(r) r^n dr \qquad (10.5)$$

For $n = 0$, Eq. (10.5) reflects the normalization condition. The first moment ($n = 1$) yields an average value of the distribution function. The second moment ($n = 2$) contains information about the broadness of the distribution function. The third moment ($n = 3$) describes the asymmetry of the distribution function. Unfortunately, these moments cannot directly be measured. However, they can be related to quantities accessible by experimental observations. For the characterization of polymers, several average values with respect to molar mass were applied (i.e., number-average molar mass (M_N), mass-average molar mass (M_M), and z-average molar mass (M_Z). In polymer thermodynamics, mostly the concept of segment-molar quantities is used, and hence the number-average segment number (r_N), the mass-average segment number (r_M), and the z-average segment number (r_Z) occur in the thermodynamic equations.

Using the distribution function (i.e., Eq. (10.3)), several average values can be defined. The number-average segment number, which is proportional to the number-average molar mass, which can be measured using the vapor pressure of solutions or osmometry, is given by the ratio of the 0 and -1 moment of the distribution function:

$$r_N = \frac{\overline{M}^{(0)}}{\overline{M}^{(-1)}} = \frac{\int_0^\infty W(r)dr}{\int_0^\infty \frac{W(r)}{r}dr} \tag{10.6}$$

Similar the mass-average segment number results in

$$r_M = \frac{\overline{M}^{(1)}}{\overline{M}^{(0)}} = \frac{\int_0^\infty rW(r)dr}{\int_0^\infty W(r)dr} \tag{10.7}$$

The mass-average segment number can be obtained by light-scattering measurements. The z-average segment number is given by

$$r_Z = \frac{\overline{M}^{(2)}}{\overline{M}^{(1)}} = \frac{\int_0^\infty r^2 W(r)dr}{\int_0^\infty rW(r)dr} \tag{10.8}$$

The z-average molar mass can be measured using an ultracentrifuge. Gel permeation chromatography allows the measurement of the complete distribution function and hence the estimation of all average molar masses.

Example 10.1

Calculation of the average segment number:

The polymer under consideration can be described with a Schulz–Flory distribution function with $r_N = 100$ and $k = 1$. For the thermodynamic modeling, the quantities r_M and r_Z must be calculated.

Solution

The Schulz–Flory distribution function is given in Eq. (10.3). For the calculation of the average values r_M and r_Z, the moments of the distribution function must be known. The moments can be computed using Eq. (10.5). Applying Eq. (10.3) to Eq. (10.5) leads to

$$n = 0 \quad \overline{M}^{(0)} = \int_0^\infty W(r)dr = \frac{k^k}{\Gamma(k)r_N}\int_0^\infty \left(\frac{r}{r_N}\right)^k \exp\left(-k\frac{r}{r_N}\right)dr$$

$$n = -1 \quad \overline{M}^{(-1)} = \int_0^\infty \frac{W(r)}{r}dr = \frac{k^k}{\Gamma(k)r_N}\int_0^\infty \left(\frac{r}{r_N}\right)^k \exp\left(-k\frac{r}{r_N}\right)\frac{1}{r}dr$$

$$n = 1 \quad \overline{M}^{(1)} = \int_0^\infty W(r)rdr = \frac{k^k}{\Gamma(k)r_N}\int_0^\infty \left(\frac{r}{r_N}\right)^k \exp\left(-k\frac{r}{r_N}\right)rdr$$

$$n = 2 \quad \overline{M}^{(2)} = \int_0^\infty W(r)r^2 dr = \frac{k^k}{\Gamma(k)r_N}\int_0^\infty \left(\frac{r}{r_N}\right)^k \exp\left(-k\frac{r}{r_N}\right)r^2 dr$$

The integrals in these equations can be solved analytically by substituting $\frac{r}{r_N} = x$ resulting in

$$n = 0 \quad \overline{M}^{(0)} = \frac{k^k}{\Gamma(k)} \int_0^\infty x^k \exp(-kx)dx$$

$$n = -1 \quad \overline{M}^{(-1)} = \frac{k^k}{\Gamma(k)r_N} \int_0^\infty x^{k-1} \exp(-kx)dx$$

$$n = 1 \quad \overline{M}^{(1)} = \frac{k^k r_N}{\Gamma(k)} \int_0^\infty x^{k+1} \exp(-kx)xdx$$

$$n = 2 \quad \overline{M}^{(2)} = \frac{k^k r_N^2}{\Gamma(k)} \int_0^\infty x^{k+2} \exp(-kx)dx$$

Using the general formula

$$\int_0^\infty x^m \exp(-ax) = \frac{\Gamma(m+1)}{a^{m+1}}$$

and the calculation rules for the Γ function

$$\Gamma(x) = (x-1)! \quad x \in N; \quad \Gamma(x+1) = x\Gamma(x)$$

leads to

$$\overline{M}^{(0)} = 1; \quad \overline{M}^{(-1)} = \frac{1}{r_N}; \quad \overline{M}^{(1)} = \frac{r_N(k+1)}{k}; \quad \overline{M}^{(2)} = \frac{r_N^2(k+2)(k+1)}{k^2}$$

$\overline{M}^{(0)} = 1$ means that the normalization condition is fulfilled. This is the analog to the statement that the sum of all mole fractions of the components in a mixture must be 1. $\overline{M}^{(1)}$ describes the broadness of the distribution function and $\overline{M}^{(2)}$ characterizes the asymmetry. Finally, the average values r_M and r_Z are given by inserting $\overline{M}^{(n)}$ into Eq. (10.7) and, respectively, into Eq. (10.8):

$$r_M = \frac{\overline{M}^{(1)}}{\overline{M}^{(0)}} = \frac{r_N(k+1)}{k} \quad r_Z = \frac{\overline{M}^{(2)}}{\overline{M}^{(1)}} = \frac{r_N^2(k+2)(k+1)k}{k^2 r_N(k+1)} = \frac{r_N(k+2)}{k}$$

Using $r_N = 100$ and $k = 1$, the final results are $r_M = 200$ and $r_Z = 300$.

In mixtures of low-molar-mass components, the structure of the components will not depend on concentration. However, the structure of the polymer depends on solvent concentration. At a certain polymer concentration (overlapping concentration c^*), entanglements of the polymer chains will occur (semidiluted polymer solution). This effect is neglected in most thermodynamic equations.

Nowadays, many polymerization reactions lead to products having very narrow molar-mass distributions, but such product samples still differ in average molar-mass values. Also, fractionation [2] can narrow the distribution function, but never produces a monodisperse polymer. As a hypothetical case, we consider a monodisperse polymer dissolved in one solvent as quasi-binary. In this chapter, first the thermodynamic models (g^E-models and equations of state (EOS)) for a quasi-binary system, consisting of a monodisperse polymer and a solvent, will be introduced, and later the influence of polydispersity on the phase behavior will be discussed. Caused by the negligible vapor pressure of polymers, the most important phase equilibrium is the liquid–liquid equilibrium. For this reason, we focus our attention on liquid–liquid phase split, where equilibrium between two phases differing in polymer concentration and in polymer molar-mass distribution (fractionation effect) is established.

If the polydispersity is neglected, the polymer solution behaves similarly to a binary system (Figure 5.67). For such a polymer solution, no fractionation effect occurs. In the equilibrium state of polymer solutions, two stable liquid phases of differing compositions are rather formed than one single liquid phase, at least over certain ranges of temperature and composition. The liquid–liquid equilibrium of a polymer solution can be seen in Figure 10.2. At temperatures above the critical solution temperature (marked with a star in Figure 10.2), the polymer solution is stable and hence one homogeneous liquid phase is present. If the temperature and composition are in the single-phase region (marked with stable in Figure 10.2), the equilibrium state is a single liquid phase. If, however, the temperature and overall composition of the mixture lie in the two-phase region (marked with meta-stable and unstable in Figure 10.2), two liquid phases will be formed. The compositions of these phases are given by the intersection of the constant temperature line (tie line in Figure 10.2) with the boundaries of the two-phase region. The stable region is separated from the metastable region by the binodal line.

For example, a polymer solution with a polymer concentration according to the left arrow in Figure 10.2 will cool down from a temperature above the binodal line to a temperature below the binodal line (T_1 in Figure 10.2). During this cooling process, the polymer solution becomes cloudy and the first droplet of the coexisting phase will be formed. For this reason, the binodal line is also called *cloud point curve*. At the temperature T_1, two coexisting phases differing in the polymer concentration will establish in equilibrium. The polymer concentrations in both phases can be read on the binodal line (or cloud point curve) at T_1. Both phases in equilibrium are connected via the tie line. The spinodal line separates

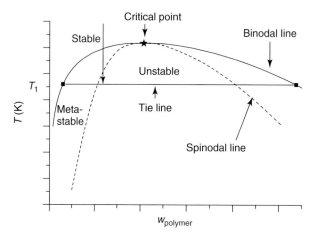

Figure 10.2 Liquid–liquid equilibrium of binary polymer solutions at constant pressure.

the metastable from the unstable region of the phase diagram. A metastable phase is stable against small fluctuations, but not against all variations, like a stable state. Simple examples are subcooled liquid or superheated vapor. The spinodal line runs through a maximum and meets the binodal line in the critical point. In order to calculate phase diagrams (Figure 10.2), a g^E-model or an EOS can be applied.

10.2
g^E-models

Similarly to the description of real phase behavior of mixtures of low-molar-mass components, mixture models based on activity coefficients can be formulated. Whereas in the case of low-molar-mass components the models describe the deviations from an ideal mixture, the models for polymer solutions account for the deviations from an ideal-athermic mixture. As a starting point for the development of a model, all segments are placed on a lattice (Figure 10.3). Polymer chains will be arranged on lattice sites of equal size, where the number of occupied lattice sites depends on the segment number r. For a quasi-binary polymer solution, all other places are occupied with solvent segments.

Using statistical thermodynamics, Flory [1] derived an expression for the entropy of mixing for this situation (Figure 10.3), leading for a polymer (B) dissolved in solvent (A) to

$$\Delta s^{id} = -R \left[\frac{x_A}{r_A} \ln x_A + \frac{x_B}{r_B} \ln x_B \right] \tag{10.9}$$

where x_B is the segment fraction of the polymer and x_A is the segment fraction of the solvent. Both segment fractions can be calculated applying Eq. (10.2), if for the general quantity z the mole fraction is used. Equation 10.9 defines the ideal-athermic mixture (or Flory–Huggins mixture, $h^E = 0$) and is used as a standard state for polymer solutions. This equation replaces the entropy of mixing of an ideal mixture for low-molar-mass components. In order to describe deviations

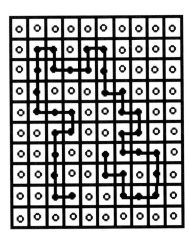

Figure 10.3 Schematic picture of the Flory–Hugging lattice.

from the ideal-athermic mixture, models for the excess enthalpy can be formulated. One well-known option is the χ-parameter concept, originally introduced by Flory [1] and Huggins [3]:

$$\frac{h^E}{RT} = x_B x_A \chi \left(T, x_B, r_B\right) \tag{10.10}$$

Considering the χ-parameter only as a function of temperature corresponds to the original Flory–Huggins theory [1]. Later, different expressions for the dependence of χ on the polymer concentration and the segment number have been suggested. The dependence of χ on the segment number r allows taking into account oligomer effects. For simplicity, it is assumed that the polymer has a sufficiently high molar mass and hence no oligomer effects occur. Combining Eq. (10.9) with Eq. (10.10) results in the excess Gibbs energy for a quasi-binary polymer solution:

$$\frac{g^E}{RT} = \frac{x_A}{r_A} \ln x_A + \frac{x_B}{r_B} \ln x_B + x_B x_A \chi \left(T, x_B\right) \tag{10.11}$$

The dependence of χ on the segment number is not taken into account, because the end group effects are neglected. Using Eq. (10.11), the Gibbs energy of the quasi-binary polymer solution is given by

$$\frac{g}{RT} = x_A \frac{g_{0A}}{RT} + x_B \frac{g_{0B}}{RT} + \frac{x_A}{r_A} \ln x_A + \frac{x_B}{r_B} \ln x_B + x_B x_A \chi \left(T, x_B\right) \tag{10.12}$$

where g_{0A} and g_{0B} represent the Gibbs energies for the pure solvent and the pure polymer, respectively.

With the help of Eq. (10.12), all thermodynamic quantities necessary for liquid–liquid equilibria calculations can be obtained. One important quantity is the segment molar chemical potential μ_i of a component i present in the mixture. This quantity can be obtained by derivation of Eq. (10.12) with respect to the segment number of moles of the considered compound. For the segment chemical potential of the polymer, it can be found with $x_A = 1 - x_B$:

$$\frac{\mu_B}{RT} = \left(\frac{\partial g/RT}{\partial n_B}\right)_{T,P,n_A} = \frac{g}{RT} + x_A \left(\frac{\partial g/RT}{\partial x_B}\right)_{T,P}$$

$$\frac{\mu_B}{RT} = \frac{g_{0B}}{RT} + \frac{\ln x_B}{r_B} + x_A \left(\frac{1}{r_B} - \frac{1}{r_A}\right) + x_A^2 \left(\chi\left(T, x_B\right) + x_B \left(\frac{\partial \chi\left(T, x_B\right)}{\partial x_B}\right)_{T,P}\right) \tag{10.13}$$

For the solvent, the segment chemical potential is

$$\frac{\mu_A}{RT} = \left(\frac{\partial g/RT}{\partial n_A}\right)_{T,P,n_B} = \frac{g}{RT} - x_B \left(\frac{\partial g/RT}{\partial x_B}\right)_{T,P}$$

$$\frac{\mu_A}{RT} = \frac{g_{0A}}{RT} + \frac{\ln x_A}{r_A} + x_B \left(\frac{1}{r_A} - \frac{1}{r_B}\right) + x_B^2 \left[\chi\left(T, x_B\right) - x_A \left(\frac{\partial \chi\left(T, x_B\right)}{\partial x_B}\right)_{T,P}\right] \tag{10.14}$$

For phase equilibrium calculations, the usual phase equilibrium conditions can be applied:

$$\mu_A^I\left(x_B^I, T\right) = \mu_A^{II}\left(x_B^{II}, T\right); \mu_B^I\left(x_B^I, T\right) = \mu_B^{II}\left(x_B^{II}, T\right) \tag{10.15}$$

Inserting Eqs. (10.13) and (10.14) into Eq. (10.15) results in

$$\ln\left(\frac{x'_A}{x''_A}\right) = r_A \rho_A; \quad \ln\left(\frac{x'_B}{x''_B}\right) = r_B \rho_B \tag{10.16}$$

where the abbreviations ρ_A and ρ_B are given by

$$\rho_A = (x''_B - x'_B)\left(\frac{1}{r_A} - \frac{1}{r_B}\right) + \left(x''^2_B - x'^2_B\right)\chi(T, x''_B)$$

$$+ \left(x'^2_B x'_A - x''^2_B x''_A\right)\left(\frac{\partial \chi(T, x'_B)}{\partial x'_B}\right)_{T,P}$$

$$\rho_B = (x''_A - x'_A)\left(\frac{1}{r_B} - \frac{1}{r_A}\right) + \left(x''^2_A - x'^2_A\right)\chi(T, x''_B)$$

$$+ \left(x''^2_A x''_B - x'^2_A x'_B\right)\left(\frac{\partial \chi(T, x'_B)}{\partial x'_B}\right)_{T,P} \tag{10.17}$$

In phase equilibrium calculations, three unknowns, namely the polymer concentration in the dilute phase x'_B, the polymer concentration in the concentrated phase x''_B, and the temperature T, occur. Specifying one of the unknowns, the two other ones can be calculated by solving Eq. (10.15) simultaneously. In order to find suitable initial values, the spinodal and/or the critical solution point can be helpful. The spinodal (see Figure 10.2) separates the metastable from the unstable region in the phase diagram and can be calculated with the help of the stability theory, where the necessary condition reads

$$\left(\frac{\partial^2 g/RT}{\partial x_B^2}\right)_{T,P} = 0 \tag{10.18}$$

If phase separation in the considered mixture at a given temperature occurs, the Gibbs energy as a function of composition can show two minima (Figure 10.4). The equilibrium criterion for a closed system at constant temperature and pressure is the minimum of the Gibbs energy of the system. For a mixture with an overall composition between the minima of G (solid line in Figure 10.4), the lowest value of the Gibbs energy of the mixture is obtained when the mixture separates into two phases. In this case, the Gibbs energy of the mixture is a linear combination of the Gibbs energies of the two coexisting phases, and it is represented by the dashed line in Figure 10.4. The inflection point (marked with stars in Figure 10.4) of the Gibbs energy as a function of composition (solid line in Figure 10.4) characterizes the boundary between the metastable and the unstable region at constant temperature in Figure 10.2. From the mathematical point of view, this means that the second derivative with respect to the composition must be zero (Eq. (10.18)).

For a quasi-binary system, the critical solution point is located in the extremum of the spinodal and hence the necessary condition is

$$\left(\frac{\partial^3 g/RT}{\partial x_B^3}\right)_{T,P} = 0 \tag{10.19}$$

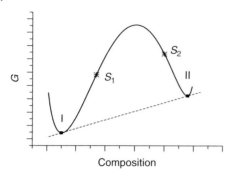

Figure 10.4 Schematic drawing of the Gibbs energy of the quasi-binary polymer solution as a function of composition at constant temperature and pressure.

An example for an analytical equation for the χ parameter is the modified semiempirical Koningsveld–Kleintjens model [2]:

$$\chi(T, x_B) = \frac{\beta(T)}{1 - \gamma x_B} \tag{10.20}$$

where $\beta(T)$ and γ are the adjustable parameters.

These parameters must be estimated using pseudo-binary data. Inserting Eq. (10.20) into Eq. (10.12) and applying Eq. (10.14) reads

$$\frac{\mu_A}{RT} = \frac{g_{0A}}{RT} + \frac{\ln x_A}{r_A} + x_B\left(\frac{1}{r_A} - \frac{1}{r_B}\right) + \frac{x_B^2 \beta(T)(1-\gamma)}{(1-\gamma x_B)^2} \tag{10.21}$$

Inserting Eq. (10.20) into Eq. (10.12) and applying Eq. (10.13) yields

$$\frac{\mu_B}{RT} = \frac{g_{0B}}{RT} + \frac{\ln x_B}{r_B} + x_A\left(\frac{1}{r_B} - \frac{1}{r_A}\right) + \frac{x_A^2 \beta(T)}{(1-\gamma x_B)^2} \tag{10.22}$$

With Eq. (10.20) for the spinodal condition (Eq. (10.18)), one obtains

$$\left(\frac{\partial^2 g/RT}{\partial x_B^2}\right)_{T,P} = \frac{1}{r_A x_A} + \frac{1}{r_B x_B} - \frac{2\beta(T)(\gamma-1)}{(1-\gamma x_B)^3} = 0 \tag{10.23}$$

Using Eq. (10.20) for the critical condition (Eq. (10.19)) results in

$$\left(\frac{\partial^3 g/RT}{\partial x_B^3}\right)_{T,P} = \frac{1}{r_A x_A^2} - \frac{1}{r_B x_B^2} + \frac{6\beta(T)(\gamma-1)\gamma}{(1-\gamma x_B)^4} = 0 \tag{10.24}$$

Combining the spinodal condition (Eq. (10.23)) with the critical condition (Eq. (10.19)) gives the opportunity to estimate the adjustable parameter γ using the critical concentration, x_B^C:

$$\gamma = \frac{1}{x_B^C + \dfrac{3\left(\dfrac{1}{r_A x_A^C} + \dfrac{1}{r_B x_B^C}\right)}{\left(\dfrac{1}{r_A (x_A^C)^2} - \dfrac{1}{r_B (x_B^C)^2}\right)}} \tag{10.25}$$

In order to determine the quantities ρ_A and ρ_B in Eq. (10.17) for phase equilibrium calculations, the chemical potentials given in Eqs. (10.21) and (10.22) must be equated, resulting in

$$\rho_A = (x_B'' - x_B') \left(\frac{1}{r_A} - \frac{1}{r_B} \right) + \beta(T)(1-\gamma) \left[\frac{x_B''^2}{(1-\gamma x_B'')^2} - \frac{x_B'^2}{(1-\gamma x_B')^2} \right] \tag{10.26}$$

and

$$\rho_B = (x_A'' - x_A') \left(\frac{1}{r_B} - \frac{1}{r_A} \right) + \beta(T) \left(\frac{x_A''^2}{(1-\gamma x_B'')^2} - \frac{x_A'^2}{(1-\gamma x_B')^2} \right) \tag{10.27}$$

The quantity ρ_A or ρ_B can be interpreted as distance from the critical solution point. Having in mind that the temperature is only involved in the $\beta(T)$ function, a simplification is possible. Rearranging both Eqs. (10.26) and (10.27) and using Eq. (10.16) allows elimination of $\beta(T)$:

$$\frac{\frac{1}{r_A} \ln\left(\frac{x_A'}{x_A''}\right) - (x_B'' - x_B')\left(\frac{1}{r_A} - \frac{1}{r_B}\right)}{(1-\gamma)\left[\frac{(x_B'')^2}{(1-\gamma x_B'')^2} - \frac{(x_B')^2}{(1-\gamma x_B')^2}\right]} = \frac{\frac{1}{r_B} \ln\left(\frac{x_B'}{x_B''}\right) - (x_A'' - x_A')\left(\frac{1}{r_B} - \frac{1}{r_A}\right)}{\left(\frac{(x_A'')^2}{(1-\gamma x_B'')^2} - \frac{(x_A')^2}{(1-\gamma x_B')^2}\right)} \tag{10.28}$$

Eq. (10.28) allows the calculation of the liquid–liquid phase split by searching the root of one simple equation. For each value of x_A', the corresponding value of x_A'' can be calculated. The phase separation temperature is given, if a detailed function of $\beta(T)$ is specified. Usually, a simple function is used:

$$\beta(T) = \beta_0 + \frac{\beta_1}{T} + \frac{\beta_2}{T^2} \tag{10.29}$$

The choice of the parameter β_i has a great impact on the resulting phase behavior. Typical phase diagrams (see Figure 5.67) can by characterized as having

1) an upper critical solution temperature (UCST),
2) a lower critical solution temperature (LCST),
3) a closed loop miscibility gap with two critical solution points, and
4) two separated miscibility gaps with two critical solution points.

In the cases (1) and (2), the critical condition (Eq. (10.19)) needs one root with respect to temperature, and hence the parameter β_2 in Eq. (10.29) can be set to zero. The sign of the parameter β_1 in Eq. (10.29) determines the type of critical solution point. If $\beta_1 < 0$, the polymer solution will have an LCST, and if $\beta_1 > 0$, the polymer solution will have an UCST. In the cases (3) and (4), the critical condition (Eq. (10.19)) needs two roots with respect to temperature and hence the parameter β_2 in Eq. (10.29) cannot be set to zero. In order to find values for these parameters β_i, the critical solution point (or critical solution points), given spinodal data or given liquid–liquid equilibria (LLE) data can be used. If experimental LLE data are available, the following strategy can be recommended:

1) estimation of the parameter γ using Eq. (10.25),
2) calculation of the phase composition using Eq. (10.28),

3) calculation of the β value using Eqs. (10.26) or (10.27), and
4) fitting $\beta(T)$ to the experimental demixing temperatures by linear regression.

If no experimental demixing data are available, the adjustable model parameter can be estimated using other experimental data. One option is the utilization of experimental spinodal data in Eq. (10.23). Spinodal data can be measured using temperature pulsed or pressure pulsed induced critical scattering.

Another option is the utilization of experimental excess enthalpy data, which can be measured calorimetrically. Alternatively, inverse gas chromatography can be useful to estimate the adjustable parameters. This method refers to the characterization of the chromatographic stationary phase (polymer) using a known amount of solvent. The stationary phase is prepared by coating an inert support with polymer and packing the coated particles into a conventional gas chromatography column. The activity of the given solvent can be related to its retention time on the column.

The model parameters can also be obtained by measurements of the vapor pressure of the polymer solution. Assuming that the vapor pressure of the polymer is zero leads to

$$\ln\left(\frac{P^s(T)}{P_A^s(T)}\right) = \frac{\ln x_A}{r_A} + x_B\left(\frac{1}{r_A} - \frac{1}{r_B}\right) + \frac{x_B^2 \beta(T)(1-\gamma)}{(1-\gamma x_B)^2} \tag{10.30}$$

where P^s is the vapor pressure of the polymer solution, $P_A^s(T)$ is the vapor pressure of the pure solvent at temperature T.

Below typical examples of all four principal types of phase diagrams are discussed. First, the classical UCST behavior is shown in Figure 10.5 for the system toluene + polymethacrylate, where the polymer has a mass average molar mass of 1770 kg/mol. Using the Koningsveld–Kleintjens model (Eq. (10.20)), an excellent agreement between the experimental data (M. Haberer and B.A. Wolf, private communication) and the model could be achieved. As mentioned above, the value β_1 has a positive sign and hence the system shows a UCST behavior, meaning that the demixing starts during cooling. Several polymers (i.e., N-isopropylacrylamide, methylcellulose, polyols, poly(vinylether), poly(N-vinylcaprolactam), and poly(N-vinylisobutyramide)) show a demixing curve with a LCST in water, where the demixing starts with heating.

An example is shown in Figure 10.6, where the polymer has a molar mass of 150 kg/mol. The LCST behavior can be modeled using a negative value for β_1 in the selected model. Example 10.2 explains the calculation procedure in detail.

Example 10.2

Calculation of the cloud point curve for the system water + methylcellulose (Figure 10.6).

The experimental cloud point curve of the system water + methylcellulose having a molar mass of 150 kg/mol shall be modeled and the parameter of the Koningsveld–Kleintjens model (Eq. (10.20)) shall be estimated.

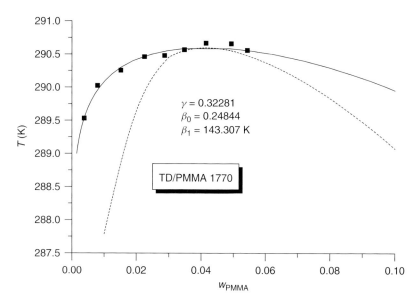

Figure 10.5 Comparison of experimental (M. Haberer and B.A. Wolf, private communication) and calculated phase equilibria (solid line: binodal and dotted line: spinodal) for the system toluene + polymethacrylate.

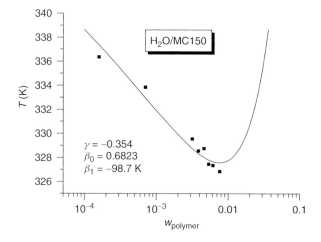

Figure 10.6 Comparison of experimental [4] and calculated phase diagram for the system water + methylcellulose.

Solution

The experimental demixing data of the system water + methylcellulose having a molar mass of 150 kg/mol are shown in Figure 10.6. They were measured using differential thermal analysis [4]. In order to model these data, the first step is the estimation of the number of segments of the solvent and the polymer. If the solvent (A) is chosen as standard segment and the molar mass as physical property L in Eq. (10.1), this leads to $r_B = 150\,000/18 = 8333.3$ and $r_A = 1$. This definition means that the mass fraction w is identical with the segment fraction x. In the literature [4], no information about the polydispersity is available. For this reason, it is necessary to assume that the polymer is monodisperse, as polydispersity effects cannot be taken into account. For the calculation of the cloud point curve, a suitable g^E-model (Eq. (10.11)) must be selected. Choosing Eq. (10.20) in combination with Eq. (10.29), the parameters γ, β_0, β_1, and β_2 must be estimated. The cloud point curve in Figure 10.6 shows an LCST behavior and hence the parameter β_2 can be set to zero. If the polymer solution is considered to be a binary system, then the critical solution point is located in the minimum of the cloud point curve. Using the data given in Figure 10.6, the critical mass fraction or the critical segment fraction is equal to $w_B^C = x_B^C = 0.00756$. Equation 10.25 can be applied to calculate the parameter γ:

$$\gamma = \cfrac{1}{0.00756 + \cfrac{3\left(\cfrac{1}{(1-0.00756)} + \cfrac{1}{8333.3 \cdot 0.00756}\right)}{\left(\cfrac{1}{(1-0.00756)^2} - \cfrac{1}{8333.3\,(0.00756)^2}\right)}} = -0.354$$

The remaining parameters in Eq. (10.29) are β_0 and β_1. These parameters can be estimated by the calculation of $\beta(T)$ for each experimental polymer concentration. Let us consider one example. Assume that the experimental segment fraction of the polymer is $x_B' = 0.00317$, and hence the segment fraction of the solvent is $x_A' = 0.99683$. First, the corresponding concentration in the second phase x_B'' must be calculated. This can be carried out with the help of Eq. (10.28) containing only one unknown, namely x_B'':

$$f(x_B'') = \cfrac{\ln\left(\cfrac{0.99683}{1-x_B''}\right) - (x_B'' - 0.00317)\left(1 - \cfrac{1}{8333.3}\right)}{(1+0.354)\left[\cfrac{(x_B'')^2}{(1+0.354 x_B'')^2} - \cfrac{(0.00317)^2}{(1+0.354 \cdot 0.00317)^2}\right]}$$
$$- \cfrac{\cfrac{1}{8333.3}\ln\left(\cfrac{0.00317}{x_B''}\right) - (1 - x_B'' - 0.99683)\left(1 - \cfrac{1}{8333.3}\right)}{\left(\cfrac{(1-x_B'')^2}{(1+0.354 x_B'')^2} - \cfrac{(0.99683)^2}{(1+0.354 \cdot 0.00317)^2}\right)} = 0$$

This equation can be solved numerically using the Regula Falsi method, which does not need analytical derivatives in contrast to the Newton method. The interval of the search for the root should be specified. The polymer concentration of the polymer-rich phase must be above the critical concentration ($x_B^C = 0.00756$). The search interval is therefore between the critical concentration $x_{B,1}'' = 0.00756$ and a value at higher polymer concentration ($x_{B,2}'' = 0.02$). Using the Regula Falsi

Table 10.1 Iteration procedure for the calculation of a cloud point.

Iteration step	$x''_{B,i-1}$	$x''_{B,i}$	$f(x''_{B,i-1})$	$f(x''_{B,i})$	$x''_{B,i+1}$
1	0.00756	0.02	−2.84E−4	5.70E−4	0.01167394
2	0.01167394	0.02	−1.96E−4	5.70E−4	0.01378491
3	0.01378491	0.02	−4.98E−5	5.70E−4	0.01427938
4	0.01427938	0.02	−9.17E−6	5.70E−4	0.01436909
5	0.01436909	0.02	−1.58E−6	5.70E−4	0.01438454
6	0.01438454	0.02	−2.70E−7	5.70E−4	0.01438718
7	0.01438718	0.02	−4.61E−8	5.70E−4	0.01438763
8	0.01438763	0.02	−7.85E−9	5.70E−4	0.0143877

method, the following iteration formula is valid:

$$x''_{B,i+1} = x''_{B,i-1} - \frac{(x''_{B,i} - x''_{B,i-1}) f(x''_{B,i-1})}{f(x''_{B,i}) - f(x''_{B,i-1})}$$

Table 10.1 shows the iteration history when the Regula Falsi method is applied.

According to Table 10.1, the solution with an accuracy better than 10^{-8} is obtained after eight iteration steps. Knowing the polymer concentration in both phases, the quantity $\beta(T)$ can be calculated using one side of Eq. (10.28). For this example, one obtains

$$\beta(T) = \frac{\ln\left(\frac{1 - 0.00317}{1 - 0.0143877}\right) - (0.0143877 - 0.00317)\left(1 - \frac{1}{8333.3}\right)}{(1 + 0.354)\left[\frac{(0.0143877)^2}{(1 + 0.354 \cdot 0.0143877)^2} - \frac{(0.00317)^2}{(1 + 0.354 \cdot 0.00317)^2}\right]}$$

$$\beta(T) = 0.3821$$

This iteration and the calculation of $\beta(T)$ has to be performed for each experimental data point given in the literature [4], resulting in Table 10.2.

Now, the parameters β_0 and β_1 in Eq. (10.29) can be estimated by plotting the calculated β-values for each data point versus the experimental reciprocal cloud-point temperature (Figure 10.7). In this figure, the point number 9 in Table 10.2 was dropped out, because this point is completely out of the range. Fitting the data points to a straight line results in the parameters $\beta_0 = 0.6823$ and $\beta_1 = -98.7$ K in Eq. (10.29). The system H_2O + MC150 shows a LCST behavior; hence, the parameter β_1 has a negative sign. Now all parameters are specified and the whole cloud-point curve (Figure 10.6), the spinodal line, and the critical solution point can be calculated.

If two critical solution points are present in the system, the parameter β_2 in Eq. (10.29) must be taken into account. An example is the system poly(ethylene glycol) + water [5], which shows a closed loop miscibility gap. The phase behavior of this system is plotted in Figure 10.8, where the molar mass of the polymer

Table 10.2 Numerical results for the cloud-point curve for the system $H_2O + MC150$.

Number	x'_B	T (K)	x''_B	$\beta(T)$
1	0.00016	336.35	0.03413	0.38974
2	0.00071	333.85	0.02532	0.38587
3	0.00317	329.55	0.01439	0.38206
4	0.00386	328.55	0.01283	0.38168
5	0.00461	328.75	0.01142	0.38139
6	0.00536	327.45	0.01023	0.38120
7	0.00614	327.35	0.00916	0.38107
8	0.00764	326.85	0.00764	0.38107
9	0.01033	352.45	0.00529	0.38121

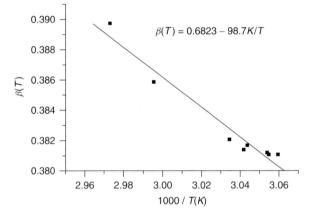

Figure 10.7 Parameter estimation for the example $H_2O + MC150$ without point 9 in Table 10.2.

is 3.35 kg/mol. In this case, demixing can be observed in the temperature range between about 420 K and 520 K, whereas outside this temperature range complete miscibility is observed.

Two critical solution points also occur if two separated miscibility gaps are present in the system, one of them connected with an UCST behavior and the other one connected with an LCST behavior. Hence, demixing can occur at heating and at cooling. The classical examples for such a complex behavior are polystyrene (PS) dissolved in different solvents (i.e., diethyl ether, *tert*-butyl acetate, cyclohexane, methyl cyclohexane (MCH), and acetone). Both critical solution points can merge if an important parameter like the molar mass or the pressure is changed [6]. An example is the system PS + MCH, where PS has a very narrow molar mass distribution with an average molar mass of 17.5 kg/mol [7]. The hour glass shape of the phase diagram can be recognized in isothermal phase diagrams, where the demixing pressure is plotted versus the concentration (Figure 10.9). At temperatures above

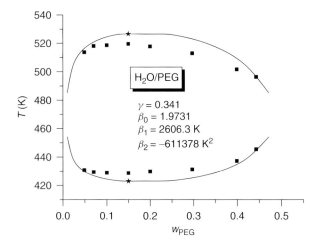

Figure 10.8 Comparison of experimental [5] and calculated phase behavior of the system poly(ethylene glycol) + water.

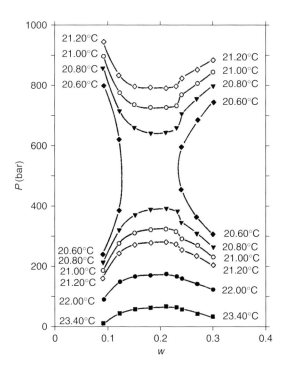

Figure 10.9 Isothermal phase behavior of the system PS + MCH at different temperatures [7].

20.6 °C, two miscibility gaps are clearly observed, where each one has a critical solution point (USCT and LCST). With increasing pressure, the binodal line connected with UCST behavior shifts to lower temperatures, whereas the critical concentration stays approximately constant. The LCST behavior shifts vice versa. At 20.8 °C, the upper and lower demixing region is separated at approximately 200 bar. Slight decrease of temperature moves the two miscibility gaps closer together until they merge and form an hour-glass-shaped two-phase region at 20.6 °C. At this temperature and below this temperature, no critical solution point can be found. The demixing region is now located at very low and at relatively high polymer concentrations. In the mass fraction range between 0.1 and 0.3, the polymer solution will be homogeneous. In case of a monodisperse polymer, first the two critical solution points merge together forming a double-critical solution point. Secondly, a new pattern of phase behavior appears (Figure 10.9). In order to model such a behavior, especially the merging process as a function of pressure, the pressure must be included in the model. One possibility to do so is to take the pressure dependence of the χ-function into account [2, 7]. It can be assumed that the parameters β_i depend on pressure via a polynomial expression:

$$\beta_i = \alpha_{i,0} + \alpha_{i,1} P + \alpha_{i,2} P^2$$

The disadvantage of this method is the large number of adjustable model parameters. An elegant solution of this problem is the replacement of the g^E-model by an EOS (see Section 10.3).

If the simplification by reducing the system to one nonlinear equation, shown in Example 10.2, is not possible, the phase equilibrium calculations can also be performed using the method suggested by Solc [8]. The basic idea of this method is to avoid the trivial solution of Eq. (10.15) by seeking a value of ρ_A or ρ_B, meaning the distance from the critical solution point, in Eq. (10.17) and calculate the concentration in both phases. Other alternative solution pathways were suggested by Horst [9]. The last method, based on the direct minimization of the Gibbs energy, is especially useful if no analytical expression for the Gibbs energy of mixing is available. This situation arises when the influence of shear stress on the phase behavior is studied. The shear can lead both to an increase or decrease of the miscibility gap, depending on the sign of the so-called storage energy. The dependence of the phase behavior on shear is much more pronounced for polymer-containing systems than for mixtures consisting of low-molar-mass components.

The extension of the lattice theory to more than two components [2] and to polymer blends [2] is straightforward.

10.3
Equations of State

Some polymers like low-density polyethylene are produced at high pressure. For the development of new production processes or for the further optimization of these processes, the involved phase equilibrium behavior must be known as a *function of pressure*. Some polymers (i.e., PS, polyurethane, and polyvinylchloride) are applied as foam for isolation purposes. In this case, the solubility of the foaming agent (e.g., carbon dioxide or *n*-pentane) as a function of pressure must be known. Using g^E-model in order to model the pressure influence, a large number of empirical parameters must be determined from experimental data. The application of an EOS is much more convenient. For the thermodynamics of low-molar-mass components, usually experimental vapor pressures as a function of temperature and experimental *PvT* data, especially saturated liquid densities, of the pure components are used to estimate the model parameters of this component. However, the vapor pressure of polymers is practically zero. The measurement of the liquid density is often also not possible, because the polymer might degrade before it melts. For these reasons, the estimation of the pure polymer parameters used in EOS is a challenging issue. Recently, methods have been developed where the parameter fitting procedure is based on *PvT* data of the considered polymer and extrapolating equations that relate the polymer parameters to those of the corresponding monomer.

For polymer systems, a lot of EOS have been developed (e.g., Flory–Orwoll–Vrij EOS [10], Sanchez–Lacombe EOS [11], Panayiotou–Vera EOS [12], lattice gas EOS [13], group-contribution lattice fluid EOS [14]). In all these equations, the pressure is introduced via empty lattice sites allowing the compressibility of the lattice.

Another approach is the extension of an EOS originally developed for low-molar-mass fluids to polymers. Some of these extensions are based on van der Waals-type equations [15]. All versions of the Statistical Association Fluid Theory (SAFT) family [16, 17], originally developed by Chapman *et al.* [18], belong to this category. An overview about the new developments can be found in literature [19, 20]. One of the most successful versions is the Perturbed-Chain-SAFT EOS (PC-SAFT EOS) [21, 22], which is based on statistical thermodynamics. Molecularly based EOS do not only provide a useful basis for deriving chemical potentials or fugacities that are needed for phase equilibrium calculations but also allow for quantifying the effect of molecular structure and interactions on bulk properties and phase behavior. In the following, we focus our attention to the PC-SAFT EOS using the version from Gross and Sadowski [21, 22].

Within this framework, it is assumed that molecules are made of a chain of hard spheres having a temperature-independent segment diameter, σ. The chains interact via a modified square-well potential $u(r)$:

$$u(r) = \begin{cases} \infty & r < (\sigma - s_1) \\ 3\varepsilon & (\sigma - s_1) \leq r < \sigma \\ -\varepsilon & \sigma \leq r < \lambda\sigma \\ 0 & r \geq \lambda\sigma \end{cases} \quad (10.31)$$

where r is the radial distance between two segments, ε denotes the depth of the potential well, and λ is the reduced well width. σ is the hard-sphere segment diameter. It is assumed that $s_1/\sigma = 0.12$ and $\lambda = 1.5$. The potential given by Eq. (10.31) is sketched in Figure 10.10.

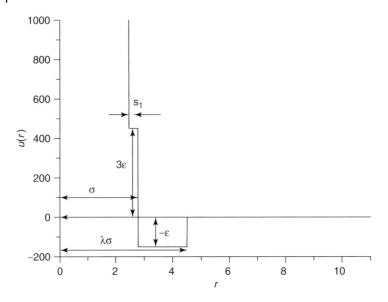

Figure 10.10 Schematic drawing of the potential $u(r)$ given in Eq. (10.31).

In the literature EOS with a varying well width (SAFT-VR versions [23]) are also available. The length of the chain is given by the parameter m. Using a g^E-model, the segment number is abbreviated with r. This quantity is calculated with the help of Eq. (10.1). In contrast to r the segment number m, used if an EOS is applied, is a pure-component parameter to be fitted to experimental data. A nonassociating molecule is characterized by three pure-component parameters: the segment diameter σ, the depth of the potential ε, and the segment number per chain m.

The interactions of molecules can be divided into a repulsive and an attractive part. For the calculation of the repulsive contribution, a reference fluid with no attractive forces is defined and the attractive interactions are treated as a perturbation of the reference system. According to the Barker–Henderson perturbation theory [21], a reference fluid with hard repulsion (Eq. (10.31)) and a temperature-dependent segment diameter d_i can be applied. For a component i, the following can be found:

$$d_i = \sigma_i \left(1 - 0.12 \exp\left(-\frac{3\varepsilon_i}{kT}\right)\right) \qquad (10.32)$$

where k is the Boltzmann constant and T is the absolute temperature.

The residual Helmholtz energy consists of the hard-chain reference, a^{hc}, and the dispersion contribution, a^{disp}. Additionally, contributions arising from polar interaction, a^{polar}, and association, a^{assoc}, can be included. The overall Helmholtz energy reads

$$a = a^{id} + a^{hc} + a^{disp} + a^{assoc} + a^{polar} \qquad (10.33)$$

where a^{id} represents the ideal gas contribution and is given by

$$\frac{a^{id}}{RT} = \ln(\rho RT) + \sum_{i=1}^{n} x_i \ln x_i \tag{10.34}$$

where ρ is the density, x_i is the mole fraction, and n is the number of components present in the mixture. The hard-chain reference contribution in Eq. (10.33) reads

$$\frac{a^{hc}}{RT} = \overline{m}\frac{a^{hs}}{RT} - \frac{1}{RT}\sum_{i=1}^{n} x_i(m_i - 1)\ln\left(g_{ii}^{hs}(\sigma_{ii})\right) \tag{10.35}$$

where $\overline{m}$ is the mean segment number in the considered mixture:

$$\overline{m} = \sum_{i=1}^{n} x_i m_i \tag{10.36}$$

The last term in Eq. (10.35) is the radial distribution function. It represents the probability of finding segment i at a given distance from segment j:

$$g_{ij}^{hs}(d_{ii}, d_{jj}) = \frac{1}{1-\zeta_3} + \left(\frac{d_{ii}d_{jj}}{d_{ii}+d_{jj}}\right)\frac{3\zeta_2}{(1-\zeta_3)^2} + \left(\frac{d_{ii}d_{jj}}{d_{ii}+d_{jj}}\right)^2\frac{2\zeta_2^2}{(1-\zeta_3)^3} \tag{10.37}$$

with ζ_k defined as

$$\zeta_k = \frac{\pi\rho N_{Av}}{6}\sum_{i=1}^{n} x_i m_i d_{ii}^k \quad k \in \{0,1,2,3\}; \eta = \zeta_3 \tag{10.38}$$

where N_{Av} is the Avogadro constant. The Helmholtz energy of the hard-sphere fluid in Eq. (10.35) is given on a per-segment basis:

$$\frac{a^{hs}}{RT} = \frac{1}{\zeta_0}\left[\frac{3\zeta_1\zeta_2}{1-\zeta_3} + \frac{\zeta_2^3}{\zeta_3(1-\zeta_3)^2} + \left(\frac{\zeta_2^3}{\zeta_3^2} - \zeta_0\right)\ln(1-\zeta_3)\right] \tag{10.39}$$

The dispersion term in Eq. (10.33) can be calculated using

$$\frac{a^{disp}}{RT} = -\pi\rho N_{Av}\left(2I_1(\eta,\overline{m})\overline{m^2\varepsilon\sigma^3} + \overline{m}C_1 I_2(\eta,\overline{m})\overline{m^2\varepsilon^2\sigma^3}\right) \tag{10.40}$$

The abbreviation C_1 in Eq. (10.40) means

$$C_1 = \left(1 + \overline{m}\frac{8\eta - 2\eta^2}{(1-\eta)^4} + (1-\overline{m})\frac{20\eta - 27\eta^2 + 12\eta^3 - 2\eta^4}{[(1-\eta)(2-\eta)]^2}\right)^{-1} \tag{10.41}$$

$\overline{m^2\varepsilon\sigma^3}$ and $\overline{m^2\varepsilon^2\sigma^3}$ stand for

$$\overline{m^2\varepsilon^l\sigma^3} = \sum_{i=1}^{n}\sum_{j=1}^{n} x_i x_j m_i m_j \left(\frac{\varepsilon_{ij}}{kT}\right)^l \sigma_{ij}^3; l = 1,2 \tag{10.42}$$

For the calculation of ε_{ij} and σ_{ij}, the conventional mixing rules are employed:

$$\varepsilon_{ij} = \sqrt{\varepsilon_{ii}\varepsilon_{jj}}\,(1-k_{ij}) \qquad \sigma_{ij} = \frac{\sigma_{ii}+\sigma_{jj}}{2} \tag{10.43}$$

where k_{ij} is the binary interaction parameter.

Table 10.3 Universal constants for the calculation of a_l in Eq. (10.45) [21].

l	a_{0l}	a_{1l}	a_{2l}
0	0.910563145	−0.30840169	−0.09061484
1	0.636128145	0.18605312	0.45278428
2	2.686134789	−2.50300473	0.59627007
3	−26.54736249	21.4197936	−1.72418291
4	97.75920878	−65.2558853	−4.13021125
5	−159.5915409	83.3186805	13.7766319
6	91.29777408	−33.7469229	−8.67284704

Table 10.4 Universal constants for the calculation of b_l in Eq. (10.45) [21].

l	b_{0l}	b_{1l}	b_{2l}
0	0.72409469	−0.57554981	0.09768831
1	2.23827919	0.69950955	−0.2557575
2	−4.00258495	3.89256734	−9.15585615
3	−21.0035768	−17.2154717	20.642076
4	26.8556414	192.672265	−38.8044301
5	206.551338	−161.826462	93.6267741
6	−355.602356	−165.207694	−29.6669056

The integrals I_1 and I_2 of the perturbation theory in Eq. (10.40) are replaced by a power series in density:

$$I_1(\eta, \overline{m}) = \sum_{l=0}^{6} a_l(\overline{m}) \eta^l \quad I_2(\eta, \overline{m}) = \sum_{l=0}^{6} b_l(\overline{m}) \eta^l \quad (10.44)$$

where the coefficients a_l and b_l depend on the chain length according to

$$a_l(\overline{m}) = a_{0l} + \frac{\overline{m}-1}{\overline{m}} a_{1l} + \frac{(\overline{m}-1)(\overline{m}-2)}{\overline{m}^2} a_{2l}$$
$$b_l(\overline{m}) = b_{0l} + \frac{\overline{m}-1}{\overline{m}} b_{1l} + \frac{(\overline{m}-1)(\overline{m}-2)}{\overline{m}^2} b_{2l} \quad (10.45)$$

The universal constants (a_{jl} and b_{jl}) in this expression are given in Tables 10.3 and 10.4.

Associating fluid mixtures contain not only monomeric molecules but also clusters of like and unlike molecules, for example, hydrogen-bonded and donor–acceptor clusters [24]. Since the effective molecular properties of the clusters (size, energy, and shape) are very different from the corresponding monomeric molecules, the bulk fluid properties also differ significantly. One way to account for such association effects is to use statistical mechanics methods, such as perturbation theory, to quantify the relationship between well-defined site–site interactions and the bulk fluid behavior [18]. Especially Wertheim's contribution [21] is suitable to provide the basis for the association term in the PC-SAFT model. Within this framework, a given molecule can have S association sites. The association sites are characterized by a noncentral potential located near the perimeter of the molecule. One can include one or more different sites on each molecule. Each of these sites has the restriction of being able to bond to only one other site. Depending on the molecule of interest, different bond scenarios can be considered. Huang and Radozs [16, 17] give some hints for the selection of the association model. Within this framework self-association and cross-association [25] can be taken into account. The associating term in Eq. (10.33) according to the Wertheim theory is given by Chapman et al. [18]

$$\frac{a^{assoc}}{RT} = \sum_i^n x_i \sum_{\alpha i}^{S_i} \left(\ln x_{\alpha i} - \frac{1}{2} x_{\alpha i} + \frac{M_\alpha}{2} \right) \quad (10.46)$$

where $x_{\alpha i}$ is the mole fraction of molecules of component i not bonded at site α. The second summation runs over all of the sites in a molecule, S_i. The values of the $x_{\alpha i}$ are obtained by the solution of the mass balance:

$$x_{\alpha i} = \frac{1}{1 + \rho N_{Av} \sum_i^n x_i \sum_{B_j}^{S_i} x_{B_j} \Delta^{A_i B_j}} \quad (10.47)$$

The association strength $\Delta^{A_i B_j}$ is the key property characterizing the association bond and depends on the segment diameter and the hard-sphere radial distribution function (Eq. (10.37)):

$$\Delta^{A_i B_j} = d_{ij}^3 g_{ij}^{hs}(d_{ij}) \kappa^{A_i B_j} \left[\exp\left(\frac{\varepsilon^{A_i B_j}}{kT}\right) - 1 \right] \quad (10.48)$$

This equation shows that the associating molecules need two more pure-component parameters, namely the association energy $\varepsilon^{A_i B_j}$, and the association volume $\kappa^{A_i B_j}$. In the case of cross-association additional mixing rules for these two parameters must be specified.

The effect of a permanent electrostatic moment of a molecule on the macroscopic behavior of the involved system can be modeled using a^{polar} in Eq. (10.33). Taking into account molecules having a dipole moment, expressions developed by Jog and Chapman [26], or by Gross and Vrabec [27] can be used. Molecules with a quadrupole moment can be described by the expression developed by Gross [28]. Vrabec and Gross [29] suggested an equation for the interaction of a dipole with a quadrupole moment. Kleiner and Gross [30] derived an equation for polarizable dipole moments. Independent of the considered molecule, the effect of a dipole or a quadrupole can be taken into account by the Pade approximation [27, 28]:

$$\frac{a^{polar}}{RT} = \frac{a_2/RT}{1 - a_3/a_2} \quad (10.49)$$

where a_2 and a_3 are the second-order and third-order perturbation terms. The first-order term is cancelled out against an angel-average portion of the anisotropic polar pair potential. For the calculation of the a_2 and a_3 values, the type of polarity must be specified. When linear and symmetric dipolar molecules are considered, a_2^{dipol} and a_3^{dipol} for the tangent sphere framework can be written as [27]

$$a_2^{\text{dipol}} = -\pi \rho N_{\text{Av}} \sum_i^n \sum_j^n x_i x_j m_i m_j \frac{\varepsilon_{ii}}{kT} \frac{\varepsilon_{jj}}{kT} \frac{\sigma_{ii}^3 \sigma_{jj}^3}{\sigma_{ij}^3} n_{\mu,i} n_{\mu,j} \mu_i^{*2} \mu_j^{*2} J_{2,ij}^{\text{dipol}} \qquad (10.50)$$

and

$$a_3^{\text{dipol}} = -\frac{\pi^2}{3} \rho^2 N_{\text{Av}}^2 \sum_i^n \sum_j^n \sum_k^n A_{ijk} \qquad (10.51)$$

$$A_{ijk} = x_i x_j x_k \frac{\varepsilon_{ii}}{kT} \frac{\varepsilon_{jj}}{kT} \frac{\varepsilon_{kk}}{kT} \frac{\sigma_{ii}^3 \sigma_{jj}^3 \sigma_{kk}^3}{\sigma_{ij} \sigma_{ik} \sigma_{jk}} n_{\mu,i} n_{\mu,j} n_{\mu,k} \mu_i^{*2} \mu_j^{*2} \mu_k^{*2} J_{3,ijk}^{\text{dipol}}$$

where μ_i^{*2} denotes the dimensionless squared dipole moment of the molecule i and $n_{\mu,i}$ is the number of dipole moments present in the molecule. The abbreviations $J_{2,ij}^{\text{dipol}}$ and $J_{3,ijk}^{\text{dipol}}$ represent integrals over the reference-fluid pair-correlation function and over the three-body correlation function, respectively. These integrals can be replaced by a power series with respect to the reduced density, η [28]:

$$J_{2,ij}^{\text{dipol}} = \sum_{n=0}^{4} \left(a_{n,ij} + b_{n,ij} \frac{\varepsilon_{ij}}{kT}\right) \eta^n \quad J_{3,ijk}^{\text{dipol}} = \sum_{n=0}^{4} c_{n,ijk} \eta^n \qquad (10.52)$$

where the coefficients $a_{n,ij}$, $b_{n,ij}$, and $c_{n,ijk}$ depend on the chain length $m_{ij} = (m_i m_j)^{1/2}$ or $m_{ijk} = (m_i m_j m_k)^{1/3}$ via a power series. The coefficients of these power series were estimated by molecular simulation and can be found in the literature [27]. Due to the results of molecular simulation, no additional adjustable parameter is necessary in order to describe polar molecules. Very similar expressions, also without any additional parameter, for molecules having a quadrupole moment can be taken from the literature [28].

The complete Helmholtz energy (Eq. (10.33)) can now be used to determine all thermodynamic properties, like the chemical potential, fugacity, and the pressure using standard thermodynamic relations.

The PC-SAFT model (Eq. (10.33)) is a theory-based EOS. In contrast to many other models, PC-SAFT explicitly accounts for the nonspherical shape of a polymer. Therefore, the model reveals predictive capabilities and allows reliable correlations as well as extrapolations for polymer systems.

The application of compressed gases for separating polymer blends, for separating polymers from solvents, and for fractioning polymers, as well as the use of supercritical gases as a continuous phase in polymer reactions needs the knowledge of the involved phase diagrams.

In this context, poly(propylene) (PP) + pentane + carbon dioxide is an interesting system. Figure 10.11 shows the phase behavior of this system, where the cloud-point pressure at constant composition is plotted against the temperature. If no CO_2 is present, the demixing pressure connected with LCST behavior increases slightly with increasing pressure. With increasing CO_2 amount, the pressure increase with temperature is much more pronounced.

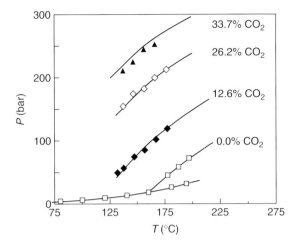

Figure 10.11 Experimental [31] and calculated [22] demixing pressure of polypropylene + n-pentane + CO_2. The initial polymer mass fraction (before the addition of CO_2) is 0.097.

In order to model this phase behavior using PC-SAFT, the pure-component parameters of the two solvents and the polymer and three binary interaction parameters must be known. The pure-component parameters of the solvents (CO_2 and pentane) can be fitted to the vapor pressure and the saturated liquid density. The identification of the pure-component parameters for the polymer is challenging and it is afflicted with a higher degree of uncertainty compared to the one for the volatile components, because no vapor pressure data, no liquid densities, and no heats of vaporization are available for polymers.

Two approaches have been established to solve the difficulties of estimating these parameters. As a first approach, pure-component parameters may be obtained by extrapolation of the pure-component parameters of a series of low-molar-mass components, in this case from the n-alkane series. However, with this method molecular effects such as entanglement, self-interactions, and shielding effects are not properly accounted for. Moreover, only those polymers for which a homologous series of low-molar-mass components exists are accessible. Alternatively, pure-component parameters can be determined by simultaneously fitting liquid density data and binary phase equilibrium data. This approach is clearly of pragmatic nature but has successfully been applied in modeling polymer systems. When experimental data for a particular polymer system is scarce or is only available in a narrow temperature range, it is recommended to assume that the segment diameter of the polymer is 4.1 Å [22]. In the first approach all binary parameters k_{ij} can be set to zero. Using this approach, the experimental data can often not be described with a sufficient accuracy. In the considered ternary system, three binary parameters must be estimated. The binary parameter describing the subsystem CO_2 and pentane can be fitted using binary vapor–liquid equilibrium data. This procedure leads to $k_{CO_2,pentane} = 0.14329$ [21]. The liquid–liquid equilibrium of the binary subsystem pentane + PP can be used to fit the corresponding binary interaction parameter ($k_{pentane,PP} = 0.0137$ [22]). The last binary interaction parameter between PP and CO_2 ($k_{PP,CO_2} = 0.177$) was obtained from the ternary mixture, adjusted to the point with the largest pressure, where 42 wt% CO_2 are present [22]. This was done because the binary system PP + CO_2 shows only a moderate sensitivity toward k_{PP,CO_2}, whereas calculations for the ternary system strongly react to changes in k_{PP,CO_2}. Using this parameter fitting procedure, the PC-SAFT describes the shift of the LCST demixing with varying CO_2 content correctly (Figure 10.11).

The gas solubilities in polymer melts play an essential role in many commercial applications of polymers, including adhesives, blends, resists, generation of foams, and spray coating and they are important in the context of nucleation, foaming, bubble formation, and growth within a polymer melt. An interesting example is the system PS + CO_2 [32]. For the study of this system, the polymer parameters were fitted only to density data taken from the literature [33], where the regarded polymer has a molar mass of 279 kg/mol, resulting in $\sigma_{PS} = 3.49$ Å, $m = M_N/29.937$, $\varepsilon/k = 292.57$ K. The pure-component parameters of CO_2 were fitted to the vapor pressure and to the saturated liquid density data [21]. The binary interaction parameter was fitted to the solubility data [34], where a polymer having a different molar mass ($M_N = 158$ kg/mol) was used. The binary interaction parameter was determined to $k_{PS,CO_2} = 0.0587$. Using these parameters, the density of the gas phase and the solubility of CO_2 as a function of pressure at different temperatures was predicted and compared to data taken from the literature [35]. The results are shown in Figure 10.12.

Example 10.3

Calculation of the pressure for the system CO_2 + PS using PC-SAFT (Figure 10.12).

A vessel contains a mixture consisting of PS (2) (molar mass: 158 kg/mol) and CO_2 (1). The liquid mole fraction is $x_1^L = 0.02$ and the liquid density is $\rho_W^L = 1007$ kg/m³. Which pressure will be built up in the vessel at 393.15 K?

Solution

In order to solve this problem, the following assumptions are made:

1) The polymer is monodisperse and hence the polydispersity is neglected.
2) At these thermodynamic conditions two phases, namely a liquid mixture and a pure gas phase, are present in the vessel.
3) For both substances no effects describing association (a^{assoc}) and polar (a^{polar}) forces in Eq. (10.33) were taken into account.
4) The pure-component parameters (m_i, σ_{ii}, and ε_{ii}/k) and the binary interaction parameter ($k_{ij} = 0.0587$) are taken from the literature [32]. The necessary pure-component parameters are collected in Table 10.5.

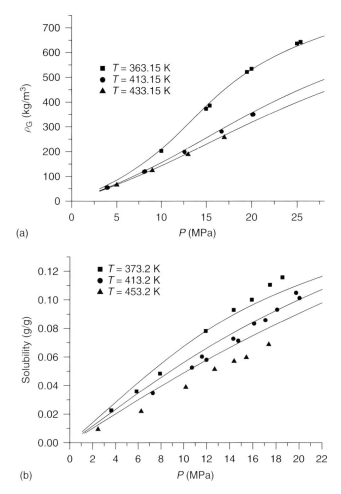

Figure 10.12 Comparison between experimental data ((a) density of the mixture [35] and (b) gas solubility [34]) and PC-SAFT modeling [32] for the system CO_2 + polystyrene (PS) with a molar mass of 158 kg/mol.

Table 10.5 PC-SAFT parameters for the system CO_2 (1) + PS 158 (2) [32].

	m_i	σ_{ii} (Å)	ε_{ii}/k (K)
CO_2	2.0729	2.7852	169.21
PS 158	5277.75	3.49	292.57

Using the pure-component parameters given in Table 10.5 and the system temperature ($T = 393.15$ K), the segment diameters of both components can be calculated with the help of Eq. (10.32):

$$d_{11} = 2.7852\,\text{Å}\left(1 - 0.12\exp\left(-\frac{3 \cdot 169.21\,\text{K}}{393.15\,\text{K}}\right)\right) = 2.6933\,\text{Å}$$

$$d_{22} = 3.49\,\text{Å}\left(1 - 0.12\exp\left(-\frac{3 \cdot 292.57\,\text{K}}{393.15\,\text{K}}\right)\right) = 3.4451\,\text{Å}$$

The segment diameters are necessary for the calculation of the quantities ζ_i defined in Eq. (10.38). The analysis of the dimension in this equation leads to the conclusion that the density ρ must be measured in mol/Å^3. The density ρ_w^L is given in kg/m^3. For the conversion of the density from kg/m^3 to mol/Å^3, the molar mass of the mixture is needed. Knowing the molar mass of the pure substances and the composition of the mixture results in

$$M = x_1 M_1 + x_2 M_2$$

$$M = 0.02 \cdot 44\,\frac{\text{g}}{\text{mol}} + 0.98 \cdot 158\,000\,\frac{\text{g}}{\text{mol}} = 154\,841\,\frac{\text{g}}{\text{mol}}$$

Using this value for the molar mass of the mixture the density of the mixture ρ^L in mol/Å^3 is

$$\rho^L = \frac{\rho_w^L}{M} = \frac{1007 \cdot 10^3\,\text{g}}{154\,841\,\text{m}^3 \frac{\text{g}}{\text{mol}}} = \frac{1007 \cdot 10^3\,\text{mol}}{10^{30} \cdot 154\,841\,\text{Å}^3} = 6.5 \cdot 10^{-30}\,\frac{\text{mol}}{\text{Å}^3}$$

The quantities ζ_i^L defined in Eq. (10.38) are given by

$$\zeta_0^L = \frac{\pi \rho^L N_{Av}}{6}(x_1 m_1 + x_2 m_2)$$

$$\zeta_0^L = \frac{\pi \cdot 6.5 \cdot 10^{-30}\,\text{mol} \cdot 6.02214 \cdot 10^{23}}{6\,\text{Å}^3\,\text{mol}}(0.02 \cdot 2.0729 + 0.98 \cdot 5277.75)$$

$$\zeta_0^L = 0.0106\,\frac{1}{\text{Å}^3}$$

$$\zeta_1^L = \frac{\pi \rho^L N_{Av}}{6}(x_1 m_1 d_{11} + x_2 m_2 d_{22})$$

$$\zeta_1^L = \frac{\pi \cdot 6.5 \cdot 10^{-30}\,\text{mol} \cdot 6.02214 \cdot 10^{23}}{6\,\text{Å}^3\,\text{mol}}$$
$$\cdot \left(0.02 \cdot 2.0729 \cdot 2.6933\,\text{Å} + 0.98 \cdot 5277.75 \cdot 3.4451\,\text{Å}\right)$$

$$\zeta_1^L = 0.03652\,\frac{1}{\text{Å}^2}$$

$$\zeta_2^L = \frac{\pi \rho^L N_{Av}}{6}(x_1 m_1 d_{11}^2 + x_2 m_2 d_{22}^2)$$

$$\zeta_2^L = \frac{\pi \cdot 6.5 \cdot 10^{-30}\,\text{mol} \cdot 6.02214 \cdot 10^{23}}{6\,\text{Å}^3\,\text{mol}}$$
$$\cdot \left(0.02 \cdot 2.0729 \cdot 2.6933^2\,\text{Å}^2 + 0.98 \cdot 5277.75 \cdot 3.4451^2\,\text{Å}^2\right)$$

$$\zeta_2^L = 0.12582 \frac{1}{\text{Å}}$$

$$\zeta_3^L = \frac{\pi \rho^L N_{Av}}{6}(x_1 m_1 d_{11}^3 + x_2 m_2 d_{22}^3)$$

$$\zeta_3^L = \frac{\pi \cdot 6.5 \cdot 10^{-30}\,\text{mol} \cdot 6.02214 \cdot 10^{23}}{6\,\text{Å}^3\,\text{mol}}$$
$$\cdot \left(0.02 \cdot 2.0729 \cdot 2.6933^3\,\text{Å}^3 + 0.98 \cdot 5277.75 \cdot 3.4451^3\,\text{Å}^3\right)$$

$$\zeta_3^L = 0.43346 = \eta$$

The next step is the calculation of the radial distribution functions of both components (g_{11}^{hs} and g_{22}^{hs}) according to Eq. (10.37):

$$g_{11}^{hs}(d_{11}) = \frac{1}{1-0.43346} + \left(\frac{2.6933^2\,\text{Å}^2}{2 \cdot 2.6933\,\text{Å}}\right)\frac{3 \cdot 0.12582}{\text{Å}(1-0.43346)^2}$$
$$+ \left(\frac{2.6933^2\,\text{Å}^2}{2 \cdot 2.6933\,\text{Å}}\right)^2 \frac{2 \cdot 0.12582^2}{\text{Å}^2(1-0.43346)^3} = 3.6679$$

$$g_{22}^{hs}(d_{22}) = \frac{1}{1-0.43346} + \left(\frac{3.4451^2\,\text{Å}^2}{2 \cdot 3.4451\,\text{Å}}\right)\frac{3 \cdot 0.12582}{\text{Å}(1-0.43346)^2}$$
$$+ \left(\frac{3.4451^2\,\text{Å}^2}{2 \cdot 3.4451\,\text{Å}}\right)^2 \frac{2 \cdot 0.12582^2}{\text{Å}^2(1-0.43346)^3} = 4.31184$$

Using the pure-component parameters given in Table 10.5 in combination with the mixing rules (Eq. (10.43)), the depth of the potential (ε_{12}/k) and the temperature-independent segment diameter (σ_{12}) of the mixture can be estimated:

$$\frac{\varepsilon_{12}}{k} = \sqrt{169.21 \cdot 292.57}\,(1 - 0.0587) = 209.44\,\text{K}$$

$$\sigma_{12} = \frac{2.7852\,\text{Å} + 3.49\,\text{Å}}{2} = 3.1376\,\text{Å}$$

According to Eq. (10.36), the mean segment number in the mixture is

$$\overline{m} = \sum_{i=1}^{n} x_i m_i = 0.02 \cdot 2.0729 + 0.98 \cdot 5277.75 = 5172$$

With Eq. (10.33), the Helmholtz energy is given. Using standard thermodynamic relations, the pressure can be obtained by

$$P = -n\left(\frac{\partial a}{\partial V}\right)_{T,x_i} = \rho^2\left(\frac{\partial a}{\partial \rho}\right)_{T,x_i} = \rho^2\left[\left(\frac{\partial a^{id}}{\partial \rho}\right)_{T,x_i} + \left(\frac{\partial a^{hc}}{\partial \rho}\right)_{T,x_i} + \left(\frac{\partial a^{disp}}{\partial \rho}\right)_{T,x_i}\right]$$
(10.53)

Therefore, the next step is the calculation of the corresponding derivatives with respect to the density. The ideal contribution in Eq. (10.53) in combination with Eq. (10.34) yields

$$P^{id} = \rho^L RT = 6.50 \cdot 10^{-30}\,\frac{\text{mol}}{\text{Å}^3}\, 8.314\,\frac{\text{J}}{\text{mol K}}\, 393.15\,\text{K}$$

$$P^{id} = 2.12 \cdot 10^{-26}\,\frac{\text{J}}{\text{Å}^3} = 0.0212\,\text{MPa}$$

The hard-sphere contribution, necessary to compute the hard-chain reference contribution, is available from Eq. (10.53), in combination with Eq. (10.39):

$$P^{hs} = \rho^L RT\overline{m}\left(\frac{\zeta_3}{1-\zeta_3} + 3\frac{\zeta_1\zeta_2}{\zeta_0(1-\zeta_3)^2} + \frac{\zeta_2^3(3-\zeta_3)}{\zeta_0(1-\zeta_3)^3}\right)$$

$$P^{hs} = 0.0212\,\text{MPa} \cdot 5172 \begin{pmatrix} \dfrac{0.43346}{1-0.43346} + 3\dfrac{0.03652\,\frac{1}{\text{Å}^2} \cdot 0.12582\,\frac{1}{\text{Å}}}{0.0106\,\frac{1}{\text{Å}^3}(1-0.43346)^2} \\ \\ + \dfrac{\left(0.12582\,\frac{1}{\text{Å}}\right)^3(3-0.43346)}{0.0106\,\frac{1}{\text{Å}^3}(1-0.43346)^3} \end{pmatrix}$$

$$P^{hs} = 822.4\,\text{MPa}$$

The hard-chain reference contribution can be determined by application of Eq. (10.35) in Eq. (10.53), leading to

$$P^{hc} = P^{hs} - RT\rho^L\left(\frac{x_1(m_1-1)}{g_{11}}\left(\frac{\partial g_{11}^{hs}}{\partial \rho}\right) + \frac{x_2(m_2-1)}{g_{22}}\left(\frac{\partial g_{22}^{hs}}{\partial \rho}\right)\right)$$

where the derivative of the radial correlation function with respect to the density is

$$\left(\frac{\partial g_{ii}^{hs}}{\partial \rho}\right) = \frac{1}{\rho^L}\left[\frac{\zeta_3}{(1-\zeta_3)^2} + d_i\left(\frac{3\zeta_2(1-\zeta_3)+6\zeta_2\zeta_3}{2(1-\zeta_3)^3}\right) + d_i^2\left(\frac{4\zeta_2^2(1-\zeta_3)+6\zeta_2^2\zeta_3}{4(1-\zeta_3)^4}\right)\right]$$

The application of these relations for both components results in

$$\rho^L \left(\frac{\partial g_{11}^{hs}}{\partial \rho}\right) = \left[\begin{array}{c} \dfrac{0.43346}{(1-0.43346)^2} + \\[2ex] 2.6933\,\text{Å} \left(\dfrac{\dfrac{3\cdot 0.12582}{\text{Å}}(1-0.43346)}{2(1-0.43346)^3\,\text{Å}} \right) \\[2ex] + \left(2.6933\,\text{Å}\right)^2 \left(\dfrac{4\left(\dfrac{0.12582}{\text{Å}}\right)^2 (1-0.43346)}{4(1-0.43346)^4} \right) \end{array}\right]$$

$$\rho^L \left(\frac{\partial g_{11}^{hs}}{\partial \rho}\right) = 6.715$$

and

$$\rho^L \left(\frac{\partial g_{22}^{hs}}{\partial \rho}\right) = \left[\begin{array}{c} \dfrac{0.43346}{(1-0.43346)^2} \\[2ex] + 3.4451\,\text{Å} \left(\dfrac{\dfrac{3\cdot 0.12582\,(1-0.43346)}{\text{Å}}}{2(1-0.43346)^3} \right) \\[2ex] + \left(3.4451\,\text{Å}\right)^2 \left(\dfrac{4\left(\dfrac{0.12582}{\text{Å}}\right)^2 (1-0.43346)}{4(1-0.43346)^4} \right) \end{array}\right]$$

$$\rho^L \left(\frac{\partial g_{22}^{hs}}{\partial \rho}\right) = 8.6968$$

10 Polymer Thermodynamics

The pressure related to the hard-chain contribution results in

$$P^{hc} = 822.4\,\text{MPa} - 0.0212\,\text{MPa}$$
$$\cdot \left(\frac{0.02\,(2.0729 - 1)}{3.6679} 6.715 + \frac{0.98\,(5277.75 - 1)}{4.31184} 8.6968 \right)$$

$$P^{hc} = 599.3\,\text{MPa}$$

The last term refers to the contribution caused by the dispersion forces (P^{disp}). This expression can be obtained using Eq. (10.40):

$$P^{disp} = -\pi RT \left(\rho^L\right)^2 N_{Av}$$
$$\cdot \left[2C_3\,(I_1 + I_{1a}) + \overline{m} C_4 \left(C_1 I_{2a} + I_2 \left(C_1 + \zeta_3 \rho^L \left(\frac{\partial C_1}{\partial \rho} \right)_{T, x_i} \right) \right) \right]$$

where the abbreviations are given either in Eqs. (10.41), (10.44) and (10.45) or in the following equations:

$$C_3 = \frac{x_1^2 m_{11}^2 \varepsilon_{11} \sigma_{11}^3}{kT} + \frac{2 x_1 x_2 m_{11} m_{22} \varepsilon_{12} \sigma_{12}^3}{kT} + \frac{x_2^2 m_{22}^2 \varepsilon_{22} \sigma_{22}^3}{kT}$$

$$C_4 = \frac{x_1^2 m_{11}^2 \varepsilon_{11}^2 \sigma_{11}^3}{(kT)^2} + \frac{2 x_1 x_2 m_{11} m_{22} \varepsilon_{12}^2 \sigma_{12}^3}{(kT)^2} + \frac{x_2^2 m_{21}^2 \varepsilon_{12}^2 \sigma_{12}^3}{(kT)^2}$$

$$I_{1a} = \sum_{i=0}^{6} i a_i \zeta_3^i \qquad I_{2a} = \sum_{i=0}^{6} i b_i \zeta_3^i$$

$$\rho^L \left(\frac{\partial C_1}{\partial \rho} \right)_{T, x_i} = -\frac{2 C_1^2 \,(p_1 + \overline{m} p_2)}{(1 - \zeta_3)^5 (2 - \zeta_3)^3}$$

$$p_1 = 20 - 64 \zeta_3 + 74 \zeta_3^2 + \zeta_3^3 (-35 + 4 \zeta_3 + \zeta_3^2)$$
$$p_2 = 12 + 96 \zeta_3 - 186 \zeta_3^2 + \zeta_3^3 (115 - 26 \zeta_3 + \zeta_3^2)$$

Except the quantities I_1, I_{1a}, I_2, and I_{2a}, all quantities can be computed immediately:

$$C_3 = \frac{0.02^2 \cdot 2.0729^2 \cdot 169.21\,\text{K}\,(2.7852\,\text{Å})^3}{393.15\,\text{K}}$$
$$+ \frac{2 \cdot 0.02 \cdot 0.98 \cdot 2.0729 \cdot 5277.75 \cdot 209.44\,\text{K}\,(3.1376\,\text{Å})^3}{393.15\,\text{K}}$$
$$+ \frac{0.98^2 \cdot 5277.75^2 \cdot 292.57\,\text{K}\,(3.49\,\text{Å})^3}{393.15\,\text{K}} = 846254896\,\text{Å}^3$$

$$C_4 = \frac{0.02^2 \cdot 2.0729^2\,(169.21\,\text{K})^2\,(2.7852\,\text{Å})^3}{(393.15\,\text{K})^2}$$
$$+ \frac{2 \cdot 0.02 \cdot 0.98 \cdot 2.0729 \cdot 5277.75\,(209.44\,\text{K})^2\,(3.1376\,\text{Å})^3}{(393.15\,\text{K})^2}$$
$$+ \frac{0.98^2 \cdot 5277.75^2\,(292.57\,\text{K})^2\,(3.49\,\text{Å})^3}{(393.15\,\text{K})^2} = 629755076\,\text{Å}^3$$

Table 10.6 Calculation of a_i and b_i values in Eq. (10.45) using Tables 10.3 and 10.4 and $\bar{m} = 5172$ [21].

i	a_i	b_i
0	0.511658801	0.24628782
1	1.274666967	2.68204432
2	0.779538269	−9.26131625
3	−6.854893287	−17.5856157
4	28.38812475	180.708728
5	−62.52032817	138.328639
6	48.88955905	−550.427807

$$p_1 = 20 - 64 \cdot 0.43346 + 74 \cdot 0.43346^2$$
$$+ 0.43346^3 \left(-35 + 4 \cdot 0.43346 + 0.43346^2\right) = 3.4683$$
$$p_2 = 12 + 96 \cdot 0.43346 - 186 \cdot 0.43346^2$$
$$+ 0.43346^3 \left(115 - 26 \cdot 0.43346 + 0.43346^2\right) = 27.1283$$

$$\rho^L \left(\frac{\partial C_1}{\partial \rho^L}\right)_{T, x_i} = -\frac{2\left(7.96 \cdot 10^{-6}\right)^2 (3.4683 + 5172 \cdot 27.1283)}{(1 - 0.43346)^5 (2 - 0.43346)^3} = -0.000079$$

For the calculation of the values I_1, I_{1a}, I_2, and I_{2a}, the universal constants occurring in Eq. (10.45) must be taken from Tables 10.3 and 10.4. Table 10.6 lists the a_i and b_i values, if $\bar{m} = 5172$ is used in Eq. (10.45).

Applying the values a_i and b_i from Table 10.6 leads to quantities I_1, I_{1a}, I_2, and I_{2a}:

$$I_1 = \sum_{i=0}^{6} a_i (0.44346)^i = 1.0221 \qquad I_2 = \sum_{i=0}^{6} b_i (0.44346)^i = 3.0817$$

$$I_{1a} = \sum_{i=0}^{6} i a_i (0.44346)^i = 0.3415 \qquad I_{2a} = \sum_{i=0}^{6} i b_i (0.44346)^i = 7.5814$$

Now, all quantities are known to calculate the pressure caused by the dispersion forces:

$$P^{\text{disp}} = -\pi \cdot 8.314 \, \frac{\text{J}}{\text{mol K}} \, 393.15 \, \text{K} \left(6.5 \cdot 10^{-30} \, \frac{\text{mol}}{\text{Å}^3}\right)^2 \, 6.02212 \cdot 10^{23} \, \frac{1}{\text{mol}}$$
$$\cdot \left(2 \cdot 846254896 \, \text{Å}^3 \, (1.0221 + 0.3415) + 5172 \cdot 629755076 \, \text{Å}^3 \right.$$
$$\left. \cdot \left(7.96 \cdot 10^{-6} \cdot 7.5814 + 3.0817 \left(7.96 \cdot 10^{-6} - 0.43346 \cdot 7.9 \cdot 10^{-5}\right)\right)\right)$$
$$= -585.4 \, \text{MPa}$$

Finally, the total pressure is accessible by summing up all contributions:

$$P = P^{\text{id}} + P^{\text{hc}} + P^{\text{disp}} = 0.0212 \, \text{MPa} + 599.3 \, \text{MPa} - 585.4 \, \text{MPa}$$
$$P = 13.92 \, \text{MPa}$$

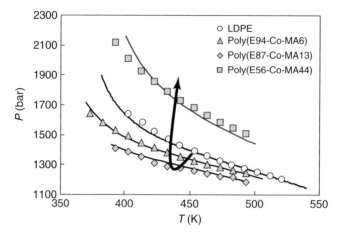

Figure 10.13 Comparison between experimental and calculated cloud-point pressures [38] for the system ethylene + poly(ethylene-co-methyl acrylate), where the polymer weight fraction is 0.05.

The SAFT EOS [36] or PC-SAFT EOS [37] can also be applied for systems with copolymers. In order to characterize the properties of copolymer systems, the PC-SAFT model was extended for hetero-nuclear hard-chain fluids, in which each segment in a chain can have different properties. In the modeling concept, a poly(ethylene-co-propylene), for example, comprises polyethylene segments and PP segments. This copolymer is on the one hand characterized by the pure-component parameters of polyethylene and polypropylene, and on the other hand, segment fractions are required, which describe the amount of different segment types in the copolymer chain, and furthermore bond fractions, which define to a certain extent the arrangement of the segments in the chain. Both quantities can be estimated on the basis of the molecular structure. Furthermore, three binary parameters have to be defined: two of them are the binary parameters of the corresponding homopolymer-solvent pairs. They are estimated from the homopolymer-solvent data. Only one additional parameter is needed which accounts for the interactions between the different segment types of different chains. This parameter has to be fitted to copolymer data.

An interesting phase behavior was observed experimentally [38] for the system poly(ethylene-co-methyl acrylate) in the solvent ethylene (Figure 10.13). Starting from low-density polyethylene, an increasing content of methyl acrylate in the copolymer shifts the demixing curve to lower pressures up to 13 mol% of methyl acrylate in the copolymer, whereas with further addition of methyl acrylate the solubility again decreases. In this case, the binary parameter is a linear function of the copolymer composition [38]. Figure 10.13 demonstrates the performance of the PC-SAFT EOS. The model is able to describe the observed dependence on temperature and molar mass as well as the nonmonotonic dependence on copolymer composition.

10.4
Influence of Polydispersity

Caused by the polymer reaction condition, polymers are not monodisperse but show a molar-mass distribution (Figure 10.1). In contrast to homopolymers, copolymers feature a two-dimensional distribution function, where the two variables are the molar mass and the chemical composition. In this chapter, the attention is focused on the influence of polydispersity on the thermodynamic properties. Due do the polydispersity, the demixing behavior becomes much more complicated for a polydisperse polymer in comparison to a monodisperse polymer as shown in Figure 10.14. The binodal curve in a quasi-binary system splits into three kinds of curves: a cloud-point curve, a shadow curve, and an infinite number of coexistence curves. The meaning of these curves becomes clear considering a cooling process. When reaching the cloud-point curve by lowering the temperature, the overall polymer content of the first droplets of the precipitated phase does not correspond to a point on the cloud-point curve but to the corresponding point on the shadow curve. When the temperature is further decreased, the two coexisting phases do not change their overall polymer content according to the cloud-point curve or to the shadow curve but rather according to the related branches of the coexistence curves. The overall polymer content of the coexisting phases is given by the intersection points of the horizontal line, at the considered temperature, with these branches. The coexistence curves are usually no closed curves but rather divided into two branches beginning at corresponding points of the cloud-point curve and of the shadow curve. Only if the composition of the initial homogeneous phase equals

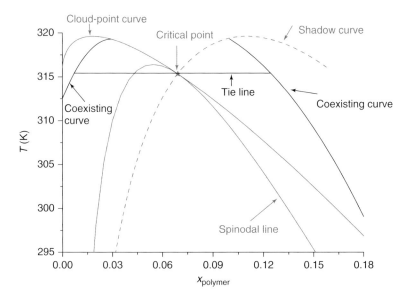

Figure 10.14 Schematic liquid–liquid phase diagram for a polydisperse polymer in a solvent [39].

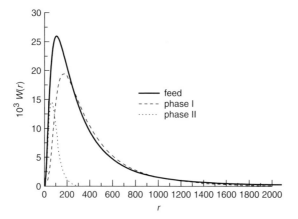

Figure 10.15 Fractionation effect.

that of the critical solution point, a closed coexistence curve is obtained whose extremum is the critical solution point. The coexistence curve can be used to construct tie lines, describing the phases which are in equilibrium. At the critical point, the cloud-point curve and shadow curve intersect. From Figure 10.14, it can be seen that for solutions of polydisperse polymers the critical solution point is not located at the extremum of the cloud-point curve or of the shadow curve as for strictly binary systems where the cloud-point curve, shadow curve, and all coexistence curves become identical.

Polymers in coexisting phases show different molar-mass distributions which are also different from that of the initial homogeneous system (Figure 10.15). This effect is called *fractionation effect* and can be used for the production of tailor-made polymers [2, 39, 40]. The phase with a lower polymer concentration contains the major part of the polymers with a lower molar mass. The cloud-point curve always corresponds to the molar-mass distribution of the initial polymer, but the first droplets of the formed coexisting new phase never do so (with the exception of the critical solution point) and, hence, they are not located on the cloud-point curve but on the shadow curve.

Owing to this polydispersity, characterization of polymers usually does not provide the number of the individual molecules or their mole fraction, mass fraction, and so on, but requires the use of continuous distribution functions or their averages. Continuous thermodynamics, developed by Rätzsch and Kehlen [39] and Cotterman *et al.* [41] can directly be applied for the calculation of thermodynamic properties including phase equilibria, because this theoretical framework is based completely on continuous distribution functions, which include all information about the polydispersity and allow a mathematically exact treatment of all related thermodynamic properties. Within this approach, a g^E-model or an EOS can be used.

An alternative to the application of continuous thermodynamics is the careful selection of pseudo-components [2]. One option for a suitable selection of the pseudocomponents is the Gauss–Hermitian quadrature.

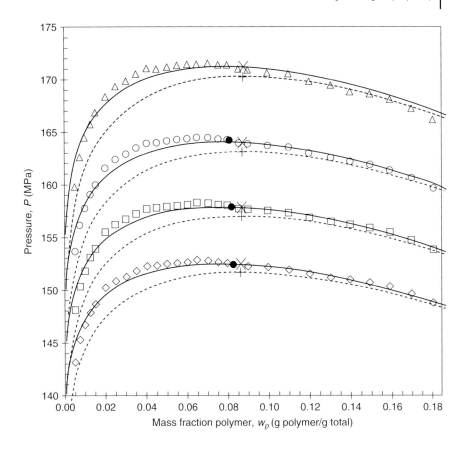

Experimental	△ 400 K	○ 410 K	□ 420 K	◇ 430 K	● critical
SL Pseudocomponents	—— 400 K	—— 410 K	—— 420 K	—— 430 K	× critical
SL Monodisperse	---- 400 K	---- 410 K	---- 420 K	---- 430 K	+ critical

Figure 10.16 Cloud-point pressures in the system LLDPE + ethylene at different temperatures (triangles $T = 400$ K, open circles $T = 410$ K, open squares $T = 420$ K, diamond $T = 430$ K) [44]. The solid lines are calculation results using Sanchez–Lacombe EOS assuming pseudo-components [44]. The dashed lines are calculation results using Sanchez–Lacombe EOS assuming monodisperse polymer.

For the calculation of the stability of a multicomponent polymer solution (spinodal line and critical solution points), the stability theory can be applied [42]. One possible consequence of the polydispersity, especially if the distribution function is bimodal, is the appearance of tri-critical solution points [2]. Suggestions for the phase equilibrium calculations of such systems can be found in the literature [2, 39, 43].

For example, the phase behavior of the system "linear low-density polyethylene (LLDPE)" + ethylene [44] is plotted in Figure 10.16. The experimental data are compared with two different modeling results, where the Sanchez–Lacombe

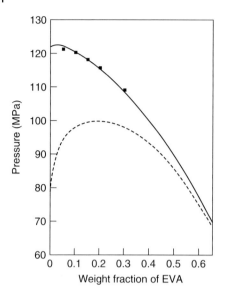

Figure 10.17 Comparison between experimental [47] and calculated [46] demixing pressure for the system poly(ethylene-co-vinyl acetate) + ethylene. The dashed line is calculated ignoring the polydispersity. The solid line is calculated taking the polydispersity via the Stockmayer distribution function into account.

EOS is used. First, the modeling was performed assuming that the polymer is monodisperse. Secondly, the modeling was repeated, but the polydispersity using pseudocomponents was taken into account. It can be clearly recognized that the consideration of the polydispersity leads to a significant improvement of the calculation, particularly in the region of diluted polymer solution.

Due to the very large number of different chemical species, the composition of polydisperse systems cannot be described by the mole fraction of the individual components, but by a continuous distribution function. In the case of statistical copolymers a two-dimensional distribution function according to the molar mass and the chemical composition must be used [45]. One classical example is the phase behavior of poly(ethylene-co-vinyl acetate) in ethylene. In Figure 10.17, two calculation results [46] are compared with experimental data. In both methods, an EOS, in this case Sako, Wu, Prausnitz-EOS [15], is used. However, in the first case the polydispersity is neglected and in the second case the polydispersity is taken into account using the two-dimension Stockmayer distribution function. Again, taking into account polydispersity permits a significantly better agreement with the experimental data.

Additional Problems

P10.1 The mole fraction of a polymer solution made from monodisperse polystyrene (PS) having a molar mass of $M_{PS} = 100\,\text{kg/mol}$ and cyclohexane (CH) having a molar mass of $M_{CH} = 64.16\,\text{g/mol}$ is $x_{PS} = 0.00001$. Calculate the segment mole fraction of PS in this polymer solution.

P10.2 The polydispersity of polyethylene can often be described with the help of the Wesslau distribution function:

$$W(r) = \frac{1}{Ar\sqrt{\pi}} \exp\left(-\frac{(\ln(r) - \ln(\bar{r}))^2}{A^2}\right)$$

Calculate the number-average and the mass-average segment number.

P10.3 In the original Flory-Huggins theory a concentration independent interaction parameter χ occurs. This interaction parameter can be estimated using the vapor pressure of the polymer solution. For the system polystyrene (PS) + n-hexane the vapor pressure was measured for two different polymer segment fractions at $T = 298.15\text{K}$. The experimental results are $P^s = 20.56\,\text{kPa}$ for $\overline{x_{PS}} = 0.1$ and $P^s = 20.4\,\text{kPa}$ for $\overline{x_{PS}} = 0.2$. The vapor pressure of the pure solvent at $T = 298.15\,\text{K}$ is $P_0^s = 20.6\,\text{kPa}$ and the segment number of PS is $r_{PS} = 100$. Can the original Flory-Huggins theory be applied for this polymer solution?

P10.4 Polystyrene can be produced by dissolving the monomer styrene in the solvent cyclohexane at $25\,°\text{C}$. During polymerization the temperature increases to $85\,°\text{C}$. The dependence of vapor pressure of cyclohexane on temperature is given by:

$$\log\left(P^s/\text{bar}\right) = A - \frac{B}{T/K + C}$$

where $A = 3.96988$, $B = 1203.526$, and $C = -50.287$.
a. Calculate the vapor pressure of cyclohexane at $25\,°\text{C}$, $50\,°\text{C}$, and $75\,°\text{C}$.
b. Plot the calculated vapor pressure of the solvent in a suitable diagram.
c. How much heat is removed by the evaporation of the solvent, if you assume cyclohexane vapor to behave as an ideal gas, the liquid volume is negligible small, and the enthalpy of vaporization does not depend on temperature?
d. Why is it desired that the polymerization temperature does not exceed $85\,°\text{C}$?
e. Discuss the influence of the polymer on the vapor pressure of cyclohexane.

P10.5 The vapor pressure of polyisobutylene (PIB) with a number-average molar mass of $M_{PIB} = 45\,\text{kg/mol}$ dissolved in benzene (B) ($M_B = 78.11\,\text{g/mol}$) was measured as function of polymer concentration at $313.15\,\text{K}$. The vapor pressure of benzene at this temperature is $24.3\,\text{kPa}$. For a polymer segment fraction of 0.4 the vapor pressure was $22.6\,\text{kPa}$.
a. Calculate the activity of the solvent for this polymer solution.
b. Calculate the Flory-Huggins interaction parameter χ.
c. Is this mixture stable, if you assume that the Flory-Huggins equation describes the system correctly?

P10.6 The LLE of a polymer solution consisting of polymer B and solvent A, should be modeled using Flory-Huggins theory, where the χ-function is given by:

$$\chi = \left(1 - a\bar{\bar{x}}_B\right)\left(1 + \frac{\beta_1}{T} + \frac{\beta_2}{T^2}\right)$$

a. Calculate the segment-based activity coefficients of the polymer and of the solvent.
b. Derive the spinodal and the critical conditions.

P10.7 Linear poly-N-isopropylacrylamide (PNIPAM) dissolved in water shows a lower critical solution temperature (LCST). For the calculation of the liquid–liquid equilibrium with the help of the Koningsveld-Kleintjens model the parameter γ and the function $\beta(T)$ are required. The experimental critical temperature is $T_c = 306.65$ K and the experimental critical segment fraction is $\overline{\overline{x^C_{PNIPAM}}} = 0.015$. The used PNIPAM was monodisperse having a segment number of $r_{PNIPAM} = 4000$. Calculate the parameter γ and $\beta(T_c)$.

P10.8 A vessel contains a mixture consisting of polystyrene ($M_{PS} = 70.3$ kg/mol) and CO_2 ($M_{CO_2} = 44.014$ g/mol). The liquid composition is $x^L_{CO_2} = 0.01$ and the liquid density is $\rho^L = 1.454 \cdot 10^{-5}$ mol/cm^3. Calculate the pressure in the vessel at $T = 373.2$ K using PC-SAFT-EOS with $k_{ij} = 0.0587$.

References

1. Flory, P.J. (1953) *Principles of Polymer Chemistry*, Cornell University Press, Ithaca.
2. Koningsveld, R., Stockmayer, W.H., and Nies, E. (2001) *Polymer Phase Diagrams*, Oxford University Press, Oxford.
3. Huggins, M.L. (1942) *J. Phys. Chem.*, **46**, 151–158.
4. Kagemoto, A., Baba, Y., and Fujishiro, R. (1972) *Makromol. Chem.*, **154**, 105–110.
5. Saeki, S., Kuwahra, N., Nakata, M., and Kaneko, M. (1976) *Polymer*, **17**, 685–689.
6. Solc, K. and Koningsveld, R. (1992) *J. Phys. Chem.*, **96**, 4056–4068.
7. Vanhee, S., Kiepen, F., Brinkmann, D., Borchard, W., Koningsveld, R., and Berghmans, H. (1994) *Macromol. Chem. Phys.*, **195**, 759–780.
8. Solc, K. (1970) *Macromolecules*, **3**, 665–673.
9. Horst, R. (1995) *Macromol. Theory Simul.*, **4**, 449–458.
10. Flory, P.J., Orwoll, R.A., and Vrij, A. (1964) *J. Am. Chem. Soc.*, **86**, 3507–3514.
11. Lacombe, R.H. and Sanchez, I.C. (1976) *J. Phys. Chem.*, **80**, 2568–2580.
12. Panayiotou, C.P. and Vera, J.H. (1982) *Polym. J.*, **14**, 681–694.
13. Kleintjens, L.A. and Koningsveld, R. (1980) *Colloid Polym. Sci.*, **258**, 711–718.
14. High, M.S. and Danner, R.P. (1990) *AIChE J.*, **36**, 1625–1632.
15. Sako, T., Wu, A.H., and Prausnitz, J.M. (1989) *J. Appl. Polym. Sci.*, **38**, 1839–1858.
16. Huang, S.H. and Radosz, M. (1990) *Ind. Eng. Chem. Res.*, **29**, 2284–2294.
17. Huang, S.H. and Radosz, M. (1991) *Ind. Eng. Chem. Res.*, **30**, 1994–2005.
18. Chapman, W.G., Gubbins, K.E., Jackson, G., and Radosz, M. (1989) *Fluid Phase Equilib.*, **52**, 31–38.
19. Müller, E.A. and Gubbins, K.E. (2001) *Ind. Eng. Chem. Res.*, **40**, 2193–2211.
20. Tan, S.P., Adidharma, H., and Radozs, M. (2008) *Ind. Eng. Chem. Res.*, **47**, 8063–8082.

21. Gross, J. and Sadowski, G. (2001) *Ind. Eng. Chem. Res.*, **40**, 1244–1260.
22. Gross, J. and Sadowski, G. (2002) *Ind. Eng. Chem. Res.*, **41**, 1084–1093.
23. Gil-Villegas, A., Galindo, A., Whitehead, P.J., Mills, S.J., Jackson, G., and Burgess, A.N. (1997) *J. Chem. Phys.*, **106**, 4168–4186.
24. Müller, E.A. and Gubbins, K.E. (2000) Associating fluids and fluid mixtures, in *Equation of State for Fluids and Fluid Mixtures* (eds J.V. Sengers, R.F. Kayser, C.J. Peters, and H.J. White Jr.), Elsevier, Amsterdam, 435–478.
25. Wolbach, J.P. and Sandler, S.L. (1998) *Ind. Eng Chem. Res.*, **37**, 2917–2928.
26. Jog, P.K. and Chapman, W.G. (1999) *Mol. Phys.*, **97**, 307–319.
27. Gross, J. and Vrabec, J. (2006) *AIChE J.*, **52**, 1194–1204.
28. Gross, J. (2005) *AIChE J.*, **51**, 2556–2558.
29. Vrabec, J. and Gross, J. (2008) *J. Phys. Chem. B*, **112**, 51–60.
30. Kleiner, M. and Gross, J. (2006) *AIChE J.*, **52**, 1951–1961.
31. Martin, T.M., Lateef, A.A., and Roberts, C.B. (1999) *Fluid Phase Equilib.*, **154**, 241–259.
32. Enders, S., Kahl, H., and Winkelmann, J. (2005) *Fluid Phase Equilib.*, **228–229**, 511–522.
33. Quach, A. and Simha, R. (1971) *J. Appl. Phys.*, **42**, 4592–4606.
34. Sato, Y., Yurugi, M., Fujiwara, K., Takishima, S., and Masuoka, H. (1996) *Fluid Phase Equilib.*, **125**, 129–138.
35. Jaeger, P.T., Eggers, R., and Baumgartl, H. (2002) *J. Supercrit. Fluids*, **24**, 203–217.
36. Hasch, B.M., Lee, S.H., and McHugh, M.A. (1993) *Fluid Phase Equilib.*, **83**, 341–348.
37. Tumakaka, F., Gross, J., and Sadowski, G. (2002) *Fluid Phase Equilib.*, **194–197**, 541–551.
38. Becker, F., Buback, M., Latz, H., Sadowski, G., and Tumakaka, F. (2004) *Fluid Phase Equilib.*, **215**, 263–282.
39. Rätzsch, M.T. and Kehlen, H. (1989) *Prog. Polym. Sci.*, **14**, 1–46.
40. Tung, L.H. (1977) *Fractionation of Synthetic Polymers Principles and Practices*, Marcel Dekker, New York.
41. Cotterman, R.L., Bender, R., and Prausnitz, J.M. (1985) *Ind. Eng. Chem. Process Des. Dev.*, **24**, 194–203.
42. Solc, K., Kleintjens, L.A., and Koningsveld, R. (1984) *Macromolecules*, **17**, 573–585.
43. Krenz, R.A. and Heidemann, R.A. (2007) *Fluid Phase Equilib.*, **262**, 217–226.
44. Trumpi, H., de Loos, T.W., Krenz, R.A., and Heidemann, R.A. (2003) *J. Supercrit. Fluids*, **27**, 205–214.
45. Rätzsch, M.T., Kehlen, H., and Borwarzik, D. (1985) *J. Macromol. Sci. Part A Chem.*, **22**, 1679–1690.
46. Browarzik, C., Browarzik, D., and Kehlen, H. (2001) *J. Supercrit. Fluids*, **20**, 73–88.
47. Wagner, P. (1983) Untersuchungen zum Hochdruckphasengleichgewicht in Mischungen aus Ethylen und (Ethylen-Vinylacetat)-Copolymeren, PhD Thesis. TH Merseburg, Germany.

11
Applications of Thermodynamics in Separation Technology

Besides pure component properties, a reliable knowledge of the phase equilibrium behavior of multicomponent systems with nonpolar, polar, supercritical compounds, electrolytes, and sometimes polymers is the prerequisite for the successful synthesis, design, and optimization of the different unit operations. In the development stage, the chemical engineer has to decide about the separation process used. Then, he has to check for existing separation problems, for example, azeotropic points in the case of distillation processes. He has to find how the separation problem can be solved, for example, with the help of an entrainer in the case of azeotropic or extractive distillation, a hybrid process, and so on. In the design step, he has to determine the number of theoretical stages or the height of the columns. Furthermore, he has to arrange the separation sequence. At the end all product specifications have to be met at minimal operating and investment costs.

For solving the tasks mentioned above, reliable experimental pure component properties (vapor pressures, densities, heat capacities, transport properties, etc.) and mixture data (phase equilibria, and excess properties) for the system considered would be most desirable. Some decades ago, a time-consuming literature search was always necessary to obtain these data. In the meantime comprehensive factual data banks, for example, the Dortmund Data Bank (DDB), NIST data bank, DIPPR data bank, and so on, have been built up. These data banks contain a great part of the worldwide available experimental data.

From the data banks mentioned before, the DIPPR data bank covers experimental pure component properties for approx. 2000 selected compounds. Additionally, recommended basic data and temperature-dependent correlation parameters for the various pure component properties are available for all compounds. When experimental pure component data were missing in the DIPPR data bank, predictive methods were used to calculate the required pure component data and to fit the required temperature-dependent correlation parameters.

In the DDB and in the NIST data bank, additionally the various mixture data (phase equilibria, excess properties, transport properties, etc.) are stored besides the pure component properties. While the NIST data bank mainly contains the data from important thermodynamic journals, in the DDB additionally an enormous amount of unpublished experimental data from private communications, PhD, MSc, BSc theses, and data from industry were collected besides the data from scientific journals published worldwide in various languages. Furthermore, a great

Chemical Thermodynamics: for Process Simulation, First Edition.
Jürgen Gmehling, Bärbel Kolbe, Michael Kleiber, and Jürgen Rarey.
© 2012 Wiley-VCH Verlag GmbH & Co. KGaA. Published 2012 by Wiley-VCH Verlag GmbH & Co. KGaA.

Table 11.1 Current status of the Dortmund Data Bank (September 2011).

Data Bank	Additional information	Data sets (data points)
Vapor–liquid equilibria	Normal boiling substances	32200
Vapor–liquid equilibria	Low boiling substances	32300
Vapor–liquid equilibria	Electrolyte systems	8700
Liquid–liquid equilibria	Mainly for organic compounds, water,	23400
Activity coefficients	At infinite dilution (in pure solvents)	(62500)
Activity coefficients	At infinite dilution (in nonelectrolyte mixtures)	1500
Gas solubilities	Nonelectrolyte systems	20200
Gas solubilities	Electrolyte systems	2200
Critical data of mixtures	Critical lines	2600
Solid–liquid equilibria	Mainly organic compounds, for example, pharmaceuticals, and so on	41000
Salt solubilities	Mainly in water	32000
Azeotropic data	Binary to quaternary systems	(53200)
Partition coefficients	Octanol–water partition coefficients	10800
Adsorption equilibria	Vapor phase	4000
Polymer mixtures	Liquid–liquid equilibria (LLE), solubility, PVT, swelling data, and so on	18200
Excess enthalpies	Heats of mixing	21000
Heat capacities of mixtures	Includes also excess heat capacities	3900
Mixture densities	Includes also excess volumes	51000
Mixture viscosities	Includes also viscosity deviations	14400
Various mixture properties	Heat conductivities, surface tensions, and so on	22600
Pure component properties	Vapor pressures, densities, heat capacities, and so on	215000

part of the unpublished data measured systematically for the development of the various predictive methods are stored in the DDB.

In Table 11.1, the present status of the DDB is shown. It can be seen that the DDB (September 2011) contains more than 695 000 data sets. These experimental data were taken from more than 60 000 references, where data for more than 33 000 compounds including salts, adsorbents, and polymers can be found. The number of data tuples (this means T, P_i^s in the case of vapor pressures; x_i, y_i, P, T in the case of vapor–liquid equilibria (VLE)) is ∼5 300 000.

About 10% of the data stored in the DDB were published in unified form in different data compilations [1–5]. Every year the number of data sets in the DDB increases by approx. 9%. Analyzing the available mixture data, it can be recognized that for example, in the case of VLE 88.3% of the data were measured for binary systems, 10.6% for ternary, and only 1.1% for quaternary and higher systems.

However, in practice not only the knowledge of the phase equilibrium behavior of binary, but also of multicomponent systems is required. This was the reason why powerful thermodynamic models were developed, which allow the reliable prediction of the phase equilibrium behavior of multicomponent systems using binary data alone. The different models (g^E-models, equations of state, and electrolyte models) have been introduced in Chapters 5 and 7, where also their advantages and disadvantages were discussed. For a reliable application of these models in the composition and temperature range of interest, special care is required when fitting the binary parameters to experimental data. It is advantageous to fit the parameters of the g^E-models or equations of state simultaneously to all kinds of available data (phase equilibria, excess properties, data in the dilute range, etc.) covering a large temperature range. Then, it is possible to describe the K-factors or separation factors in the whole composition and a large temperature range with the required accuracy.

If no binary experimental data are available, powerful predictive models (group contribution methods and group contribution equations of state) can be applied today to reliably predict the missing pure component properties (see Chapter 3) and the phase equilibrium behavior (see Chapters 5 and 7). These predicted mixture data can be used, for example, to fit the missing binary parameters for a multicomponent system.

From the remarks above, it can be concluded that a sophisticated software tool for process synthesis and design should have direct access to all worldwide available pure component and mixture data. At the same time, it should use recommended basic data (T_c, P_c, v_c, ω, T_m, T_{tr}, etc.) and reliable temperature (pressure)-dependent parameters to describe the pure component properties at the given temperature (pressure). Additionally, for the most important mixtures, recommended model parameters for g^E-models, equations of state, and electrolyte models should be provided. Furthermore, a software package should be available which allows the prediction of the required pure component properties and phase equilibria.

A sophisticated software tool for process synthesis was realized in connection with the DDB. The rough structure of this software tool, called Dortmund Data Bank Software Package (DDBSP), is shown in Figure 11.1. There is a direct access to all experimental pure component and mixture data stored in the DDB. Then, the user has direct access to the recommended basic pure component data, for example, T_c, P_c, v_c, ω, T_m, Δh_m, and so on, and recommended parameters to describe the pure component properties, such as vapor pressure, density, viscosity, and so on, as a function of temperature and for a few properties also as a function of pressure. For the mixture data, in particular phase equilibria, the user will find recommended parameters for the various g^E-models or equations of state. If no experimental pure component properties are available, a software package called *Artist* allows the user to estimate all kinds of pure component properties at the given conditions only by drawing the molecule or calling the structures of a large amount of compounds directly from a structure data bank. For the prediction of phase equilibria, the user can directly apply the different group contribution methods, like UNIFAC, modified UNIFAC or group contribution equations of

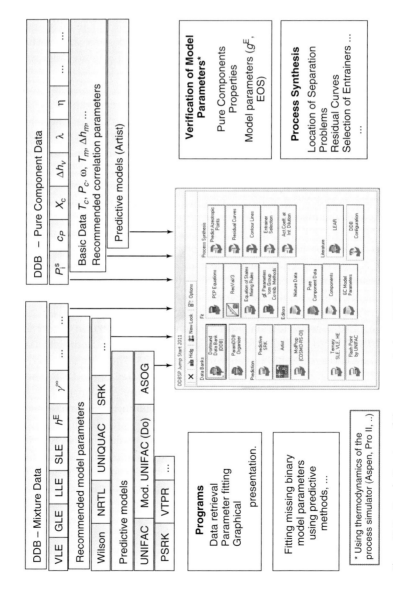

Figure 11.1 Rough Structure of the program Package DDBSP.

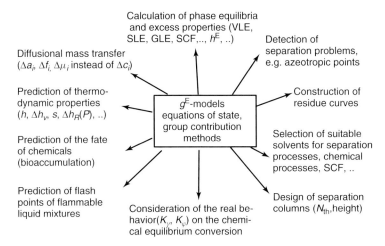

Figure 11.2 Fields of application of thermodynamic models (g^E-models, equations of state, and group contribution methods) for process development.

state, such as Predictive Soave–Redlich–Kwong (PSRK) and Volume-Translated Peng–Robinson (VTPR), or electrolyte models (LIQUAC and LIFAC).[1]

The reliable knowledge of the pure component properties and mixture data in connection with a suitable software tool allows the processing of a large number of important tasks of industrial interest. Some of them, for which, in particular, the knowledge of the phase equilibrium behavior – which means activity coefficients or fugacity coefficients – is required, are shown in Figure 11.2. Besides the calculation of the various phase equilibria, separation problems can be identified, residue curves can be constructed, the most suitable solvents for the different separation processes or chemical reactions can be selected, and the number of theoretical stages or the height of a separation column can be determined. The influence of the real behavior on the chemical equilibrium conversion can be taken into account, as well as the flash point of a flammable liquid mixture and the fate of persistent chemicals in the environment can be estimated. Furthermore, the influence of the pressure on the enthalpy of reaction can be calculated, more precise standard thermodynamic properties can be derived from experimental enthalpy data and chemical equilibria, and more suitable potential differences can be applied for the calculation of transport processes.

Besides the applications mentioned above and shown in Figure 11.2, the DDB and the integrated software package can be applied to a large number of other tasks, for example, the verification of pure component properties and models prior to process simulation, the selection of suitable working fluids for thermodynamic

---

1) Additionally there is the possibility to apply quantum-chemical methods for the calculation of phase equilibria, such as COSMO-RS (Ol) [6], for which the required σ-profiles for approx. 4500 compounds are also directly available in the Dortmund Data Bank.

cycles such as organic Rankine cycles (ORC processes), refrigeration cycles, air conditioning, the calculation of Joule–Thomson coefficients, and so on.

In this chapter, a few important examples of industrial interest in the field of separation processes are discussed. In the case of the design of distillation processes, the quality of the thermodynamic model parameters has a great influence on the reliability of the results. Therefore, the parameters used for the calculation of pure component properties, phase equilibria, and so on, should be checked carefully prior to process simulation.

11.1
Verification of Model Parameters Prior to Process Simulation

In most commercial process simulators, model parameters for pure component properties and binary parameters can be found for a large number of compounds and binary systems. However, the simulator providers repeatedly warn in their software documentations and user manuals that these default parameters should be applied only after careful examination by the company's thermodynamic experts prior to process simulation. For verification of the model parameters again, a large factual data bank like the DDB is the ideal tool. The DDB allows checking all the parameters used for the description of the pure component properties as a function of temperature and of the binary parameters of a multicomponent system by access to the experimental data stored. On the basis of the results for the different pure component properties and phase equilibria, excess enthalpies, activity coefficients at infinite dilution, separation factors, and so on, the experienced chemical engineer can decide whether all the data and parameters are sufficiently reliable for process simulation.

11.1.1
Verification of Pure Component Parameters

With a sophisticated software package, the verification of the default pure component model parameters can be performed fast and easily. Often the results can be judged graphically. Hopefully, in most cases the user will find good agreement between the calculated and the experimental data stored in the factual data bank. But sometimes also poor results may be obtained. As an example, the dynamic viscosity of hexafluorobenzene is shown in Figure 11.3. Deviations larger than 200% between calculated and experimental data are obtained. The reason for the poor results is that the parameters were fitted to predicted data and not to the data available in the literature. The deviation is caused by the fact that the chosen predictive method leads to poor dynamic viscosities for fluorine compounds.[2]

---

2) In the last year the parameters were refitted using available experimental data. But of course poor results occur also for other pure component properties and compounds. The best option is the use of agreeing experimental data measured in different laboratories.

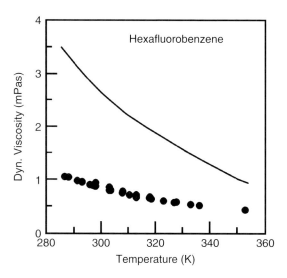

Figure 11.3 Experimental [7] and calculated dynamic viscosities for hexafluorobenzene in the temperature range from 280 to 360 K.

11.1.2
Verification of g^E-Model Parameters

Besides the pure component parameters, in particular the mixture parameters, for example of a g^E-model or an equation of state, should be checked carefully prior to process simulation. The procedure is shown in Figure 11.4 for the binary system acetone–cyclohexane, which may be one of the binary key systems of a multicomponent mixture. From the results shown in Figure 11.4, it can be concluded that the VLE behavior of the binary system can be reliably described in the temperature range 0–50 °C with the Wilson parameters used. But from the poor h^E-results, it seems that an extrapolation to higher or lower temperature may be dangerous, as already can be seen from the solid–liquid equilibrium (SLE) results of the eutectic system in the temperature range 0 to −100 °C and also from the incorrect temperature dependence of the calculated azeotropic data.

Poor parameters of course lead to a questionable number of theoretical stages of a distillation column when solving the balance (MESH) equations [8]. This means that the height or the number of theoretical stages of the distillation column for a given separation problem depends on the g^E-model and the parameters chosen. This is shown in Figure 11.5 for the system acetone–water using the Wilson, NRTL, and UNIQUAC parameters fitted only to VLE data. At a first glance at the y–x diagram, it seems that the VLE behavior is described within the same accuracy by all g^E-models chosen. But when the number of theoretical stages for the given separation problem is calculated, surprisingly for the Wilson model more theoretical stages than for the NRTL equation are required. The lowest number of stages is required for the UNIQUAC model. The reason is that the separation factor at infinite dilution for the different models is quite different. While a separation

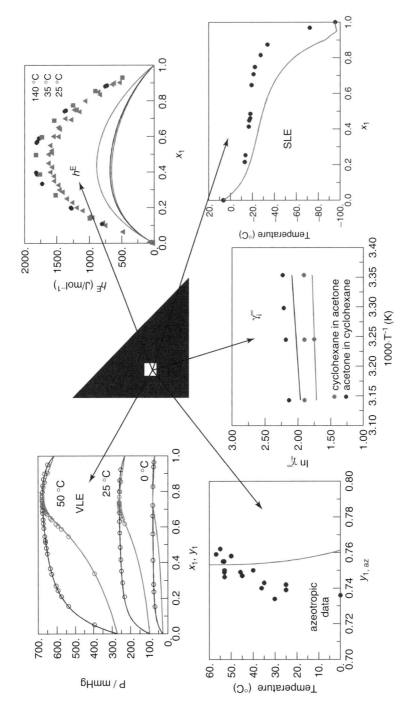

Figure 11.4 Verification of Wilson parameters, with the help of experimental data stored in the Dortmund Data Bank prior to process simulation, exemplarily shown for the binary system acetone (1)–cyclohexane (2). ■, ··· 140 °C, ♦, – – 35 °C, ▲, – – 25 °C.

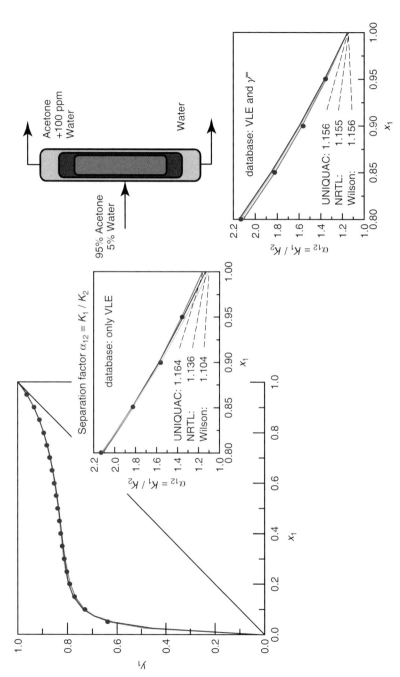

Figure 11.5 Separation factors obtained for the system acetone (1)–water (2) at atmospheric pressure using parameters fitted either to VLE or fitted simultaneously to VLE and activity coefficients at infinite dilution.

factor α_{12}^∞ of 1.104 is calculated for the Wilson equation, a separation factor α_{12}^∞ of 1.164 is obtained by the UNIQUAC equation. In the case of positive deviation from Raoult's law – as in our case – the largest separation effort is usually required at the top of the column. The separation effort mainly depends on the difference $\alpha_{12}^\infty - 1$. This means, in particular, that the separation factors α_{12}^∞ at the top of the column have a great influence on the number of required stages for a given separation problem. When different values are calculated with the different g^E-models, the calculated height of the column depends on the g^E-model.

This situation can only be improved when a larger database is used for fitting the g^E-model parameters and the correct description of the most relevant data is enforced by adequate weighting during regression. Besides VLE data, also reliable activity coefficients at infinite dilution should be used. As can be seen from Figure 11.5, nearly identical separation factors ($\alpha_{12}^\infty = 1.156$) are obtained for the g^E-models considered using an extended database with both VLE data and activity coefficients at infinite dilution. This leads to the same number of theoretical stages for the three g^E-models for the given separation problem.

To get an idea about the quality of the default parameters stored in the different process simulators, results for the system acetone–water are shown in Figures 11.6 and 11.7. From Figure 11.6, it can be seen that the separation factor strongly depends on the g^E-model chosen. Using the UNIQUAC equation, less stages are calculated for the distillation process than for the NRTL equation, since higher separation factors are calculated using the UNIQUAC equation. With the default Wilson parameters at atmospheric pressure, even azeotropic behavior for the nonazeotropic system acetone–water is incorrectly calculated.

Using the default parameters of a second simulator, again a different number of stages is required for the separation of the system acetone–water at atmospheric pressure. While with the NRTL and UNIQUAC parameters similar separation factors are obtained, surprisingly high separation factors are calculated using the default Wilson parameters. As can be seen from Figure 11.7 instead of a separation factor of approximately 1.2 at infinite dilution – using the NRTL and the UNIQUAC model – approximately a value of 3.0 is calculated using the default Wilson parameters. This means that for the separation of the system acetone–water, a distillation column with only a few stages would be sufficient following the Wilson equation.[3]

These examples demonstrate the importance of reliable model parameters for process simulation. All the binary parameters of a multicomponent system used for process simulation should provide a reliable description of the phase equilibrium behavior of the binary system in the whole composition range (including the dilute region) and temperature range covered. Prior to process simulation at least the interaction parameters of the critical binary systems (separation factor not far away from unity, occurrence of a miscibility gap, etc.) and further key systems should be checked carefully with the help of the experimental data stored in a factual data bank, such as DDB. If no reliable data are available, the measurement of the phase equilibrium behavior of the key systems is strongly recommended.

3) A further example with poor default parameters is given in Appendix F.

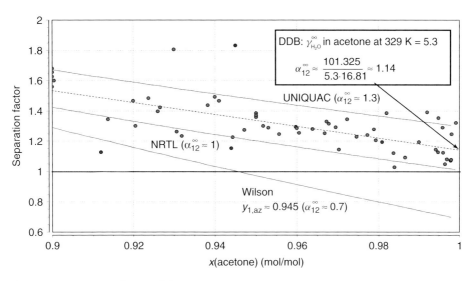

Figure 11.6 Separation factors of the system acetone (1)–water (2) at 101.325 kPa calculated using the default parameters for the g^E-models (Wilson, NRTL, and UNIQUAC) delivered with process simulator 1: - - - - - mean separation factor α_{12} determined from the scattering VLE.

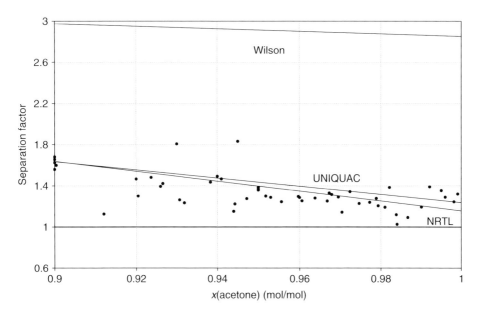

Figure 11.7 Separation factors obtained for the system acetone (1)–water (2) at atmospheric pressure using the default parameters delivered with process simulator 2.

Example 11.1

Calculate the separation factor for the system acetone (1)–water (2) at infinite dilution ($x_1 = 1$) (top of the column) at the normal boiling point of acetone ($T_b = 56.27°C$). For the calculation, the following value of the activity coefficient at infinite dilution $\gamma_2^\infty = 5.3$ (measured at 55.3 °C by Bergmann and Eckert [9]) should be applied.

Antoine parameters for water ($\log P_i^s \text{ (mmHg)} = A - B/(\vartheta\,(°C) + C)$)

Compound	A	B	C
Water	8.07131	1730.63	233.426

Solution

The separation factor at infinite dilution in pure acetone can directly be calculated (see Chapter 5.4). Since the vapor pressure of acetone at the normal point is 760 mmHg, only the vapor pressure of water at 56.27 °C is required additionally:

$$\log P_{H_2O}^s = 8.07131 - \frac{1730.63}{56.27 + 233.426} = 2.0973$$

$$P_{H_2O}^s = 125.13 \text{ mm Hg}$$

Using this vapor pressure, the separation factor of the system acetone–water at the top of a column at atmospheric pressure is

$$\alpha_{12}^\infty = \frac{760}{5.3 \cdot 125.13} = 1.146$$

The calculated separation factor is shown in Figure 11.6 together with the experimental values derived from the VLE data published by different authors, the calculated results using the default model parameters from different g^E-models, and the mean separation factor (dashed line) calculated from the scattering experimental VLE data. It can be seen that the most reliable separation factor is obtained using the activity coefficient at infinite dilution for fitting the binary parameters.

In most cases not binary but multicomponent systems have to be separated. Sometimes an additional component is needed as an entrainer e.g. for the separation by extractive distillation. As an example selected separation factors α_{12} for the system benzene (1)–cyclohexane (2)–NMP (3) calculated using modified UNIFAC and default UNIQUAC parameters from a simulator are shown in Figures 11.8 and 11.9, respectively. As benzene and cyclohexane form an azeotrope, the main task of the entrainer NMP is to shift the separation factor between benzene and cyclohexane as far from unity as possible; this means $\alpha_{12} \gg 1$ or $\alpha_{12} \ll 1$. In practice, typical entrainer concentrations of 50–80 mol% are employed to achieve satisfying separation factors. A higher entrainer concentration usually improves the separation factor.

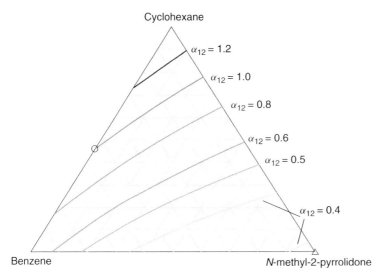

Figure 11.8 Calculated separation factors α_{12} at atmospheric pressure for the system benzene (1)–cyclohexane (2) in the presence of NMP (3) using modified UNIFAC.

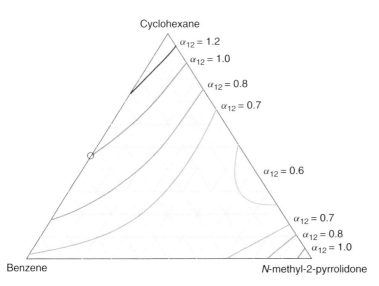

Figure 11.9 Calculated separation factors α_{12} at atmospheric pressure for the system benzene (1)–cyclohexane (2) in the presence of NMP (3) using default binary UNIQUAC parameters from simulator 1.

In Figure 11.8, the predicted separation factors between benzene (1) and cyclohexane (2) using modified UNIFAC are shown for the ternary system with NMP as entrainer at atmospheric pressure. It can be seen that the separation factor α_{12} of the azeotropic system benzene–cyclohexane always shows values smaller than one at mole fractions of NMP greater than about 0.25. With increasing concentrations of NMP, the separation factor α_{12} further decreases monotonically. This means that the higher boiling component cyclohexane becomes more volatile than benzene in the presence of NMP. With increasing NMP concentration, this effect is more pronounced, which is in agreement with the experimental findings.

In Figure 11.9, the results are shown using the default UNIQUAC parameters delivered with a process simulator. The calculated separation factors look totally different. While the separation factor α_{12} firstly also decrease with increasing NMP concentrations, surprisingly for $x_3 > 0.4$ the opposite behavior is observed. At very high NMP concentration even separation factors of one are calculated. This is in contrast to the behavior observed experimentally and calculated using modified UNIFAC or with the help of reliable g^E-model parameters.

The wrong description of the real behavior of the ternary system results from the fact that not all binary systems are described correctly using the default UNIQUAC parameters. The results for the two binary systems with NMP are shown in Figure 11.10. While satisfying results are obtained for the system benzene–NMP, poor results are observed for the binary system cyclohexane–NMP. In particular, bad separation factors at infinite dilution of cyclohexane (α_{23}^∞) are calculated. Since the separation factor α_{12} can be calculated from the separation factor α_{13} and α_{23}, poor results are obtained for the separation factor α_{12}:

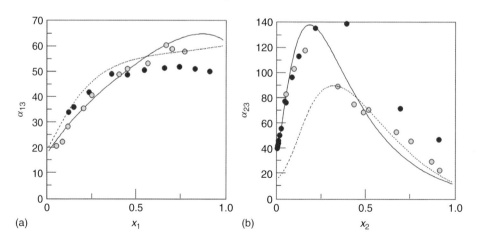

Figure 11.10 Experimental [7] and calculated separation factors at atmospheric pressure for the binary systems benzene (1)–NMP (3) (a) and cyclohexane (2)–NMP (3) (b) - - - using default UNIQUAC parameters from simulator 1, ⎯⎯ using modified UNIFAC, ● ○ experimental data from different references.

$$\alpha_{12} = \frac{y_1/x_1}{y_2/x_2} = \frac{\alpha_{13}}{\alpha_{23}} = \frac{(y_1/x_1)/(y_3/x_3)}{(y_2/x_2)/(y_3/x_3)}$$

The examples shown for the binary system acetone–water and the ternary system benzene–cyclohexane–NMP demonstrate that the parameters used for process simulation have to be checked carefully prior to process simulation using experimental data from different sources stored in factual data banks. The verification of the binary parameters is of particular importance for systems where the separation factor shows values not far from unity. In comprehensive factual data banks, for example, the DDB, sophisticated software packages are integrated to allow the verification of all pure component properties and mixture parameters prior to process simulation. Suitable for the verification of the pure component properties and mixture parameters are in particular graphical presentations, as exemplarily shown in Figures 11.3 and 11.4.

11.2
Investigation of Azeotropic Points in Multicomponent Systems

For process design, in particular, the knowledge of separation problems is of great importance. For distillation processes, the main problems arise from the occurrence of binary, ternary, and quaternary azeotropes. This leads to points, boundary lines, or boundary surfaces, which cannot be crossed by ordinary distillation. Azeotropic behavior can be determined with the help of thermodynamic models (g^E-models and equations of state) using binary parameters or predictive methods (group contribution methods and group contribution equations of state). Furthermore, a direct access to a factual data bank with experimental azeotropic data can help to determine the binary and higher azeotropic points for given conditions (temperature and pressure). Prior to process simulation it is important to verify that the model parameters used reproduce the azeotropic points and at the same time not falsely predict nonexistent azeotropes like in the case of the system acetone–water shown in Figure 11.6.

Example 11.2

Check with the help of modified UNIFAC, whether the homogeneous quaternary system benzene (1)–cyclohexane (2)–acetone (3)–ethanol (4) shows binary, ternary, or quaternary azeotropes at atmospheric pressure.

Solution

The azeotropic points in homogeneous systems can be determined with the procedures introduced in Chapter 5, for example, in binary systems by looking for a composition for which the separation factor α_{12} shows a value of unity – or more general – the following objective function F shows a value of zero for binary or

Table 11.2 Comparison of the predicted azeotropic data for the quaternary system benzene (1)–cyclohexane (2)–acetone (3)–ethanol (4) at atmospheric pressure with the mean value of the experimental azeotropic data stored in the Dortmund Data Bank [5, 7].

System	Predicted (modified UNIFAC (Do))				Experimental			
	Type of azeotrope	ϑ (°C)	$y_{1,\,az}$	$y_{2,\,az}$	Type of azeotrope	ϑ (°C)	$y_{1,\,az}$	$y_{2,\,az}$
1–2	homPmax	77.5	0.543	–	homPmax	77.6	0.543	–
1–3	none	–	–	–	none	–	–	–
1–4	homPmax	68.0	0.537	–	homPmax	67.9	0.552	–
2–3	homPmax	54.3	–	0.221	homPmax	53.2	–	0.248
2–4	homPmax	65.3	–	0.545	homPmax	64.8	–	0.553
3–4	none	–	–	–	none	–	–	–
1–2–3	none	–	–	–	none	–	–	–
1–2–4	homPmax	65.1	0.126	0.441	homPmax	64.9	0.113	0.462
1–3–4	none	–	–	–	none	–	–	–
2–3–4	none	–	–	–	none	–	–	–
1–2–3–4	none	–	–	–	n.a.[a]	–	–	–

[a] n.a. not available.

multicomponent systems:

$$F = \sum_j \sum_i \left| \alpha_{ij} - 1 \right| = 0$$

The azeotropic points calculated using modified UNIFAC with the help of the Mathcad program are listed in Table 11.2 together with the mean values of the experimental azeotropic points stored in the DDB. It can be seen that all existing binary and ternary azeotropic points for the quaternary system were predicted using modified UNIFAC, and that at the same time the predicted temperature and composition is in good agreement with the experimental findings.

Instead of a g^E-model of course also an equation of state can be applied to calculate or predict the required separation factors. This is demonstrated in Example 11.3.

Example 11.3

With the help of the group contribution equation of state VTPR, it should be checked whether the quaternary system carbon dioxide (1)–ethane (2)–hydrogen sulfide (3)–propane (4) shows binary, ternary, or quaternary azeotropes at 266.5 K.

Solution

For the determination of the azeotropic points again, the Mathcad program can be used. The results are shown in Table 11.3 together with the experimental data

Table 11.3 Comparison of the predicted azeotropic data for the quaternary system carbon dioxide (1)–ethane (2)–hydrogen sulfide (3)–propane (4) at 266.5 K using VTPR with the mean value of the experimental azeotropic data stored in the Dortmund Data Bank [5, 7].

System	Predicted using VTPR				Experimental			
	Type	P (bar)	$y_{2,az}$	$y_{3,az}$	Type	P (bar)	$y_{2,az}$	$y_{3,az}$
1–2	homPmax	33.36	0.3127	–	homPmax	33.27	0.33	–
1–3	none	–	–	–	none	–	–	–
1–4	none	–	–	–	none	–	–	–
2–3	homPmax	20.59	0.9097	0.0903	homPmax	20.68	0.896	0.104
2–4	none	–	–	–	none	–	–	–
3–4	homPmax	8.97	–	0.8301	homPmax	n.a.	–	0.83
1–2–3	none	–	–	–	n.a.[a]	–	–	–
1–2–4	none	–	–	–	n.a.	–	–	–
1–3–4	none	–	–	–	n.a.	–	–	–
2–3–4	none	–	–	–	n.a.	–	–	–
1–2–3–4	none	–	–	–	n.a.	–	–	–

[a]n.a. not available.

stored in the DDB. For three binary systems, azeotropic behavior is calculated. No ternary or quaternary azeotropic point is found using the group contribution equation of state VTPR. A comparison with the experimental data stored in the DDB shows that this is in agreement with the experimental findings.

The results of Examples 11.2 and 11.3 show that today even predictive models can be applied successfully to find the binary and higher azeotropes of a multicomponent system. With the development of the group contribution equations of state like PSRK and VTPR, the range of applicability was extended to compounds which are not covered by group contributions methods such as UNIFAC or modified UNIFAC.

11.3
Residue Curves, Distillation Boundaries, and Distillation Regions

For a better understanding of the separation of ternary or quaternary systems by distillation, the construction of residue curves is quite helpful. That is of particular importance when binary or higher azeotropes occur in the system to be separated. Residue curves describe the change of the liquid composition of a mixture L in a still during open evaporation.

Residue curves can be determined experimentally by open evaporation ensuring exactly one equilibrium stage. During evaporation, the less volatile compounds are enriched in the still. With an increasing amount of high boilers in the still, the

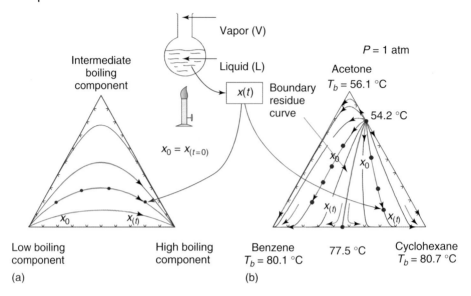

Figure 11.11 Composition as a function of time for two ternary systems in the case of open distillation at atmospheric pressure. Ideal ternary system with a low, intermediate and high-boiling compound (a). Ternary system acetone–benzene–cyclohexane (b).

temperature will increase. This is shown in Figure 11.11 for two different systems at atmospheric pressure, where the arrows point to the direction of increasing temperature. While in Figure 11.11a an ideal ternary system is considered, in Figure 11.11b the ternary system acetone–benzene–cyclohexane with two binary azeotropic points was chosen. As can be recognized from the change of the liquid composition of the ternary system, the high boiler is enriched with time. Associated with the enrichment of the high boiler in the still, the temperature increases which is shown by the direction of the arrows. Finally, the high-boiling compound remains in the still as typical bottom product of a typical distillation process. In the example, on the right-hand side surprisingly either benzene or cyclohexane remains in the still as high-boiling compound. The remaining high boiler depends on the initial feed composition. This means two distillation regions are formed, which are separated by the distillation boundary. While starting from an initial composition in the distillation region on the left hand side of Figure 11.11b benzene is obtained as bottom product, cyclohexane is found as bottom product starting with an initial composition on the right hand side of Figure 11.11b.

The residue curves can also be calculated by numerical integration of the differential equation for open evaporation (Rayleigh equation) if the vapor–liquid equilibrium behavior is known (see Figure 11.11) [8]. Starting from the material balance of component i:

$$y_i \, dV = -d(L \, x_i) = -x_i \, dL - L \, dx_i \tag{11.1}$$

one obtains after substitution of dL by $-dV$:

$$\frac{dx_i}{dL/L} = y_i - x_i \tag{11.2}$$

for $i = 1, \ldots, n-1$.

By introducing a nonlinear dimensionless timescale ξ:

$$\frac{dx_i}{d\xi} = \frac{dx_i}{dL/L} = y_i - x_i \tag{11.3}$$

For a given composition, Eq. (11.3) can be integrated numerically with positive and negative $\Delta\xi$.

End points of the residue curves are so-called singular[4] points. These are the pure components or binary and higher azeotropes. In topology analysis, these singular points ($dx_{i,\text{singular}}/d\xi = 0$) are called stable nodes in case of a local temperature maximum in the region considered and unstable nodes in case of a local temperature minimum in the isobaric case. In the isothermal case, stable nodes show a pressure minimum and unstable nodes a pressure maximum. The remaining pure components and azeotropic points show saddle points. Here, the residue curves move toward and then away from the saddle. All these singular points determine the topology of the system. The determination of the bottom and top product of a distillation column, of the boundary distillation lines, and the distillation regions requires the knowledge of the boiling points or vapor pressures of the pure components and of all azeotropic points of the system to be separated. The topology depends on the number, type, and position of the singular points.

Example 11.4

Calculate two isothermal residue curves and the composition of the end points for the system acetone (1)–benzene (2)–cyclohexane (3) assuming ideal vapor phase behavior at 70 °C with the help of the Wilson equation starting from the following initial compositions:

(a) $x_1 = 0.25$; $x_2 = 0.25$; $x_3 = 0.5$
(b) $x_1 = 0.4$; $x_2 = 0.4$; $x_3 = 0.2$

by numerical integration of Eq. (11.3) with positive and negative $\Delta\xi$.

Molar volumes:

$v_1 = 74.04 \text{ cm}^3 \text{ mol}^{-1}$; $v_2 = 89.40 \text{ cm}^3 \text{ mol}^{-1}$; $v_3 = 108.75 \text{ cm}^3 \text{ mol}^{-1}$

Vapor pressures at 70 °C:

$P_1^s = 159.28 \text{ kPa}$; $P_2^s = 73.36 \text{ kPa}$; $P_3^s = 72.38 \text{ kPa}$.

---

[4] At a singular point (pure component, binary, ternary, or quaternary azeotrope) the bubble and dew point line meet.

Recommended Wilson parameters $\Delta\lambda_{ij}$ (cal/mol) = $a_{ij} + b_{ij}T + c_{ij}T^2$:

	a_{ij} (cal/mol)	b_{ij} (cal/mol K)	c_{ij} (cal/mol K^2)
Acetone–benzene	−113.717	2.52916	−0.0035364
	201.963	−1.55162	0.0026447
Acetone–cyclohexane	3109.21	−10.6218	0.013757
	1670.65	−4.51887	0.0012266
Benzene–cyclohexane	1558.96	−8.23833	0.0128559
	−203.702	3.21397	−0.0075245

Solution

With the help of the temperature-dependent recommended Wilson parameters $\Delta\lambda_{ij}$ at 70 °C, one obtains (see also Mathcad Solution 11.04):

$$\Delta\lambda_{12} = -113.717 + 2.52916 \cdot 343.15 - 0.0035364 \cdot 343.15^2$$
$$= 337.75 \text{ cal/mol}.$$

In the same way, all other interaction parameters $\Delta\lambda_{ij}$ can be determined:

$$\Delta\lambda_{21} = -19.055 \text{ cal/mol},$$

$$\Delta\lambda_{13} = 1084.25 \text{ cal/mol}; \quad \Delta\lambda_{31} = 264.44 \text{ cal/mol}$$

$$\Delta\lambda_{23} = 245.78 \text{ cal/mol}; \quad \Delta\lambda_{32} = 13.151 \text{ cal/mol}.$$

Using these values and the molar volumes, the Wilson parameters Λ_{ij} can be calculated:

$$\Lambda_{ij} = \frac{v_j}{v_i} \exp\left[-\frac{\Delta\lambda_{ij}}{RT}\right]$$

$$\Lambda_{12} = \frac{89.40}{74.04} \exp\left[\frac{337.75}{1.98721 \cdot 343.15}\right] = 0.7358$$

In the same way, one obtains

$$\Lambda_{21} = 0.8517$$

$$\Lambda_{13} = 0.2995; \quad \Lambda_{31} = 0.4620$$

$$\Lambda_{23} = 0.8482; \quad \Lambda_{32} = 0.8064$$

Using these Wilson parameters Λ_{ij} the activity coefficients can be determined. For the calculation of the activity coefficient of component 1, the following equation holds:

$$\ln\gamma_1 = -\ln(x_1\Lambda_{11} + x_2\Lambda_{12} + x_3\Lambda_{13}) + 1 - \frac{x_1\Lambda_{11}}{x_1\Lambda_{11} + x_2\Lambda_{12} + x_3\Lambda_{13}}$$
$$- \frac{x_2\Lambda_{21}}{x_1\Lambda_{21} + x_2\Lambda_{22} + x_3\Lambda_{23}} - \frac{x_3\Lambda_{31}}{x_1\Lambda_{31} + x_2\Lambda_{32} + x_3\Lambda_{33}}$$

11.3 Residue Curves, Distillation Boundaries, and Distillation Regions

For a composition of $x_1 = 0.25$, $x_2 = 0.25$, and $x_3 = 0.5$, one gets

$$\ln \gamma_1 = -\ln(0.25 \cdot 1 + 0.25 \cdot 0.7358 + 0.5 \cdot 0.2995) + 1$$
$$-\frac{0.25 \cdot 1}{0.25 \cdot 1 + 0.25 \cdot 0.7358 + 0.5 \cdot 0.2995}$$
$$-\frac{0.25 \cdot 0.8517}{0.25 \cdot 0.8517 + 0.25 \cdot 1 + 0.5 \cdot 0.8482}$$
$$-\frac{0.5 \cdot 0.4620}{0.25 \cdot 0.4620 + 0.25 \cdot 0.8064 + 0.5 \cdot 1} = 0.5874$$

$$\gamma_1 = 1.7992$$

For the other two components, the following activity coefficients are obtained:

$$\gamma_2 = 1.0299 \quad \gamma_3 = 1.2495$$

With the help of the activity coefficients and the given vapor pressures, the partial pressures, the total pressure, and the mole fraction in the vapor phase at 70 °C can be calculated:

$$P = \sum p_i = \sum x_i \cdot \gamma_i \cdot P_i^s$$

$$P = 0.25 \cdot 1.7992 \cdot 159.28 + 0.25 \cdot 1.0299 \cdot 73.36$$
$$+ 0.5 \cdot 1.2495 \cdot 72.38 = 135.75 \text{ kPa}$$

The ratio of the partial to the total pressure is identical with the vapor phase mole fraction:

$$y_1 = \frac{0.25 \cdot 1.7992 \cdot 159.28}{135.75} = 0.5278$$

For the other two components, one obtains

$$y_2 = 0.1391 \quad y_3 = 0.3331$$

By numerical integration of Eq. (11.2), the residue curves can be calculated. Starting from the chosen composition, the composition change is:

$$x_1^{(1)} = x_1^{(0)} + \left(y_1^{(0)} - x_1^{(0)}\right) \Delta \xi$$

The integration can then be performed with positive and negative $\Delta \xi$ steps.

For a fixed step width of $\Delta \xi = 0.02$, one obtains

$$x_1^{(1)} = 0.25 + (0.5278 - 0.25) \cdot 0.02 = 0.2556$$

In the same way, one gets a value of $x_2 = 0.2478$ and for $x_3 = 1 - x_1 - x_2 = 0.4967$. For $\Delta \xi = -0.02$, the integration leads to $x_1 = 0.2444$, $x_2 = 0.2522$, and $x_3 = 0.5033$.

11 Applications of Thermodynamics in Separation Technology

Table 11.4 Compositions in the liquid and vapor phase and pressure for the different steps of the forward and backward integration.

	x_1	x_2	x_3	y_1	y_2	y_3	P
	0.7692	0.0000	0.2308	0.7692	0.0000	0.2308	172.00
...	...	...	...	...	...	...	...
0.10	0.2775	0.2390	0.4836	0.5474	0.1295	0.3231	138.89
0.08	0.2720	0.2412	0.4868	0.5437	0.1313	0.3250	138.29
0.06	0.2666	0.2434	0.4901	0.5399	0.1332	0.3269	137.68
0.04	0.2611	0.2456	0.4933	0.5359	0.1351	0.3289	137.05
0.02	0.2556	0.2478	0.4967	0.5319	0.1371	0.3310	136.41
0	0.25	0.25	0.50	0.5278	0.1391	0.3331	135.75
−0.02	0.2444	0.2522	0.5033	0.5235	0.1412	0.3353	135.07
−0.04	0.2389	0.2544	0.5067	0.5191	0.1433	0.3375	134.38
−0.06	0.2333	0.2567	0.5101	0.5146	0.1455	0.3399	133.67
−0.08	0.2276	0.2589	0.5135	0.5100	0.1477	0.3423	132.94
−0.10	0.2220	0.2611	0.5169	0.5052	0.1500	0.3448	132.19
...	...	...	...	...	...	...	...
...	0.00	0.00	1.0000	0.00	0.00	1.0000	72.38

For the next integration steps, the vapor phase composition has to be calculated again with the help of the Wilson equation using the new liquid composition.[5] In Table 11.4, the compositions in the liquid and vapor phase are given together with the system pressures. It can be seen that the end points of the integration are the azeotropic point of acetone (1) with cyclohexane (3) and the pure component cyclohexane (3), which are the possible top and bottom products of the distillation column.

If $x_1 = 0.4$ and $x_2 = 0.4$ are used as starting composition not cyclohexane but benzene is obtained as typical bottom product after integration besides the binary azeotrope as low-boiling compound at the top of the column.

Distillation boundaries always exist if there is more than one origin or terminus of residue curves in the system, which means more than one feasible bottom or top product. Typical residue curves for the system benzene–cyclohexane–acetone are shown in Figure 11.11. The most important lines are the distillation boundaries, which cannot be crossed by distillation. While the residue curves are connections between the high-boiling compound (stable node) and the low-boiling compound (unstable node), the distillation boundaries are connecting lines between the saddle point and the stable or the unstable node, respectively.

Residue curve maps can be roughly constructed if only the boiling points or vapor pressures of the singular points are available. This is shown in Figures 11.12

---

5) In most cases the calculations are required for isobaric conditions, often 1 atm. In this case of course the temperature has to be determined iteratively for every integration step.

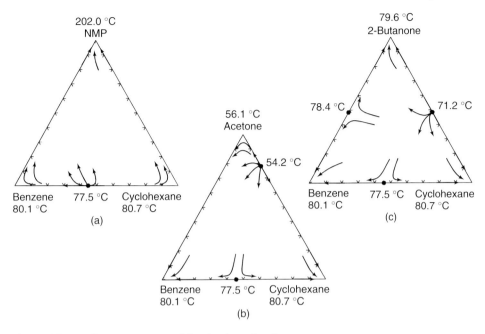

Figure 11.12 Boiling temperatures of the singular points for different ternary systems with benzene and cyclohexane at atmospheric pressure.

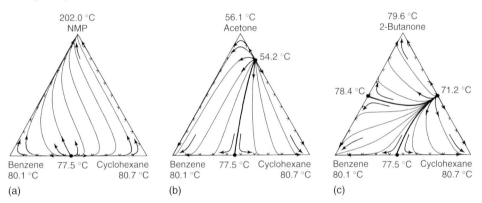

Figure 11.13 Residue curves and distillation boundaries for different ternary systems with benzene and cyclohexane at atmospheric pressure.

and 11.13 for three ternary systems consisting of the binary azeotropic system benzene–cyclohexane with NMP, acetone, and 2-butanone as third compound. In Figure 11.12, for all singular points the boiling points are given. While in the ternary system with NMP only one binary azeotrope between benzene and cyclohexane with a boiling point of 77.5 °C exists, in the ternary system with acetone an additional azeotrope between cyclohexane–acetone (boiling point 54.2 °C) is found. With

2-butanone two further azeotropes with boiling points of 71.2 °C and 78.4 °C are introduced. Looking at the boiling points of the singular points, directly the course of the temperature can be drawn and the type of singular point can be assigned. While for the ternary system with NMP only one stable node is found, for the system with acetone two stable nodes and for the system with 2-butanone three stable nodes are found. In all three cases besides saddle points, one unstable node is observed. Since a stable node is the typical bottom product of a distillation

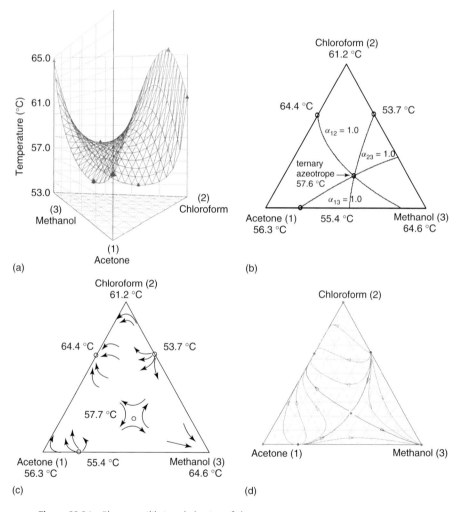

Figure 11.14 Phase equilibrium behavior of the ternary system acetone–chloroform–methanol at atmospheric pressure calculated using modified UNIFAC. (a) Tx-behavior; (b) lines of constant separation factors ($\alpha_{12} = 1$, $\alpha_{13} = 1$, $\alpha_{23} = 1$); (c) nodes and saddles of the different singular points; (d) residue curves and distillation boundaries.

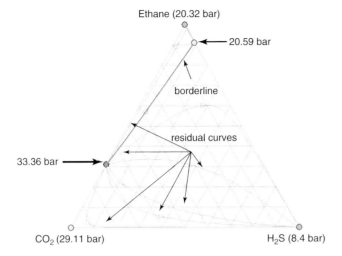

Figure 11.15 Calculated residue and distillation boundary for the system carbon dioxide–hydrogen sulfide–ethane at 266.5 K using VTPR.

column, the bottom product of the column must depend on the feed composition as already discussed before. If more than one stable node is formed, distillation boundaries exist which cannot be crossed by ordinary distillation. The distillation boundary runs from the saddle point to the stable or unstable node, so that for the system with acetone two distillation regions and for the system with 2-butanone three distillation regions exist (see Figure 11.13).

For the system acetone–chloroform–methanol, even four distillation regions are found. This is illustrated in Figure 11.14, where besides the Tx-behavior, the lines with separation factors equal to one ($\alpha_{ij} = 1$), the stable and unstable nodes and saddles, the residue curves, and the distillation boundaries at atmospheric pressure are shown.

Instead of a g^E-model or a group contribution method like modified UNIFAC also an equation of state or a group contribution equation of state can be used for the calculation of residue curves and distillation boundaries. In Figure 11.15, the results are shown for the ternary system carbon dioxide–hydrogen sulfide–ethane at 266.5 K using VTPR. As can be seen, two binary azeotropes and one distillation boundary is observed.

11.4
Selection of Entrainers for Azeotropic and Extractive Distillation

For the separation of azeotropic systems by distillation, different techniques can be applied. There is the possibility that the azeotropic behavior disappears at lower or higher pressure. This means that the separation could be performed in a column running at lower or higher pressure. Furthermore, a strong pressure dependence of the azeotropic composition can be used to separate the azeotropic system by pressure swing distillation, this means in two columns running at different pressures. Heterogeneous azeotropic systems can be separated with the help of two columns and a decanter. Furthermore, hybrid processes, such as distillation and a membrane separation process or distillation and adsorption, can be applied to separate azeotropic systems, for example, ethanol–water.

In most cases, special distillation processes such as azeotropic or extractive distillation are applied to separate azeotropic systems by distillation, where a suitable solvent is added. While in the case of azeotropic distillation a solvent is required which forms a lower boiling azeotropic point, in the case of extractive distillation a high-boiling selective solvent is used which changes the separation factor in a way that it becomes distinctly different from unity. Both processes are shown in Figure 11.16 together with the column configuration.

While for azeotropic distillation the knowledge of the azeotropic points and of the miscibility gap is most important, for the selection of solvents (entrainers) for extractive distillation the knowledge of the influence of the entrainer on the separation factor is required.

Looking at the simplified Eq. (5.18) for the separation factor, the task of the high-boiling entrainer can directly be recognized:

$$\alpha_{12} = \frac{\gamma_1 P_1^s}{\gamma_2 P_2^s}$$

The entrainer should alter the separation factor in a way that the separation factor becomes different from unity. Since the entrainer has no influence on the pure component vapor pressures, the entrainer has to shift the ratio of the activity coefficients selectively. Although in practice a concentration of 50–80% entrainer is used in the columns, for the selection of a selective entrainer in the first step the selectivity at infinite dilution S_{12}^∞ is used.

$$S_{12}^\infty = \frac{\gamma_1^\infty}{\gamma_2^\infty} \gg 1 \, (\ll 1) \tag{11.4}$$

This means that the knowledge about activity coefficients at infinite dilution is required. Azeotropic points and activity coefficients at infinite dilution can be obtained either with the help of predictive thermodynamic models (UNIFAC, modified UNIFAC, PSRK, and VTPR) or by access to the information (azeotropic information and activity coefficients at infinite dilution) stored in a factual data bank, for example, the DDB (DDB) [7]. Sophisticated software packages for searching for a suitable solvent were developed [10]. The flow diagrams are shown in Figure 11.17. In Figure 11.17a, the procedure is shown for the application of predictive thermodynamic models. In this case, all 33 000 components of the DDB can be considered as possible solvents for azeotropic or extractive distillation. The flow diagram in Figure 11.17b shows the procedure by access to the Dortmund Data Bank, where more than 53 200 azeotropic data points and more than 62 500 activity coefficients at infinite dilution can be used to select the most suitable solvent from the thermodynamic point of view. Both methods supplement each other. The use of predictive models allows calculating the required properties also for systems for

11.4 Selection of Entraziners for Azeotropic and Extractive Distillation | 513

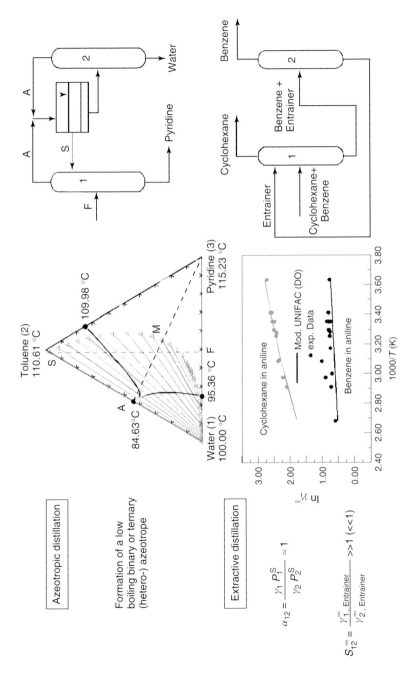

Figure 11.16 Separation of azeotropic systems by azeotropic (feed F: pyridine–water, solvent: toluene) and extractive distillation (feed: benzene-cyclohexane).

514 | *11 Applications of Thermodynamics in Separation Technology*

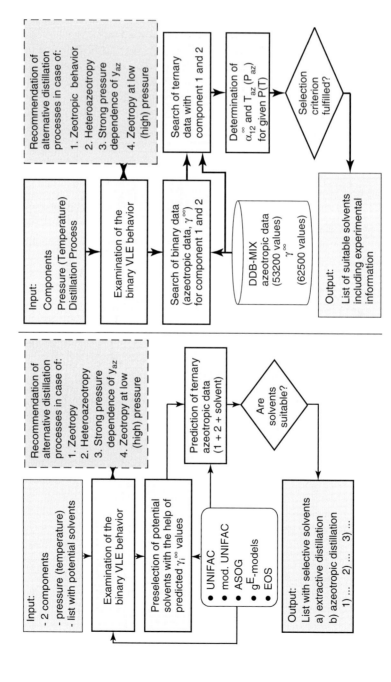

Figure 11.17 Flow diagram for the selection of suitable solvents for azeotropic or extractive distillation by direct access to the Dortmund Data Bank or the application of predictive methods.

which no experimental data exist. The access to the factual data bank also allows finding solvents which cannot be found with the predictive models.

To run the program "Process Synthesis [7]", the user first has to specify the binary system to be separated and the temperature or pressure of interest. Then, the program analyzes the VLE behavior and advises the user if the system can be separated without the use of a solvent, for example, when the system shows nonazeotropic behavior or when it forms a heterogeneous azeotrope which can be separated in two columns with a decanter. The program will also remind the user, if the binary system can be separated by pressure swing distillation or that the azeotropic behavior disappears at lower or higher pressure. If the user decides to proceed with the search for an entrainer, first the activity coefficients at infinite dilution of the components to be separated are calculated for all solvents selected. From these data, information about the selectivity and at the same time about azeotropic behavior of the binary system using the criterion mentioned in Section 5.8 is available. Then, the program package analyzes the ternary system to look for ternary azeotropes, which can be useful for azeotropic distillation but should be avoided in the case of extractive distillation. At the end, the user will get a list with suitable solvents for azeotropic and extractive distillation. The procedure by direct access to a factual data bank is shown on the right-hand side of Figure 11.17. It is similar, but in this procedure the required data (experimental azeotropic data and activity coefficients at infinite dilution) are directly taken from the DDB.

Typical results for the selection of suitable solvents for azeotropic distillation for the separation of the azeotropic system ethanol – water and the system acetic acid – water at atmospheric pressure are given in Table 11.5. It can be seen that the predicted temperatures and compositions of the heterogeneous ternary azeotropes in the case of the system ethanol–water and of the binary heterogeneous azeotropes organic compound–water in the case of acetic acid–water using modified UNIFAC are in good agreement with the experimental data stored in the DDB. While mainly hydrocarbons are selected for the formation of the ternary heterogeneous azeotrope with ethanol and water, heterogeneous binary azeotropes of water with esters, ethers and ketones, are suitable for the separation of acetic acid–water by azeotropic distillation.

The selected solvents for the separation of the azeotropic system benzene–cyclohexane (chosen as a representative system for the separation of aromatics from aliphatics) by extractive distillation at 353.15 K via direct access to the Dortmund Data Bank and predicted using modified UNIFAC are given in Table 11.6. In this table, the separation factors at infinite dilution are sorted by the value of the separation factor at infinite dilution. It can be seen that by direct access to the DDB different ionic liquids are found, which cannot be detected by modified UNIFAC since up to now the group interaction parameters for these ionic liquids are not available. As can be seen, the ionic liquids show a higher selectivity than the organic compounds. For the organic compounds, the predicted and the experimental separation factors at infinite dilution are in good agreement. As well, other thermodynamic criteria are important for solvent selection. For example, a miscibility gap is unwanted in extractive distillation, because it may lead

Table 11.5 Selected entrainers for the separation of ethanol (1)–water (2) and water (1)–acetic acid (2) by azeotropic distillation.

Entrainer (3)	DDB			Modified UNIFAC		
	$\gamma_{1,az}$	$\gamma_{2,az}$	T_{az} (K)	$\gamma_{1,az}$	$\gamma_{2,az}$	T_{az} (K)
Ethanol (1)–water (2)						
Cyclohexane	0.312	0.164	335.25	0.293	0.160	336.3
Methylcyclohexane	0.423	0.236	342.80	0.422	0.213	343.7
Hexane	0.168	0.210	329.15	0.243	0.121	330.2
Benzene	0.228	0.233	338.00	0.264	0.208	338.0
Toluene	0.432	0.326	347.55	0.456	0.286	347.3
Ethyl acetate	0.125	0.276	343.45	0.095	0.294	343.2
Water (1)– acetic acid (2)						
Butyl acetate	0.712	–	363.8	0.717	–	364.2
Dibutyl ether	0.768	–	367.5	0.796	–	366.9
3-Pentanone	0.475	–	356.3	0.540	–	357.1

Table 11.6 Experimental and predicted separation factors at infinite dilution at 353 K using mod. UNIFAC for selected entrainers for the separation of the system cyclohexane (1)–benzene (2) by extractive distillation.

Dortmund Data Bank		Modified UNIFAC	
Entrainer (3)	α_{12}^∞	Entrainer (3)	α_{12}^∞
MMIM methylsulfate[a]	12.6	Adiponitrile	8.7
EMIM ethylsulfate	12.0	Sulfolane	8.2
Sulfolane	7.2	N-Methylpyrrolidone (NMP)	4.9
Adiponitrile	6.4	Diethylenglycol	4.6
N-Formylmorpholine (NFM)	5.6	Aniline	4.2
Diethylenglycol	5.3	Propylencarbonate	4.2
Propylencarbonate	5.1	Furfural	4.1
N-Methylpyrrolidone (NMP)	4.8	Dimethylformamide (DMF)	3.7

[a] Ionic liquids: MMIM methylsulfate = Methylmethylimidazolium methylsulfate; EMIM ethylsulfate = Ethylmethylimidazolium ethylsulfate.

to distribution problems inside the column. Additional criteria are the capacity of the solvent or the ease of stripping.[6]

6) Besides the phase equilibrium behavior (azeotropy, activity coefficients) other properties such as flammability, toxicity, availability, price, thermal and chemical stability, etc., are of similar importance for the selection of suitable solvents for separation processes. Therefore time consuming tests are required, before a new solvent is applied in practice.

Separation factors at infinite dilution show the largest deviation from ideal behavior, but at infinite dilution the necessary amount of solvent would also be infinite. In practice, extractive distillation processes are performed with entrainer compositions between 50 and 80 mol%. After the selection of the best suited entrainer, its amount has to be chosen in a way which allows the most economical separation. How the separation factor is altered by different amounts of entrainer is shown in Example 11.5.

Example 11.5

With the help of modified UNIFAC it should be checked how the separation factor α_{12} of the binary system benzene (1) and cyclohexane (2) is altered in the presence of 50 and 80 mol% NMP(3) at 80 °C.

Solution

The solution is performed in the Mathcad-File 11.05. For the whole composition range, the mole fractions in the vapor phase and the separation factors as a function of the mole fraction in the liquid phase on a solvent-free basis are shown in Figure 11.18. From the figures, it can be recognized that the azeotropic binary system benzene–cyclohexane can be separated by extractive distillation, since cyclohexane in comparison to benzene shows higher activity coefficients; this means cyclohexane becomes more volatile in the presence of NMP and is obtained at the top of the extractive distillation column. The deviation of the separation factor from a value of one strongly depends on the concentration of the entrainer (see Figure 11.18). The higher the concentration of the selective solvent, the stronger

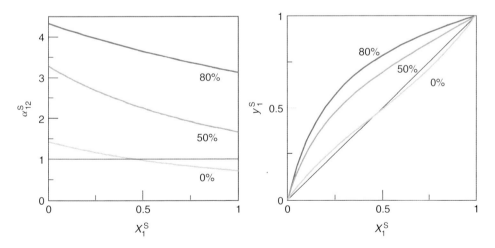

Figure 11.18 Calculated separation factors and y–x behavior of the system benzene (1)–cyclohexane (2) (solvent-free basis) using modified UNIFAC at 80 °C in the presence of NMP (3) ($x_3 = 0$ mol%; $x_3 = 50$ mol%; $x_3 = 80$ mol%).

the deviation of the separation factor α_{12} from a value of one and the smaller the number of stages required for the separation of the azeotropic system by extractive distillation.

11.5
Selection of Solvents for Other Separation Processes

In Section 11.4, it was shown how suitable solvents can be selected with the help of powerful predictive thermodynamic models or direct access to the DDB using a sophisticated software package. A similar procedure for the selection of suitable solvents was also realized for other separation processes, such as physical absorption, extraction, solution crystallization, supercritical extraction, and so on. In the case of absorption processes or supercritical extraction instead of a g^E-model, for example, modified UNIFAC, of course an equation of state such as PSRK or VTPR has to be used. For the separation processes mentioned above instead of azeotropic data or activity coefficients at infinite dilution, now gas solubility data, liquid–liquid equilibrium data, distribution coefficients, solid–liquid equilibrium data or VLE data with supercritical compounds are required and can be accessed from the DDB.

For the removal of the sour gases from natural gas, the experimental and predicted Henry constants in methanol using the group contribution equation of

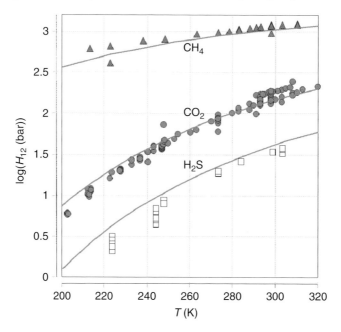

Figure 11.19 Experimental [7] and predicted Henry constants using PSRK for the gases methane, carbon dioxide, and hydrogen sulfide in methanol.

state PSRK are shown in Figure 11.19. It can be seen that there is good agreement between the experimental and predicted Henry constants. Furthermore from the diagram, it can be recognized that the Henry constant of methane in methanol is distinctly higher than the Henry constant of CO_2 and H_2S. Since the vapor pressure of the solvent must be low to avoid solvent losses in the case of methanol (Rectisol process), one has to work at very low temperatures. But from the diagram, it seems that at very low temperatures at the same time higher selectivities are achieved. In practice, the absorption is run at approx. $-40\,°C$.

11.6
Examination of the Applicability of Extractive Distillation for the Separation of Aliphatics from Aromatics

Aromatic compounds such as benzene, toluene, or the different xylenes are mainly produced by the hydrogenated C_{5+}-stream (pyrolysis gasoline) of a steam cracker. Besides the aromatics, this stream contains the different aliphatics and naphthenes. There is the question if all the aromatics (C_6–C_{12}) can be separated from the other C_6–C_{12} compounds by extractive distillation using for example sulfolane as entrainer. Simplifying, it is assumed that the aliphatics only consist of n-alkanes (n-hexane–n-dodecane). A temperature of $80\,°C$ is chosen. The separation problem and the column configuration is shown in Figure 11.20.

The most critical separation problem is the separation of all the n-alkanes from the most volatile aromatic compound, namely benzene. To be able to achieve the separation shown in Figure 11.20, the separation factors of all n-alkanes with

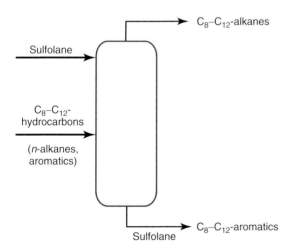

Figure 11.20 Column configuration for the separation of aliphatics from aromatics by extractive distillation using sulfolane as entrainer.

respect to benzene:

$$\alpha_{ij} = \frac{\gamma_{n\text{-alkane}} \cdot P^s_{n\text{-alkane}}}{\gamma_{benzene} \cdot P^s_{benzene}}$$

must be higher than unity, to obtain even the high-boiling compounds such as undecane and dodecane as top product (distillate). The separation factors at infinite dilution can be obtained from vapor pressure data and predicted or experimental activity coefficients at infinite dilution. The following vapor pressures at 80 °C can be used to calculate the separation factors.

Vapor pressures (kPa):

Benzene	Hexane	Heptane	Octane	Nonane	Decane	Undecane	Dodecane
101	142.4	57.1	23.4	9.7	4.07	1.71	0.74

Using modified UNIFAC, the following activity coefficients at infinite dilution in sulfolane are predicted with Mathcad at 80 °C:

Benzene	Hexane	Heptane	Octane	Nonane	Decane	Undecane	Dodecane
2.39	31.0	43.1	59.8	82.9	114	158	218

In the DDB [7] and in the data compilation [4], the following activity coefficients at infinite dilution in sulfolane can be found for a temperature of 80 °C.

Benzene	Hexane	Heptane	Octane	Nonane	Decane	Undecane	Dodecane
2.38	35	40	66	88	112	142	182

Using these values, the following separation factors for the n-alkanes with respect to benzene at infinite dilution α_{12}^∞ are calculated.

	Hexane	Heptane	Octane	Nonane	Decane	Undecane	Dodecane
Modified UNIFAC	18.3	10.2	5.8	3.33	1.92	1.13	0.67
Experimental	20.7	9.5	6.4	3.55	1.90	1.01	0.56

In Figure 11.21, the ratios P_1^s/P_2^s, the experimental and predicted ratios of $\gamma_1^\infty/\gamma_2^\infty$, and the separation factors at infinite dilution α_{12}^∞ are shown. It can be seen

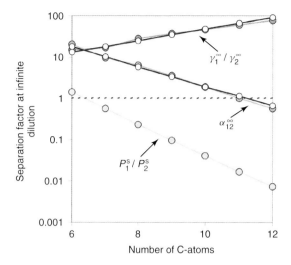

Figure 11.21 Ratio of the vapor pressures and ratio of the predicted and experimental activity coefficients and separation factors in sulfolane (3) at infinite dilution.

that the predicted values are in good agreement with the experimental findings. Furthermore, it can be seen that at infinite dilution a separation of benzene from C_6–C_{10}-aliphatics would be possible, since the separation factors of n-hexane to n-decane are significantly larger than unity.

Since a separation at infinite dilution is not of practical interest, additionally the separation factors for a sulfolane mole fraction of 0.8 should be calculated using modified UNIFAC. The Mathcad-File can be used to predict the activity coefficients of benzene and the n-alkanes in a mixture of 80 mol% sulfolane, 10 mol% benzene, and 10 mol% n-alkane. The predicted activity coefficients using modified UNIFAC and the separation factors are given in the table below. Now lower activity coefficients are predicted, and the activity coefficient of benzene also depends on the n-alkane considered. From the activity coefficients and the vapor pressures again, the separation factors can be calculated. The ratios of the activity coefficients and the ratio of the vapor pressures are shown together with the separation factor α_{12} in Figure 11.22.

	Hexane	Heptane	Octane	Nonane	Decane	Undecane	Dodecane
γ_{alkane}	13.07	15.38	17.91	20.65	23.58	26.69	29.96
$\gamma_{benzene}$	1.72	1.68	1.64	1.60	1.57	1.54	1.51
α_{12}	10.7	5.17	2.53	1.24	0.61	0.29	0.15

From Figure 11.22, it can be concluded that the separation of aliphatics from aromatics by extractive distillation using sulfolane as entrainer is only possible

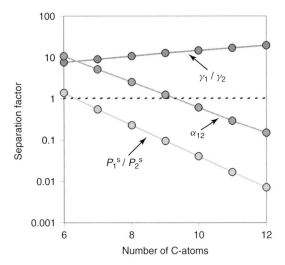

Figure 11.22 Ratio of the vapor pressures $P^s_{alkane}/P^s_{benzene}$, predicted ratio of the activity coefficients $\gamma_{alkane}/\gamma_{benzene}$, and separation factors at 80 °C in sulfolane (3) with a mole fraction $x_3 = 0.8$.

for a C_6-C_8 stream. This is mainly caused by the low vapor pressures of the higher n-alkanes. This means that before extractive distillation can be applied for the aliphatic/aromatic separation, a C_6-C_8-cut has to be produced from the hydrocarbon feed.

In the case of extraction processes, the vapor pressures do not influence the distribution coefficients. They only depend on the activity coefficients. Therefore, it seems that instead of extractive distillation, an extraction process could be an alternative for the separation considered.

Additional Problems

P11.1 For the separation of butadiene-1,3 from the C4-fraction of the steam cracker by extractive distillation various entrainers are used. Calculate the separation factor between butadiene-1,3(1) ($x_1 = 0.25$) and 1-butene(2) ($x_2 = 0.25$) at a temperature of 50 °C using the modified UNIFAC model to get an impression about the selectivity of the following entrainers (3): NMP, Acetonitrile, DMF ($x_3 = 0.5$).

P11.2 Check with the help of the Wilson equation how the separation factor α_{12} between benzene (1) and cyclohexane (2) is altered in the presence of 50 and 70 mol% NMP (3) at 60 °C.

Wilson parameters, molar volumes, and vapor pressures of NMP(3) at 60 °C:

$$\Delta\lambda_{13} = -205.34 \text{ K} \quad \Delta\lambda_{31} = 414.16 \text{ K}$$
$$\Delta\lambda_{23} = 321.30 \text{ K} \quad \Delta\lambda_{32} = 436.34 \text{ K}$$

$$v_3 = 96.44 \text{ cm}^3 \text{ mol}^{-1} \quad P_3^s = 0.47 \text{ kPa}$$

The additionally required Wilson parameters and pure component properties are given in Table 5.10 and in Appendix A, respectively.

P11.3 Calculate the separation factor for the system acetone (1)–water (2) for an acetone concentration of $x_1 = 0.95$ at 50 °C and at 1 atm with the help of the NRTL equation assuming ideal vapor phase behavior. Binary NRTL parameters can be found in Table 5.9. All other properties are given in Appendix A.

P11.4 In the free DDBSP Explorer Version, search for all mixture data for the system benzene–water. Due to very limited miscibility in the liquid phase, the system shows heteroazeotropic behavior. Estimate the azeotropic composition using the vapor pressure correlations given in Appendix A and compare the results to the experimental data.

References

1. Gmehling, J., Onken, U. et al. (1977) *Vapor-Liquid Equilibrium Data Collection*, DECHEMA Chemistry Data Series, 37 parts, Vol. **I**, DECHEMA, Frankfurt.
2. Sørensen, J.M., Arlt, W., Macedo, E., and Rasmussen, P. (1979) *Liquid–Liquid Equilibrium Data Collection*, DECHEMA Chemistry Data Series, 4 parts, DECHEMA, Frankfurt.
3. Gmehling, J., Holderbaum, T., Christensen, C., Rasmussen, P., and Weidlich, U. (1984) *Heats of Mixing Data Collection*, DECHEMA Chemistry Data Series, 4 parts, Vol. **III**, DECHEMA, Frankfurt.
4. Gmehling, J., Menke, J. et al. (1986) *Activity Coefficients at Infinite Dilution*, DECHEMA Chemistry Data Series, 6 parts, Vol. **IX**, DECHEMA, Frankfurt.
5. Gmehling, J., Menke, J., Krafczyk, J., and Fischer, K. (2004) *Azeotropic Data*, 3 parts, Wiley-VCH Verlag GmbH, Weinheim.
6. Grensemann, H. and Gmehling, J. (2005) *Ind. Eng. Chem. Res.*, **44**, 1610–1624.
7. Dortmund Data Bank www.ddbst.com, January 2011.
8. Baerns, M., Behr, A., Brehm, A., Gmehling, J., Hofmann, H., Onken, U., and Renken, A. (2006) *Technische Chemie*, Wiley-VCH Verlag GmbH, Weinheim.
9. Bergmann, D.L. and Eckert, C.A. (1991) *Fluid Phase Equilib.*, **63**, 141–150.
10. Gmehling, J. and Möllmann, C. (1998) *Ind. Eng. Chem. Res.*, **37**, 3112–3123.

12
Enthalpy of Reaction and Chemical Equilibria

The knowledge of the maximum conversion and the enthalpy of reaction is, besides the reliable knowledge of the reaction rate as a function of the various variables, such as temperature, pressure, and composition, a prerequisite for the design of chemical reactors.

When, for example, the reversible exothermal oxidation of sulfur dioxide to sulfur trioxide is considered:

$$SO_2 + 1/2\, O_2 \rightleftarrows SO_3$$

kinetic measurements show that even in the presence of catalysts, temperatures above 300 °C are required for the reaction to start. With a further increase of the temperature, a strongly exponential increase of the reaction rate is observed. From the kinetic point of view, a high temperature should be chosen for the realization of the chemical reaction. However, a temperature increase leads to a decrease in maximum conversion in the case of exothermal reversible reactions. For the optimization of the reaction step, the maximum conversion and at the same time the enthalpy of reaction as a function of temperature and pressure have to be known besides the reaction rate. For the determination of the maximum conversion and of the enthalpy of reaction as a function of temperature and pressure, thermodynamics can be applied. Using thermodynamics, the maximum conversion as well as the enthalpy of reaction can be calculated starting from tabulated data without time-consuming experimental measurements.

Sometimes these results already allow a decision whether the realization of a planned chemical process is economically reasonable. For example, during process development, it can be decided if the planned project should be stopped or an alternative process should be applied, when from the economic point of view a minimum conversion of 35% is required, but only a conversion of 10% can theoretically be achieved. In an alternative process, the reaction could be directly combined with the separation step, as in the case of reactive distillation, membrane reactors, and so on, to increase the chemical conversion.

Chemical Thermodynamics: for Process Simulation, First Edition.
Jürgen Gmehling, Bärbel Kolbe, Michael Kleiber, and Jürgen Rarey.
© 2012 Wiley-VCH Verlag GmbH & Co. KGaA. Published 2012 by Wiley-VCH Verlag GmbH & Co. KGaA.

12.1
Enthalpy of Reaction

The value and the sign (exothermal and endothermic) of the enthalpy of reaction do not only depend on temperature but also on pressure. The tabulation of the enthalpies of reaction for all the possible chemical reactions as a function of temperature and pressure is not possible. However, the effort can be drastically reduced if the reactants and the products are considered in their standard states. As standard state usually the pure component in one of the possible states (liquid, solid, and hypothetical ideal gas) at 25 °C and 1 atm (101.325 kPa) is chosen. In process simulators usually the hypothetical ideal gas state is used.

But also at fixed standard states, a large number of possible standard reaction enthalpies remain. Since enthalpies are state functions, this means that the values do not depend on the reaction path, enthalpies of reaction can be calculated using Hess' law. Following the law of Hess, the enthalpy change for the overall reaction, obtained by adding up the individual reaction steps, is the sum of the enthalpy changes associated with each of the individual chemical reactions. Thus, a chemical reaction can be derived by a linear combination of other reactions, for example, by a linear combination of the standard reactions of formation of the compounds involved. The standard reaction of formation is defined as a reaction, where 1 mol of the compound in a given aggregate state is formed at standard state conditions (25 °C, 1 atm) from the elements in their stable form, as for example:

$$C(S) + O_2(G) \rightarrow CO_2(G) \qquad \Delta h_R^0 = \Delta h_f^0 = -393.50 \text{ kJ/mol} \qquad (12.1)$$

$$H_2(G) + \tfrac{1}{2}O_2(G) \rightarrow H_2O(L) \qquad \Delta h_R^0 = \Delta h_f^0 = -286.04 \text{ kJ/mol} \qquad (12.2)$$

$$H_2(G) + \tfrac{1}{2}O_2(G) \rightarrow H_2O(G) \qquad \Delta h_R^0 = \Delta h_f^0 = -241.82 \text{ kJ/mol} \qquad (12.3)$$

$$H_2(G) \rightarrow H_2(G) \qquad \Delta h_R^0 = \Delta h_f^0 = 0 \text{ kJ/mol} \qquad (12.4)$$

The value of the standard enthalpies of formation of the elements in the stable form at standard conditions is set to 0 by definition.[1] This means that the standard enthalpy of reaction Δh_R^0 of the formation reaction is identical with the standard enthalpy of formation Δh_f^0. From the definitions above, it can easily be understood that the enthalpy of formation of an element in the stable form has a value of 0 not only at the standard temperature 25 °C, but also at other temperatures.

Only a few standard enthalpies of formation can be determined experimentally, for example, for carbon dioxide and water. However, the values for other compounds can be derived from the various enthalpies of reaction measured, for example, enthalpies of combustion, enthalpies of hydrogenation, and so on, using Hess' law.

---

1) In case that different modifications of the element exist, the stable form has to be defined: For example in the case of carbon the more stable graphite and not the less stable diamond was chosen. In the case of sulfur the rhombic form is used.

12.1 Enthalpy of Reaction

Using the standard enthalpies of formation of the compounds involved, the standard enthalpy of reaction can be determined for every reaction:

$$\Delta h_R^0 = \Sigma v_i \Delta h_{f,i}^0 \qquad (12.5)$$

where v_i = stoichiometric coefficient (positive for the products and negative for the reactants) as shown below for the production of acetic acid from methanol and carbon monoxide:

$$CH_3OH(L) \rightarrow C(S) + 2\,H_2(G) + 1/2\,O_2(G) \qquad -\Delta h_{f,CH_3OH(L)}^0$$
$$CO(G) \rightarrow C(S) + 1/2\,O_2(G) \qquad -\Delta h_{f,CO(G)}^0$$
$$2\,C(S) + 2H_2(G) + O_2(G) \rightarrow CH_3COOH(L) \qquad \Delta h_{f,CH_3COOH(L)}^0$$

$$CH_3OH(L) + CO(G) \rightarrow CH_3COOH(L) \qquad \Delta h_R^0$$

For a large number of compounds, the standard enthalpies of formation at 25 °C were determined and can be found in factual data banks and data compilations, for example, [1–3].

For selected compounds, the values in the ideal gas state at 25 °C are given in Appendix A. If the values are not available, approximate values in the ideal gas state can be predicted with the help of group contribution methods (see Section 3.1.5). Values in the ideal gas state are convenient as they represent values for isolated molecules.

12.1.1
Temperature Dependence

Usually chemical reactions are not carried out at standard temperature and pressure. In most cases, only the influence of the temperature on the enthalpy of reaction has to be considered. But for gas-phase reactions at high pressures, the influence of the pressure on the enthalpy of reaction is not negligible (see Example 12.2).

For the calculation of the standard enthalpy of reaction at a temperature different from 25 °C (298.15 K), Kirchhoff's law can be applied:

$$\Delta h_R^0(T) = \Delta h_R^0(T_0) + \int_{T_0}^{T} \Sigma v_i c_{Pi}^0 dT \qquad (12.6)$$

This means that the standard enthalpy of reaction at another temperature can be calculated when the heat capacities c_{Pi}^0 in the standard state (liquid, solid, and hypothetical ideal gas) of the compounds involved are known. The temperature dependence of the heat capacities can be described, for example, by a polynomial of the following form:

$$c_P^0 = a + bT + cT^2 + dT^3 \qquad (12.7)$$

For the calculation of the standard enthalpy of reaction at a temperature different from 298.15 K using Eq. (12.6), the sum of the heat capacities multiplied by the stoichiometric coefficients over all the reactants and products is required. These values can be calculated using the parameters a, b, c and d for the individual compounds:

$$\Delta a = \sum_i v_i a_i \quad \Delta b = \sum_i v_i b_i \quad \Delta c = \sum_i v_i c_i \quad \Delta d = \sum_i v_i d_i \qquad (12.8)$$

With the help of the values obtained for Δa, Δb, Δc, and Δd, the standard enthalpy of reaction for other temperatures can be calculated:

$$\Delta h_R^0(T) = \Delta h_R^0(T_0) + \Delta a (T - T_0) + \frac{\Delta b}{2} \left(T^2 - T_0^2\right)$$
$$+ \frac{\Delta c}{3} \left(T^3 - T_0^3\right) + \frac{\Delta d}{4} \left(T^4 - T_0^4\right) \qquad (12.9)$$

For the calculation only the parameters of the polynomial (Eq. 12.7) are required. An alternative way is shown in Chapter 6.

Example 12.1

With the help of the standard thermodynamic properties and heat capacities given in Appendix G, calculate the enthalpy of reaction of the ammonia synthesis in the ideal gas state at 450 °C:

$$1/2 N_2 + 3/2 H_2 \rightleftharpoons NH_3$$

Solution

For a temperature of 25 °C, the standard enthalpy of reaction (Eq. (12.5)) is obtained from the tabulated values of the standard enthalpies of formation Δh_f^0:

$$\Delta h_R^0 = \sum v_i \Delta h_{f,i}^0 = -45.773 - 1/2 \cdot 0 - 3/2 \cdot 0 = -45.773 \text{ kJ/mol}$$

For each of the parameters, the following Δ values are obtained using Eq. (12.8):

$$\Delta a = 27.296 - \frac{1}{2} \cdot 31.128 - \frac{3}{2} \cdot 27.124 = -28.954 \text{ J/mol K}$$

$$\Delta b = \left[23.815 - \frac{1}{2} \cdot (-13.556) - \frac{3}{2} \cdot 9.267\right] 10^{-3} = 16.692 \cdot 10^{-3} \text{ J/mol K}^2$$

$$\Delta c = \left[17.062 - \frac{1}{2} \cdot 26.777 - \frac{3}{2} \cdot (-13.799)\right] 10^{-6} = 24.372 \cdot 10^{-6} \text{ J/mol K}^3$$

$$\Delta d = \left[-11.84 - \frac{1}{2} \cdot (-11.673) - \frac{3}{2} \cdot 7.64\right] 10^{-9} = -17.464 \cdot 10^{-9} \text{ J/mol K}^4$$

Using these values, the standard enthalpy of reaction at any temperature can be determined. Thus, according to Eq. (12.9), the following value is obtained for the standard enthalpy of reaction at 450 °C:

$$\Delta h_R^0 (723.15\,\text{K}) = -45773 - 28.954 \cdot (723.15 - 298.15)$$
$$+ \frac{16.692}{2} \cdot 10^{-3} \left(723.15^2 - 298.15^2\right)$$
$$+ \frac{24.372}{3} \cdot 10^{-6} \left(723.15^3 - 298.15^3\right)$$
$$- \frac{17.464}{4} \cdot 10^{-9} \left(723.15^4 - 298.15^4\right)$$

$$\Delta h_R^0 (723.15\,\text{K}) = -52758\,\text{J/mol}.$$

12.1.2
Consideration of the Real Gas Behavior on the Enthalpy of Reaction

Besides the temperature, the pressure influences the value of the enthalpy of reaction. While this effect is negligible in the case of reactions in the liquid phase, considerable effects are observed for gas-phase reactions at high pressures. The deviations from ideal gas behavior can directly be taken into account with the help of equations of state, introduced in Section 2.2.1 using residual enthalpies $(h - h^{id})_{T,P}$ (see Table 2.2), which describe the enthalpy difference between the real state and the ideal gas state at a given temperature and pressure. Using residual enthalpies of the reactants and products, the enthalpy of reaction in the real state at given temperature and pressure can be calculated if the standard enthalpy of reaction in the ideal gas state at this temperature is known. An expression for the calculation of the enthalpy of reaction for a given pressure and temperature can be derived from the thermodynamic cycle shown in Figure 12.1:

$$\Delta h_R(T, P) = \Delta h_R^0 (T, P^0) + \Sigma v_i \left(h - h^{id}\right)_{T,P,i} \qquad (12.10)$$

In the case of the ammonia synthesis, the enthalpy of ammonia decreases since attractive forces dominate, whereas for both highly supercritical compounds N_2 and H_2, the forces are mainly of repulsive nature.

This means that for the calculation of the pressure effect on the enthalpy of reaction only the residual enthalpies of the compounds involved have to be calculated with an appropriate equation of state.

Example 12.2
Calculate the enthalpy of reaction for the ammonia synthesis at 450 °C and a pressure of 600 atm using the value of the standard enthalpy of reaction at 450 °C in the ideal gas state calculated in Example 12.1. For the calculation of the residual enthalpies, the group contribution equation of state volume-translated Peng–Robinson (VTPR) should be applied. The required VTPR parameters are given in Appendix K.

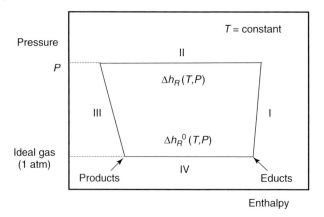

Figure 12.1 Thermodynamic cycle for the calculation of the enthalpy of reaction for a selected pressure P and temperature T from the enthalpy of reaction in the ideal gas state at temperature T for an exothermal reaction.

Solution

Using the Mathcad-File 12.02, the following residual enthalpies are obtained at 600 atm:

Compound	$(h - h^{id})_{450°C, 600\,atm}$, (J/mol)
N_2	815
H_2	1152
NH_3	−4896

With these values, the following enthalpy of reaction at 600 atm is obtained:

$$\Delta h_R (450\,°C, 600\,atm) = -52758 - 0.5 \cdot 815 - 1.5 \cdot 1152 - 4896$$
$$= -59790 \text{ J/mol}$$

From the results, it can be seen that the pressure has a great influence on the enthalpy of reaction. While a value of $\Delta h_R = -52758$ J/mol is calculated for the ideal gas state, a much higher exothermal value of −59790 J/mol is obtained at 600 atm, which is in good agreement with the value 59650 J/mol obtained using the equations of Gillespie and Beatty [4] used in Ullmann [5].[2] This value is more than

---

2) While at 450 °C the enthalpies of reaction as function of pressure reported in Ullmann [5] are in qualitative agreement with the values calculated using, for example, the VTPR or a reference equation of state, the values listed for low temperatures (e.g., 0 °C) are in total disagreement with thermodynamics.

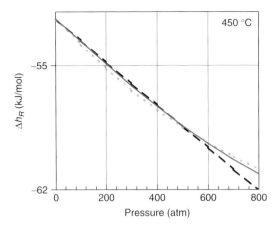

Figure 12.2 Enthalpy of reaction of the ammonia synthesis at 450 °C as a function of pressure: - - - - VTPR; --- Gillespie; —— reference equation of state.

10% higher than the one obtained from the standard thermodynamic properties in the ideal gas state and has to be taken into account for the design of reactors, heat exchangers, and so on, of ammonia plants. Of course, the same is true for other gas phase reactions carried out at high pressures. For pressures up to 800 atm, the calculated enthalpies of reaction for the ammonia synthesis are compared with the values of the Gillespie correlation and the values calculated using a reference equation of state in Figure 12.2.

12.2
Chemical Equilibrium

In addition to the determination of enthalpies of reaction as a function of temperature and pressure, thermodynamics allows us to calculate the equilibrium conversion for single or complex reversible reactions at given conditions (temperature, pressure, and initial composition).

At constant temperature and pressure, chemical equilibrium is reached when the Gibbs energy shows a minimum. To describe the change of the Gibbs energy with temperature, pressure, and composition, the following fundamental equation can be applied:

$$dG = -SdT + VdP + \sum \mu_i dn_i \tag{12.11}$$

where μ_i is the chemical potential of component i, and n_i is the number of moles of component i.

At constant temperature and pressure, Eq. (12.11) simplifies to

$$dG = \sum \mu_i dn_i \tag{12.12}$$

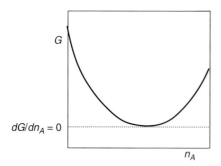

Figure 12.3 Typical curve of the Gibbs energy for a reacting mixture.

When, for example, the following reaction is considered:

$$A + B \rightleftharpoons 2C \tag{12.13}$$

the change in the number of moles for the reaction shown above is

$$-dn_A = -dn_B = \frac{1}{2}dn_C$$

This means that Eq. (12.12) can be rearranged in the following way:

$$dG = (-\mu_A - \mu_B + 2\mu_C)dn_A \tag{12.14}$$

As can be seen from Eq. (12.12), the value of the molar Gibbs energy changes with the composition of the reaction mixture. At chemical equilibrium, the molar Gibbs energy shows a minimum, as shown in Figure 12.3, where the change of the Gibbs energy with the change of the number of moles is zero:

$$\frac{dG}{dn_A} = 2\mu_C - \mu_A - \mu_B = 0$$

or in general form:

$$\sum_i \nu_i \mu_i = 0 \tag{12.15}$$

where the chemical potential μ_i of component i can be described by the chemical potential in the standard state μ_i^0 and the fugacities in the real state f_i and in the standard state f_i^0:

$$\mu_i = \mu_i^0 + RT \ln \frac{f_i}{f_i^0} \tag{12.16}$$

Combination of Eqs. (12.15) and (12.16) gives

$$\sum_i \nu_i \mu_i^0 = -RT \sum_i \ln \left(\frac{f_i}{f_i^0}\right)^{\nu_i} \tag{12.17}$$

The chemical potential in the standard state μ_i^0 is identical with the molar Gibbs energy of formation $\Delta g_{f,i}^0$. Thus, the part on the left-hand side of Eq. (12.17) is identical with the standard Gibbs energy of reaction Δg_R^0. The part

$$K = \Pi \left(\frac{f_i}{f_i^0}\right)^{\nu_i} \tag{12.18}$$

is called the chemical equilibrium constant K and can be calculated from the standard Gibbs energy of reaction:

$$\Delta g_R^0 = \sum_i v_i \Delta g_{f,i}^0 = -RT \sum_i \ln\left(\frac{f_i}{f_i^0}\right)^{v_i} = -RT \ln K \qquad (12.19)$$

For the reaction

$$A + B \rightleftharpoons 2C$$

the following equilibrium constant is obtained:

$$\Delta g_R^0 = 2\Delta g_{f,C}^0 - \Delta g_{f,A}^0 - \Delta g_{f,B}^0 = -RT \ln K = -RT \ln \frac{\left(\frac{f_C}{f_C^0}\right)^2}{\left(\frac{f_A}{f_A^0}\right)\left(\frac{f_B}{f_B^0}\right)} \qquad (12.20)$$

This means that the value of the equilibrium constant at 25 °C can directly be calculated from the tabulated standard Gibbs energies of formation. For a large number of compounds, the required standard Gibbs energies of formation at 25 °C and 1 atm can be found in factual data banks or data compilations for the different states of aggregation (liquid, solid, and hypothetical ideal gas state) [3]. In the various process simulators, usually the hypothetical ideal gas state is used.

The difference of the Gibbs energies of formation at 25 °C in the different states (solid, liquid, and hypothetical ideal gas) is only caused by the different fugacities in the standard state. The values can directly be converted. The standard fugacities of pure liquids and solids correspond approximately to the vapor respectively sublimation pressure at 25 °C. The standard fugacity of the hypothetical ideal gas is 1 atm. The difference of the standard Gibbs energies of formation between the liquid and the hypothetical ideal gas state is therefore approximately $RT \ln (P_i^s(25\,°C)/1\,atm)$. The values of the standard Gibbs energies of formation of the elements in their stable form are defined as 0.

If no standard Gibbs energies for a given compound can be found, the value in the ideal gas state can be estimated with the help of group contribution methods (see Section 3.1.5).

With the help of the tabulated standard Gibbs energies of formation at 25 °C, only the equilibrium constants at 25 °C can be determined. However, in most cases the equilibrium behavior is required at a temperature different from 25 °C. For the calculation of the equilibrium constant at a temperature different from 25 °C, the van't Hoff equation can be applied:

$$\frac{d \ln K}{dT} = \frac{\Delta h_R^0}{RT^2} \qquad (12.21)$$

If the standard enthalpy of reaction Δh_R^0 can be considered as constant in the temperature range covered, Eq. (12.21) can directly be integrated to obtain the equilibrium constant at the desired temperature T:

$$\ln K(T) = \ln K(T_0) - \frac{\Delta h_R^0}{R}\left(\frac{1}{T} - \frac{1}{T_0}\right) \qquad (12.22)$$

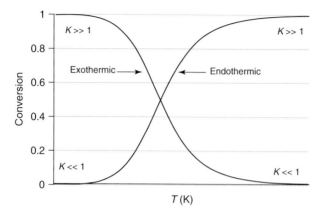

Figure 12.4 Qualitative trend of the equilibrium conversion X for reversible exothermic and endothermic reactions.

In case that the standard enthalpy of reaction Δh_R^0 is not constant in the temperature range considered, the temperature dependence can be described using Kirchhoff's law (Eq. (12.6)), taking into account the temperature-dependent heat capacities of the compounds involved. After integration, the following relation is obtained to calculate the equilibrium constant at the desired temperature:

$$\ln K_T = \ln K_{T_0} + \frac{\Delta a}{R} \ln \frac{T}{T_0} + \frac{\Delta b}{2R}(T - T_0) + \frac{\Delta c}{6R}(T^2 - T_0^2) + \frac{\Delta d}{12R}(T^3 - T_0^3)$$
$$+ \frac{1}{R}\left(-\Delta h_{R_{T_0}} + \Delta a T_0 + \frac{\Delta b}{2}T_0^2 + \frac{\Delta c}{3}T_0^3 + \frac{\Delta d}{4}T_0^4\right)\left(\frac{1}{T} - \frac{1}{T_0}\right) \quad (12.23)$$

Since the heat capacities in the ideal gas state are usually better known than the heat capacities of the liquids or the solids, even for reactions in the liquid phase it is often advantageous to use the values for the ideal gas for the determination of the equilibrium constant. This choice has no influence on the results, if the correct value of the standard fugacity f_i° (1 atm instead of $\approx P_i^s$) is used for the calculation of the equilibrium conversion.

In the case of exothermic reactions (negative values of the standard enthalpy of reaction), the equilibrium constant will decrease with increasing temperature. This means that also the maximum conversion will decrease, as for example, in the case of the SO_3 and the NH_3 production. The opposite is true in the case of endothermic reactions, for example, in the case of the production of synthesis gas (CO and H_2) by the steam reforming process. Qualitatively, this behavior is shown in Figure 12.4. While for endothermic reactions, high temperatures have to be realized (see Examples 12.6 and 12.8) to get high conversions. For exothermic reactions low temperatures are needed, although it must be taken into account that at low temperatures the reaction rates are slow.

Example 12.3

Determine the equilibrium constant for the ammonia synthesis at 450 °C with the help of the standard thermodynamic properties given in Appendix A and [2] and determine the equilibrium conversion at 600 atm assuming ideal behavior of the gas phase. The molar ratio of nitrogen to hydrogen should be 1/3.

a. For the calculation, it should be assumed that the standard enthalpy of reaction is constant in the temperature range covered.
b. The temperature dependence of the standard enthalpy of reaction is taken into account.

Solution

a. Using the tabulated values of the standard Gibbs energies of formation in the ideal gas state ($f_i^0 = 1$ atm) for a temperature of 25 °C, for the standard Gibbs energy of reaction one obtains

$$\Delta g_R^0 = \sum v_i \Delta g_{f,i}^0 = -16.150 - 1/2\,(0) - 3/2\,(0) = -16.150 \text{ kJ/mol}$$

Using this value for the standard Gibbs energy of reaction, the chemical equilibrium constant K at 25 °C can be calculated directly:

$$K = \exp\left(-\frac{\Delta g_R^0}{RT}\right) = \exp\left(\frac{16150}{8.31433 \cdot 298.15}\right) = 675.15$$

If the standard enthalpy of reaction is constant in the temperature range considered, Eq. (12.22) can directly be used for the calculation of the equilibrium constant at a temperature T. Thus, for a temperature of 450 °C (723.15 K), one obtains (Δh_R^0 from Example 12.1)

$$\ln K\,(723.15\text{ K}) = \ln 675.15 + \frac{45773}{8.31433}\left(\frac{1}{723.15} - \frac{1}{298.15}\right) = -4.337$$

$$K\,(723.15\text{ K}) = 0.01308$$

With the help of the equilibrium constant, the partial pressures or vapor phase mole fractions of the compounds involved can be determined. Starting from (12.19) and assuming ideal gas phase behavior, the following equation can be applied:

$$K = \frac{p_{NH_3}}{p_{N_2}^{0.5} p_{H_2}^{1.5}}\left(\frac{1}{1\text{ atm}}\right)^{-1} = \frac{y_{NH_3}}{y_{N_2}^{0.5} y_{H_2}^{1.5} P}\left(\frac{1}{1\text{ atm}}\right)^{-1}$$

The equilibrium conversion can be determined by a material balance. The equilibrium conversion is the maximum conversion X_{max}, which can be achieved for the number of moles present at the beginning.
By definition, the conversion X_i is

$$X_i \equiv \frac{n_{i0} - n_{ie}}{n_{i0}}$$

with
n_{i0} the initial number of moles of component i
n_{ie} the number of moles of component i in chemical equilibrium.

For the case that the conversion is related to nitrogen, the number of moles of all other components can be expressed as follows:

$$n_{N_2} = n_{N_20} \left(1 - X_{N_2}\right)$$
$$n_{H_2} = 3 n_{N_20} \left(1 - X_{N_2}\right)$$
$$n_{NH_3} = 2 n_{N_20} \cdot X_{N_2}$$

The total number of moles $n_T = \sum n_i$ is

$$n_T = n_{N_20} \left(4 - 2 X_{N_2}\right)$$

This means that all mole fractions $y_i = n_i/n_T$ can be represented by the initial number of moles of nitrogen and the conversion X_{N_2}. Consequently, the equilibrium constant results in

$$K = \frac{2 X_{N_2} \left(4 - 2 X_{N_2}\right)}{\left(1 - X_{N_2}\right)^{0.5} 3^{1.5} \left(1 - X_{N_2}\right)^{1.5} P} = \frac{2 X_{N_2} \left(4 - 2 X_{N_2}\right)}{3^{1.5} \left(1 - X_{N_2}\right)^2 P}$$

Introducing the auxiliary quantity K'

$$K' = \frac{3^{1.5} P K}{4}$$

the conversion X_{N_2} can be determined for every temperature and every pressure using the following relation:

$$X_{N_2} = 1 - \frac{1}{\sqrt{1 + K'}}$$

For a temperature of 450 °C ($K = 0.01308$) and a pressure of 600 atm, the following conversion is obtained:

$$K' = \frac{3^{1.5} \cdot 600 \cdot 0.01308}{4} = 10.195$$

$$X_{N_2} = 1 - \frac{1}{\sqrt{1 + 10.195}} = 0.7011$$

From the conversion, the equilibrium composition can be calculated. For nitrogen, we obtain

$$y_{N_2} = \frac{n_{N_2}}{n_T} = \frac{1 - X_{N_2}}{4 - 2 X_{N_2}} = \frac{1 - 0.7011}{4 - 2 \cdot 0.7011} = 0.1151$$

For the other mole fractions, the following values are determined:

$$y_{H_2} = 3 y_{N_2} = 0.3452$$
$$y_{NH_3} = 1 - 4 y_{N_2} = 0.5397$$

In the same way, conversions at other temperatures and other pressures can be calculated. The results at 25 and 450 °C and different pressures are listed in Table 12.1.

From Table 12.1, it can be recognized that low temperatures and high pressures are advantageous for the ammonia synthesis. Nevertheless, temperatures around 450 °C are used in chemical industry. The reason is that despite improved

Table 12.1 Equilibrium conversion and composition of the NH_3-synthesis at different conditions assuming ideal gas phase behavior.

ϑ (°C)	P (atm)	X_{N_2}	y_{N_2}	y_{H_2}	y_{NH_3}
25	1	0.9663	0.0163	0.0490	0.9347
25	100	0.9966	0.0017	0.0050	0.9933
450	100	0.3913	0.1892	0.5676	0.2432
450	600	0.7011	0.1151	0.3452	0.5397

catalysts the reaction rate is too slow at low temperatures, so that the conversion is much lower than the equilibrium conversion for a limited residence time. Only at higher temperatures, the reaction rate is fast enough to achieve a conversion close to chemical equilibrium. For an exothermal reaction, this is shown qualitatively in Figure 12.5.

b. To get more reliable results, the temperature dependence of the standard enthalpy of reaction Δh_R^0 has to be considered. Therefore, the heat capacities of the components involved in the reaction have to be taken into account as additional information.

As shown in Example 12.1, the absolute value of the enthalpy of reaction increases with increasing temperature. Therefore, the more precise calculation will lead to a smaller equilibrium constant and thus a smaller conversion than the calculation with a constant standard enthalpy of reaction at 25 °C. Thus, according to Eq. (12.23), the following equilibrium constant is obtained for a temperature of 450 °C,

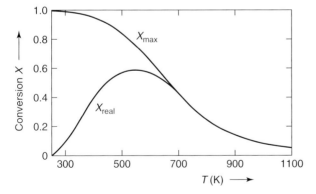

Figure 12.5 Equilibrium conversion (X_{max}) and conversion for a limited residence time X_{real} for an exothermal reversible reaction as a function of the absolute temperature.

when taking the coefficients Δa, Δb, Δc, and Δd from Example 12.1:

$$\ln\frac{K(723.15\text{ K})}{K(298.15\text{ K})} = -3.482\ln\frac{723.15}{298.15} + 1.0038\cdot 10^{-3}(723.15 - 298.15)$$
$$+ 4.8855\cdot 10^{-7}(723.15^2 - 298.15^2)$$
$$- 1.7504\cdot 10^{-10}(723.15^3 - 298.15^3)$$
$$+ \frac{1}{8.31433}(45773 - 28.954\cdot 298.15 + 8.346\cdot 10^{-3}\cdot 298.15^2$$
$$+ 8.124\cdot 10^{-6}\cdot 298.15^3 - 4.366\cdot 10^{-9}\cdot 298.15^4)$$
$$\cdot\left(\frac{1}{723.15}\cdot\frac{1}{298.15}\right)$$

$$\ln\frac{K(723.15\text{ K})}{K(298.15\text{ K})} = -11.5321$$

$$K(723.15\text{ K}) = 6.62\cdot 10^{-3}$$

Still assuming ideal gas phase behavior, the following conversion and mole fractions are obtained for the chemical equilibrium at 600 atm:

$X_{N_2} = 0.5971$

$y_{N_2} = 0.1436$

$y_{H_2} = 0.4308$

$y_{NH_3} = 0.4256$

From the results, it can be seen that the simplified calculation leads to distinctly different results. While the less precise calculation provides a conversion of 0.7011, a conversion of 0.5971 is obtained by taking into account the temperature dependence of the standard enthalpy of reaction.

The conversion for a large temperature and pressure range calculated by the method discussed above is also shown in Figure 12.6. However, taking into account the temperature dependence of the equilibrium constant correctly, but assuming ideal gas behavior ($K_\varphi = 1$), can still lead to large errors at high pressure, as can be seen in Figure 12.6 by a comparison with the experimental results.

At low pressures, the differences between the calculation assuming ideal gas behavior and taking into account the deviation from ideal gas behavior can be neglected. For a consideration of the real behavior, a reliable representation of the PvT-behavior of the reacting compounds is required.

By definition, the fugacity in the gas phase (see Chapter 3) can be described with the help of fugacity coefficients:

$$f_i = p_i\varphi_i = y_i\varphi_i P \tag{12.24}$$

Using this relation, the equilibrium constant K can be written in the following form (standard fugacity $f_i^0 = 1$ atm):

$$K = K_P K_\varphi\left(\frac{1}{1\text{ atm}}\right)^{\Sigma v_i} = \Pi p_i^{v_i}\Pi\varphi_i^{v_i}\left(\frac{1}{1\text{ atm}}\right)^{\Sigma v_i} \tag{12.25}$$

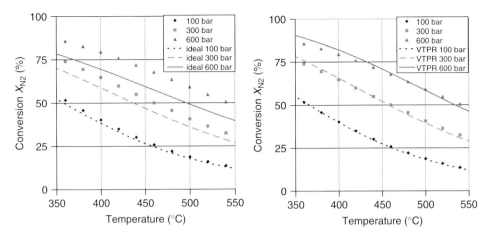

Figure 12.6 Experimental and calculated equilibrium conversions as a function of pressure and temperature of the ammonia synthesis assuming ideal (left hand side) and real behavior (right hand side).

Equation (12.25) can also be written in the following form:

$$K = K_y K_\varphi \left(\frac{P}{1 \text{ atm}}\right)^{\Sigma v_i} = \Pi y_i^{v_i} \Pi \varphi_i^{v_i} \left(\frac{P}{1 \text{ atm}}\right)^{\Sigma v_i} \quad (12.26)$$

This means that the equilibrium constant for the reversible reaction shown before ($\Sigma v_i = 0$) can also be described as follows:

$$K = \frac{p_C^2}{p_A p_B} \frac{\varphi_C^2}{\varphi_A \varphi_B} = \frac{y_C^2}{y_A y_B} \frac{\varphi_C^2}{\varphi_A \varphi_B} \quad (12.27)$$

For reactions where the number of moles changes, it is important that the same unit is used for the partial pressures as for the standard fugacities (atm).

In the case of ideal gases ($\varphi_i = 1$) K_φ shows a value of 1; this means $K = K_P$ for the considered reversible reaction. In this case, the partial pressures p_i can also be substituted by concentrations c_i:

$$p_i = \frac{n_i RT}{V} = c_i RT$$

Using concentrations, the following equilibrium constant is obtained:

$$K_P = K_c (RT)^{\Sigma v_i}$$

where

$$K_c = \Pi c_i^{v_i}$$

When Eq. (12.26) is considered, it can be seen how the pressure influences the equilibrium conversion. In the case of gas phase reactions, the pressure has an immense influence on the equilibrium conversion, if the number of moles increases or decreases ($\Sigma v_i \neq 0$), since the value of the equilibrium constant K_y

depends on the pressure. So for all gas-phase reactions, a pressure increase will cause a conversion increase, when the number of moles is reduced ($\Sigma \nu_i < 0$) as for example in the case of the ammonia synthesis. If the number of moles increases during the reaction ($\Sigma \nu_i > 0$), a conversion decrease is observed with increasing pressure. But also in the case of constant number of moles, the more and more real behavior ($\varphi_i \neq 1$) of the gas phase with increasing pressure has an influence on the equilibrium conversion. Depending on the values of K_φ, the equilibrium conversion will increase or decrease. For K_φ values smaller than 1, the real behavior leads to an equilibrium conversion higher than that in the case of ideal behavior as for the ammonia synthesis (see for Example 12.4). With K_φ values greater than 1, a lower conversion than that in the ideal case is obtained.

Example 12.4

Calculate the equilibrium conversion of the ammonia synthesis at 450 °C and 600 atm taking into account the real behavior of the gas phase for stoichiometric amounts of nitrogen and hydrogen ($N_2/H_2 = 1/3$):

a. using the Soave–Redlich Kwong (SRK) equation of state;
b. using the group contribution equations of state Predictive Soave–Redlich Kwong (PSRK) and VTPR.

The required pure component properties T_c, P_c, v_c, and ω together with the parameters for the Twu α-function are given in Appendices A and K. The volume translation parameter c should be calculated using Eq. (2.179).

Simplifying, in the case of the SRK equation of state, it should be assumed that the influence of the binary parameters k_{ij} can be neglected, that is, $k_{ij} = 0$.

PSRK group interaction parameters:

1	2	a_{12} (K)	a_{21} (K)	b_{12}	b_{21}
N_2	H_2	77.701	247.42	0.	0.
N_2	NH_3	1900.	973.45	3.415	−0.03016
H_2	NH_3	1893.1	1339.	11.556	0.14026

VTPR group interaction parameters:

1	2	a_{12} (K)	a_{21} (K)
N_2	H_2	85.862	78.387
N_2	NH_3	689.28	841.26
H_2	NH_3	1473.9	1031.7

Solution

With the above parameters and assumptions, for each concentration the fugacity coefficients can be determined for the given pressure and temperature. The calculation is performed in the same way as shown in Chapter 5 for the system N_2-CH_4. However, it is simpler, since only the fugacity coefficients in the gas phase are required. Nevertheless, the calculation has to be performed iteratively, because the fugacity coefficients depend on composition.

With the critical data, acentric factors, and the binary parameters, the pure component and mixture parameters have to be calculated for a temperature of 723.15 K. As initial composition, the mole fractions determined assuming ideal gas behavior are used and the parameters required for the calculation of the fugacity coefficients calculated.

In the next step, the volume at 723.15 K and 600 atm can be determined using the mixture parameters a and b.

a. For a volume of 0.11583 dm^3/mol, a pressure of 600 atm is obtained with the SRK equation of state. With the calculated volume, the fugacity coefficients can be calculated. For nitrogen, the following fugacity coefficient is obtained for the given temperature, pressure, and composition (see Mathcad-file 12.04).

$$\varphi_{N_2}^V = 1.3472$$

In the same way, the fugacity coefficients for hydrogen and ammonia can be calculated:

$$\varphi_{H_2} = 1.2356 \qquad \varphi_{NH_3} = 1.0445$$

which can then be used to calculate K_φ and to determine a new value for K_P as shown below:

$$K_\varphi = \frac{1.0445}{1.3472^{0.5} \cdot 1.2356^{1.5}} = 0.6552$$

$$K_P = \frac{K}{K_\varphi} = \frac{6.62 \cdot 10^{-3}}{0.6552} = 10.1 \cdot 10^{-3}$$

which can be used to obtain more reliable mole fractions. From the new K_P, the following conversion and mole fractions are obtained:

$$X_{N_2} = 0.6644$$
$$y_{N_2} = 0.1257$$
$$y_{H_2} = 0.3770$$
$$y_{NH_3} = 0.4973$$

After a few iterations, chemical equilibrium is achieved:

$$X_{N_2} = 0.6711$$

$$y_{N_2} = 0.1238 \qquad \varphi_{N_2} = 1.3681$$
$$y_{H_2} = 0.3713 \qquad \varphi_{H_2} = 1.2535$$
$$y_{NH_3} = 0.5049 \qquad \varphi_{NH_3} = 1.0276$$

$$K_P = 6.62 \cdot 10^{-3} \, \frac{1.3681^{0.5} \cdot 1.2535^{1.5}}{1.0276} = 10.58 \cdot 10^{-3}$$

b. For the PSRK and the VTPR group contribution equations of state, the following results are obtained at chemical equilibrium at 450 °C and 600 atm:

	PSRK	VTPR
φ_{N_2}	1.6267	1.4451
φ_{H_2}	1.3970	1.4198
φ_{NH_3}	1.0659	0.9445
K_P	$13.08 \cdot 10^{-3}$	$14.25 \cdot 10^{-3}$
X_{N_2}	0.701	0.713
y_{N_2}	0.1150	0.1116
y_{H_2}	0.3451	0.3348
y_{NH_3}	0.5398	0.5535

At 600 atm, a mole fraction of 0.536 was observed for NH_3 experimentally [4, 5]. For a larger temperature and pressure range, the calculated conversions assuming ideal and real behavior using Mathcad are shown in Figure 12.6 together with the experimental values reported in Ullmann [5]. From the results, the improvements obtained when the real behavior is taken into account can clearly be recognized.

The same conclusions can be drawn from Figure 12.7, where the experimental equilibrium constants K_P at 450 °C are shown as a function of pressure. Similar to the results shown in Figure 12.6, reliable values for K_P are obtained for VTPR and PSRK, whereas the deviation from ideal behavior is only qualitatively described by the SRK equation of state with $k_{ij} = 0$.[3] In case of ideal gas behavior of course no influence of the pressure on the value of K_P is calculated.

In the case of the ammonia synthesis, the equilibrium conversion is increased by the pressure not only because of the reduction of the number of moles (Le Chatelier principle), but also because the real behavior in the gas phase leads to K_φ values smaller than 1. This is in agreement with experimental results. While an equilibrium constant K_P of $6.59 \cdot 10^{-3}$ atm^{-1} was measured at 1 atm, a value of

---
3) But also the use of binary parameters did not improve the results.

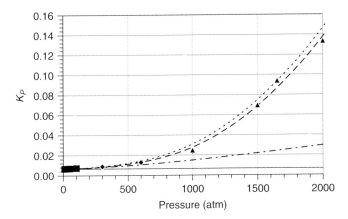

Figure 12.7 Experimental and calculated equilibrium constants K_P as a function of pressure: ■ [6]; ▲ [7]; ♦ [8]; ---- VTPR; — — PSRK; —·— SRK; —— ideal.

$23.28 \cdot 10^{-3}$ atm^{-1} was experimentally determined at 1000 atm [6]. At a pressure of 3500 atm, even a value of 1.075 for K_P was measured by Winchester and Dodge [7]. This value is by a factor 150 higher than at atmospheric pressure and shows the strong contribution of the real behavior on the chemical equilibrium conversion.

As shown in Example 12.4, the required fugacity coefficients and the value for K_φ can be calculated with the help of equations of state. For the whole pressure range, the influence of the real behavior on the equilibrium constant K_P and on the conversion can be seen from Figure 12.7. At the same time, it can be seen that modern equations of state like PSRK and VTPR are able to predict the correct influence of the gas phase reality, this means K_φ as a function of the pressure. Only the correct consideration of the fugacity coefficients permits a reliable calculation of the equilibrium composition for reversible gas-phase reactions at high pressure starting from tabulated standard thermodynamic properties.

For the calculation of the chemical equilibrium conversion of reversible reactions in the liquid phase, the corresponding fugacities are required. For the description of fugacities in the liquid phase, in most cases activity coefficients γ_i are used (see Section 5.2):

$$f_i = x_i \gamma_i f_i^0 = a_i f_i^0 \approx x_i \gamma_i P_i^s \tag{12.28}$$

When the standard thermodynamic properties in the liquid state ($f_i^0 \approx P_i^s$) are used for the calculation of the equilibrium constant K, the standard fugacity cancels out[4]

---

4) In reality the standard fugacities used are slightly different because of the different pressures. For the calculation of the fugacity f_i it is the system pressure in contrast to the pressure 1 atm in the standard state of the thermodynamic property).

and the following expression is obtained for the equilibrium constant:

$$K = K_a = \Pi a_i^{v_i} = K_x K_\gamma = \Pi x_i^{v_i} \Pi \gamma_i^{v_i} \tag{12.29}$$

In the case of an ideal mixture, K_γ is equal to 1; this means $K = K_x$. The deviation from ideal behavior in the liquid phase can be much stronger than that in the gas phase because of the small distance between the molecules. For a correct calculation of the equilibrium composition, the deviation from ideal behavior, this means K_γ, has to be taken into account with the help of g^E-models or group contribution methods (see Section 5.3 or 5.12).

For the reversible reaction considered,

$$A + B \rightleftarrows 2C$$

the following relation is obtained:

$$K = \frac{a_C^2}{a_A a_B} = \frac{x_C^2}{x_A x_B} \frac{\gamma_C^2}{\gamma_A \gamma_B} \tag{12.30}$$

Substituting the mole fraction x_i by the concentration c_i (mol/dm³) and the total molar concentration $c = \Sigma c_i$, a relation with the often used equilibrium constant K_c can be derived:

$$K = K_c K_\gamma \left(\frac{1}{c}\right)^{\Sigma v_i} \tag{12.31}$$

Example 12.5

Calculate the equilibrium conversion for the following reaction in the temperature range from 25 to 90 °C:

$$\text{methanol} + i\text{-butene} \rightleftarrows \text{MTBE}$$

in the liquid phase, starting from the following initial amounts of methanol and i-butene:

$$n_{\text{MeOH}} = 1.06 \text{ kmol}; \quad n_{i\text{-butene}} = 1 \text{ kmol}$$

a. assuming ideal behavior;
b. taking into account the real behavior with the help of the Wilson equation and the modified UNIFAC method;
c. for a solution in inert n-butane.

Standard thermodynamic properties in the ideal gas state:

Compound	Δg_f^0 (kJ/mol)	Δh_f^0 (kJ/mol)
methanol	−162.21	−200.8
i-butene	58.04	−17.09
MTBE	−117.42	−283.31

Molar volumes and Antoine constants $\log P_i^s \text{ (kPa)} = A - B/(\vartheta + C)$ [9]:

Compound	v (cm³/mol)	A	B	C
methanol (1)	40.73	7.20587	1582.27	239.726
i-butene (2)	94.46	5.96625	923.20	240.00
MTBE (3)	119.0	6.25487	1265.40	242.50
n-butane (4)	100.39	6.35575	1175.58	271.1

Wilson parameters (K) fitted to VLE data [9]:

i\j	methanol (1)	i-butene (2)	MTBE (3)	n-butane (4)
methanol (1)	0	1218.5	592.5	1362.4
i-butene (2)	57.14	0	463.5	36.96
MTBE (3)	−102.8	−302.95	0	53.18
n-butane (4)	168.4	−21.14	−53.28	0

Solution

For the calculation of the chemical equilibrium constant at 25 °C, only the standard Gibbs energies of formation are required:

$$-RT \ln K = \Delta g_R^0 = \sum_i v_i \Delta g_{f,i}^0 = -117.42 - (-162.21 + 58.04)$$

$$= -13.25 \text{ kJ/mol}$$

$$K(298.15 \text{ K}) = e^{-\frac{\Delta g_R^0}{RT}} = e^{13250/(8.31433 \cdot 298.15)} = 209.57$$

For the determination of the equilibrium constant at other temperatures, the standard enthalpy of reaction is needed. This value can be calculated using the tabulated standard enthalpies of formation:

$$\Delta h_R^0 = \sum_i v_i \Delta h_{f,i}^0 = -283.31 - (-200.8 - 17.09) = -65.42 \text{ kJ/mol}$$

With the assumption that the standard enthalpy of reaction is constant, the equilibrium constant at a given temperature can be calculated using Kirchhoff's law (Eq. (12.6)). As an example, the following chemical equilibrium constant is obtained at 80 °C (353.15 K):

$$\ln K(353.15 \text{ K}) = \ln K_{T_0} - \frac{\Delta h_R^0}{R}\left(\frac{1}{T} - \frac{1}{T_0}\right)$$

$$= \ln 209.57 + \frac{65420}{8.31433}\left(\frac{1}{353.15} - \frac{1}{298.15}\right)$$

$$= 1.235$$

$$K(353.15 \text{ K}) = 3.438$$

12 Enthalpy of Reaction and Chemical Equilibria

The chemical equilibrium constant for the reaction considered is defined as follows:

$$K = \frac{f_{MTBE}/f^0_{MTBE}}{\frac{f_{MeOH}}{f^0_{MeOH}} \cdot \frac{f_{i\text{-butene}}}{f^0_{i\text{-butene}}}}$$

Since the standard thermodynamic properties for the ideal gas state were chosen, all standard fugacities f_i^0 are equal to 1 atm. The fugacities in the liquid phase can be expressed by (Eq. 12.28):

$$f_i^L \approx x_i \gamma_i P_i^s$$

This means that the chemical equilibrium constant can also be expressed in the following way:

$$K = \frac{x_{MTBE}}{x_{MeOH} \cdot x_{i\text{-butene}}} \cdot \frac{\gamma_{MTBE}}{\gamma_{MeOH} \cdot \gamma_{i\text{-butene}}} \cdot \frac{P^s_{MTBE}}{P^s_{MeOH} \cdot P^s_{i\text{-butene}}} \cdot 1\ \text{atm}$$
$$= K_x \cdot K_\gamma \cdot K_{ps} \cdot 1\ \text{atm}$$

Using the Antoine constants given above, one obtains for the vapor pressures and K_{ps}:

Temperature	P^s_{MTBE} (kPa)	P^s_{MeOH} (kPa)	$P^s_{i\text{-butene}}$ (kPa)	$K_{ps} \cdot 101.325$ (kPa)
25 °C	33.45	16.94	303.7	0.6588
80 °C	214.36	180.73	1205.7	0.09968

a. Assuming ideal behavior ($K_\gamma = 1$), the following values for K_x are obtained:

$$25\ °\text{C}: K_x = \frac{K}{K_{ps} \cdot 1\ \text{atm}} = \frac{209.57}{0.6588} = 318.1$$

$$80\ °\text{C}: K_x = \frac{K}{K_{ps} \cdot 1\ \text{atm}} = \frac{3.438}{0.09968} = 34.49$$

From these values, the chemical equilibrium conversion can be calculated:

$$n_{MeOH} = 1.06 - \Delta n\ \text{kmol} \quad x_{MeOH} = (1.06 - \Delta n)/\sum n_i$$
$$n_{i\text{-butene}} = 1.00 - \Delta n\ \text{kmol} \quad x_{i\text{-butene}} = (1.00 - \Delta n)/\sum n_i$$
$$n_{MTBE} = \Delta n\ \text{kmol} \quad x_{MTBE} = \Delta n/\sum n_i$$
$$\sum n_i = 2.06 - \Delta n\ \text{kmol}$$

Now the value for Δn has to be determined, which fulfills the chemical equilibrium constant:

$$K_x = \frac{\Delta n \cdot (2.06 - \Delta n)}{(1.06 - \Delta n) \cdot (1.00 - \Delta n)}$$

This can be done by solving the resulting quadratic equation. While for 25 °C a value of 0.9650 kmol is obtained, a value of 0.8545 kmol is found at 80 °C. The conversion of i-butene at 25 and 80 °C is calculated as follows:

$$X_{i\text{-butene}} = \Delta n / n_{0,i\text{-butene}}$$

For the whole temperature range, the conversions assuming ideal behavior are shown in Figure 12.8 (lower line) together with the experimental values. It can be seen that the agreement is not very good.

b. The deviations are caused by the real behavior of the ternary system, which can be taken into account using g^E-models or group contribution methods such as UNIFAC. For this purpose, an iterative procedure has to be applied. To determine K_γ using one of the above-mentioned models, the composition has to be specified. Starting point for the iterative procedure are the mole fractions obtained assuming ideal behavior (part a of this example). This means for a temperature of 80 °C:

$$x_{\text{MeOH}} = \frac{1.06 - 0.8545}{2.06 - 0.8545} = 0.1705$$

$$x_{i\text{-butene}} = \frac{1 - 0.8545}{2.06 - 0.8545} = 0.1207$$

$$x_{\text{MTBE}} = \frac{0.8545}{2.06 - 0.8545} = 0.7088$$

Now for this composition the activity coefficients for the three compounds are calculated using the Wilson equation.

The calculations are performed for a temperature of 80 °C. All required pure component data (molar volumes) and interaction parameters are given in the

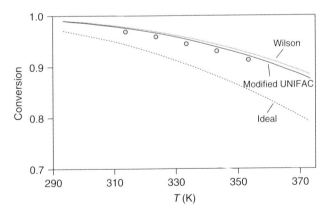

Figure 12.8 Experimental and predicted equilibrium conversion for the MTBE synthesis in the liquid phase as a function of temperature (initial number of moles: methanol 1.06 kmol and i-butene 1 kmol): • experimental.

table above. First of all, the Λ_{ij}'s have to be calculated:

$$\Lambda_{12} = \frac{v_2}{v_1}\exp\left(-\frac{\Delta\lambda_{12}}{T}\right) = \frac{94.46}{40.73}\exp\left(-\frac{1218.5}{353.15}\right) = 0.0736$$

$$\Lambda_{21} = \frac{v_1}{v_2}\exp\left(-\frac{\Delta\lambda_{21}}{T}\right) = \frac{40.73}{94.46}\exp\left(-\frac{57.14}{353.15}\right) = 0.3667$$

In the same way, the other Λ_{ij}'s are obtained:

$$\Lambda_{13} = 0.5457 \qquad \Lambda_{31} = 0.4580 \qquad \Lambda_{23} = 0.3390 \qquad \Lambda_{32} = 1.8718$$

Now the activity coefficients for the selected composition ($x_1 = 0.1705$, $x_2 = 0.1207$, $x_3 = 0.7088$) can be calculated. For methanol (1), the following equation is valid:

$$\ln \gamma_1 = -\ln(x_1\Lambda_{11} + x_2\Lambda_{12} + x_3\Lambda_{13}) + 1 - \frac{x_1\Lambda_{11}}{x_1\Lambda_{11} + x_2\Lambda_{12} + x_3\Lambda_{13}}$$

$$- \frac{x_2\Lambda_{21}}{x_1\Lambda_{21} + x_2\Lambda_{22} + x_3\Lambda_{23}} - \frac{x_3\Lambda_{31}}{x_1\Lambda_{31} + x_2\Lambda_{32} + x_3\Lambda_{33}}$$

resulting in the activity coefficient for methanol:

$$\ln \gamma_1 = -\ln(0.1705 + 0.1207 \cdot 0.0736 + 0.7088 \cdot 0.5457)$$

$$+ 1 - \frac{0.1705}{0.1705 + 0.1207 \cdot 0.0736 + 0.7080 \cdot 0.5457}$$

$$- \frac{0.1207 \cdot 0.3667}{0.1705 \cdot 0.3667 + 0.1207 + 0.7088 \cdot 0.3390}$$

$$- \frac{0.7088 \cdot 0.4580}{0.1705 \cdot 0.4580 + 0.1207 \cdot 1.8718 + 0.7088}$$

$$= 0.8427$$

$$\gamma_1 = 2.323$$

In a similar way, the activity coefficients for the other two components can be calculated. The results are $\gamma_{i\text{-butene}} = 1.272$ and $\gamma_{\text{MTBE}} = 1.026$. This means for K_γ:

$$K_\gamma = \frac{\gamma_3}{\gamma_1 \cdot \gamma_2} = \frac{1.026}{2.323 \cdot 1.272} = 0.347$$

Since K_γ is smaller than 1, it leads to an increase of K_x and thus of the chemical conversion.

$$K_x = \frac{K}{K_\gamma \cdot K_{ps} \cdot 1 \text{ atm}} = \frac{3.438}{0.347 \cdot 0.09968} = 99.40$$

With the equilibrium compositions obtained now, new activity coefficients can be calculated. After a few iterations an i-butene conversion of 0.9266 is achieved, which leads to the following equilibrium composition, activity coefficients, and value for K_γ:

$$x_1 = 0.1177; \; x_2 = 0.0648; \; x_3 = 0.8178$$

$$\gamma_1 = 2.506; \; \gamma_2 = 1.258; \; \gamma_3 = 1.013$$

$$K_\gamma = 0.3213$$

for which the equilibrium constant K is fulfilled:

$$K = K_x K_\gamma K_{Ps} = \frac{0.8178}{0.1177 \cdot 0.0648} \cdot \frac{1.013}{2.506 \cdot 1.258} \cdot 0.09968$$

$$= 107.22 \cdot 0.3213 = 3.434$$

Using modified UNIFAC, an i-butene conversion of 0.9211 is obtained, which results in the following equilibrium composition, activity coefficients and K_γ (see Mathcad file 12.05).

$$x_1 = 0.1236; \; x_2 = 0.0710; \; x_3 = 0.8054$$

$$\gamma_1 = 2.567; \; \gamma_2 = 1.050; \; \gamma_3 = 1.013$$

$$K_\gamma = 0.3758$$

$$K = K_x K_\gamma K_{Ps} = \frac{0.8054}{0.1236 \cdot 0.0710} \cdot \frac{1.013}{2.567 \cdot 1.050} \cdot 0.09968$$

$$= 91.77 \cdot 0.3758 = 3.438$$

The results for the temperature range from 290 to 370 K are shown in Figure 12.8 together with the experimentally determined conversions. It can be seen that higher conversions than for the ideal case are reached and that at least qualitative agreement with the experimental findings is obtained, when the real behavior is taken into account [10]. Higher conversions compared to the ideal conversion are always achieved when K_γ becomes smaller than unity.

c. In practice, instead of i-butene the C_4-raffinate[5] is used to produce MTBE. It is the question how the additional compounds influence the equilibrium composition. For simplicity, it may be assumed that besides i-butene only n-butane is present as an inert solvent and it can be checked how the equilibrium conversion is affected by the presence of n-butane:

$$n_{MeOH} = 1.06 \text{ kmol}; \; n_{i\text{-butene}} = 1 \text{ kmol}; \; n_{n\text{-butane}} = 4 \text{ kmol}$$

[5] C_4-raffinate is the butadiene free C_4-stream of a naphtha cracker. It contains all the C_4-hydrocarbons with the exception of butadiene.

First, Δn has to be determined again. Starting from the initial numbers of moles, the chemical equilibrium can be described as follows:

$$n_{MeOH} = 1.06 - \Delta n \text{ kmol}; \quad x_{MeOH} = (1.06 - \Delta n)/\sum n_i$$

$$n_{i\text{-butene}} = 1.00 - \Delta n \text{ kmol}; \quad x_{i\text{-butene}} = (1.00 - \Delta n)/\sum n_i$$

$$n_{MTBE} = \Delta n \text{ kmol}; \quad x_{MTBE} = \Delta n/\sum n_i$$

$$n_{butane} = 4 \text{ kmol}; \quad x_{butane} = 4/\sum n_i$$

$$\sum n_i = 6.06 - \Delta n \text{ kmol}$$

This means that for a temperature of 353.15 K, the following equation has to be solved:

$$K_x = \frac{\Delta n \cdot (6.06 - \Delta n)}{(1.06 - \Delta n) \cdot (1.00 - \Delta n)} = 34.49$$

By solving the resulting quadratic equation, a value of $\Delta n = 0.6990$ is obtained. From Δn, the number of moles of the other compounds and thus the composition can be calculated directly. The values are given in Table 12.2. Taking into account the real behavior, an iterative procedure is used to obtain the conversion for which the following relation is fulfilled:

$$K_x K_\gamma = \frac{K}{K_{ps} \cdot 1 \text{ atm}} = 34.49$$

The procedure is identical to the procedure for the ternary system. But now the solvent (n-butane) influences the activity coefficients of the different compounds. The solution is given in the Mathcad-file. The resulting conversion, the activity coefficients, and the composition in equilibrium using the Wilson or the modified UNIFAC method are given in Table 12.2. In Figure 12.9, the equilibrium conversion assuming ideal and real behavior (Wilson, modified UNIFAC) is shown for the temperature range from 290 to 370 K. It can be recognized that very different results are obtained compared to the results assuming an ideal mixture.

In the presence of n-butane, K_γ becomes smaller than in the ternary system because of the increasing activity coefficient of methanol in the presence of butane, which leads to an increase of K_x and the equilibrium conversion. But this effect is compensated by the dilution of the reaction mixture, which leads to a decrease of the equilibrium conversion following the principle of LeChatelier, since for the reaction considered Σv_i is negative.

In the example chosen, n-butane is a main component of the C_4-stream (raffinate). But for other reversible reactions, a solvent can be chosen which has favorable impact on the reaction. In this case, the right selection of the solvent can lead to a distinct increase of the conversion, which is important for economical reasons.

Table 12.2 Calculated *i*-butene conversions assuming ideal behavior, respectively, taking into account the real behavior using the Wilson model or modified UNIFAC.

$T = 353.15$ K	Ideal	Wilson	Modified UNIFAC
$X_{i\text{-butene}}$	0.6990	0.9157	0.9060
x_1	0.0673	0.0280	0.0299
x_2	0.0561	0.0164	0.0182
x_3	0.1304	0.1780	0.1758
x_4	0.7462	0.7776	0.7761
γ_1	1.000	10.22	9.83
γ_2	1.000	0.986	0.992
γ_3	1.000	0.897	1.043
γ_4	1.000	1.020	1.029
K_γ	1.000	0.089	0.107

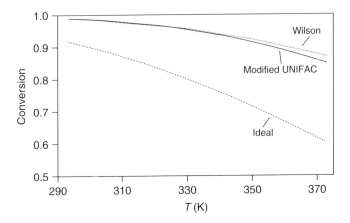

Figure 12.9 Experimental and predicted equilibrium conversion of *i*-butene for the MTBE synthesis in the liquid phase as a function of temperature (initial number of moles: methanol 1.06 kmol, *i*-butene 1 kmol, *n*-butane 4 kmol).

12.3
Multiple Chemical Reaction Equilibria

So far, only single reversible chemical reactions were considered. In practice, often multiple reversible chemical reactions take place, for example in the case of coal gasification or the steam-reforming process. During the calculation only the set of independent reactions has to be considered, this means only the reactions which cannot be obtained by linear combinations from the already selected reversible chemical reactions.

For the calculation of the equilibrium composition in the case of multiple reversible reactions different procedures can be applied. Either the compositions are determined for which all equilibrium constants are satisfied, or it is searched for the minimum of the Gibbs energy. In the first procedure, all equilibrium constants at the considered temperature of the various reversible chemical reactions have to be known. For the second procedure only the Gibbs energies at the considered temperature and the elemental formulas of the compounds involved are required. The second procedure has great advantages, when the occurring reactions are not known or when a large number of components participating in the reactions has to be considered. If the real behavior should be taken into account, the parameters for a chosen g^E-model, equation of state, or predictive method for the calculation of the real behavior (fugacity coefficients or activity coefficients) have to be given additionally.

12.3.1
Relaxation Method

In the relaxation method, the equilibrium constants of the different simultaneous reversible reactions are regarded independently. The number of reactors must therefore be identical with the number of independent reversible reactions, as shown in Figure 12.10. In the procedure, it is assumed that in reactor i only reaction i and in reactor n only reaction n, and so on, takes place.

Starting from the initial number of moles in reactor 1, the number of moles n_i has to be determined for which the equilibrium constant K_1 of the first reaction is fulfilled. The number of moles leaving the first reactor is used to calculate the equilibrium composition for the second reversible reaction; this means the moles of the compounds involved to fulfill the equilibrium constant K_2 for the second reaction. In the same way, the number of moles for all n reactions (reactors) is determined. The mixture leaving reactor n is then recycled to the first reactor. Then, the calculations with the new numbers of moles are continued until no further change is observed for the number of moles and all equilibrium constants are fulfilled.

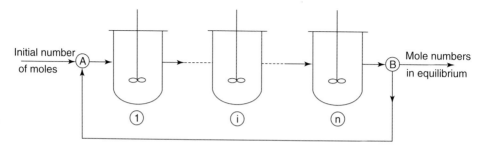

Figure 12.10 Schematic presentation of the relaxation method for the determination of the chemical equilibrium composition for n reversible chemical reactions.

12.3 Multiple Chemical Reaction Equilibria

Example 12.6

For the production of synthesis gas with the help of the steam-reforming process, the equilibrium composition at a temperature of 1100 K and a pressure of 40 atm should be determined starting from 1 mol of methane and 3.2 mol of water (steam), taking into account the following independent chemical reactions:

$$CH_4 + H_2O \rightleftharpoons CO + 3H_2$$

$$CO + H_2O \rightleftharpoons CO_2 + H_2$$

Ideal gas behavior should be assumed.

Standard thermodynamic properties in the ideal gas state at 25 °C and 1 atm [2].

Compound	Δh_f^0 (kJ/mol)	Δg_f^0 (kJ/mol)
CH_4	−74.85	−50.835
H_2O	−241.82	−228.59
CO	−110.54	−137.27
H_2	0	0
CO_2	−393.50	−394.38

Ideal gas heat capacities $c_p^{id} = a + bT + cT^2 + dT^3$ (J/mol K) [2].

Compound	a	$b \cdot 10^3$	$c \cdot 10^6$	$d \cdot 10^9$
CH_4	19.238	52.09	11.966	−11.309
H_2O	32.220	1.9225	10.548	−3.594
CO	30.848	−12.84	27.87	−12.71
H_2	27.124	9.267	−13.799	7.64
CO_2	19.78	73.39	−55.98	17.14

Solution

Using the standard properties given for the ideal gas state, the following standard enthalpies of reaction, standard Gibbs energies of reaction, and equilibrium constants are obtained for the two reversible reactions at 25 °C:

Reaction 1:

$$\Delta h_{R1}^0 = -110.54 + 3 \cdot 0 + 74.85 + 241.82 = 206.13 \text{ kJ/mol}$$

$$\Delta g_{R1}^0 = -137.27 + 3 \cdot 0 + 50.835 + 228.59 = 142.155 \text{ kJ/mol}$$

$$K_1 = \exp\left(\frac{-142155}{8.31433 \cdot 298.15}\right) = 1.24 \cdot 10^{-25}$$

Reaction 2:

$$\Delta h_{R2}^0 = -393.50 + 0 + 110.54 + 241.82 = -41.14 \text{ kJ/mol}$$
$$\Delta g_{R2}^0 = -394.38 + 0 + 137.27 + 228.59 = -28.52 \text{ kJ/mol}$$
$$K_2 = \exp\left(\frac{28250}{8.31433 \cdot 298.15}\right) = 9.92 \cdot 10^4$$

While the first reaction is strongly endothermic, the second reaction is exothermic at 25 °C. From the values of the equilibrium constants K_i, it can be recognized that at 25 °C the equilibrium of reaction 1 is far on the reactant side while for the second reaction it is on the product side.

Starting from the equilibrium constants at 298 K, the equilibrium constants at 1100 K can be determined with the help of the van't Hoff equation assuming constant standard enthalpies of reaction. At 1100 K, the following values for the equilibrium constants are obtained:

$$\ln K_1 (1100 \text{ K}) = \ln\left(1.24 \cdot 10^{-25}\right) - \frac{206130}{8.31433}\left(\frac{1}{1100} - \frac{1}{298.15}\right) = 3.2654$$
$$K_1 = 26.19$$

$$\ln K_2 (1100 \text{ K}) = \ln\left(9.92 \cdot 10^4\right) + \frac{41140}{8.31433}\left(\frac{1}{1100} - \frac{1}{298.15}\right) = -0.5928$$
$$K_2 = 0.5528$$

In reality, the enthalpy of reaction is strongly temperature-dependent. When the temperature dependence of the standard enthalpy of reaction with the help of the molar heat capacities is taken into account (see Eq. 12.6), the following values for the enthalpies of reaction, Gibbs energies of reaction, and equilibrium constants are obtained (see Example 12.8):

$$\Delta h_{R1}^0 = 226.432 \text{ kJ/mol} \quad \Delta g_{R1}^0 = -52.1975 \text{ kJ/mol} \quad K_1 = 301 \text{ at } 1100 \text{ K}$$

and for the second reaction:

$$\Delta h_{R2}^0 = -33.7385 \text{ kJ/mol} \quad \Delta g_{R2}^0 = 0.5618 \text{ kJ/mol} \quad K_2 = 0.94 \text{ at } 1100 \text{ K}$$

It can be seen that the equilibrium constants are distinctly different than the ones calculated assuming constant standard enthalpies of reaction in the temperature range covered.

Both equilibrium constants have to be matched. In the relaxation method, first only reaction 1 is considered. If Δn_1 is the number of moles of CH_4 that need to react to reach equilibrium, the number of moles in equilibrium of each component can be calculated via

$$CH_4: n_{CH_4} = 1 - \Delta n_1 \quad H_2O: n_{H_2O} = 3.2 - \Delta n_1$$
$$CO: \quad n_{CO} = \Delta n_1 \quad \quad H_2: n_{H_2} = 3 \cdot \Delta n_1$$
$$\sum n_i: 4.2 + 2\Delta n_1$$

12.3 Multiple Chemical Reaction Equilibria

Using these values, the equilibrium constant of the first reaction can be written as follows:

$$K_1 = 301 = \frac{p_{CO} p_{H_2}^3}{p_{CH_4} p_{H_2O}} = \frac{y_{CO} y_{H_2}^3 P^2}{y_{CH_4} y_{H_2O}} = \frac{n_{CO} n_{H_2}^3 P^2}{n_{CH_4} n_{H_2O} (\Sigma n_i)^2}$$

$$= \frac{\Delta n_1 (3\Delta n_1)^3 P^2}{(1 - \Delta n_1)(3.2 - \Delta n_1)(4.2 + 2\Delta n_1)^2}$$

For the determination of the equilibrium composition, only the value for Δn_1 has to be determined. As this equation cannot easily be solved analytically, a numerical method is suggested. As objective function F for the minimum search, either the absolute or the least square deviation can be used:

The determination of the minimum of the objective function by changing the values of Δn_1 can be performed iteratively or with the help of appropriate solvers (e.g., one-dimensional search routines, such as the method of the golden section or the Fibonacci search [11]).

$$F = \left(301 - \frac{\Delta n_1 (3\Delta n_1)^3 P^2}{(1 - \Delta n_1)(3.2 - \Delta n_1)(4.2 + 2\Delta n_1)^2}\right)^2 = 0$$

For a value of $\Delta n_1 = 0.65606$, the equilibrium constant K_1 is fulfilled. Using this value, the following numbers of moles are leaving the first reactor:

$n_{CH_4} = 0.34394 \quad n_{H_2O} = 2.54394$

$n_{CO} = 0.65606 \quad n_{H_2} = 1.96818$

This mixture enters the second reactor, where only the second reaction is considered; this means that now the second equilibrium constant K_2 has to be fulfilled:

$n_{CH_4} = 0.34394 \quad\quad n_{H_2O} = 2.54394 - \Delta n_2$

$n_{CO} = 0.65606 - \Delta n_2 \quad n_{H_2} = 1.96818 + \Delta n_2$

$n_{CO_2} = \Delta n_2$

$$0.94 = \frac{p_{CO_2} p_{H_2}}{p_{CO} p_{H_2O}} = \frac{y_{CO_2} y_{H_2}}{y_{CO} y_{H_2O}} = \frac{n_{CO_2} n_{H_2}}{n_{CO} n_{H_2O}} = \frac{\Delta n_2 \cdot (1.96818 + \Delta n_2)}{(0.65606 - \Delta n_2)(2.54394 - \Delta n_2)}$$

The equilibrium constant K_2 is fulfilled for $\Delta n_2 = 0.31415$. This means the following numbers of moles are obtained behind the second reactor:

$n_{CH_4} = 0.34394 \quad n_{H_2O} = 2.22979$

$n_{CO} = 0.34191 \quad n_{H_2} = 2.28233$

$n_{CO_2} = 0.31415$

This stream is now recycled to the first reactor, where with the new number of moles Δn_1 has now to be calculated again to fulfill equilibrium constant K_1, where a much smaller value $\Delta n_1 = 7.9773 \cdot 10^{-3}$ is obtained. After a few steps, no further change of the number of moles is obtained. For the first four steps, the numbers

12 Enthalpy of Reaction and Chemical Equilibria

Table 12.3 Calculated mole numbers for the different steps of the relaxation method.

Step	Reactor	n_{CH_4}	n_{H_2O}	n_{CO}	n_{H_2}	n_{CO_2}	Δn_i	Σn_i
–		1.00000	3.20000	0.00000	0.00000	0.00000	–	4.20000
1	1	0.34394	2.54394	0.65606	1.96818	0.00000	0.65606	5.51212
1	2	0.34394	2.22979	0.34191	2.28233	0.31415	0.31415	5.51212
2	1	0.33596	2.22181	0.34990	2.30627	0.31415	0.00798	5.52808
2	2	0.33596	2.22051	0.34859	2.30757	0.31545	0.00131	5.52808
3	1	0.33581	2.22035	0.34874	2.30803	0.31545	0.00015	5.52838
3	2	0.33581	2.22033	0.34872	2.30806	0.31548	0.00002	5.52838
4	1	0.33580	2.22033	0.34872	2.30807	0.31548	<0.00001	5.52840
4	2	0.33580	2.22033	0.34872	2.30807	0.31548	<0.00001	5.52840

of moles are given in Table 12.3. It can be recognized that already after four steps only very small changes of the mole numbers are observed.

Using the number of moles, the equilibrium composition can be calculated. For a temperature of 1100 K and 40 atm the following mole fractions are obtained:

$y_{CH_4} = 0.06074$
$y_{H_2O} = 0.40162$
$y_{CO} = 0.06308$
$y_{H_2} = 0.41749$
$y_{CO_2} = 0.05707$

For a larger temperature range, the resulting equilibrium concentrations at 40 atm are shown in Figure 12.12. The highest hydrogen concentrations are observed at the highest temperature, as expected for endothermic reactions.

12.3.2
Gibbs Energy Minimization

In Section 12.2, expressions for the calculation of the chemical equilibrium constant were derived (Eqs. (12.18–12.23)), starting from the knowledge that the Gibbs energy shows a minimum. So, it should also be possible to skip the auxiliary chemical equilibrium constant K and to determine the minimum of the Gibbs energy directly

$$G = \sum_i \mu_i n_i \tag{12.32}$$

at given temperature and pressure.

Using the standard Gibbs energies in the ideal gas state, this means $f_i^0 = 1$ atm and assuming ideal gas behavior, the required fugacities f_i (atm) can be expressed by the partial pressures p_i (atm):

$$\mu_i = \mu_i^0 + RT \ln p_i = \Delta g_{f,i}^0 + RT \ln p_i = \Delta g_{f,i}^0 + RT \ln \frac{n_i}{\sum_j n_j} P \tag{12.33}$$

This means that Eq. (12.32) can be replaced by the following relation to calculate the total Gibbs energy at a given temperature T:

$$G(T) = \sum_i n_i \left(\Delta g^0_{f,i}(T) + RT \ln \frac{n_i P/(1 \text{ atm})}{\sum_j n_j} \right)$$

In chemical equilibrium, the total Gibbs energy should show a minimum. For a simple reversible reaction, this can be shown graphically (see Example 12.7).

In the case of multiple reaction equilibria, the number of moles of all reactive compounds in chemical equilibrium can be determined with the help of nonlinear regression methods [11]. But at the same time, the element balance has to be satisfied; this means the amount of carbon, hydrogen, oxygen, nitrogen has to be the same before and after the reaction. This can either be taken into account with the help of Lagrange multipliers or using penalty functions [11], as shown in Example 12.8.

Example 12.7

Determine the molar Gibbs energy as a function of composition and the equilibrium concentration for the isomerization reaction:

$$n\text{-butane} \rightleftarrows i\text{-butane}$$

for temperatures of 298.15 K and 400 K and a pressure of 1 atm, assuming ideal gas behavior.

Standard Gibbs energies and enthalpies of formation in the ideal gas state.

Component	Δg^0_f (298.15 K, 1 atm)	Δh^0_f (298.15 K, 1 atm)
i-Butane (1)	−20 878 J/mol	−134 510 J/mol
n-Butane (2)	−17 154 J/mol	−126 150 J/mol

Solution

The Gibbs energy according to Eq. (12.33) is as follows:

$$G = \sum_i n_i \mu^0_i + \sum_i n_i RT \ln y_i \, P/(1 \text{ atm})$$

Starting with 1 mol of reactants, the number of moles can be replaced by the mole fractions, so that at 1 atm the molar Gibbs energy can be calculated using the following expression:

$$g = y_1 \Delta g^0_{f,1} + y_2 \Delta g^0_{f,2} + RT(y_1 \ln y_1 + y_2 \ln y_2)$$

For a mole fraction $y_1 = 0.1$, the following value for the molar Gibbs energy g is obtained at $T = 298.15$ K:

$$g = 0.1 \cdot (-20878) + 0.9 \cdot (-17154) + 8.31433 \cdot 298.15$$
$$\cdot (0.1 \cdot \ln 0.1 + 0.9 \cdot \ln 0.9)$$
$$g = -18332 \text{ J/mol}.$$

For other compositions, the values of the molar Gibbs energy can be calculated similarly. The values of the molar Gibbs energy for the whole composition range are shown in Figure 12.11. The minimum is reached at a concentration of $y_1 = 0.818$ at 298.15 K, which is the equilibrium composition.

The minimum of the molar Gibbs energy g can also be calculated by differentiation of the expression for the molar Gibbs energy $(y_2 = 1 - y_1)$:

$$\frac{dg}{dy_1} = \Delta g^0_{f,1} - \Delta g^0_{f,2} + RT \ln \frac{y_1}{y_2} = 0$$

$$\frac{y_1}{y_2} = \exp\left(-\frac{\left(\Delta g^0_{f,1} - \Delta g^0_{f,2}\right)}{RT}\right) = \exp\left(-\frac{-20878 + 17154}{8.31433 \cdot 298.15}\right) = 4.492$$

$$y_1 = 0.818$$

The equilibrium conversion can of course also be determined in the classical way via the equilibrium constant:

$$\Delta g^0_R = -20878 + 17154 = -3724 \text{J/mol}$$

$$K = \exp\left(-\frac{\Delta g^0_R}{RT}\right) = \exp\left(\frac{3724}{8.31433 \cdot 298.15}\right) = 4.492$$

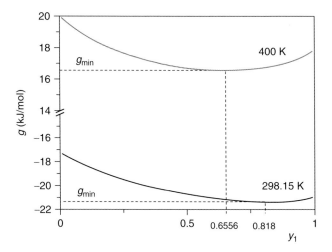

Figure 12.11 Composition dependence of the molar Gibbs energy at 298.15 and 400 K for the system i-butane (1)–n-butane (2).

Assuming ideal behavior of the gas phase, the equilibrium constant K can directly be replaced by K_y (see Eq. 12.28), since $(\sum v_i = 0)$, that is,

$$K = K_y = \frac{y_1}{y_2} = 4.492$$

$$y_1 = 0.818$$

For a temperature of 400 K, the Gibbs energies of formation at this temperature have to be calculated. Assuming that the standard enthalpy of formation is constant in the temperature range covered, the Gibbs energies of formation for i-butane at 400 K can be calculated by integrating the van't Hoff equation:

$$\Delta g_f (T, 1 \text{ atm}) = T \cdot \left[\frac{\Delta g_f^0 (T_0, 1 \text{ atm})}{T_0} + \Delta h_f^0 (T_0, 1 \text{ atm}) \cdot \left(\frac{1}{T} - \frac{1}{T_0} \right) \right]$$

For i-butane, the following Gibbs energy of formation is obtained:

$$\Delta g_{f,1} (400\text{K}, 1 \text{ atm}) = 400\text{K} \cdot \left[\frac{-20878}{298.15} + -134510 \cdot \left(\frac{1}{400} - \frac{1}{298.15} \right) \right]$$

$$= 17939 \frac{\text{J}}{\text{mol}}$$

Similarly, for n-butane, the following value is obtained:

$$\Delta g_{f,2} (400\text{K}, 1 \text{ atm}) = 20080 \frac{\text{J}}{\text{mol}}$$

Now the Gibbs energy at 400 K as a function of composition can be calculated. For a mole fraction $y_1 = 0.1$, one obtains

$$g = y_1 \Delta g_{f,1}^0 + y_2 \Delta g_{f,2}^0 + RT \left(y_1 \ln y_1 + y_2 \ln y_2 \right)$$

$$g = 0.1 \cdot 17939 + 0.9 \cdot 20080 + 8.31433 \cdot 400 \cdot \left(0.1 \cdot \ln 0.1 + 0.9 \cdot \ln 0.9 \right)$$

$$= 18785 \text{ J/mol}$$

For other compositions, the values of the molar Gibbs energy can be calculated in a similar way. The values of the molar Gibbs energy for the whole composition range are shown in Figure 12.11. While at 298.15 K, the minimum is found at a mole fraction of $y_1 = 0.818$, for the exothermal reaction at 400 K the equilibrium composition is moved to a lower mole fraction of $y_1 = 0.6556$.

Example 12.8

Determine the equilibrium composition for the reversible reactions investigated in Example 12.6 at 1100 K and 40 atm for 1 mol of methane and 3.2 mol of water by minimization of the Gibbs energy.

Polynomial coefficients for the calculation of molar heat capacities (J/mol. K) (see Eq. 12.7) as a function of temperature for different elements.

	a	$b \cdot 10^3$	$c \cdot 10^6$	$d \cdot 10^9$
H_2	27.124	9.267	−13.799	7.640
O_2	28.087	−0.0042	17.447	−10.644
C (graphite)	−5.416	58.981	−43.559	11.604

Solution

For the calculation, the Gibbs energies of formation of all compounds at 1100 K are required. For an element such as hydrogen, the value for the Gibbs energy of formation is zero at any temperature (Eq. (12.3)). For the other compounds, the temperature dependence has to be calculated by integrating the van't Hoff equation and Kirchhoff's law. For example the Gibbs energy of formation for CO_2 can be calculated taking into account the molar heat capacities of CO_2 and of the elements O_2 and C forming this compound, in this case O_2 in the ideal gas state and carbon in the solid form (graphite) as a function of temperature.

With the help of these values the Gibbs energy of formation of CO_2 at 1100 K is determined to $\Delta g_f^{id} = -395966$ J/mol.

For CO_2 and the other compounds involved, the values at 1100 K are given in the table below together with the literature values.

Calculated and tabulated standard Gibbs energies of formation at 1100 K and 1 atm in the ideal gas state.

	CH_4	H_2O	CO	CO_2
Δg_f^{id} (kJ/mol) calc	29.947	−187.139	−209.389	−395.966
Δg_f^{id} (kJ/mol) [1]	30.358	−187.052	−209.084	−395.984

As can be seen, the temperature dependence of the standard Gibbs energy of formation can be calculated if the temperature dependencies of the molar heat capacities of the compound considered and of the elements are known.

In process simulators often a different way for calculating the required molar Gibbs energies of the compounds at the desired temperature is chosen. The molar Gibbs energies are calculated starting from the standard Gibbs energies of formation at 298.15 K and integration over the temperature range from 298.15 K to the desired temperature. This has the advantage that the heat capacities for the elements not involved in the reversible reactions are not required.

Instead only the molar heat capacities of the compounds considered are required. At 1100 K, the following value is obtained for H_2:

$$\frac{g^{id}(T, 1 \text{ atm})}{T} = \frac{g^{id}(T^0, 1 \text{ atm})}{T^0} - \int_{T^0}^{T} \frac{h^{id}(T, 1 \text{ atm})}{T^2} dT$$

$$g^{id}_{H_2}(1100\text{K}, 1 \text{ atm}) = -18.545 \text{ kJ/mol}$$

The molar Gibbs energies for all five compounds in the ideal gas state are summarized in the table below. Using these values instead of the Gibbs energies of formation, the same results are obtained, since the contributions of the elements not involved cancel.

Of course both methods provide the same standard Gibbs energies of reaction and thus the same equilibrium constants as shown below.

Gibbs energies for the compounds involved at 1100 K and 1 atm in the ideal gas state.

	H_2	CH_4	H_2O	CO	CO_2
g^{id} (kJ/mol)	−18.545	−16.078	−215.567	−228.207	−424.667

Gibbs energies of reaction calculated using the Gibbs energies in the ideal gas state:

$$\Delta g^{id}_{R,1} = 3 \cdot (-18.545) - 228.207 + 16.078 + 215.567 = -52.197 \text{ kJ/mol}$$

$$\Delta g^{id}_{R,2} = -424.667 - 18.545 + 228.207 + 215.567 = 0.562 \text{ kJ/mol}$$

Gibbs energies of reaction calculated using the Gibbs energies of formation in the ideal gas state:

$$\Delta g^{id}_{R,1} = 3 \cdot 0 - 209.389 - 29.947 + 187.139 = -52.197 \text{ kJ/mol}$$

$$\Delta g^{id}_{R,2} = -395.966 + 0 + 209.389 + 187.139 = 0.562 \text{ kJ/mol}$$

To satisfy the element balance for the compounds involved, that is, methane, water, hydrogen, carbon monoxide, and carbon dioxide in chemical equilibrium, the following amounts of carbon, hydrogen, and oxygen should be found starting from the given initial amounts of methane (1 mol) and water (3.2 mol):

$$n_C = 1 \cdot 1 = 1 \text{ mol}$$
$$n_H = 1 \cdot 4 + 3.2 \cdot 2 = 10.4 \text{ mol}$$
$$n_O = 3.2 \cdot 1 = 3.2 \text{ mol}$$

so that the following objective function can be used. The quadratic parts are the penalty functions, which contribute to the objective function, if the material balances for carbon, hydrogen, and oxygen are not fulfilled:

$$F = \sum_{i}^{5} n_i \left(\Delta g_{f,i}^0(T) + RT \ln \frac{n_i P/(1 \text{ atm})}{\sum_{j} n_j} \right) + (1 - n_{CH_4} - n_{CO} - n_{CO_2})^2$$

$$+ (10.4 - 4n_{CH_4} - 2n_{H_2O})^2 + (3.2 - n_{H_2O} - n_{CO} - 2n_{CO_2})^2$$

$$\stackrel{!}{=} \text{minimum}$$

For the initial number of moles, only the reactants have to be considered and the element balance is automatically fulfilled. This means that the penalty function delivers no contribution to the objective function. With the initial number of moles for methane and water, the following value for the objective function F is obtained:

$$F = G = \sum_{i}^{5} n_i \left(\Delta g_{f,i}^0(T) + RT \ln \frac{n_i P/(1 \text{ atm})}{\sum_{j} n_j} \right)$$

$$= 1 \cdot \left(29947 + 8.31433 \cdot 1100 \ln \frac{1 \cdot 40}{4.2} \right)$$

$$+ 3.2 \cdot \left((-187139) + 8.31433 \cdot 1100 \ln \frac{3.2 \cdot 40}{4.2} \right)$$

$$= -448283.3 \text{ J}$$

When it is assumed that 0.1 mol CH_4 and 0.1 mol H_2O react to CO and H_2, the following number of moles and the resulting value for the Gibbs energy are obtained:

$n_{CH_4} = 0.9$ mol, $n_{H_2O} = 3.1$ mol, $n_{CO} = 0.1$ mol, $n_{H_2} = 0.3$ mol,

$n_{CO_2} = 0$ mol $\quad G = -459492.4$ J

When additionally 0.02 mol of CO react with water to CO_2 and hydrogen, the following values are obtained:

$n_{CH_4} = 0.9$ mol, $n_{H_2O} = 3.08$ mol, $n_{CO} = 0.08$ mol, $n_{H_2} = 0.32$ mol,

$n_{CO_2} = 0.02$ mol $\quad G = -460359.5$ J

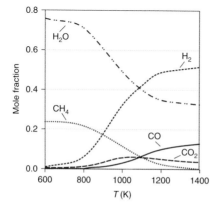

Figure 12.12 Mole fractions of CH_4, H_2O, CO, H_2, and CO_2 in chemical equilibrium at 40 atm as a function of temperature.

It can be seen that the value for the Gibbs energy decreases when the reactions take place. The task of the software package for minimization of the Gibbs energy is to determine the numbers of moles n_i of all five compounds for which the Gibbs energy shows a minimum where at the same time the element balance is fulfilled. To obtain the minimum of G, the number of moles is varied with the help of a nonlinear regression method. For 1100 K and a pressure of 40 bar, the minimum value for the Gibbs energy is obtained for the following numbers of moles and corresponding mole fractions:

$$G = -479515.8 \text{ J}$$
$$n_{CH_4} = 0.33580 \text{ mol} \quad y_{CH_4} = 0.06074$$
$$n_{H_2O} = 2.22032 \text{ mol} \quad y_{H_2O} = 0.40162$$
$$n_{CO} = 0.34872 \text{ mol} \quad y_{CO} = 0.06308$$
$$n_{H_2} = 2.30807 \text{ mol} \quad y_{H_2} = 0.41749$$
$$n_{CO_2} = 0.31548 \text{ mol} \quad y_{CO_2} = 0.05707$$

Of course the result is identical with the result of the relaxation method discussed in Example 12.6. For other temperatures, the equilibrium composition is shown in Figure 12.12 for a pressure of 40 atm. The iterative solution is shown in the Mathcad file 12.08.

Additional Problems

P12.1 Calculate the equilibrium composition of the TAME (tertiary amyl methyl ether) synthesis in the liquid phase using equimolar amounts of methanol (MeOH) and 2-methyl-2-butene (2M2B) at 25 °C and 80 °C

$$\text{MeOH} + \text{2M2B} \rightleftarrows \text{TAME}$$

with the help of the standard thermodynamic properties for the ideal gas state given below
a. without a solvent
b. in the presence of n-pentane with $n_{MeOH}/n_{pentane} = 1:1$ and $n_{MeOH}/n_{pentane} = 1:4$.
Calculations should be performed
1) assuming ideal behavior; that is, $\gamma_i = 1$.
2) taking into account the real behavior.
and
a. using the Wilson equation
b. using modified UNIFAC.

	Δh_f^0 (kJ/mol)	Δg_f^0 (kJ/mol)	Antoine constants		
			A	B	C
MeOH	−201.16	−162.5	8.08097	1582.27	239.7
2M2B	−42.58	59.7	6.92322	1099.07	233.317
TAME	−298.7	−108.05	6.85141	1208.39	217.900

Standard thermodynamic properties in the ideal gas state and Antoine constants (log P_i^s (mm Hg) = $A - B/(\vartheta(°C) + C)$).

	a (J/(mol K))	b (J/(mol K^2))	c (J/(mol K^3))	d (J/(mol K^4))
MeOH	21.137	$7.088 \cdot 10^{-2}$	$2.582 \cdot 10^{-5}$	$-2.850 \cdot 10^{-8}$
2M2B	11.81	$3.509 \cdot 10^{-1}$	$-1.117 \cdot 10^{-4}$	$-5.807 \cdot 10^{-9}$
TAME	36.391	$4.911 \cdot 10^{-1}$	$2.760 \cdot 10^{-5}$	$-1.287 \cdot 10^{-7}$

Parameters for the description of the molar heat capacities c_P for the ideal gas as a function of temperature ($c_P = a + bT + cT^2 + dT^3$).

Compound	Molar volume (cm^3/mol)	Wilson parameters (cal/mol)			
i\j		MeOH	2M2B	TAME	Pentane
MeOH	40.73	–	2090.97	1467.13	2027.39
2M2B	105.9	287.62	–	408.06	−115.94
TAME	133.45	−354.79	−280.62	–	91.02
Pentane	116.11	616.61	285.64	85.82	–

Molar volumes and Wilson parameters for the compounds considered.

P12.2 For the TAME synthesis the influence of the solvents benzene, and acetone on the chemical equilibrium conversion in the liquid phase should be calculated with the help of the modified UNIFAC model at 80 °C. The initial mixture should consist of equimolar amounts of methanol (MeOH), 2-methyl-2-butene (2M2B), and solvent. All required properties and parameters can be found in problem P12.1 and in the Appendices A and I.

P12.3 Ethanol can be produced from a feed of about 60 vol% ethylene and 40 vol% water. The mixture reacts over a phosphoric acid catalyst at 300 °C and 60 bar. Calculate the equilibrium constant at 25 °C and 300 °C for this reaction. Calculate the effect of the real vapor phase behavior on the equilibrium constant K_P and equilibrium composition at 300 °C as function of pressure up to 100 bar using VTPR.

The standard thermodynamic data of formation and the ideal gas heat capacity correlations can be found in Appendix A.

P12.4 At high temperatures, ethane dissociates via the reaction (ethane cracker)

$$C_2H_6 \rightleftarrows C_2H_4 + H_2$$

Calculate the equilibrium concentrations at $T = 800$ K and $P = 2$ bar. All required data can be found in Appendix A.

P12.5 Calculate the optimal feed ratio of nitrogen and hydrogen for the ammonia synthesis at 450 °C and a pressure of 600 atm using the parameters given in Examples 12.1–12.4 using
   a. ideal gas behavior
   b. taking into account the real behavior using the VTPR group contribution equation of state.

P12.6 The dissociative gas and important rocket propellant dinitrogen tetroxide is in rapid equilibrium with nitrogen dioxide following the reaction:

$$N_2O_4 \rightleftarrows 2\,NO_2$$

At a total pressure of 1 bar, the equilibrium concentrations of N_2O_4 in the gas phase were reported to be 0.6349 (30 °C), 0.4088 (50 °C), 0.2113 (70 °C), and 0.09321 (90 °C). Calculate the difference between the heat capacity of the mixture and a mixture of ideal gases between 20 °C and 100 °C. Assume that only the chemical reaction contributes to the non-ideal behavior.

P12.7 Calculate the equilibrium of water formation from the elements in their most stable state for a stoichiometric feed of oxygen and hydrogen at $P = 0.01$ Pa and $T = 2000$ K using the data from Appendix A and assuming ideal gas phase behavior:

$$H_2 + 0.5\,O_2 \rightleftarrows H_2O$$

P12.8 In Example 12.6 the equilibrium conversion for the steam reforming process was calculated for a water/methane ratio of 2.7 and a pressure of 1 atm as a function of temperature. Please check how the equilibrium conversion is changed, if:
   a. the water/methane ratio is changed to 2
   b. the reaction is performed at 5 atm.

References

1. Lide, D.R. (ed.) (2004) *Handbook of Chemistry and Physics*, 85th edn, CRC Press, Boca Raton, FL.
2. Pedley, J.B. (1994) *Thermochemical Data and Structures of Organic Compounds*, Part 1, TRC, College Station.
3. Stull, D.R., Westrum, E.F., and Sinke, G.C. (1969) *The Chemical Thermodynamics of Organic Compounds*, John Wiley & Sons, Inc., New York.
4. Gillespie, L.J. and Beatty, J.A. (1930) The thermodynamic treatment of chemical

equilibria in systems composed of real gases. *Phys. Rev.*, **36**, 1008.
5. *Ullmann's Encyclopedia of Industrial Chemistry*, 5th edn, (1997), Wiley-VCH, Weinheim.
6. Larson, A.T. and Dodge, R.L. (1923) The ammonia equilibrium. *J. Am. Chem. Soc.*, **45**, 2918–2930.
7. Winchester, L.J. and Dodge, B.F. (1956) The chemical equilibrium of the ammonia synthesis reaction at high temperatures and extreme pressures. *AIChE J.*, **2**, 431–436.
8. Larson, A.T. (1924) The ammonia equilibrium at high pressures. *J. Am. Chem. Soc.*, **46**, 367–372.
9. Dortmund Data Bank *www.ddbst.com*, January 2011
10. Izquierdo, J.F., Cunnill, F., Vila, M., Tejero, J., and Iborra, M. (1992) Equilibrium Constants for Methyl tert-Butyl Ether Liquid-Phase Synthesis. *J. Chem. Eng. Data*, **37**, 339–343.
11. Hoffmann, U. and Hofmann, H. (1971) *Einführung in die Optimierung*, Verlag Chemie, Weinheim.

13
Special Applications

In the previous chapters, phase equilibria as well as chemical equilibria have been thoroughly discussed. For the development of a process model, it is the usual way to describe the chemical reactions first, and then focus on the phase equilibria for the separation process. However, there are some substances where special models have to be developed, as phase equilibria and chemical reactions have a strong influence on each other. Furthermore, components occur which do not exist in the pure form so that pure component properties cannot be assigned in the usual way.

The strategy for the development of a process model mentioned above cannot be applied in these cases. In this chapter, two examples for special process models are presented.

13.1
Formaldehyde Solutions

Formaldehyde is one of the most important basic chemical compounds [1]. Aqueous formaldehyde solutions are used as disinfection and conservation agents. It is an educt for the production of resins from phenol, urea, and melamin. Water-free formaldehyde is the raw material for polyoxymethylene, a dimensionally stable polymer.

Formaldehyde is a low-boiling substance with a normal boiling point of approx. 254 K. It is not stable in its pure form, so it usually occurs in aqueous or methanolic solutions. Mixtures of formaldehyde and water or alcohols are not binary solutions in the usual sense, as formaldehyde reacts with both of them to a wide variety of species which are not stable as pure compounds themselves. Therefore, the standard procedure for building up a thermodynamic model by setting up the pure component properties and the binary interaction parameters fails in this case. The formaldehyde–water–methanol system is a good example for a reactive phase equilibrium, where a special model has to be developed. This has been done by the group of Maurer [2–6].

The reaction of formaldehyde with water leads to methylene glycol:

$$CH_2O + H_2O \rightleftharpoons HO-CH_2-OH \qquad (13.1)$$

Chemical Thermodynamics: for Process Simulation, First Edition.
Jürgen Gmehling, Bärbel Kolbe, Michael Kleiber, and Jürgen Rarey.
© 2012 Wiley-VCH Verlag GmbH & Co. KGaA. Published 2012 by Wiley-VCH Verlag GmbH & Co. KGaA.

which does, as mentioned, not exist as a pure substance. Further addition of formaldehyde molecules results in the homologous series of methylene glycols according to the reactions

$$HO-(CH_2O)_{i-1}H + HOCH_2OH \rightleftharpoons HO-(CH_2O)_iH + H_2O \qquad (13.2)$$

Similarly, the reactions with alcohols are

$$CH_2O + R-OH \rightleftharpoons HO-CH_2O-R \qquad (13.3)$$

where the products are called *hemiformals*, which are again not stable as pure substances. As well, hemiformal chains can be formed according to

$$HO-(CH_2O)_{i-1}-R + HO-CH_2O-R \rightleftharpoons HO-(CH_2O)_i-R + R-OH \qquad (13.4)$$

The most important hemiformals are the ones formed by the reaction of formaldehyde with methanol ($R = CH_3$ in reaction (13.3)). In the following section, the higher methylene glycols and hemiformals are named MG_i and HF_i, respectively.

These chemical reactions have a significant influence on the physical properties of the solution and on the phase equilibria. The formaldehyde, which would normally be expected preferably in the vapor phase, behaves now like a component with a similar vapor pressure as water and methanol. To create a model, the reactions listed above have to be taken into account, at least to an extent where significant amounts of the higher methylene glycols or hemiformals are formed.

The particular difficulty in the Maurer model is that the species created in these reactions do not exist in pure form. Therefore, their vapor pressures and their binary parameters with other components cannot be determined in the usual way. Moreover, methylene glycols and hemiformals can exist simultaneously only in ternary and higher mixtures of water (W), formaldehyde (FA), and methanol (MeOH).

To reduce the number of adjustable parameters, the UNIFAC method (see Section 5.9.2) with special groups for the species occurring in formaldehyde mixtures has been introduced. The groups and their size and surface parameters are given in Table 13.1. For the reactions mentioned above, the mass action laws come into operation. For methylene glycols and hemiformals, the equilibrium constants are given by

$$K_{MG1} = \frac{(x_{MG1}\gamma_{MG1})}{(x_{FA}\gamma_{FA})(x_W\gamma_W)} \qquad (13.5)$$

and

$$K_{HF1} = \frac{(x_{HF1}\gamma_{HF1})}{(x_{FA}\gamma_{FA})(x_{MeOH}\gamma_{MeOH})} \qquad (13.6)$$

The equilibrium constants for the liquid phase are related to the equilibrium constants for the vapor phase via

$$K_{MG1} = K_{MG1,gas}\frac{P^s_{FA}P^s_W}{P^s_{MG1}P^0} \qquad (13.7)$$

Table 13.1 Main groups of the Maurer model with their volume and surface area parameters [6].

No	Main group	R	Q	Incrementation example
1	CH_2O	0.9183	0.780	Formaldehyde: 1 CH_2O
2	H_2O	0.9200	1.400	Water: 1 H_2O
3	$HO(CH_2O)H$	2.6744	2.940	Methylene glycol 1 : 1 $HO(CH_2O)H$
4	OH	1.0000	1.200	Methylene glycol 2 : 1 CH_2O, 2 OH, 1 CH_2
5	CH_2	0.6744	0.540	Methylene glycol 2 : 1 CH_2O, 2 OH, 1 CH_2
6	CH_3O	1.1459	1.088	Hemiformal 1 : 1 CH_3O, 1 CH_2OH
7	CH_3OH	1.4311	1.432	Methanol: 1 CH_3OH
8	CH_2OH	1.2044	1.124	Hemiformal 1 : 1 CH_3O, 1 CH_2OH

and

$$K_{HF1} = K_{HF1,gas} \frac{P^s_{FA} P^s_{MeOH}}{P^s_{HF1} P^0} \tag{13.8}$$

where $P^0 = 101325$ Pa and

$$K_{MG1,gas} = \frac{\gamma_{MG1}}{\gamma_{FA}\gamma_w} \frac{P^0}{P} \tag{13.9}$$

$$K_{HF1,gas} = \frac{\gamma_{HF1}}{\gamma_{FA}\gamma_{MeOH}} \frac{P^0}{P} \tag{13.10}$$

assuming an ideal vapor phase. Similarly, for the higher methylene glycols and hemiformals, the chemical equilibria are described by

$$K_{MG,i} = \frac{(x_{MG,i}\,\gamma_{MG,i})(x_W\,\gamma_W)}{(x_{MG1}\,\gamma_{MG1})(x_{MG,i-1}\,\gamma_{MG,i-1})} \tag{13.11}$$

and

$$K_{MG,i} = \frac{(x_{HF,i}\,\gamma_{HF,i})(x_{MeOH}\,\gamma_{MeOH})}{(x_{HF1}\,\gamma_{HF1})(x_{HF,i-1}\,\gamma_{HF,i-1})} \tag{13.12}$$

The equilibrium constants, the UNIFAC interaction parameters, and the vapor pressure parameters of the components which do not exist in pure form have to be fitted simultaneously to phase equilibrium data and reaction equilibrium data obtained from spectroscopic measurements for the systems formaldehyde–water, formaldehyde–methanol, and the ternary system formaldehyde–water–methanol, for which large amounts of data are available. To keep the significance of the particular parameters, a number of simplifying assumptions are made.

- The vapor pressures of the higher methylene glycols and hemiformals MG_i and HF_i, with $i \geq 2$, are set to zero. Therefore, they do not occur in the vapor phase according to the model.
- The reaction constants for the formation of higher methylene glycols with $i \geq 3$, and for the formation of higher hemiformals with $i \geq 2$, are identical.

Table 13.2 Group interaction parameters a_{nm} (K) for the UNIFAC approach in the Maurer model [6].

n/m	1	2	3	4	5	6	7	8
1	–	867.8	189.2	237.7	83.4	0	238.4	238.4
2	−254.5	–	189.2	−229.1	300.0	−219.3	289.6	451.64 −114100/(T/K)
3	59.2	−191.8	–	−229.1	300.0	−142.4	289.6	289.6
4	28.1	353.5	353.5	–	156.4	112.8	−137.1	−137.1
5	251.5	1318.0	1318.0	986.5	–	447.8	697.2	697.2
6	0	423.8	774.8	1164.8	273.0	–	238.4	238.4
7	−128.6	−181.0	−181.0	249.1	16.5	−128.6	–	0
8	−128.6	−1018.57 +329900/(T/K)	−181	249.1	16.5	−128.6	0	–

Table 13.3 Chemical equilibrium constants in the Maurer model [6]: $\ln K = A + B/(T/K) + C(T/K) + D(T/K)^2$.

	A	B	C	D
$K_{MG1,gas}$	−16.984	5233.2	–	–
K_{MG2}	0.005019	834.5	–	–
$K_{MGn},\ n>2$	0.01312	542.1	–	–
$K_{HF1,gas}$	56.364	0	−0.2509	0.0002758
$K_{HF2,gas}$	−0.6291	−407.7	–	–
$K_{HFn},\ n>2$	−0.5018	−526.2	–	–

- The vapor phase is considered to be ideal.
- Several UNIFAC interaction parameter pairs are set to zero or supposed to be identical.

The final parameter sets used in the Maurer model are given in Tables 13.2–13.4.

For practical applications, the resulting distribution of the particular methylene glycols and hemiformals is not interesting for the mass balance. It has to be transformed into overall concentrations of formaldehyde, methanol, and water. This is done by applying the formal reactions:

$$MG_i \longrightarrow iCH_2O + H_2O \tag{13.13}$$

and

$$HF_i \longrightarrow iCH_2O + CH_3OH \tag{13.14}$$

for the final mass balance.

Figures 13.1 and 13.2 give a qualitative overview on the reactive systems formaldehyde–water and formaldehyde–methanol by applying the Maurer model.

Table 13.4 Antoine coefficients used in the Maurer model [6]: $\ln(P^s/\text{kPa}) = A + B/[T/K + C]$.

Component	A	B	C
Formaldehyde	14.4625	−2204.13	−30.00
Water	16.2886	−3816.44	−46.13
Methanol	16.5725	−3626.55	−34.29
Methylene glycol	17.4364	−4762.07	−51.21
Hemiformal	19.5736	−5646.71	0.00

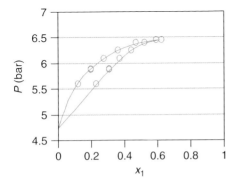

Figure 13.1 The system formaldehyde (1)–water (2) at $T = 423$ K. (Data from Kuhnert [6].)

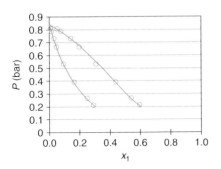

Figure 13.2 The system formaldehyde (1)–methanol (2) at $T = 333.1$ K. (Data from Kogan and Ogorodnikov [7].)

The expected curvature with formaldehyde as a low-boiling substance is not obtained. Instead, the formaldehyde behaves like a substance with a similar boiling point as water and methanol. For formaldehyde–water, an azeotrope is obtained, which vanishes at high temperatures. The presentations are interrupted at $x_{FA} = 0.6$, as for higher concentrations solid formation is likely to occur. In this area, there are no data points available; thus, the application of the model is not justified. The Maurer model has been successfully applied to several process steps in the POM (polyoxymethylene) production. For this purpose, additional inert components (e.g., trioxane and methylal) have been introduced into the model [3], as well as by-product reactions like the Canizzarro reaction:

$$2CH_2O + H_2O \rightleftharpoons HCOOH + CH_3OH \tag{13.15}$$

or the Tischenko reaction [6]:

$$2CH_2O \rightleftharpoons HCOOCH_3 \tag{13.16}$$

where formaldehyde is transformed to methyl formate.

The enthalpies of the methylene glycols and hemiformals can be evaluated by taking into account the enthalpies of the particular formation reactions. For MG_1 and HF_1, it can be written as

$$h_{MG1}^{id}(T) = h_{FA}^{id}(T) + h_{W}^{id}(T) + \Delta h_{R,MG1}^{id}(T) \tag{13.17}$$

$$h_{HF1}^{id}(T) = h_{FA}^{id}(T) + h_{MeOH}^{id}(T) + \Delta h_{R,HF1}^{id}(T) \tag{13.18}$$

where the enthalpies of reaction are obtained from the van't Hoff equation

$$\Delta h_R^{id} = RT^2 \left(\frac{d \ln K}{dT} \right) \tag{13.19}$$

To obtain the liquid enthalpy, the enthalpy of vaporization is subtracted:

$$h_{MG1}^{L} = h_{MG1}^{id} - \Delta h_{V,MG1} \tag{13.20}$$

$$h_{HF1}^{L} = h_{HF1}^{id} - \Delta h_{V,HF1} \tag{13.21}$$

The enthalpies of vaporization can be evaluated with the Clausius–Clapeyron equation (see Section 2.5.3), using the vapor pressure equations given in Table 13.4. The vapor phase is treated as an ideal gas, this means the vapor phase reality is neglected.

As the vapor pressures of the higher methylene glycols and hemiformals had been set to zero, only their liquid enthalpies are of interest. They can be determined by

$$h_{MG,i}^{L}(T) = h_{MG,i-1}^{L}(T) + h_{MG1}^{L}(T) - h_{W}^{L}(T) + \Delta h_{R,MG,i}^{L} \tag{13.22}$$

$$h_{HF,i}^{L}(T) = h_{HF,i-1}^{L}(T) + h_{HF1}^{L}(T) - h_{MeOH}^{L}(T) + \Delta h_{R,HF,i}^{L} \tag{13.23}$$

where the Δh_R^L are obtained via the van't Hoff equation analogously to Eq. (13.19). The procedure refers to the enthalpy route A described in Section 6.2. As a derivative of a roughly determined enthalpy of reaction is involved and an ideal gas phase is used, it cannot be expected that the results especially for c_p^L are very exact. Instead, they must be interpreted as an estimation.

In many cases, the influence of reaction kinetics is significant, especially at low temperatures. On the other hand, the evaluation of the reaction kinetics is difficult because of the number of reactions to be regarded. Moreover, any kinetics must be written with activities and not with concentrations. Otherwise, the results will not be consistent, as the reaction progresses do not move toward the equilibrium state.

In recent works, even a promising application of the slow reaction kinetics has been shown [8]. At low temperatures and low residence times e.g. in an evaporator, the reaction kinetics are too slow to reach equilibrium. This is advantageous, if the formaldehyde content of the vapor should be as low as possible, which is the case for the enrichment of trioxane by evaporation in the POM process.

The reverse reactions of the MG- and HF-chains to form MG_1 and molecular formaldehyde are too slow to reach equilibrium; thus, less formaldehyde is evaporated. Different to common absorption processes, the absorption of formaldehyde containing vapors should take place at relatively high temperatures to obtain a sufficient reaction rate in the liquid phase which is necessary for the absorption.

13.2
Vapor Phase Association

The equations of state described in Section 2.6 take into account that at high pressures the intermolecular forces are no more negligible and the deviations from the ideal gas law become larger. At ambient pressure, the ideal gas law is at least a sufficient approximation to evaluate the PvT behavior.

This is not valid for a certain number of components, for example some organic acids and hydrogen fluoride (HF). Table 13.5 shows the compressibility factors of the pure component vapor phase at the normal boiling point, which deviate significantly from the ideal gas law $z = Pv/RT = 1$.

According to the association model, it is assumed that these substances form associates in the vapor phase, which themselves behave like an ideal gas. In practical applications, only the occurrence of dimers is taken into account, which is sufficient with respect to the relatively poor data situation. There are two exceptions: for acetic acid, much more experimental data are available, and it makes sense to make the model more flexible and accurate. Just for this purpose, tetramers have been introduced as species, although it has been shown that they actually do not occur to a significant extent [9, 10]. As well, for HF quite a lot of data exist. In the vapor phase, a number of different species have been detected. Most often, hexamers occur [11]. The PvT behavior can be successfully described

Table 13.5 Compressibility factors of associating substances at the normal boiling point.

	Normal boiling point (K)	z
Acetic acid	391	0.6
Formic acid	374	0.65
Hydrogen fluoride	293	0.33

by assuming that there are only monomers and hexamers. The consideration of other species does not result in a better representation.

The character of the association is that of a chemical reaction in the vapor phase. Each reaction

$$n(i) \rightleftarrows (i)_n$$

can be described with the law of mass action:

$$K_n = \frac{z_n}{z_1^n (P/P_0)^{n-1}} \tag{13.24}$$

where K is the equilibrium constant (see Chapter 12) and z_n denotes for the true concentration of the associate of the degree n, this means the concentration of the species that consists of n molecules ($n = 1$, $K_1 = 1$). P_0 is an arbitrary reference pressure; in this context, it is chosen as

$$P_0 = 1 \text{Pa} \tag{13.25}$$

The equilibrium constant is related to the Gibbs energy of reaction (see Chapter 12) via

$$\Delta g_{R,n}^0 = \Delta h_{R,n}^0 - T \Delta s_{R,n}^0 = -RT \ln K_n \tag{13.26}$$

Therefore, the equilibrium constant can be written as

$$\ln K_n = \frac{\Delta s_{R,n}^0}{R} - \frac{\Delta h_{R,n}^0}{RT} \tag{13.27}$$

Assuming that $\Delta h_{R,n}^0$ and $\Delta s_{R,n}^0$ are independent of temperature, the correlation function

$$\ln K_n = A_n + \frac{B_n}{T} \tag{13.28}$$

with $A_n = \Delta s_{R,n}^0/R$ and $B_n = -\Delta h_{R,n}^0/R$ is obtained. As the total number of species in the vapor phase is reduced by the association reaction, $\Delta s_{R,n}^0$ is always negative, that is, $A < 0$. Furthermore, the reaction is always exothermic, which means $\Delta h_{R,n}^0$ is negative and therefore $B > 0$. Values of the association constants for the most important associating substances are given in Table 13.6.

To obtain the true concentrations z_n, the mass balance in the vapor phase has to be evaluated. The sum of the concentrations of the particular species must be 1:

$$\sum_n z_n = \sum_n K_n z_1^n (P/P_0)^{n-1} = 1 \tag{13.29}$$

which results in a nonlinear equation for the monomer concentration z_1. If only dimers are regarded, a quadratic equation

$$K_2 z_1^2 (P/P_0) + z_1 - 1 = 0 \tag{13.30}$$

is obtained, which can be directly solved to

$$z_1 = \frac{\sqrt{1 + 4K_2(P/P_0)} - 1}{2K_2(P/P_0)} \tag{13.31}$$

Table 13.6 Association constants for the most important associating substances.

Substance	Formula	A_2	B_2/K	A_4	B_4/K	A_6	B_6/K
		Dimerization		Tetramerization		Hexamerization	
Formic acid	HCOOH	−29.607	7 108	−	−	−	−
Acetic acid	CH_3COOH	−29.129	7 335	−78.232	17 268	−	−
Chloroacetic acid	$CH_2ClCOOH$	−43.568	13 035	−	−	−	−
Dichloroacetic acid	$CHCl_2COOH$	−38.134	10 636	−	−	−	−
Propionic acid	CH_3-CH_2-COOH	−29.791	7 821	−	−	−	−
Acrylic acid	$CH_2=CH-COOH$	−33.989	9 384	−	−	−	−
Butyric acid	C_3H_7COOH	−28.232	7 090	−	−	−	−
Benzoic acid	C_6H_5-COOH	−25.467	6 012	−	−	−	−
Hydrogen fluoride	HF	−	−	−	−	−119.306	18 735

For acetic acid and HF, where tetramers or hexamers are regarded, the equation must be solved numerically, which is, however, pretty easy, as $0 < z_1 < 1$. Having calculated z_1, the compressibility factor z can be determined as the ratio between the number of associates and the number of molecules in the vapor phase:

$$z = \frac{Pv}{RT} = \frac{\sum_n z_n}{\sum_n n z_n} = \frac{1}{\sum_n n z_n} \quad (13.32)$$

The specific volume is then obtained via

$$v = z\frac{RT}{P} \quad (13.33)$$

As for all chemical reactions, the enthalpy changes are considerable and cannot be neglected. According to Eq. (13.27 and 13.28), the enthalpy of reaction is given by

$$\Delta h^0_{R,n} = -RB_n \quad (13.34)$$

For each molecule in an association complex of degree n, the contribution to $(h - h^{id})$ is equal to $\Delta h^0_{R,n}/n$. On the other hand, the fraction of molecules being integrated in a complex of degree n is equal to $nz_n/\sum_n nz_n$. Thus, $(h - h^{id})$ can be calculated by

$$(h - h^{id}) = \sum_n \frac{nz_n}{\sum_n nz_n}(\Delta h^0_{R,n}/n) = \sum_n \frac{-RB_n z_n}{\sum_n nz_n} = -zR\sum_n B_n z_n \quad (13.35)$$

For comparison with experimental data, it is useful to have an expression for c_p^V. With

$$h^V(T,P) = \Delta h^0_f + \int_{T_0}^{T} c_p^{id} dT + (h - h^{id})(T,P) \quad (13.36)$$

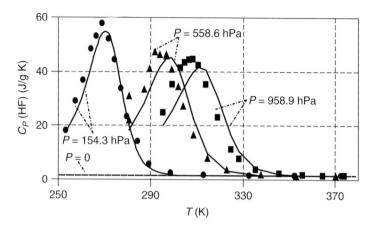

Figure 13.3 Specific heat capacity of gaseous HF as a function of temperature. (Data from Franck and Meyer [12].)

we obtain

$$c_P^V = c_P^{id} + \left(\frac{d(h-h^{id})}{dT}\right)_P \tag{13.37}$$

Due to the complexity of Eq. (13.35), it is recommended to evaluate the differential quotient in Eq. (13.37) numerically.

In Figure 13.3, the unusual curvature of the specific heat capacity of gaseous HF as a function of temperature is shown with the pressure as a parameter. For $P = 0$, it is an almost constant function with $c_P^V = c_P^{id} \approx 1.46 \, J/(g/K)$. For finite pressures, there is a strong influence of the association reactions, and a peak is formed. For the curve on the left-hand side, the top of the peak is approx. 40 times larger than c_P^{id}. With increasing pressure, the peak becomes less steep, and it is transferred to higher temperatures. The simple association model can reproduce this complicated shape quite well; however, the agreement is not exact.

The curvature itself is easy to understand: at the boiling point, the vapor consists of hexamers to a large extent. Additional heat supply leads to splitting of hexamers, as it favors the endothermic reaction. The splitting has the character of a chemical reaction; therefore, its energy demand is extremely high. Therefore, c_P^V increases drastically, as the heat is mainly used for the splitting and not for a temperature increase. Finally, with increasing temperature, the number of hexamers to be splitted decreases; thus, the heat capacity passes a maximum and comes down to the normal value of c_P^{id}. The experimental data shown in Figure 13.3 are taken from Franck and Meyer [12]. For organic acids, the curvature of c_P^V is similar but less dramatic as it is for HF (Figure 13.4).

As well, the curvature of the enthalpy of vaporization as a function of temperature is different from the usual ones shown in Figure 3.12. Figure 13.5 shows the plot for HF and acetic acid. Both of them show a maximum value. For HF, the

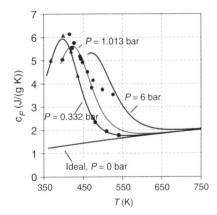

Figure 13.4 c_p^V of acetic acid vapor as a function of temperature.

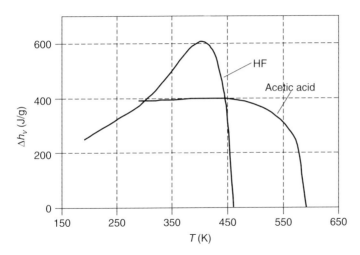

Figure 13.5 Enthalpy of vaporization as a function of temperature for HF and acetic acid.

shape is so extraordinary that only Eq. (3.62) can yield acceptable correlation results.

The parameters to describe the equilibrium constant A and B can be fitted to data for vapor density and c_p^V. Due to the steep shape of the c_p^V function, relatively small errors cause large contributions to the objective function. Therefore, weighting factors should be introduced to make sure that both types of data are regarded in the evaluation of the objective function. If Route A is used for the enthalpy description (Section 6.2), it is strongly recommended to integrate data for c_p^L as well [13]. In this case, the association constants and the coefficients of the correlation of the enthalpy of vaporization are fitted simultaneously to the objective function

$$F = w_1 \sum_i \left(\frac{\Delta h_{vi,calc} - \Delta h_{vi,exp}}{\Delta h_{vi,exp}}\right)^2 + w_2 \sum_i \left(\frac{c^L_{Pi,calc} - c^L_{Pi,exp}}{c^L_{Pi,exp}}\right)^2$$

$$+ w_3 \sum_i \left(\frac{c^V_{Pi,calc} - c^V_{Pi,exp}}{c^V_{Pi,exp}}\right)^2 + w_4 \sum_i \left(\frac{v^V_{i,calc} - v^V_{i,exp}}{v^V_{i,exp}}\right)^2 \quad (13.38)$$

It should be mentioned that there is no chance to estimate association constants. Even qualitative estimations are often dangerous. For example, trichloroacetic acid does not show association behavior [14], although it is more polar than acetic acid, monochloroacetic acid, and dichloroacetic acid.

Example 13.1

Calculate the specific vapor volume of acetic acid at $P = 0.6486$ bar and $T = 383.15$ K.

Solution

From Table 13.6, we get

$$K_2 = \exp\left(-29.129 + \frac{7335}{383.15}\right) = 4.608 \cdot 10^{-5}$$

$$K_4 = \exp\left(-78.232 + \frac{17268}{383.15}\right) = 3.956 \cdot 10^{-15}$$

Eq. (13.29) yields

$$z_1 + K_2 z_1^2 \left(\frac{P}{P_0}\right) + K_4 z_1^4 \left(\frac{P}{P_0}\right)^3 = 1$$

$$\Rightarrow f(z_1) = z_1 + 2.98896 z_1^2 + 1.07947 z_1^4 - 1 = 0$$

Starting with an estimated value of $z_1 = 0.5$, the following iteration is obtained according to the Newton method:

$$z_{1,i+1} = z_{1,i} - \frac{f(z_{1,i})}{f'(z_{1,i})}$$

with

$$f'(z_1) = 1 + 5.97792 z_1 + 3.23841 z_1^3 = 0$$

one gets

Step	$z_{1,i}$	$f(z_{1,i})$	$f'(z_{1,i})$	$z_{1,i+1}$
1	0.5	0.3147	4.3938	0.4284
2	0.4284	0.0133	3.8156	0.4249
3	0.4249	−0.00029	3.7884	0.4250
4	0.4250	0.0001	3.7892	0.4250

With $z_1 = 0.425$, Eq. (13.24) yields $z_2 = 0.5399$ and $z_4 = 0.0352$. Thus, the compressibility factor is

$$z = \frac{Pv}{RT} = \frac{1}{\sum_n n z_n} = \frac{1}{z_1 + 2z_2 + 4z_4} = 0.6077$$

The specific vapor volume is

$$v = \frac{zRT}{P} = 29.846 \; \frac{m^3}{kmol}$$

According to Barton and Hsu [15], the experimental value is $v = 30.0 \; m^3/kmol$.

As well as for pure components, the association model can be applied to mixtures of associating compounds and to mixtures of associating and nonassociating substances. Similarly, the concentrations of the true species can be calculated from material balances. As more than one component occurs, the indices system must be revised for K and z: In the following section, the first index denotes for the component, and the second one is the degree of association, for example, z_{12} ... true concentration of dimers of component 1.

For mixtures of organic acids, co-dimers formed by the reaction

$$(i) + (j) \rightleftarrows (ij)$$

have to be introduced as a new species. The corresponding law of mass action is

$$K_{Mij} = \frac{z_{Mij}}{z_{i1} \, z_{j1} (P/P_0)} \tag{13.39}$$

For K_{Mij}, Prausnitz et al. [16] recommend the relationship

$$K_{Mij} = 2\sqrt{K_{i2} K_{j2}} \tag{13.40}$$

It has to be distinguished between the true concentrations z_{in} of the particular species and the apparent concentrations y_i, which represent the stoichiometric mole fractions of the components occurring in the mixture.

To obtain the true concentrations z_{in}, a system of equations much more complicated than for pure components must be solved. Again, the true monomer concentrations z_{i1} are the key concentrations. After they have been determined, the concentrations of the other species can be derived by the equilibrium conditions Eqs (13.24) and (13.39). For each component, a true monomer concentration exists, no matter if it associates or not. Therefore, one equation must be available for each component. The first equation of the system is a simple mass balance. The sum of the true concentrations of all species is 1:

$$\sum_i \sum_n z_{in} + \sum_i \sum_{j>i} z_{Mij} = 1 \tag{13.41}$$

For each component $i > 1$, an equation can be set up in the following way: the ratio between the numbers of monomers of the particular component and the first component is equal to the ratio of the apparent concentrations:

$$\frac{\sum\limits_{n} nz_{in} + \sum\limits_{j\neq i} z_{Mij}}{\sum\limits_{n} nz_{1n} + \sum\limits_{j\neq 1} z_{M1j}} = \frac{y_i}{y_1} \tag{13.42}$$

Inserting Eqs. (13.24) and (13.39), we obtain

$$\sum_i \sum_n K_{in} z_{i1}^n (P/P_0)^{n-1} + \sum_i \sum_{j>i} K_{Mij} z_{i1} z_{j1} (P/P_0) = 1 \tag{13.43}$$

and

$$\frac{\sum\limits_{n} n K_{in} z_{i1}^n (P/P_0)^{n-1} + \sum\limits_{j\neq i} K_{Mij} z_{i1} z_{j1}(P/P_0)}{\sum\limits_{n} n K_{1n} z_{11}^n (P/P_0)^{n-1} + \sum\limits_{j\neq 1} K_{M1j} z_{11} z_{j1}(P/P_0)} = \frac{y_i}{y_1} \tag{13.44}$$

where the values for K_{i1} must formally be set to unity.

Thus, a system of nonlinear equations is available to evaluate the true monomer concentrations z_{i1}, which can be solved with standard methods [17]. The compressibility factor is equal to the ratio between the number of associates and the number of monomers in the associates:

$$z = \frac{Pv}{RT} = \frac{\sum\limits_i \sum\limits_n z_{in} + \sum\limits_i \sum\limits_{j>1} z_{Mij}}{\sum\limits_i \sum\limits_n nz_{in} + 2\sum\limits_i \sum\limits_{j>1} z_{Mij}} \tag{13.45}$$

Using Eq. (13.41), it can be written as

$$z = \frac{Pv}{RT} = \frac{1}{\sum\limits_i \sum\limits_n nz_{in} + 2\sum\limits_i \sum\limits_{j>1} z_{Mij}} \tag{13.46}$$

Similarly, the residual part of the vapor enthalpy $(h - h^{id})$ can be derived. Using N as the total number of the particular species, it is possible to set up the ratios

$$\frac{N_{in}}{N} = \frac{nz_{in}}{\sum\limits_i \sum\limits_n nz_{in} + 2\sum\limits_i \sum\limits_{j>1} z_{Mij}} \tag{13.47}$$

and

$$\frac{N_{Mij}}{N} = \frac{2z_{Mij}}{\sum\limits_i \sum\limits_n nz_{in} + 2\sum\limits_i \sum\limits_{j>1} z_{Mij}} \tag{13.48}$$

With the help of the van't Hoff equation

$$\Delta h_R^0 = RT^2 \left(\frac{d\ln K}{dT}\right) \tag{13.49}$$

the enthalpies of reaction

$$\Delta h_{in} = -RB_{in} \tag{13.50}$$

and

$$\Delta h_{Mij} = -\frac{R}{2}(B_{i2} + B_{j2}) \tag{13.51}$$

are obtained.

For each molecule in an associate of the degree n ($n > 1$), the contribution to the enthalpy correction is $\Delta h_{in}/n$ or $\Delta h_{Mij}/2$, respectively. Setting $\Delta h_{i1} = 0$, the enthalpy correction is

$$(h - h^{id}) = \frac{\sum_i \sum_n N_{in} \Delta h_{in}/n + \sum_i \sum_{j>i} N_{Mij} \Delta h_{Mij}/2}{N} \tag{13.52}$$

Using Eqs. (13.47) and (13.48), $(h - h^{id})$ can be expressed as

$$(h - h^{id}) = \frac{\sum_i \sum_n z_{in} \Delta h_{in} + \sum_i \sum_{j>i} z_{Mij} \Delta h_{Mij}}{\sum_i \sum_n n z_{in} + 2\sum_i \sum_{j>i} z_{Mij}}$$

$$= z \left(\sum_i \sum_n z_{in} \Delta h_{in} + \sum_i \sum_{j>i} z_{Mij} \Delta h_{Mij} \right) \tag{13.53}$$

The specific heat capacity can again be calculated according to Eq. (13.37).

For the calculation of vapor–liquid equilibria, an analytical expression for the fugacity coefficient is required. It can be derived according to Prausnitz et al. [16].

With the material balances for each component

$$N_i = \sum_n n N_{in} + \sum_i \sum_{j \neq i} N_{Mij} \tag{13.54}$$

and the equilibrium conditions for the chemical potentials

$$\mu_{in} = n\mu_{i1} \tag{13.55}$$

$$\mu_{Mij} = \mu_{i1} + \mu_{j1} \tag{13.56}$$

the total differential of the Gibbs energy at constant temperature and pressure is set up in two ways. First, it is formulated without consideration of the true species:

$$dG = \sum_i \mu_i dN_i \tag{13.57}$$

In the second way, the existence of the particular true species is regarded:

$$dG = \sum_i \sum_n \mu_{in} dN_{in} + \sum_i \sum_{j>i} \mu_{Mij} dN_{Mij} \tag{13.58}$$

Substituting the chemical potentials from Eqs. (13.55) and (13.56) in Eq. (13.58), we obtain

$$dG = \sum_i \sum_n n\mu_{i1} dN_{in} + \sum_i \sum_{j>i} (\mu_{i1} + \mu_{j1}) dN_{Mij}$$

$$= \sum_i \sum_n n\mu_{i1} dN_{in} + \sum_i \sum_{j>i} \mu_{i1} dN_{Mij} + \sum_i \sum_{j>i} \mu_{j1} dN_{Mij}$$

$$= \sum_i \sum_n n\mu_{i1} dN_{in} + \sum_i \sum_{j>i} \mu_{i1} dN_{Mij} + \sum_i \sum_{j<i} \mu_{i1} dN_{Mij}$$

$$= \sum_i \sum_n n\mu_{i1} dN_{in} + \sum_i \sum_{j\neq i} \mu_{i1} dN_{Mij}$$

$$= \sum_i \mu_{i1} \left[\sum_n n\, dN_{in} + \sum_{j\neq i} dN_{Mij} \right] \tag{13.59}$$

Differentiating Eq. (13.54)

$$dN_i = \sum_n n\, dN_{in} + \sum_i \sum_{j\neq i} dN_{Mij} \tag{13.60}$$

and inserting it into Eq. (13.59), the result is

$$dG = \sum_i \mu_{i1} dN_i \tag{13.61}$$

The comparison between Eqs. (13.61) and (13.57) yields

$$\mu_i = \mu_{i1} \tag{13.62}$$

From the stoichiometric point of view, μ_i is the chemical potential of a nonideal gas. Therefore, it can be assigned with a fugacity coefficient

$$\mu_i = g_i^{\text{pure}}(T, P_0) + RT \ln \frac{y_i P \varphi_i}{f_i^0} \tag{13.63}$$

From the true component point of view, the evaluation of the chemical potential can be performed by treating the mixture as an ideal gas:

$$\mu_{i1} = g_{i1}^{\text{pure}}(T, P_0) + RT \ln \frac{z_{i1} P}{f_{i1}^0} \tag{13.64}$$

as the particular species are treated as ideal gases. As for $g^{\text{pure}}(T, P^0)$ and f^0 the same conditions are referred to, the fugacity coefficient is simply obtained as

$$\varphi_i = \frac{z_{i1}}{y_i} \tag{13.65}$$

The association model has been successfully applied in many cases for mixtures containing organic acids or HF. The most important system where it comes into operation is acetic acid/water. Figure 13.6 shows the phase diagram and the representation of the NRTL/association model. For comparison, the fit with NRTL and ideal gas phase is shown as well, demonstrating the considerable quality

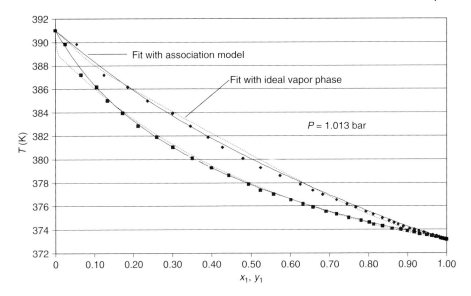

Figure 13.6 Phase diagram for the water (1)/acetic acid (2) system. (Data from Hui et al. [18].)

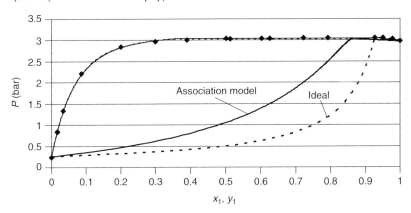

Figure 13.7 Phase diagram for the R22 (1)/HF (2) system. (Data from Wilson et al. [19].)

difference. The same comparison is made for the R22/HF system in Figure 13.7. In this case, only PTx data are available. The fit using the ideal gas model has the same quality as the fit with the association model. However, the vapor concentrations that have not been measured are significantly different.

Nevertheless, a model is missing that combines the association model with an equation of state that takes into account the intermolecular forces at high pressures. There are several applications where such a model would be useful [20].

A number of attempts have been made [11, 21–33]; however, none of them fulfills the particular demands of a process simulator, that is, extension to multicomponent mixtures, proved enthalpy description, and derivation of a fugacity coefficient.

Example 13.2

Calculate the specific vapor volume of a mixture of 40 mol% formic acid (1) and 60 mol% nitrogen (2) at $T = 400$ K and $P = 1$ bar.

Solution

According to Table 13.6, the dimerization constant can be calculated to be

$$K_2 = \exp\left(-29.607 + \frac{7108}{400}\right) = 7.23196 \cdot 10^{-6}$$

Eqs. (13.43) and (13.44) yield

$$z_{11} + K_2 z_{11}^2 (P/P_0) + z_{21} = 1$$

$$\frac{z_{11} + 2K_2 z_{11}^2 (P/P_0)}{z_{21}} = \frac{y_1}{y_2}$$

$$\rightarrow z_{11} + 0.7232 z_{11}^2 + z_{21} = 1$$

$$z_{11} + 1.4464 z_{11}^2 - \frac{2}{3} z_{21} = 0$$

$$\rightarrow 2.5 z_{11} + 2.8928 z_{11}^2 = 1$$

$$\rightarrow z_{11}^2 + 0.8642 z_{11} - 0.3457 = 0$$

$$\rightarrow z_{11} = -0.4321 + \sqrt{0.1867 + 0.3457} = 0.2976$$

From Eq. (13.24), we get

$$z_{12} = K_2 z_{11}^2 (P/P_0) = 0.064$$

and

$$z_{21} = 1 - z_{11} - z_{12} = 0.6384$$

Using Eq. (13.46), we obtain

$$z = \frac{Pv}{RT} = \frac{1}{z_{11} + 2z_{12} + z_{21}} = 0.9398$$

and

$$v = \frac{zRT}{P} = 31.257 \; \frac{m^3}{kmol}$$

Example 13.3

Calculate the fugacity coefficients φ_i^V and φ_i^S for the system water (1)–acetic acid (2) at $\vartheta = 80\,°C$ with the help of the association model ($y_1 = 0.35$, $P = 35.57$ kPa). The required association parameters can be taken from Table 13.6.

Solution

To be able to calculate the fugacity coefficient φ_i^V, first the equilibrium constants K_2 and K_4 have to be determined at $\vartheta = 80\,°C$. Then, the vapor phase mole fraction of the monomeric acetic acid and for water has to be calculated:

$$\ln K_{22} = -29.129 + \frac{7335\text{ K}}{353.15\text{ K}} = -8.3588 \Rightarrow K_{22} = 2.3433 \cdot 10^{-4}$$

$$\ln K_{24} = -78.232 + \frac{17268\text{ K}}{353.15\text{ K}} = -29.3349 \Rightarrow K_{24} = 1.8197 \cdot 10^{-13}$$

Using Eqs. (13.24), (13.41) and (13.42), we get

$$K_{22} = \frac{z_{22}}{z_{21}^2 (P/P_0)^1}$$

$$K_{24} = \frac{z_{24}}{z_{21}^4 (P/P_0)^3}$$

$$z_{11} + z_{21} + K_{22}z_{21}^2(P/P_0) + K_{24}z_{21}^4(P/P_0)^3 = 1 \quad (13.66)$$

and

$$\frac{z_{21} + 2K_{22}z_{21}^2(P/P_0) + 4K_{24}z_{21}^4(P/P_0)^3}{z_{11}} = \frac{y_2}{y_1} \quad (13.67)$$

With $P = 35.57$ kPa, $P_0 = 1$ Pa, and the values for K_{22} and K_{24} obtained above, we get two equations for z_{11} and z_{21}:

$$z_{11} + z_{21} + 8.3351 z_{21}^2 + 8.18939 z_{21}^4 = 1$$

$$z_{21} + 16.6702 z_{21}^2 + 32.7576 z_{21}^4 = 1.8571 z_{11}$$

Eliminating z_{11} yields

$$z_{21} + 16.6702 z_{21}^2 + 32.7576 z_{21}^4$$
$$= 1.8571 - 1.8571 z_{21} - 15.4791 z_{21}^2 - 15.2085 z_{21}^4$$

and

$$47.9661 z_{21}^4 + 32.1493 z_{21}^2 + 2.8571 z_{21} - 1.8571 = 0$$

This yields $z_{21} = 0.1955$ and, subsequently, $z_{11} = 0.4740$.

Using the true mole fractions, the fugacity coefficients can be obtained:

$$\varphi_1^V = \frac{z_{11}}{y_1} = \frac{0.4740}{0.35} = 1.354$$

$$\varphi_2^V = \frac{z_{21}}{y_2} = \frac{0.1955}{0.65} = 0.301$$

For pure acetic acid, Eq. (13.43) reads

$$z_{21} + K_{22}z_{21}^2(P_2^s/P_0) + K_{24}z_{21}^4(P_2^s/P_0)^3 = 1$$

With $P_2^s = 27.590$ kPa at $\vartheta = 80\,°C$, we obtain

$$z_{21} + 6.46516 z_{21}^2 + 3.82169 z_{21}^4 - 1 = 0$$

Table 13.7 VLE data for the system water (1)–acetic acid (2) at 80 °C [34, 35].

x_1	y_1	P (kPa)
0.0000	0.0000	27.918
0.0776	0.1480	31.731
0.1185	0.2040	32.811
0.1778	0.2810	34.157
0.2392	0.3500	35.570
0.3511	0.4590	37.864
0.4724	0.5800	40.343
0.6080	0.7100	42.290
0.6915	0.7820	43.596
0.7648	0.8400	44.556
0.8908	0.9240	45.916
1.0000	1.0000	47.343

giving

$$z_{21} = \varphi_2^s = 0.3161$$

Example 13.4

Calculate the activity coefficients for the system water (1)–acetic acid (2) at $\vartheta = 80\,°C$ for the compositions given in Table 13.7, using Eq. 5.14. The Poynting factor shall be neglected.

Solution

This should be shown for the data point at $x_1 = 0.2392$, since for this data point the values for the fugacity coefficients have already been calculated in Example 13.3. However, it makes sense to use the experimental vapor pressure of the authors to obtain a more reliable representation (see Appendix F), which causes a small change to $\varphi_2^s = 0.3144$. φ_1^s can be set to 1, as pure water does not show strong association in the vapor phase. The experimental activity coefficients can directly be calculated using Eq. (5.14):

$$\gamma_1 = \frac{y_1 \varphi_1^V P}{x_1 \varphi_1^s P_1^s} = \frac{0.35 \cdot 1.354 \cdot 35.57\,\text{kPa}}{0.2392 \cdot 1 \cdot 47.343\,\text{kPa}} = 1.4885$$

$$\gamma_2 = \frac{y_2 \varphi_2^V P}{x_2 \varphi_2^s P_2^s} = \frac{0.65 \cdot 0.301 \cdot 35.57\,\text{kPa}}{0.7608 \cdot 0.3144 \cdot 27.918\,\text{kPa}} = 1.042$$

Assuming ideal vapor phase behavior, different activity coefficients are calculated:

$$\gamma_1(\text{id}) = \frac{y_1 P}{x_1 P_1^s} = \frac{0.35 \cdot 35.57\,\text{kPa}}{0.2392 \cdot 47.343\,\text{kPa}} = 1.0993$$

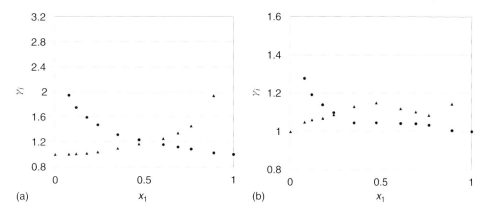

Figure 13.8 Activity coefficients calculated for the system water (1)–acetic acid (2) at $\vartheta = 80\,°C$ taking into account (a) real and (b) ideal vapor phase behavior.

$$\gamma_2(\text{id}) = \frac{y_2 P}{x_2 P_2^s} = \frac{0.65 \cdot 35.57\,\text{kPa}}{0.7608 \cdot 27.918\,\text{kPa}} = 1.0885$$

For all data points, the activity coefficients assuming ideal and real vapor phase behavior are shown in Figure 13.8. It can be recognized that the activity coefficients taking into account nonideal vapor phase behavior show the typical curvature, whereas the ideal calculation results in a very strange curve shape. This is absolutely important for the modeling of the VLE behavior of this system; the dewatering of acetic acid via distillation is quite a difficult task and requires a considerable amount of energy.

Additional Problems

P13.1 Calculate the compressibility factor of formic acid vapor at 420 K and 0.5 bar using the chemical theory. The required dimerization and tetramerization constants can be calculated from the parameters given in Table 13.6.

P13.2 Estimate the heat capacity of acetic acid along the vapor-liquid coexistence curve for both phases using the vapor pressure and ideal gas heat capacity correlations given in Appendix A and the dimerization and tetramerization constants from Table 13.6.

P13.3 Estimate the entropy and enthalpy change upon isothermal compression of propionic acid from 0.01 to 1 atm at a temperature of 145 °C using the chemical theory with dimerization constants given in Table 13.6.

P13.4 Estimate the change of the degree of dimerization along the vapor pressure curve for formic acid and propionic acid using the dimerization constants and a simplified vapor pressure equation ($\log(P^s/1\,\text{bar}) = A + B/T$). Calculate the values of the parameters A and B from the boiling temperatures at 0.5 bar and 2 bar.

Boiling Points:

	0.5 bar	2 bar
Formic Acid	78.4 °C	125 °C
Propionic Acid	119.7 °C	164.7 °C

Dimerization constants can be calculated from the parameters given in Table 13.6.

P13.5 Calculate the saturated vapor pressure and fugacity along the vapor-liquid coexistence curve for benzene, water and acetic acid between 25 °C and the critical temperature of the components. For the real vapor phase behavior, use the virial equation truncated after the second virial coefficient in case of water and benzene and the chemical theory in case of acetic acid. Discuss the results. Inside which temperature range are the results reliable? Do the calculations lead to under- or overprediction of the fugacity outside the reliable temperature range?

Vapor pressure equation coefficients and second virial coefficient correlations as function of temperature are given in Appendix A. Dimerization and tetramerization constant parameters are given in Table 13.6.

P13.6 In case of mixtures of components which do not strongly or differently associate in the vapor phase, the separation factor α_{ij} can be approximated by $\alpha_{ij} = P_i^s/P_j^s \cdot \gamma_i/\gamma_j$. In other cases the ratio of the real gas factors has to be taken into account. Calculate the ratio of the pure component vapor pressures, the activity coefficients (calculated using the UNIQUAC model) and the real gas factors for the system water (1)–acetic acid (2) at 80 °C as function of concentration. Discuss the results.

Vapor pressure equation constants can be found in Appendix A. Calculate the dimerization and tetramerization constant of acetic acid using the parameters given in Table 13.6.

UNIQUAC parameters: $\Delta u_{12} = 12.0164$ cal/mol $\quad \Delta u_{21} = 68.3212$ cal/mol
$r_1 = 0.9200 \quad\quad\quad q_1 = 1.4000$
$r_2 = 2.2024 \quad\quad\quad q_2 = 2.0720$

P13.7 Search for experimental heats of vaporization data for acetic acid in the free DDBSP Explorer Version. Interpret the behavior as function of temperature. Compare the values to those of a non-associating component of similar molecular weight (acetone).

P13.8 In the free DDBSP Explorer Version, search for all data for the system water–acetic acid and regress these data simultaneously using the Three-Suffix Margules equation and binary parameters with quadratic temperature dependence. The Three-Suffix Margules g^E-model is very seldom used today but is the only simultaneous regression model available in the free DDBSP Explorer Version. In the x–y diagram, the equilibrium curves are nearly parallel to the diagonal line at mole fractions of water greater than 0.7. What is the reason for this strange behavior?

P13.9 Calculate the fugacity coefficients of both components in the system acetic acid (1)–water (2) at T = 393.15 K and P = 1 bar for a mole fraction $y_1 = 0.5$. Use the association model and the corresponding constants from Table 13.6.

References

1. Walker, J.F. (1964) *Formaldehyde*, Reinhold Publishing Corporation, New York.
2. Maurer, G. (1986) *AIChE J.*, **32** (6), 932–948.
3. Albert, M., Garcia, B.C., Kuhnert, C., Peschla, R., and Maurer, G. (2000) *AIChE J.*, **46** (8), 1676–1687.
4. Hasse, H. (1990) Dampf-Flüssigkeits-Gleichgewichte, Enthalpien und Reaktionskinetik in formaldehydhaltigen Mischungen. Thesis. University Kaiserslautern, Germany.
5. Albert, M. (1999) Thermodynamische Eigenschaften formaldehydhaltiger Mischungen. Thesis. University Kaiserslautern, Germany.
6. Kuhnert, C. (2004) Dampf-Flüssigkeits-Gleichgewichte in mehrkomponentigen formaldehydhaltigen Systemen. Thesis. University Kaiserslautern, Germany.
7. Kogan, L.V. and Ogorodnikov, S.K. (1980) *Zh. Prikl. Khim.*, **53** (1), 115–119.
8. Schilling, K., Sohn, M., Ströfer, E., and Hasse, H. (2003) *Chem. Ing. Tech.*, **75** (3), 240–244.
9. Ritter, H.L. and Simons, J.H. (1945) *J. Am. Chem. Soc.*, **67**, 757–762.
10. Büttner, R. and Maurer, G. (1983) *Ber. Bunsen-Ges. Phys. Chem.*, **87**, 877–882.
11. Anderko, A. and Prausnitz, J.M. (1994) *Fluid Phase Equilib.*, **95**, 59–71.
12. Franck, E.U. and Meyer, F. (1959) *Z. Elektrochem.*, **63** (5), 571–582.
13. Kleiber, M. (2003) *Ind. Eng. Chem. Res.*, **42**, 2007–2014.
14. Landee, F.A. and Johns, I.B. (1941) *J. Am. Chem. Soc.*, **63**, 2891–2895.
15. Barton, J.R. and Hsu, C.C. (1969) *J. Chem. Eng. Data*, **14** (2), 184–187.
16. Prausnitz, J.M., Lichtenthaler, R.N., and Azevedo, E.G. (1986) *Molecular Thermodynamics of Fluid Phase Equilibria*, Prentice-Hall.
17. Engeln-Müllges, G. and Reutter, F. (1988) *Formelsammlung zur Numerischen Mathematik mit Standard-FORTRAN 77 Programmen*, 6th edn, BI-Wissenschaftsverlag, Mannheim, Wien, Zürich.
18. Hui, F. et al. (1986) *J. Chem. Eng. (China)*, **6**, 56.
19. Wilson, L.C., Wilding, W.V., and Wilson, G.M. (1989) *AIChE Symp. Ser.*, **85** (271), 51.
20. Chem Systems Inc. (1997) Vinyl Acetate. Chem Systems Rep. 96/97-5.
21. Anderko, A. (1989) *Fluid Phase Equilib.*, **45**, 39–67.
22. Anderko, A. (1989) *Chem. Eng. Sci.*, **44** (3), 713–725.
23. Anderko, A. (1992) *Fluid Phase Equilib.*, **75**, 89–103.
24. Derawi, S.O., Zeuthen, J., Michelsen, M.L., Stenby, E.H., and Kontogeorgis, G.M. (2004) *Fluid Phase Equilib.*, **225**, 107–113.
25. Kontogeorgis, G.M., Voutsas, E.C., Yakoumis, I.V., and Tassios, D.P. (1996) *Ind. Eng. Chem. Res.*, **35**, 4310–4318.
26. Visco, D.P., Kofke, D.A., and Singh, R.R. (1997) *AIChE J.*, **43** (9), 2381–2384.
27. Chai Kao, C.-P., Paulaitis, M.E., Sweany, G.A., and Yokozeki, M. (1995) *Fluid Phase Equilib.*, **108**, 27–46.

28. Huang, S.H. and Radosz, M. (1990) *Ind. Eng. Chem. Res.*, **29**, 2284–2294.
29. Chapman, W.G., Gubbins, K.E., Jackson, G., and Radosz, M. (1990) *Ind. Eng. Chem. Res.*, **29**, 1709–1721.
30. Fu, Y.-H. and Sandler, S.I. (1995) *Ind. Eng. Chem. Res.*, **34**, 1897–1909.
31. Gmehling, J., Liu, D.D., and Prausnitz, J.M. (1979) *Chem. Eng. Sci.*, **34**, 951–958.
32. Grenzheuser, P. and Gmehling, J. (1986) *Fluid Phase Equilib.*, **25**, 1–29.
33. Pang, J. and Peng, D.-Y. (2006) *Fluid Phase Equlib.*, **241**, 31–40.
34. Dortmund Data Bank *www.ddbst.com*. January 2011.
35. Lazeeva, M.S. and Markuzin, N.P. (1973) *J. Appl. Chem. USSR*, **46** (2), 373–376.

14
Practical Applications

In process simulation, there are a few basic unit operations that are simple to set up but require profound knowledge of thermodynamics to exactly match the intended application in the plant. In the following chapter, some of them are discussed.

14.1
Flash

The flash is the most widely used unit operation block in process simulation. A flash means that a stream is transformed to a certain equilibrium state defined by two of the variables P, T, enthalpy h, or vapor fraction q. There are several kinds of equipment which can be represented by a flash calculation.

- **Vapor–liquid separator**: The stream is led into a drum where vapor and liquid separate. The pressure P and the heat duty (adiabatic operation, $\dot{Q}=0$) are given; thus, it is a P–h flash.
- **Heat exchanger**: The outlet of a heat exchanger is also defined by a flash specification. Usually, the pressure drop and thus the outlet pressure is given. As the second coordinate, the outlet temperature or the vapor fraction (total condenser: $q=0$; total evaporator: $q=1$) are given. If the heat exchanger is part of a network, the heat duty can often be specified.
- **Throttle valve**: A throttle valve reduces the pressure of a gas at adiabatic conditions ($\dot{Q}=0$); thus, it is a P–h flash. The resulting temperature is determined by the Joule–Thomson coefficient.
- **Mixer**: In a mixer block, two or more streams are merged. The outlet pressure is the lowest pressure of the inlet streams. Normally, the mixing is also considered to be adiabatic; thus, the total enthalpy remains constant ($\dot{Q}=0$).

The standard state-of-the-art steady-state process simulation is based on stream flows, which do not change with time. Therefore, flash calculations with other variables like constant specific volume or entropy are not provided. For the design of vessels and safety valves, the so-called isochoric flash is important. It uses the specific volume as the second fixed variable and determines the pressure which

Chemical Thermodynamics: for Process Simulation, First Edition.
Jürgen Gmehling, Bärbel Kolbe, Michael Kleiber, and Jürgen Rarey.
© 2012 Wiley-VCH Verlag GmbH & Co. KGaA. Published 2012 by Wiley-VCH Verlag GmbH & Co. KGaA.

builds up in a closed vessel and the conditions at that state. For the energy balance, it must be taken into account that in a closed vessel the internal energy is decisive instead of the enthalpy:

$$Q_{12} = U_2 - U_1 = H_2 - H_1 - (P_2 - P_1) V \tag{14.1}$$

Example 14.1

An evacuated vessel ($V = 1$ m^3) is filled with 900 kg water and 1 kg nitrogen. The temperature is $T_1 = 300$ K. What pressure is built up? What is the necessary heat duty to increase the temperature to $T_2 = 400$ K? What pressure is built up then? The Henry coefficient of nitrogen in water in the temperature range of interest can be evaluated by

$$\ln\left(H_{N_2-H_2O}/\text{Pa}\right) = 176.507 - \frac{8432.77}{T/K} - 21.558 \ln(T/K) - 0.00844 \frac{T}{K}$$

Solution

Using Henry's law and the Peng–Robinson equation of state for the vapor phase in a commercial simulator, we first estimate the pressure and then check whether it was correct by adding up the volumes of liquid and vapor and comparing it with the vessel volume. Making first guesses and applying the Regula Falsi, we get the following results for $T_1 = 300$ K (Table 14.1):

The commercial simulator reports a total enthalpy of $H_1 = -14265.5$ MJ. At $T_2 = 400$ K, we obtain the results shown in Table 14.2. In this case, the commercial simulator reports a total enthalpy of $H_2 = -13888$ MJ. Thus, the required heat duty can be calculated to be

$$Q_{12} = U_2 - U_1 = H_2 - H_1 - (P_2 - P_1) V$$
$$= -13888 \text{ MJ} + 14265.5 \text{ MJ} - (24.839 - 8.0897) \text{ bar} \cdot 1 \text{ m}^3$$
$$= 377.5 \text{ MJ} - 1.675 \text{ MJ} = 375.8 \text{ MJ}$$

In commercial simulators, this iterative procedure can be accomplished automatically by using a specification/variation paragraph.

The PV term is usually small compared with the enthalpy changes. The unexpected high pressure obtained in the calculation above must be interpreted.

Table 14.1 Iterative solution for the pressure in the vessel at $T_1 = 300$ K.

P (bar)	V^L (m^3)	V^V (m^3)	($V^L + V^V$) (m^3)
1	0.9086	0.9032	1.8118
10	0.9036	0.0751	0.9787
8	0.9035	0.0975	1.0010
8.0897	0.9035	0.0962	0.9997

Table 14.2 Iterative solution for the pressure in the vessel at $T_2 = 400$ K.

P (bar)	V^L (m³)	V^V (m³)	$(V^L + V^V)$ (m³)
10	0.9604	0.1431	1.1035
15	0.9607	0.0804	1.0410
30	0.9613	0.0288	0.9901
27.083	0.9612	0.0339	0.9951
24.106	0.9611	0.0405	1.0016
24.839	0.9611	0.0387	0.9999

The vapor pressure of water is built up to approx. 2.5 bar, which does not explain the result. In this case, the expansion of the liquid with temperature is important. At $T_2 = 400$ K, the liquid density of water changes from 996.5 to 937.2 kg/m³. Referring to 900 kg water, its volume increases from 0.903 to 0.961 m³. This means that approx. 60% of the space for the vapor phase left at $T_1 = 300$ K is consumed by the additional volume required by the liquid, giving a corresponding pressure increase, which is, on the other hand, mitigated by the increased solubility of nitrogen in water due to the elevated pressure.

14.2
Joule–Thomson Effect

When a gas flows through a valve or an orifice in a pipe, there will be a pressure loss (Figure 14.1). Considering that no heat exchange takes place and that differences in velocity and height at the end of the control volume are negligible for the energy balance, the First Law can be formulated as

$$h_2 = h_1 \qquad (14.2)$$

Nevertheless, a change in temperature can occur. Behind the orifice, vortices are formed, and two effects compete: the flow involving friction of the vortices causes a temperature rise, and the pressure loss causes a lowering of the temperature, as energy is necessary to overcome the attractive forces between the molecules.[1]

The Joule–Thomson coefficient $(\partial T/\partial P)_h$ determines which of these two effects dominates. It can be evaluated as follows.

The total differential of the specific enthalpy is

$$dh = \left(\frac{\partial h}{\partial T}\right)_P dT + \left(\frac{\partial h}{\partial P}\right)_T dP \stackrel{!}{=} 0 \qquad (14.3)$$

[1] It must be emphasized that Eq. (14.2) refers to the cross-sections 1 and 2 as shown in Figure 14.1, which are far away from the orifice. It is assumed that the velocities in cross-sections 1 and 2 do not significantly change, even though an expansion has taken place and the cross-section areas might have changed. This assumption is usually valid. In the orifice itself (index *), the velocity is increased according to the continuity equation, and the First Law reads $h_1 + \frac{w_1^2}{2} = h^* + \frac{w^{*2}}{2}$.

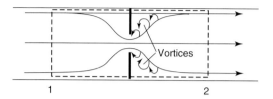

Figure 14.1 Illustration of the flow in an orifice of a pipe.

With $(\partial h/\partial T)_P = c_P$, one gets

$$\left(\frac{\partial T}{\partial P}\right)_h = -\frac{(\partial h/\partial P)_T}{c_P} \tag{14.4}$$

In Appendix C, the following relation is derived:

$$\left(\frac{\partial h}{\partial P}\right)_T = v - T\left(\frac{\partial v}{\partial T}\right)_P \tag{14.5}$$

Combining Eqs. (14.4) and (14.5), the Joule–Thomson coefficient becomes

$$\left(\frac{\partial T}{\partial P}\right)_h = \frac{T(\partial v/\partial T)_P - v}{c_P} \tag{14.6}$$

For ideal gases, the Joule–Thomson coefficient is

$$\left(\frac{\partial T}{\partial P}\right)_h = \frac{T(R/P) - v}{c_P} = 0 \tag{14.7}$$

and there is no temperature change by the Joule–Thomson effect.

The sign of the Joule–Thomson coefficient in Eq. (14.6) depends on the temperature. At the inversion temperature, $(\partial T/\partial P)_h = 0$. Below the inversion temperature, the temperature of the gas is lowered by throttling (Joule–Thomson effect). For air, the inversion temperature is $T_{\text{inv}} = 760$ K. For hydrogen ($T_{\text{inv}} \approx 200$ K) and helium ($T_{\text{inv}} \approx 40$ K), it is considerably lower. The Joule–Thomson effect is used for liquefaction of these three gases with the Linde process; because of the low inversion temperatures of hydrogen and helium, a precooling with liquefied air or, liquefied hydrogen, respectively, is necessary. For process simulation, the Joule–Thomson coefficient plays a minor role, as the extent of the Joule–Thomson effect can be more easily calculated as a simple change of conditions with an appropriate equation of state.

Example 14.2

Nitrogen ($T_1 = 300$ K, $P_1 = 200$ bar) is expanded to $P_2 = 1$ bar by a throttle valve. Calculate the outlet temperature using

1) a high-precision equation of state
2) the Peng–Robinson equation of state

For c_P^{id}, the equation given in Appendix A can be used.

Solution

(1) Using Eq. (2.113) and the coefficients listed in Appendix B, the specific enthalpy at $T_1 = 300$ K and $P_1 = 200$ bar is determined to be $h_1 = 279.11$ J/g. The outlet temperature T_2 at $h_2 = h_1$ and $P_2 = 1$ bar is calculated to be $T_2 = 269.19$ K.

(2) Using the Peng–Robinson equation (2.166) and the corresponding expression for the enthalpy departure of a pure fluid [1], we get

$$h_1 = \Delta h_f^0 + \int_{298.15 \text{ K}}^{300 \text{ K}} c_P^{id} dT + \left(h - h^{id}\right)(300 \text{ K}, 200 \text{ bar})$$

$$= 0 + 1.92345 \, \frac{\text{J}}{\text{g}} - 17.816 \, \frac{\text{J}}{\text{g}} = -15.893 \, \frac{\text{J}}{\text{g}}$$

To obtain a good first guess for the iteration procedure, the enthalpy departure at $P_2 = 1$ bar is neglected in the first step. With $c_P^{id} \approx 1$ J/(g K), we get

$$h_2 = -15.893 \, \frac{\text{J}}{\text{g}} = c_P^{id} \left(T_2 - 298.15 \text{ K}\right) \Rightarrow T_2 = 282.257 \text{ K}$$

For a check, we insert the guess for T_2^* into the enthalpy equation:

$$h_2^* = \Delta h_f^0 + \int_{298.15 \text{ K}}^{282.257 \text{ K}} c_P^{id} dT + (h - h^{id})(282.257 \text{ K}, 1 \text{ bar})$$

$$= 0 - 16.5219 \, \frac{\text{J}}{\text{g}} - 0.195 \, \frac{\text{J}}{\text{g}} = -16.717 \, \frac{\text{J}}{\text{g}}$$

The next guess is $T_2^* = 283.5$ K:

$$h_2^* = \Delta h_f^0 + \int_{298.15 \text{ K}}^{283.5 \text{ K}} c_P^{id} dT + (h - h^{id})(283.5 \text{ K}, 1 \text{ bar})$$

$$= 0 - 15.2298 \, \frac{\text{J}}{\text{g}} - 0.192 \, \frac{\text{J}}{\text{g}} = -15.4218 \, \frac{\text{J}}{\text{g}}$$

Applying the Regula Falsi, we get the solution $T_2 = 283.05$ K:

$$h_2^* = \Delta h_f^0 + \int_{298.15 \text{ K}}^{283.05 \text{ K}} c_P^{id} dT + (h - h^{id})(283.05 \text{ K}, 1 \text{ bar})$$

$$= 0 - 15.6976 \, \frac{\text{J}}{\text{g}} - 0.193 \, \frac{\text{J}}{\text{g}} = -15.891 \, \frac{\text{J}}{\text{g}}.$$

In this case, the results from the high-precision equation of state and Peng–Robinson differ by around 14 K. It can often be seen that caloric properties calculated with the Peng–Robinson equation are not very exact, and the equation has not been designed for very high pressures. Therefore, its poor result can be understood.

14.3 Adiabatic Compression and Expansion

Compressors, pumps, fans, and other fluid flow engines for pressure elevation are widely applied in industry for the transport of fluids or for establishing a certain pressure to carry out a reaction or a separation. In process simulation, compressors for gases cannot be regarded as a simple flash, as they cannot be specified by two outlet variables. Instead, more information about the course of the change of state is necessary. For most types of compressors, it can be assumed that they are adiabatic, that means the heat exchange with the environment does not play a major role.

In the following section, the calculation route is illustrated by the adiabatic compression of a vapor. The changes in kinetic energy can be neglected in the energy balance. The calculation is divided into the reversible adiabatic calculation and the integration of losses.

1) **Reversible calculation.** The reversible case characterizes the process that requires the lowest power consumption. It is specified by the outlet pressure P_2 at constant entropy. According to the Second Law, the outlet temperature is calculated by the isentropic condition

$$s_2(T_{2\text{rev}}, P_2) = s_1(T_1, P_1) \tag{14.8}$$

In case of an ideal gas, Eq. (14.8) leads to

$$s_0(T_0, P_0) + \int_{T_0}^{T_{2\text{rev}}} c_p^{id} \frac{dT}{T} - R \ln \frac{P_2}{P_0}$$

$$= s_0(T_0, P_0) + \int_{T_0}^{T_1} c_p^{id} \frac{dT}{T} - R \ln \frac{P_1}{P_0} \tag{14.9}$$

with s_0 as the reference enthalpy, giving

$$\int_{T_1}^{T_{2\text{rev}}} c_p^{id} \frac{dT}{T} = R \ln \frac{P_2}{P_1} \tag{14.10}$$

For manual approximations, a temperature-independent c_p^{id} can be assumed, which results in

$$c_p^{id} \ln \frac{T_{2\text{rev}}}{T_1} = R \ln \frac{P_2}{P_1} \tag{14.11}$$

With the abbreviation $\kappa = c_p^{id}/c_v^{id}$, we get

$$\frac{R}{c_p^{id}} = \frac{c_p^{id} - c_v^{id}}{c_p^{id}} = 1 - \frac{1}{\kappa} = \frac{\kappa - 1}{\kappa} \tag{14.12}$$

and we obtain the well-known formula

$$\frac{T_{2\text{rev}}}{T_1} = \left(\frac{P_2}{P_1}\right)^{\frac{\kappa-1}{\kappa}} \tag{14.13}$$

In process simulation calculations, Eq. (14.8) is directly evaluated with the corresponding equation of state to determine $T_{2\text{rev}}$.

The specific technical work for the reversible case is given by

$$W_{t12\text{rev}} = h_{2\text{rev}}(T_{2\text{rev}}, P_2) - h_1(T_1, P_1) \tag{14.14}$$

2) **Integration of losses.** The actual specific technical work required is calculated with the isentropic and mechanical efficiency:

$$W_{t12} = \frac{W_{t12\text{rev}}}{\eta_{\text{th}}\eta_{\text{mech}}} \tag{14.15}$$

The isentropic efficiency η_{th} is an empirical factor which summarizes all the effects about the irreversibility of the process. $\eta_{\text{th}} = 0.8$ is often a reasonable approximation. It decreases with increasing pressure ratio. η_{mech} is the efficiency of the energy transformation of the compressor engine, which is not related to the process flow. For large drives, $\eta_{\text{mech}} = 0.95$ is often a reasonable approach.

The outlet conditions of the flow are then calculated backward via

$$h_2(T_2, P_2) = h_1(T_1, P_1) + W_{t12\text{rev}}/\eta_{\text{th}} \tag{14.16}$$

Knowing h_2, the outlet temperature $T_2 = f(h_2, P_2)$ can be calculated by an iterative procedure.

Example 14.3

In a heat pump, the refrigerant R22 ($CHClF_2$) is fed at $\dot{m} = 100$ kg/h into the compressor at $T_1 = 300$ K and $P_1 = 1$ bar. It is adiabatically compressed to $P_2 = 5$ bar. The isentropic efficiency is $\eta_{\text{th}} = 0.7$ and $\eta_{\text{mech}} = 0.9$. What power is required and what is the outlet temperature?

Solution

1) **Reversible calculation**

   Using the Peng–Robinson equation of state, we calculate the specific entropy of stream 1 to be $s_1 = 1.9876$ J/(g K) and $h_1 = 430.451$ J/g. In the reversible case, it is $s_{2\text{rev}} = s_1$. Thus, the state variables required are $P_2 = 5$ bar and $s_{2\text{rev}} = 1.9876$ J/(g K). The simulator evaluates $T_{2\text{rev}} = 377.37$ K and $h_{2\text{rev}} = 481.649$ J/g. The power required is

$$P_{12\text{rev}} = \dot{m}\, w_{t12} = \dot{m}(h_{2\text{rev}} - h_1) = 100 \text{ kg/h} \cdot (481.649 - 430.451) \text{ J/g}$$
$$= 5120 \text{ kJ/h} = 1.422 \text{ kW}$$

2) **Integration of losses**

   The actual power can be evaluated with the efficiency

$$P_{12} = P_{12\text{rev}}/\eta_{\text{th}}\eta_{\text{mech}} = 1.422 \text{ kW}/(0.7 \cdot 0.9) = 2.257 \text{ kW}$$

   This corresponds to a specific enthalpy

$$h_2 = h_1 + (P_{12\text{rev}}/\eta_{\text{th}})/\dot{m} = 430.451 \text{ J/g} + \frac{1.422 \text{ kW}}{0.7 \cdot 100 \text{ kg/h}} = 503.582 \text{ J/g}$$

   For the fixed variables $h_2 = 503.582$ J/g and $P_2 = 5$ bar, the outlet temperature can now be determined to be $T_2 = 405.68$ K.

Example 14.4

A refrigeration unit is used to cool down a brine to $\vartheta_{br} = 15\,°C$. The removed heat can be released to cooling water, which has to be returned with $\vartheta_{cw} = 45\,°C$. For the heat transfer, a temperature difference of $\Delta T = 5\,K$ shall be provided.

For a first guess, the compressor of the refrigeration unit is supposed to operate in a reversible and adiabatic, this means isentropic way. Figure 14.2 shows a scheme of the arrangement.

Make a first guess whether propane or n-butane is more appropriate as refrigerant.

Solution

The vapor pressures of the two substances at $\vartheta_1 = 10\,°C$ and $\vartheta_2 = 50\,°C$ are as follows:

$$P^s_{propane}(\vartheta_1 = 10\,°C) = 6.367\,bar \quad P^s_{butane}(\vartheta_1 = 10\,°C) = 1.485\,bar$$
$$P^s_{propane}(\vartheta_2 = 50\,°C) = 17.137\,bar \quad P^s_{butane}(\vartheta_2 = 50\,°C) = 4.958\,bar$$

In both cases, the pressure ratios are well below the capability of a one-stage compressor, which is approximately $(P_2/P_1)_{max} = 6\text{–}8$. As well, the magnitude of the vapor pressures does not cause a mechanical strength problem in both cases. However, concerning the operation of the compressor, there is a decisive difference.

In both cases, the compressor is fed with saturated vapor at $\vartheta_1 = 10\,°C$. Using a high-precision equation of state (Eq. (2.113)), we can calculate the state of the compressor outlet stream for the vapor pressure at 50 °C.

For propane,

$$s_2 = s_{2rev} = s_1 = s^V(\vartheta_1 = 10\,°C) = -0.4876\,\frac{J}{g\,K}$$

$$\vartheta_2 = \vartheta\left(P_2 = 17.137\,bar, s_2 = -0.4876\,\frac{J}{g\,K}\right) = 54.03\,°C$$

The compressor outlet is a superheated vapor.

For n-butane,

$$s_2 = s_{2rev} = s_1 = s^V(\vartheta_1 = 10\,°C) = -0.1590\,\frac{J}{g\,K}$$

As

$$s^L(P_2 = 4.958\,bar) = -1.1623\,\frac{J}{g\,K}$$

and

$$s^V(P_2 = 4.958\,bar) = -0.1293\,\frac{J}{g\,K}$$

meaning $s^L(P_2) < s_2 < s^V(P_2)$, we will end up in the two-phase region at $P_2 = 4.958\,bar$, and $\vartheta_2 = \vartheta_s(P_2) = 50\,°C$. From

$$s_2 = q \cdot s^V(50\,°C) + (1-q) \cdot s^L(50\,°C)$$

14.3 Adiabatic Compression and Expansion

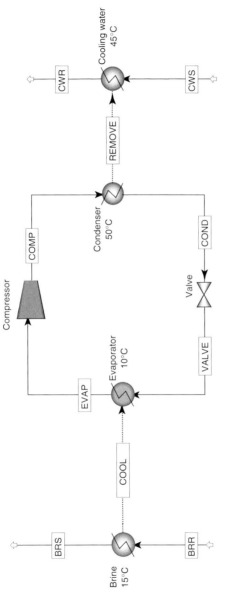

Figure 14.2 Scheme of a refrigeration unit.

one can calculate the vapor fraction q:

$$q = \frac{s_2 - s^L}{s^V - s^L} = \frac{-0.1590 - (-1.1623)}{-0.1293 - (-1.1623)} = 0.9712$$

This means that after compression approx. 3% of the vapor has condensed during compression. Droplets will be formed inside the compressor, which will rapidly lead to erosion. For propane, the whole compression procedure takes place in the vapor phase. Thus, propane can be regarded as an appropriate refrigerant, whereas the use of n-butane does not make much sense, although it is possible that due to an irreversible compression we actually can end up in the vapor region.

Getting into the two-phase region during the compression becomes the more probable, the larger the refrigerant molecule is. This can be made clear when the ideal gas behavior is taken as an approximation. In this case, according to Eq. (14.13), the compressor outlet temperature of the saturated vapor for the reversible case is

$$\frac{T_{2rev}}{T_1} = \left(\frac{P_2}{P_1}\right)^{\frac{\kappa-1}{\kappa}}$$

The temperature T_2 must be greater than the dew point temperature at pressure P_2. Otherwise, the outlet stream will end up in the two-phase region. T_2 depends on the exponent $(\kappa - 1)/\kappa$, which can be written according to Eq. (14.12) as

$$\frac{\kappa - 1}{\kappa} = \frac{R}{c_p^{id}}$$

T_2 decreases with increasing molar c_p^{id} and vice versa. On the other hand, the molar c_p^{id} increases with the size of the molecule, as more degrees of freedom (see Section 3.2.4) are available. For n-butane in the example, the exponent is already too small, so that T_{2rev} calculated with Eq. (14.13) is smaller than the dew point temperature of $P = P_2$, leading to a state in the two-phase region.

14.4
Pressure Relief

In process industry, a considerable amount of work is done to assign limiting operation temperatures and pressures to the particular pieces of equipment. To save costs, there is a tendency to keep the overdesign as low as possible. In a so-called HAZOP study[2], it is discussed how to deal with conditions exceeding the design presumptions. There are organizational and process control measures to maintain safe plant operation; however, the last stage of protection for a piece of equipment is often the pressure relief through a safety valve or a rupture disk.

Possible causes for overpressures are increased heat supply by fire, a failure in the heating system, a breakdown of the cooling system, or an exothermic chemical

---

2) HAZOP = Hazard and Operability Study.

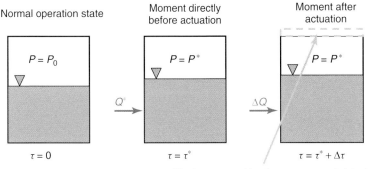

Figure 14.3 Steps for the pressure relief calculation.

reaction, as well as leakages. For each failure case, an estimated scenario giving overpressure and temperatures is calculated so that the most suitable safety relief device can be chosen. Thermodynamic models have a considerable influence on the calculation of safety valves, both for the calculation of the scenario and for the design of the safety device.

For the calculation of the mass flow to be removed from a vessel in a pressure relief process, three steps are applied (Figure 14.3).

1) Evaluation of the state directly before actuation of the safety device.
2) Calculation of the mass flow in the moment of actuation.
3) Tracking the process over a relevant period of time.

First, starting from the normal operation case at $\tau = 0$, the state at $\tau = \tau^*$ is determined where the actuation pressure P^* of the safety device is reached, which usually corresponds to the design pressure of the equipment. In this way, the thermophysical properties necessary for the design of the safety device are obtained. In a second step, a differential time step $\Delta \tau$ at constant pressure $P = P^*$ is regarded. The new volume ΔV of the content of the equipment is calculated; the difference between this volume and the volume of the vessel has to be removed through the safety valve to prevent a further increase in pressure:

$$\dot{m} = \rho^V \frac{\Delta V}{\Delta \tau} \tag{14.17}$$

The calculation has to be repeated for several time steps, as it is not guaranteed that the largest mass flow occurs immediately after actuation.

For the simple case of a pure boiling substance in the vessel far away from the critical point, the evaluation of Eq. (14.17) gives

$$\dot{m} = \frac{\dot{Q}}{\Delta h_v} \tag{14.18}$$

which is often applied even for mixtures, where a constant isobaric enthalpy of vaporization does not make sense. This is demonstrated in Example 14.5.

Example 14.5

In a vessel ($V = 2$ m^3) 1000 kg of a liquid mixture of water (1) (10 mass%) and propanediol-1,2 (2) (90 mass%) is stored under its own vapor pressure and the vessel is protected by a rupture disk actuating at $P^* = 3$ bar. The heat flux responsible for building up the overpressure shall be 1000 kW. The binary parameters for the NRTL equation (Eq. (C.249)) are given to be

$$A_{12} = 0.1078 \qquad B_{12} = 62.0818$$
$$A_{21} = -0.2811 \qquad B_{21} = -1.4101$$
$$\alpha_{12} = 0.3$$

The vapor phase is assumed to be ideal. Determine the mass flow to be relieved.

Solution

As a first step, the state is evaluated where the mixture builds up the pressure to actuate the rupture disk. For this purpose, an isochoric flash is performed. The temperature at $P^* = 3$ bar is estimated, and with P and T as coordinates the volume of the content of the vessel is calculated. The temperature is varied until the calculated volume is equal to the vessel volume. The iteration history is shown in Table 14.3.

At this state, the rupture disk actuates. Choosing $\Delta\tau = 1$ s, the heat

$$Q = 1000 \text{ kW} \cdot 1 \text{ s} = 1000 \text{ kJ}$$

is added to the system. Assuming that the rupture disk is designed correctly, the pressure in the vessel stays at $P^* = 3$ bar. The new state is then evaluated to be

$$P^* = 3 \text{ bar}, \quad m_L = 997.57 \text{ kg}, \quad \rho^L = 901.36 \text{ kg/m}^3$$
$$\vartheta = 172.46 \,°\text{C}, \quad m_V = 2.43 \text{ kg}, \quad \rho^V = 2.115 \text{ kg/m}^3$$

The mole fractions are $x_1 = 0.32$ (9.9 wt%) and $y_1 = 0.86$ (59 wt%).

Table 14.3 Iteration history for Example 14.5.

P (bar)	ϑ (°C)	m_L (kg)	m_V (kg)	ρ^L (kg/m^3)	ρ^V (kg/m^3)	V^L (m^3)	V^V (m^3)	V_{total} (m^3)
3	175	976.81	23.19	898.7	2.19	1.087	10.589	11.676
3	173	993.16	6.84	900.8	2.13	1.103	3.211	4.314
3	172.5	997.22	2.78	901.3	2.12	1.106	1.311	2.418
3	172.4	998.03	1.97	901.4	2.11	1.107	0.934	2.041
3	172.39	998.12	1.88	901.4	2.11	1.107	0.891	1.998

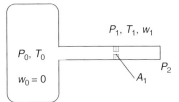

Figure 14.4 Flow through actuated rupture disk.

Thus, one obtains $V_L = 1.107 \text{ m}^3$ and $V_V = 1.148 \text{ m}^3$, giving $V_{total} = 2.255 \text{ m}^3$. Applying Eq. (14.17) yields

$$\dot{m} = \rho^V \frac{\Delta V}{\Delta \tau} = 2.115 \frac{\text{kg}}{\text{m}^3} \frac{0.255 \text{ m}^3}{1 \text{ s}} = 0.54 \text{ kg/s} = 1940 \text{ kg/h}$$

For the application of Eq. (14.18), an averaging procedure for the enthalpies of vaporization of the pure components is required. Using a linear mixing rule and the feed composition, the result would be

$$\Delta h_v = 0.1 \cdot 2041 \text{ J/g} + 0.9 \cdot 734 \text{ J/g} = 864.7 \text{ J/g}$$

giving a significantly different mass flow of 4163 kg/h. This example clearly demonstrates the necessity of a flash calculation for the pressure relief. In this example, Eq. (14.17) takes into account that mainly water is evaporated and that the temperature of the remaining liquid is slightly increased, which has often a considerable contribution to the energy balance in spite of the small temperature elevation.

To determine the maximum possible flow through an actuated pressure relief device, use of the phenomenon that a maximum mass flux exists which only depends on the state in the vessel (see Figure 14.4) is made. For a given pressure P_0 in the vessel, the mass flow is determined by the size of the narrowest cross-flow area A_1 in the line. When the outlet pressure P_2 is decreased, the mass flow increases. However, if P_2 falls below a certain value, no further increase of the mass flow takes place (Figure 14.5). The maximum mass flow is called the *critical mass flow* $\dot{m}_{crit}$.

If the content of the vessel is gaseous, it can be shown that the maximum possible velocity w_1 is the speed of sound (see Appendix C). Nevertheless, its determination is not trivial, as it refers to P_1 and T_1 in the narrowest cross flow area, which are unknown. For the determination of the maximum mass flux, the following equations have to be fulfilled:

1) **Speed of sound in the narrowest cross-flow area:**

$$w_1 = w^*(T_1, P_1) \quad (14.19)$$

2) **No losses (friction, etc.), that is, isentropic condition:**

$$s_1(T_1, P_1) = s(T_0, P_0) \quad (14.20)$$

3) **First Law:**

$$h_1(T_1, P_1) + \frac{1}{2}w_1^2 = h_0(T_0, P_0) + \frac{1}{2}w_0^2 \quad (14.21)$$

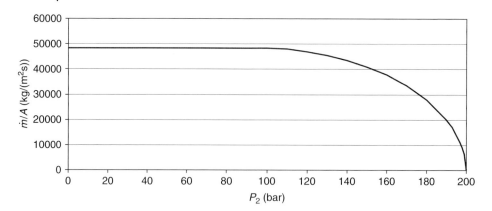

Figure 14.5 Mass flux for a given temperature and pressure as a function of the outlet pressure. Example: Nitrogen at $T_1 = 300$ K and $P_1 = 200$ bar.

These are three equations for the three unknowns T_1, P_1, and w_1. The maximum mass flow can then be determined by the equation of continuity

$$\dot{m} = \frac{w_1 A_1}{v_1} \tag{14.22}$$

If the ideal gas law is used as an equation of state, the system of equations (14.19)–(14.22) has an explicit solution (Appendix C, G2):

$$\frac{\dot{m}_{\text{crit}}}{A_1} = \sqrt{\frac{2P_0}{v_0} \frac{\kappa}{\kappa - 1} \left[\left(\frac{P_1}{P_0}\right)^{\frac{2}{\kappa}} - \left(\frac{P_1}{P_0}\right)^{\frac{\kappa+1}{\kappa}} \right]} \tag{14.23}$$

with

$$\frac{P_1}{P_0} = \left(\frac{2}{\kappa + 1}\right)^{\frac{\kappa}{\kappa - 1}} \tag{14.24}$$

In practical applications, a correction factor has to be introduced to take into account friction losses and the influence of geometry [2].

Example 14.6

A rupture disk protects a vessel containing gaseous ethylene at $P_0 = 200$ bar and $T_0 = 400$ K. During malfunction, a mass flow of 5 t/h has to be released by opening a valve (Figure 14.4). Calculate the minimum possible cross flow area A_1 of the valve. Use

1) the ideal gas equation of state
2) a high-precision equation of state.

Solution

1) Ideal gas
   Using $c_p^{id}(400\text{ K}) = 1.889$ J/(g K) and 28.052 g/mol for the molar mass of ethylene one obtains

   $$\kappa(400\text{ K}) = c_p^{id}(400\text{ K})/(c_p^{id}(400\text{ K}) - R) = 1.1861$$

   $$v_0 = \frac{RT_0}{P_0} = \frac{8314.3}{28.052}\frac{\text{J}}{\text{kg K}}\frac{400\text{ K}}{200 \cdot 10^5\text{ Pa}} = 0.005928\ \frac{\text{m}^3}{\text{kg}}$$

   $$\frac{P_1}{P_0} = \left(\frac{2}{1.1861 + 1}\right)^{\frac{1.1861}{1.1861-1}} = 0.5672$$

   $$\frac{\dot{m}_{\text{crit}}}{A_1} = \sqrt{\frac{2 \cdot 200 \cdot 10^5\text{ Pa}}{0.005928\text{ m}^3/\text{kg}}\frac{1.1861}{1.1861-1}\left[(0.5672)^{\frac{2}{1.1861}} - (0.5672)^{\frac{1.1861+1}{1.1861}}\right]}$$

   $$= 37512\ \frac{\text{kg}}{\text{m}^2\text{s}}$$

   giving

   $$A_1 = \frac{\dot{m}}{(\dot{m}_{\text{crit}}/A_1)} = \frac{5000\text{ kg m}^2\text{s}}{3600\text{s} \cdot 37512\text{ kg}} = 37\text{ mm}^2$$

2) High-precision equation of state.
   The maximum flow is obtained for the reversible adiabatic, that is, isentropic state change. However, for the calculation of this process, the speed of sound has to be evaluated at the conditions in the cross-flow area (index 1) of the valve. For this purpose, an iterative procedure is necessary. The necessary steps are as follows:
   1) Set entropy $s_1 = s_0$
   2) Estimate P_1
      Calculate $T_1 = f(P_1, s_1)$. For ideal gases, we get $\frac{T_1}{T_0} = \left(\frac{P_1}{P_0}\right)^{\frac{\kappa-1}{\kappa}}$
   3) Calculate $w_1 = w^*(T_1, P_1)$
      Check First Law: $h_0 + \frac{1}{2}w_0^2 = h_1 + \frac{1}{2}w_1^2$
   4) If check First Law fails, go to step 2.
   5) Calculate $v_1 = f(T_1, P_1)$
      Evaluate $A_1 = \frac{\dot{m}v_1}{w_1}$

   Using Eq. (2.113) and the coefficients from Appendix B, the iteration history is shown in Table 14.4, giving $z_1 = 0.6497$, $v_1 = 0.006517$ m^3/kg, and $A_1 = 29$ mm^2. In this case, the result for the ideal gas calculation is conservative; however, the opposite can happen if $z_0 > 1$.

   The final result has to be multiplied by an empirical factor considering friction and the influence of geometry [2].

Table 14.4 Iteration history for Example 14.6.

Step	P_1 (bar)	T_1 (K)	w^* (m/s)	w_1 (m/s) from First Law	w^*/w_1
1	100	351.52	310.89	326.25	0.9529
2	110	358.46	321.63	305.62	1.0524
3	104.73	354.90	315.94	316.49	0.9983
4	104.90	355.01	316.12	316.14	0.9999
5	104.91	355.02	316.13	316.13	1.0000

14.5
Limitations of Equilibrium Thermodynamics

In process simulation, one must be aware that equilibrium thermodynamics is not the only decisive issue. Often mass transfer or reaction kinetics must be taken into account to get results that ensure an appropriate equipment design. Special literature is available in these areas [3, 4].

Typical examples are as follows:

- **Partial condensation:** If noncondensable components occur in a stream, it can happen that the condensation is no longer only determined by the heat transfer between condensate film and cooling agent. Instead, the mass transfer of the particular components on the vapor phase side is decisive. The thermodynamic equilibrium still plays a major role as it determines the partial pressure differences between the bulk and the phase boundary.

Heat and mass transfer have an influence on each other, which can be taken into account by the Ackermann [5] correction factor E for the heat transfer stream

$$\dot{q} = \alpha^V E \left(T^V - T^L \right) \text{ with } E = \frac{\frac{(\dot{n}/A)c_P^V}{\alpha^V}}{1 - \exp\left(-\frac{(\dot{n}/A)c_P^V}{\alpha^V}\right)} \quad (14.25)$$

where $\dot{n}/A$ is the mole flow density and α is the heat transfer coefficient. The whole set of equations and a detailed description of the procedure are given in the literature [6, 7]. A highly iterative system of equations results, which is currently not covered by commercial simulation and design programs. Instead, more or less empirical corrections [8, 9] are provided.

Correlations for the mass transfer coefficient β can be obtained from the corresponding correlations for the heat transfer coefficients, taken from the literature for the particular geometrical constraints [10]. The diffusion coefficients necessary for the determination of β can be calculated as described in Section 3.3.6.

Beyond the mass transfer problem, it has to be considered if aerosol formation plays a major role [11].

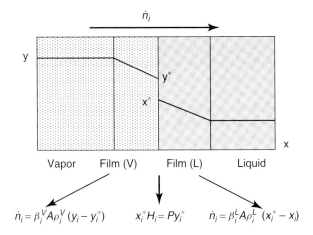

Figure 14.6 Illustration of the two-film theory.

- **Absorption processes:** A standard application for absorption processes is the removal of dilute components from a gas stream by dissolving them in a liquid. Often it is advantageous to work preferably at low temperatures in order to increase the solubility of the gases. Similar to the partial condensation, it is not appropriate for the design to assume that vapor and liquid reach the thermodynamic equilibrium in a certain volume element, for example, on a tray in a column. An equilibrium model would even yield qualitatively wrong results; a design engineer would get the impression that an increase of the mass flow of the absorptive agent will improve the performance of the equipment in any case. However, the performance may be limited by the available mass transfer area, which does not increase with the liquid mass flow.

  Again, a mass transfer calculation is necessary. Figure 14.6 illustrates the two-film theory, which is applied in commercial simulation programs also for multicomponent systems. For simplification, an ideal vapor phase is assumed. Therefore, the driving force for the mass transfer, which is normally given by the difference of the chemical potentials, is reduced to a simple concentration difference.

  It is assumed that the mass transfer occurs in the layers on both sides of the phase boundary. According to the model, the mass transfer is caused by molecular diffusion. Therefore, a linear concentration profile is built up in the boundary layers. At the phase boundary itself, thermodynamic equilibrium is reached. More or less empirical correlations for the mass transfer coefficients β and for the effective phase interface area are usually provided by the simulation program for different column internals (random packings, structured packings, and trays). The necessary diffusion coefficients can again be calculated according to Section 3.3.6.

  In most cases, the mass transfer resistance in one of the two layers dominates. For high solubilities (e.g., ammonia in water or hydrogen chloride in water), the

gas-phase resistance dominates, and for low solubilities (e.g., oxygen in water or carbon dioxide in water), it is the liquid-phase resistance.
- **Reaction kinetics:** In most chemical processes, chemical reactions do not reach the chemical equilibrium. Instead, reaction kinetics are decisive for the yield, the chemical equilibrium calculations are mainly used to determine the maximum yield. Also for the design of reactive distillation processes and chemical absorption processes, the reaction kinetics must be known. As well, the Maurer model for formaldehyde (see Section 13.1) must be regarded as a reactive distillation and especially at low temperatures, the equilibrium is by far not reached.

Process simulators normally offer a standard framework for the formulation of kinetic equations, which must be filled in by the user. To describe the kinetic modeling in detail goes far beyond the scope of this book, but some difficulties in establishing a reliable reaction kinetics should be mentioned here.

- For sufficiently long residence times, the calculations using reaction kinetics must yield the results obtained with equilibrium calculations. This is not the case if the reaction kinetics are not formulated in terms of chemical potentials, that is, with activities and fugacities as concentration measures instead of the concentration itself.
- For heterogeneous reactions (e.g., reactions at a solid catalyst or chemical absorption), the mass transfer plays a major role, which must be considered.
- Reaction kinetics cannot be predicted. The true reaction mechanism is known only for a few cases. Instead, often empirical overall reaction rates are used, which may not be true, when extrapolated to changed conditions.

Additional Problems

P14.1 Calculate the Joule-Thomson coefficient of nitrogen at a temperature of 150 K and a pressure of 10 atm using
  a. the virial equation truncated after the second term using a second virial coefficient estimated via the Tsonopoulos method.
  b. the Soave-Redlich-Kwong equation of state.
  All required parameters are given in Appendix A.

P14.2 Oxygen at 25 °C and a pressure of 1 bar is compressed to 10 bar. Calculate the required specific work and the temperature of the compressed gas assuming isentropic compression
  a. using ideal gas law
  b. using the Soave-Redlich-Kwong equation of state.
  All required physical property parameters are given in Appendix A.

P14.3 A liquid water stream (2000 kg/h at 25 °C and 1 bar) is compressed to 100 bar in a cooled pump. The process can be assumed to be reversible and isothermal. Calculate the required work and heat duty of the cooling system. The thermal expansion coefficient $\alpha = 0.207 \cdot 10^{-3}$ K^{-1} and the compressibility coefficient $\chi = 0.46 \cdot 10^{-4}$ bar^{-1} can be assumed to be constant over the relevant temperature and pressure range. The molar volume of liquid water at feed conditions is 18.07 cm^3/mol.

P14.4 A feed stream of 1600 kg/h of methane is adiabatically mixed with 170 kg/h of n-dodecane. Both streams are at 160 °C and a pressure of 20 bar, the mixing is isobaric. Calculate the temperature of the stream leaving the mixer using the Soave-Redlich-Kwong equation of state with a binary parameter of $k_{12} = 0$. Explain the result.

All physical property parameters for methane can be found in Appendix A. The required values for n-dodecane are: $T_c = 658.8$ K, $P_c = 1809.7$ kPa, $\omega = 0.562$, $c_p^{id} = 379.8$ J/mol K.

P14.5 The refrigerant R134a is compressed from $\vartheta_1 = 5$ °C, saturated vapor, to $P_2 = 10$ bar. The isentropic efficiency of the compressor is $\eta_{th} = 0.7$. The mechanical efficiency is $\eta_{mech} = 0.95$. Calculate the power of the compressor. The mass flow is 3000 kg/h. Use a high-precision equation of state.

P14.6 In a 50 m^3 vessel, liquid ammonia at $\vartheta_1 = 50$ °C is stored at $P_1 = 100$ bar. Due to a vessel failure, the ammonia is collected in a backup vessel. Which is the necessary volume of the backup vessel, if $P_2 = 10$ bar must not be exceeded?

P14.7 In a Linde plant, nitrogen at -104 °C, 240 bar is let down to $P = 1$ bar through a valve. How much liquid nitrogen is produced?

P14.8 A mixed stream consisting of 1 kmol/h CO_2 and 1 kmol/h O_2 is compressed from $T_1 = 290$ K, $P_1 = 1$ bar to $P_2 = 5$ bar. Calculate the compressor power for
a. adiabatic compression
b. isothermal compression.

The mixture should be regarded as an ideal gas. The compression should be assumed to be reversible in both cases.

P14.9 In an LDPE (Low Density Polyethylene) plant, ethylene is expanded from $P_0 = 3000$ bar, $T_0 = 600$ K to $P_1 = 300$ bar by a throttle valve. By a second throttle valve, it is expanded to environmental pressure $P_2 = 1$ bar. Calculate the temperatures T_1 and T_2 by using a high-precision equation of state. The velocity terms in the First Law should be neglected.

P14.10 In vacuum distillation columns, the leakage of ambient air into the column is always a problem and might lead to an explosive atmosphere in the condenser. How does the leakage rate rise if the column operating pressure is lowered from $P_1 = 400$ mbar to $P_2 = 100$ mbar? The ambient pressure shall be $P_{amb} = 1.013$ bar.

P14.11 A vessel (1 m^3) containing 500 kg propylene at $\vartheta = 30$ °C is exposed to sun radiation. What is the initial pressure? The safety valves of the vessel actuate at $P = 60$ bar. Use a high-precision equation of state to calculate the respective temperature.

P14.12 Oxygen ($\vartheta = 25\,°C$, $\dot{m} = 250$ kg/h) is compressed adiabatically from $P_1 = 1$ bar to $P_2 = 10$ bar. Calculate the power of the compressor and the outlet temperature of the gas using
   a. the ideal gas law
   b. the Soave-Redlich-Kwong equation of state.
   The isentropic efficiency of the compressor is $\eta_{th} = 0.75$. The mechanical efficiency is $\eta_{mech} = 0.95$.

P14.13 A thermodynamic cycle is operated with water at the following conditions:
   1) Isobaric heating to $P_1 = 100$ bar, $\vartheta_1 = 350\,°C$
   2) Reversible and adiabatic expansion of the vapor in a turbine to $P_2 = 1$ bar.
   3) Isobaric condensation.
   4) Isentropic compression of the liquid in a pump to $P_4 = 100$ bar.
   Calculate the thermal efficiency of the process defined by
   $$\eta_{th} = \frac{-(P_{12} + P_{34})}{\dot{Q}_{41}}$$
   Use the high-precision equation of state.

P14.14 A refrigerator is operated with R12 (dichlorodifluoromethane). The particular steps of the compression cycle are:
   1) Isobaric condensation without subcooling at $\vartheta_1 = 30\,°C$.
   2) Adiabatic pressure relief by a throttle valve to $P_2 = P^s(-20\,°C)$.
   3) Complete isobaric evaporation of the refrigerant at $\vartheta_2 = \vartheta_3 = -20\,°C$ without superheating
   4) Isentropic compression of the saturated vapor to $P_4 = P^s(30\,°C)$.
   Calculate the process data for the steps 1–4 using the Peng-Robinson equation of state.

P14.15 Calculate the Joule-Thomson coefficient for methane at $T = 300$ K and $P = 30$ bar using the Peng-Robinson equation of state. The critical data and the acentric factor can be taken from Appendix A.

P14.16 Gaseous ammonia (100 °C, 5 bar) is compressed to $P_2 = 10$ bar. The thermal efficiency is $\eta_{th} = 0.8$, the mechanical efficiency is $\eta_{mech} = 0.9$. Calculate the specific compressor duty and the state properties at the compressor outlet.

References

1. IK-CAPE Thermodynamics-Package for CAPE-Applications, www.dechema.de/dechema_media/Downloads/Informationssysteme/IK_CAPE_Equations.pdf (accessed March 2002).
2. Brodhagen, A. and Schmidt, F. (2006) VDI-Wärmeatlas, 10th edn, Part Lbd. Springer, Berlin.
3. Bird, R.B., Stewart, W.E., and Lightfoot, E.N. (2002) Transport Phenomena, 2nd edn, John Wiley & Sons, Inc., New York.
4. Baerns, M., Behr, A., Brehm, A., Gmehling, J., Hofmann, H., Onken, U., and Renken, A. (2006) Technische Chemie, Wiley-VCH, Weinheim.

5. Ackermann, G. (1937) *Wärmeübertragung und molekulare Stoffübertragung im gleichen Feld bei großen Temperatur- und Partialdruckdifferenzen*, VDI-Forschungsheft 382, Ausg. B, Bd. **8**.
6. Fullarton, D. and Schlünder, E.-U. (2006) *VDI-Wärmeatlas*, 10th edn, Part Jba. Springer, Berlin.
7. Burghardt, A. (2006) *VDI-Wärmeatlas*, 10th edn, Part Jbb. Springer, Berlin.
8. Bell, K.J. and Ghaly, M.A. (1973) *AIChE Symp. Ser.*, **69** (131), 72–79.
9. Colburn, A.P. and Drew, T.B. (1937) *Trans. Am. Inst. Chem. Eng.*, **33**, 197–215.
10. (2010) *VDI Heat Atlas*, 2nd edn, Springer-Verlag, Berlin/Heidelberg.
11. Schaber, K. and Schnerr, G.H. (2010) *VDI Heat Atlas, Section M11*, 2nd edn, Springer-Verlag, Berlin.

15
Introduction to the Collection of Example Problems

In order to enable the reader to gain practical experiences with the different topics discussed in this textbook, a number of examples is available free for download via the Internet. These include many of the examples found in the book and a collection of additional problems, which we plan to further update in the future. While the location of the files may change in the future, the material itself or a link to the material will always be available on *www.ddbst.com*.

15.1
Mathcad Examples

To construct a part of the example calculations, the well-known and widely available mathematical software Mathcad was used. Major reasons for selecting this software are its simplicity and intuitiveness. In addition, the Mathcad format very closely resembles the way that equations and calculations are usually presented in textbooks. This makes the examples easy to read and understand. An additional advantage is that values in Mathcad usually carry their correct units and unit conversions are handled automatically. In some of the examples, symbolic manipulations using the integrated MAPLE software are demonstrated, which allow for example, re-arranging, simplifying, and solving mathematical expressions as well as symbolic differentiation and integration. A typical Mathcad document is shown in Figure 15.1.

For a broader comparison and user experiences with Mathcad and other mathematical software like MatLab, Mathematica, and so on, as well as introductions to use Mathcad, consult the Internet. A free 30-day-trial of Mathcad is available from *www.ptc.com*. For readers who do not have access to Mathcad, the files will also be made available in other formats (XPS, MathML, etc.).

Besides the examples prepared for this textbook, a multitude of further calculations from different other authors using Mathcad can be found online, either in form of single files or complete books written in Mathcad.

In addition, PTC, the provider of Mathcad, maintains a massive library of example calculations within their Mathcad Resource Center.

614 | 15 Introduction to the Collection of Example Problems

Calculation of the interaction parameters $\Delta\lambda$ as function of temperature:

$$\Delta\lambda(T) := a_\lambda + b_\lambda \cdot T \qquad \Delta\lambda(323.15\ K) = \begin{pmatrix} 0 & -80.513 \\ 523.416 & 0 \end{pmatrix} \frac{cal}{mol}$$

Calculation of the interaction parameters $\Delta\Lambda$ as function of temperature:

$$\Delta\Lambda(T, i, j) := \frac{v_j}{v_i} \cdot e^{\left[\frac{(-\Delta\lambda(T)_{i,j})}{R_{gas} \cdot T}\right]} \qquad \Delta\Lambda(323.15\ K, 1, 2) = 0.624$$

Calculation of the activity coefficient as function of temperature and composition:

$$\gamma(T, x) := \begin{vmatrix} \text{for } i \in 1..n_{comp} \\ \gamma_i \leftarrow \exp\left[-\ln\left(\sum_j x_j \cdot \Delta\Lambda(T, i, j)\right) + 1 - \sum_k \left(\frac{x_k \cdot \Delta\Lambda(T, k, i)}{\sum_j x_j \cdot \Delta\Lambda(T, k, j)}\right)\right] \\ \gamma \end{vmatrix}$$

In order to plot the ratios of the vapor pressures and the ratio of the activity coefficients, the following functions are defined:

$$Ps_{ratio}(T) := \frac{Ps(T, 2)}{Ps(T, 1)} \qquad x(x_1) := \begin{pmatrix} x_1 \\ 1 - x_1 \end{pmatrix} \qquad \gamma_{ratio}(T, x_1) := \frac{\gamma(T, x(x_1))_1}{\gamma(T, x(x_1))_2}$$

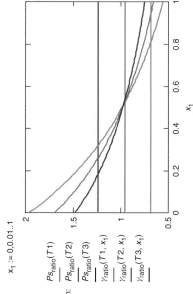

The values of both functions as function of composition are shown for the three different temperatures below. The azeotropic composition can be found at the intersection of same colored lines.

$x_1 := 0, 0.01..1$

$\overline{Ps_{ratio}(T1)}$
$\overline{Ps_{ratio}(T2)}$
$\overline{Ps_{ratio}(T3)}$
$\overline{\gamma_{ratio}(T1, x_1)}$
$\overline{\gamma_{ratio}(T2, x_1)}$
$\overline{\gamma_{ratio}(T3, x_1)}$

Figure 15.1 Excerpt from a typical Mathcad document from the example collection.[1]

As a first introduction to the software, consult the texts under "Help – Resource Center – Overview and Tutorials" in Mathcad. Beside this, a multitude of further introductory texts is available on the Internet.

Examples in this collection use both the file formats for Mathcad Version 2001 and more up to date versions. Any version of Mathcad higher than Version 2001 is sufficient to solve the problems.

The complexity of the examples ranges from simple calculations to very extended and complex documents. Usually, the files also contain more or less trivial intermediate calculation steps in order to assist the mathematically less-trained reader. The authors hope that these examples not only help the reader to better follow the details of the calculations but that they also serve as a convenient starting point for own modifications and extensions.

15.2
Examples Using the Dortmund Data Bank (DDB) and the Integrated Software Package DDBSP

In addition to the Mathcad files, numerous further problems are included that require the Dortmund Data Bank (DDB) and the integrated software package DDBSP. In order to enable anybody to follow these examples, DDBST GmbH has provided a free Explorer Version of their products that can be downloaded from *www.ddbst.com*. The free version, although restricted in functionality, is sufficient for most of the problems given on the website. It contains basic constants and model parameters as well as several thousand sets of experimental data for 30 common components and their mixtures. The components have been selected in order to provide a sufficient amount of reliable experimental data and at the same time cover a large range of the phenomena discussed in the textbook.

Using the integrated software package DDBSP, physical properties from the data bank can be searched, plotted, and regressed via a variety of thermodynamic models (pure component property correlations as well as g^E-models and equations of state for the real mixture behavior). Simultaneous regression and analysis of phase equilibrium data and excess properties of subcritical nonelectrolyte mixtures can be performed using the software Recval/3. In order to estimate a large variety of pure component properties from molecular structure, the program package Artist is included. The functionality of DDBSP has been restricted in the free Explorer version, only a subset of the regression and prediction models is included and they may only be used for a certain range of components. Process synthesis tools were to a large part excluded, especially where they require a larger number of components and experimental data like in the case of entrainer selection. The Explorer version also does not allow us to build up private data banks and thus excludes the various data bank editor programs. For an extended functionality in

---

[1] "05.13 Azeotropic Points of the System Acetone – Methanol.mcd."

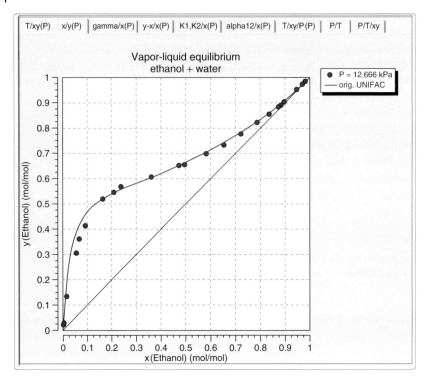

Figure 15.2 Experimental VLE dataset and curve calculated using UNIFAC as presented by DDBSP.

a university or a nonprofit research institution, special low-cost single user and classroom Educational versions are available (see *www.ddbst.com*). The Professional Educational version includes basic data for all the nearly 33000 DDB components and allows us to build up in-house data banks to store and process for example, own measurements.

A typical graphical output of a vapor–liquid equilibrium dataset together with the UNIFAC prediction as presented by DDBSP is shown in Figure 15.2.

15.3
Examples Using Microsoft Excel and Microsoft Office VBA

Microsoft Excel and Visual Basic for Application (VBA) are also frequently used for scientific calculations. They are convenient to use and generally available but suffer from the fact that the calculations performed in Basic code or the Excel sheet are not easy to read and understand by users other than the developer. Several examples employ Excel and/or VBA and demonstrate applications where they offer great advantages due to their flexibility and the ability to easily interface to COM components of other software products. For the less-trained or infrequent programmer, VBA is definitely considered the language of choice and its similarity to the science and engineering language FORTRAN is a clear advantage. From VBA, calculations implemented in Mathcad documents or many functionalities in DDBSP can be easily called as *COM* functions. VBA is familiar to many engineers as it is also able to programmatically employ most functionality including the thermodynamic engines of all major process simulators.

Appendix A
Pure Component Parameters

General data for selected compounds.

Substance	Formula	T_c (K)	P_c (bar)	V_c (cm³/mol)	M (g/mol)	ω	T_b (K)	T_m (K)	Δh_m (J/g)	Δh_f^0 (J/mol)	Δg_f^0 (J/mol)
Water	H_2O	647.096	220.640	55.947	18.015	0.3443	373.13[a]	273.15	333.1	−241820	−228590
Ammonia	NH_3	405.500	113.592	75.768	17.031	0.2560	239.82	195.45	332.2	−45773	−16150
Hydrogen chloride	HCl	324.650	83.100	81.016	36.461	0.1314	188.08	158.95	54.9	−92310	−95300
Chlorine	Cl_2	416.958	79.911	122.927	70.906	0.0874	239.17	172.15	90.3	0	0
Nitrogen	N_2	126.192	33.958	89.416	28.014	0.0372	77.36	63.15	25.7	0	0
Oxygen	O_2	154.599	50.464	74.948	31.999	0.0221	90.19	54.35	13.9	0	0
Hydrogen	H_2	33.190	13.150	66.938	2.016	−0.2187	20.38	13.95	58.1	0	0
Sulfur dioxide	SO_2	430.643	78.757	122.087	64.062	0.2552	263.13	200.00	115.5	−296850	−300150
Carbon monoxide	CO	132.860	34.982	92.088	28.010	0.0503	81.64	68.15	30.0	−110540	−137270
Carbon dioxide	CO_2	304.128	73.773	94.117	44.009	0.2236	–	216.55	204.9	−393500	−394380
Methane	CH_4	190.564	45.992	98.629	16.043	0.0114	111.67	90.65	58.7	−74850	−50835
Ethane	C_2H_6	305.322	48.722	145.843	30.070	0.0995	184.57	90.35	95.1	−84680	−32930
Propane	C_3H_8	369.825	42.477	200.004	44.097	0.1524	231.03	85.45	79.9	−103840	−23470
n-Butane	C_4H_{10}	425.125	37.960	254.930	58.124	0.2008	272.66	134.85	80.2	−126150	−17154
Isobutane	C_4H_{10}	407.810	36.290	257.756	58.124	0.1835	261.40	113.55	78.1	−134510	−20878
n-Pentane	C_5H_{12}	469.659	33.689	306.766	72.151	0.2517	309.21	143.40	116.4	−146760	−8813
n-Hexane	C_6H_{14}	507.795	30.416	386.753	86.178	0.3002	341.87	177.85	151.8	−167190	−250
Cyclohexane	C_6H_{12}	553.600	40.750	308.263	84.162	0.2092	353.86	279.65	32.6	−123300	31910
n-Heptane	C_7H_{16}	541.226	27.738	445.551	100.205	0.3460	371.53	182.55	140.2	−187650	8165
n-Octane	C_8H_{18}	569.570	25.067	501.835	114.231	0.3943	398.78	216.35	181.6	−208750	16000
Ethylene	$CH_2=CH_2$	282.350	50.418	130.947	28.054	0.0866	169.38	104.05	119.4	52300	68110
Propylene	C_3H_6	365.570	46.646	188.376	42.081	0.1408	225.46	87.85	71.4	20420	62720

Appendix A Pure Component Parameters

Substance	Formula	T_c (K)	P_c (bar)	V_c (cm³/mol)	M (g/mol)	ω	T_b (K)	T_m (K)	Δh_m (J/g)	Δh_f^0 (J/mol)	Δg_f^0 (J/mol)
1-Butene	C_4H_8	419.290	40.057	235.802	56.108	0.1919	266.84	87.85	68.6	−540	70270
Methanol	CH_3OH	513.380	82.159	113.828	32.042	0.5625	337.63	175.50	100.3	−201160	−162500
Ethanol	C_2H_5OH	513.900	61.480	166.917	46.069	0.6441	351.41	159.05	107.0	−234800	−168280
n-Propanol	C_3H_7OH	536.750	51.750	218.990	60.096	0.6211	370.31	146.95	89.4	−255200	−159900
n-Butanol	C_4H_9OH	563.050	44.230	274.996	74.123	0.5905	390.90	183.85	126.4	−274600	−150300
Ethylene glycol	$C_2H_4(OH)_2$	719.150	82.000	190.983	62.068	0.5129	470.22	260.15	160.4	−392200	−301800
Isopropanol	C_3H_7OH	508.250	47.620	220.011	60.096	0.6631	355.36	185.25	90.0	−272700	−173470
Acetic acid	CH_3COOH	591.950	57.860	179.676	60.052	0.4630	391.04	289.85	195.3	−434830	−376680
Methyl acetate	$CH_3\text{–}COO\text{–}CH_3$	506.550	47.500	228.015	74.079	0.3306	330.08	175.15	107.6	−411900	−324200
Ethyl acetate	$CH_3\text{–}COO\text{–}C_2H_5$	523.200	38.301	285.992	88.106	0.3606	350.27	189.65	118.9	−442910	−327390
Vinyl acetate	$C_4H_6O_2$	519.150	39.580	269.978	86.090	0.3526	345.86	180.35	62.4	−314900	−227900
MTBE	$C_5H_{12}O$	497.150	34.300	328.976	88.150	0.2662	328.29	164.55	86.2	−283500	−117500
Acetone	C_3H_6O	508.100	46.924	212.299	58.080	0.3064	329.23	178.45	99.4	−215700	−151300
Benzene	C_6H_6	562.014	49.010	255.044	78.114	0.2103	353.24	278.65	126.3	82930	129660
Toluene	$C_6H_5\text{–}CH_3$	591.749	41.263	315.422	92.141	0.2657	383.75	178.15	72.0	50170	122200
p-Xylene	C_8H_{10}	616.250	35.110	377.958	106.168	0.3218	411.51	286.45	161.2	18030	121400
R23	CHF_3	299.747	48.690	133.115	70.013	0.2584	191.09	117.95	58.0	−697050	−662610
R32	CH_2F_2	351.255	57.826	122.696	52.023	0.2770	221.50	137.05	99.2	−452710	−425160
R12	CCl_2F_2	385.120	41.361	214.006	120.913	0.1795	243.39	115.15	34.2	−481160	−442250
R22	$CHClF_2$	369.280	49.885	166.282	86.468	0.2207	232.35	115.75	47.7	−481600	−450500
R134a	$CF_3\text{–}CH_2F$	374.212	40.593	199.299	102.030	0.3268	247.08	172.15	63.5	−895790	−826900
R125	$CF_3\text{–}CHF_2$	340.229	37.039	212.131	120.020	0.3017	225.04	170.15	47.0	−1105000	−1030000

[a] According to ITS-90.

Vapor pressure correlations for selected compounds.

Equation: $\ln \dfrac{P^s}{P_c} = \dfrac{1}{T_r}\left[A(1-T_r) + B(1-T_r)^{1.5} + C(1-T_r)^{2.5} + D(1-T_r)^5\right]$

Substance	A	B	C	D	Range of validation		Verification point	
					T_{min} (K)	T_{max}	T (K)	P^s (bar)
Water	−7.870154	1.906774	−2.310330	−2.063390	274	T_c	393.15	1.9859
Ammonia	−7.303825	1.649953	−2.021615	−1.960295	196	T_c	275.15	4.6237
Hydrogen chloride	−6.634574	1.068097	0.043272	−4.443638	159	T_c	243.15	10.7137
Chlorine	−6.442452	1.492841	−1.225096	−2.015398	174	T_c	273.15	3.6877
Nitrogen	−6.123680	1.260610	−0.760446	−1.794726	65	T_c	103.15	9.6198
Oxygen	−6.051465	1.234823	−0.628118	−1.614180	64	T_c	123.15	12.2153
Hydrogen	−4.836839	0.943915	0.763880	−0.467794	15	T_c	18.05	0.4651
Sulfur dioxide	−7.278016	1.726871	−2.371926	−2.708750	203	T_c	303.15	4.6239
Carbon monoxide	−6.194175	1.319639	−0.943212	−2.001545	68	T_c	101.15	5.9215
Carbon dioxide	−7.026565	1.527245	−2.246311	−2.630030	218	T_c	253.15	19.6972
Methane	−6.024057	1.268690	−0.570278	−1.375360	93	T_c	173.15	26.0390
Ethane	−6.461277	1.353559	−1.043360	−2.044654	123	T_c	243.15	10.6417
Propane	−6.715816	1.387038	−1.311343	−2.563166	160	T_c	260.15	3.1220
n-Butane	−7.084375	1.789499	−1.994745	−2.325711	180	T_c	280.15	1.3356
Isobutane	−6.907184	1.578704	−1.803286	−2.427186	180	T_c	300.15	3.7173
n-Pentane	−7.363978	1.943238	−2.471007	−2.349144	213	T_c	343.15	2.8355
n-Hexane	−7.611393	2.007192	−2.744096	−2.825408	233	T_c	303.15	0.2498
Cyclohexane	−7.009954	1.575242	−1.968888	−3.260142	283	T_c	373.15	1.7462
n-Heptane	−7.754461	1.847289	−2.802496	−3.625021	263	T_c	373.15	1.0619

(continued overleaf)

(continued).

Substance	A	B	C	D	Range of validation		Verification point	
					T_{min} (K)	T_{max}	T (K)	P^s (bar)
n-Octane	-8.010138	1.984728	-3.259138	-4.003580	283	T_c	343.15	0.1583
Ethylene	-6.412447	1.452364	-1.239075	-1.996814	110	T_c	260.15	30.1310
Propylene	-6.666610	1.436075	-1.395381	-2.466962	140	T_c	280.15	7.1796
1-Butene	-7.078279	1.876157	-2.019940	-2.651169	183	T_c	293.15	2.5462
Methanol	-8.726980	1.450050	-2.771770	-0.723874	233	T_c	343.15	1.2545
Ethanol	-8.338015	0.087185	-3.305779	-0.259857	253	T_c	353.15	1.0845
n-Propanol	-8.606755	2.173634	-8.046856	3.691766	290	T_c	364.18	0.7990
n-Butanol	-8.330857	2.054134	-8.175663	0.190068	275	T_c	418.85	2.4959
Ethylene glycol	-7.807002	0.915488	-4.927491	-1.926135	333	T_c	453.75	0.6057
Isopropanol	-8.439864	1.149220	-6.938392	0.615959	276	T_c	373.70	2.0101
Acetic acid	-9.345421	3.784980	-3.602383	-1.553062	290	T_c	364.64	0.4227
Methyl acetate	-8.574704	4.224264	-5.368114	-0.827557	227	T_c	332.15	1.0881
Ethyl acetate	-7.897191	2.167535	-3.523305	-3.107122	242	T_c	330.15	0.4977
Vinyl acetate	-7.554439	1.364168	-2.656813	-2.993477	252	T_c	352.25	1.2471
Methyl-tert-butyl ether	-7.565301	2.577840	-3.917256	-0.671678	217	T_c	374.55	3.7435
Acetone	-7.670734	1.965917	-2.445437	-2.899873	213	T_c	353.15	2.1544
Benzene	-7.114987	1.841409	-2.254156	-3.147450	283	T_c	373.15	1.8014
Toluene	-7.498890	2.084280	-2.556275	-2.860135	273	T_c	333.15	0.1853
p-Xylene	-7.671695	1.812883	-2.387960	-3.456690	286	T_c	373.15	0.3206
R23	-7.285947	1.636279	-2.135618	-2.481506	158	T_c	263.15	18.8963
R32	-7.470600	1.756960	-2.015692	-2.619790	158	T_c	263.15	5.8274
R12	-6.919773	1.622716	-1.734359	-2.589279	168	T_c	263.15	2.1879
R22	-7.092406	1.619996	-2.013516	-2.729725	158	T_c	263.15	3.5500
R134a	-7.649281	1.798190	-2.660067	-3.320921	173	T_c	263.15	2.0063
R125	-7.537003	1.837339	-2.675330	-2.787620	173	T_c	283.15	9.0868

The required critical data T_c and P_c are given in the Table "General data for selected compounds".

Liquid density correlations for selected compounds.

Equation: $\dfrac{\rho^L}{\text{kg/m}^3} = \dfrac{\rho_c}{\text{kg/m}^3} + \left[A\left(1 - \dfrac{T}{T_c}\right)^{0.35} + B\left(1 - \dfrac{T}{T_c}\right)^{2/3} + C\left(1 - \dfrac{T}{T_c}\right) + D\left(1 - \dfrac{T}{T_c}\right)^{4/3} \right]$

Substance	ρ_c (kg/m³)	A	B	C	D	Range of validation		Verification point	
						T_{min} (K)	T_{max}	T (K)	ρ^L (kg/m³)
Water	322.00	see Eq. 3.52				274	T_c	373.15	958.20
Ammonia	224.78	533.0864	−39.1990	271.4070	−72.5196	196	T_c	310.15	584.33
Hydrogen chloride	450.05	640.0083	462.2040	−180.0889	273.3080	159	T_c	246.15	1016.46
Chlorine	576.81	908.9017	948.0483	−1353.2687	1093.5252	173	T_c	273.15	1467.02
Nitrogen	313.30	471.2169	492.8646	−561.5971	391.1454	68	T_c	123.15	471.54
Oxygen	426.95	748.3728	396.2376	−416.2389	372.6904	58	T_c	123.15	952.52
Hydrogen	30.12	52.8201	2.7120	6.2764	−4.0288	15.0	T_c	18.05	73.30
Sulfur dioxide	524.72	1026.4147	287.9415	−59.1833	242.7178	203	T_c	273.15	1435.62
Carbon monoxide	304.17	571.9312	−67.0962	387.1504	−121.7320	68	T_c	113.15	626.53
Carbon dioxide	467.60	897.8433	170.0655	169.2649	37.6341	218	T_c	253.15	1031.67
Methane	162.66	267.8594	129.3958	−73.6070	69.9714	93	T_c	133.15	388.78
Ethane	206.18	339.3617	278.2759	−326.5676	246.4990	93	T_c	213.15	505.90
Propane	220.48	372.1807	329.3106	−439.7196	331.6824	90	T_c	280.15	518.73
n-Butane	228.00	418.6984	246.8435	−317.6268	274.8872	140	T_c	280.15	592.99
Isobutane	225.50	383.5780	363.7638	−483.8143	353.4964	120	T_c	280.15	572.34
n-Pentane	235.20	331.1728	680.8988	−965.2019	602.3687	153	T_c	293.15	625.55
n-Hexane	222.82	537.4149	87.8736	−283.5449	344.6594	183	T_c	293.15	659.38
Cyclohexane	273.02	370.9435	865.3421	−1291.8141	834.1977	283	T_c	353.15	720.11
n-Heptane	224.90	308.6265	1070.9890	−1663.6827	989.8326	183	T_c	293.15	683.72
n-Octane	227.63	314.9330	1030.8629	−1576.0010	939.7454	223	T_c	293.15	702.33

(continued overleaf)

(continued).

Substance	ρ_c (kg/m³)	A	B	C	D	Range of validation		Verification point	
						T_{min} (K)	T_{max}	T (K)	ρ^L (kg/m³)
Ethylene	214.24	364.8832	208.8422	−198.1499	185.2595	120	T_c	280.15	287.91
Propylene	223.39	369.7124	440.5713	−608.9573	438.0698	90	T_c	280.15	533.55
1-Butene	237.95	374.9006	532.7375	−823.2356	585.6584	93	T_c	283.15	605.90
Methanol	281.49	164.7427	2257.8517	−3545.8257	1929.8087	183	T_c	273.15	810.52
Ethanol	276.00	748.6190	−412.3645	776.4385	−436.6754	253	T_c	343.15	744.74
n-Propanol	274.42	816.2710	−549.2099	696.9837	−232.0819	147	T_c	290.55	806.76
n-Butanol	269.54	777.2536	−446.8411	578.8807	−172.9552	185	T_c	284.15	816.58
Ethylene glycol	324.99	1305.6439	−1374.3218	1690.8786	−664.8168	261	T_c	381.15	1048.42
Isopropanol	273.15	865.9179	−744.0946	975.9034	−381.1280	186	T_c	304.25	776.23
Acetic acid	334.22	925.3877	−312.7976	340.0648	29.7201	290	T_c	337.55	1000.46
Methyl acetate	324.89	735.4603	−131.4940	371.8328	−53.9224	176	T_c	297.25	928.75
Ethyl acetate	308.07	660.3749	8.8513	207.1687	1.5101	207	T_c	277.45	918.19
Vinyl acetate	318.88	752.0380	−204.4664	468.8042	−107.5550	181	T_c	287.35	939.17
Methyl-tert-butyl ether	267.95	615.1648	−332.9179	716.5664	−284.8751	165	T_c	304.55	729.60
Acetone	273.58	548.0439	205.2643	−197.7406	250.6303	183	T_c	313.15	767.89
Benzene	306.28	502.4341	531.5958	−663.9853	469.5977	283	T_c	343.15	825.92
Toluene	292.12	439.5835	839.1558	−1234.8445	797.8741	183	T_c	313.15	848.53
p-Xylene	280.90	660.6612	−279.7990	602.9859	−192.8244	287	T_c	321.15	836.72
R23	525.96	640.0829	2166.2973	−3069.2655	1827.1404	158	T_c	283.15	941.76
R32	424.00	895.2757	224.1921	−50.9230	229.6594	138	T_c	273.15	1054.86
R12	565.00	914.8334	994.9288	−1324.2838	961.0109	123	T_c	273.15	1395.42
R22	520.01	904.0763	874.3000	−1176.1556	880.9367	123	T_c	283.15	1246.61
R134a	511.94	1043.9503	239.7177	−102.7633	287.7073	173	T_c	283.15	1260.67
R125	565.78	800.1882	1917.3007	−2752.5653	1696.7682	173	T_c	283.15	1272.60

The required critical temperature T_c is given in the Table "General data for selected compounds".

Enthalpy of vaporization correlations for selected compounds.

Equation: $\Delta h_v = RT_c(A\tau^{1/3} + B\tau^{2/3} + C\tau + D\tau^2 + E\tau^6)$, $\tau = 1 - T/T_c$

Substance	A	B	C	D	E	Range of validation		Verification point	
						T_{min} (K)	T_{max} (K)	T (K)	Δh_v (J/g)
Water	6.853064	7.437940	−2.937398	−3.282184	8.396833	273	T_c	323.15	2380.25
Ammonia	5.744770	7.282878	−2.428749	−2.261923	2.909393	196	T_c	300.15	1157.70
Hydrogen chloride	7.669223	0.211221	0.917107	−1.245487	4.526673	159	T_c	246.15	370.87
Chlorine	5.113960	5.494794	−1.639730	−2.193617	4.356455	173	T_c	273.15	267.42
Nitrogen	5.063172	5.518154	−2.645913	−1.981109	5.333368	65	T_c	93.15	174.93
Oxygen	4.969617	5.305319	−2.426090	−2.151263	3.466285	58	T_c	123.15	167.73
Hydrogen	4.115256	2.511597	−1.539154	−2.578672	2.165454	15	T_c	20.05	461.39
Sulfur dioxide	6.431380	6.405860	−2.656015	−0.900937	6.502211	203	T_c	323.15	328.16
Carbon monoxide	5.365160	4.615630	−1.502620	−2.360346	7.214583	68	T_c	101.15	181.89
Carbon dioxide	6.285898	5.640077	−1.240625	−2.040365	26.542058	218	T_c	253.15	282.44
Methane	4.990555	5.035151	−2.283393	−2.460933	4.378278	93	T_c	183.15	214.96
Ethane	5.240570	7.195872	−4.635360	−0.641593	2.271410	93	T_c	243.15	388.62
Propane	5.532218	7.865983	−5.298339	0.075567	2.154822	90	T_c	280.15	364.59
n-Butane	5.894590	7.877690	−5.041876	−0.151283	3.790784	140	T_c	280.15	379.00
Isobutane	5.985358	6.870170	−3.957940	−0.356806	2.957430	120	T_c	280.15	347.69
n-Pentane	5.752590	9.973145	−6.896600	0.534831	4.463225	153	T_c	273.15	385.00
n-Hexane	5.823448	11.201235	−8.071440	1.348380	3.456936	183	T_c	343.15	333.92
Cyclohexane	3.437910	14.061510	−8.730995	0.671730	0.025579	283	T_c	294.15	394.34
n-Heptane	3.316640	21.992820	−18.808160	5.534334	2.931020	183	T_c	303.15	362.02
n-Octane	4.464230	19.783260	−16.839300	5.360490	3.956600	223	T_c	303.15	361.66

(continued overleaf)

(continued).

Substance	A	B	C	D	E	Range of validation		Verification point	
						T_{min} (K)	T_{max} (K)	T (K)	Δh_V (J/g)
Ethylene	5.143747	6.934186	−4.268831	−0.584890	3.213931	110	T_C	250.15	303.81
Propylene	4.863014	9.287801	−6.283872	0.380303	1.325099	88	T_C	280.15	366.30
1-Butene	5.495498	9.245589	−6.920092	1.448008	2.307289	93	T_C	303.15	354.54
Methanol	5.465579	15.616850	−7.676416	−4.926600	6.334842	183	T_C	373.15	1020.27
Ethanol	14.687649	−15.271194	26.062303	−20.049661	15.816495	253	T_C	293.15	925.91
n-Propanol	5.890074	16.292544	−5.777913	−2.767181	−8.575518	293	T_C	343.65	740.74
n-Butanol	3.925226	18.801993	−5.348588	−2.937749	−0.181003	224	T_C	343.85	654.47
Ethylene glycol	7.079169	8.721527	1.013498	−5.214016	4.441808	333	T_C	333.15	1027.05
Isopropanol	13.846539	−16.693668	32.098404	−19.900215	−9.894509	253	T_C	304.25	757.07
Acetic acid	6.686640	15.014483	−22.086618	3.077698	17.354972	290	T_C	369.35	395.64
Methyl acetate	6.398833	13.125464	−12.779956	5.864096	−9.168706	227	T_C	297.25	439.04
Ethyl acetate	8.568302	3.691585	−0.614594	−0.635779	0.817636	190	T_C	330.15	381.82
Vinyl acetate	7.959075	5.923838	−2.473449	−1.445393	4.013308	273	T_C	336.35	377.62
Methyl-tert-butyl ether	7.677125	3.101996	0.388899	−1.672830	0.763138	217	T_C	339.55	311.02
Acetone	5.731751	9.174230	−4.934225	0.048998	3.735669	183	T_C	373.15	448.73
Benzene	5.007470	10.690810	−7.316719	1.140714	6.786710	283	T_C	333.15	408.31
Toluene	4.607790	13.962160	−10.579148	2.112462	4.284860	183	T_C	323.15	398.25
p-Xylene	8.707394	1.081836	2.571094	−3.435021	9.405508	287	T_C	373.15	361.13
R23	5.346016	10.065090	−6.445266	0.855815	0.100753	158	T_C	253.15	170.98
R32	6.302860	6.634830	−2.530184	−1.546727	4.892836	138	T_C	253.15	344.12
R12	5.672474	8.015016	−5.135777	−0.186518	4.616427	123	T_C	253.15	162.07
R22	5.941889	7.847445	−4.736810	0.118054	4.060834	123	T_C	253.15	220.00
R134a	7.258160	4.942105	−0.155284	−2.761780	12.455860	173	T_C	273.15	198.76
R125	5.543650	11.761440	−8.102561	0.932260	5.177319	173	T_C	273.15	133.16

The required critical temperature T_C and the molar mass M are given in the Table "General data for selected compounds".

Liquid heat capacity correlations for selected compounds.

Equation: $c_P^L = R\left(\dfrac{A}{\tau} + B + C\tau + D\tau^2 + E\tau^3 + F\tau^4\right)$, $\tau = 1 - \dfrac{T}{T_c}$

Substance	A	B	C	D	E	F	Range of validation		Verification point	
							T_{min} (K)	T_{max} (K)	T (K)	c_P^L (J/(g K))
Water	0.255980	12.545950	−31.408960	97.766500	−145.423600	87.018500	273	586	343.15	4.1875
Ammonia	0.518380	8.111677	−3.922537	16.080840	−18.078723	−4.265080	205	355	265.15	4.5730
Hydrogen chloride	–	9.205142	−3.518062	–	–	–	159	185	182.85	1.7487
Chlorine	0.348448	9.581768	−35.823396	198.970547	−456.440234	365.477124	173	383	273.15	0.9791
Nitrogen	0.468003	5.859125	−3.294471	10.949390	−10.931129	3.521050	68	113	103.15	2.4109
Oxygen	0.491044	4.971189	3.165818	−15.730230	35.129358	−24.450400	58	138	103.15	1.7546
Hydrogen	0.412752	1.939203	−0.676502	2.161263	−9.684044	14.410197	16	28	18.75	11.2238
Sulfur dioxide	0.492030	9.882450	−5.736710	12.991290	−7.898405	4.251653	203	413	303.15	1.4033
Carbon monoxide	0.434226	6.852270	−7.826240	25.693570	−39.219300	31.506260	68	125	101.15	2.3261
Carbon dioxide	0.500361	8.830377	−3.868837	10.324592	0.675891	1.845183	218	283	253.15	2.1654
Methane	0.407112	7.061365	−11.142546	37.881423	−60.945893	38.024820	93	168	123.15	3.5803
Ethane	0.428053	10.459037	−18.182884	51.711500	−75.884470	44.500540	93	283	213.15	2.5942
Propane	0.558680	12.802352	−5.738203	−9.551125	26.702477	−15.775134	81	330	280.15	2.5415
n-Butane	0.494221	19.195325	−11.646233	−13.063580	37.552637	−18.508970	140	400	280.15	2.3449
Isobutane	0.497330	18.773678	−14.552160	−0.254646	19.359040	−15.881925	120	380	280.15	2.3207
n-Pentane	0.493157	25.403260	−18.571150	−8.455922	27.383110	−5.186790	153	443	273.15	2.2100
n-Hexane	0.503827	30.695832	−16.437834	−28.145485	50.312710	−18.044650	183	463	333.15	2.4093
Cyclohexane	0.452602	29.040500	−17.260650	−69.508970	254.381420	−307.047120	283	523	298.65	1.8145
n-Heptane	0.274131	39.690828	−38.327579	48.535630	−116.903726	115.685562	183	523	393.15	2.6629

(continued overleaf)

(continued).

Substance	A	B	C	D	E	F	Range of validation		Verification point	
							T_{min} (K)	T_{max} (K)	T (K)	c_p^L (J/(g K))
n-Octane	0.615215	42.753600	−27.390185	8.228510	−58.652690	79.116170	223	533	293.15	2.2091
Ethylene	0.475969	7.565091	−3.361502	−1.602014	24.541197	−22.072075	110	260	250.15	3.3689
Propylene	0.498817	12.324942	−10.168120	11.351221	−9.765026	10.914720	93	350	280.15	2.4918
1-Butene	0.474628	17.444100	−16.144424	18.563360	−20.820925	16.654776	93	383	243.15	2.0809
Methanol	0.612632	13.195540	−5.208870	−45.762120	91.190270	−44.456180	183	463	323.15	2.7082
Ethanol	0.503568	22.442022	−36.783235	160.373246	−466.432673	396.027992	253	473	343.15	3.0966
n-Propanol	0.313992	42.598117	−78.401260	30.612206	40.102166	−16.588491	147	400	293.45	2.3626
n-Butanol	1.203976	50.595314	−91.331488	37.897784	28.225069	−2.351873	185	391	303.95	2.4513
Ethylene glycol	−0.106462	32.547745	−27.000190	22.166665	−43.223418	21.074136	261	493	284.65	2.3342
Isopropanol	−0.028011	37.272721	−60.212901	145.141653	−415.612074	383.532117	185	480	293.85	2.5706
Acetic acid	0.129079	23.227878	−19.035099	11.172422	−32.209650	32.222624	290	413	332.23	2.1868
Methyl acetate	−3.270153	54.948876	−110.920230	73.069065	86.152790	−91.367990	177	373	329.15	2.0263
Ethyl acetate	1.558660	24.189499	−26.451003	27.644553	−25.493382	29.857034	190	350	316.35	1.9958
Vinyl acetate	0.532294	28.920121	−26.329666	18.066371	−41.778918	48.998543	260	389	266.35	1.8743
Methyl-tert-butyl ether	0.681223	28.641641	−26.276132	26.982673	−34.177133	22.912098	165	328	293.85	2.1107
Acetone	0.440183	17.664600	−8.866356	−11.726743	34.192440	−17.424056	183	503	333.15	2.2404
Benzene	0.497842	22.393654	−15.407940	3.391770	−25.948450	39.510594	283	523	343.15	1.8693
Toluene	0.472576	28.776500	−29.760240	48.793090	−113.292550	94.204740	183	563	313.15	1.7496
p-Xylene	−0.298962	48.971447	−144.539120	451.279840	−811.386690	551.296903	287	567	335.95	1.8151
R23	0.583673	8.825460	3.660254	−28.715384	71.336899	−57.631705	158	288	263.15	1.6319
R32	0.482781	9.736560	−7.746770	18.994675	−22.694329	16.232080	138	343	293.15	1.8863
R12	0.469917	14.390100	−17.006910	53.677190	−104.511075	78.215790	123	373	293.15	0.9757
R22	0.509972	11.154020	−3.730131	−1.359015	11.325878	−5.456086	123	353	293.15	1.2395
R134a	0.509614	17.066100	−15.834770	39.390760	−69.727496	53.455900	173	368	293.15	1.4053
R125	0.619806	15.473500	2.397560	−32.903800	33.244360	22.729920	173	333	293.15	1.3682

The required critical temperature T_c and the molar mass M are given in the Table "General data for selected compounds".

Ideal gas heat capacity correlations for selected compounds.

Equations:

$$\frac{c_P^{id}}{R} = B + (C - B)\gamma^2 \left[1 + (\gamma - 1)(D + E\gamma + F\gamma^2 + G\gamma^3 + H\gamma^4)\right], \quad \gamma = \frac{T}{A + T} \quad (A.1)$$

$$\frac{c_P^{id}}{J/(kmol\ K)} = A + B\left(\frac{C/T}{\sinh(C/T)}\right)^2 + D\left(\frac{E/T}{\cosh(E/T)}\right)^2 \quad (A.2)$$

Substance	Number of equation	A	B	C	D
Water	A.2	33484.75	9275.30	1218.48	20241.42
Ammonia	A.2	34083.18	26087.00	990.77	33100.02
Hydrogen chloride	A.2	29141.68	8602.70	2038.41	99556.02
Chlorine	A.2	29197.65	8502.80	405.49	−3253.99
Nitrogen	A.2	29108.79	8526.28	1678.41	66784.83
Oxygen	A.2	29116.90	10437.46	2565.44	9338.84
Hydrogen	A.2	20463.61	9591.19	216.79	16524.83
Sulfur dioxide	A.1	904.53351	3.98118	5.66358	3.21967
Carbon monoxide	A.2	29104.60	8665.81	3043.79	8336.18
Carbon dioxide	A.1	188.27392	3.50426	8.29771	4.71883
Methane	A.2	33356.35	45763.85	1025.42	48699.01
Ethane	A.1	903.41135	4.48148	11.69046	8.47923
Propane	A.1	1222.85277	4.63428	6.17777	−31.84476
n-Butane	A.1	668.64918	8.90810	14.24670	41.04664
Isobutane	A.1	2084.48334	5.07542	7.06198	−264.30218
n-Pentane	A.1	1074.74180	8.97762	11.92509	31.16797
n-Hexane	A.1	918.01459	11.29386	18.05869	29.33969
Cyclohexane	A.1	671.99039	4.04945	5.43698	67.17766
n-Heptane	A.1	751.21484	14.78843	17.82902	121.15456
n-Octane	A.1	580.69298	22.78850	32.50664	73.77537
Ethylene	A.1	100.67955	3.46632	18.86798	−1.58496
Propylene	A.1	2039.57238	4.27246	6.58039	−84.96332
1-Butene	A.1	444.42742	4.00215	4.77657	−103.78383
Methanol	A.1	72.32165	−43.80600	18.04246	−8.74148
Ethanol	A.1	1165.86482	4.70209	9.77865	−1.17688
n-Propanol	A.2	61723.88	203342.08	757.27	−96727.69
n-Butanol	A.2	74802.09	167505.02	−711.59	130880.37
Ethylene glycol	A.1	316.79019	3.05241	3.58464	−152.76283
Isopropanol	A.1	313.10693	8.96218	9.41683	702.94470
Acetic acid	A.1	589.88417	5.10642	5.52776	211.48957
Methyl acetate	A.2	54664.40	179099.95	595.63	−106514.52
Ethyl acetate	A.2	99847.16	210798.73	−947.39	−52228.85
Vinyl acetate	A.1	339.00453	10.85084	11.20264	1116.09686
Methyl-tert-butyl ether	A.2	100524.03	206853.45	−807.44	137641.68
Acetone	A.2	58083.25	101389.15	−761.09	84650.28
Benzene	A.1	560.93643	4.00088	6.20811	22.19301
Toluene	A.1	1250.94941	3.99957	10.73366	−19.78047
p-Xylene	A.1	1299.91280	4.43091	6.06047	−229.73910
R23	A.1	766.08865	4.00055	4.56857	57.65285
R32	A.1	889.23598	4.39579	9.60005	15.90687
R12	A.1	722.38297	3.13821	3.33217	−818.72651
R22	A.1	1036.65937	3.60012	4.64243	−89.43873
R134a	A.1	1108.39974	4.00203	6.01963	−138.01236
R125	A.1	863.90053	6.11410	6.41788	−301.84855

(continued overleaf)

(continued).

E	F	G	H	Range of validation		Verification point	
				T_{min} (K)	T_{max} (K)	T(K)	c_p^{id} (J/(g K))
2919.59	–	–	–	278	1273	373.15	1.8909
2905.60	–	–	–	196	1500	394.75	2.2598
8582.33	–	–	–	50	1773	355.25	0.7996
3892.43	–	–	–	173	1123	273.15	0.4721
10672.63	–	–	–	73	1773	313.15	1.0399
1149.97	–	–	–	63	1773	1200	1.1150
3955.44	–	–	–	24	2273	513.15	14.6353
−88.88144	186.92401	−121.55529	0.0	203	1773	343.15	0.6483
1531.93	–	–	–	73	1773	313.15	1.0407
−11.75109	13.94035	−6.73002	0.0	223	1273	373.15	0.9158
2664.97	–	–	–	93	1823	293.15	2.2073
−77.02151	122.97656	−74.05999	0.0	123	1500	273.15	1.6518
−487.58918	1216.90986	−972.09252	0.0	123	1500	373.15	2.0080
−258.18297	411.82384	−258.68803	0.0	163	1500	373.15	2.0314
−47.27861	2309.95342	−3524.85868	0.0	143	1223	373.15	2.0137
−592.50351	1201.64991	−830.32720	0.0	183	1673	323.15	1.7761
−307.72814	556.38463	−356.72160	0.0	193	1773	313.15	1.7239
−795.75587	1090.93493	−701.78714	0.0	283	1273	373.15	1.6512
−843.82591	1360.01383	−824.87302	0.0	183	1773	303.15	1.6734
−348.56358	493.58782	−270.29141	0.0	233	1773	303.15	1.6663
6.37008	0.00005	−0.00003	0.0	143	1273	473.15	2.1337
−389.76058	2181.03718	−2786.71808	0.0	100	1210	280.15	1.4656
269.01469	−545.25186	27.15311	0.0	93	1273	293.15	1.4991
10.37780	−0.33156	−1.05900	0.0	183	2023	303.15	1.3840
−135.76765	322.75959	−247.73488	0.0	100	1500	373.15	1.6635
880.13	–	–	–	200	1773	473.65	2.0484
2223.47	–	–	–	200	1773	473.65	2.1413
212.94465	84.60646	−662.81134	0.0	100	1000	373.15	1.5568
−3295.08198	5345.62488	−3435.86066	0.0	100	1500	573.15	2.3965
−1404.10912	2057.41805	−1317.64721	0.0	50	1195	373.15	1.2540
660.06	–	–	–	298	1773	424.55	1.4729
1203.35	–	–	–	268	1273	473.65	1.7668
−5117.97374	8086.03265	−5022.37492	0.0	100	1500	394.75	1.4167
2358.16	–	–	–	268	1500	336.85	1.5946
2186.59	–	–	–	183	1773	313.15	1.3247
−159.24845	−48.97058	155.54132	0.0	200	1023	373.15	1.3487
−218.24713	677.57410	−595.41549	0.0	183	1273	313.15	1.1858
−186.84354	1156.78350	−1164.53180	0.0	200	1500	336.85	1.3369
−767.99181	1432.42397	−923.07484	0.0	163	1773	363.15	0.8223
−111.31449	190.11067	−112.69276	0.0	143	1773	373.15	0.9440
1463.34239	−979.86258	−188.05965	0.0	123	1773	373.15	0.6677
22.60799	204.58845	−241.99955	0.0	123	1773	373.15	0.7332
334.51996	−417.38313	−4.62768	0.0	173	1773	373.15	0.9632
−423.46781	1665.31109	−2612.24748	0.0	173	1773	373.15	0.8932

The required molar mass M is given in the Table "General data for selected compounds".

Liquid viscosity correlations for selected compounds.

$$\text{Equation:} \quad \frac{\eta^L}{\text{Pa s}} = E \cdot \exp\left[A \left(\frac{C-T}{T-D}\right)^{1/3} + B \left(\frac{C-T}{T-D}\right)^{4/3} \right]$$

Substance	A	B	C	D	10^8 E	Range of validation		Verification point	
						T_{min} (K)	T_{max} (K)	T (K)	η^L (mPa s)
Water	1.023096	0.669483	640.884	124.8677	4694.832	274	638	333.15	0.4645
Ammonia	2.081580	0.637466	409.229	29.3262	2290.330	196	400	278.65	0.1605
Hydrogen chloride	5.973708	−0.692313	446.332	99.6543	55.655	233	313	233.15	0.1647
Chlorine	0.485630	1.344777	439.888	−109.3797	17135.830	173	416	223.15	0.5582
Nitrogen	2.238583	0.018882	127.141	36.1391	1343.698	68	123	113.15	0.0478
Oxygen	2.148550	0.000532	157.422	36.1141	1839.315	73	148	113.15	0.1098
Hydrogen	1.525476	0.000001	33.246	7.4900	285.192	14	33	18.2	0.0157
Sulfur dioxide	12.383610	10.060090	598.154	−1525.6990	12.008	225	400	280.25	0.3347
Carbon monoxide	1.037410	1.391890	169.937	−43.9656	2893.372	68	131	88.15	0.1460
Carbon dioxide	2.701205	4.821786	304.972	−219.7839	2965.800	218	298	221.15	0.2374
Methane	2.313648	0.668030	200.211	−19.1784	1039.192	93	178	123.15	0.0920
Ethane	2.435910	0.097102	309.734	33.4667	1554.180	103	293	213.15	0.1175
Propane	2.563440	0.161368	372.533	38.0331	1751.140	90	350	280.15	0.1175
n-Butane	2.629043	0.237324	428.065	37.0486	1801.085	140	420	272.65	0.2035
Isobutane	2.794999	0.373353	419.617	28.0546	1567.570	130	390	250.15	0.2615
n-Pentane	2.803312	0.250738	474.455	32.8562	1652.600	163	463	273.15	0.2830
n-Hexane	0.654192	0.364412	902.902	18.8312	4603.184	175	343	256.15	0.4596
Cyclohexane	2.889560	0.000853	4884.230	115.5702	17.677	280	553	303.15	0.8214
n-Heptane	2.960020	0.247153	553.942	62.1719	1424.064	203	513	263.15	0.6070

(continued overleaf)

(continued).

Substance	A	B	C	D	10^8 E	Range of validation		Verification point	
						T_{min} (K)	T_{max} (K)	T (K)	η^L (mPa s)
n-Octane	2.870880	0.354830	577.534	48.9830	1721.664	223	553	283.15	0.6176
Ethylene	2.098284	0.091540	281.461	47.5527	2107.291	110	270	250.15	0.0655
Propylene	1.413955	0.360563	712.791	−16.7129	1180.602	88	320	283.35	0.1039
1-Butene	1.122814	1.464498	431.564	−78.5456	4506.029	120	413	259.15	0.2009
Methanol	3.748710	0.213107	546.888	82.9003	818.340	176	483	303.65	0.5010
Ethanol	6.168237	−0.001200	740.289	90.7653	38.125	169	516	326.35	0.6496
n-Propanol	6.140716	0.001411	571.434	133.6983	135.184	217	370	276.25	3.4021
n-Butanol	4.364716	1.005752	743.418	27.0658	211.208	190	562	327.35	1.2962
Ethylene glycol	−0.443561	1.318848	895.794	79.3616	20317.761	261	450	320.15	7.6071
Isopropanol	0.536067	1.881026	1143.342	−77.3655	386.920	188	363	293.45	2.3078
Acetic acid	2.230867	0.715487	527.102	77.2313	5567.007	288	393	327.15	0.7514
Methyl acetate	2.072517	2.239494	489.756	−176.4315	3968.760	250	425	277.65	0.4462
Ethyl acetate	2.233189	5.921434	608.986	−574.6203	1981.913	220	473	286.65	0.4893
Vinyl acetate	−0.076134	1.103327	668.129	−13.9030	10998.540	225	346	275.75	0.5284
Methyl-tert-butyl ether	1.416944	0.905146	712.916	−42.5044	2260.808	180	450	293.65	0.3503
Acetone	1.895877	0.337109	766.008	21.4982	1613.371	190	333	298.15	0.3046
Benzene	2.486500	2.199610	787.253	−255.5922	855.545	279	545	303.15	0.5633
Toluene	2.954350	−0.000010	1047.179	137.2255	397.694	221	588	275.55	0.7500
p-Xylene	0.809643	1.371902	638.509	−40.4018	6772.298	282	613	319.75	0.4728
R23	1.720363	0.385773	288.676	44.7211	5570.377	118	285	276.25	0.1074
R32	2.575043	−0.035320	358.978	117.8077	2372.017	137	287	279.35	0.1789
R12	2.886860	−0.000216	380.360	53.1287	2553.078	183	373	253.15	0.3056
R22	2.506270	0.001628	369.199	55.9635	2948.260	178	363	273.15	0.1991
R134a	2.589129	0.000033	374.968	127.0868	2718.736	173	373	243.15	0.4051
R125	2.913370	0.211906	3895.012	−209.8739	3.090	216	298	293.13	0.1581

Vapor viscosity correlations for selected compounds at low pressures.

$$\text{Equation:} \quad \frac{\eta^{id}}{\text{Pa s}} = \frac{AT^B}{1 + CT^{-1} + DT^{-2}}$$

Substance	10^6 A	B	C	D	Range of validation		Verification point	
					T_{min} (K)	T_{max} (K)	T (K)	η^{id} (μPa s)
Water	0.501246	0.709247	869.465599	−90063.891	278	1173	353.15	11.7342
Ammonia	1.113854	0.573628	695.164458	−39937.499	196	1000	280.15	9.4972
Hydrogen chloride	2.001380	0.494814	414.835270	−8538.403	200	1000	284.25	13.9219
Chlorine	0.164926	0.801668	42.635734	3223.277	200	1000	273.15	12.3467
Nitrogen	0.847662	0.574033	75.437536	56.771	73	1773	313.15	18.4896
Oxygen	1.102259	0.563255	96.450180	−3.515	54	1500	358.65	23.8665
Hydrogen	0.169104	0.692485	−7.634394	467.120	78	3000	328.25	9.5254
Sulfur dioxide	4.465290	0.379352	600.546520	−469.440	198	1000	493.25	21.1826
Carbon monoxide	0.734306	0.588574	52.318660	1018.822	68	1473	379.15	21.1283
Carbon dioxide	4.719875	0.373279	512.686300	−6119.961	223	1473	233.15	11.7023
Methane	1.119178	0.493234	214.627200	−3952.087	93	1000	233.15	8.9131
Ethane	0.612179	0.567691	199.802139	799.607	103	973	233.15	7.2233
Propane	0.173966	0.734798	143.207060	−7147.859	85	1000	313.15	8.5718
n-Butane	0.075828	0.837082	67.618677	−2141.762	143	963	313.15	7.7972
Isobutane	0.121656	0.767652	92.147580	−2356.256	131	1000	284.25	7.1853
n-Pentane	0.208109	0.692504	209.355310	−9032.643	153	1173	323.15	7.2872
n-Hexane	0.829617	0.524075	548.537420	−19506.100	183	1223	283.15	5.9364
Cyclohexane	1.707204	0.444162	797.904850	−55646.900	280	900	377.65	8.7486
n-Heptane	0.893462	0.523746	800.554090	−50014.750	213	973	303.15	5.7538
n-Octane	0.647995	0.558004	678.981600	−35397.490	223	1273	303.15	5.5056

(continued overleaf)

(continued).

Substance	10^6 A	B	C	D	Range of validation		Verification point	
					T_{min} (K)	T_{max} (K)	T (K)	η^{id} (μPa s)
Ethylene	1.503552	0.456140	288.342422	73.362	170	1000	300.55	10.3537
Propylene	0.921458	0.514534	301.722140	−179.984	88	1000	375.95	10.8115
1-Butene	1.590010	0.435591	427.677300	1432.659	175	1000	348.65	9.0969
Methanol	0.477915	0.641076	284.838034	−3230.713	240	1000	400.05	13.1568
Ethanol	0.106484	0.806176	53.133947	−19.624	200	1000	284.25	8.5323
n-Propanol	1.842372	0.446996	622.575600	1735.926	200	1000	284.25	7.1686
n-Butanol	1.261930	0.473919	510.297350	−143.068	185	1000	313.25	7.3169
Ethylene glycol	0.149939	0.770025	161.702296	−7162.270	260	1000	376.95	10.4782
Isopropanol	0.197482	0.724489	176.811793	13.775	185	1000	313.95	8.1363
Acetic acid	2.181432	0.479850	1126.897790	−76689.777	290	800	343.55	9.9011
Methyl acetate	2.515403	0.453254	1231.787070	−82656.705	250	800	336.85	8.9530
Ethyl acetate	1.845324	0.424328	522.471640	−2316.919	190	1000	360.25	9.2229
Vinyl acetate	1.698675	0.447738	583.233050	−17183.346	180	1000	352.95	9.3404
Methyl-tert-butyl ether	0.168450	0.724727	119.588355	−538.542	165	1000	340.45	8.5561
Acetone	0.060442	0.894053	137.302579	−10508.844	178	1000	308.15	7.6028
Benzene	33.944850	0.179320	3840.605800	−114442.650	279	1000	353	8.8669
Toluene	26.356700	0.089851	1411.952400	49235.071	178	1000	351.15	8.2336
p-Xylene	1.297141	0.436899	446.549129	−1248.050	286	1000	323.95	6.8501
R23	0.588485	0.620799	76.103710	10940.869	191	1000	318.75	15.6563
R32	2.505242	0.445997	489.840880	−10877.245	173	513	303.15	12.8280
R12	2.043170	0.483200	633.356640	−49455.239	250	575	284.25	11.9751
R22	2.496580	0.445339	480.172300	−10278.585	173	513	293.15	12.4428
R134a	2.535768	0.434922	519.195800	−11741.298	173	493	293.15	11.3870
R125	2.322973	0.450581	419.744700	−7324.273	173	453	293.15	12.8005

Liquid thermal conductivity correlations for selected compounds.

Equations:
$$\frac{\lambda^L}{W/(Km)} = A(1 + B\tau^{1/3} + C\tau^{2/3} + D\tau), \quad \tau = 1 - \frac{T}{T_c} \quad (A.3)$$

$$\frac{\lambda^L}{W/(Km)} = A + B\frac{T}{K} + C\left(\frac{T}{K}\right)^2 + D\left(\frac{T}{K}\right)^3 + E\left(\frac{T}{K}\right)^4 \quad (A.4)$$

Substance	Number of equation	A	B	C	D	E	Range of validation T_{min} (K)	T_{max} (K)	Verification point T (K)	λ^L (W/(Km))
Water	A.4	−2.41488460	0.024516	−7.312129·10⁻⁵	9.949215·10⁻⁸	−5.373019·10⁻¹¹	275	623	403.15	0.6843
Ammonia	A.3	0.45812600	−3.407603	7.363882	−3.110926	—	203	383	303.15	0.4592
Hydrogen chloride	A.3	0.12208099	—	—	5.589949	—	173	323	275.75	0.2249
Chlorine	A.3	0.08364446	−2.244378	6.000759	−1.768471	—	173	410	273.15	0.1478
Nitrogen	A.3	0.03723700	0.819379	1.546830	2.956170	—	73	123	113.15	0.0756
Oxygen	A.3	0.04849099	0.662250	−0.157473	4.180248	—	73	148	133.15	0.0912
Hydrogen	A.3	0.31257102	−4.480248	8.852520	−5.476372	—	14	31	22.05	0.1012
Sulfur dioxide	A.3	0.11179000	0.049672	−0.083144	2.454405	—	198	400	304.15	0.1920
Carbon monoxide	A.3	0.04654000	0.258036	−0.495500	5.394290	—	68	125	83.15	0.1372
Carbon dioxide	A.3	0.11094000	−2.381636	4.902130	0.184516	—	218	293	273.15	0.1082
Methane	A.3	0.09222700	−1.466680	2.951520	1.069580	—	93	188	143.15	0.1394
Ethane	A.3	0.03263560	5.689980	−11.845846	15.965704	—	103	303	243.15	0.1142
Propane	A.3	0.04821980	1.815034	−4.456460	7.138120	—	85	360	290.15	0.0976
n-Butane	A.3	0.03328196	6.417034	−14.997547	15.443150	—	135	420	358.15	0.0840
Isobutane	A.3	0.01668096	15.527922	−36.061721	35.162423	—	203	373	293.15	0.0931
n-Pentane	A.3	0.01128515	26.817246	−55.053399	47.844687	—	193	443	293.15	0.1090
n-Hexane	A.3	0.04551534	—	—	3.941700	—	178	370	309.35	0.1156

(continued overleaf)

(continued).

Substance	Number of equation	A	B	C	D	E	Range of validation		Verification point	
							T_{min} (K)	T_{max} (K)	T (K)	λ^L (W/(Km))
Cyclohexane	A.3	−0.01629480	−36.933160	73.652630	−52.004130	–	280	354	287.45	0.1260
n-Heptane	A.3	0.00554613	30.528260	−38.792760	45.698080	–	190	370	260	0.1343
n-Octane	A.3	0.00609920	28.137870	−38.823160	45.582000	–	216	399	273.95	0.1354
Ethylene	A.3	0.00553668	76.084404	−153.166503	150.029799	–	111	270	230.15	0.1239
Propylene	A.3	0.00393223	83.538953	−142.763856	128.355015	–	253	330	313.15	0.0944
1-Butene	A.3	0.06673001	–	–	2.200248	–	88	267	248.05	0.1267
Methanol	A.3	0.13951266	–	–	1.033317	–	176	338	298.15	0.2000
Ethanol	A.3	0.11124054	–	–	1.217548	–	160	353	312.25	0.1644
n-Propanol	A.3	0.10464898	–	–	1.104913	–	201	535	372.15	0.1401
n-Butanol	A.3	0.09905170	–	–	1.156703	–	185	553	303.95	0.1518
Ethylene glycol	A.3	0.00135174	−1170.50402	3868.88659	−2632.07239	–	260	593	326.55	0.2592
Isopropanol	A.3	0.06591520	5.104116	−12.120860	9.634190	–	186	507	403.15	0.1168
Acetic acid	A.3	0.10546986	–	–	1.028583	–	292	573	300.45	0.1589
Methyl acetate	A.3	0.06649329	–	–	3.175936	–	176	493	286.25	0.1583
Ethyl acetate	A.3	0.00008142	4414.597964	−7292.83987	6016.90670	–	190	493	282.55	0.1491
Vinyl acetate	A.3	0.07232333	–	–	2.541833	–	181	513	313.35	0.1452
Methyl-tert-butyl ether	A.3	0.03620530	2.747481	−3.446200	5.939223	–	165	328	311.15	0.1235
Acetone	A.3	0.07089878	–	–	3.058789	–	179	343	299.85	0.1598
Benzene	A.3	0.06389367	–	–	2.610350	–	279	413	363.55	0.1228
Toluene	A.3	−0.01200600	−32.702200	55.625500	−42.111000	–	178	475	287.45	0.1339
p-Xylene	A.3	0.06531592	2.109022	−6.476927	6.704619	–	287	413	333.05	0.1209
R23	A.3	0.06626834	–	–	2.521323	–	118	243	216.75	0.1125
R32	A.3	0.09911900	0.076417	3.712623	−1.460831	–	163	348	263.15	0.2139
R12	A.3	0.03202120	0.066322	−0.117768	5.320465	–	150	363	273.45	0.0812
R22	A.3	0.06064400	−1.049875	2.115769	1.076857	–	163	363	263.15	0.0933
R134a	A.3	0.07364600	−1.756262	2.633737	1.096548	–	173	373	293.15	0.0834
R125	A.3	0.06652500	−2.316396	5.016910	−1.260304	–	173	333	298.15	0.0622

The required critical temperature T_c is given in the Table "General data for selected compounds."

Appendix A Pure Component Parameters | 637

Vapor thermal conductivity correlations for selected compounds at low pressures.

Equation: $\dfrac{\lambda^{\mathrm{id}}}{\mathrm{W/(Km)}} = \dfrac{\sqrt{T_r}}{A + \dfrac{B}{T_r} + \dfrac{C}{T_r^2} + \dfrac{D}{T_r^3}}$

Substance	A	B	C	D	Range of validation		Verification point	
					$T_{\min}$ (K)	$T_{\max}$ (K)	T (K)	λ^{id} (W/(Km))
Water	2.000796	14.169449	5.237322	−2.013030	278	1073	353.15	0.0223
Ammonia	1.771596	25.648577	−2.263180	1.120586	203	1000	279.75	0.0221
Hydrogen chloride	21.127674	59.560815	−21.349900	3.611350	159	1000	291.75	0.0144
Chlorine	35.164546	37.254610	6.665306	−1.775010	200	1000	373.15	0.0114
Nitrogen	30.516440	100.761278	−84.097324	35.705221	93	1223	323.15	0.0270
Oxygen	25.097426	69.251077	−35.326230	10.740183	80	1495	284.4	0.0251
Hydrogen	9.355729	86.604301	−163.421012	81.498903	22	1600	271.15	0.1619
Sulfur dioxide	27.058610	−3.315050	54.097200	−15.764240	250	900	386.85	0.0138
Carbon monoxide	32.987680	73.919420	−28.849020	4.398120	70	1500	973.15	0.0636
Carbon dioxide	18.890387	11.895399	40.897183	−13.035620	223	973	373.15	0.0227
Methane	0.826598	71.374130	−29.034600	6.107060	93	1000	273.15	0.0310
Ethane	3.233358	19.029493	27.805380	−4.717205	113	1000	313.35	0.0231
Propane	1.351410	19.983850	20.026000	−3.553930	203	1000	352.55	0.0243
n-Butane	3.215504	10.073633	21.826312	−3.621102	203	1000	387.45	0.0267
Isobutane	9.038150	−1.395520	32.309630	−5.577840	203	1000	273.15	0.0135
n-Pentane	−1.175174	22.560610	10.125880	−1.653540	273	1000	273.15	0.0129
n-Hexane	−1.521446	21.520796	9.432466	−1.141610	339	1000	478.25	0.0317
Cyclohexane	−1.927981	16.082202	6.833899	2.209833	323	1000	467.15	0.0302
n-Heptane	0.059077	15.934620	11.446140	−1.120901	273	1000	373.95	0.0190
n-Octane	−1.585376	18.857260	9.328904	−1.062463	339	1000	443.35	0.0247

(*continued overleaf*)

(continued).

Substance	A	B	C	D	Range of validation		Verification point	
					T_{min} (K)	T_{max} (K)	T (K)	λ^{id} (W/(Km))
Ethylene	0.284146	33.661050	29.235217	−10.510851	169	900	323.25	0.0238
Propylene	3.226603	20.193072	19.143530	−2.027584	225	1000	306.95	0.0180
1-Butene	−1.734943	28.508830	10.986120	−1.536220	225	800	376.35	0.0228
Methanol	−2.097847	17.979169	12.828005	−3.546304	273	1000	324.38	0.0179
Ethanol	−2.259996	21.881454	5.488168	−0.304888	293	1000	293.15	0.0147
n-Propanol	−1.658351	19.595786	6.903138	−0.755881	372	720	372.15	0.0215
n-Butanol	−22.982020	69.128385	−28.209030	5.678522	370	800	573.65	0.0438
Ethylene glycol	−2.986535	18.296988	3.589649	−0.321265	470	1000	470.45	0.0251
Isopropanol	0.670345	9.782420	20.438520	−5.081330	304	1000	473.15	0.0339
Acetic acid	0.704143	7.249838	22.437444	−4.751947	295	687	294.75	0.0105
Methyl acetate	−3.526070	24.686506	11.814704	−1.450865	273	1000	330.05	0.0142
Ethyl acetate	−1.481710	15.774236	24.322500	−5.597174	273	1000	311.45	0.0115
Vinyl acetate	−2.302838	18.466591	16.708090	−2.785195	346	1000	345.65	0.0152
Methyl-tert-butyl ether	5.121987	17.774514	6.213344	0.976908	273	1000	349.55	0.0183
Acetone	−1.729843	20.390480	15.370780	−1.971064	293	1000	364.75	0.0166
Benzene	1.014925	17.781470	9.661990	−0.409016	339	1000	478.25	0.0267
Toluene	0.631015	17.186500	6.139255	0.177187	384	1000	383.75	0.0190
p-Xylene	−1.261201	36.201340	−26.431401	14.075927	320	1000	391.55	0.0177
R23	21.179034	15.915758	49.016112	−13.336371	191	1000	318.75	0.0151
R32	−7.018654	77.642855	−5.084861	−1.550638	163	473	283.15	0.0114
R12	1.049970	67.165700	3.157079	−0.843364	243	700	363.55	0.0130
R22	3.086424	39.018330	32.276130	−8.032932	173	473	313.15	0.0114
R134a	−1.346631	67.782530	−34.251530	19.287077	193	1000	313.15	0.0144
R125	15.887050	−7.776550	75.408970	−25.482380	225	1000	333.15	0.0166

The required critical temperature T_c is given in the Table "General data for selected compounds."

Appendix A Pure Component Parameters

Surface tension correlations for selected compounds.

Equation: $\dfrac{\sigma}{\text{N/m}} = A(1 - T_r)^{B + CT_r + DT_r^2 + ET_r^3}$

Substance	A	B	C	D	E	Range of validation		Verification point	
						T_{min} (K)	T_{max} (K)	T (K)	σ (mN/m)
Water	0.158990	1.800132	−1.178740	−0.410700	0.964067	273	608	292.85	73.8843
Ammonia	0.103258	1.320142	−0.300843	0.350729	−0.154149	196	395	239.65	34.2903
Hydrogen chloride	0.085166	1.296700	—	—	—	159	316	246.15	13.5142
Chlorine	0.069660	1.290711	−0.595300	0.689530	−0.297926	173	404	273.15	21.3050
Nitrogen	0.029512	1.278290	0.044647	−0.157933	0.089708	63	123	112.95	1.7387
Oxygen	0.037848	1.200264	−0.019308	0.080429	−0.050479	54	149	73.15	17.4984
Hydrogen	0.005631	1.201187	0.024380	−0.439855	0.303610	14	32	18.05	2.3136
Sulfur dioxide	0.088735	1.329490	−0.475261	0.585308	−0.258743	198	406	295.85	22.1365
Carbon monoxide	0.040422	2.716037	−3.185805	2.537085	−0.839567	68	123	85.15	8.7400
Carbon dioxide	0.081563	1.527091	−0.836249	0.924520	−0.348195	216	299	239.65	11.3432
Methane	0.033795	0.837809	0.524994	−0.287428	0.048288	91	185	153.75	5.5421
Ethane	0.048339	1.181172	−0.012966	0.082275	−0.055848	90	294	282.65	2.1598
Propane	0.050936	1.225863	−0.031551	0.052927	−0.027713	85	355	280.05	9.0520
n-Butane	0.051625	1.214678	−0.063855	0.120511	−0.053450	135	355	273.30	14.8635
Isobutane	0.058431	2.186481	−2.691774	3.097050	−1.323847	114	377	283.95	11.4621
n-Pentane	0.052049	1.208440	−0.014871	0.019269	−0.008941	143	435	280.85	17.3662
n-Hexane	0.054521	1.260288	—	—	—	175	456	299.35	17.7507
Cyclohexane	0.065637	1.266746	—	—	—	273	524	332.15	20.5630
n-Heptane	0.054835	1.269376	—	—	—	183	521	314.35	18.1866
n-Octane	0.053060	1.241357	—	—	—	216	532	290.55	21.8806

(continued overleaf)

(continued).

Substance	A	B	C	D	E	Range of validation		Verification point	
						T_{min} (K)	T_{max} (K)	T (K)	σ (mN/m)
Ethylene	0.052923	1.278061	–	–	–	104	273	272.95	0.6841
Propylene	0.053598	1.213030	–	–	–	88	351	307.15	5.7954
1-Butene	0.056424	1.349660	−0.561878	0.987326	−0.561644	88	402	297.55	12.0642
Methanol	0.071115	2.452836	−1.961439	−0.637268	1.248386	226	467	285.25	23.2116
Ethanol	0.063735	2.466253	−3.218905	1.914451	−0.100171	273	479	285.25	23.1181
n-Propanol	0.044017	0.781202	–	–	–	153	473	323.15	21.4290
n-Butanol	0.049406	0.925929	–	–	–	273	523	323.15	22.4239
Ethylene glycol	0.075776	1.134073	−0.434579	−1.570191	2.334810	263	471	453.15	33.4920
Isopropanol	0.046935	0.908095	–	–	–	287	353	323.15	18.7562
Acetic acid	0.057531	1.077013	–	–	–	290	576	353.45	21.6121
Methyl acetate	0.062040	1.045756	–	–	–	175	489	279.85	26.7625
Ethyl acetate	0.061747	1.332643	−0.604840	0.690144	−0.292236	190	506	294.95	23.6160
Vinyl acetate	0.065003	1.218115	–	–	–	180	501	305.15	22.0854
Methyl-tert-butyl ether	0.059863	1.236617	–	–	–	165	480	304.55	18.5303
Acetone	0.070273	1.215138	–	–	–	178	480	298.00	24.0301
Benzene	0.070259	1.177885	0.028702	0.078840	−0.056076	279	547	368.15	19.2833
Toluene	0.067024	1.246125	0.024241	−0.035845	0.009383	178	570	308.75	26.6487
p-Xylene	0.064561	1.268807	–	–	–	286	599	373.15	19.8339
R23	0.062289	1.570226	−0.795218	0.739061	−0.251189	118	280	261.15	4.5902
R32	0.077760	1.411980	−0.405818	0.513851	−0.242771	137	329	295.15	7.2997
R12	0.065030	1.746093	−0.990751	0.852173	−0.289736	243	370	280.65	10.9640
R22	0.065425	1.549562	−0.847518	0.922886	−0.374168	116	353	293.15	8.7897
R134a	0.058065	1.400084	−0.682828	0.922975	−0.408788	172	364	293.00	8.7995
R125	0.080614	2.465560	−1.588870	0.445522	0.057276	233	328	283.15	5.6915

The required critical temperature T_c is given in the Table "General data for selected compounds."

Appendix B
Coefficients for High Precision Equations of State

Table B.1 Coefficients for nonpolar fluids (Eq. (2.113)).

Coefficients	Methane 1	Ethane 2	Propane 3	n-Butane 4	n-Pentane 5
n_1	0.89269676	0.97628068	1.0403973	1.0626277	1.0968643
n_2	−2.5438282	−2.6905251	−2.8318404	−2.8620952	−2.9988888
n_3	0.64980978	0.73498222	0.8439381	0.88738233	0.99516887
n_4	0.020793471	−0.035366206	−0.076559592	−0.12570581	−0.16170709
n_5	0.070189104	0.084692031	0.094697373	0.10286309	0.1133446
n_6	0.00023700378	0.00024154594	0.00024796475	0.00025358041	0.00026760595
n_7	0.16653334	0.23964954	0.2774376	0.323252	0.40979882
n_8	−0.043855669	−0.042780093	−0.043846001	−0.037950761	−0.040876423
n_9	−0.1572678	−0.22308832	−0.26991065	−0.32534802	−0.38169482
n_{10}	−0.035311675	−0.051799954	−0.069313413	−0.079050969	−0.10931957
n_{11}	−0.029570024	−0.027178426	−0.029632146	−0.020636721	−0.032073223
n_{12}	0.014019842	0.011246305	0.014040127	0.0057053809	0.016877016

Coefficients	n-Hexane 6	n-Heptane 7	n-Octane 8	Argon 9	Oxygen 10
n_1	1.0553238	1.0543748	1.0722545	0.85095715	0.88878286
n_2	−2.6120616	−2.6500682	−2.4632951	−2.4003223	−2.4879433
n_3	0.76613883	0.81730048	0.65386674	0.54127841	0.59750191
n_4	−0.29770321	−0.30451391	−0.36324974	0.016919771	0.0096501817
n_5	0.11879908	0.12253869	0.1271327	0.068825965	0.071970429
n_6	0.00027922861	0.00027266473	0.00030713573	0.00021428033	0.00022337443
n_7	0.4634759	0.49865826	0.52656857	0.17429895	0.18558686
n_8	0.011433197	−0.00071432815	0.019362863	−0.033654496	−0.038129368
n_9	−0.48256969	−0.54236896	−0.58939427	−0.135268	−0.15352245
n_{10}	−0.093750559	−0.13801822	−0.14069964	−0.016387351	−0.026726815
n_{11}	−0.0067273247	−0.0061595287	−0.0078966331	−0.024987667	−0.025675299
n_{12}	−0.0051141584	0.0004860251	0.0033036598	0.0088769205	0.0095714302

(continued overleaf)

Table B.1 (continued).

Coefficients	Nitrogen 11	Ethylene 12	Isobutane 13	Cyclohexane 14	SF$_6$ 15
n_1	0.92296567	0.9096223	1.0429332	1.0232354	1.2279403
n_2	−2.5575012	−2.4641015	−2.8184273	−2.9204964	−3.3035623
n_3	0.64482463	0.56175311	0.86176232	1.073663	1.2094019
n_4	0.01083102	−0.019688013	−0.10613619	−0.19573985	−0.12316
n_5	0.073924167	0.078831145	0.098615749	0.12228111	0.11044657
n_6	0.00023532962	0.00021478776	0.00023948209	0.00028943321	0.00032952153
n_7	0.18024854	0.23151337	0.30330005	0.27231767	0.27017629
n_8	−0.045660299	−0.037804454	−0.041598156	−0.04483332	−0.062910351
n_9	−0.1552106	−0.20122739	−0.29991937	−0.38253334	−0.3182889
n_{10}	−0.03811149	−0.044960157	−0.080369343	−0.089835333	−0.099557419
n_{11}	−0.031962422	−0.02834296	−0.029761373	−0.024874965	−0.036909694
n_{12}	0.015513532	0.012652824	0.01305963	0.010836132	0.019136427

Coefficients	Carbon monoxide 16	Carbonyl sulfide 17	n-Decane 18	Hydrogen sulfide 19	Isopentane 20
n_1	0.90554	0.94374	1.0461	0.87641	1.0963
n_2	−2.4515	−2.5348	−2.4807	−2.0367	−3.0402
n_3	0.53149	0.59058	0.74372	0.21634	1.0317
n_4	0.024173	−0.021488	−0.52579	−0.050199	−0.1541
n_5	0.072156	0.082083	0.15315	0.066994	0.11535
n_6	0.00018818	0.00024689	0.00032865	0.00019076	0.00029809
n_7	0.19405	0.21226	0.84178	0.20227	0.39571
n_8	−0.043268	−0.041251	0.055424	−0.0045348	−0.045881
n_9	−0.12778	−0.22333	−0.73555	−0.2223	−0.35804
n_{10}	−0.027896	−0.050828	−0.18507	−0.034714	−0.10107
n_{11}	−0.034154	−0.028333	−0.020775	−0.014885	−0.035484
n_{12}	0.016329	0.016983	0.012335	0.0074154	0.018156

Coefficients	Neopentane 21	Isohexane 22	Krypton 23	n-Nonane 24	Toluene 25
n_1	1.1136	1.1027	0.83561	1.1151	0.96464
n_2	−3.1792	−2.9699	−2.3725	−2.702	−2.7855
n_3	1.1411	1.0295	0.54567	0.83416	0.86712
n_4	−0.10467	−0.21238	0.014361	−0.38828	−0.1886
n_5	0.11754	0.11897	0.066502	0.1376	0.11804
n_6	0.00034058	0.00027738	0.0001931	0.00028185	0.00025181
n_7	0.29553	0.40103	0.16818	0.62037	0.57196
n_8	−0.074765	−0.034238	−0.033133	0.015847	−0.029287
n_9	−0.31474	−0.43584	−0.15008	−0.61726	−0.43351
n_{10}	−0.099401	−0.11693	−0.022897	−0.15043	−0.1254
n_{11}	−0.039569	−0.019262	−0.021454	−0.012982	−0.028207
n_{12}	0.023177	0.0080783	0.0069397	0.0044325	0.014076

Table B.1 (continued).

Coefficients	Xenon 26	R116 27
n_1	0.83115	1.1632
n_2	−2.3553	−2.8123
n_3	0.53904	0.77202
n_4	0.014382	−0.14331
n_5	0.066309	0.10227
n_6	0.00019649	0.00024629
n_7	0.14996	0.30893
n_8	−0.035319	−0.028499
n_9	−0.15929	−0.30343
n_{10}	−0.027521	−0.068793
n_{11}	−0.023305	−0.027218
n_{12}	0.0086941	0.010665

Table B.2 Coefficients for polar fluids (Eq. (2.114)).

Coefficients	R11 1	R12 2	R22 3	R32 4
n_1	1.0656383	1.0557228	0.96268924	0.92876414
n_2	−3.2495206	−3.3312001	−2.5275103	−2.4673952
n_3	0.87823894	1.0197244	0.31308745	0.40129043
n_4	0.087611569	0.084155115	0.072432837	0.055101049
n_5	0.00029950049	0.00028520742	0.00021930233	0.00011559754
n_6	0.42896949	0.39625057	0.33294864	−0.25209758
n_7	0.70828452	0.63995721	0.63201229	0.42091879
n_8	−0.017391823	−0.021423411	−0.0032787841	0.0037071833
n_9	−0.37626522	−0.36249173	−0.33680834	−0.10308607
n_{10}	0.011605284	0.001934199	−0.022749022	−0.11592089
n_{11}	−0.089550567	−0.092993833	−0.087867308	−0.044350855
n_{12}	−0.030063991	−0.024876461	−0.021108145	−0.012788805

Coefficients	R113 5	R123 6	R125 7	R134a 8
n_1	1.0519071	1.116973	1.1290996	1.0663189
n_2	−2.8724742	−3.074593	−2.8349269	−2.449597
n_3	0.41983153	0.51063873	0.29968733	0.044645718
n_4	0.087107788	0.094478812	0.087282204	0.075656884
n_5	0.00024105194	0.00029532752	0.00026347747	0.00020652089
n_6	0.70738262	0.66974438	0.61056963	0.42006912
n_7	0.93513411	0.96438575	0.90073581	0.76739111
n_8	−0.0096713512	−0.014865424	−0.0068788457	0.0017897427
n_9	−0.52595315	−0.49221959	−0.44211186	−0.36219746
n_{10}	0.022691984	−0.022831038	−0.035041493	−0.06780937
n_{11}	−0.14556325	−0.1407486	−0.1269863	−0.10616419
n_{12}	−0.02741995	−0.025117301	−0.025185874	−0.018185791

(continued overleaf)

Table B.2 (continued).

Coefficients	R143a 9	R152a 10	Carbon dioxide 11	Ammonia 12
n_1	1.0306886	0.95702326	0.89875108	0.7302272
n_2	−2.9497307	−2.3707196	−2.1281985	−1.1879116
n_3	0.6943523	0.18748463	−0.06819032	−0.68319136
n_4	0.071552102	0.063800843	0.076355306	0.040028683
n_5	0.00019155982	0.00016625977	0.00022053253	0.000090801215
n_6	0.079764936	0.082208165	0.41541823	−0.056216175
n_7	0.56859424	0.57243518	0.71335657	0.44935601
n_8	−0.0090946566	0.0039476701	0.00030354234	0.029897121
n_9	−0.24199452	−0.23848654	−0.36643143	−0.18181684
n_{10}	−0.070610813	−0.080711618	−0.0014407781	−0.09841666
n_{11}	−0.075041709	−0.073103558	−0.089166707	−0.055083744
n_{12}	−0.016411241	−0.015538724	−0.023699887	−0.0088983219

Coefficients	Acetone 13	Nitrous oxide 14	Sulfur dioxide 15	R141b 16
n_1	0.90041	0.88045	0.93061	1.1469
n_2	−2.1267	−2.4235	−1.9528	−3.6799
n_3	−0.083409	0.38237	−0.17467	1.3469
n_4	0.065683	0.068917	0.061524	0.083329
n_5	0.00016527	0.00020367	0.00017711	0.00025137
n_6	−0.039663	0.13122	0.21615	0.3272
n_7	0.72085	0.46032	0.51353	0.46946
n_8	0.0092318	−0.0036985	0.010419	−0.029829
n_9	−0.17217	−0.23263	−0.25286	−0.31621
n_{10}	−0.14961	−0.00042859	−0.05472	−0.026219
n_{11}	−0.076124	−0.04281	−0.059856	−0.078043
n_{12}	−0.018166	−0.023038	−0.016523	−0.020498

Coefficients	R142b 17	R218 18	R245fa 19
n_1	1.0038	1.327	1.2904
n_2	−2.7662	−3.8433	−3.2154
n_3	0.42921	0.922	0.50693
n_4	0.081363	0.1136	0.093148
n_5	0.00024174	0.00036195	0.00027638
n_6	0.48246	1.1001	0.71458
n_7	0.75542	1.1896	0.87252
n_8	−0.00743	−0.025147	−0.015077
n_9	−0.4146	−0.65923	−0.40645
n_{10}	−0.016558	−0.027969	−0.11701
n_{11}	−0.10644	−0.1833	−0.13062
n_{12}	−0.021704	−0.02163	−0.022952

Appendix C
Useful Derivations

The following derivations are performed in this appendix:

- **A. Differential Relationships**
  - A1. Relationship between $(\partial s/\partial T)_P$ and $(\partial s/\partial T)_v$
  - A2. Expressions for $(\partial u/\partial v)_T$ and $(\partial s/\partial v)_T$
  - A3. c_P and c_v as Derivatives of the Specific Entropy
  - A4. Relationship between c_P and c_v
  - A5. Expression for $(\partial h/\partial P)T$
  - A6. Expression for $(\partial s/\partial P)T$
  - A7. Expression for $[\partial(g/RT)/\partial T]_P$ and van't Hoff Equation
  - A8. General Expression for c_v
  - A9. Expression for $(\partial P/\partial v)_T$
  - A10. Cardano's Formula.
- **B. General Derivations**
  - B1. Derivation of the Kelvin Equation
  - B2. Equivalence of Chemical Potential μ and Gibbs Energy g for a Pure Substance
  - B3. Phase Equilibrium Condition for a Pure Substance
  - B4. Relationship between Partial Molar Property and State Variable (Euler Theorem)
  - B5. Chemical Potential in Mixtures
  - B6. Relationship between Second Virial Coefficients of Leiden and Berlin Form
  - B7. Derivation of Expressions for the Speed of Sound for Ideal and Real Gases
  - B8. Activity of the Solvent in an Electrolyte Solution
  - B9. Temperature Dependence of the Azeotropic Composition.
- **C. Caloric Properties**
  - C1. $(s-s^{id})_{T,P}$
  - C2. $(h-h^{id})_{T,P}$
  - C3. $(g-g^{id})_{T,P}$.
- **D. Fugacity Coefficients**
  - D1. Fugacity Coefficient for a Pressure-Explicit Equation of State
  - D2. Fugacity Coefficient of the Virial Equation (Leiden Form)
  - D3. Fugacity Coefficient of the Virial Equation (Berlin Form)
  - D4. Fugacity Coefficient of the Soave–Redlich–Kwong Equation of State

- D5. Fugacity Coefficient of the Predictive Soave–Redlich–Kwong (PSRK) Equation of State.
- E. **Activity Coefficients**
  - E1. Derivation of the Wilson Equation
  - E2. Notation of the Wilson, NRTL, and UNIQUAC Equations in Process Simulation Programs
  - E3. Inability of the Wilson Equation to Describe a Miscibility Gap.
- F. **Expressions for Caloric Properties**
  - F1. $(h - h^{id})$ for Soave–Redlich–Kwong Equation of State
  - F2. $(s - s^{id})$ for Soave–Redlich–Kwong Equation of State
  - F3. $(g - g^{id})$ for Soave–Redlich–Kwong Equation of State
  - F4. Antiderivatives of c_P^{id} Correlations.
- G. **Pressure Relief**
  - G1. Speed of Sound as Maximum Velocity with Constant Cross-Flow Area
  - G2. Maximum Mass Flux of an Ideal Gas.

A1. Relationship between $(\partial s/\partial T)_P$ and $(\partial s/\partial T)_v$

The total differential of the specific entropy can be written in two ways:

$$ds = \left(\frac{\partial s}{\partial T}\right)_v dT + \left(\frac{\partial s}{\partial v}\right)_T dv = \left(\frac{\partial s}{\partial T}\right)_P dT + \left(\frac{\partial s}{\partial P}\right)_T dP \tag{C.1}$$

Furthermore, we can write

$$dv = \left(\frac{\partial v}{\partial T}\right)_P dT + \left(\frac{\partial v}{\partial P}\right)_T dP \tag{C.2}$$

and we obtain

$$ds = \left[\left(\frac{\partial s}{\partial T}\right)_v + \left(\frac{\partial s}{\partial v}\right)_T \left(\frac{\partial v}{\partial T}\right)_P\right] dT + \left(\frac{\partial s}{\partial v}\right)_T \left(\frac{\partial v}{\partial P}\right)_T dP \tag{C.3}$$

The comparison between Eqs. (C.1) and (C.3) yields

$$\left(\frac{\partial s}{\partial T}\right)_P = \left(\frac{\partial s}{\partial T}\right)_v + \left(\frac{\partial s}{\partial v}\right)_T \left(\frac{\partial v}{\partial T}\right)_P \tag{C.4}$$

A2. Expressions for $(\partial u/\partial v)_T$ and $(\partial s/\partial v)_T$

In the relationship

$$du = Tds - Pdv \tag{C.5}$$

ds can be replaced by the total differential set up in Eq. (C.1):

$$du = T\left(\frac{\partial s}{\partial T}\right)_v dT + \left(T\left(\frac{\partial s}{\partial v}\right)_T - P\right) dv \tag{C.6}$$

A comparison with the total differential of the specific internal energy

$$du = \left(\frac{\partial u}{\partial T}\right)_v dT + \left(\frac{\partial u}{\partial v}\right)_T dv \qquad (C.7)$$

yields

$$c_v = \left(\frac{\partial u}{\partial T}\right)_v = T\left(\frac{\partial s}{\partial T}\right)_v \qquad (C.8)$$

and

$$\left(\frac{\partial u}{\partial v}\right)_T = T\left(\frac{\partial s}{\partial v}\right)_T - P \qquad (C.9)$$

or, respectively,

$$\left(\frac{\partial s}{\partial v}\right)_T = \frac{1}{T}\left(\frac{\partial u}{\partial v}\right)_T + \frac{P}{T} \qquad (C.10)$$

$$\left(\frac{\partial s}{\partial T}\right)_v = \frac{1}{T}\left(\frac{\partial u}{\partial T}\right)_v \qquad (C.11)$$

The mixed quadratic derivatives of the specific entropy must be identical according to the law of Schwarz:

$$\left(\frac{\partial^2 s}{\partial v \partial T}\right) = -\frac{1}{T^2}\left(\frac{\partial u}{\partial v}\right)_T + \frac{1}{T}\left(\frac{\partial^2 u}{\partial v \partial T}\right) - \frac{P}{T^2} + \frac{1}{T}\left(\frac{\partial P}{\partial T}\right)_v$$

$$\left(\frac{\partial^2 s}{\partial T \partial v}\right) = \frac{1}{T}\left(\frac{\partial^2 u}{\partial T \partial v}\right) \qquad (C.12)$$

After rearranging, we get

$$-\frac{1}{T^2}\left(\frac{\partial u}{\partial v}\right)_T - \frac{P}{T^2} + \frac{1}{T}\left(\frac{\partial P}{\partial T}\right)_v = 0 \qquad (C.13)$$

and therefore

$$-\left(\frac{\partial u}{\partial v}\right)_T - P + T\left(\frac{\partial P}{\partial T}\right)_v = 0 \qquad (C.14)$$

which is equivalent to

$$\left(\frac{\partial u}{\partial v}\right)_T = T\left(\frac{\partial P}{\partial T}\right)_v - P \qquad (C.15)$$

or, using Eq. (C.10),

$$\left(\frac{\partial s}{\partial v}\right)_T = \left(\frac{\partial P}{\partial T}\right)_v \qquad (C.16)$$

A3. c_P and c_V as Derivatives of the Specific Entropy

In Eq. (C.8), it has been shown that c_v is related to the temperature derivative of the specific entropy. Analogously, a similar expression for c_P can be evaluated:
With

$$dh = Tds + vdP \qquad (C.17)$$

and the total differential of the specific entropy

$$ds = \left(\frac{\partial s}{\partial T}\right)_P dT + \left(\frac{\partial s}{\partial P}\right)_T dP \qquad (C.18)$$

we get

$$dh = T\left(\frac{\partial s}{\partial T}\right)_P dT + \left[T\left(\frac{\partial s}{\partial P}\right)_T + v\right] dP \qquad (C.19)$$

Comparison with the total differential of the specific enthalpy

$$dh = \left(\frac{\partial h}{\partial T}\right)_P dT + \left(\frac{\partial h}{\partial P}\right)_T dP \qquad (C.20)$$

gives

$$c_P = \left(\frac{\partial h}{\partial T}\right)_P = T\left(\frac{\partial s}{\partial T}\right)_P \qquad (C.21)$$

A4. Relationship between c_P and c_v

Starting with Eq. (C.4)

$$\left(\frac{\partial s}{\partial T}\right)_P = \left(\frac{\partial s}{\partial T}\right)_v + \left(\frac{\partial s}{\partial v}\right)_T \left(\frac{\partial v}{\partial T}\right)_P \qquad (C.22)$$

and multiplication with T

$$T\left(\frac{\partial s}{\partial T}\right)_P = T\left(\frac{\partial s}{\partial T}\right)_v + T\left(\frac{\partial s}{\partial v}\right)_T \left(\frac{\partial v}{\partial T}\right)_P \qquad (C.23)$$

we get with the help of Eqs. (C.21) and (C.8)

$$c_P = c_v + T\left(\frac{\partial s}{\partial v}\right)_T \left(\frac{\partial v}{\partial T}\right)_P \qquad (C.24)$$

With particular expressions for $(\partial s/\partial v)_T$ (Eqs. (C.10) and (C.16)) one gets

$$c_P = c_v + \left[\left(\frac{\partial u}{\partial v}\right)_T + P\right]\left(\frac{\partial v}{\partial T}\right)_P \qquad (C.25)$$

or, respectively,

$$c_P = c_v + T\left(\frac{\partial P}{\partial T}\right)_v \left(\frac{\partial v}{\partial T}\right)_P \qquad (C.26)$$

For a pressure-explicit equation of state, the term $(\partial v/\partial T)_P$ is not very useful. Setting up the total differential of the pressure

$$dP = \left(\frac{\partial P}{\partial T}\right)_v dT + \left(\frac{\partial P}{\partial v}\right)_T dv \qquad (C.27)$$

we get for the isobaric case with $dP = 0$

$$0 = \left(\frac{\partial P}{\partial T}\right)_v dT + \left(\frac{\partial P}{\partial v}\right)_T dv \qquad (C.28)$$

and therefore

$$\left(\frac{\partial v}{\partial T}\right)_P = -\frac{\left(\frac{\partial P}{\partial T}\right)_v}{\left(\frac{\partial P}{\partial v}\right)_T} \tag{C.29}$$

After inserting Eq. (C.29) into Eq. (C.26), we get

$$c_P = c_v - T\frac{\left(\frac{\partial P}{\partial T}\right)_v^2}{\left(\frac{\partial P}{\partial v}\right)_T} \tag{C.30}$$

For a volume-explicit equation of state, an analogous expression can be derived, starting from Eq. (C.26). In this case, the term $(\partial P/\partial T)_v$ must be replaced. Setting up the total differential of the specific volume

$$dv = \left(\frac{\partial v}{\partial T}\right)_P dT + \left(\frac{\partial v}{\partial P}\right)_T dP \tag{C.31}$$

we get for the isochoric case with $dv = 0$

$$0 = \left(\frac{\partial v}{\partial T}\right)_P dT + \left(\frac{\partial v}{\partial P}\right)_T dP \tag{C.32}$$

and therefore

$$\left(\frac{\partial P}{\partial T}\right)_v = -\frac{\left(\frac{\partial v}{\partial T}\right)_P}{\left(\frac{\partial v}{\partial P}\right)_T} \tag{C.33}$$

After inserting Eq. (C.33) into Eq. (C.26), we get

$$c_P = c_v - T\frac{\left(\frac{\partial v}{\partial T}\right)_P^2}{\left(\frac{\partial v}{\partial P}\right)_T} \tag{C.34}$$

A5. Expression for $(\partial h/\partial P)_T$

The total differentials of the specific enthalpy and the specific entropy are

$$dh = \left(\frac{\partial h}{\partial T}\right)_P dT + \left(\frac{\partial h}{\partial P}\right)_T dP \tag{C.35}$$

$$ds = \left(\frac{\partial s}{\partial T}\right)_P dT + \left(\frac{\partial s}{\partial P}\right)_T dP \tag{C.36}$$

We can compare it with the fundamental equation

$$dh = Tds + vdP \tag{C.37}$$

Combining Eqs. (C.36) and (C.37), we get

$$dh = T\left(\frac{\partial s}{\partial T}\right)_P dT + \left[T\left(\frac{\partial s}{\partial P}\right)_T + v\right] dP \tag{C.38}$$

Comparison of the coefficients in Eqs. (C.35) and (C.38) yields

$$\left(\frac{\partial h}{\partial T}\right)_P = T\left(\frac{\partial s}{\partial T}\right)_P \Rightarrow \left(\frac{\partial s}{\partial T}\right)_P = \frac{1}{T}\left(\frac{\partial h}{\partial T}\right)_P \tag{C.39}$$

$$\left(\frac{\partial h}{\partial P}\right)_T = T\left(\frac{\partial s}{\partial P}\right)_T + v \Rightarrow \left(\frac{\partial s}{\partial P}\right)_T = \frac{1}{T}\left(\frac{\partial h}{\partial P}\right)_T - \frac{v}{T} \tag{C.40}$$

The mixed-quadratic derivatives must be equal, that is,

$$\frac{\partial^2 s}{\partial T \partial P} = \frac{\partial^2 s}{\partial P \partial T} \tag{C.41}$$

Thus, we get

$$\frac{1}{T}\frac{\partial^2 h}{\partial T \partial P} = -\frac{1}{T^2}\left(\frac{\partial h}{\partial P}\right)_T + \frac{1}{T}\frac{\partial^2 h}{\partial P \partial T} + \frac{v}{T^2} - \frac{1}{T}\left(\frac{\partial v}{\partial T}\right)_P, \tag{C.42}$$

and finally, after simplification:

$$\left(\frac{\partial h}{\partial P}\right)_T = v - T\left(\frac{\partial v}{\partial T}\right)_P \tag{C.43}$$

A6. Expression for $(\partial s/\partial P)_T$

Using the fundamental equation and the total differential for the specific enthalpy, one can write

$$Tds = dh - vdP = \left(\frac{\partial h}{\partial T}\right)_P dT + \left(\frac{\partial h}{\partial P}\right)_T dP - vdP \tag{C.44}$$

Inserting Eq. (C.43), one obtains

$$Tds = \left(\frac{\partial h}{\partial T}\right)_P dT + vdP - T\left(\frac{\partial v}{\partial T}\right)_P dP - vdP$$

$$= \left(\frac{\partial h}{\partial T}\right)_P dT - T\left(\frac{\partial v}{\partial T}\right)_P dP \tag{C.45}$$

which can easily be transformed into

$$\left(\frac{\partial s}{\partial P}\right)_T = -\left(\frac{\partial v}{\partial T}\right)_P \tag{C.46}$$

A7. Expression for $[\partial(g/RT)/\partial T]_P$ and van't Hoff Equation

From the definition of the specific Gibbs energy g one gets

$$\frac{g}{RT} = \frac{h - Ts}{RT} \tag{C.47}$$

and subsequently

$$\left(\frac{\partial(g/RT)}{\partial T}\right)_P = \frac{RT\left(\frac{\partial h}{\partial T}\right)_P - RTs - RT^2\left(\frac{\partial s}{\partial T}\right)_P - Rh + RTs}{R^2 T^2}$$

$$= \frac{1}{RT}\left(\frac{\partial h}{\partial T}\right)_P - \frac{1}{R}\left(\frac{\partial s}{\partial T}\right)_P - \frac{h}{RT^2} \tag{C.48}$$

Replacing the differential quotients with the specific isobaric heat capacity according to Eq. (C.21), we obtain

$$\left(\frac{\partial(g/RT)}{\partial T}\right)_P = \frac{1}{RT}c_P - \frac{1}{RT}c_P - \frac{h}{RT^2} = -\frac{h}{RT^2} \tag{C.49}$$

Analogously, an expression for the excess Gibbs energy can be derived:

$$\left(\frac{\partial(g^E/RT)}{\partial T}\right)_P = -\frac{h^E}{RT^2} \tag{C.50}$$

Moreover, the van't Hoff equation for the temperature dependence of the equilibrium constant of chemical reactions

$$\frac{d \ln K(T)}{dT} = \frac{\Delta h_R^0}{RT^2} \tag{C.51}$$

is a direct conclusion from Eq. (C.49), as

$$\ln K(T) = -\frac{\Delta g_R^0}{RT} \tag{12.19}$$

A8. General Expression for c_V

The total differential of the internal energy is

$$du = \left(\frac{\partial u}{\partial T}\right)_v dT + \left(\frac{\partial u}{\partial v}\right)_T dv \tag{C.52}$$

Starting from a point in the ideal gas state $T_0, v_0 \to \infty$, we can perform the integration to get the internal energy at an arbitrary state as follows:

$$u(T, v) = u(T_0, v_0) + \int_{T_0}^{T} c_v^{id} dT + \int_{\infty}^{v} \left(\frac{\partial u}{\partial v}\right)_T dv \tag{C.53}$$

which is, according to Eq. (C.15), equivalent to

$$u(T, v) = u(T_0, v_0) + \int_{T_0}^{T} c_v^{id} dT + \int_{\infty}^{v} \left[T\left(\frac{\partial P}{\partial T}\right)_v - P\right] dv \tag{C.54}$$

A general expression for c_v can be obtained by differentiation with respect to T:

$$c_v(T,v) = \left(\frac{\partial u}{\partial T}\right)_v = c_v^{id}(T) + \int_\infty^v \left[\left(\frac{\partial P}{\partial T}\right)_v + T\left(\frac{\partial^2 P}{\partial T^2}\right)_v - \left(\frac{\partial P}{\partial T}\right)_v\right]dv$$

$$= c_v^{id}(T) + T\int_\infty^v \left(\frac{\partial^2 P}{\partial T^2}\right)_v dv \quad (C.55)$$

A9. Expression for $(\partial P/\partial v)_T$

The total differentials of pressure and temperature can be written as

$$dP = \left(\frac{\partial P}{\partial v}\right)_T dv + \left(\frac{\partial P}{\partial T}\right)_v dT \quad (C.56)$$

$$dT = \left(\frac{\partial T}{\partial v}\right)_P dv + \left(\frac{\partial T}{\partial P}\right)_v dP \quad (C.57)$$

Eliminating dT yields

$$dP = \left(\frac{\partial P}{\partial v}\right)_T dv + \left(\frac{\partial P}{\partial T}\right)_v \left(\frac{\partial T}{\partial v}\right)_P dv + \left(\frac{\partial P}{\partial T}\right)_v \left(\frac{\partial T}{\partial P}\right)_v dP \quad (C.58)$$

As the terms with dP cancel out, one can set up the relationship

$$\left(\frac{\partial P}{\partial v}\right)_T = -\left(\frac{\partial P}{\partial T}\right)_v \left(\frac{\partial T}{\partial v}\right)_P = -\frac{\left(\frac{\partial P}{\partial T}\right)_v}{\left(\frac{\partial v}{\partial T}\right)_P} \quad (C.59)$$

A10. Cardano's Formula

A cubic equation of state can be transformed into the form

$$z^3 + Uz^2 + Sz + T = 0 \quad (C.60)$$

with z as the compressibility factor. Using Cardanos formula [1], this type of equation can be solved analytically.

With the abbreviations

$$P = \frac{3S - U^2}{3} \quad Q = \frac{2U^3}{27} - \frac{US}{3} + T \quad (C.61)$$

the discriminant can be determined to be

$$D = \left(\frac{P}{3}\right)^3 + \left(\frac{Q}{2}\right)^2 \quad (C.62)$$

For $D > 0$, the equation of state has one real solution:

$$z = \left[\sqrt{D} - \frac{Q}{2}\right]^{1/3} - \frac{P}{3\left[\sqrt{D} - \frac{Q}{2}\right]^{1/3}} - \frac{U}{3} \quad (C.63)$$

For $D < 0$, there are three real solutions. With the abbreviations

$$\Theta = \sqrt{-\frac{P^3}{27}} \quad \text{and} \quad \Phi = \arccos\left(\frac{-Q}{2\Theta}\right) \tag{C.64}$$

they can be written as

$$z_1 = 2\Theta^{1/3}\cos\left(\frac{\Phi}{3}\right) - \frac{U}{3} \tag{C.65}$$

$$z_2 = 2\Theta^{1/3}\cos\left(\frac{\Phi}{3} + \frac{2\pi}{3}\right) - \frac{U}{3} \tag{C.66}$$

$$z_3 = 2\Theta^{1/3}\cos\left(\frac{\Phi}{3} + \frac{4\pi}{3}\right) - \frac{U}{3} \tag{C.67}$$

The largest and the smallest of the three values correspond to the vapor and to the liquid solution, respectively. The middle one has no physical meaning.

B1. Derivation of the Kelvin Equation

The Kelvin equation describes the dependence of the vapor pressure from the droplet size. Its derivation is based on the equilibrium of forces on a droplet, which is visualized in Figure C.1.

The force balance is given by

$$\left(P^L - P^V\right)\frac{\pi d^2}{4} = \sigma \pi d \tag{C.68}$$

giving

$$P^L - P^V = \frac{4\sigma}{d} \tag{C.69}$$

for the pressure difference between inside and outside the droplet.

For a small droplet of a pure substance in a vapor volume, the chemical equilibrium can be written as

$$\mu^V\left(P^V, T\right) = \mu^L\left(P^L, T\right) \tag{C.70}$$

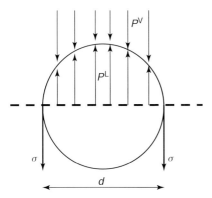

Figure C.1 Equilibrium of forces on a droplet.

To eliminate P^L, we can make use of the relationship

$$\mu^L(P^L, T) = \mu^L(P^V, T) + \int_{P^V}^{P^L} v^L dP \tag{C.71}$$

Assuming that $v^L \neq f(P)$ and introducing Eq. (C.69), we get

$$\mu^V(P^V, T) = \mu^L(P^V, T) + \frac{4v^L \sigma}{d} \tag{C.72}$$

With the phase equilibrium condition for even surfaces

$$\mu^V(P^s, T) = \mu^L(P^s, T) \tag{C.73}$$

and the relationships

$$\mu^V(P^V, T) - \mu^V(P^s, T) = RT \ln \frac{P^V}{P^s} \tag{C.74}$$

$$\mu^L(P^V, T) - \mu^L(P^s, T) = v^L(P^V - P^s) \tag{C.75}$$

one obtains

$$\mu^V(P^V, T) - RT \ln \frac{P^V}{P^s} = \mu^L(P^V, T) - v^L(P^V - P^s) \tag{C.76}$$

where the vapor phase is assumed to be ideal. Rearranging gives

$$RT \ln \frac{P^V}{P^s} = \mu^V(P^V, T) - \mu^L(P^V, T) + v^L(P^V - P^s) \tag{C.77}$$

Combining Eqs. (C.72) and (C.77), one gets

$$RT \ln \frac{P^V}{P^s} = \frac{4v^L \sigma}{d} + v^L(P^V - P^s) \tag{C.78}$$

Neglecting the second term and writing $\rho^L = 1/v^L$, the Kelvin equation

$$\ln \frac{P^V}{P^s} = -\frac{4\sigma}{\rho^L RT d} \tag{C.79}$$

is obtained.

B2. Equivalence of Chemical Potential μ and Gibbs Energy g for a Pure Substance

The fundamental equation of a pure substance in a closed system is given by

$$TdS = dU + PdV \tag{C.80}$$

In an open system, the mass transfer across the border of the system has to be taken into account additionally. Assigning the contribution of the mass transfer with an arbitrarily introduced quantity μ as the chemical potential, we get

$$dU = TdS - PdV + \mu dn \tag{C.81}$$

with n as the number of moles. From Eq. (C.81) we get:

$$ndu + udn = nTds + Tsdn - nPdv - Pvdn + \mu dn \tag{C.82}$$

Rearrangement yields

$$n(du - Tds + Pdv) = dn(Ts - u - Pv + \mu) \tag{C.83}$$

or, respectively,

$$0 = [\mu - (u + Pv - Ts)] dn \tag{C.84}$$

Thus, for a pure substance the chemical potential can be determined to be

$$\mu = u + Pv - Ts = h - Ts = g \tag{C.85}$$

and the chemical potential μ is equivalent to the Gibbs energy g.

B3. Phase Equilibrium Condition for a Pure Substance

In the following section, the vapor–liquid phase equilibrium condition is derived for two different arrangements:

1) **Fully closed system**
   In a fully closed system, the constraints are according to Figure C.2.:

$$dV = 0 = dV^L + dV^V \tag{C.86}$$

$$dS = 0 = dS^L + dS^V \tag{C.87}$$

and

$$dn = 0 = dn^L + dn^V \tag{C.88}$$

The fundamental equation for the system is

$$dU = TdS - PdV = 0 = dU^L + dU^V \tag{C.89}$$

For both phases, mass transfer across the border between the phases is allowed. Introducing the chemical potential μ (see Eq. (C.81)), we obtain for both phases

$$dU^L = T^L dS^L - P^L dV^L + \mu^L dn^L \tag{C.90}$$

$$dU^V = T^V dS^V - P^V dV^V + \mu^V dn^V \tag{C.91}$$

Figure C.2 Phase equilibrium in a fully closed system.

Appendix C Useful Derivations

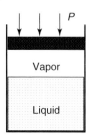

Figure C.3 Phase equilibrium for a closed system.

Adding up Eqs. (C.90) and (C.91) and taking into account the constraints (C.86), (C.87), and (C.89), the result is

$$0 = (T^L dS^L + T^V dS^V) - (P^L dv^L + P^V dv^V) + (\mu^L dn^L + \mu^V dn^V) \tag{C.92}$$

or

$$0 = (T^L - T^V)dS^L - (P^L - P^V)dv^L + (\mu^L - \mu^V)dn^L \tag{C.93}$$

As the differentials are not zero due to the constraints, the conclusions are

$$T^L = T^V \tag{C.94}$$

$$P^L = P^V \tag{C.95}$$

and

$$\mu^L = \mu^V \tag{C.96}$$

which is equivalent to

$$g^L = g^V \tag{C.97}$$

for a pure substance (Eq. (C.85)).

2) **Closed system**

   For a closed system which is in mechanical equilibrium with the environmental pressure, it is possible that heat passes the system border (Figure C.3.). As long as a pure substance is considered, the temperature does not vary when heat is transferred to the system (see Section 2.4.4). In this case, the constraints are

$$dP = dP^L = dP^V = 0 \tag{C.98}$$

$$dT = dT^L = dT^V = 0 \tag{C.99}$$

$$dn = 0 = dn^L + dn^V \tag{C.100}$$

For the entire system, the fundamental equation is

$$dU = TdS - PdV \tag{C.101}$$

The differential of the Gibbs energy is given by

$$dG = dU + PdV + VdP - TdS - SdT \tag{C.102}$$

Inserting Eq. (C.101) into Eq. (C.102), we get

$$dG = VdP - SdT = 0 = dG^L + dG^V \qquad (C.103)$$

Applying Eqs. (C.101) and (C.102) to the particular phases and allowing mass transfer between the phases, we obtain

$$dU^L = dG^L - P^L dV^L - V^L dP^L + T^L dS^L + S^L dT^L$$
$$= T^L dS^L - P^L dV^L + \mu^L dn^L \qquad (C.104)$$

and, respectively,

$$dU^V = dG^V - P^V dV^V - V^V dP^V + T^V dS^V + S^V dT^V$$
$$= T^V dS^V - P^V dV^V + \mu^V dn^V \qquad (C.105)$$

Adding up Eqs. (C.104) and (C.105) and regarding the constraints (C.98)–(C.103), the result is

$$0 = \left(\mu^L - \mu^V\right) dn^L \qquad (C.106)$$

giving again

$$\mu^L = \mu^V \qquad (C.107)$$

or, as a pure substance is regarded,

$$g^L = g^V \qquad (C.108)$$

B4. Relationship between Partial Molar Property and State Variable (Euler Theorem)

The well-known relationship between a state variable and its partial molar property

$$M = n_T m = \sum_i n_i \overline{m}_i \quad \text{or}$$

$$m = \sum_i x_i \overline{m}_i \qquad (C.109)$$

can be shown with the Euler theorem for homogeneous functions. For an arbitrary state variable, we can write

$$M(T, P, \lambda n_1, \lambda n_2, \ldots, \lambda n_n) = \lambda M(T, P, n_1, n_2, \ldots, n_n) \qquad (C.110)$$

Differentiating M with respect to λ gives

$$\frac{\partial M}{\partial (\lambda n_1)} \frac{\partial (\lambda n_1)}{\partial \lambda}\bigg|_{T,P} + \frac{\partial M}{\partial (\lambda n_2)} \frac{\partial (\lambda n_2)}{\partial \lambda}\bigg|_{T,P} + \cdots + \frac{\partial M}{\partial (\lambda n_n)} \frac{\partial (\lambda n_n)}{\partial \lambda}\bigg|_{T,P}$$
$$= M(T, P, n_1, n_2, \ldots, n_n) \qquad (C.111)$$

and therefore

$$M(T, P, n_1, n_2, \ldots, n_n) = n_1 \frac{\partial M}{\partial (\lambda n_1)}\bigg|_{T,P} + n_2 \frac{\partial M}{\partial (\lambda n_2)}\bigg|_{T,P}$$
$$+ \cdots + n_n \frac{\partial M}{\partial (\lambda n_n)}\bigg|_{T,P} \qquad (C.112)$$

Eq. (C.112) is valid for any λ. Therefore, the case $\lambda = 1$ is included, and we get

$$M(T, P, n_1, n_2, \ldots, n_n) = n_T m = \sum_i n_i \overline{m}_i \tag{C.113}$$

or, respectively,

$$m = \sum_i x_i \overline{m}_i \tag{C.114}$$

B5. Chemical Potential in Mixtures

For mixtures, the definition of the entropy (2.6) can be extended to be

$$TdS = dU + PdV + T \sum_i \left(\frac{\partial S}{\partial n_i}\right)_{U,V,n_{j \neq i}} dn_i \tag{C.115}$$

Inserting the total differentials

$$dH = dU + PdV + VdP \tag{C.116}$$
$$dA = dU - TdS - SdT \tag{C.117}$$
$$dG = dH - TdS - SdT \tag{C.118}$$

yields

$$dU = TdS - PdV - T \sum_i \left(\frac{\partial S}{\partial n_i}\right)_{U,V,n_{j \neq i}} dn_i \tag{C.119}$$

$$dH = TdS + VdP - T \sum_i \left(\frac{\partial S}{\partial n_i}\right)_{U,V,n_{j \neq i}} dn_i \tag{C.120}$$

$$dA = -SdT - PdV - T \sum_i \left(\frac{\partial S}{\partial n_i}\right)_{U,V,n_{j \neq i}} dn_i \tag{C.121}$$

$$dG = -SdT + VdP - T \sum_i \left(\frac{\partial S}{\partial n_i}\right)_{U,V,n_{j \neq i}} dn_i \tag{C.122}$$

Comparing to the approaches

$$dU = TdS - PdV + \sum_i \left(\frac{\partial U}{\partial n_i}\right)_{S,V,n_{j \neq i}} dn_i \tag{C.123}$$

$$dH = TdS + VdP + \sum_i \left(\frac{\partial H}{\partial n_i}\right)_{S,P,n_{j \neq i}} dn_i \tag{C.124}$$

$$dA = -SdT - PdV + \sum_i \left(\frac{\partial A}{\partial n_i}\right)_{T,V,n_{j \neq i}} dn_i \tag{C.125}$$

$$dG = -SdT + VdP + \sum_i \left(\frac{\partial G}{\partial n_i}\right)_{T,P,n_{j \neq i}} dn_i \tag{C.126}$$

the chemical potential can be set equal to

$$\mu_i = -T\left(\frac{\partial S}{\partial n_i}\right)_{U,V,n_{j\neq i}} = \left(\frac{\partial G}{\partial n_i}\right)_{T,P,n_{j\neq i}} = \left(\frac{\partial U}{\partial n_i}\right)_{S,V,n_{j\neq i}}$$
$$= \left(\frac{\partial H}{\partial n_i}\right)_{S,P,n_{j\neq i}} = \left(\frac{\partial A}{\partial n_i}\right)_{T,V,n_{j\neq i}} \tag{C.127}$$

B6. Relationship between Second Virial Coefficients of Leiden and Berlin Form

The Leiden form of the virial equation is given by

$$z = 1 + B(T)\rho + C(T)\rho^2 + \cdots \tag{C.128}$$

in contrast to the Berlin form

$$z = 1 + B'(T)P + C'(T)P^2 + \cdots \tag{C.129}$$

Multiplying both equations with ρRT yields

$$P = \rho RT + B(T)\rho^2 RT + C(T)\rho^3 RT + \cdots \tag{C.130}$$

and

$$P = \rho RT + B'(T)P\rho RT + C'(T)P^2 \rho RT + \cdots \tag{C.131}$$

Inserting Eq. (C.130) into (C.131) gives

$$P = \rho RT + B'(T)(\rho RT + B(T)\rho^2 RT + C(T)\rho^3 RT + \cdots)\rho RT$$
$$\quad + C'(T)(\rho RT + B(T)\rho^2 RT + \cdots)^2 \rho RT + \cdots$$
$$= \rho RT + B'(T)\rho^2 R^2 T^2 + \cdots \tag{C.132}$$

After comparing the coefficients in Eq. (C.132) and (C.130), we obtain

$$B(T)RT = B'(T)R^2 T^2 \tag{C.133}$$

and, subsequently,

$$B'(T) = \frac{B(T)}{RT} \tag{C.134}$$

B7. Derivation of Expressions for the Speed of Sound for Ideal and Real Gases

The general formula for the calculation of the speed of sound is [2]

$$(w^*)^2 = -v^2 \left(\frac{\partial P}{\partial v}\right)_s \tag{C.135}$$

The differential quotient in Eq. (3.94) can be evaluated as follows:
For $s = $ const., we get $ds = 0$ and

$$Tds = dh - vdP = 0 \tag{C.136}$$

Setting up the total differential of the enthalpy, we get

$$\left(\frac{\partial h}{\partial T}\right)_P dT + \left(\frac{\partial h}{\partial P}\right)_T dP - v dP = 0 \qquad (C.137)$$

For an ideal gas, the equation of state

$$Pv = RT \qquad (C.138)$$

can be transferred into

$$P dv + v dP = R dT \qquad (C.139)$$

and

$$\frac{dv}{v} + \frac{dP}{P} = \frac{dT}{T} \qquad (C.140)$$

As for an ideal gas $(\partial h/\partial P)_T = 0$, the combination of Eqs. (C.137) and (C.140) leads to

$$c_P T \left(\frac{dv}{v} + \frac{dP}{P}\right) - v dP = 0 \qquad (C.141)$$

Using Eq. (C.138) and the abbreviation $\kappa = c_P^{id}/c_v^{id}$, we get

$$\left(\frac{\partial P}{\partial v}\right)_s = -\kappa \frac{P}{v} \qquad (C.142)$$

and finally

$$w^{*id} = \sqrt{\kappa RT} \qquad (C.143)$$

For a nonideal gas, $(\partial h/\partial P)_T$ does not vanish. An analogous equation can be derived by setting up a relationship between dP and dv. At constant entropy, we can write

$$ds = \left(\frac{\partial s}{\partial T}\right)_v dT + \left(\frac{\partial s}{\partial v}\right)_T dv = \frac{c_v}{T} dT + \left(\frac{\partial P}{\partial T}\right)_v dv \stackrel{!}{=} 0$$

$$ds = \left(\frac{\partial s}{\partial T}\right)_P dT + \left(\frac{\partial s}{\partial P}\right)_T dP = \frac{c_P}{T} dT - \left(\frac{\partial v}{\partial T}\right)_P dP \stackrel{!}{=} 0 \qquad (C.144)$$

The particular partial derivatives of the specific entropy are explained in Sections C-A2 and C-A6, respectively.

Eliminating the temperature terms, we can summarize to

$$\frac{c_v}{c_P}\left(\frac{\partial v}{\partial T}\right)_P dP + \left(\frac{\partial P}{\partial T}\right)_v dv = 0 \qquad (C.145)$$

giving according to the rules of partial derivatives [3]:

$$\left(\frac{\partial P}{\partial v}\right)_s = \frac{c_P \left(\frac{\partial P}{\partial T}\right)_v}{c_v \left(\frac{\partial v}{\partial T}\right)_P} = \frac{c_P}{c_v}\left(\frac{\partial P}{\partial v}\right)_T \qquad (C.146)$$

where the rearrangement of the partial derivatives is explained in detail in Section C-A9.

Therefore, the analytical expression for the speed of sound is

$$(w^*)^2 = -v^2 \left(\frac{\partial P}{\partial v}\right)_T \frac{c_P}{c_v} \tag{C.147}$$

Using the relationship [3] (see Section C-A4)

$$c_P = c_v + T \left(\frac{\partial P}{\partial T}\right)_v \left(\frac{\partial v}{\partial T}\right)_P \tag{C.148}$$

and Eq. (C.146), the speed of sound can as well be expressed as

$$(w^*)^2 = v^2 \left[\frac{T}{c_v}\left(\frac{\partial P}{\partial T}\right)_v^2 - \left(\frac{\partial P}{\partial v}\right)_T\right] \tag{C.149}$$

For c_v, the relationship

$$c_v = c_v^{id} + T \int_\infty^v \left(\frac{\partial^2 P}{\partial T^2}\right)_v dv \tag{C.150}$$

can be used (see Section C-A8).

B8. Activity of the Solvent in an Electrolyte Solution

The relationship between the activities of the electrolyte components and the activity of the solvent is the Gibbs–Duhem equation at constant temperature and pressure:

$$x_{\text{solv}} d \ln a_{\text{solv}} + \sum_{i \neq \text{solv}} x_i d \ln a_i^{*m} = 0 \tag{C.151}$$

The molalities

$$m_i = \frac{n_i}{m_{\text{solv}}} = \frac{n_i}{n_{\text{solv}} M_{\text{solv}}} \tag{C.152}$$

can be introduced by

$$\frac{x_i}{x_{\text{solv}}} = \frac{n_i}{n_T} \frac{n_T}{n_{\text{solv}}} = m_i \frac{n_{\text{solv}} M_{\text{solv}}}{n_i} = m_i M_{\text{solv}} \tag{C.153}$$

with

$$a_i^{*m} = m_i \gamma_i^{*m} \tag{C.154}$$

Eq. (C.151) gives

$$d \ln a_{\text{solv}} = -M_{\text{solv}} \sum_{i \neq \text{solv}} m_i d \ln (m_i \gamma_i^{*m}) \tag{C.155}$$

The total differential of the activities of the ions can be expressed as

$$d\ln(m_i\gamma_i^{*m}) = \sum_{k\ne solv} \frac{\partial \ln(m_i\gamma_i^{*m})}{\partial m_k} dm_k$$

$$= \sum_{k\ne i,\,solv} \frac{\partial \ln(m_i\gamma_i^{*m})}{\partial m_k} dm_k + \frac{\partial \ln(m_i\gamma_i^{*m})}{\partial m_i} dm_i$$

$$= \sum_{k\ne i,\,solv} \left[\frac{1}{m_i\gamma_i^{*m}} m_i \frac{\partial \gamma_i^{*m}}{\partial m_k}\right] dm_k + \left[\frac{1}{m_i\gamma_i^{*m}}\left(\gamma_i^{*m} + m_i \frac{\partial \gamma_i^{*m}}{\partial m_i}\right)\right] dm_i$$

$$= \sum_{k\ne i,\,solv} \left[\frac{\partial \ln \gamma_i^{*m}}{\partial m_k}\right] dm_k + \left[\frac{1}{m_i} + \frac{\partial \ln \gamma_i^{*m}}{\partial m_i}\right] dm_i$$

$$= \sum_{k\ne solv} \left[\frac{\partial \ln \gamma_i^{*m}}{\partial m_k}\right] dm_k + \frac{dm_i}{m_i} \qquad (C.156)$$

Combining Eqs. (C.155) and (C.156) gives

$$d\ln a_{solv} = -M_{solv} \sum_{i\ne solv} dm_i - M_{solv} \sum_{i\ne solv} m_i \sum_{k\ne solv} \left[\frac{\partial \ln \gamma_i^{*m}}{\partial m_k}\right] dm_k \qquad (C.157)$$

The integration of the second term is carried out in the following way: Starting from the pure solvent with all the $m_{i\ne solv} = 0$, the particular ions are successively added to the solution. In this way, only the activities of the species already present in the solution are affected, that is, all the species $i \le k$. After rearranging the sequence of the summation, the result of the integration is

$$\ln a_{solv} = -M_{solv} \sum_{i\ne solv} m_i - M_{solv} \sum_{k\ne solv} \int_0^{m_k} \sum_{i\le k} m_i \left[\frac{\partial \ln \gamma_i^{*m}}{\partial m_k}\right] dm_k \qquad (C.158)$$

This derivation clearly demonstrates the advantage of the molality concept, as during the integration the molality of one component can be varied from 0 to m_k without affecting the molalities of the other species.

B9. Temperature Dependence of the Azeotropic Composition

In [4], an equation is derived which describes how the azeotropic composition of a binary mixture depends on the temperature. Starting with the phase equilibrium for both components

$$x_1\gamma_1 P_1^s = y_1 P$$
$$x_2\gamma_2 P_2^s = y_2 P \qquad (C.159)$$

where an ideal vapor phase is assumed, and taking into account that $x_i = y_i$ at the azeotropic point, one gets

$$\gamma_1 P_1^s = \gamma_2 P_2^s \qquad (C.160)$$

and therefore

$$\ln \frac{\gamma_1 P_1^s}{\gamma_2 P_2^s} = 0 = \ln \gamma_1 - \ln \gamma_2 + \ln \frac{P_1^s}{P_2^s} \tag{C.161}$$

As the azeotropic composition itself is a function of temperature, differentiating with respect to T gives

$$\left(\frac{\partial \ln \gamma_1}{\partial T}\right)_{x_1} + \left(\frac{\partial \ln \gamma_1}{\partial x_1}\right)_T \left(\frac{\partial x_1}{\partial T}\right)_{az} - \left(\frac{\partial \ln \gamma_2}{\partial T}\right)_{x_1} - \left(\frac{\partial \ln \gamma_2}{\partial x_1}\right)_T \left(\frac{\partial x_1}{\partial T}\right)_{az}$$
$$+ \frac{1}{P_1^s}\frac{dP_1^s}{dT} - \frac{1}{P_2^s}\frac{dP_2^s}{dT} = 0 \tag{C.162}$$

The factors $(\partial \ln \gamma_i/\partial x_1)_T$ can be determined from Eq. (C.159), which can be written in logarithmic form as well:

$$\ln x_1 + \ln \gamma_1 = \ln \frac{P}{P_1^s} + \ln y_1$$

$$\ln x_2 + \ln \gamma_2 = \ln \frac{P}{P_2^s} + \ln y_2 \tag{C.163}$$

The derivations of Eqs. (C.163) with respect to x_1 are

$$\left(\frac{\partial \ln \gamma_1}{\partial x_1}\right)_T = \frac{1}{P}\left(\frac{\partial P}{\partial x_1}\right)_T + \frac{1}{y_1}\left(\frac{\partial y_1}{\partial x_1}\right)_T - \frac{1}{x_1}$$
$$\left(\frac{\partial \ln \gamma_2}{\partial x_1}\right)_T = \frac{1}{P}\left(\frac{\partial P}{\partial x_1}\right)_T - \frac{1}{y_2}\left(\frac{\partial y_1}{\partial x_1}\right)_T + \frac{1}{x_2} \tag{C.164}$$

At the azeotropic point, the constraints

$$y_i = x_i \tag{C.165}$$

and

$$\left(\frac{\partial P}{\partial x_i}\right)_{T,az} = 0 \tag{C.166}$$

are valid. Thus, Eqs. (C.164) turn into

$$\left(\frac{\partial \ln \gamma_1}{\partial x_1}\right)_T = \frac{1}{x_1}\left(\left(\frac{\partial y_1}{\partial x_1}\right)_T - 1\right)$$
$$\left(\frac{\partial \ln \gamma_2}{\partial x_1}\right)_T = -\frac{1}{x_2}\left(\left(\frac{\partial y_1}{\partial x_1}\right)_T - 1\right) \tag{C.167}$$

Inserting Eqs. (C.167) into Eq. (C.162) yields

$$\left(\frac{\partial \ln \gamma_1}{\partial T}\right)_{x_1} + \frac{1}{x_1}\left(\left(\frac{\partial y_1}{\partial x_1}\right)_T - 1\right)\left(\frac{\partial x_1}{\partial T}\right)_{az} - \left(\frac{\partial \ln \gamma_2}{\partial T}\right)_{x_1}$$
$$+ \frac{1}{x_2}\left(\left(\frac{\partial y_1}{\partial x_1}\right)_T - 1\right)\left(\frac{\partial x_1}{\partial T}\right)_{az} + \frac{1}{P_1^s}\frac{dP_1^s}{dT} - \frac{1}{P_2^s}\frac{dP_2^s}{dT} = 0 \tag{C.168}$$

Solving for the temperature dependence of the azeotropic composition gives

$$\left(\frac{\partial x_1}{\partial T}\right)_{az} = \frac{\frac{1}{P_2^s}\frac{dP_2^s}{dT} - \frac{1}{P_1^s}\frac{dP_1^s}{dT} - \left(\frac{\partial \ln \gamma_1}{\partial T}\right)_{x_1} + \left(\frac{\partial \ln \gamma_2}{\partial T}\right)_{x_1}}{\left(\frac{1}{x_1} + \frac{1}{x_2}\right)\left(\left(\frac{\partial y_1}{\partial x_1}\right)_T - 1\right)} \tag{C.169}$$

Using the conversion

$$\frac{1}{x_1} + \frac{1}{x_2} = \frac{x_2 + x_1}{x_1 x_2} = \frac{1}{x_1 x_2} \qquad \text{(C.170)}$$

the Clausius–Clapeyron equation (2.86) with the approximation $(v^V - v^L) = RT/P$ valid for ideal gases

$$\Delta h_{vi} = T(v^V - v^L)\frac{dP_i^s}{dT} \approx RT^2 \frac{1}{P_i^s} \frac{dP_i^s}{dT} \qquad \text{(C.171)}$$

and the relationship for partial molar excess enthalpy (5.26)

$$\overline{h_i^E} = R\frac{\partial \ln \gamma_i}{\partial (1/T)} = -RT^2 \frac{\partial \ln \gamma_i}{\partial T} \qquad \text{(C.172)}$$

one obtains

$$\left(\frac{\partial x_1}{\partial T}\right)_{az} = x_{1az}x_{2az} \frac{\Delta h_{v1} - \Delta h_{v2} + \overline{h_2^E} - \overline{h_1^E}}{RT^2\left(1 - \left(\frac{\partial \gamma_1}{\partial x_1}\right)_T\right)} \qquad \text{(C.173)}$$

C1. $(s - s^{id})_{T,P}$

According to Eq. (2.51), we had the expression

$$\left(s - s^{id}\right)_{T,P} = \left(s - s^{id}\right)_{T,P \to 0} + \int_0^P \left[\left(\frac{\partial s}{\partial P}\right)_T - \left(\frac{\partial s^{id}}{\partial P}\right)_T\right] dP \qquad (2.51)$$

Rearranging, using Eq. (C.16) and defining $s_0(T_0,P_0)$ as a reference point at ideal gas conditions yields

$$\left(s - s^{id}\right)_{T,P} = s(T, P) - s^{id}(T, P)$$

$$= s(T, v(P)) - s^{id}(T, v(P)) + s^{id}(T, v(P)) - s^{id}(T, P)$$

$$= \int_\infty^v \left[\left(\frac{\partial P}{\partial T}\right)_v - \frac{R}{v}\right] dv + s_0 + \int_{T_0}^T \frac{c_v^{id} dT}{T} + R \ln \frac{v(P)}{v_0} - s_0$$

$$- \int_{T_0}^T \frac{c_p^{id} dT}{T} + R \ln \frac{P}{P_0}$$

$$= \int_\infty^v \left[\left(\frac{\partial P}{\partial T}\right)_v - \frac{R}{v}\right] dv - R \ln \frac{T}{T_0} + R \ln \frac{v(P)}{v_0} + R \ln \frac{P}{P_0}$$

$$= \int_\infty^v \left[\left(\frac{\partial P}{\partial T}\right)_v - \frac{R}{v}\right] dv + R \ln \left(\frac{v(P)P}{RT} \frac{RT_0}{v_0 P_0}\right)$$

$$= \int_\infty^v \left[\left(\frac{\partial P}{\partial T}\right)_v - \frac{R}{v}\right] dv + R \ln z \qquad \text{(C.174)}$$

C2. $(h-h^{id})_{T,P}$

The enthalpy can be related to the internal energy according to its definition:

$$(h - h^{id})_{T,P} = (u - u^{id})_{T,P} + Pv - RT$$

$$= \int_\infty^v \left(\frac{\partial u}{\partial v}\right)_T dv + Pv - RT \tag{C.175}$$

Using Eq. (C.15), we get

$$(h - h^{id})_{T,P} = \int_\infty^v \left[T\left(\frac{\partial P}{\partial T}\right)_v - P\right] dv + Pv - RT \tag{C.176}$$

C3. $(g-g^{id})_{T,P}$

Again, $(g-g^{id})_{T,P}$ can be calculated according to its definition:

$$(g - g^{id})_{T,P} = (h - h^{id})_{T,P} - T(s - s^{id})_{T,P}$$

$$= \int_\infty^v \left[T\left(\frac{\partial P}{\partial T}\right)_v - P - T\left(\frac{\partial P}{\partial T}\right)_v + \frac{RT}{v}\right] dv$$

$$+ Pv - RT - RT \ln z$$

$$= \int_\infty^v \left[\frac{RT}{v} - P\right] dv + RT(z - 1 - \ln z) \tag{C.177}$$

D1. Fugacity Coefficient for a Pressure-Explicit Equation of State

Starting with Eq. (4.60)

$$\bar{g}_i = g_i^{\text{pure}}(T, P^0) + RT \ln \frac{f_i}{f_i^0} \tag{4.60}$$

and the relationships

$$\varphi_i = \frac{f_i}{z_i P} \tag{C.178}$$

and

$$\mu_i = \bar{g}_i \tag{C.179}$$

one can write

$$RT \ln \varphi_i = \bar{g}_i - g_i^{\text{pure}}(T, P^0) - RT \ln \frac{z_i P}{f_i^0} \tag{C.180}$$

For an ideal gas, it is

$$\bar{g}_i^{id}(T, P, z_i) = g_i^{\text{pure}}(T, P^0) + RT \ln \frac{z_i P}{f_i^0} \tag{C.181}$$

which yields

$$RT \ln \varphi_i = \bar{g}_i(T, P, z_i) - \bar{g}_i^{id}(T, P, z_i)$$

$$= \mu_i(T, P, z_i) - \mu_i^{id}(T, P, z_i) \tag{C.182}$$

Introducing the fundamental Eq. (C.127), one can relate the fugacity coefficient to the Helmholtz energy:

$$RT \ln \varphi_i = \frac{\partial}{\partial n_i} \left[A - A^{id} \right]_{T,V,n_{j \neq i}} \tag{C.183}$$

The difference in brackets can be expressed by

$$\left(A - A^{id} \right) = \int_{V^{id}}^{V} \left(\frac{\partial A}{\partial V} \right)_{T,n_i} dV = -\int_{V^{id}}^{V} P \, dV \tag{C.184}$$

The integral can be split:

$$\int_{V^{id}}^{V} P \, dV = \int_{\infty}^{V} P \, dV - \int_{\infty}^{V^{id}} P \, dV = \int_{\infty}^{V} P \, dV - \int_{\infty}^{V^{id}} \frac{n_T RT}{V} dV \tag{C.185}$$

as the ideal gas equation of state is valid for $V > V^{id}$. n_T is the total number of moles. Rearranging gives

$$\begin{aligned}
\int_{V^{id}}^{V} P \, dV &= \int_{\infty}^{V} \left(P - \frac{n_T RT}{V} \right) dV + \int_{\infty}^{V} \frac{n_T RT}{V} dV - \int_{\infty}^{V^{id}} \frac{n_T RT}{V} dV \\
&= \int_{\infty}^{V} \left(P - \frac{n_T RT}{V} \right) dV + \int_{V^{id}}^{V} \frac{n_T RT}{V} dV \\
&= \int_{\infty}^{V} \left(P - \frac{n_T RT}{V} \right) dV + n_T RT \ln \frac{V}{V^{id}} \\
&= \int_{\infty}^{V} \left(P - \frac{n_T RT}{V} \right) dV + n_T RT \ln z
\end{aligned} \tag{C.186}$$

As

$$\left(\frac{\partial n_T}{\partial n_i} \right)_{T,V,n_{j \neq i}} = 1 \tag{C.187}$$

combining Eqs. (C.183), (C.184), and (C.186) gives

$$RT \ln \varphi_i = \int_{V}^{\infty} \left[\left(\frac{\partial P}{\partial n_i} \right)_{T,V,n_{j \neq i}} - \frac{RT}{V} \right] dV - RT \ln z \tag{C.188}$$

D2. Fugacity Coefficient of the Virial Equation (Leiden Form)

The relationship for the fugacity coefficient of a component in mixtures is given according to Eq. (4.64):

$$\ln \varphi_k = \frac{1}{RT} \int_{\infty}^{V} \left[\frac{RT}{V} - \left(\frac{\partial P}{\partial n_k} \right)_{T,V,n_j} \right] dV - \ln z \tag{4.64}$$

If the virial equation in the Leiden form is truncated after the second term, we get according to Eq. (2.88):

$$z = 1 + \frac{B}{V} \tag{2.84}$$

or
$$P = \frac{RT}{V} + \frac{BRT}{V^2} \tag{C.189}$$

Introducing n_T as the total number of moles, we can write
$$P = \frac{n_T RT}{V} + \frac{n_T^2 BRT}{V^2} \tag{C.190}$$

and therefore
$$\left(\frac{\partial P}{\partial n_k}\right)_{T,V,n_j} = \frac{RT}{V} + \frac{RT}{V^2}\left[2Bn_T + n_T^2\left(\frac{\partial B}{\partial n_k}\right)_{T,n_j}\right] \tag{C.191}$$

To obtain an expression for $(\partial B/\partial n_k)$, it has to be referred to the mixing rule (4.89)
$$B = \sum_i \sum_j y_i y_j B_{ij} = \sum_i \sum_j \frac{n_i n_j}{n_T^2} B_{ij} \tag{C.192}$$

It is useful to split the sum in Eq. (C.192) into four parts, that is, for the cases $i = k, j = k$, $i = k, j \neq k$, $i \neq k, j = k$ and $i \neq k, j \neq k$. We get
$$B = \frac{n_k^2}{n_T^2} B_{kk} + \sum_{j \neq k} \frac{n_k n_j}{n_T^2} B_{kj} + \sum_{i \neq k} \frac{n_i n_k}{n_T^2} B_{ik} + \sum_{i \neq k} \sum_{j \neq k} \frac{n_i n_j}{n_T^2} B_{ij} \tag{C.193}$$

which can be summarized to
$$B = \frac{n_k^2}{n_T^2} B_{kk} + 2\sum_{i \neq k} \frac{n_i n_k}{n_T^2} B_{ik} + \sum_{i \neq k} \sum_{j \neq k} \frac{n_i n_j}{n_T^2} B_{ij} \tag{C.194}$$

as $B_{ij} = B_{ji}$ is valid. Now the differentiation can be performed to obtain an expression for $n_T(\partial B/\partial n_k)$:

$$\begin{aligned}
n_T\left(\frac{\partial B}{\partial n_k}\right)_{T,n_j} &= n_T \frac{2n_k n_T^2 - 2n_k^2 n_T}{n_T^4} B_{kk} + 2n_T \sum_{i \neq k} \frac{n_i n_T^2 - 2n_i n_k n_T}{n_T^4} B_{ik} \\
&\quad + n_T \sum_{i \neq k}\sum_{j \neq k} \frac{-2n_i n_j n_T}{n_T^4} B_{ij} \\
&= 2y_k B_{kk} - 2y_k^2 B_{kk} + 2\sum_{i \neq k} y_i B_{ik} - 4\sum_{i \neq k} y_i y_k B_{ik} - 2\sum_{i \neq k}\sum_{j \neq k} y_i y_j B_{ij} \\
&= -2y_k^2 B_{kk} + 2\sum_i y_i B_{ik} - 2\sum_{i \neq k} y_i y_k B_{ik} - 2\sum_{i \neq k}\sum_j y_i y_j B_{ij} \\
&= 2\sum_i y_i B_{ik} - 2\sum_i y_i y_k B_{ik} - 2\sum_{i \neq k}\sum_j y_i y_j B_{ij} \\
&= 2\sum_i y_i B_{ik} - 2\sum_j y_k y_j B_{kj} - 2\sum_{i \neq k}\sum_j y_i y_j B_{ij} \\
&= 2\sum_i y_i B_{ik} - 2\sum_i \sum_j y_i y_j B_{ij} \\
&= 2\sum_i y_i B_{ik} - 2B \tag{C.195}
\end{aligned}$$

after making use of Eq. (4.89).

Inserting Eqs. (C.195) and (C.191) into Eq. (4.64) yields

$$\ln \varphi_k = \int_V^\infty \frac{2n_T \sum_i y_i B_{ik}}{V^2} dV - \ln z = \left(2n_T \sum_i y_i B_{ik}\right) \int_V^\infty \frac{dV}{V^2} - \ln z$$

$$= \frac{2}{V} \sum_i y_i B_{ik} - \ln z \tag{C.196}$$

D3. Fugacity Coefficient of the Virial Equation (Berlin Form)

The Berlin form of the virial equation, cut after the first term, is given by

$$z = 1 + B'(T) P \tag{C.197}$$

or

$$v = \frac{RT}{P} + B(T) \tag{C.198}$$

when Eq. (C.134) is applied. The mixing rule for B is

$$B = \sum_i \sum_j y_i y_j B_{ij} \tag{4.89}$$

Starting with the equation for the fugacity coefficient

$$\ln \varphi_k = \frac{1}{RT} \int_0^P \left[\left(\frac{\partial V}{\partial n_k}\right)_{T,P,n_j} - \frac{RT}{P}\right] dP \tag{4.63}$$

one gets

$$V = n_T \frac{RT}{P} + n_T B \tag{C.199}$$

and

$$\left(\frac{\partial V}{\partial n_k}\right)_{T,P,n_j} = \frac{RT}{P} + B + n_T \left(\frac{\partial B}{\partial n_k}\right)_{T,n_j} \tag{C.200}$$

The last term in Eq. (C.200) has already been evaluated in Eq. (C.195). Inserting the result into Eq. (4.63), we get

$$\ln \varphi_k = \frac{1}{RT} \int_0^P \left[\frac{RT}{P} + B + 2\sum_i y_i B_{ik} - 2B - \frac{RT}{P}\right] dP$$

$$= \frac{1}{RT} \int_0^P \left[2\sum_i y_i B_{ik} - B\right] dP$$

$$= \left[2\sum_i y_i B_{ik} - B\right] \frac{P}{RT} \tag{C.201}$$

D4. Fugacity Coefficient of the Soave–Redlich–Kwong Equation of State

The Soave–Redlich–Kwong equation is given by

$$P = \frac{RT}{v-b} - \frac{a}{v(v+b)} \tag{C.202}$$

with the mixing rules

$$a = \sum_i \sum_j y_i y_j a_{ij} = \sum_i \sum_j \frac{n_i n_j}{n_T^2} a_{ij} \tag{C.203}$$

$$b = \sum_i y_i b_i = \sum_i \frac{n_i}{n_T} b_i \tag{C.204}$$

Using the formula for the fugacity coefficient

$$\ln \varphi_k = \frac{1}{RT} \int_\infty^V \left[\frac{RT}{V} - \left(\frac{\partial P}{\partial n_k}\right)_{T,V,n_j} \right] dV - \ln z \tag{4.64}$$

we obtain

$$P = \frac{n_T RT}{V - n_T b} - \frac{n_T^2 a}{V(V + n_T b)} \tag{C.205}$$

or

$$P = \frac{n_T RT}{V - \sum_i n_i b_i} - \frac{\sum_i \sum_j n_i n_j a_{ij}}{V\left(V + \sum_i n_i b_i\right)} \tag{C.206}$$

The term containing the a_{ij} is the most difficult one for differentiation. We can split it with respect to the component k to

$$\sum_i \sum_j n_i n_j a_{ij} = n_k^2 a_{kk} + \sum_{j \neq k} n_k n_j a_{kj} + \sum_{i \neq k} n_i n_k a_{ik} + \sum_{i \neq k} \sum_{j \neq k} n_i n_j a_{ij} \tag{C.207}$$

and differentiate

$$\frac{\partial}{\partial n_k} \left(\sum_i \sum_j n_i n_j a_{ij} \right)_{T,V,n_i} = 2 n_k a_{kk} + \sum_{j \neq k} n_j a_{kj} + \sum_{i \neq k} n_i a_{ik}$$

$$= 2 \sum_i n_i a_{ik} \tag{C.208}$$

as $a_{ik} = a_{ki}$. Therefore, we can write

$$\left(\frac{\partial P}{\partial n_k}\right)_{T,V,n_i} = \frac{RT\left(V - \sum_i n_i b_i\right) + n_T RT b_k}{\left(V - \sum_i n_i b_i\right)^2}$$

$$- \frac{2V\left(V + \sum_i n_i b_i\right)\sum_i n_i a_{ik} - V b_k \sum_i \sum_j n_i n_j a_{ij}}{V^2 \left(V + \sum_i n_i b_i\right)^2}$$

$$= \frac{RT}{V - n_T b} + \frac{n_T RT b_k}{(V - n_T b)^2} - \frac{2\sum_i n_i a_{ik}}{V(V + n_T b)}$$

$$+ \frac{b_k n_T^2 a}{V(V + n_T b)^2} \tag{C.209}$$

and the fugacity coefficient is

$$\ln \varphi_k = -\ln \frac{Pv}{RT} - \int_\infty^V \left[\frac{1}{V - n_T b} + \frac{n_T b_k}{(V - n_T b)^2} - \frac{2\sum_i n_i a_{ik}}{RTV(V + n_T b)}\right.$$

$$\left. + \frac{b_k n_T^2 a}{RTV(V + n_T b)^2} - \frac{1}{V}\right] dV$$

$$= -\ln \frac{Pv}{RT} - \int_\infty^V \frac{dV}{V - n_T b} - \int_\infty^V \frac{n_T b_k dV}{(V - n_T b)^2}$$

$$+ \int_\infty^V \left[\frac{2\sum_i n_i a_{ik}}{RTV(V + n_T b)}\right] dV$$

$$- \int_\infty^V \left[\frac{b_k n_T^2 a}{RTV(V + n_T b)^2}\right] dV + \int_\infty^V \frac{dV}{V} \tag{C.210}$$

The particular integrals can be solved as follows:

$$\int_\infty^V \frac{dV}{V} - \int_\infty^V \frac{dV}{V - n_T b} = \lim_{V^+ \to \infty} \left[\ln \frac{V}{V^+} - \ln \frac{V - n_T b}{V^+ - n_T b}\right]$$

$$= \ln \frac{V}{V - n_T b} = \ln \frac{v}{v - b} \tag{C.211}$$

$$-\int_\infty^V \frac{n_T b_k dV}{(V - n_T b)^2} = -n_T b_k \left[-\frac{1}{V - n_T b}\right]_\infty^V = \frac{n_T b_k}{V - n_T b} = \frac{b_k}{v - b}$$

$$\tag{C.212}$$

$$\int_\infty^V \left[\frac{2\sum_i n_i a_{ik}}{RTV(V+n_T b)}\right] dV = \frac{2\sum_i n_i a_{ik}}{RT}\left[-\frac{1}{n_T b}\ln\frac{V+n_T b}{V}\right]_\infty^V$$

$$= \frac{-2\sum_i n_i a_{ik}}{RT n_T b}\ln\frac{V+n_T b}{V}$$

$$= \frac{-2\sum_i y_i a_{ik}}{RTb}\ln\frac{v+b}{v} \quad (C.213)$$

$$-\int_\infty^V \left[\frac{b_k n_T^2 a}{RTV(V+n_T b)^2}\right] dV = -\frac{b_k n_T^2 a}{RT}\left[-\frac{1}{(n_T b)^2}\left(\ln\frac{V+n_T b}{V} - \frac{n_T b}{V+n_T b}\right)\right]_\infty^V$$

$$= \frac{b_k n_T^2 a}{RT n_T^2 b^2}\left[\ln\frac{V+n_T b}{V} - \frac{n_T b}{V+n_T b}\right]$$

$$= \frac{b_k a}{RTb^2}\left[\ln\frac{v+b}{v} - \frac{b}{v+b}\right] \quad (C.214)$$

Thus, Eq. (C.210) can be summarized to

$$\ln\varphi_k = -\ln\frac{Pv}{RT} + \ln\frac{v}{v-b} + \frac{b_k}{v-b} - \frac{2\sum_i y_i a_{ik}}{RTb}\ln\frac{v+b}{v}$$

$$+ \frac{b_k a}{RTb^2}\left[\ln\frac{v+b}{v} - \frac{b}{v+b}\right] \quad (C.215)$$

D5. Fugacity Coefficient of the PSRK Equation of State

The PSRK equation of state is given by

$$P = \frac{RT}{v-b} - \frac{a}{v(v+b)} \quad (C.216)$$

with the pure component parameters

$$a_i = 0.42748\frac{R^2 T_{ci}^2}{P_{ci}}f(T) \quad (C.217)$$

$$f(T) = \left[1 + c_1(1 - T_r^{0.5}) + c_2(1 - T_r^{0.5})^2 + c_3(1 - T_r^{0.5})^3\right]^2 \quad \text{for} \quad T_r \leq 1$$
$$f(T) = \left[1 + c_1(1 - T_r^{0.5})\right]^2 \quad \text{for} \quad T_r > 1$$
$$(C.218)$$

$$b_i = 0.08664\frac{RT_{ci}}{P_{ci}} \quad (C.219)$$

The corresponding mixing rules are

$$b = \sum_i x_i b_i \quad (C.220)$$

$$\frac{a}{bRT} = \frac{g^E}{ART} + \sum_i x_i \frac{a_i}{b_i RT} + \frac{1}{A} \sum_i x_i \ln \frac{b}{b_i} \qquad (C.221)$$

with $A = -0.64663$ and g^E taken from an arbitrary g^E model.
The fugacity coefficient can be calculated via

$$\ln \varphi_i = -\ln \frac{Pv}{RT} - \frac{1}{RT} \int_\infty^V \left[\left(\frac{\partial P}{\partial n_i}\right)_{T,V,n_{k \neq i}} - \frac{RT}{V} \right] dV \qquad (C.222)$$

With n_T as the total number of moles and Eq. (C.220), we get

$$P = \frac{n_T RT}{V - \sum_k n_k b_k} - \frac{a n_T^2}{V(V + \sum_k n_k b_k)} \qquad (C.223)$$

and

$$\left(\frac{\partial P}{\partial n_i}\right)_{T,V,n_{k \neq i}} - \frac{RT}{V}$$

$$= \frac{RT\left(V - \sum_k n_k b_k\right) + b_i n_T RT}{\left(V - \sum_k n_k b_k\right)^2}$$

$$- \frac{\left(2 a n_T + n_T^2 \left(\frac{\partial a}{\partial n_i}\right)_{T,V,n_{k \neq i}}\right) V \left(V + \sum_k n_k b_k\right) - a n_T^2 V b_i}{V^2 \left(V + \sum_k n_k b_k\right)^2} - \frac{RT}{V}$$

$$= \frac{RT}{\left(V - \sum_k n_k b_k\right)} + \frac{b_i n_T RT}{\left(V - \sum_k n_k b_k\right)^2}$$

$$- \frac{\left(2 a n_T + n_T^2 \left(\frac{\partial a}{\partial n_i}\right)_{T,V,n_{k \neq i}}\right)}{V\left(V + \sum_k n_k b_k\right)} + \frac{a n_T^2 b_i}{V\left(V + \sum_k n_k b_k\right)^2} - \frac{RT}{V} \qquad (C.224)$$

The corresponding integral is

$$\int_{V^+ \to \infty}^V \left[\left(\frac{\partial P}{\partial n_i}\right)_{T,V,n_{k \neq i}} - \frac{RT}{V} \right] dV =$$

$$= \lim_{V^+ \to \infty} \left\{ RT \ln \left(\frac{V - \sum_k n_k b_k}{V^+ - \sum_k n_k b_k} \right) - \frac{b_i n_T RT}{V - \sum_k n_k b_k} + \frac{b_i n_T RT}{V^+ - \sum_k n_k b_k} \right\}$$

$$+ \lim_{V^+ \to \infty} \left\{ \begin{array}{c} \dfrac{an_T^2 b_i}{\left(\sum_k n_k b_k\right)^2} \ln \dfrac{V}{V^+} - \dfrac{an_T^2 b_i}{\left(\sum_k n_k b_k\right)^2} \ln \left(\dfrac{V + \sum_k n_k b_k}{V^+ + \sum_k n_k b_k}\right) \\ + \dfrac{an_T^2 b_i}{\sum_k n_k b_k} \left(\dfrac{1}{V + \sum_k n_k b_k} - \dfrac{1}{V^+ + \sum_k n_k b_k}\right) \end{array} \right\}$$

$$+ \lim_{V^+ \to \infty} \left\{ \begin{array}{c} -\dfrac{\left(2an_T + n_T^2 \left(\frac{\partial a}{\partial n_i}\right)_{T,V,n_{k \neq i}}\right)}{\sum_k n_k b_k} \ln \dfrac{V}{V^+} \\ + \dfrac{\left(2an_T + n_T^2 \left(\frac{\partial a}{\partial n_i}\right)_{T,V,n_{k \neq i}}\right)}{\sum_k n_k b_k} \ln \dfrac{V + \sum_k n_k b_k}{V^+ + \sum_k n_k b_k} - RT \ln \dfrac{V}{V^+} \end{array} \right\}$$

$$= RT \ln \left(\dfrac{v-b}{v}\right) - \dfrac{b_i RT}{v-b} + \dfrac{ab_i}{b^2} \ln \left(\dfrac{v}{v+b}\right)$$

$$+ \dfrac{ab_i}{b(v+b)} + \dfrac{\left(2a + n_T \left(\frac{\partial a}{\partial n_i}\right)_{T,V,n_{k \neq i}}\right)}{b} \ln \dfrac{v+b}{v} \tag{C.225}$$

and the fugacity coefficient gives

$$\ln \varphi_i = \ln \left(\dfrac{RT}{P(v-b)}\right) + \dfrac{b_i}{v-b} - \dfrac{ab_i}{RTb(v+b)}$$

$$- \dfrac{\left(2ab - ab_i + n_T b \left(\frac{\partial a}{\partial n_i}\right)_{T,V,n_{k \neq i}}\right)}{b^2 RT} \ln \dfrac{v+b}{v}$$

$$= \dfrac{b_i}{b}\left(\dfrac{Pv}{RT} - 1\right) - \ln \dfrac{P(v-b)}{RT}$$

$$- \dfrac{\left(2ab - ab_i + n_T b \left(\frac{\partial a}{\partial n_i}\right)_{T,V,n_{k \neq i}}\right)}{b^2 RT} \ln \dfrac{v+b}{v} \tag{C.226}$$

The term in brackets containing $(\partial a/\partial n_i)$ can be calculated as follows:
Rearranging Eq. (C.221) gives

$$a = \dfrac{bg^E}{A} + b \sum_i x_i \dfrac{a_i}{b_i} + \dfrac{bRT}{A} \sum_i x_i \ln \dfrac{b}{b_i} \tag{C.227}$$

and the derivation is

$$\left(\frac{\partial a}{\partial n_i}\right)_{T,V,n_{k\neq i}} = \frac{g^E}{A}\left(\frac{\partial b}{\partial n_i}\right)_{T,V,n_{k\neq i}} + \frac{b}{A}\left(\frac{\partial g^E}{\partial n_i}\right)_{T,V,n_{k\neq i}}$$
$$+ \left(\frac{\partial b}{\partial n_i}\right)_{T,V,n_{k\neq i}} \sum_k x_k \frac{a_k}{b_k} + b\frac{\partial}{\partial n_i}\left(\sum_k x_k \frac{a_k}{b_k}\right)_{T,V,n_{k\neq i}}$$
$$+ \frac{RT}{A}\left[\left(\frac{\partial b}{\partial n_i}\right)_{T,V,n_{k\neq i}} \sum_k x_k \ln\frac{b}{b_k} + b\frac{\partial}{\partial n_i}\left(\sum_k x_k \ln\frac{b}{b_k}\right)_{T,V,n_{k\neq i}}\right]$$
(C.228)

An arbitrary quantity with the structure

$$Y = \sum_k x_k Y_k \tag{C.229}$$

can be derived as follows:

$$\left(\frac{\partial Y}{\partial n_i}\right)_{T,V,n_{k\neq i}} = \frac{\partial}{\partial n_i}\left(\frac{\sum_k n_k Y_k}{n_T}\right)_{T,V,n_{k\neq i}} = \frac{n_T Y_i - \sum_k n_k Y_k}{n_T^2} = \frac{Y_i - Y}{n_T}$$
(C.230)

This means that

$$\left(\frac{\partial b}{\partial n_i}\right)_{T,V,n_{k\neq i}} = \frac{b_i - b}{n_T} \tag{C.231}$$

$$\frac{\partial}{\partial n_i}\left(\sum_k x_k \frac{a_k}{b_k}\right)_{T,V,n_{k\neq i}} = \frac{\frac{a_i}{b_i} - \sum_k x_k \frac{a_k}{b_k}}{n_T} \tag{C.232}$$

and

$$\left(\frac{\partial g^E}{\partial n_i}\right)_{T,V,n_{k\neq i}} = \frac{RT \ln \gamma_i - g^E}{n_T} \tag{C.233}$$

More complicated is

$$\frac{\partial}{\partial n_i}\left(\sum_k x_k \ln \frac{b}{b_k}\right)_{T,V,n_{k\neq i}}$$
$$= \frac{\partial}{\partial n_i}\left(\frac{\sum_k n_k \ln\frac{b}{b_k}}{n_T}\right)_{T,T,n_{k\neq i}} = \frac{\partial}{\partial n_i}\left(\frac{n_i \ln\frac{b}{b_i} + \sum_{k\neq i} n_k \ln\frac{b}{b_k}}{n_T}\right)$$

$$= \frac{n_T \ln \frac{b}{b_i} + n_T n_i \frac{b_i}{b} \frac{1}{b_i} \frac{b_i-b}{n_T} + n_T \sum_{k \neq i} n_k \frac{b_k}{b} \frac{1}{b_k} \frac{b_i-b}{n_T} - n_i \ln \frac{b}{b_i} - \sum_{k \neq i} n_k \ln \frac{b}{b_k}}{n_T^2}$$

$$= \frac{\ln \frac{b}{b_i} + \frac{b_i}{b} - 1 - \sum_k x_k \ln \frac{b}{b_k}}{n_T} \qquad (C.234)$$

Summarizing gives

$$2ab - ab_i + n_T b \left(\frac{\partial a}{\partial n_i}\right)_{T,V,n_{k \neq i}}$$

$$= \frac{2b^2 g^E}{A} + 2b^2 \sum_i x_i \frac{a_i}{b_i} + \frac{2b^2 RT}{A} \sum_i x_i \ln \frac{b}{b_i}$$

$$- \frac{bb_i g^E}{A} - bb_i \sum_i x_i \frac{a_i}{b_i} - \frac{bb_i RT}{A} \sum_i x_i \ln \frac{b}{b_i} + \frac{bg^E}{A}(b_i - b)$$

$$+ \frac{b^2}{A}(RT \ln \gamma_i - g^E) + (b_i - b)b \sum_k x_k \frac{a_k}{b_k} + b^2 \frac{a_i}{b_i} - b^2 \sum_k x_k \frac{a_k}{b_k}$$

$$+ \frac{RTb}{A}\left[(b_i - b)\sum_k x_k \ln \frac{b}{b_k} + b\left(\ln \frac{b}{b_i} + \frac{b_i}{b} - 1 - \sum_k x_k \ln \frac{b}{b_k}\right)\right]$$

$$= \frac{RTb^2}{A} \ln \gamma_i + b^2 \frac{a_i}{b_i} + \frac{RTb^2}{A}\left(\ln \frac{b}{b_i} + \frac{b_i}{b} - 1\right) \qquad (C.235)$$

and

$$\frac{2ab - ab_i + n_T b \left(\frac{\partial a}{\partial n_i}\right)_{T,V,n_{k \neq i}}}{b^2 RT} = \frac{1}{A} \ln \gamma_i + \frac{a}{RTb_i} + \frac{1}{A}\left(\ln \frac{b}{b_i} + \frac{b_i}{b} - 1\right) \qquad (C.236)$$

E1. Derivation of the Wilson Equation

Figure C.4. shows two examples for possible local distributions of molecules of the species 1 and 2 in a binary mixture:

It can be seen that the concentrations of the species 1 and 2 can be different in the areas around a molecule of type 1 and around a molecule of type 2, although the overall concentration in the mixture remains constant. Taking this into account, the excess Gibbs energy can according to Wilson [5] be assumed to be

$$\frac{g^E}{RT} = \sum_i x_i \ln \frac{\xi_i}{x_i} \qquad (C.237)$$

where ξ_i denotes for the so-called local concentration

$$\xi_i = \frac{x_{ii} V_i}{\sum_j x_{ji} V_j} \qquad (C.238)$$

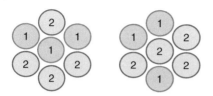

Figure C.4 Possible molecule distributions in a binary mixture.

with x_{ji} as the local mole fraction of species j in the area around species i.

Inserting Eq. (C.238) into Eq. (C.237) gives

$$\frac{g^E}{RT} = \sum_i x_i \ln \frac{\frac{x_{ii} V_i}{\sum_j x_{ji} V_j}}{x_i} \tag{C.239}$$

Then, it is assumed that the local concentrations and the overall concentrations are related by Boltzmann factors:

$$\frac{x_{ji}}{x_{ii}} = \frac{x_j}{x_i} \frac{\exp\left(-\lambda_{ji}/(RT)\right)}{\exp\left(-\lambda_{ii}/(RT)\right)} \tag{C.240}$$

Combining Eqs. (C.239) and (C.240), we obtain

$$\frac{g^E}{RT} = \sum_i x_i \ln \frac{x_{ii} V_i}{x_i \sum_j x_{ii} V_j \frac{x_j}{x_i} \frac{\exp\left(-\lambda_{ji}/(RT)\right)}{\exp\left(-\lambda_{ii}/(RT)\right)}}$$

$$= \sum_i x_i \ln \frac{V_i}{\sum_j V_j x_j \frac{\exp\left(-\lambda_{ji}/(RT)\right)}{\exp\left(-\lambda_{ii}/(RT)\right)}}$$

$$= -\sum_i x_i \ln \sum_j x_j \frac{V_j}{V_i} \exp\left(-\frac{(\lambda_{ji} - \lambda_{ii})}{RT}\right) \tag{C.241}$$

Setting $\lambda_{ij} = \lambda_{ji}$, the abbreviation

$$\Lambda_{ij} = \frac{V_j}{V_i} \exp\left(-\frac{(\lambda_{ij} - \lambda_{ii})}{RT}\right) = \frac{V_j}{V_i} \exp\left(-\frac{\Delta \lambda_{ij}}{RT}\right) \tag{C.242}$$

can be introduced, where the $\Delta \lambda_{ij}$ are the interaction parameters. The expression for the excess Gibbs energy becomes

$$\frac{g^E}{RT} = -\sum_i x_i \ln \sum_j x_j \Lambda_{ij} \tag{C.243}$$

For the binary mixture, Eq. (C.243) gives

$$\frac{g^E}{RT} = -x_1 \ln (x_1 + x_2 \Lambda_{12}) - x_2 \ln (x_1 \Lambda_{21} + x_2) \tag{C.244}$$

The activity coefficients can be calculated via Eq. (4.85)

$$\ln \gamma_i = \frac{\partial \left(n_T \frac{g^E}{RT}\right)}{\partial n_i}\bigg|_{T,P,n_{j\neq i}} = \frac{\partial}{\partial n_i}\left[-\sum_k n_k \ln \sum_j \frac{n_j}{n_T}\Lambda_{kj}\right]_{T,P,n_{j\neq i}}$$

$$= -\ln \sum_j x_j \Lambda_{ij} - n_i \frac{\partial}{\partial n_i}\left(\ln \sum_j \frac{n_j}{n_T}\Lambda_{ij}\right)_{T,P,n_{j\neq i}}$$

$$+ \frac{\partial}{\partial n_i}\left[-\sum_{k\neq i} n_k \ln \sum_j \frac{n_j}{n_T}\Lambda_{kj}\right]_{T,P,n_{j\neq i}}$$

$$= -\ln \sum_j x_j \Lambda_{ij} - \frac{n_i}{\sum_j \frac{n_j}{n_T}\Lambda_{ij}} \frac{\partial}{\partial n_i}\left(\sum_j \frac{n_j}{n_T}\Lambda_{ij}\right)_{T,P,n_{j\neq i}}$$

$$- \sum_{k\neq i} \frac{n_k}{\sum_j \frac{n_j}{n_T}\Lambda_{kj}} \frac{\partial}{\partial n_i}\left(\sum_j \frac{n_j}{n_T}\Lambda_{kj}\right)_{T,P,n_{j\neq i}}$$

$$= -\ln \sum_j x_j \Lambda_{ij} - \sum_k \frac{n_k}{\sum_j \frac{n_j}{n_T}\Lambda_{kj}} \frac{\partial}{\partial n_i}\left(\sum_j \frac{n_j}{n_T}\Lambda_{kj}\right)_{T,P,n_{j\neq i}}$$

$$= -\ln \sum_j x_j \Lambda_{ij} - \sum_k \frac{n_k}{\sum_j \frac{n_j}{n_T}\Lambda_{kj}} \left(\frac{n_T - n_i}{n_T^2}\Lambda_{ki} - \sum_{j\neq i}\frac{n_j}{n_T^2}\Lambda_{kj}\right)$$

$$= -\ln \sum_j x_j \Lambda_{ij} - \sum_k \frac{1}{\sum_j x_j \Lambda_{kj}}\left(x_k \Lambda_{ki} - \sum_j x_j x_k \Lambda_{kj}\right)$$

$$-\ln \sum_j x_j \Lambda_{ij} - \sum_k \frac{x_k \Lambda_{ki}}{\sum_j x_j \Lambda_{kj}} + \sum_k \frac{\sum_j x_j x_k \Lambda_{kj}}{\sum_j x_j \Lambda_{kj}}$$

$$-\ln \sum_j x_j \Lambda_{ij} - \sum_k \frac{x_k \Lambda_{ki}}{\sum_j x_j \Lambda_{kj}} + \sum_k \frac{x_k \sum_j x_j \Lambda_{kj}}{\sum_j x_j \Lambda_{kj}}$$

$$-\ln \sum_j x_j \Lambda_{ij} - \sum_k \frac{x_k \Lambda_{ki}}{\sum_j x_j \Lambda_{kj}} + 1 \quad \text{(C.245)}$$

In a similar way, the UNIQUAC equation is derived in [6]. The NRTL equation is explained in [7].

E2. Notation of the Wilson, NRTL, and UNIQUAC Equations in Process Simulation Programs

For the use in a process simulator, Eq. (C.242) is not appropriate as an expression for the Wilson equation, as the specific volumes v_i have an influence on the activity coefficients. When the binary parameters $\Delta\lambda_{ij}$ are stored, they are related to the specific volumes that were used in the parameter regression run. If the pure component values are changed, maybe due to an improved data situation, the binary parameters would have to be refitted, which is usually not considered by the user. In process simulation programs, the Wilson equation is therefore written in a different way which avoids these disadvantages. Starting with Eq. (C.242), we can write

$$\begin{aligned}
\Lambda_{ij} &= \frac{v_j}{v_i} \exp\left(-\frac{\Delta\lambda_{ij}}{RT}\right) \\
&= \exp\left(\ln\frac{v_j}{v_i} + \frac{-\Delta\lambda_{ij}/R}{T}\right) \\
&= \exp\left(A_{ij} + \frac{B_{ij}}{T}\right) \\
&= \exp\left(\frac{A_{ij}T + B_{ij}}{T}\right)
\end{aligned} \qquad (C.246)$$

A further advantage is that the gas constant R is no more involved in the parameters; therefore, the parameters do not depend on the units or the value used for R. As well, the parameters are always written with a positive sign in the equation. It is worth mentioning that $\ln v_j/v_i$ represents the linear temperature dependence of the binary interaction parameter.

In case temperature-dependent parameters are necessary, the A parameters are fitted to experimental data as well, making the equation completely independent of the specific volume.

In case temperature-independent parameters are used, only the B parameters are fitted. As worked out in Eq. (C.246), the A parameters can be set to $A_{ij} = \ln(v_j/v_i)$ and stored in this way. Without loss of quality, the A_{ij} can also be set to 0, before the B parameters are adjusted.

Further temperature-dependent terms can be added, for example,

$$\Lambda_{ij} = \exp\left(A_{ij} + \frac{B_{ij}}{T} + C_{ij}\ln\frac{T}{K} + D_{ij}T\right) \qquad (C.247)$$

However, there are only a few cases where it is justified to use them (see Appendix F). Similarly, the terms containing the interaction parameters in the NRTL and the UNIQUAC equation are written in the forms

$$\tau_{ij} = \exp\left(A_{ij} + \frac{B_{ij}}{T} + E_{ij}\ln\frac{T}{K} + F_{ij}T\right) \qquad (C.248)$$

or, respectively,

$$\tau_{ij} = \exp\left(A_{ij} + \frac{B_{ij}}{T} + C_{ij}\ln\frac{T}{K} + D_{ij}T + E_{ij}T^2\right) \qquad (C.249)$$

In some cases, even the α parameter in the NRTL equation is made temperature-dependent. The notation in the process simulators is less consequent in this case, as

$$\alpha_{ij} = C_{ij} + D_{ij}(T - 273.15 K) \tag{C.250}$$

Again, the temperature-dependent terms should be used with caution; a sufficient amount of data should be available when they are used.

E3. Inability of the Wilson Equation to Describe a Miscibility Gap

The Wilson equation never indicates a miscibility gap. This can be shown using the condition (5.80):

$$\left(\frac{\partial^2 \Delta g}{\partial x_1^2}\right)_{T,P} < 0 \tag{C.251}$$

with

$$\Delta g = \Delta g^{id} + g^E \tag{C.252}$$

Considering the case of a binary mixture, the necessary condition for the occurrence of a miscibility gap would be

$$\frac{\partial^2}{\partial x_1^2}[x_1 \ln x_1 + x_2 \ln x_2 - x_1 \ln(x_1 + x_2\Lambda_{12}) - x_2 \ln(x_1\Lambda_{21} + x_2)] \stackrel{!}{<} 0 \tag{C.253}$$

One gets

$$\frac{\partial}{\partial x_1}[x_1 \ln x_1 + x_2 \ln x_2 - x_1 \ln(x_1 + x_2\Lambda_{12}) - x_2 \ln(x_1\Lambda_{21} + x_2)]_{T,P}$$

$$= \frac{\partial}{\partial x_1}[x_1 \ln x_1 + (1 - x_1) \ln(1 - x_1) - x_1 \ln(x_1 + \Lambda_{12} - x_1\Lambda_{12})$$
$$\quad - (1 - x_1) \ln(x_1\Lambda_{21} + 1 - x_1)]_{T,P}$$

$$= \ln x_1 + 1 - \frac{1}{1 - x_1} - \ln(1 - x_1) + \frac{x_1}{1 - x_1} - \ln(x_1 + \Lambda_{12} - x_1\Lambda_{12})$$
$$\quad - \frac{x_1(1 - \Lambda_{12})}{x_1 + \Lambda_{12} - x_1\Lambda_{12}}$$
$$\quad - \frac{\Lambda_{21} - 1}{x_1\Lambda_{21} + 1 - x_1} + \ln(x_1\Lambda_{21} + 1 - x_1) + \frac{x_1(\Lambda_{21} - 1)}{x_1\Lambda_{21} + 1 - x_1}$$

$$= \ln x_1 - \ln(1 - x_1) - \ln(x_1 + \Lambda_{12} - x_1\Lambda_{12}) - 1 + \frac{\Lambda_{12}}{x_1 + \Lambda_{12} - x_1\Lambda_{12}}$$
$$\quad + \ln(x_1\Lambda_{21} + 1 - x_1) + 1 - \frac{\Lambda_{21}}{x_1\Lambda_{21} + 1 - x_1}$$

$$= \ln x_1 - \ln(1 - x_1) - \ln(x_1 + \Lambda_{12} - x_1\Lambda_{12}) + \frac{\Lambda_{12}}{x_1 + \Lambda_{12} - x_1\Lambda_{12}}$$
$$\quad + \ln(x_1\Lambda_{21} + 1 - x_1) - \frac{\Lambda_{21}}{x_1\Lambda_{21} + 1 - x_1} \tag{C.254}$$

and

$$\frac{\partial^2}{\partial x_1^2}[x_1 \ln x_1 + x_2 \ln x_2 - x_1 \ln (x_1 + x_2\Lambda_{12}) - x_2 \ln (x_1\Lambda_{21} + x_2)]$$

$$= \frac{1}{x_1} + \frac{1}{1-x_1} - \frac{1-\Lambda_{12}}{x_1+\Lambda_{12}-x_1\Lambda_{12}} - \frac{\Lambda_{12}(1-\Lambda_{12})}{(x_1+\Lambda_{12}-x_1\Lambda_{12})^2}$$

$$+ \frac{\Lambda_{21}-1}{x_1\Lambda_{21}+1-x_1} + \frac{\Lambda_{21}(\Lambda_{21}-1)}{(x_1\Lambda_{21}+1-x_1)^2}$$

$$= \frac{1}{x_1} + \frac{1}{x_2} - \frac{(1-\Lambda_{12})(x_1+\Lambda_{12}-x_1\Lambda_{12})}{(x_1+\Lambda_{12}-x_1\Lambda_{12})^2} - \frac{\Lambda_{12}(1-\Lambda_{12})}{(x_1+\Lambda_{12}-x_1\Lambda_{12})^2}$$

$$+ \frac{(\Lambda_{21}-1)(x_1\Lambda_{21}+1-x_1)}{(x_1\Lambda_{21}+1-x_1)^2} + \frac{\Lambda_{21}(\Lambda_{21}-1)}{(x_1\Lambda_{21}+1-x_1)^2}$$

$$= \frac{1}{x_1} + \frac{1}{x_2} - \frac{(1-\Lambda_{12})(x_1+2\Lambda_{12}-x_1\Lambda_{12})}{(x_1+\Lambda_{12}-x_1\Lambda_{12})^2}$$

$$+ \frac{(\Lambda_{21}-1)(x_1\Lambda_{21}+\Lambda_{21}+1-x_1)}{(x_1\Lambda_{21}+1-x_1)^2}$$

$$= \left[\frac{1}{x_1} - \frac{x_1+2x_2\Lambda_{12}-(2-x_1)\Lambda_{12}^2}{(x_1+x_2\Lambda_{12})^2}\right]$$

$$+ \left[\frac{1}{x_2} - \frac{x_2+2x_1\Lambda_{21}-(2-x_2)\Lambda_{21}^2}{(x_1\Lambda_{21}+x_2)^2}\right] \tag{C.255}$$

In order to judge which sign the resulting term has, it makes sense to examine the two symmetric terms in rectangular brackets separately. We focus on the first term in brackets.

First, it is shown that it increases in a strictly monotonic way with Λ_{12}, independently from x_1:

$$\frac{d}{d\Lambda_{12}}\left[\frac{1}{x_1} - \frac{x_1+2x_2\Lambda_{12}-(2-x_1)\Lambda_{12}^2}{(x_1+x_2\Lambda_{12})^2}\right]_{x_1,x_2}$$

$$= -\frac{(2x_2-(4-2x_1)\Lambda_{12})(x_1+x_2\Lambda_{12})^2 - 2x_2(x_1+x_2\Lambda_{12})[x_1+2x_2\Lambda_{12}-(2-x_1)\Lambda_{12}^2]}{(x_1+x_2\Lambda_{12})^4}$$

$$= -\frac{(2x_2-(4-2x_1)\Lambda_{12})(x_1+x_2\Lambda_{12}) - 2x_2[x_1+2x_2\Lambda_{12}-(2-x_1)\Lambda_{12}^2]}{(x_1+x_2\Lambda_{12})^3} \tag{C.256}$$

$$= -\frac{2x_1x_2 + 2x_2^2\Lambda_{12} -4x_1\Lambda_{12} + 2x_1^2\Lambda_{12} -4x_2\Lambda_{12}^2 + 2x_1x_2\Lambda_{12}^2 -2x_1x_2 -4x_2^2\Lambda_{12} + 4x_2\Lambda_{12}^2 -2x_1x_2\Lambda_{12}^2}{(x_1+x_2\Lambda_{12})^3}$$

$$= -\frac{-4x_1\Lambda_{12} + 2x_1^2\Lambda_{12} - 2x_2^2\Lambda_{12}}{(x_1+x_2\Lambda_{12})^3}$$

$$= \frac{\Lambda_{12}[2x_1(2-x_1) + 2x_2^2]}{(x_1+x_2\Lambda_{12})^3}$$

As Λ_{12} is always positive according to Eqs. (C.242) or (C.247) and $0 \leq x_1, x_2 \leq 1$, it turns out that the terms in brackets strictly increase with increasing Λ_{12}. Therefore, the minimum value occurs at $\Lambda_{12} \rightarrow 0$, which cannot be reached due to the curvature of the exponential function for Λ_{12}. The same considerations can be made for the second term in brackets in Eq. (C.255), indicating that the minimum value occurs for $\Lambda_{21} \rightarrow 0$. Therefore the minimum value of the second derivative of Δg is calculated to be ($\Lambda_{12} = \Lambda_{21}$)

$$\frac{\partial^2}{\partial x_1^2} [x_1 \ln x_1 + x_2 \ln x_2 - x_1 \ln(x_1 + x_2\Lambda_{12}) - x_2 \ln(x_1\Lambda_{21} + x_2)]$$
$$= \left[\frac{1}{x_1} - \frac{x_1}{x_1^2}\right] + \left[\frac{1}{x_2} - \frac{x_2}{x_2^2}\right] = 0 \qquad (C.257)$$

which means that the condition for the occurrence of a miscibility gap given in Eq. (C.251) can never be fulfilled.

More restrictions for the Wilson equation and their reasons can be found in [8]

F1. ($h-h^{id}$) for Soave–Redlich–Kwong Equation of State

The Soave–Redlich–Kwong equation is given by

$$P = \frac{RT}{v-b} - \frac{a(T)}{v(v+b)} \qquad (C.258)$$

with the mixing rules

$$a = \sum_i \sum_j y_i y_j a_{ij} \qquad (C.259)$$

$$b = \sum_i y_i b_i \qquad (C.260)$$

and

$$a_{ij} = \sqrt{a_i a_j}(1 - k_{ij}) \qquad (C.261)$$

$$a_i(T) = 0.42748 \frac{R^2 T_{ci}^2}{P_{ci}} \alpha_i(T) \qquad (C.262)$$

with

$$\alpha_i(T) = \left[1 + (0.48 + 1.574\omega_i - 0.176\omega_i^2)(1 - T_{ri}^{0.5})\right]^2$$
$$= \left[1 + m_i(1 - T_{ri}^{0.5})\right]^2 \qquad (C.263)$$

$$b_i = 0.08664 \frac{RT_{ci}}{P_{ci}} \qquad (C.264)$$

In Section C2, a calculation formula for $(h - h^{id})_{T,P}$ has been derived:

$$\left(h - h^{id}\right)_{T,P} = \int_\infty^v \left[T\left(\frac{\partial P}{\partial T}\right)_v - P\right] dv + Pv - RT \qquad (C.265)$$

The evaluation of Eq. (C.265) can be performed as follows:

$$\left(\frac{\partial P}{\partial T}\right)_v = \frac{R}{v-b} - \frac{da/dT}{v(v+b)} \tag{C.266}$$

$$\frac{da}{dT} = \sum_i \sum_j y_i y_j \frac{da_{ij}}{dT} = \sum_i \sum_j y_i y_j (1-k_{ij}) \frac{1}{2\sqrt{a_i a_j}} \left(a_i \frac{da_j}{dT} + a_j \frac{da_i}{dT}\right) \tag{C.267}$$

$$\frac{da_i}{dT} = 0.42748 \frac{R^2 T_{ci}^2}{P_{ci}} \frac{d\alpha_i(T)}{dT} = \frac{a_i(T)}{\alpha_i(T)} \frac{d\alpha_i(T)}{dT} \tag{C.268}$$

with

$$\frac{d\alpha_i(T)}{dT} = -2\left[1 + m_i\left(1 - T_{ri}^{0.5}\right)\right] \cdot m_i \frac{1}{2T_{ri}^{0.5}} \frac{1}{T_{ci}} = -\frac{1}{\sqrt{T}} \frac{m_i \sqrt{\alpha_i}}{\sqrt{T_{ci}}} \tag{C.269}$$

giving

$$\frac{da_i}{dT} = -\frac{1}{\sqrt{T}} \frac{a_i m_i}{\sqrt{\alpha_i T_{ci}}} \tag{C.270}$$

Summarizing yields

$$\frac{da}{dT} = -\frac{1}{\sqrt{T}} \sum_i \sum_j y_i y_j (1-k_{ij}) \frac{1}{2\sqrt{a_i a_j}} \left(a_i \frac{a_j m_j}{\sqrt{\alpha_j T_{cj}}} + a_j \frac{a_i m_i}{\sqrt{\alpha_i T_{ci}}}\right)$$

$$= -\frac{1}{\sqrt{T}} \sum_i \sum_j y_i y_j \frac{a_{ij}}{2} \left(\frac{m_j}{\sqrt{\alpha_j T_{cj}}} + \frac{m_i}{\sqrt{\alpha_i T_{ci}}}\right) \tag{C.271}$$

which will be abbreviated by

$$\frac{da}{dT} = -\frac{X}{2} \tag{C.272}$$

Thus, the integrand in Eq. (C.265) yields

$$T\left(\frac{\partial P}{\partial T}\right)_v - P = \frac{RT}{v-b} + \frac{TX}{2v(v+b)} - \frac{RT}{v-b} + \frac{a}{v(v+b)}$$

$$= \frac{a + \frac{XT}{2}}{v(v+b)} \tag{C.273}$$

and the integral can be evaluated to be

$$\left(h - h^{id}\right)_{T,P} = \left(a + \frac{XT}{2}\right) \int_\infty^v \frac{dv}{v(v+b)} + Pv - RT$$

$$= -\frac{1}{b}\left(a + \frac{XT}{2}\right) \ln \frac{v+b}{v} + Pv - RT \tag{C.274}$$

F2. $(s-s^{id})$ for Soave–Redlich–Kwong Equation of State

According to Section C1, the residual part of the specific entropy is

$$\left(s - s^{id}\right)_{T,P} = \int_{\infty}^{v} \left[\left(\frac{\partial P}{\partial T}\right)_v - \frac{R}{v}\right] dv + R \ln z \tag{C.275}$$

Using Eqs. (C.266) and (C.272), the integrand can be written as

$$\left(\frac{\partial P}{\partial T}\right)_v - \frac{R}{v} = \frac{R}{v-b} + \frac{X}{2v(v+b)} - \frac{R}{v} \tag{C.276}$$

and the result is

$$\left(s - s^{id}\right)_{T,P} = \int_{\infty}^{v} \left[\frac{R}{v-b} + \frac{X}{2v(v+b)} - \frac{R}{v}\right] dv + R \ln z$$

$$= R \ln \frac{v-b}{v} - \frac{X}{2b} \ln \frac{v+b}{v} + R \ln \frac{Pv}{RT}$$

$$= R \ln \frac{P(v-b)}{RT} - \frac{X}{2b} \ln \frac{v+b}{v} \tag{C.277}$$

F3. $(g-g^{id})$ for Soave–Redlich–Kwong Equation of State

Again, $g - g^{id}$ can be evaluated from the terms obtained for the enthalpy and the entropy.

From Eqs. (C.274) and (C.277) we obtain

$$\left(g - g^{id}\right)_{T,P} = \left(h - h^{id}\right)_{T,P} - T\left(s - s^{id}\right)_{T,P}$$

$$= -\frac{1}{b}\left(a + \frac{XT}{2}\right) \ln \frac{v+b}{v} + Pv - RT$$

$$\quad - RT \ln \frac{P(v-b)}{RT} + \frac{XT}{2b} \ln \frac{v+b}{v}$$

$$= -RT \ln \frac{P(v-b)}{RT} - \frac{a}{b} \ln \frac{v+b}{v} + Pv - RT \tag{C.278}$$

F4. Antiderivatives of c_p^{id} Correlations

The antiderivatives for the particular correlations for the calculation of enthalpies and entropies are

For Eq. (3.77):

$$c_p^{id} = A + B\left(\frac{C/T}{\sinh(C/T)}\right)^2 + D\left(\frac{E/T}{\cosh(E/T)}\right)^2$$

$$\int c_p^{id} dT = AT + BC \coth\left(\frac{C}{T}\right) - DE \tanh\left(\frac{E}{T}\right) + \text{const.} \tag{C.279}$$

$$\int \frac{c_p^{id}}{T} dT = A \ln \frac{T}{K} + B \left[\frac{C}{T} \coth\left(\frac{C}{T}\right) - \ln \sinh\left(\frac{C}{T}\right) \right] -$$
$$- D \left[\frac{E}{T} \tanh\left(\frac{E}{T}\right) - \ln \cosh\left(\frac{E}{T}\right) \right] + \text{const.} \quad \text{(C.280)}$$

For Eq. (3.78):
$$\frac{c_p^{id}}{R} = B + (C - B)\,y^2 \left[1 + (y - 1)(D + Ey + Fy^2 + Gy^3 + Hy^4) \right]$$
with $y = \frac{T}{A+T}$

$$\int c_p^{id} dT = RBT + R(C - B) \cdot \left[T - (D + E + F + G + H + 2)A \ln \frac{A+T}{K} \right.$$
$$- (2D + 3E + 4F + 5G + 6H + 1) \frac{A^2}{A+T}$$
$$+ (D + 3E + 6F + 10G + 15H) \frac{A^3}{2(A+T)^2}$$
$$- (E + 4F + 10G + 20H) \frac{A^4}{3(A+T)^3}$$
$$+ (F + 5G + 15H) \frac{A^5}{4(A+T)^4}$$
$$- (G + 6H) \frac{A^6}{5(A+T)^5}$$
$$\left. + H \frac{A^7}{6(A+T)^6} \right] + \text{const.} \quad \text{(C.281)}$$

$$\int \frac{c_p^{id}}{T} dT = RB \ln \frac{T}{K}$$
$$+ R(C - B) \cdot \left[\ln \frac{A+T}{K} + (1 + D + E + F + G + H) \frac{A}{A+T} \right.$$
$$- \left(\frac{D}{2} + E + \frac{3F}{2} + 2G + \frac{5}{2} H \right) \frac{A^2}{(A+T)^2}$$
$$+ \left(\frac{E}{3} + F + 2G + \frac{10}{3} H \right) \frac{A^3}{(A+T)^3}$$
$$- \left(\frac{F}{4} + G + \frac{5}{2} H \right) \frac{A^4}{(A+T^4)}$$
$$+ \left(\frac{G}{5} + H \right) \frac{A^5}{(A+T)^5}$$
$$\left. - \frac{H}{6} \frac{A^6}{(A+T)^6} \right] + \text{const.} \quad \text{(C.282)}$$

For Eq. (3.80):
$$c_p^{id} = A + BT + CT^2 + DT^3$$

$$\int c_p^{id} dT = AT + \frac{B}{2}T^2 + \frac{C}{3}T^3 + \frac{D}{4}T^4 + \text{const.} \tag{C.283}$$

$$\int \frac{c_p^{id}}{T} dT = A \ln \frac{T}{K} + BT + \frac{C}{2}T^2 + \frac{D}{3}T^3 + \text{const.} \tag{C.284}$$

G1. Speed of Sound as Maximum Velocity in an Adiabatic Pipe with Constant Cross-Flow Area

In an adiabatic pipe with constant cross-flow area, there are three equations which define the maximum velocity that can be developed. They can be written in the differential form:

1) **First Law**

$$dh + w\,dw = 0 \tag{C.285}$$

2) **Maximum mass flux**

$$d\left(\frac{\dot{m}}{A}\right) = 0 \tag{C.286}$$

3) **Isentropic change of state**

$$T\,ds = dh - v\,dP = 0 \tag{C.287}$$

As $\dot{m} = \dfrac{wA}{v}$, Eq. (C.286) yields

$$d\left(\frac{\dot{m}}{A}\right) = d\left(\frac{w}{v}\right) = \frac{dw}{v} - \frac{w\,dv}{v^2} = 0 \tag{C.288}$$

Combining Eqs. (C.285), (C.287), and (C.288), one obtains

$$dw = -\frac{dh}{w} = -\frac{v\,dP}{w} = \frac{w}{v}dv \tag{C.289}$$

and

$$w = \sqrt{-v^2 \left(\frac{\partial P}{\partial v}\right)_s} = w^* \tag{C.290}$$

which is identical to Eq. (3.98).

G2. Maximum Mass Flux of an Ideal Gas

To calculate the maximum mass flux of an ideal gas, it is assumed that the gas is stored in a vessel. A valve is opened for pressure relief. In the narrowest cross flow area, the speed of sound is the maximum possible velocity, as shown in G1. Index 0 denotes for the state in the vessel, index 1 for the state in the narrowest cross-flow area. For an ideal gas with a constant c_P^{id}, Eqs. (C.285), (C.290), and (C.285) read in their integrated form

1) **First law**

$$c_P^{id}(T_1 - T_0) + \frac{1}{2}(w_1^2 - w_0^2) = 0 \tag{C.291}$$

2) **Speed of sound**

$$w_1 = w^*(T_1) = \sqrt{\kappa R T_1} \tag{C.292}$$

3) **Isentropic change of state**

$$\frac{T_1}{T_0} = \left(\frac{P_1}{P_0}\right)^{\frac{\kappa-1}{\kappa}} \tag{C.293}$$

Rearranging Eq. (C.291) and combining with Eq. (C.292) yields

$$\frac{1}{2}\kappa R T_1 = c_P^{id}(T_0 - T_1) \tag{C.294}$$

if w_0 is set to zero. Solving for T_1/T_0, and using $\kappa = (c_P^{id})/(c_P^{id} - R)$ the result is

$$\frac{T_1}{T_0} = \frac{c_P^{id}}{\frac{1}{2}\kappa R + c_P^{id}} = \frac{\frac{c_P^{id}}{R}}{\frac{1}{2}\kappa + \frac{c_P^{id}}{R}}$$

$$= \frac{\frac{\kappa}{\kappa-1}}{\frac{\kappa}{2} + \frac{\kappa}{\kappa-1}} = \frac{2\kappa}{\kappa(\kappa-1) + 2\kappa} = \frac{2}{\kappa+1} \tag{C.295}$$

Introducing Eq. (C.278) into Eq. (C.280) gives

$$\frac{P_1}{P_0} = \left(\frac{2}{\kappa+1}\right)^{\frac{\kappa}{\kappa-1}} \tag{C.296}$$

Integrating the ideal gas equation of state and the First Law (Eq. (C.276)), the maximum possible mass flux can be set up as

$$\left(\frac{\dot{m}}{A}\right)_{max} = \frac{w_1}{v_1}$$

$$= \frac{P_1}{RT_1}\sqrt{2c_P^{id}(T_0 - T_1)}$$

$$= \sqrt{\frac{2P_1^2 c_P^{id}}{R^2 T_1^2}(T_0 - T_1)}$$

$$= \sqrt{\frac{2P_0}{v_0} \frac{RT_0}{P_0^2} \frac{P_1^2 c_P^{\text{id}}}{R^2 T_1^2} (T_0 - T_1)}$$

$$= \sqrt{\frac{2P_0}{v_0} \frac{\kappa}{\kappa - 1} \left(\frac{P_1^2 T_0^2}{P_0^2 T_1^2} - \frac{P_1^2 T_0}{P_0^2 T_1} \right)}$$

$$= \sqrt{\frac{2P_0}{v_0} \frac{\kappa}{\kappa - 1} \left[\left(\frac{P_1}{P_0}\right)^2 \left(\frac{P_1}{P_0}\right)^{\frac{1-\kappa}{\kappa} \cdot 2} - \left(\frac{P_1}{P_0}\right)^2 \left(\frac{P_1}{P_0}\right)^{\frac{1-\kappa}{\kappa}} \right]}$$

$$= \sqrt{\frac{2P_0}{v_0} \frac{\kappa}{\kappa - 1} \left[\left(\frac{P_1}{P_0}\right)^{\frac{2}{\kappa}} - \left(\frac{P_1}{P_0}\right)^{\frac{\kappa+1}{\kappa}} \right]} \qquad (C.297)$$

References

1. Bronstein, I.N. and Semendjajev, K.A. (1981) *Taschenbuch der Mathematik*, 21 Auflage, Verlag Harri Deutsch, Thun und Frankfurt a.M.
2. Baehr, H.D. (1992) *Thermodynamik*, 8. Auflage, Springer, Berlin.
3. Falk, G. and Ruppel, W. (1976) *Energie und Entropie*, Springer, Berlin.
4. Novak, J.P., Matous, J., and Vonka, P. (1991) *Collect. Czech Chem. Commun.*, **56**, 745–749.
5. Wilson, G.M. (1964) *J. Am. Chem. Soc.*, **20** (1), 127–130.
6. Maurer, G. (1978) *Fluid Phase Equilib.*, **2**, 91–99.
7. Renon, H. and Prausnitz, J.M. (1968) *AIChE J.*, **14** (1), 135–144.
8. Pentermann, W. (1983) *VT Verfahrenstech.*, **17** (2), 89–90.

Appendix D
Standard Thermodynamic Properties for Selected Electrolyte Compounds

Compound	$\Delta h_f^0(*,m)$ (kJ/mol)	$\Delta g_f^0(*,m)$ (kJ/mol)	$c_P^0(*,m)$ (J/(mol K))
CO_2 (aq)	−413.8	−385.98	−
SO_2 (aq)	−323.16	−300.503	194.8
H_2S (aq)	−39.7	−27.83	−
HF (aq)	−320.08	−296.82	−
NH_3 (aq)	−80.29	−26.5	−
Hg (aq)	21.776	37.244	154.2
H^+ (aq)	0	0	0
Ca^{2+} (aq)	−542.83	−553.58	−
K^+ (aq)	−252.38	−283.27	21.8
Na^+ (aq)	−240.12	−261.905	46.4
NH_4^+ (aq)	−132.51	−79.31	79.9
Cu^{2+} (aq)	65.78	65.7	−
Zn^{2+} (aq)	−152.84	−147.16	−
Hg^{2+} (aq)	170.16	164.703	−
OH^- (aq)	−229.994	−157.244	−148.8
Cl^- (aq)	−167.159	−131.228	−136.4
F^- (aq)	−332.63	−278.79	−106.7
HCO_3^- (aq)	−691.99	−586.77	−
CO_3^{2-} (aq)	−677.14	−527.81	−
SO_3^{2-} (aq)	−630.44	−486.092	−263.9
HSO_3^- (aq)	−626.79	−527.032	−1.9
HSO_4^- (aq)	−887.34	−755.91	−84.0
SO_4^{2-} (aq)	−909.27	−744.53	−293.0
HS^- (aq)	−17.6	12.08	−
S^{2-} (aq)	33.1	85.8	−
NO_2^- (aq)	−104.6	−32.2	−97.5
NO_3^- (aq)	−205	−108.74	−86.6

	$\Delta h_f^0(L)$ (kJ/mol)	$\Delta g_f^0(L)$ (kJ/mol)	$c_P^0(L)$ (J/(mol K))
H_2O	−285.83	−237.129	75.291

Data from: Luckas/Krissmann: Thermodynamik der Elektrolytlösungen, Springer, Berlin, Heidelberg, 2001.

Appendix E
Regression Technique for Pure Component Data

To perform a regression means to adjust a correlation function to given data points in a way that the representation of the data points is as good as possible. This is achieved by minimizing an objective function F. Consider m to be a general pure component property, this objective function is usually defined as

$$F = \sum_i w_i \left(\frac{m_{calc,i} - m_{exp,i}}{m_{exp,i}} \right)^2 \overset{!}{\to} \min \qquad (E.1)$$

where w_i is a weighting factor for a particular data point.[1]

The summation of the squares of the deviations has the target to eliminate the influence of the sign of the deviations and to emphasize the points with large deviations. The weighting factor can sort out obviously poor data points or assign a special weight to individual data points, for example, according to their experimental uncertainties. Many other forms for the objective function are possible, for example, the sum of relative or absolute deviations.

In a regression, three kinds of problems can come up:

1) The chosen correlation might not be capable to reproduce the data satisfactorily with the required accuracy. In this case, a better one, maybe with a larger number of adjustable parameters, has to be chosen. Currently, there are correlations available for each property which are finally capable enough to reproduce any reasonable data for a specified component in the temperature range of interest.
2) Some data points or whole data sets can be wrong or not accurate enough. In this case, they should be removed from the database, for example, by setting their weighting factor to 0. To decide which data points have to be removed, the following procedure has proved to be successful:
   a. Simultaneous regression of all data to get a quick overview on the particular deviations. If it turns out that they are all consistent with each other, only single outliers have to be removed.

---

1) The following considerations are general ones, independent from the objective function.

b. If the data situation could not be clarified, each data set should be fitted on its own to check its internal consistency. The weight of inconsistent data sources should be set to zero.
c. Find combinations of data sets which can be correlated simultaneously. Do not forget to remove single outlier data points.
3) Extrapolation might not be reasonable, which is often the case for polynomials. Therefore, a graphical representation showing the whole temperature range of interest should be checked to obtain a reasonable curvature.

The vapor pressure regression is the most important and most difficult one. From the triple point to the critical point, the vapor pressure range covers several orders of magnitude. Therefore, only a logarithmic diagram can visualize the curvature of the vapor pressure in the whole temperature range. On the other hand, a logarithmic diagram cannot be used to detect small differences between correlated and experimental values that correspond to the high accuracy demand of vapor pressures ($\ll 1\%$ deviation). Figure E.1 gives an example. The logarithmic diagram seems to indicate good agreement between measured data points and

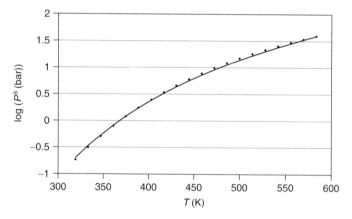

Figure E.1 Logarithmic vapor pressure curve.

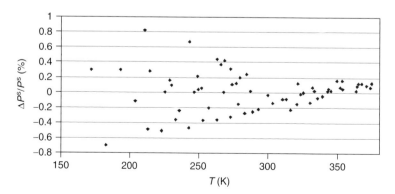

Figure E.2 Deviation plot for a good vapor pressure fit (Wagner equation).

calculated vapor pressure curve. However, the mean deviation is larger than 5% and therefore not acceptable. Instead, a deviation plot can be used for this purpose.

Figure E.2 shows the deviation plot for the vapor pressure curve of R134a which could be fitted reasonably well with the Wagner equation. The deviations are scattered around the zero line, and a systematic error cannot be detected. The absolute maximum deviation is ∼0.8%, already in the low vapor pressure region where it can be tolerated. The average deviation is much lower (0.26%). In comparison to that, the situation in Figure E.3 is worse. Although the same data have been used, the plot results in a sinusoidal line, which indicates that the Antoine correlation used is not flexible enough. This could be tolerated if the deviations were small, however, in this case, the average deviation is too large (∼0.63%). Thus, a more flexible correlation is recommended (e.g., the Wagner equation). Another option would be to assign a smaller validation range to the fitted equation.

Finally, Figure E.4 shows a plot for toluene where measured data from different authors do not agree. In this case, an improved correlation cannot solve the

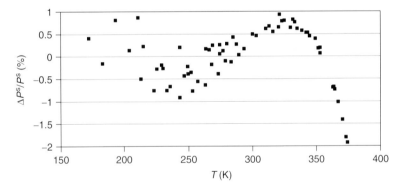

Figure E.3 Deviation plot for the vapor pressure regression; correlation is not flexible enough (Antoine equation).

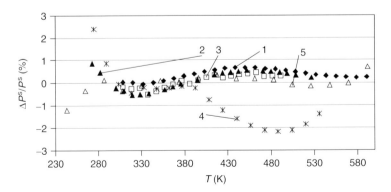

Figure E.4 Deviation plot for several data sources that disagree.

problem. Instead, the data points from reference 4 should be removed from the database by setting their weighting factor to 0.

To obtain a reliable correlation, care should be taken that the ratio between the number of data points and the number of adjustable parameters is reasonable. At least, a ratio of 2 should be aspired.

Appendix F
Regression Techniques for Binary Parameters

The results of simulation models are strongly affected by the particular assumptions made in the plant model and by the model parameters. This is the reason why these parameters should be fitted to reliable data describing simplified systems, for example, binary mixtures. Regressing parameters to observations on the complex target system is usually not feasible, as the influence of the assumptions made on the performance of the equipment (e.g., tray efficiency in distillation) is not sufficiently isolated from the influence of the model parameters. The experimental data for the isolated subsystems are the key for a successful process model. They contain all the information about the process model, the model parameters are only a representation. If new data become available, the model can be upgraded by extending the database and perform a new regression.

In a process simulation of a chemical plant, usually 20 ... 40 components are involved. When a process model is developed, the adjustment of the binary parameters is the most time-consuming part. For example, if a system with 20 components is considered, a total of $20 \cdot 19/2 = 190$ binary parameter sets have to be fitted. To avoid diluting one's effectiveness, it should be taken into account that not all of them have to be necessarily determined. Usually, many of the by-products only occur at very low concentrations. Therefore, only their binary parameters with the main components are of interest; their binary parameters among themselves have hardly any influence on the results. Commercial simulation programs provide integrated parameter tables. In many cases, their accuracy is sufficient, but the user should be aware that the information about their range of applicability, the quantity of the data the binary parameters are based on, and the quality of the regression is lost; not to mention that sometimes the given parameters are simply erroneous. The following example illustrates the consequences which might occur if default binary parameters are used without further check.

Example F.1

For the binary system 1,1,2-trichloroethane (1)–water (2), the following NRTL parameters are listed as built-in default parameters for the form

$$\tau_{ij} = A_{ij} + \frac{B_{ij}}{T/K} + E_{ij} \ln(T/K) + F_{ij}(T/K)$$

Chemical Thermodynamics: for Process Simulation, First Edition.
Jürgen Gmehling, Bärbel Kolbe, Michael Kleiber, and Jürgen Rarey.
© 2012 Wiley-VCH Verlag GmbH & Co. KGaA. Published 2012 by Wiley-VCH Verlag GmbH & Co. KGaA.

$$A_{12} = 670.9628 \qquad A_{21} = -108.7561$$
$$B_{12} = -26907.7813 \qquad B_{21} = 4489.9561$$
$$E_{12} = -101.3913 \qquad E_{21} = 17.4607$$
$$F_{12} = 0 \qquad F_{21} = 0$$
$$\alpha_{12} = 0.2$$

The parameters are indicated to be valid in the temperature range $0\,°C < \vartheta < 55\,°C$.

In a process simulation, they play a decisive role for the mass balance in a decanter, operating at $\vartheta_1 = 50\,°C$, and in a waste water stripper, operating at $P_2 = 2.3$ bar, corresponding to a temperature $\vartheta_2 = 125\,°C$. The task of the stripper is to take the organic components overhead to simplify the biological waste water treatment. Check whether the parameters can be used in the process simulation. If not, replace them by appropriate ones.

Solution

The parameters have a limited range of validity which does not cover the operating temperature of the stripper. They have to be extrapolated beyond this range, which is nothing unusual. However, the parameter figures are suspicious, because of their unusually high values. It is quite normal that for a strongly nonideal system temperature-dependent parameters must be used; however, six parameters to cover just a range of 55 K are probably not justified. This impression is confirmed in the Pxy diagrams at $\vartheta_1 = 50\,°C$ and $\vartheta_2 = 125\,°C$ (Figure F.1).

As can be seen, the parameters work as expected at $\vartheta_1 = 50\,°C$, where a large miscibility gap is exhibited. This is expected for a system consisting of water and a nonpolar substance like 1,1,2-trichloroethane. Surprisingly, at

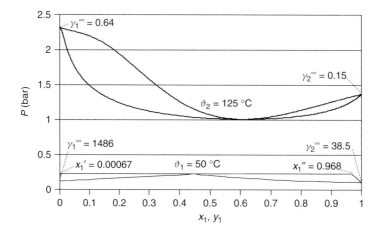

Figure F.1 Pxy diagrams of the system 1,1,2-trichloroethane (1)–water (2) at $\vartheta_1 = 50\,°C$ and $\vartheta_2 = 125\,°C$ on the basis of the default parameters.

Table F.1 LLE data of the system 1,1,2-trichloroethane (1)–water (2).

T/K	x_1' (mol/mol)	x_1'' (mol/mol)
273.15	0.0006293	0.994841
282.35	0.0005952	0.9927901
292.85	–	0.991038
304.45	0.0006211	0.988492
314.15	0.0006551	0.986753
323.75	0.0007028	0.983867
333.65	0.0006742	0.980565
344.15	0.0007533	0.977496
354.85	0.0008939	0.974514
363.95	0.0009554	0.9698601

Data from Stephenson [1].

$\vartheta_2 = 125\,°C$, the miscibility gap has vanished (see Figure F.1). Instead, large negative deviations from Raoult's law and a pressure-minimum azeotrope are observed. This can hardly be believed. A negative deviation from Raoult's law means that the 1,1,2-trichloroethane will leave the stripper at the bottom of the column, this means the stripper will not work. Together with the suspicion that the parameters are strange, a revision of the model should take place.

This can be done by fitting the parameters to the liquid–liquid equilibria (LLE) data of Stephenson [1], which are listed in Table F.1.

The LLE data can be fitted with the NRTL equation, giving the parameters

$$A_{12} = -3.2453 \qquad A_{21} = 5.9389$$
$$B_{12} = 1839.4651 \qquad B_{21} = -61.7085$$
$$E_{12} = 0 \qquad E_{21} = 0$$
$$F_{12} = 0 \qquad F_{21} = 0$$
$$\alpha_{12} = 0.2$$

Having a look at Figure F.2, the advantages of the new parameters become obvious. It is now clear what the parameters are based on. Still, the application in the stripper is beyond their range of validity. Nevertheless, the extrapolation looks much more reasonable, and the parameters are assured up to approx. 90 °C. The expected behavior showing large miscibility gaps both in the decanter and in the stripper is now calculated. Moreover, the solubility of water in 1,1,2-trichloroethane at $\vartheta = 50\,°C$ turns out to be only half the value calculated with the default parameters, which should cover this case. In fact, with the new parameters, the behavior of the 1,1,2-trichloroethane in the process can be adequately described.

It must be emphasized that a case like this is not an error of the simulation program. The default parameters are suggestions: it is the responsibility of the user

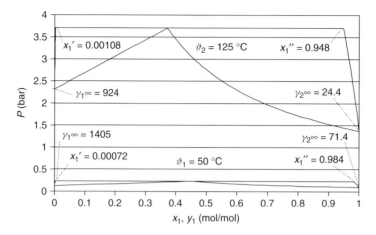

Figure F.2 Pxy diagrams of the system 1,1,2-trichloroethane (1)–water (2) at $\vartheta_1 = 50\,°C$ and $\vartheta_2 = 125\,°C$ on the basis of the adjusted parameters.

to clarify whether they can be used or not. Especially in this case, the given range of applicability clearly indicated that there might be a problem for the stripper calculation.

Thus, it is strongly recommended that at least the decisive binary parameters should be checked carefully to have a defined accuracy and application range. If literature data are not available, it is worth carrying out own measurements. Predictive methods like modified UNIFAC or VTPR often provide good results; however, with estimated binary parameters there will ever remain imponderabilities in the design of the equipment. For binary parameters which are not decisive or which are only used for preliminary work, predictive methods are a useful tool. They can easily be combined with other models by using them for generating artificial data sets, which can then be regressed again to obtain the missing parameters for the process model.

There are various types of datasets:

- **TPxy data** For VLE points, the whole information is available. If a miscibility gap occurs, the vapor concentrations have the largest uncertainties.
- **TPx data** The vapor concentration is not measured. The data are often measured with a static equilibrium cell.
  Usually, the liquid concentration is determined by feeding certain liquid amounts of the degassed compounds into the evacuated cell. Especially for high pressures, low feed amounts, and large volatility differences between the two components, the amounts of substance in the vapor phase for each component have to be taken into account when calculating the true composition in the liquid phase by an iterative procedure. For this purpose the total volume of the static cell must be known.

- **Txy data** This data type is typical for gas chromatographic headspace measurements. One should be aware that for this type of measurement, only the vapor concentration is experimentally evaluated. If the pressure is given additionally, it has been obtained by a calculation with the help of a model.
- **y-values** Old data sources often give calculated values for the activity coefficients. If possible, the original data points should be used.
- **Activity coefficients at infinite dilution** are extremely important for the design of distillation columns, as most of the stages are necessary for the removal of the last traces of the impurities.
- **LLE data** For LLE, the concentrations of both phases or only of one phase (cloud-point method) are given. In the latter case, binary parameters can be fitted if cloud points on both sides are available. After adjusting the parameters, one should be aware that they do not necessarily describe multicomponent LLEs correctly. If possible, ternary LLE should be integrated into the database to make the parameters more significant.
- **h^E data** Binary parameters can also be fitted to excess enthalpy data. As stand-alone datasets, h^E data are not appropriate, as the prediction of vapor–liquid equilibria from h^E data does not work with the models. Thus, h^E data are only useful as supplementary information. As h^E is related to the temperature dependence of the activity coefficients, a good h^E fit helps to achieve a reliable temperature extrapolation outside the temperature range covered by the data.

The excess enthalpy itself is usually not very important in process simulation, as it is in most cases negligible in comparison to enthalpies of vaporization. There are certain well-known exceptions (ammonia–water and acetaldehyde–water).

For each data type, there are several alternatives for the objective function. For VLE data, the usually applied function is

$$F = \sum_i \left[\left(\frac{P_{i,\text{calc}} - P_{i,\text{exp}}}{P_{i,\text{exp}}} \right)^2 + w \left(y_{1i,\text{calc}} - y_{1i,\text{exp}} \right)^2 \right] \quad (F.1)$$

This function has proved to give satisfactory results even if the weighting factor w is simply set to unity. It can be applied both to isothermal and isobaric datasets. Alternatively, the function

$$F = \sum_i \left[\left(\frac{T_{i,\text{calc}} - T_{i,\text{exp}}}{T_{i,\text{exp}}} \right)^2 + w \left(y_{1i,\text{calc}} - y_{1i,\text{exp}} \right)^2 \right] \quad (F.2)$$

is often used for isobaric datasets. However, one must take into account that with $w = 1$, a 1% deviation in the vapor fraction has the same contribution to the objective function as for instance a large deviation of 3.5 K in the temperature at $T = 350$ K. Therefore, for $w = 1$ mainly the vapor concentrations are fitted. For measurements by headspace gas chromatography, where only the vapor concentrations are experimentally determined, the first term can simply be left out:

$$F = \sum_i \left(y_{1i,\text{calc}} - y_{1i,\text{exp}} \right)^2 \quad (F.3)$$

Considering the vapor fraction term in Eqs. (F.1–F.3), one might wonder whether the absolute differences were taken for the objective function instead of a term like

$$F = \sum_i \left(\frac{y_{1i,\text{calc}} - y_{1i,\text{exp}}}{y_{1i,\text{exp}}} \right)^2 \tag{F.4}$$

In fact, this would be a bad choice. The advantage of the absolute differences is that all data points and both components are treated in the same way. In Eq. (F.4), the points at low concentrations of component 1 have a higher weight than the ones at high concentration. Therefore, essentially only one side of the binary mixture is fitted, and it is hardly possible that γ_2^∞ is represented correctly. For a correct representation of both sides, a symmetric objective function

$$F = \sum_i \left[\left(\frac{y_{1i,\text{calc}} - y_{1i,\text{exp}}}{y_{1i,\text{exp}}} \right)^2 + \left(\frac{y_{2i,\text{calc}} - y_{2i,\text{exp}}}{y_{2i,\text{exp}}} \right)^2 \right] \tag{F.5}$$

must be applied. With this function, both components are treated in the same way, but the dilute regions are emphasized. The results are often significantly different to those obtained with Eq. (F.3). The choice of the objective function depends on the targets of the user; there is no correct or incorrect objective function.

Similarly, there are several options for the parameter adjustment of LLE, for instance

$$F = \sum_i \left[\left(x'_{1i,\text{calc}} - x'_{1i,\text{exp}} \right)^2 + \left(x''_{1i,\text{calc}} - x''_{1i,\text{exp}} \right)^2 \right] \tag{F.6}$$

and

$$F = \sum_i \left[\left(\frac{x'_{1i,\text{calc}} - x'_{1i,\text{exp}}}{x'_{1i,\text{exp}}} \right)^2 + \left(\frac{x''_{1i,\text{calc}} - x''_{1i,\text{exp}}}{x''_{1i,\text{exp}}} \right)^2 + \left(\frac{x'_{2i,\text{calc}} - x'_{2i,\text{exp}}}{x'_{2i,\text{exp}}} \right)^2 + \left(\frac{x''_{2i,\text{calc}} - x''_{2i,\text{exp}}}{x''_{2i,\text{exp}}} \right)^2 \right] \tag{F.7}$$

Both functions treat the two components in the same way. However, Eq. (F.6) is only appropriate if the mutual solubilities have the same order of magnitude. For illustration, consider a miscibility gap in the concentration range 0.001–0.6. An error $\Delta x = 0.01$ would be excellent for x''_1 but disastrous for $x'_1 (\Delta x \gg x'_1)$. Nevertheless, their contribution to the objective function is identical, and the fitting procedure might have problems to yield a good representation of the left-hand side. Equation (F.7) has less problems in these cases.

For direct fitting of activity coefficient data, the two options are

$$F = \sum_i \left[\left(\gamma_{1i,\text{calc}} - \gamma_{1i,\text{exp}} \right)^2 + \left(\gamma_{2i,\text{calc}} - \gamma_{2i,\text{exp}} \right)^2 \right] \tag{F.8}$$

and

$$F = \sum_i \left[\left(\frac{\gamma_{1i,\text{calc}} - \gamma_{1i,\text{exp}}}{\gamma_{1i,\text{exp}}} \right)^2 + \left(\frac{\gamma_{2i,\text{calc}} - \gamma_{2i,\text{exp}}}{\gamma_{2i,\text{exp}}} \right)^2 \right] \tag{F.9}$$

Again, Eq. (F.8) is only useful if the activity coefficients have the same order of magnitude. Otherwise, only the contributions of high γ-values are significant.

For h^E data, the favorite function is

$$F = \sum_i \left(\frac{h^E_{i,\text{calc}} - h^E_{i,\text{exp}}}{|h^E_{\max}|} \right)^2 \qquad (F.10)$$

$|h^E_{\max}|$ is the largest absolute value in the dataset.

If several data types occur, the corresponding objective functions are added up to an overall objective function, which has then to be minimized. This is like comparing apples and oranges; the particular functions contribute to the overall objective function in a different way. For instance, contributions of Eq. (F.7) are usually much larger than those of the other quantities. Therefore, it is strongly recommended to think about the regression strategy first before just generating parameters. As an example, LLE data could first be regressed separately and then be regarded in the database as simple activity coefficient data.

The problem of arbitrary objective functions can be overcome with the so-called maximum-likelihood method [2]. With this method, the objective function is,

$$F = \sum_i \left[\left(\frac{T_{\text{true},i} - T_{\text{exp},i}}{\sigma_T} \right)^2 + \left(\frac{x_{1,\text{true},i} - x_{1,\text{exp},i}}{\sigma_x} \right)^2 \right.$$
$$\left. + \left(\frac{P_{\text{true},i} - P_{\text{exp},i}}{\sigma_P} \right)^2 + \left(\frac{y_{1,\text{true},i} - y_{1,\text{exp},i}}{\sigma_y} \right)^2 \right] \qquad (F.11)$$

with σ as standard deviation of the particular quantity, which corresponds to the experimental uncertainty. In the maximum likelihood procedure, the independent variables for phase equilibrium calculation (T and x) are varied as well to obtain the so-called true values, as the measured values are never exact. The true values for the dependent variables P and y are related to them via the phase equilibrium condition. Thus, one will get many more variables in the objective function, and the minimization effort increases rapidly. Although it seems to be the most reasonable way to formulate an objective function, it is doubtful whether the additional effort is really justified.

The evaluation of the regression results should be done with a table of deviations as well as with a graphical representation. Both are necessary to get a sufficient impression about the quality of the dataset and the regression. The deviations should be within a certain range. For $TPxy$ data, the average deviations

$$\overline{\Delta P/P} = \sqrt{\frac{1}{n} \sum_{i=1}^{n} \left(\frac{P_{i,\text{calc}} - P_{i,\text{exp}}}{P_{i,\text{exp}}} \right)^2} \approx 1\%$$

and

$$\overline{\Delta y} = \sqrt{\frac{1}{n} \sum_{i=1}^{n} \left(y_{i,\text{calc}} - y_{i,\text{exp}} \right)^2} \approx 0.5\%$$

can be regarded as reasonable. If the deviations are higher, the reason should be evaluated. For other data, a rule of thumb for acceptable deviations cannot be set up as easily. For γ^∞-values, larger deviations like 5% could be accepted. For other data, the deviations must be regarded in context with the data themselves. For LLE data, it is very important that at least the order of magnitude of the mutual solubilities is well represented in the whole temperature range that is considered.

For an evaluation of a regression, the following items might be useful.

- Are there data points in the list which are represented a good deal worse than the others? If so, these outliers should be removed from the database.
- Are the deviations uniformly distributed around the zero line? If yes, it can be assumed that the errors are statistical ones. Furthermore, if they are within the expected uncertainty, the regression can be taken as a confirmation of the thermodynamic consistency.
- If all or a large majority of the data show positive or negative deviations from the zero line, systematic errors, a weakness of the chosen model (e.g., wrong description of the vapor phase nonidealities), or a failure of the regression program can be the reason. The parameter fit must not be accepted.
- For systems with large miscibility gaps, pressure and boiling temperature outside the miscibility gap increase very steeply with concentration. Although the deviations in P and y might be large, the graphical representation can still be acceptable, as long as the miscibility gap is described correctly.

If several datasets have to be regressed simultaneously, it is recommended to fit them first separately to detect outliers and check their internal thermodynamic consistency. If a dataset cannot be regressed with a sufficient quality, it should be removed from the database, as it might spoil the fit of the other datasets. Afterward, it can still happen that different internally consistent datasets do not fit together. In this case, a clear judgment of the situation is not possible. To obtain at least a solution, one could try to find combinations of datasets which fit together to enhance the probability of the chosen solution. Furthermore, there should be a tendency to prefer recently evaluated data to older data due to the improvement of the measurement techniques.

Before accepting regressed parameters, a graphical representation should be checked for plausibility. Figure F.3 shows the normal, monotonic curvature of the activity coefficient with respect to concentration for constant temperature. The highest values occur at infinite dilution. Other curvatures have to be checked carefully (e.g., possible outliers or bad choice of the objective function), although errors are not always involved. Two possible reasons for strange γ-values are as follows.

- The local composition models Wilson, NRTL, and UNIQUAC have some difficulties with small nonidealities ($\gamma^\infty \approx 0.9$–$1.1$) Therefore, curvatures like those shown in Figure F.4 can be accepted.
- For isobaric datasets, it is not unusual that the γ-plot shows a maximum, as the activity coefficient additionally depends on temperature, which is not constant for widely boiling mixtures (see Figure 5.32).

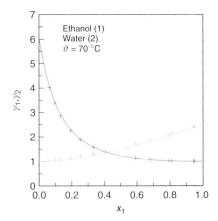

Figure F.3 Normal, monotonic γ plot as a function of concentration. (Data from Mertl [3].)

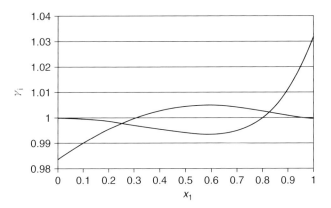

Figure F.4 Unexpected γ-curvature for small nonidealities.

- It is strange if one γ^∞-value is less than unity and one is larger than unity. The dataset should be checked again in this case. Otherwise, the values obtained can simply be correct like in Figure F.5, where the system benzene–hexafluorobenzene has in fact a double azeotrope.

Concerning their order of magnitude, activity coefficients cover a wide range. $\gamma < 1$ occurs less often than $\gamma > 1$. For γ-values less than 0.1, chemical reactions are probably involved (e.g., formaldehyde–water or HCl–water). In these cases, an activity coefficient model on its own is not appropriate; instead, the particular reactions must be taken into account.

For $\gamma > 1$, values up to 10^5 have been observed (e.g., p-xylene–water, Figure F.6). Sometimes, regressions yield γ^∞-values like 10^{10}. In these cases, the regression has to be carefully checked. Due to numerical difficulties, it should be avoided to use the corresponding parameters in process simulations.

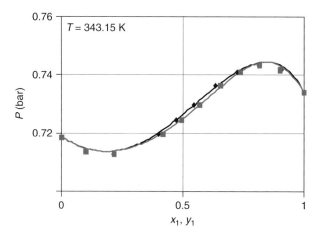

Figure F.5 Double azeotropy for the system benzene (1)–hexafluorobenzene (2) at $T = 343.15$ K. (Data from Gaw and Swinton [4].)

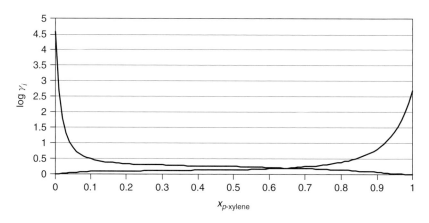

Figure F.6 Extraordinarily high γ-values for the system p-xylene-water at $T = 303.15$ K.

When binary data are regressed, first temperature-independent parameters should be used. If the fit does not satisfy, the use of additional parameters can be taken into account. First, a linear temperature dependence described by the parameter A_{ij} can be introduced:

$$\tau_{ij} = \exp\left[\frac{A_{ij}T + B_{ij}}{T}\right] = \exp\left[A_{ij} + \frac{B_{ij}}{T}\right] \tag{F.12}$$

One has to take care that the additional parameters are significant. This is usually the case for the following data situations:

- at least two isothermal datasets with sufficient temperature difference (approx. 30 … 50 K);

- an isobaric dataset with many data points and a wide temperature spread (~50–70 K). One has to keep in mind that this situation is not desirable, see Section 5.4;
- LLE data for different temperatures. Often, a relatively small temperature spread is sufficient to justify temperature-dependent parameters;
- a VLE dataset supplemented by h^E data.

A reasonable range for the A_{ij} parameters is $-6 < A_{ij} < 6$. Values beyond this range can be fully correct; however, before application one should carefully check their extrapolation behavior. If the use of temperature-dependent parameters does not yield significant improvements, they should be cancelled again.

If the regression results are still not satisfactory, a further pair of parameters can be considered, for example, a quadratic temperature dependence:

$$\tau_{ij} = \exp\left[\frac{A_{ij}T + B_{ij} + C_{ij}T^2}{T}\right] = \exp\left[A_{ij} + \frac{B_{ij}}{T} + C_{ij}T\right] \tag{F.13}$$

In this case, the extrapolation behavior is even more critical. There are no rules for the range of the parameters, and the fitting procedure is often difficult. It is recommended to fit B_{ij} and A_{ij} first. Again, the additional parameters should be cancelled if they do not yield significant improvements.

For the NRTL equation, there is also the opportunity to adjust the α-value or make it even temperature-dependent. In the latter case, α is written as

$$\alpha_{ij} = C_{ij} + D_{ij}(T - 273.15 \text{ K}) \tag{F.14}$$

Fitting α can improve the regression results significantly. There are some systems where the adjustment of α is even necessary. An example is the system methyl acetate–water, which exhibits a complicated isothermal behavior with a miscibility gap and a homogeneous azeotrope (Figure F.7).

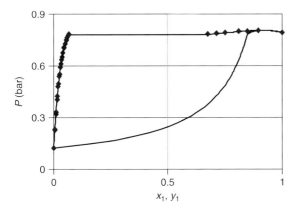

Figure F.7 Isothermal VLE behavior of the system methyl acetate (1)–water (2) at 323.15 K. (Data from Loehe et al. [5].)

Nevertheless, one should be careful with the fitting of the α parameter. For multicomponent systems, experience shows that the application of NRTL in process simulation is somewhat easier if all the α-values are close to each other. The larger the α, the more probable are multiple solutions for the LLE (see below).

If data at low concentrations of one component occur ($x_1 < 0.05$, $x_1 > 0.95$), the use of vapor pressure shifting should be considered to get the activity coefficients at infinite dilution correctly. Figure F.8 illustrates an example for the system water–trioxane at $T = 338.15$ K. The solid line represents the fit with vapor pressure shifting, whereas it has not been used for the fit represented by the dashed line.

For the dashed line, the regression program has used the correct vapor pressure equation which has in this case been carefully fitted to high precision data. However, the vapor pressure for trioxane measured by the author shows a deviation of approx. 6%. This can hardly be regarded as an acceptable accuracy; nevertheless, data for this system are relatively scarce and any available dataset should be utilized in some way. Moreover, the deviation in the trioxane vapor pressure measurement is likely to be caused by traces of water, which has low influence on the other data points. With the disagreement in the pure component data, the regression produces a wrong slope of the bubble point line at low water concentrations, as it only tries to minimize the deviations in the objective function between experimental and calculated data points. A wrong slope in this region means that the γ^∞-values of water as the dilute component will be poorly represented. The result is $\gamma^\infty = 5.1$. The alternative is that the vapor pressure measured by the authors is used for the regression (see Fig. 5.29).

For this purpose, the values obtained from the vapor pressure equations for both components are multiplied with a factor which is specific for each dataset. If the Antoine equation is used, this can be achieved by simply changing the parameter A in Eq. 3.30. The solid line in Figure F.8 shows that the correct slope of the bubble point line can be represented. The water activity coefficient at infinite

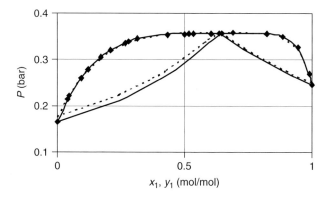

Figure F.8 Fit of the system water (1)–trioxane (2) at $T = 338.15$ K with (solid line) and without (dashed line) vapor pressure shifting. (Data from Brandani et al. [6].)

dilution is determined to be $\gamma^\infty = 5.9$. There is a significant difference between the two cases for the dew-point line. This example demonstrates that the vapor pressure shifting is an essential tool to obtain binary parameters that represent the correct γ^∞-values. In process simulation, the original vapor pressure equation will of course be used again. The whole dataset should be cancelled if the differences between the experimental vapor pressures and a well-assured vapor pressure equation are too large (approx. 2 ... 3%, if alternative datasets are available). If a VLE dataset does not contain the pure component values, the situation becomes more difficult. There are users who even fit the correction factor for the vapor pressure; however, a general recommendation cannot be given in this case.

If a miscibility gap occurs, the Gibbs energy of mixing should be checked. The equilibrium condition

$$(x_i \gamma_i)' = (x_i \gamma_i)'' \tag{F.15}$$

is only a necessary, not a sufficient condition for the LLE. The sufficient condition is that the Gibbs energy must be at its minimum. The thermodynamic fundamentals are explained in Section 6.4 [7]. The recipe for checking the sufficient condition is as follows.

- Draw an isothermal plot of the function

$$\frac{\Delta g}{RT} = x_1 \ln(x_1 \gamma_1) + x_2 \ln(x_2 \gamma_2) \tag{F.16}$$

- If there is a miscibility gap, there are two points with a common tangent which is lower than the curve itself.
- Check if there are tangents which are lower than the one evaluated with Eq. (F.15).

The following diagrams shall illustrate the procedure. Figure F.9 shows the Δg-plot for the system methyl acetate–water after fitting the parameters. There is a miscibility gap between $0.062 < x_1 < 0.686$, which is confirmed by the tangent shown in the plot. There is no other common tangent possible; thus, the parameters can be accepted.

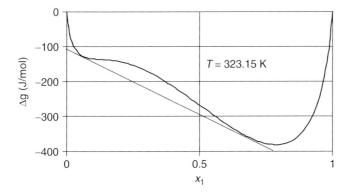

Figure F.9 Δg-plot for the system methyl acetate (1)–water (2).

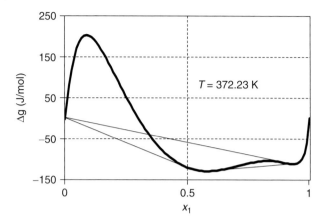

Figure F.10 Δg-plot for the system methyl benzoate (1)–water (2) with inappropriate parameters.

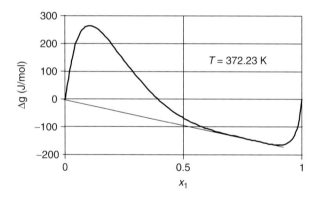

Figure F.11 Δg-plot for the system methyl benzoate (1)–water (2) with corrected parameters.

The case is different for the fit shown in Figure F.10, which refers to the system methyl benzoate–water. With the parameters from a regression, a Δg-plot with three possible tangents is obtained. In process simulation, the calculated LLE would depend on the starting value of the iteration. In fact, the parameters indicate that there are two miscibility gaps, as the two shorter tangents are lower than the long one, which represents the true miscibility gap. For correction, the α-value of the NRTL equation was modified and fixed in the regression. The new parameters have a slightly worse value of the objective function, but a stable solution for the LLE (Figure F.11).

A lot has been written about consistency tests [2, 8–10]. The Redlich–Kister test has already been described in Section 5.4.1. However, for testing the consistency, it has the serious disadvantage that the pressure cancels out. At low pressures, where an ideal gas phase can be assumed, it can be written for each data point:

$$\frac{\gamma_1}{\gamma_2} = \frac{\gamma_1 x_2 P_2^s}{\gamma_2 x_1 P_1^s} \neq f(P) \tag{F.17}$$

This means that the pressure, which is the quantity that can usually be measured with an excellent accuracy, has no influence on the consistency checked with the Redlich–Kister method.

Alternatively, the simple check of the regression results can be taken as a consistency test, as each of the equations for the activity coefficients fulfills the Gibbs–Duhem equation

$$x_1 d \ln \gamma_1 + x_2 d \ln \gamma_2 = 0 \tag{F.18}$$

Therefore, if data can be fitted with a sufficient accuracy, they can be regarded as consistent. Furthermore, outliers can directly be detected and removed from the database. One should always be careful if chemical reactions can occur in the system. In this case, a model describing only phase equilibria is not sufficient.

The decision, which of the three activity coefficient models Wilson, NRTL, or UNIQUAC should be taken, is usually not a matter of quality, if the parameters are fitted to data and not taken from a parameter bank. Essentially, no significant differences will be detected between the three equations.

In fact, this question hardly occurs in practical applications. For the choice of the model, it is decisive that all systems occurring in the project can be satisfactorily described. The following items must be taken into account.

- The Wilson equation is not appropriate for LLE. Currently, the process simulators do not offer an electrolyte extension. If both LLE and electrolytes do not occur, there are no concerns about using the Wilson equation.
- The NRTL equation can describe LLE and has an electrolyte extension, although the quality in both applications is limited. Nevertheless, it can at least be applied to most of the problems. Often, the third parameter is helpful for describing difficult systems.
- UNIQUAC can describe miscibility gaps, but is less flexible than NRTL. There are electrolyte extensions (LIQUAC, see Section 7.3.4); however, they are hardly available in process simulators.
- If decisive components are supercritical, the use of an equation of state for both phases (φ–φ-approach, for instance VTPR) must be considered.

References

1. Stephenson, R.M. (1992) *J. Chem. Eng. Data*, **37** (1), 80–95.
2. Anderson, T.F., Abrams, D.S., and Grens, E.A. (1978) *AIChE J.*, **24** (1), 20–29.
3. Mertl, I. (1972) *Collect. Czech. Chem. Commun.*, **37**, 366.
4. Gaw, W.J. and Swinton, F.L. (1968) *Trans. Faraday Soc.*, **64**, 2023.
5. Loehe, J.R., van Ness, H.C., and Abbott, M.M. (1983) *J. Chem. Eng. Data*, **28**, 405.
6. Brandani, S., Brandani, V., and Flammini, D. (1994) *J. Chem. Eng. Data*, **39**, 184–185.

7. Bowrey, R.G. and Marek, C.J. (1971) *Br. Chem. Eng.*, **16** (1), 57–58.
8. Bertucco, A., Barolo, M., and Elvassore, N. (1997) *AIChE J.*, **43** (2), 547–554.
9. Jackson, P.L. and Wilsak, R.A. (1995) *Fluid Phase Equilib.*, **103**, 155–197.
10. Miklos, D., Kemeny, S., Almasy, G., and Kollar-Hunek, K. (1995) *Fluid Phase Equilib.*, **110**, 89–113.

Appendix G
Ideal Gas Heat Capacity Polynomial Coefficients for Selected Compounds

Equation: $\dfrac{c_P^{id}}{\text{J/(mol K)}} = A + B\left(\dfrac{T}{K}\right) + C\left(\dfrac{T}{K}\right)^2 + D\left(\dfrac{T}{K}\right)^3$

Compound	Formula	M (g/mol)	A	$10^3 B$	$10^6 C$	$10^9 D$	Verification point T (K)	c_P^{id} (J/(g K))
Hydrogen	H_2	2.016	27.124	9.267	−13.799	7.640	513.15	14.5229
Water	H_2O	18.015	32.220	1.923	10.548	−3.594	373.15	1.8995
Ammonia	NH_3	17.031	27.296	23.815	17.062	−11.840	394.75	2.2681
Nitrogen dioxide	NO_2	46.005	24.216	48.324	−20.794	0.293	584.25	0.9871
Nitrogen	N_2	28.014	31.128	−13.556	26.777	−11.673	313.15	1.0406
Oxygen	O_2	31.999	28.087	−0.004	17.447	−10.644	313.15	0.9210
Sulfur dioxide	SO_2	64.062	23.836	66.940	−49.580	13.270	343.15	0.6479
Dichlorodifluoromethane	CCl_2F_2	120.913	31.577	178.110	−150.750	43.388	373.15	0.6559
Carbon monoxide	CO	28.010	30.848	−12.840	27.870	−12.710	313.15	1.0414
Carbon dioxide	CO_2	44.009	19.780	73.390	−55.980	17.140	373.15	0.9148
Methane	CH_4	16.043	19.238	52.090	11.966	−11.309	293.15	2.1973
Methanol	CH_3OH	32.042	21.137	70.880	25.820	−28.500	303.15	1.3795
Ethylene	$CH_2=CH_2$	28.054	3.803	156.480	−83.430	17.540	473.15	2.1752
Ethylene oxide	C_2H_4O	44.053	−7.514	222.100	−125.560	25.900	298.15	1.0948
Acetic acid	CH_3COOH	60.052	4.837	254.680	−175.180	49.454	373.15	1.2997
Ethane	C_2H_6	30.070	5.406	177.980	−69.330	−1.916	273.15	1.6232
Ethanol	C_2H_5OH	46.069	9.008	213.900	−83.846	1.372	373.15	1.6762
Propylene	$CH_2=CH-CH_3$	42.081	3.707	234.380	−115.940	22.030	280.15	1.4437
Propane	C_3H_8	44.097	−4.222	306.050	−158.530	32.120	373.15	2.0313
Ethyl acetate	$CH_3-COO-C_2H_5$	88.106	7.2298	406.900	−209.030	28.526	473.65	1.7717
n-Butane	C_4H_{10}	58.124	9.481	331.100	−110.750	−2.820	373.15	2.0209
Isobutane	$CH_3-CH(CH_3)-CH_3$	58.124	−1.389	384.500	−184.470	28.932	373.15	2.0285
Benzene	C_6H_6	78.114	−33.890	474.040	−301.490	71.250	373.15	1.3406
n-Hexane	C_6H_{14}	86.178	−4.410	581.570	−311.660	64.890	313.15	1.7306

Parameters from: Gmehling, J., Kolbe, B., Thermodynamik, **2**, Auflage, VCH, Weinheim, 1992.

Chemical Thermodynamics: for Process Simulation, First Edition.
Jürgen Gmehling, Bärbel Kolbe, Michael Kleiber, and Jürgen Rarey.
© 2012 Wiley-VCH Verlag GmbH & Co. KGaA. Published 2012 by Wiley-VCH Verlag GmbH & Co. KGaA.

Appendix H
UNIFAC Parameters

Table H.1 Relative van der Waals surface area and volume parameters for the main groups 1–9.

Number	Subgroup	Number	Main group	R_k	Q_k
1	CH_3	1	CH_2	0.9011	0.8480
2	CH_2			0.6744	0.5400
3	CH			0.4469	0.2280
4	C			0.2195	0.0000
5	$CH_2=CH$	2	$C=C$	1.3454	1.1760
6	$CH=CH$			1.1167	0.8670
7	$CH_2=C$			1.1173	0.9880
8	$CH=C$			0.8886	0.6760
70	$C=C$			0.6605	0.4850
9	ACH	3	ACH	0.5313	0.4000
10	AC			0.3652	0.1200
11	$ACCH_3$	4	$ACCH_2$	1.2663	0.9680
12	$ACCH_2$			1.0396	0.6600
13	ACCH			0.8121	0.3480
14	OH	5	OH	1.0000	1.2000
15	CH_3OH	6	CH_3OH	1.4311	1.4320
16	H_2O	7	H_2O	0.9200	1.4000
17	ACOH	8	ACOH	0.8952	0.6800
18	CH_3CO	9	CH_2CO	1.6724	1.4880
19	CH_2CO			1.4457	1.1800

Table H.2 UNIFAC group interaction parameters a_{nm} (K) for the main groups 1–9.

n\m		1	2	3	4	5	6	7	8	9
1	CH_2	0	86.02	61.13	76.5	986.5	697.2	1318.	1333.	476.4
2	C=C	−35.36	0	38.81	74.15	524.1	787.6	270.6	526.1	182.6
3	ACH	−11.12	3.446	0	167	636.1	637.35	903.8	1329	25.77
4	$ACCH_2$	−69.7	−113.6	−146.8	0	803.2	603.25	5695	884.9	−52.1
5	OH	156.4	457	89.6	25.82	0	−137.1	353.5	−259.7	84
6	CH_3OH	16.51	−12.52	−50	−44.5	249.1	0	−180.95	−101.7	23.39
7	H_2O	300	496.1	362.3	377.6	−229.1	289.6	0	324.5	−195.4
8	ACOH	275.8	217.5	25.34	244.2	−451.6	−265.2	−601.8	0	−356.1
9	CH_2CO	26.76	42.92	140.1	365.8	164.5	108.65	472.5	−133.1	0

Further parameters were published in the following reference

Hansen, H.K., Schiller, M., Fredenslund, Aa., Gmehling, J., and Rasmussen, P. (1991) *Ind. Eng. Chem. Res.*, **30**, 2352–2355.

The following internet site: *www.ddbst.com* contains the entire list of the published UNIFAC group interaction parameters. The revised and extended UNIFAC group interaction parameter matrix is only available for the sponsors of the UNIFAC consortium (*www.unifac.org*).

Appendix I
Modified UNIFAC Parameters

Table I.1 Relative van der Waals surface area and volume parameters for the main groups 1–9.

Number	Subgroup	Number	Main group	R_k	Q_k
1	CH_3	1	CH_2	0.6325	1.0608
2	CH_2			0.6325	0.7081
3	CH			0.6325	0.3554
4	C			0.6325	0.0000
5	$CH_2=CH$	2	C=C	1.2832	1.6016
6	CH=CH			1.2832	1.2489
7	$CH_2=C$			1.2832	1.2489
8	CH=C			1.2832	0.8962
70	C=C			1.2832	0.4582
9	ACH	3	ACH	0.3763	0.4321
10	AC			0.3763	0.2113
11	$ACCH_3$	4	$ACCH_2$	0.9100	0.9490
12	$ACCH_2$			0.9100	0.7962
13	ACCH			0.9100	0.3769
14	OH (primary)	5	OH	1.2302	0.8927
81	OH (secondary)			1.0630	0.8663
82	OH (tertiary)			0.6895	0.8345
15	CH_3OH	6	CH_3OH	0.8585	0.9938
16	H_2O	7	H_2O	1.7334	2.4561
17	ACOH	8	ACOH	1.0800	0.9750
18	CH_3CO	9	CH_2CO	1.7048	1.6700
19	CH_2CO			1.7048	1.5542

Group Interaction Parameters (Modified UNIFAC)

Table I.2 UNIFAC group interaction parameters a_{nm} (K) for the main groups 1–9.

n\m	Main group	1	2	3	4	5	6	7	8	9
1	CH_2	0	189.66	114.2	7.339	2777	2409.4	1391.3	1381.0	433.6
2	C=C	−95.418	0	174.1	117.3	2649	−628.07	778.3	1207	179.8
3	ACH	16.07	−157.2	0	139.2	3972.	1604.3	792	1356	146.2
4	$ACCH_2$	47.2	−113.1	−45.330	0	3989	436.21	1050.2	1375	1001
5	OH	1606	1566	3049	2673	0	346.31	−801.9	83.91	−250
6	CH_3OH	82.593	−96.297	13.733	145.54	−1218.2	0	−328.5	−867	86.439
7	H_2O	−17.253	−1301	332.30	24.144	1460	−524.3	0	−2686	190.5
8	ACOH	1987	191.60	2340	1825	465.4	265.5	148.40	0	−145.2
9	CH_2CO	199	91.811	−57.53	−146.6	653.3	394.78	770.60	−666.80	0

Table I.3 UNIFAC group interaction parameters b_{nm} for the main groups 1–9.

n\m	Main group	1	2	3	4	5	6	7	8	9
1	CH_2	0	−0.2723	0.0933	−0.4538	−4.674	−3.0099	−3.6156	−0.9977	0.1473
2	C=C	0.061708	0	−0.5886	−0.8552	−6.508	10	0.1482	−1.955	0.69911
3	ACH	−0.2998	0.61660	0	−0.65	−13.16	−2.0299	−1.726	−2.118	−1.237
4	$ACCH_2$	0.3575	1.172	0.4223	0	−14.09	1.9094	−1.9939	−1.702	−1.871
5	OH	−4.746	−5.809	−12.77	−5.765	0	−2.4583	3.824	−1.262	2.857
6	CH_3OH	−0.48575	0.6304	−0.11768	−0.48799	9.7928	0	1.0823	−1.258	−0.465
7	H_2O	0.8389	4.072	1.158	1.6504	−8.673	4.6065	0	19.44	−3.669
8	ACOH	−4.615	0.4936	−5.043	−3.743	−1.841	−2.905	−2.757	0	−0.738
9	CH_2CO	−0.8709	−0.71715	1.212	0.2419	−1.412	−0.36048	−0.5873	1.918	0

Table I.4 UNIFAC group interaction parameters c_{nm} (K^{-1}) for the main groups 1–9.

n\m	Main group	1	2	3	4	5	6	7	8	9
1	CH_2	0	0	0	0	1.551e−03	0	1.144e−03	0	0
2	C=C	0	0	0	0	4.822e−03	−1.497e−02	0	0	0
3	ACH	0	0	0	0	1.208e−02	0	0	0	4.237e−03
4	$ACCH_2$	0	0	0	0	1.530e−02	0	0	0	2.390e−04
5	OH	9.181e−04	5.197e−03	1.435e−02	−3.320e−04	0	2.929e−03	−7.514e−03	0	−6.022e−03
6	CH_3OH	0	−1.800e−03	0	0	−1.616e−02	0	−2.2e−03	2.998e−02	0
7	H_2O	9.021e−04	0	0	0	1.641e−02	−4.000e−03	0	−2.702e−02	8.838e−03
8	ACOH	0	0	0	0	0	2.283e−03	2.329e−03	0	0
9	CH_2CO	0	0	−3.715e−03	1.133e−04	9.540e−04	0	−3.252e−03	0	0

Further parameters were published in the following references

Lohmann, J. and Gmehling, J. (2001) *J. Chem. Eng. Jpn.*, **34**, 43–54.

Lohmann, J., Joh, R., and Gmehling, J. (2001) *Ind. Eng. Chem. Res.*, **40**, 957–964.

Gmehling, J., Li, J., and Schiller, M. (1993) *Ind. Eng. Chem. Res.*, **32**, 178.

Gmehling, J., Lohmann, J., Jakob, A., Li, J., and Joh, R. (1998) *Ind. Eng. Chem. Res.*, **37**, 4876–4882.

Gmehling, J., Wittig, R., Lohmann, J., and Joh, R. (2002) *Ind. Eng. Chem. Res.*, **41**, 1678–1688.

Jakob, A., Grensemann, H., Lohmann, J., and Gmehling, J. (2006) *Ind. Eng. Chem. Res.*, **45**, 7924–7933.

Nebig, S. and Gmehling, J. (2011) *Fluid Phase Equilib.*, **302**, 220–225.

Weidlich, U. and Gmehling, J. (1987) *Ind. Eng. Chem. Res.*, **26**, 1372.

Wittig, R., Lohmann, J., and Gmehling, J. (2003) *AIChE J.*, **49** (2), 530–537.

Wittig, R., Lohmann, J., Joh, R., Horstmann, S., and Gmehling, J. (2001) *Ind. Eng. Chem. Res.*, **40**, 5831–5838.

The following internet site *www.ddbst.com* contains the entire list of the published modified UNIFAC group interaction parameters.

The revised and extended group interaction parameter matrix of modified UNIFAC is only available for the sponsors of the UNIFAC consortium (*www.unifac.org*).

Appendix J
PSRK Parameters

For most of the structural groups the group interaction parameters of the group contribution equation of state are identical with the UNIFAC group interaction parameters. Therefore only van der Waals properties of a few selected gases are given below.

Table J.1 Relative van der Waals surface area and volume parameters for selected main groups.

Number	Subgroup	Number	Main group	R_k	Q_k
117	CO_2	56	CO_2	1.3000	0.9820
118	CH_4	57	CH_4	1.1292	1.1240
119	O_2	58	O_2	0.7330	0.8490
116	Ar	59	Ar	1.1770	1.1160
115	N_2	60	N_2	0.8560	0.9300
114	H_2S	61	H_2S	1.2350	1.2020
113	H_2	62	H_2	0.4160	0.5710
120	D_2	–	–	0.3700	0.5270
112	CO	63	CO	0.7110	0.8280

Table J.2 PSRK group interaction parameters a_{nm} (K) for a few selected gases (structural groups 56–63).

n\m	Main group	56	57	58	59	60	61	62	63
56	CO_2	0	73.563	208.14	568.2	−580.82	78.98	838.06	161.54
57	CH_4	196.16	0	–	17.425	64.108	511.99	253.92	62.419
58	O_2	32.043	–	0	−7.7202	−23.358	–	–	–
59	Ar	−201.60	11.868	32.631	0	11.986	–	190.87	64.100
60	N_2	694.28	11.865	44.349	2.1214	0	862.84	77.701	25.060
61	H_2S	114.96	278.1	–	–	648.2	0	628	665.70
62	H_2	3048.9	128.55	–	4196.2	247.42	137	0	863.18
63	CO	4.2038	1.6233	–	−18.703	6.423	116.97	494.67	0

Table J.3 PSRK group interaction parameters b_{nm} for the structural groups 56–63.

n\m	Main group	56	57	58	59	60	61	62	63
56	CO_2	0	0	0	0	3.6997	0	−1.0158	0
57	CH_4	0	0	–	0	0	−1.1761	0	0
58	O_2	0	–	0	0	0	–	–	–
59	Ar	0	0	0	0	0	–	0	0
60	N_2	−3.0173	0	0	0	0	−2.1569	0	−0.77261
61	H_2S	0	−0.23	–	–	−0.30072	0	0	0
62	H_2	−10.247	0	–	0	0	0	0	−12.309
63	CO	0	0	–	0	0.57946	0	−8.1869	0

Table J.4 PSRK group interaction parameters c_{nm} (K^{-1}) for the main structural groups 56–63.

n\m	Main group	56	57	58	59	60	61	62	63
56	CO_2	0	0	0	0	0	0	0	0
57	CH_4	0	0	–	0	0	0	0	0
58	O_2	0	–	0	0	0	–	–	–
59	Ar	0	0	0	0	0	–	0	0
60	N2	0	0	0	0	0	0	0	0
61	H_2S	0	0	–	–	0	0	0	0
62	H_2	0	0	–	0	0	0	0	4.632e−02
63	CO	0	0	–	0	0	0	4.718e−02	0

Further parameters were published in the following references

Fischer, K. and Gmehling, J. (1996) *Fluid Phase Equilib.*, **121**, 185–206.

Gmehling, J., Li, J., and Fischer, K. (1997) *Fluid Phase Equilib.*, **141**, 113–127.

Holderbaum, T. and Gmehling, J. (1991) *Fluid Phase Equilib.*, **70**, 251–265.

Horstmann, S., Fischer, K., and Gmehling, J. (2000) *Fluid Phase Equilib.*, **167**, 173–186.

Horstmann, S., Jabloniec, A., Krafczyk, J., Fischer, K., and Gmehling, J. (2005) *Fluid Phase Equilib.*, **227**, 157–164.

The following internet site: *www.ddbst.com* contains the entire list of the published PSRK group interaction parameters. The revised and extended group interaction parameter matrix of PSRK is only available for the sponsors of the UNIFAC consortium (*www.unifac.org*).

Appendix K
VTPR Parameters

Table K.1 Group surface areas for selected subgroups of the VTPR GC-EOS [1].

Number	Main group	Number	Subgroup	Q_k
1	CH_2	[1]	CH_3	1.2958
		[2]	CH_2	0.9471
		[3]	CH	0.2629
		[4]	C	0
3	ACH	[9]	ACH	0.4972
		[10]	AC	0.1885
9	CH_2CO	[18]	CH_3CO	1.448
		[19]	CH_2CO	1.18
42	cy-CH_2	[78]	cy-CH_2	0.8635
		[79]	cy-CH	0.1071
		[80]	cy-C	0
151	CO_2	[306]	CO_2	0.982
152	CH_4	[307]	CH_4	1.124
155	N_2	[304]	N_2	0.93

Chemical Thermodynamics: for Process Simulation, First Edition.
Jürgen Gmehling, Bärbel Kolbe, Michael Kleiber, and Jürgen Rarey.
© 2012 Wiley-VCH Verlag GmbH & Co. KGaA. Published 2012 by Wiley-VCH Verlag GmbH & Co. KGaA.

Table K.2 Selected group interaction parameters for the VTPR GC-EOS [1].

Main group n	Main group m	a_{nm} (K)	b_{nm}	c_{nm} (K^{-1})	a_{mn} (K)	b_{mn}	c_{mn} (K^{-1})
1	3	54.2589	0.2882		35.4832	−0.36933	
1	9	425.3118	0.68790	−3.07810E-04	284.2495	−1.77310	1.63580E-03
1	42	−27.2894	−0.06410	1.13850E-04	60.4525	−0.04770	7.33000E-06
1	151	403.1138	−0.19990	−6.68000E-05	204.8302	−1.30960	1.19670E-03
1	152	66.2551	0.01350		−23.3722	−0.08440	
1	155	282.5594	−0.41093	3.12489E-03	26.9233	−0.51246	−7.85810E-04
3	42	42.8927	−0.5895	5.34900E-04	91.002	0.26959	2.70370E-05
9	42	180.2850	−1.06370	6.82760E-04	547.6216	0.03250	1.79820E-04
151	152	99.0671			176.6828		
151	155	184.1479			80.8253		
152	155	53.3084			12.1806		

Table K.3 L,M,N-parameter for the Twu-α function (Eq. (2.173)), critical data, translation parameters and acentric factors for selected compounds [2].

Component	L	M	N	T_c (K)	P_c (kPa)	c_i (cm^3/mol)	ω
Heptane	0.87995	0.917159	0.971917	540.3	2734	−1.18	0.3457
CO_2	0.821045	0.84233	0.801334	304.2	7376	1.59	0.2252
Tetrachloromethane	0.706852	0.882672	0.867511	556.4	4560	4.81	0.194
Dichloromethane	1.04931	0.788845	0.594894	510	6355	1.58	0.2027
Water	0.44132	−0.8734	1.7599	647.3	22048	−4.40	0.344
Benzene	0.652522	0.832782	0.923292	562.1	4894	2.85	0.212
Cyclohexane	0.99168	0.921466	0.657534	553.8	4080	6.42	0.213
Ethane	0.21225	0.87204	1.701	305.4	4884	4.55	0.098
m-Xylene	0.807679	0.938647	1.04714	617	3546	−2.49	0.331
Toluene	1.21115	1.06277	0.601298	591.7	4114	0.82	0.257
Acetone	1.15623	1.05341	0.690726	508.1	4701	−10.24	0.309
n-Butane	1.16264	1.07189	0.553013	425.2	3800	4.89	0.193
2-Methylpentane	1.95756	2.11206	0.254845	497.7	3040	4.76	0.279
Octane	0.94534	0.897616	0.968495	568.8	2495	−3.81	0.394
Dodecane	1.05832	0.863245	1.06679	658.8	1810	−21.96	0.562
Octacosane	1.34201	0.815227	1.30126	845.4	727.5	−254.65	1.195
Eicosane	1.41753	1.00998	1.11291	767	1070	−93.89	0.8805
Hexane	1.09414	1.00138	0.723981	507.4	3014	1.28	0.2975
Propane	0.773132	0.9124	0.726325	369.95	4245	4.88	0.152
Methylisopropylketone	0.42506	0.81569	1.57975	552.8	3850	−3.40	0.3244
Pentane	0.420675	0.830283	1.41573	469.7	3369	3.52	0.251
2-Pentanone	0.759252	0.84014	1.04804	561.19	3672	−6.94	0.3396
Hexadecane	1.14769	0.863263	1.16627	722.4	1401	−48.86	0.742
2-Butanone	0.744647	0.88926	1.08045	535.6	4154	−8.32	0.329
Nitrogen	0.3295	0.88275	1.05208	126.2	3394	4.05	0.04
Methane	0.94543	1.24525	0.42415	190.6	4600	4.09	0.008
Tetradecane	1.16443	0.903594	1.08532	691.8	1572	−41.49	0.679

References

1. Schmid, B. and Gmehling, J. (2012) *Fluid Phase Equilib.*, in press.
2. Dortmund Data Bank (*www.ddbst.com*).

Index

a

α-function
– generalized
– – Soave–Redlich–Kwong 48
– – Peng–Robinson 48
– Mathias–Copeman 53
– Twu 53–57
absorption process 259–261, 606–607
acentric factor 46, 71ff
activity
– and activity coefficient 161–162, 188, 193–197, 675–680
– – Bromley extension 376–377
– – Debye–Hückel limiting law 374–376
– – electrolyte-NRTL model 378–387
– – LIQUAC model 387–395
– – Mean Spherical Approximation (MSA) model 396
– – Pitzer model 377–378
– – pressure dependence 200–201
– – temperature dependence 200–201
adiabatic compression and expansion 595–600
advanced cubic equations of state 52–58
Aly–Lee equation 106
Antoine constants 546
Antoine equation 83–84
apparent permeability constant 444
applications
– practical
– – adiabatic compression and expansion 595–600
– – equilibrium thermodynamics limitations 606–608
– – flash 591–593
– – Joule–Thomson effect 593–595
– – pressure relief 600–606
– special

– – formaldehyde solutions 567–573
– – vapor phase association 573–587
ARTIST software package 489
ASOG method 293
azeotropic behavior occurrence, conditions for 248–259
azeotropic points 197
– investigation in multicomponent systems 257–259, 501–503
– temperature dependence of azeotropic composition 243, 251–258, 307, 321, 662–664

b

Bancroft point 255
Barker–Henderson perturbation theory 464
Berlin form of virial equation 659
Born term 381, 396
Boyle curve 39
Brock–Bird–Miller equation 134
Bromley extension 376–377

c

caloric properties 10–14, 333, 664–665, 681–685
– caloric equations of state
– – entropy 336–337
– – Helmholtz energy and Gibbs energy 337–339
– – internal energy and enthalpy 333–336
– chemical reactions 354–361
– enthalpy descriptions in process simulation programs 339–346
– G-minimization technique 361–364
Canizzarro reaction 572
Cardano's formula 42, 652–653

Chemical Thermodynamics: for Process Simulation, First Edition.
Jürgen Gmehling, Bärbel Kolbe, Michael Kleiber, and Jürgen Rarey.
© 2012 Wiley-VCH Verlag GmbH & Co. KGaA. Published 2012 by Wiley-VCH Verlag GmbH & Co. KGaA.

chemical equilibrium 531–551
– influence of the real behavior 538–552
– multiple 551–552
– – Gibbs energy minimization 556–563
– – relaxation method 552–556
chemical potential 154
– equivalence of, for pure compounds 654–655
– mixtures 658–659
– polymer solutions 452–454
Chung equation 130
Clausius–Clapeyron equation 26, 27, 56, 75, 99, 100, 216, 236, 347, 572
cloud point curve 450, 479
complex electrolyte systems 400–401
correlation 80–81
corresponding-states principle 46–47
COSTALD method 67, 95
Coulomb's law 368
Cox charts 86
critical condensation point 182
critical locus 183
critical mass flow 603
critical opalescence 8
cross virial coefficients 163
cryoscopic constant 418
cubic equations of state 40–45, 164–174

d

Dalton's law 155
Debye–Hückel limiting law 374–376
Debye–Hückel term 381
degree of freedom 103
departure functions 19
differential relationships 646–653
diffusion coefficients 136–140
dissociation equilibrium 396–398
Dortmund Data Bank (DDB) 615–616
– current status of 488
Dortmund Data Bank Software Package (DDBSP) 489, 490, 615–616

e

electrolyte-NRTL model 378–387
electrolyte solutions 365–369
– activity coefficient models for
– – Bromley extension 376–377
– – Debye–Hückel limiting law 374–376
– – electrolyte-NRTL model 378–387
– – LIQUAC model 387–395
– – Mean Spherical Approximation(MSA) model 396
– – Pitzer model 377–378
– complex electrolyte systems 400–401

– dissociation equilibrium 396–398
– salt influence on vapor-liquid equilibrium behavior 398–399
– solvent activity in 661–662
– thermodynamics of 369–374
enantiotropy 406
enthalpy 11, 12, 15, 74–80
– descriptions, in process simulation programs 339–340
– – equation of state 346–354
– internal energy 333–336
– vaporization 97–102
enthalpy of reaction 525–527
– real gas behavior consideration on 529–531
– temperature dependence 527–529
entrainer selection for azeotropic and extractive distillation 511–518
entropy 11
equations of state 27
– advanced cubic 52–58
– application to mixtures 162–163
– – cubic equations of state 164–174
– – virial equation 163–164
– caloric
– – entropy 336–337
– – Helmholtz energy and Gibbs energy 337–339
– – internal energy and enthalpy 333–336
– coefficients for high precision 641–644
– cubic 40–45
– fugacity coefficient
– – for pressure-explicit 665–666
– – of PSRK 671–675
– – of Soave–Redlich–Kwong 669–671
– – of virial equation (Berlin form) 668
– – of virial equation (Leiden form) 666–668
– generalized 45–52
– group contribution 317–326
– high precision 32–39
– polymer thermodynamics 462–478
– predictive Soave–Redlich–Kwong 312–317
– process simulation programs 346–354
– solubility of gases in liquids 270–271
– van der Waals 40
– vapor–liquid equilibrium calculation using 235–240
– – fitted binary parameter of cubic equations of state 240–249
– virial equation 27–32

– VTPR group contribution equation of state   317–326
equilibrium thermodynamics limitations   606–608
Euler theorem   657–658
eutectic systems, and solid–liquid equilibrium (SLE)   410–419

f

Fenske equation   230
Fick's first law   137
flash   591–593
Flory–Huggins equation   207–209, 451–452
– of PSRK g^E mixing rule   172
fluid systems, phase equilibrium in   177–186
– application of activity coefficient models   193–197
– conditions for the occurence of azeotropic behavior   248–259
– g^E model parameters fitting   216–221
– – recommended model parameters   231–235
– – VLE data check for thermodynamic consistency   221–231
– liquid–liquid equilibrium   273–286
– – calculation using the K-factor method   282–286
– – diagrams   274, 276
– – isoactivity criterion   273, 275
– – pressure dependence   288–289
– – temperature dependence of ternary systems   286–288
– predictive models   289–290
– – group contribution methods   292–293
– – predictive Soave–Redlich–Kwong equation of state   312–317
– – regular solution theory   290–291
– – UNIFAC method   293–312
– – VTPR group contribution equation of state   317–326
– solubility of gases in liquids   259–260
– – calculation using equations of state   270–271
– – calculation using Henry constants   261–270
– – prediction   271–273
– thermodynamic fundamentals   186–193
– vapor–liquid equilibrium calculation, using equation of state   235–240
– – binary parameter fitting of cubic equations of state   240–249
– vapor–liquid equlibria using g^E models   197–216
formaldehyde solutions   567–573
freezing point depression   417–419
fugacity
– and fugacity coefficient   19–23, 188
– – for pressure-explicit equation of state   665–666
– – of PSRK equation of state   671–675
– – of Soave–Redlich–Kwong equation of state   669–671
– – of virial equation (Berlin form)   668
– – of virial equation (Leiden form)   666–668
– in mixtures   159
– – ideal   159–160
– – phase equilibrium   160–161
Fuller method   137

g

Gauss–Hermitian quadrature   480–481
g^E mixing rule   165–172, 241
g^E models   197–216, 451–462
– fitting of model parameters   216–230
– recommended model parameters   231–235
generalized equations of state   45–52
Gibbs–Duhem equation   140, 153–154, 162, 372, 709
Gibbs energy   11, 21, 26, 197–198, 213–214, 277–278, 337–339, 357, 360, 362, 372, 377, 409, 452, 453, 531–533, 554, 655
– minimization   361–364, 556–563
– standard enthalpy and   77–80
Gibbs–Helmholtz equation   201, 216, 227, 228, 254, 374
Gibbs phase rule   42
group contribution methods   292–293
group interactions   70
Guldberg rule   66

h

Helmholtz energy   11, 114, 337–339, 465
Henry's law   262
Henry constant   261–268
high precision equation of state   32–39, 641–643
Hoffmann–Florin equation   86–87
Huron–Vidal-g^E mixing rule   166, 168

i

ideal gases   14–15, 27
- derivation of expressions for the speed of sound   659–660
- heat capacity   102–109
- – polynomial coefficients for selected compounds   711
- ideal mixture of   154–156
- maximum mass flux   685–687

ideal mixture
- of ideal gases   154–156
- of real fluids   156–157

interaction parameters   380
internal energy, of system   10
inverse gas chromatography   456
inverse reduced viscosity   121
isofugacity condition   187
isothermal compressibility factor   17

j

Jamieson equation   126
Joback method   66–67, 87, 88, *108*
- estimation formula   72
- group contributions for   68
Joule–Thomson effect   593–595
Joule–Thomson inversion curve   39

k

Kelvin equation   92–93
- derivation of   653–654
Kirchhoff's law   527, 534, 546
Kirchhoff equation   83, 116
Koningsveld–Kleintjens model   454–456
Krichevsky–Kasarnovsky equation   267

l

Lee–Kesler approach   162
Lee–Kesler–Plöcker equation   163
Leiden form of virial equation   659
Lewis–Randall rule   160
LIFAC method   317
Li method   128
linear low-density polyethylene (LLDPE)   481
LIQUAC model   387–395, 429, 430
liquid density   94–97
liquid heat capacity   109–113
liquid–liquid equilibrium (LLE)   179, 212, 273–286
- polymer solutions   449–462, 479–482
- pressure dependence   288
- temperature dependence of ternary   286–288

liquid thermal conductivity   125–130
liquid viscosity   114–120
local composition   207
lower critical solution temperature (LCST)   275, 462
Lucas equation   118–125

m

Mathcad examples   613–615
Mathias–Copeman α-function   53
Maurer model   568–572, 608
Maxwell's equal area construction   43
Maxwell relations   13, 18
mean ion activity coefficient   372
Mean Spherical Approximation (MSA) model   396
membrane processes
- osmosis   439–442
- pervaporation   443–444
Microsoft Excel   616–617
Microsoft Office Visual Basic for Application   616–617
Missenard method   128
mixture properties   147
- activity and activity coefficient   161–162
- application of equation of state to mixtures   162–163
- – cubic equations of state   164–174
- – virial equation   163–164
- excess properties   157–158
- fugacity in mixtures   159
- – of ideal mixture   159–160
- – phase equilibrium   160–161
- Gibbs–Duhem equation   153–154
- ideal mixture
- – of ideal gases   154–156
- – of real fluids   156–157
- partial molar properties   149–152
- property change of mixing and   148–149
molality   366
monotropy   406
multiple chemical equilibrium   551–552
- Gibbs energy minimization   556–563
- relaxation method   552–556

n

Nernst distribution coefficients   276
NRTL model   213, 378–387, 678–679, 709
- recommended model parameters   233
Nußelt number   126

o

opposite lever arms, law of   180
osmosis   439–442
osmotic coefficient   373ff
osmotic pressure   439, 442

p

pair parameters   380
partial condensation   606
partial molar Gibbs energy   155, 156, 160
Peng–Robinson equation   48, 99, 101, 167–169, 339, 352
– generalized α-function   48
permanent gases   8–9
Perturbed-Chain-SAFT (PC-SAFT) model   467–468
Perturbed-Chain-SAFT EOS (PC-SAFT EOS)   463
pervaporation   443–444
phase equilibrium   23–27, 160–161. See also fluid systems, phase equilibrium in
– for closed system   656–657
– for fully closed system   655–656
Pitzer model   377–378
Planck–Einstein function   104
polydispersity   450, 479–482
polymer thermodynamics   445–451
– g^E models   451–462
– equation of state   462–478
– polydispersity influence   479–482
polymorphism   406
Porter equation   152, 198, 262, 421
Poynting factor (Poy$_i$)   189, 432, 433
Prandtl number   126
Prausnitz–Shair method   271
predictive models   289–290
– group contribution methods   292–293
– predictive Soave–Redlich–Kwong equation of state   312–317
– regular solution theory   290–291
– UNIFAC method   293–312
– VTPR group contribution equation of state   317–326
predictive Soave–Redlich–Kwong (PSRK)   271, 312–317
– difference between VTPR and   317
pressure relief   600–606, 685–687
process simulation programs   1–2
– enthalpy descriptions in   339–340
– – equation of state   346–354
– model parameter verification   492

– – g^E model parameter verification   493–501
– – pure component parameter verification   492–493
– notation of equations in   678–679
PT-graph   9
PSRK   166, 169–172, 671–675
– parameters   721–723
pure component parameters   619–640
pure component properties, correlation and estimation of   65
– characteristic physical property constants   65–66
– – acentric factor   71–72
– – critical data   66–71
– – melting point and enthalpy of fusion   74–77
– – normal boiling point   72–74
– – standard enthalpy and standard Gibbs energy of formation   77–80
– correlation and estimation of transport properties   114
– – diffusion coefficients   136–140
– – liquid thermal conductivity   125–130
– – liquid viscosity   114–120
– – surface tension   133–136
– – vapor thermal conductivity   130–133
– – vapor viscosity   120–125
– temperature-dependent properties   80–81
– – enthalpy of vaporization   97–102
– – ideal gas heat capacity   102–109
– – liquid density   94–97
– – liquid heat capacity   109–113
– – speed of sound   113–114
– – vapor pressure   82–94
pure components, PvT behavior of   5
– caloric properties   10–14
– equations of state   27
– – advanced cubic   52–58
– – cubic   40–45
– – generalized   45–52
– – high precision   32–39
– – virial equation   27–32
– ideal gases   14–15
– real fluids
– – auxiliary functions   16–17
– – fugacity and fugacity coefficient   19–23
– – phase equilibrium   23–27
– – residual functions   17–19
PvT-diagram   6
Pxy-diagram   179–180, 181–182

r

Rackett equation 94
Raoult's law 197, 206
Rarey/Moller method 89–91
Rarey/Nannoolal methods 69–70, 72, 74, 116
reaction kinetics 608
real fluids
- auxiliary functions 16–17
- fugacity and fugacity coefficient 19–23
- ideal mixture of 156–157
- phase equilibrium 23–27
- residual functions 17–19
real gas, expression derivation for speed of sound 660–661
Rectisol process 262
Redlich–Kister expansion 198, 199
Redlich–Kister test 221–223, 708–709
Redlich–Kwong equation 44
refrigeration arrangement, scheme of 599
regression technique
- for binary parameters 695–709
- for pure component data 691–694
Regula Falsi method 458–459
regular solution theory 272–273, 290–291
relaxation method 552–556
residual curves 503–511
- boundary residual curves 185–186, 503–511
retrograde condensation 183
reverse osmosis 439
Roy–Thodos method 131

s

salting out 398
salt solubility 427–432
Sato–Riedel equation 127
saturated liquid 23
saturated vapor 23
Schulz–Flory distribution 446
Schwarz theorem 13
separation technology, thermodynamic application in 487–492
- azeotropic points in multicomponent systems 501–503
- entrainer selection for azeotropic and extractive distillation 511–518
- extractive distillation applicability examination, for the separation of aliphatics from aromatics 519–522

- model parameter verification, prior to process simulation 492
- – g^E model parameter verification 493–501
- – pure component parameter verification 492–493
- residue curves, distillation boundaries, and distillation regions 503–511
- solvent selection, for separation processes 518–519
Simon–Glatzel equation 75, 77, 427
- pressure dependence of melting point description using 76
Simplex–Nelder–Mead method 218
Soave–Redlich–Kwong equation 47–48, 101, 165–169, 339, 681–685
- generalized α-function 48
solid–liquid equilibrium (SLE) 179, 405, 408–410
- diagrams 407
- with intermolecular compound formation in solid state 424–426
- pressure dependence of 427
- salt solubility 427–432
- of simple eutectic systems 410–419
- – freezing point depression 417–419
- solubility of solids in supercritical fluids 432–434
- of systems with solid solutions
- – ideal systems 419–420
- – nonideal systems 420–423
solubility, of gases in liquids 259–260
- calculation using equation of state 270–271
- calculation using Henry constants 261–270
- prediction 271–273
solution of groups concept 292–293
solvation 368
speed of sound 105, 113–114
- expression derivation, for ideal and real gases 659–661
- as maximum velocity in adiabatic pipe with constant cross-flow area 685
specific heat capacity 15, 102ff
state variable and partial molar property, relationship between 657–658
Statistical Association Fluid Theory (SAFT) 463
Stiel–Thodos equation 132
supercritical fluids, solubility of solids in 432–434
surface tension 133–136

t

Tait equation   96
ternary azeotrope   185
thermal expansion coefficient   17
thermodynamics
– first Law   10
– second Law   11
– third Law   12
throttle valve   591
Twu-α-function   53–57
Txy-diagram   180
Tyn/Calus equation   139

u

UNIFAC   293–300, 515, 521, 568–570
– modified   300–*310*, 412
– – parameters   *715–718*
– parameters   *713–714*
– weaknesses of group contribution methods   309–312
UNIQUAC equation   207, 213, *233*, 279, 283, 294, 496, 500, 678, 709
– recommended model parameters   231
upper critical solution temperature (UCST)   275, 462

v

van't Hoff equation, application of   200, 269, 428, 533, 572, 651
van der Waals equation of state   40
van Velzen method   116
vapor fraction   24
vapor–liquid equilibrium (VLE)   179, 193, *196*, *217*, *220*, *226*
– calculation, using equation of state   235–240
– – binary parameter fitting of cubic equations of state   240–249
– data check for thermodynamic consistency   221–231
– – recommended g^E model parameters   231–235
– diagrams   196
– salt influence on   398–399
– using g^E models   197–216
vapor–liquid separator   591
vapor phase association   343, 573–587
vapor pressure   82–94
vapor thermal conductivity   130–133
vapor viscosity   120–125
Vignes correlation   140
virial equation   27, 163–164
– estimation of second virial coefficients
– – Tsonopoulos method   30
– – Hayden–O'Connell method   30
Visual Basic for Application   616–617
volume-translated Peng–Robinson equation (VTPR)   55, 67, 170
– group contribution equation of state   317–326
– parameters   *725–727*

w

Wagner equation   81, 84–85
Walden rule   75
Wassiljewa mixing rule   132
Watson equation   98, 133
Wilke mixing rule   124, 132
Wilson equation   209–210, 213, 278, 548, 678, 709
– derivation   675–677
– inability, to describe miscibility gaps   679–681
– recommended model parameters   232
Wood's metal   406

z

Zeno line   39